ASTRONOMY
AT PLAY IN THE COSMOS

ADAM FRANK
UNIVERSITY OF ROCHESTER

WITH CONTRIBUTIONS BY

JEFF BARY
COLGATE UNIVERSITY

CAROL LATTA
ROCHESTER ACADEMY OF SCIENCE

W. W. NORTON
NEW YORK • LONDON

W. W. Norton & Company has been independent since its founding in 1923, when William Warder Norton and Mary D. Herter Norton first published lectures delivered at the People's Institute, the adult education division of New York City's Cooper Union. The firm soon expanded its program beyond the Institute, publishing books by celebrated academics from America and abroad. By midcentury, the two major pillars of Norton's publishing program—trade books and college texts—were firmly established. In the 1950s, the Norton family transferred control of the company to its employees, and today—with a staff of four hundred and a comparable number of trade, college, and professional titles published each year—W. W. Norton & Company stands as the largest and oldest publishing house owned wholly by its employees.

Editor: Erik Fahlgren
Project Editor: Christine D'Antonio
Developmental Editor: Beth Ammerman
Assistant Editor: Arielle Holstein
Manuscript Editor: Stephanie Hiebert
Managing Editor, College: Marian Johnson
Managing Editor, College Digital Media: Kim Yi
Production Manager: Eric Pier-Hocking
Media Editor: Rob Bellinger
Associate Media Editor: Julia Sammaritano
Media Project Editor: Marcus Van Harpen and Liz Vogt
Media Editorial Assistant: Ruth Bolster
Marketing Manager, Astronomy: Stacy Loyal
Design Director: Rubina Yeh
Book Designer: DeMarinis Design LLC
Photo Editor: Stephanie Romeo
Photo Research: Jane Miller
Permissions Manager: Megan Jackson
Composition: Lachina
Illustrations: Imagineering Media Services, Inc., and Lachina
Manufacturing: Transcontinental Interglobe, Inc.
The text of this book is composed in Adobe Caslon with the display set in Gotham.

Permission to use copyrighted material is included on page C-1.

ISBN 978-0-393-93522-6

W. W. Norton & Company, Inc., 500 Fifth Avenue, New York, NY 10110-0017

wwnorton.com

W. W. Norton & Company Ltd., Castle House, 75/76 Wells Street, London W1T 3QT

2 3 4 5 6 7 8 9 0

To my parents Ingrid Frank and George Richardson. Thank you for teaching me that wonder is the highest virtue and pursuing wonder the highest aspiration.

CONTENTS

PART I
Bedrock

PART III

A Galaxy of Stars

PART IV
A Universe of Galaxies

In my introductory astronomy course, I have two overarching goals. First, I want my students, who are almost exclusively not science majors, to be scientifically literate. I believe it's now imperative for more of our society's citizens to have at least a basic understanding of science and physical concepts. Just as important, people must have an understanding of how science works and be both curious and critical about what is presented to them concerning science. My great hope is that throughout their lives, my students will turn up the volume when a science story is reported on the news or click the link when one appears on their computer screen. This brings me to my second goal: I want my students to *like* science. Actually, I want them to love science, but I will be satisfied if they tell their roommates or relatives that their astronomy course was one of the best classes they took in college.

In my experience, students *are* interested in astronomy. But I have seen that, in general, they don't read their introductory astronomy textbooks much anymore. These students, many of whom are humanities and social science majors, do like to read. They carry around novels and critical works and start many of their questions with, "I read that…" The problem is that they just don't like to read textbooks, choosing instead to get their information from a buffet of often mismatched sources (usually from the Web) that don't integrate well with the class. It's not only a problem, but also a missed opportunity. The textbook for an introductory astronomy course can be both engaging and critical to a student's success. This is the main reason I wanted to write a book that *doesn't read like a textbook*. I wanted to transform students' inherent interest in astronomy into the imaginative concentration required to understand the details of the science.

In keeping with this goal, one of the most important distinctions between *At Play in the Cosmos* and other textbooks is the use of more than one voice to explicate the science in each chapter. I began my career as a popular-science writer during graduate school. While working on my PhD thesis in computational astrophysics, I also began writing for small magazines such as *Exploratorium Quarterly*.

Later I had the chance to reach a wider audience with regular contributions to national publications such as *Discover* and *Astronomy* magazines. Thus, during the same time that I was being trained as a scientist, I also was extremely lucky to receive training from some of the best science writers and editors in the country. They taught me the importance of "narrative drive" in writing to nonscientists about science. Whether the topic is the birth of solar systems or the birth of the Universe, there has to be something driving the reader to turn to the next page. The stories of science become easier for nonscientists to access when they are part of a broader human narrative: difficulties faced and overcome, people willing to endure long hours in service of an inner drive to know—these are the gateways that make the science behind the story interesting and worthy of time and concentration.

As I went on to author two books, cofound National Public Radio's 13.7 Cosmos and Culture blog, and become a regular contributor to the *New York Times,* I took these lessons with me, along with the conviction that communicating the excitement of science to the broadest audience is essential in the era we inhabit.

With that experience in mind, in writing *At Play in the Cosmos* I have made extensive use of interviews with working scientists. Each chapter includes interviews with two scientists of various ages and at various career levels (the first chapter has only one interviewee—the Nobel Prize–winning chemist and author Roald Hoffmann). The interviews serve two functions. First, they provide the all-important human element giving readers a face and story to accompany the science they're learning. Second, the use of interviews fulfills one of the cardinal rules of good science writing: let someone else tell the story. In writing for popular audiences, writers use the voice of the scientist-interviewee to fill in explanatory details. This device, mixed with more standard exposition, enlivens the writing by changing cadence and tone. For student readers faced with a dizzying array of facts and details to digest, this device can mean the difference between falling asleep and falling in love with a topic.

In this book I have tried to use the fundamental skills I learned as a magazine writer—such as effective transitions and the proper balance of detail and narrative—to improve readability. In addition to including interviews with scientists having a diversity of backgrounds, the text at times uses historical examples of researchers and their stories to tell the science. For example, the introduction to extragalactic studies begins with an account of the Great Debate in 1920. The discussion of white dwarf mass limits also tells the story of the rift between Chandrasekhar and his mentor Eddington after the latter surprisingly and publicly attacked the relevant physics. And the chapter on stellar death (13) begins with Stan Woosley (one of our interviewees) remembering how his family vacation was cut short by news of supernova 1987A.

At the same time, from my 25 years of experience teaching intro astronomy, I know the crucial function that a textbook serves as a study tool and reference. To this end, I developed several in-chapter features that make this book a key resource for all students.

- Each chapter begins with a set of learning goals keyed to that chapter's sections. I have tried to provide goals that span Bloom's taxonomy and require that students not only know the basics but also can explain a concept to a classmate. The chapter summary is organized by section and serves as a useful bookend.

- Each chapter provides integrated section summaries so that students can quickly review the concepts covered in each section.

- Each chapter ends with an overarching summary so that students can review, in a few paragraphs, the points covered in the learning goals.

- To support these features, a running glossary provides definitions of boldface terms in the margins for easy reference.

Being mathematically literate is an important part of being scientifically literate. In astronomy, seeing how the math works often helps students better understand a concept or work through a problem. Recognizing that students come to the class with different levels of mathematical background (and anxiety), I treat most quantitative aspects of topics *separately* in the "Going Further" boxes. Instructors thus can adjust how mathematical their courses are. In the "Going Further" boxes I take care to explain each step so that students understand why and how the math is being applied.

At the end of each chapter, students will find three categories of questions and problems to help them check their understanding. "Narrow It Down" contains multiple-choice questions that range from fact-based questions to ranking exercises. "To the Point" contains open-ended, qualitative questions that can be used to spark a discussion in class. "Going Further" contains quantitative problems that ask students to use the skills they learned in the "Going Further" boxes.

As I wrote each chapter, I also considered how the concepts in the book could be animated or simulated to make them even clearer. An NSF Career Award in 1997 allowed me to begin developing interactives for my classes. That experience showed how powerful digital tools can be for introductory astronomy, but only if they are clearly and cleanly integrated into the textbook and the rest of the class work. For this textbook, Jeff Bary of Colgate University and I worked together not only to identify the concepts chosen for the 50 interactives, but also to develop the storyboards that have gone into each of these online tools. Identified by an icon in the text and ebook, these simulation-based interactives can be used in class, recitations, or labs and allow the students to "play" with ideas first encountered in the text. These same "smart" graphics appear again in modified form as interactive assessment questions in Norton's online tutorial and homework system, Smartwork5.

The 18 chapters of *At Play in the Cosmos* are organized with the understanding that certain topics are more effective at holding student interest than others (extraterrestrial life, black holes, the Big Bang, etc.). The course is designed to develop a cadence whereby these popular topics come at the end of each section, enabling instructors to reignite students' excitement. The treatment of life in the Universe comes at the end of the planet section rather than in the book's final chapter (which sometimes is never reached because of time constraints). A chapter on black holes concludes the stellar evolution section, and cosmology concludes the section on galaxies and large-scale structure. Each of these chapters is written to present a broad and engaging view that speaks both to the current state of the science and to the questions that students bring to the topics.

The teaching and learning program to accompany the book will set the bar for this course in terms of quality of the material and innovation.

Ancillaries for Students

At Play in the Cosmos: The Video Game

Backed by academic research on how students learn through play, the video game *At Play in the Cosmos* is being developed by Dr. Kurt Squire's Learning Games Network, with authorial input and guidance from Adam Frank and Jeff Bary. The game challenges students to apply what they've learned and to learn more, by flying challenging missions and confronting problems in astronomy—such as finding habitable exoplanets. There is one multipart level for each of the four main parts of the text. Instructors can have students play the game before or after class, in the classroom or the lab. Or students can play whenever they want outside of class.

The video game reports student diagnostic data to a grade book, enabling instructors to assess students' engagement and progress.

Interactive Simulations

Jeff Bary, *Colgate University*

Nearly 50 easy-to-use and tablet-friendly Interactive Simulations enable students to play with physical relationships that are key to the study of astronomy. The Simulations use art from both the textbook and the video game. They are incorporated into the video game, the ebook, and Smartwork5 online assessment. The Simulations are also available for professors to use in the classroom.

smartwork5

Juan Cabanela, *Minnesota State University-Moorhead*
Christopher Claysmith, *Chemeketa Community College*
William Dirienzo, *University of Wisconsin-Cheboygan*
Violet Mager, *Penn State Wilkes-Barre*

Smartwork5 is Norton's tablet-compatible homework system. Over 900 Smartwork5 questions support *At Play in the Cosmos*—all with answer-specific feedback, hints, and ebook links. Questions include ranking and sorting tasks, selected end-of-chapter problems (both multiple-choice and algorithmic numeric entry), labeling exercises based on book art, and guided inquiry activities based on the Interactive Simulations. In addition, "Process of Science" assignments help students apply the scientific method to important questions in astronomy, challenging them to think like scientists.

Smartwork5 provides rich diagnostic data on student performance, and it can integrate with campus Learning Management Systems (LMS).

Learning Astronomy by Doing Astronomy *Workbook*: Collaborative Lecture Activities

Stacy Palen, *Weber State University*
Ana Larson, *University of Washington*

Students learn best by doing. Devising, writing, testing, and revising suitable in-class activities that use real astronomical data, illuminate astronomical concepts, and ask probing questions that encourage students to confront misconceptions can be challenging and time-consuming. In this workbook, the authors draw on their experience teaching thousands of students in many different types of courses (large in-class, small in-class, hybrid, online, flipped, and so on) to present 30 field-tested activities that can be used in any classroom today. The activities have been designed to require no special software, materials, or equipment, and to take no more than 50 minutes to complete.

An instructor's manual, as well as PowerPoint versions of the pre- and postactivity "clicker" questions, is available at Norton's Instructor's Site.

Starry Night Planetarium Software (College Version) and Norton *Starry Night Workbook*

Steven Desch, *Guilford Technical Community College*
Michael Marks, *Bristol Community College*

Starry Night is a realistic, user-friendly planetarium simulation program designed to enable students in urban areas to perform observational activities on a computer screen. Norton's unique accompanying workbook offers observation assignments that guide students' virtual explorations and help them apply what they've learned from reading assignments in the text.

Ancillaries for Instructors

Instructor's Manual

Gerceida Jones, *New York University*
Elisha Polomski, *St. Cloud State University*

The Instructor's Manual contains brief chapter overviews, suggestions for using the text and Interactive Simulations in the classroom, and a list of all corresponding *Learning Astronomy by Doing Astronomy Workbook* activities for each chapter. The manual also contains worked solutions for all end-of-chapter problems and answers to exercises in the Norton *Starry Night Workbook*.

In addition, the Instructor's Manual provides support for integrating the video game into your course.

Test Bank

Steven Furlanetto, *University of California–Los Angeles*

The Test Bank assesses a common set of learning objectives consistent with the textbook and Smartwork5 online homework, and provides over 1,200 multiple-choice and short-answer questions. Every chapter consists of six question levels classified according to Bloom's taxonomy, and questions are further classified by section and difficulty, making it easy to construct meaningful and diagnostic tests and quizzes.

PowerPoint Lecture Slides

Michael Frey, *Cypress College*
Gerceida Jones, *New York University*

These ready-made lecture slides integrate selected art from the text, "clicker" questions, and links to the Interactive Simulations. Designed with accompanying lecture outlines, these lecture slides are fully editable and are available in Microsoft PowerPoint format.

Norton also provides an update service—quarterly PowerPoint presentations on engaging new topics—that enables instructors to cover new developments in astronomy only weeks after they occur.

Norton Instructor's Resource Site

This web resource contains the following resources to download:

- Test Bank, available in ExamView, Word, and PDF formats
- Instructor's Manual, in PDF format
- PowerPoint lecture slides with lecture notes
- All art and tables in JPEG and PPT formats
- Coursepacks, available in BlackBoard, Angel, Desire2Learn, Moodle, and Canvas formats

Coursepacks

Norton's Coursepacks feature a variety of activities and assessment materials, including multiple-choice quizzes, flash cards, and the Test Bank, at no extra cost. In addition, pre- and postactivity quizzes for the *Learning Astronomy by Doing Astronomy Workbook* are available. Coursepacks are available in BlackBoard, Angel, Desire2Learn, Moodle, and Canvas LMS formats.

Instructor's Resource USB Drive

This USB drive contains the Test Bank and all instructor resources, including PowerPoint lecture slides, labeled and unlabeled art slides, the Instructor's Manual with solutions, offline versions of the Interactive Simulations, and all art and photos from the text.

Acknowledgments

This book is the product of seven years of work that could not have been completed without the combined effort of a team of highly creative professionals. I am forever indebted to Carol Latta for her tireless work and boundless energy for all things astronomical in assembling the book via questions, summaries, images, and other features. It was a great pleasure to work with Jeff Bary, whose insights sharpened the text and whose creative work has made the interactives so powerful. Beth Ammerman's guidance as editor was essential; her remarkable ability to hold the near infinite details of an intro textbook helped us keep both the big picture and the task at hand in view. Jane Miller and Stephanie Romeo were invaluable help in obtaining the many beautiful images that grace the pages. Robert Bellinger brought an infectious sense of enthusiasm to his work leading the development of the many digital components of the textbook. Finally, I am very grateful to have had Erik Fahlgren as the lead at W. W. Norton for this project. Erik's vision, dedication to excellence, perseverance, and good humor throughout the long work of creating this book were essential to its completion. In general, the entire team at Norton, including Arielle Holstein, was a pleasure to work with.

It's worth more than just a note that without the kindness and support of my wife, Alana Cahoon, this project would not have been as much fun, because nothing is as much fun without her.

Many thanks to the instructors who provided feedback and helped shape this book, as well as the instructors who class-tested an early version of Chapter 5 with their students.

Reviewers

William Bagnuolo, *Georgia State University*

Becky Baker, *Missouri State University*

Celso Batalha, *Evergreen Valley College*

Elisabeth Benchich, *University of North Carolina at Charlotte*

Jeffrey Bodart, *Chipola College*

David Bradford, *State University of New York at Canton*

David Branning, *Trinity College*

Juan E. Cabanela, *Minnesota State University–Moorhead*

Todd Carriero, *City College of San Francisco*

Clinton Case, *Truckee Meadows Community College*

Damian Christian, *California State University–Northridge*

Fred Ciesla, *University of Chicago*

Juliana Constantinescu, *University of Wisconsin–Whitewater*

James Cooney, *University of Central Florida*

Michael Corwin, *University of North Carolina at Charlotte*

Santo D'Agostino, *Brock University*

Kate Dellenbusch, *Bowling Green State University*

Alessandra Di Credico, *Northeastern University*

Jess Dowdy, *Abilene Christian University*

Donald Farnelli, *Rowan University*

Tim Farris, *Volunteer State Community College*

Andrew Fittingoff, *Laney College*

Ian Freedman, *Dutchess Community College*

Michael Frey, *Cypress College*

Thor Garber, *Pensacola State College*

Christopher Gay, *Santa Fe College*

Christopher Gerardy, *University of North Carolina at Charlotte*

Nancy Gerber, *San Francisco State University*

Clint Harper, *Moorpark College*

Paul Heckert, *Western Carolina University*

Michael Hood, *Mt. San Antonio College*

Olencka Hubickyj-Cabot, *San Jose State University*

Gerceida Jones, *New York University*

Robert D. Joseph, *University of Hawaii*

Kishor Kapale, *Western Illinois University*

Julia Kregenow, *Pennsylvania State University*

Monika Kress, *San Jose State University*

Danilo Marchesini, *Tufts University*

Brian Mazur, *University of Mississippi*

Suzanne Metlay, *Western Governors University*

Wouter Montfrooij, *University of Missouri*

Thomas Nelson, *University of Minnesota*

Christopher Palma, *Pennsylvania State University*

Jon Pedicino, *College of the Redwoods*

Ylva Pihlström, *University of New Mexico*

John Pratte, *Arkansas State University*

Claude Pruneau, *Wayne State University*

Richard Rand, *University of New Mexico*

Michael Reid, *University of Toronto*

Ronald Revere, *Coastal Carolina University*

Andreas Riemann, *Western Washington University*

Lawrence Rudnick, *University of Minnesota*

Eric Schlegel, *University of Texas at San Antonio*

Sara Schultz, *Minnesota State University–Moorhead*

Kathy Shan, *University of Toledo*

John-David Smith, *University of Toledo*

Kelli Spangler, *Montgomery County Community College*

Edward Stander, *State University of New York at Cobleskill*

Winfield Sylvester, *Queensborough Community College*

Brian Thomas, *Washburn University*

Bruce Twarog, *University of Kansas*

Geoffrey Tweedale, *University of Wisconsin–Milwaukee*

Christopher Tycner, *Central Michigan University*

Robert Tyson, *University of North Carolina at Charlotte*

Arun Venkatachar, *Ohio University*

Saeqa Vrtilek, *Northeastern University*

G. Scott Watson, *Syracuse University*

Paul Wiida, *Georgia State University*

Derek Wills, *University of Texas*

Jie Zhang, *George Mason University*

Class Testers

Esteban Araya, *Western Illinois University*

Raymond Benge, *Tarrant County College*

Matthew Bobrowsky, *University of Maryland*

David Bradford, *State University of New York at Canton*

Juan E. Cabanela, *Minnesota State University–Moorhead*

Todd Carriero, *City College of San Francisco*

Erik Christensen, *South Florida State College*

Aaron Clevenson, *Lone Star College–Montgomery*

James Dickinson, *Clackamas Community College*

Ethan Dolle, *Northern Arizona University*

Jess Dowdy, *Abilene Christian University*

Andrew Fittingoff, *Laney College*

Michael Frey, *Cypress College*

Jeffrey Gillis-Davis, *University of Hawaii*

Gerceida Jones, *New York University*

Kishor Kapale, *Western Illinois University*

Frank Pyzcki, *County College of Morris*

Ronald Revere, *Coastal Carolina University*

Brian Schuft, *North Carolina Agricultural and Technical State University*

Christine Staver, *County College of Morris*

INTERACTIVES

ABOUT THE AUTHOR

Adam Frank

Adam Frank is a physicist, astronomer, and writer. His research focuses on computational astrophysics with an emphasis on star formation and late stages of stellar evolution. His popular writing has focused on issues of science in its cultural context including issues of science and religion and the role of technology in the human experience of time. He is a cofounder of National Public Radio's 13.7 Cosmos and Culture Blog.

ASTRONOMY

AT PLAY IN THE COSMOS

GETTING STARTED

Science, Astronomy, and Being Human

In this chapter you will learn about the history and human context of astronomical science. After reading through each section, you should be able to do the following:

1.1 Miles Apart and Years Between
Trace the long history of humanity's study of astronomy and how it has been applied across cultures and eras.

1.2 Very, Very Old and Really, Really Big
Conceptualize the size scales discussed in astronomy and express them in scientific notation.

1.3 Contents of the Cosmos
List the major categories of structure in the Universe and compare their sizes.

1.4 Why Science?
Explain how the scientific method is used to increase our understanding about the Universe and how science differs from pseudoscience.

1.5 Science, Politics, and the Human Prospect on a Changing Planet
Give examples of ways in which an understanding of science is necessary for informed political participation in our high-technology era.

The night sky and the Milky Way as seen above red sandstone formations at Arches National Park in Utah. For as long as human beings have existed, the sky has been a source of inspiration and wonder.

1

1.1
Miles Apart and Years Between

Newgrange, Ireland. The entrance is barely wide enough for a grown man to pass through. Unlike the bright Irish summer day behind us, with its impossibly blue sky and iridescent green hills, the path ahead is dark, cool, and damp. The ceiling hangs low and the walls press in close. We thread single file down a narrow passageway, a corridor that was constructed 5,000 years ago, a full half millennium before the Egyptian pyramids were finished. The passage opens into a low circular chamber set into the stone. As my Irish astronomer colleagues and I circle the chamber to make room for our guide, the silence of five millennia settles on us like a layer of ash.

Until the day before I had never heard of Newgrange, a prehistoric monument 50 miles from Dublin (**Figure 1.1**). I was just an American astrophysicist who had come to Ireland to collaborate with Tom Ray, an Irish astronomer who studies how stars like the Sun form. But Ray, who moonlights as an "archaeoastronomer," is often called upon to verify the astronomical orientation of Ireland's many archaeological sites. He had recently completed work at Newgrange. In a pub the night before, Ray suggested we visit this massive window into our collective astronomical past.

The guide at Newgrange waits for a moment as we settle down in the central chamber, and then begins. "Some time around fifty centuries ago, this great monument was built," he intones in a rich baritone. "Without aid of horse or wheel, the 26-ton slabs of granite around you were dragged across many miles and set in place where you see them. We do not know what rites and rituals were performed here. We do not know the meaning of the spiral patterns you see incised on the walls. This, however, we do know: One morning each year, in the very heart of winter, on the day of the winter solstice, the rising Sun aligns perfectly with the passageway you just traversed. On that morning, and a few others before and after, a beam of sunlight from the rising Sun pierces the 24-meter-long hallway and falls here in the central chamber. The light within this black chamber flares for a mere 14 minutes. Then it passes. The Sun rises above the hill, the alignment is broken, and the chamber remains dark for another year. We do not know much about the long-forgotten builders of this incredible place, but the alignment of the hallway with the winter solstice tells us they were, to some degree, astronomers."

Socorro, New Mexico. The road disappears over the sand-colored hills and bright-blue sky. It's just eleven o'clock in the morning and the desert is already unnerving me. I have been driving for more than 2 hours through the uplands of central New Mexico and have passed only two human outposts. There is no one out here for hundreds of miles in any direction. Still, this seems like a fitting location for civilization's most advanced signal post to the stars. Lonely tufts of grass roll by as my rental car noses up and begins a long climb along the edge of a dry plateau. Half an hour later, I crest the final rise and reach the Very Large Array, or VLA (**Figure 1.2**).

Stretching 10 miles into the distance, they stand with their giant heads upturned, all silent, all still, all facing the same patch of sky. The array is composed of 27 radio telescopes, each the size of an eight-story building and supporting a radio dish weighing 230 tons. Special railroad engines periodically lift the towering radio detectors and glide them into different configurations to optimize their sensitivity. Sensitivity is what the VLA is all about. All that steel and all the miles of interconnecting wires, all the human effort and human genius that went into their design—all of it serves a single purpose: to catch faint radio signals from the far ends of the Universe. The VLA's mission is to serve the astronomers who come here in hopes of converting pinprick pulses of electromagnetic

FIGURE 1.1
Newgrange
The 5,000-year-old monument at Newgrange, Ireland, provides direct evidence of the astronomical interests and knowledge of its ancient builders.

FIGURE 1.2 Very Large Array
The Very Large Array (VLA) in New Mexico consists of 27 radio telescopes in a Y-shaped configuration. The data from the individual antennas are combined electronically to give the resolution of a much larger telescope. The VLA studies celestial objects and phenomena that emit radio waves, such as jets coming from the regions around black holes and clouds of gas where stars form.

radiation caught by the telescopes into a narrative of the Universe: stories that tell us where we are, how we got here, and where we're going.

Science and the Human Journey

Newgrange and the Very Large Array are separated by more than 7,000 kilometers of Earth's surface and 5,000 years of human history. Still, despite all that space and all that time, they share an essentially human ambition. Each site stands as a stark reminder that human beings have always been fervent star watchers. The night sky has been the gateway to our

greatest mysteries, igniting in us a seemingly innate and creative curiosity. This boundless inquisitiveness seems to be one of the great gifts that evolution bestowed on us. We do not have sharp claws. We do not have wings to fly. What we do have is an intelligence that takes the world in and then has the audacity to ask it questions.

Astronomy—the study of celestial objects such as moons, planets, stars, nebulas, and galaxies—is now a modern science, but science in its modern form is part of an old game we play as we try to understand and control the world around us (**Figure 1.3**). The game can be deadly serious when the answers we seek determine our survival—as they did for early farmers who needed to read the stars to decide when to begin planting. It can also be creative, and even playful, as it was for the 17th- and 18th-century astronomers delighting in applying newly developed laws of physics to the movement of celestial objects.

From our very beginnings as modern humans 50,000 or more years ago, the nightly theater of stars, the Moon, and the planets has invited us to ask enduring questions of who we are, what we are, and how we got here. Our personal questions about life (what should I do with my time?), love (whom should I marry?), and meaning (why is life so hard sometimes?) are all, ultimately, framed against the background of our place in the **Universe**—the total of space, time, matter, and energy. Astronomy asks big, exhilarating questions, and from its perspective almost every aspect of life and the Universe can be explored.

This book and the class that you're starting are about the modern story of astronomy. It is a story of planets, stars, and galaxies, and the Universe as a

interactive
Scale of the Universe

astronomy The study of celestial objects such as moons, planets, stars, nebulas, and galaxies.

Universe The total of space, time, matter, and energy.

FIGURE 1.3 The Cosmological Perspective
The photo on the left shows Earth "rising" from the perspective of the Moon, as seen by *Apollo* astronauts. On the right is a view of the Milky Way as seen from Earth. In the middle is a market in a busy city. Everyday life is embedded in a cosmic context, but how often do we stop to think about our place in the Universe?

whole. It is a story of origins and endings. It is a story of forces and their capacity to shape all that we see and experience.

Other stories of the cosmos have come before the one you will encounter in this book. The great creation myths of cultures that preceded ours were attempts to understand the Universe and find humanity's proper place within it. Similarly, modern astronomy forces us to step back and think on the largest scale about *context*—our context and that of everything else in the Universe. What sets our modern story apart from the earlier great works of imagination is the process of science, developed and refined over centuries. Now, in the early years of the 21st century, we live in a world saturated with the fruits of science—and sometimes the poisons of the technologies derived from science. Dealing with the great challenges that face the human species over the coming decades will require that you have some understanding of how science works—how science knows what it claims to know and how that knowledge is evaluated. This book is therefore about both astronomy and the process of science in general.

Astronomy is a gateway, a window that lets us see both where we came from and where we might be headed. It shows us what we know and what we don't know, and the play of ideas between those two domains. Like all other sciences, astronomy shows us how much we can learn about the world in which we find ourselves, if only we're willing to pay attention, ask questions, and be vigilant in our search for answers.

SECTION SUMMARY
Science and the Human Journey

● As shown by ancient astronomical structures like Newgrange, human beings have always been star watchers, using their observations for both survival and creative purposes.

● Astronomy helps us confront our individual existence in the context of the Universe.

1.2
Very, Very Old and Really, Really Big

One of the most difficult aspects of learning the modern story of astronomy is the sheer size and age of the Universe. The numbers that astronomers use when accounting for astronomical distances and timescales can make your head spin. Somehow you have to get a feeling for what these numbers mean before you truly grasp what astronomy can tell you about the cosmos you live in.

Numbers and Human Time

Let's start with a relatively modest number: the timescale of modern human evolution. Archaeologists tell us that the species *Homo sapiens* (us) has been around for less than a million years. They also tell us that only in the last 70,000–50,000 years or so have we become "modern" in the sense that our bodies and our culture (the product of our minds) first came to resemble what they are today. Sometime during that period, human beings started burying their dead, wearing clothes, making musical instruments, and painting images of animals on cave walls.

Fifty thousand years is a long time. But really, how long? How can you wrap your mind around that number? How are we supposed to understand it as we understand a span of time that we've all experienced, such as "3 weeks" or "a month and half"?

One way to comprehend this timescale might be to think about human "generations." You and all your friends of about the same age make up one generation. Your parents make up another. Their parents (your grandparents) are another generation still. Your great-grandparents are another generation earlier. So if we take the approximate age at which humans give birth, we have a unit for measuring generations. We could even give this unit a name, the "grandma," which might help it make even more sense, since we want to count the number of "great-great-greats" between you and your distant ancestors.

Let's start by saying that, on average, people have children at 25 years of age. That means every 25 years, the next generation is born. If 25 years = 1 generation, how long has it been since the dawn of modern humankind? Divide 50,000 by 25 and you get 2,000. Thus, there have been only 2,000 generations (2,000 "grandmas") since the dawn of human culture. That means just 2,000 parents, grandparents, great-grandparents, great-great grandparents, and so on, linking you to the small tribes of early, ice-age nomadic humans who hunted wild game and foraged for edible plants. Think of it this way: in just 2,000 generations, we have invented all of civilization and transformed the entire planet (**Figure 1.4**).

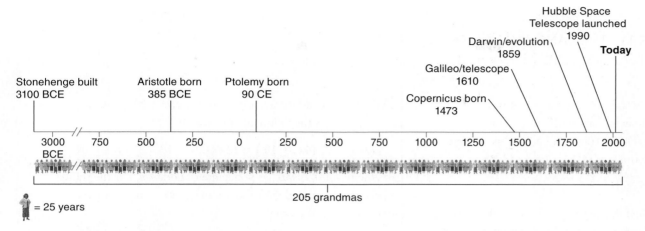

FIGURE 1.4   History Time Line in Units of Grandmas
We can measure back in time to historical events using time units of "grandmas," or generations, estimated at 25 years. Here we see the number of grandmas back to famous events in history related to science in general and astronomy in particular.

SECTION SUMMARY
Numbers and Human Time
- Human beings, in our current ("modern") form, have existed for 50,000 years, or about 2,000 generations—a very small amount of time in astronomical terms.

Scientific Notation

The numbers involved in astronomy get much larger than 50,000. The age of Earth, for example, is about 5 billion years. With numbers of this size, we have not only the problem of grasping what the number means, but also the problem of writing it down. Scientists have come up with a special way of writing very large or very small numbers, called **scientific notation**. For example, in scientific notation the number 5 billion is written as 5×10^9, which means 5 with nine zeros after it: 5,000,000,000. And 5 million, which is 1,000 times smaller than 5 billion, is written as 5×10^6, which means 5 with six zeros after it. In a similar fashion, the number 0.005 would be written as 5×10^{-3}, where the negative exponent tells us how many places to the right of the decimal place the digit 5 should be shifted. But knowing how to write numbers in scientific notation doesn't help us get a feel for how much larger, say, 5×10^7 is than 5×10^4. To understand this difference, let's think about something near and dear to our hearts: money.

The first column of the table in **Figure 1.5** shows different amounts of money in dollars, using the usual representation for numbers. The second column gives the same amounts of cash represented in scientific notation. The third column shows what that much

money actually looks like in each case. The fourth column gives a rough estimate of what you can buy with that amount of money. Notice as we move down the second column that the amount of money is jumping by one power of 10 in each row (1 dollar = 1×10^0, 10 dollars = 1×10^1, 100 dollars = 1×10^2). Each

scientific notation A way to write very large or very small numbers using the form $a \times 10^b$, where the coefficient a is any real number and the exponent b is an integer.

$1	1×10^0		
$10	1×10^1		
$100	1×10^2		
$1,000	1×10^3		
$10,000	1×10^4		
$100,000	1×10^5		
$1,000,000	1×10^6		

FIGURE 1.5   Scientific Notation
Using powers of 10 (also called scientific notation) is illustrated here by a series of purchases, each of which is one power of 10 greater than the one before it.

GOING FURTHER 1.1 THE LANGUAGE OF THE COSMOS
Working with Scientific Notation

Since most quantities in astronomy are expressed in scientific notation, it's important to know how to work with these kinds of numbers. When the number 245,000,000,000 (245 billion) is written in scientific notation, it looks like this:

$$2.45 \times 10^{11}$$

The first number, 2.45, is called the *coefficient*. The second number (without its exponent) is called the *base*. In scientific notation, the base is always 10. The number 11 above and to the right of the base is referred to as the *exponent* or the *power of 10*.

MULTIPLICATION. Let's start with multiplying two numbers in scientific notation. To do so, just multiply the two coefficients and add the exponents. For example,

$$(3.0 \times 10^7) \times (2.0 \times 10^5) = (3.0 \times 2.0) \times 10^{(7+5)} = 6.0 \times 10^{12}$$

What do you do if the product of the two coefficients is more than 10? Consider the following expression:

$$(4.0 \times 10^7) \times (3.0 \times 10^5) = (4.0 \times 3.0) \times 10^{(7+5)} = 12.0 \times 10^{12}$$

There is nothing wrong with the result 12.0×10^{12}, but in general, the coefficients in numbers expressed in scientific notation should be less than 10 and greater than or equal to 1. To meet that requirement, in this example you would shift the decimal place to the left and add one to the exponent:

$$12.0 \times 10^{12} = (1.2 \times 10^1) \times 10^{12} = 1.2 \times 10^{(1+12)} = 1.2 \times 10^{13}$$

DIVISION. Dividing two numbers in scientific notation is similar. Divide the two coefficients and subtract the exponents. Here is an example:

$$(6.0 \times 10^4)/(2.0 \times 10^3) = (6.0/2.0) \times 10^{(4-3)} = 3.0 \times 10^1$$

What do you do if the quotient of the two coefficients is less than 10? Consider the following expression:

$$(2.0 \times 10^9)/(4.0 \times 10^5) = (2.0/4.0) \times 10^{(9-5)} = 0.5 \times 10^4$$

Again, the result 0.5×10^4 is not standard notation. For the standard form, you would instead shift the decimal place to the right and subtract one from the exponent:

$$0.5 \times 10^4 = (5.0 \times 10^{-1}) \times 10^4 = 5.0 \times 10^{(4-1)} = 5.0 \times 10^3$$

ADDITION AND SUBTRACTION. When adding or subtracting numbers written in scientific notation, you will directly add or subtract the coefficients only if the exponent is the same for both numbers. This means that the equation

$$2.5 \times 10^3 - 1.5 \times 10^3$$

is the same as

$$2,500 - 1,500$$

which is equal to 1,000 or 1.0×10^3.

If you have to add or subtract two numbers with different exponents, you will need to manipulate them to make their exponents the same. Say, for example, someone asked you to add these two numbers:

$$(2.0 \times 10^3) + (1.0 \times 10^2)$$

This is just 2,000 + 100, which you know equals 2,100. In scientific notation, 2,100 is written as 2.1×10^3. If you wanted to add these numbers in scientific notation, you would need to represent the second number, 100, in thousands: 0.1×10^3 instead of 1.0×10^2. Now the equation becomes

$$(2.0 \times 10^3) + (0.1 \times 10^3) = 2.1 \times 10^3$$

jump in the power of 10 (the exponent) is called an increase in the **order of magnitude**.

Starting with the first row, for $1 ($1 \times 10^0$) you might be able to buy a candy bar. The second row shows that for $10 ($1 \times 10^1$) you can buy a ticket to a first-run movie. Jumping up another power of 10 to $100 ($1 \times 10^2$) makes things more interesting. With 1×10^2 dollars you could, for example, buy a fancy dinner for two at a good restaurant in Paris. Notice that a jump of just two powers of 10 (or two orders of magnitude) took us from a candy bar to fine dining. Of course, the dinner might really cost $150 or even $250, but what interests us is the fact that the meal costs hundreds of dollars rather than tens of dollars. That is what we mean by order of magnitude.

Let's keep going. How did you end up in Paris to begin with? You must have flown there, and that cost will take us up another power of 10. The cost of two tickets to Paris from the United States will be in the thousands of dollars ($1,000, or 1×10^3).

order of magnitude The size of a physical quantity in powers of 10.

And suppose you're having so much fun in Paris that you decide to stay for a while. That means you will need something to drive around. So you rent a luxury sports car for 2 months, at a cost of about $10,000 ($1 \times 10^4$). You also need a place to stay, so you rent a beautiful five-bedroom apartment overlooking the Seine River and the Notre-Dame Cathedral. That takes you up another power of 10, to $100,000 ($1 \times 10^5$). Finally, you decide you're never returning home, so you purchase the apartment for the grand price of somewhere in the millions of dollars ($1,000,000 or 1×10^6).

Do you see how this works? With just six jumps in powers of 10, we went from buying a candy bar to owning an apartment in one of the world's most expensive cities. We can all wrap our minds around the difference between a dollar and a million dollars. Scientific notation gives a better idea of what those numbers represent and prepares you to digest the staggering range of scales that make up the modern story of astronomy. **Going Further 1.1** (on p. 8) gives more details of how to work with scientific notation.

SECTION SUMMARY
Scientific Notation
- Numbers in astronomy are far larger than those we encounter in everyday life, and scientific notation is very useful for managing those numbers: 5×10^9 equals 5 billion or 5,000,000,000 (5 followed by nine zeros).
- In six orders of magnitude (increases of one in the power of 10), we can go from buying a candy bar (10^0, or $1) to buying an apartment in Paris (10^6, or $1 million).

1.3
Contents of the Cosmos

Stories need actors. The actors in the modern story of astronomy are the contents of the Universe: planets, stars, and galaxies. It has taken astronomers and physicists about 400 years to develop an inventory of the kinds of things that fill the cosmos. In all likelihood, that inventory is not complete, and there is more to be discovered; that is what makes astronomy so exciting.

The Largest Players

Since most of this book explores how the contents of the Universe came to be and how they interact, a quick survey of the major players in this cosmic drama is necessary to give a sense of scale. How much bigger is a planet than a person? How much smaller is an atom than a galaxy? Using your familiarity with powers of 10 and scientific notation, you can now get a first glimpse of the mind-boggling range of size scales that astronomy covers. In what follows, the letter L will denote the approximate size of something in meters. Let's start with something familiar and work our way up in scale.

YOU. Human beings are clearly one kind of "thing" in the Universe (even if we are unlikely to be universal). In terms of scale, the average human is somewhere between 1 and 2 meters in height, so $L \approx 10^0$ meters. (The symbol $\approx$ means "approximately equal to.")

PLANETS. The first major players among cosmic structures are planets. **Planets** come in at least two basic types: either smaller spheres of rock like Earth ($L \approx 10^7$ meters; **Figure 1.6**) or larger spheres of gas (perhaps with a rocky core) like Jupiter ($L \approx 10^8$ meters). Note that there are at least seven jumps in the power of 10 between the size of a human being and the size of a planet. Planets are, for the most part, always found orbiting stars.

STARS. Stars are the next important astronomical actors in terms of size. **Stars** are balls of gas that produce energy by means of nuclear fusion in their cores.

planet A body that orbits a star and is massive enough for self-gravity to have pulled it into a spherical shape and to have cleared smaller objects from the neighborhood of its orbit.

star A sphere of gas, such as the Sun, that produces energy via nuclear fusion in its core.

10^7 m

FIGURE 1.6 Size Scale of Planets
Earth. $L \approx 10^7$ meters.

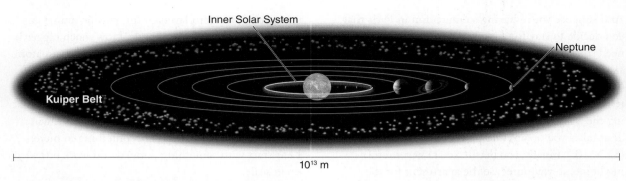

FIGURE 1.8 **Size Scale of Planetary Systems**
Our Solar System. $L \approx 10^{13}$ meters.

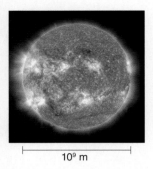

FIGURE 1.7
Size Scale of Stars
Our star, the Sun.
$L \approx 10^9$ meters.

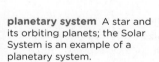

planetary system A star and its orbiting planets; the Solar System is an example of a planetary system.

galaxy Millions to billions of gravitationally bound stars. Galaxies also include gas, dust, and dark matter.

galaxy cluster A collection of hundreds or thousands of galaxies bound together by gravity.

supercluster A collection of galaxy clusters; the largest type of object in the Universe.

astronomical unit (AU) The average distance from Earth to the Sun: 1.5×10^{11} meters; used as a standard of measure for Solar System objects.

light-year (ly) The distance that light travels in a year: 9.5×10^{15} meters; used as a standard of measure for large distances.

Our Sun is an average-sized star, at $L \approx 10^9$ meters (**Figure 1.7**). Stars come in a wide range of sizes, from 100 times smaller to 1,000 times larger than the Sun.

PLANETARY SYSTEMS. A star and the planets that orbit it together compose a **planetary system**. Until the 1990s, astronomers did not know whether our system of the Sun and eight orbiting planets was the only one of its kind in the entire cosmos. Now we have enough evidence to infer that billions of other planetary systems exist in our galaxy. There are different ways to define the outer boundaries of a planetary system, but for now we can take $L \approx 10^{13}$ meters as the typical size (**Figure 1.8**). Notice that we have jumped 13 powers of 10 from the size of a human, and we haven't even left our cosmic neighborhood yet.

GALAXIES. The distances between stars are immense. The Sun's nearest neighbor lies more than 1,000 times farther away than the orbit of Neptune, the outermost planet in our Solar System. That means the typical distance between stars is about $L \approx 10^{16}$ meters. But stars do not float around randomly in empty space. All stars are born, live, and die in vast "cities" of Suns called **galaxies**. Our galaxy, the Milky Way, contains 400 billion stars and stretches across $L \approx 10^{21}$ meters (**Figure 1.9**). Notice the jump in scale from the size of planetary systems to the size of galaxies. Stars (and their planetary systems) are the building blocks of galaxies, but they are hundreds of millions of times (eight powers of 10) smaller.

GALAXY CLUSTERS. In general, galaxies form together in groups and, on larger scales, **galaxy clusters**. A typical galaxy cluster can have 1,000 galaxies in it and stretch across $L \approx 10^{23}$ meters (**Figure 1.10**).

SUPERCLUSTERS. At the top end of the size scale of cosmic structures are so-called superclusters of galaxies. A **supercluster** is a collection of galaxy clusters, and these behemoths extend across $L \approx 10^{24}$ meters of space (**Figure 1.11**). That makes them the largest "things" in the cosmos. Superclusters are still in the process of forming, as the gravity in these structures continues to pull material together from one end of the cosmos to the other.

It's worth noting here some of the units of size that will be used often in this book. The first, called an **astronomical unit**, is the average distance from the Sun to Earth (1 AU = 1.5×10^{11} meters). For larger distances, astronomers use the distance that light travels in a year—called, intuitively, the **light-year**

FIGURE 1.9 Size Scale of Galaxies
The Andromeda Galaxy, which is similar in size and shape to our Milky Way Galaxy. $L \approx 10^{21}$ meters.

FIGURE 1.10 Size Scale of Galaxy Clusters
Galaxy cluster. $L \approx 10^{23}$ meters.

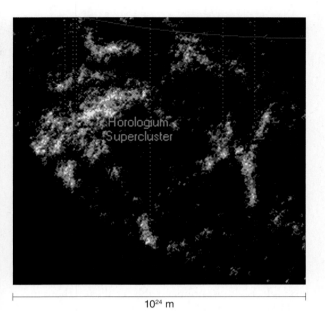

$$10^{24} \text{ m}$$

FIGURE 1.11 Size Scale of Superclusters
Galaxy supercluster. Every dot in this image is a galaxy.
$L \approx 10^{24}$ meters.

(1 ly = 9.5×10^{15} meters). Among astronomers, however, the preferred unit for large distances is the **parsec (pc)**, which is equal to 3.26 light-years.

SECTION SUMMARY
The Largest Players
- In the last 400 years, astronomers and physicists have developed an inventory of much of the contents of the Universe.
- A census of major cosmic structures, in order of increasing size, includes planets, stars, planetary systems, galaxies, galaxy clusters, and superclusters.

The Smallest Players

From humans to galaxy superclusters, we've been traveling upward in size scale. But if we move down in scale and investigate the microscopic world, what do we find in terms of the contents of the cosmos?

SINGLE-CELLED ORGANISMS. While single-celled organisms may or may not be a universal cosmic structure, they still represent one of the first important steps "down" in size relative to humans. An amoeba has a typical size of 0.2 millimeters (mm) or $L \approx 10^{-4}$ meters.

DUST. **Dust** particles are the smallest bits of solid matter in the Universe and are the starting point for building planets. The typical dust particle has a size scale of a millionth of a meter, or $L \approx 10^{-6}$ meters (**Figure 1.12**). These dust particles are composed of billions of atoms of elements such as carbon and silicon.

ATOMS. **Atoms** are the fundamental building blocks of all normal (non–dark) matter in the Universe. Atoms have a typical size scale of a billionth of a meter, or $L \approx 10^{-10}$ meters (**Figure 1.13**). Each of the different elements in the Universe, from hydrogen to uranium, has a distinct kind of atom.

SUBATOMIC PARTICLES. Atoms are made of even smaller particles—electrons, protons, and neutrons—which are called **subatomic particles**. A proton, for

FIGURE 1.12
Size Scale of Dust
Dust particle. $L \approx 10^{-6}$ meters.

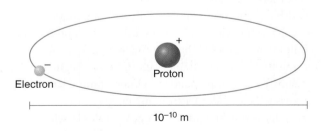

$$10^{-10} \text{ m}$$

FIGURE 1.13 Size Scale of Atoms
Hydrogen atom (the Bohr model is shown).
$L \approx 10^{-10}$ meters.

parsec (pc) The preferred unit for large distances, equal to 3.26 light-years.

dust The smallest bits of solid matter in the Universe, composed of atoms such as carbon and silicon.

atom The fundamental building block of matter, consisting of a dense central nucleus surrounded by a cloud of negatively charged electrons.

subatomic particle Any particle smaller than an atom. Subatomic particles include protons, neutrons, and electrons.

FIGURE 1.14
**Size Scale of
Subatomic Particles**
Proton (which itself is
composed of three quarks).
$L \approx 10^{-15}$ meters.

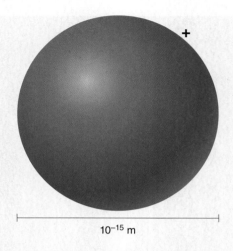

10^{-15} m

example, is almost unimaginably small, at a thousand million millionth of a meter, or $L \approx 10^{-15}$ meters (**Figure 1.14**). Some subatomic particles, such as the proton and neutron, are made of even smaller particles, called *quarks*.

All of these actors that make up the contents of the cosmos, from quarks to superclusters, are discussed in detail in the coming chapters. Note, too, that there are other important actors, such as electromagnetic fields and the newly discovered, very mysterious dark matter (a form of mass that interacts only weakly with the protons and neutrons that make up our bodies) and dark energy (a form of energy that is pushing the Universe apart). We will get to these in time.

Take a moment to reflect on the range of scales we have covered. When we were thinking about money, just six jumps took us from owning a candy bar to owning an apartment in Paris. In thinking about the contents of the cosmos, we traveled from the largest cosmic scales (superclusters, at 10^{24} meters) through the size of human beings (10^{0} meters) down to the size of a proton (10^{-15} meters), covering 39 powers of 10, or 39 orders of magnitude! The movement from the width of a proton (a million-billionth of a single meter) to the width of a supercluster (a million billion billion meters) seems beyond the imagination. Yet, using tools like scientific notation, we can begin to quantify and comprehend the almost infinite landscape across which modern astronomy ranges.

As astonishing as this range of scales is, what is equally amazing is the fact that even 150 years ago (6 "grandmas" ago), the story of this astronomical landscape could not have been told. Humans had yet to develop the tools to investigate on these size scales, so *their* stories were of a Universe much, much smaller than the one we now know to exist. The expansion of our vision of the Universe and all the actors it contains is a story that is just as important to understand as the specifics of that astronomical vision. Science itself makes such a view possible. But what is science, where did it come from, and what does it require of us?

SECTION SUMMARY
The Smallest Players
- All normal (non–dark) matter is made up of atoms, which are themselves made up of smaller, subatomic particles.
- The range in scale of all the objects in the cosmos takes us 39 orders of magnitude, from $L \approx 10^{-15}$ meters to $L \approx 10^{24}$ meters.
- There are other players in the Universe, such as electromagnetic fields, dark matter, and dark energy.

1.4
Why Science?

"Curiosity. That is where it all started," says Roald Hoffmann, a chemist at Cornell University. "If you want to understand science and its importance in human society, you have to begin at the beginning—with the curiosity we are born with." Hoffmann knows what he's talking about, for he has seen the best and worst that human society can produce. A survivor of the Nazi death camps in World War II, he went on to become a world-famous chemist, winning the Nobel Prize for his work on fast chemical reactions. Throughout his life he has been thinking and writing about the way science fits into all aspects of human culture, from poetry to politics to philosophy. In Hoffmann's view, you cannot understand what science is, or fully understand its promise or potential dangers from the technologies it can produce, without understanding where it comes from in the human mind.

Seeds of Science

"Our prehistoric ancestors were, in their way, scientists," says Hoffmann. "They were chemists before there ever was such a thing as chemistry. They were astronomers before there was astronomy." According to Hoffmann, all early cultures showed an intense interest and curiosity in the way the world worked. Along with that curiosity, these early cultures tried to use what was around them to transform their world. One example is the creation of fabric dyes like indigo. As Hoffmann explains, the blue dye indigo has been a constant in human societies for thousands of years. "To make indigo," he says, "you must learn to crush up certain kinds of snail shells or certain plants and then carry forward a pretty complicated process of extraction. This process was discovered in many ways over and over again across history, and in those multiple rediscoveries you can see the interplay between curiosity and craftsmanship. People notice certain shells or plants holding this rich blue color, and then they try to extract that color" (**Figure 1.15**).

The origins of astronomy were no different, explains Hoffmann. People found patterns in the night sky that could help them control their lives. "Early societies watched the sky," says Hoffmann. "They saw the regular cycles of Sun, Moon, and constellations. In time, and with trial and error, they found they could navigate using the constellations or determine when it was time to plant from the position of the rising Sun. As much as the night sky brought early people a sense of mystery and wonder, paying attention to the sky also led them to a deeper mastery of the world around them." Humans began with a fundamental curiosity about the world in which they found themselves and, just as important, they began with a desire to transform that world.

It took many thousands of years and the particular genius of many individuals and cultures before the curious craftsmanship of early societies developed into the practice of science that we recognize today. "It was a transformation that happened in fits and starts," says Hoffmann. "The ancient Greek cultures (500–300 BCE) played a big role. So did Arabic astronomers a thousand years later, as well as the scholars of the Renaissance in the 1400s and 1500s." Slowly, people learned how to focus their curiosity into a specific way of asking questions.

What we call modern science grew out of beginnings forged in Europe 500 years ago. Science and the technologies it produces now dominate the world. Because you live in a scientific culture, you will live longer, enjoy better health, travel farther, experience more, and have access to a wider array of information than virtually every one of the ancestors whose genes you carry. Because of science and the technologies that can grow out of it, the world you

FIGURE 1.15
Science through the Ages
The discovery of indigo, a fabric dye, has occurred many times in history and is one example of humans applying their curiosity and developing technologies in an attempt to transform their world.

inhabit also faces dangers and challenges that did not exist in the past. But what exactly is science, and what sets it apart from other human endeavors?

SECTION SUMMARY
Seeds of Science
- Our prehistoric ancestors were scientists in that they were curious and intent on finding ways to make useful changes in their world.
- Science as we know it today developed in stages, beginning with the ancient Greeks and continuing through Renaissance and later scholars.
- Science and technology now dominate the world, making it different in nearly every way from the world of our ancestors.

What Science Is and Is Not

"Lots of people think science is just a pile of facts," says Hoffmann. "But that isn't it at all. Science is a process more than anything else. It's a way of asking questions and finding answers that can be verified and repeated."

THE SCIENTIFIC METHOD. You may have already learned about the **scientific method**, which often begins when you make an observation about a particular phenomenon. Then you come up with a **hypothesis**

scientific method A body of techniques for acquiring new knowledge that includes systematic observation, experimentation, and theorizing.

hypothesis (pl. hypotheses) A proposed explanation of a phenomenon that must be rigorously tested by experiments.

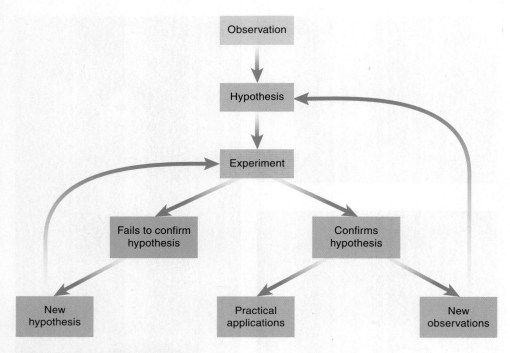

FIGURE 1.16 Scientific Method
The scientific method is the formal process by which science proceeds. It demands that scientists rigorously test hypotheses and discard or modify those that don't consistently predict results. In reality, the process is more subtle and more complicated, and factors such as accidents and surprises can play an important role.

to explain that phenomenon. Then, according to the standard account of the scientific method, you design an experiment to test that hypothesis. If the experiment fails, you have disproved your hypothesis. You have to go back to the drawing board, modify your hypothesis or invent a new one, and start again (**Figure 1.16**).

Take a concrete example: Imagine that you're watching leaves fall from a tree. You notice that some leaves tumble over and over as they fall. Others just drop straight down. That is your observation. Then you think, "I bet the leaves that tumble are wider than the ones that drop straight down because the wider ones present more air resistance." That is your hypothesis. So you take some paper and cut out 10 leaf shapes ranging from very wide to very narrow and drop them from a second-story window, counting the number of times each one tumbles on the way down. That is your experiment. Finally, you look at the data you collected to see whether the width of the leaf really did affect how much it tumbled.

This story of the scientific method is neat and reasonable, but in real life things are usually a bit more complicated. In fact, many of the greatest discoveries in human history came from accidents in the lab, and only afterward did scientists test the findings through experiments. Penicillin, perhaps the most important medical breakthrough ever, was discovered by mistake. In 1928, biologist Alexander Fleming noticed that the bacteria he needed for certain experiments would not grow on petri dishes that had mold on them (**Figure 1.17**). Fleming reasoned that the mold was somehow inhibiting the growth of the bacteria, and he recognized that what he was observing needed more investigation. He dropped the experimental program he was planning and turned his attention to isolating the factor in the mold that was killing the bacteria. Fleming's moment of recognition led to the development of antibiotics for medical use.

Many typical descriptions of the scientific method do not account for accidents, intuition, or flashes of insight that make up some of science's most interesting stories of discovery. What really makes science work is a broader account of the scientific method that not only includes rigorously testing a hypothesis against experiments but also, and most important, requires a willingness to adapt or discard any hypothesis that doesn't stand up to experimental investigation. By using experiments to test hypotheses about the way the world works, scientists achieve a degree of

FIGURE 1.17 Alexander Fleming
Alexander Fleming's discovery of penicillin is an example of discovery by accident, when the scientific method is applied after a surprising observation has turned an entire research program in a new direction.

certainty about their claims and help us gain reliable knowledge about the world.

One important point is the meaning of the word **theory** in science. Sometimes the terms *hypothesis* and *theory* can be used interchangeably. This occurs when scientists are speaking of a very specific observation or set of observations they wish to describe. We can think of cases like these as theories with a lowercase *t*, like a detective's "theory" about who committed a crime. This kind of theory can be overturned with just one or two counterexamples, meaning one or two experimental results that disprove the theory.

When scientists speak of Einstein's theory of relativity or the theory of evolution, they mean something different. We can think of these as theories with a capital *T*. They are well-developed bodies of knowledge that have been tested many times in many different ways over many decades or even over many centuries. A single new experimental result can't overturn the theory of relativity, because it has already been subjected to so much scrutiny. Instead, that new result would itself be strongly scrutinized for mistakes in how the experiment was set up or analyzed. These "capital-*T* theories" are still subject to revision or to being overturned, but given their success over so much time, it takes considerable effort to show that they're wrong. The rare cases when a major theory is overturned are therefore considered scientific revolutions.

SCIENCE VERSUS PSEUDOSCIENCE. Sometimes it can be hard to tell what is science and what is not. People make claims about the world all the time, usually with the aim of getting you to part with your money. They will tell you that a vitamin supplement will give you 50 percent more energy or that a cure for acne works 75 percent of the time. It's important to know whether these kinds of claims are valid—to separate science from **pseudoscience**.

An example of pseudoscience is a belief or practice that people claim is scientific, or that is made to look scientific, but that has not been developed through the methodology and standards discussed in this chapter. In general, pseudoscientific claims lack the supporting evidence or the plausibility associated with ideas that are backed by scientific knowledge. In other words, the proponents of a pseudoscientific claim haven't tested their ideas, or their ideas have failed experimentally but the proponents don't want to give up their cherished beliefs.

Roald Hoffmann

Nobel Prize–winning chemists don't ordinarily publish books of poetry, but then again, little about Roald Hoffmann's life could be called ordinary.

Hoffmann was born in 1937 in Złoczów, Poland, just before World War II started. "In 1941 darkness descended, and the annihilation of Polish Jewry began," explains Hoffmann, recounting his family's history. "We went to a ghetto, then a labor camp. My father smuggled my mother and me out of the camp in early 1943, and for the remainder of the war we were hidden by a good Ukrainian in the attic of a schoolhouse in a nearby village." Hoffmann's father continued organizing breakout attempts until he was caught and killed in 1943. "Most of the rest of my family suffered a similar fate."

In 1949 Hoffmann, with his mother and stepfather, emigrated to New York. "I learned English, my sixth language at this point, quite quickly," he explains. "I went to the great Stuyvesant High School, one of New York's selective science schools. Among my classmates were not only future scientists but lawyers, historians, writers." In 1962, with a PhD in chemistry from Harvard, Hoffmann joined the faculty of Cornell University, where he conducted pioneering work using a mix of computation, experimentation, and general models ("applied theoretical chemistry" is the way Hoffmann describes it). This research earned him the Nobel Prize in 1981.

But science has always been just one way that Hoffmann explores the world. His first play, *Oxygen*, written with Carl Djerassi, premiered in the United States in 2001 and had productions in London, Germany, Korea, Japan, and Toronto. Since then, Hoffmann has penned two new plays and three volumes of poetry, often exploring the boundaries between scientific and artistic ways of understanding. He has written extensively on science, art, and religion. This interest in different fields may seem contradictory, but to Hoffmann it seems natural: "Both the scientific and the artist ways of knowing are important (and there are others out there). Each contributes to our understanding of the world within and around us."

One of the longest-lasting pseudosciences, astrology, actually predates the origins of modern science. **Astrology** is the belief that the position of the planets and constellations at the time of your birth affects the future course of your life. It's a very old belief, and much of the earliest history of astronomy is tightly woven with astrological forecasting. In recent years, many highly detailed, exacting statistical studies have been done to look for correspondences between astrological predictions and reality. These studies have always shown that astrology's predictions are no better than random guesses. As much as people want to believe otherwise, astrology simply has no predictive capability. The daily horoscopes that some of us read are, in general, so vague that you could read the

theory An explanation for a phenomenon that has undergone extensive testing and is generally accepted to be accurate.

pseudoscience A claim, belief, or practice that is presented as scientific but does not adhere to a valid scientific method, lacks supporting evidence or plausibility, and/or cannot be reliably tested.

astrology The belief that there is a relationship between astronomical phenomena and events in the human world.

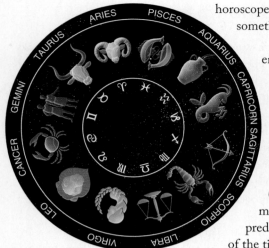

FIGURE 1.18
Astrological Zodiac
Astrology is a pseudoscience that purports to link constellation and planet positions to events in human lives.

horoscope for any of the astrological signs and find something related to your life (**Figure 1.18**).

What makes astrology a pseudoscience is its very explicit claim that a link exists between a set of phenomena (the exact positions of planets on the sky at the exact time you were born) and the future behavior of another set of phenomena (your life). Astrology appears to use precise data (planet positions) to make precise predictions (the tall, attractive stranger you will meet tonight). The problem is that these predictions do not turn out to be true most of the time—unlike, for example, astronomy's predictions of the motions of planets. Of course, many practitioners of astrology will say that it doesn't make exact predictions; instead, it predicts trends in a person's life. Don't be fooled. This is really just a way of wiggling out of the uncomfortable position of making a completely wrong prediction. In a real science, if your predictions are wrong even 10 percent of the time (one out of every 10 predictions), it's time to look for a new hypothesis.

Science's ability to predict how the world behaves, and then to help humans build powerful technologies from those predictions, is the reason it has gained such wide acceptance over the last five centuries. What separates science from pseudoscience is the willingness of scientists, and of societies that accept science, to allow the world to speak for itself, as well as the open-mindedness to give up even our most cherished beliefs about the world if they turn out to be wrong.

SECTION SUMMARY
What Science Is and Is Not

- Scientific discovery usually results from rigorous testing of a hypothesis with experiments and requires a willingness to discard or adapt a hypothesis that doesn't pass these tests.
- A pseudoscience, such as astrology, makes claims that lack supporting evidence or plausibility.
- Science enables people to predict how the world will behave and to build technologies from the predictions.

Science and Religion

How far does the reach of science extend? Can science explain all aspects of human experience? What about death? What about the existence of a god or gods?

These are deep questions that go beyond the practice of science itself and touch on the domains of philosophy and religion.

Taking the long view, however, many scholars would argue that science and religion may have shared a common link tens of thousands of years ago. "I think that science grows out of the same roots as religion in part," says Hoffmann. "The curiosity and desire to gain some control of the world I talked about before are also a part of early religious rites and rituals." As an example, Hoffmann points to his field, chemistry: "There was a relationship in early cultures between metallurgists 'winning' metal from the ores they found and that culture's religious practices. In Africa, there were all kinds of religious rituals and purification rituals by which the metalworker had to prepare himself for the process of getting the metal pure from an ore."

Over time, of course, the practices that would become science and religion separated. During the last 500 years, as science has matured it has sometimes come into conflict with some religious groups and their institutions. When science in its modern form emerged in western Europe, the Catholic Church held considerable political power. Discoveries in astronomy were seen by many church leaders to be in conflict with their interpretation of scripture and were branded as heresy. Later, Darwin's evolutionary theory was seen by some people to be in conflict with their religious beliefs about the origin of humanity. The public conflict between science and religion continues right up to the present day, as court cases continue to be raised in the United States concerning the validity of teaching evolution or cosmology rather than religious accounts in public schools (**Figure 1.19**).

The debate over the reach of science and the extent of its overlap with the domains of religion will likely continue for many years. If carried forward with tolerance and an open mind, it raises issues of the greatest scope and importance to human beings— even if, in many cases, no definitive answer can be reached. What cannot be doubted, however, is the ability of science to show us features of the physical Universe that we could not have seen without it. Through that perspective we can gain a deeper appreciation for the wonder and grandeur of the world in which we find ourselves. Science helps us see the many "miracles" hiding in plain sight before us every day: the beauty of birdsong, the wonder in the germination of seeds, and the power in the rising of the Sun. The scientific endeavor is founded on the

FIGURE 1.19 Scopes Monkey Trial
In 1925, Tennessee high school teacher John Scopes was arrested for teaching evolution in his classes. The famous trial that followed challenged limitations on teaching Darwin's evolution of species versus the biblical account of humankind's origin.

premise that nature and the Universe are comprehensible. And while we may have made more than 500 years of modern scientific progress, our fundamental conviction in nature's comprehensibility remains audacious.

SECTION SUMMARY
Science and Religion

- During the last 500 years, science has matured and sometimes has come into conflict with religious hierarchies; astronomy and biology have been two areas of such conflict.

- Evolution has been a prominent topic of conflict, embraced by some religions but also refuted by subgroups even within those same religions.

1.5
Science, Politics, and the Human Prospect on a Changing Planet

Every human generation thinks it's special. Every generation thinks it's on the cusp of history, a time when the world will either enter a bright new era or begin a decline into something darker. Europeans in the 1700s were sure they were at the beginning of a new Age of Reason. Aztecs in the 1100s thought themselves entering an era of decline after a golden

age of the past. Remarkably, the generations we belong to may be the first ones to be entirely correct in the belief that we are special. The reason for our uniqueness is inextricably tied to science.

Effects of Science and Technology

You were born into a world saturated with the fruits of science and its technologies. You were also born into a world challenged by science and its technologies. You already know some of the benefits of science: longer, healthier lives; fast, continent-spanning transportation; instant communication; the accessibility of all human knowledge through an Internet search. The challenges that science and technology bring us are also familiar to anyone who glances at the news.

Climate change tops the list of difficulties we face because of modern technologies. There is a consensus among scientists that over the last 100 years, our rapidly expanding use of fossil fuels to power everything from transportation to increased crop yields has altered the state of Earth's climate (its long-term weather patterns; **Figure 1.20**). The consequences of climate change are likely to include increases in global temperatures, changes in rainfall patterns, changes in biological habitats, and rises in sea levels. How large these changes will be over the next century is unknown; predictions span from mild to overwhelming.

Along with climate change, humanity faces challenges in the form of resource depletion in various manifestations. We are using up the things we need to live faster than we can replace them (and in some cases, we cannot replace them). Even if climate change were not a threat, most scientists would agree that the cheap, easy-to-reach oil we use to power our society will become exhausted sometime in the coming century. Other resources are threatened as well. Large-scale development fueled by increasing technological capacity has led to the exhaustion of many, if not most, of the world's fisheries, for example. Along with vanishing resources, society must contend with pollution of the resources we do have. Direct challenges from the products of science also continue to exist in the form of nuclear weapons and newer fears like biological terrorism.

In the 2002 book *Our Final Hour*, Sir Martin Rees, astronomer royal of England, reviewed a variety of threats facing the human race to measure our chances of survival. In a sobering calculation, Rees

climate change A significant, long-term change in the statistical distribution of weather patterns.

A

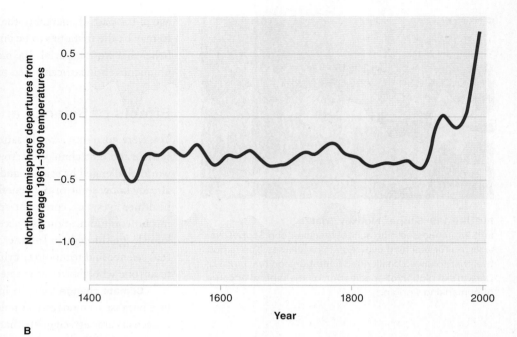

B

FIGURE 1.20 Climate Change
Climate scientists agree that our use of fossil fuels has pushed Earth's climate into unstable territory and that climate change poses a great danger to our current civilization. (A) Many scientists attribute rapid glacial melting to global warming. These two photos show recession of the Briksdal glacier in Norway in recent years. (B) This graph shows the average temperature of Earth's Northern Hemisphere going back 600 years. The temperatures at early times are reconstructed from a variety of methods, and the wide band shows the uncertainty in the temperature values. Notice how the uncertainty gets smaller the closer we get to the present era.

gave humanity 50/50 odds of making it, intact, to the next century. While many people would argue with Rees's odds, the idea that we face a critical epoch in human history has become widely accepted. That sobering reality, dependent so much on science and technology, is why you need to know what science is and how it works, regardless of your interest in any particular field of science.

As a member of a democratic society, you will be asked to make critical choices about the future. Almost all of those choices will involve science and technology in some way. From climate change to genetically modified crops, in the years ahead, all of us will have many, many opportunities to vote on issues shaped by science and technology. You may express your opinions by casting actual votes in an election, or by choosing which products you spend your money on. In some cases you may be faced with a parade of "experts" who will tell you what to believe about the science of climate change or the science of the new "green" car you're thinking of buying. By understanding how science works, how it reaches its conclusions, and how we can understand the limits of those conclusions, you

will be able to make informed decisions about your own future. The dominance of science and technology in our lives brings us great benefits, but it also comes with some very real responsibilities. At this strange and critical moment in history, we neglect those responsibilities at our own very real peril.

We human beings have always shown great resourcefulness and a capacity to attend to and solve the problems that face us. Science has been part of that resourcefulness in the past and will continue to be in the future.

SECTION SUMMARY
Effects of Science and Technology

- Science and technology offer benefits and challenges for the current generation of human beings.

- Scientific consensus is that fossil fuel use is changing Earth's climate, and climate change creates the most challenging difficulties facing humanity.

- Resource depletion and the threat of nuclear weapons and biological terrorism pose major challenges.

Astronomy and the Cosmic Perspective

If it is true that we're living in a critical historical moment driven by science and technology, then where should a person begin learning about science and how it works? With astronomy, of course!

Astronomy is one of the oldest sciences, dating far back into the millennia before written history. Astronomy also played a key role in the long march to the development of modern science, as new theories of the Solar System directly challenged the power structure of western Europe 500 years ago. But astronomy is also one of the newest sciences, pushing our understanding of distances and ages in ways that strain and invigorate our imaginations. At the same time, modern astronomy—with its stories of newly discovered planets orbiting far-flung stars, and galaxies colliding in 100-million-year-long slow-motion ballets—offers us more than just an understanding of how science works. It also offers a much-needed view of who we are, what we are, and how we got here.

In learning astronomy's stories—learning what we know about the Universe and how we have come to know it—you will gain something of great value: perspective. We may be small creatures in a very large cosmos, living short lives compared with those of stars and galaxies, but with the perspective we can gain from astronomy, we can come to understand our Universe. Not bad for a bunch of wingless, clawless mammals.

SECTION SUMMARY
Astronomy and the Cosmic Perspective
- We live in a critical historical moment driven by science and technology.
- Astronomy, as one of the oldest and yet newest sciences, provides perspective on humans' place in the cosmos and is a starting point for learning about science and how it works.

●→ chapter summary

1.1 Miles Apart and Years Between

The earliest human beings sought to understand the Universe around them and their place in it. This endeavor continues today, manifesting itself in many ways, including modern science and, in particular, the study of astronomy.

1.2 Very, Very Old and Really, Really Big

One useful way to understand human time spans is to think in terms of generations (25 years per generation). At least 50,000 years, spanning 2,000 generations, have passed since humans evolved into their culturally and biologically "modern" form. Scientific notation provides a concise way to write very large and very small numbers using powers of 10. Differences between numbers in their powers of 10 are called orders of magnitude.

1.3 Contents of the Cosmos

A brief survey of the main categories of structures in the Universe and their sizes includes human beings (10^0 meters), planets (10^7–10^8 meters), stars (10^9 meters; the typical distance between stars is 10^{16} meters), planetary systems (10^{13} meters), galaxies (10^{21} meters), galaxy clusters (10^{23} meters), and galaxy superclusters (10^{24} meters). On scales smaller than human beings are single-celled organisms (10^{-4} meters), dust grains (10^{-6} meters), atoms (10^{-10} meters), and protons (10^{-15} meters). In astronomy, numbers with the same order of magnitude are often considered roughly equivalent.

1.4 Why Science?

Science appears to be a natural result of our native human curiosity, a quality that leads us to explore and investigate. Discoveries made in this way typically have practical applications that enable humans to gain some control over (or at least increased understanding of) our environment. The modern scientific process asks questions and seeks answers that can be verified and repeated through experiments. Scientists formulate a hypothesis to explain a phenomenon and then design an experiment to test it. Often the hypothesis is disproved and discarded, but if it is not, further experiments may refine understanding of and strengthen support for the hypothesis. Pseudosciences, such as astrology, claim to explain the causes and effects of phenomena but have not been subjected to rigorous testing to verify their claims. True scientific claims are predictive; that is, the expected outcome is achieved each time it is tested, given the statistical uncertainties.

1.5 Science, Politics, and the Human Prospect on a Changing Planet

Our world today faces many challenges, some created by science and the resulting technology, and some likely to be solved through that science and technology. Climate change is foremost among those challenges. Each citizen of the world must understand the process of science enough to comprehend the challenges it presents and their potential solutions.

●→ questions and problems

Narrow It Down: Multiple-Choice Questions

1. Scientific notation is
    a. the use of Greek symbols to express scientific concepts.
    b. convenient shorthand for expressing very large or very small numbers.
    c. footnotes in scientific papers.
    d. the documentation of experimental results.
    e. formulas that express mathematical relationships.

2. Which of the following is the most correctly written in standard scientific notation?
    a. 0.014×10^{28} d. 14.0×10^{25}
    b. 0.14×10^{27} e. 140×10^{24}
    c. 1.4×10^{26}

3. Which of the following is equivalent to 9,400,000?
    a. 9.4×10^{5} d. 0.94×10^{8}
    b. 9.4×10^{6} e. 0.94×10^{9}
    c. 9.4×10^{7}

4. One number is said to be an "order of magnitude" larger than another number if
    a. it is 2 times larger.   d. it is 10 times larger.
    b. it is 3 times larger.   e. it is 100 times larger.
    c. it is 5 times larger.

5. Consider a number (x) that is 4 orders of magnitude larger than another (y). Which accurately describes the value of x compared to y?
    a. It is 4 times larger.   d. It is 10,000 times larger.
    b. It is 10 times larger.   e. It is 400 times larger.
    c. It is 1,000 times larger.

6. How many orders of magnitude larger is the Sun ($L_{Sun} \approx 10^{9}$ meters) than a terrestrial planet like Earth ($L_{Earth} \approx 10^{7}$ meters)?
    a. 0   b. 1   c. 2   d. 5   e. 10

7. By how many orders of magnitude does the gas giant planet Jupiter ($L_{J} \approx 10^{8}$ meters) exceed the size of human beings ($L_{HB} \approx 10^{0}$ meters)?
    a. 2   b. 4   c. 6   d. 8   e. 10

8. How many orders of magnitude are there between the diameter of the Sun ($L_{Sun} \approx 10^{9}$ meters) and the typical distance between stars ($L \approx 10^{16}$ meters)?
    a. 7   b. 10   c. 15   d. 20   e. 100

9. How many powers of 10 smaller than a human being ($L_{HB} \approx 10^{0}$ meters) is a typical grain of dust ($L_{DG} \approx 10^{-6}$ meters)?
    a. 2   b. 3   c. 6   d. 9   e. 15

10. To build a monument with a corridor that would catch the rays of light from the rising Sun on the winter solstice, the builders of Newgrange had to point the opening of their corridor in which general direction?
    a. northeast   d. southwest
    b. southeast   e. The direction didn't matter.
    c. northwest

11. Compared to the size of a galaxy cluster ($L \approx 10^{23}$ meters), a supercluster ($L \approx 10^{24}$ meters) is _____. Choose all that apply.
    a. twice as large
    b. 1 order of magnitude larger
    c. 10 times as large
    d. 24 times larger
    e. 1 power of 10 larger

12. Which of the following describe(s) areas where informed citizens need to understand how science works? Choose all that apply.
    a. Understanding the arguments for and against the need for human intervention in climate control.
    b. Weighing the positions for and against hydraulic fracking for obtaining natural gas.
    c. Discerning the accuracy of statements regarding the depletion of natural resources.
    d. Deciding whether children should be inoculated against various diseases
    e. Deciding which album by Led Zeppelin was the band's best.

13. How many generations of humans ("grandmas") were there between 1900 and 2000?
    a. 10   b. 4   c. 3   d. 5   e. 2

14. Comparing the size of a human ($L_{HB} \approx 10^{0}$ meters) to the size of an atom ($L \approx 10^{-10}$ meters),
    a. the human is 9 times bigger.
    b. the human is 10 times bigger.
    c. the atom is 9 orders of magnitude smaller.
    d. the human is a billion times bigger.
    e. the atom is 10 powers of 10 smaller.

15. One of the earliest practical uses of astronomy was the timing of crop planting by
    a. lunar eclipses.
    b. the appearance of comets.
    c. the appearance of specific planets.
    d. the appearance of specific constellations.
    e. solar eclipses.

16. Which of the following characteristics of astrology is/are consistent with defining it as a pseudoscience? Choose all that apply.
    a. Its advocates take a perceived correlation and claim that it shows a causal relationship.
    b. Its proponents would not give up their beliefs, even if experiments showed that those beliefs were incorrect.
    c. Its advocates assert that there are underlying principles but fail to rigorously test their validity.
    d. Its hypotheses have not been supported by experimentation.
    e. Its predictions are so broad that they fit almost any possible outcome.

17. Evolution of life on Earth has been a controversial topic, despite being well supported by scientific investigation and the resulting evidence. Which of the following is/are reasonable conclusions to be drawn from this example about science and religion in our culture? Choose all that apply.
    a. Religion and science cannot coexist.
    b. Religion may influence some people's acceptance of scientific conclusions.
    c. Scientific ideas and religious ideas have rarely conflicted over time.
    d. Some religious constituencies may refute the validity of scientific conclusions that are inconsistent with their teachings.
    e. Scientists cannot hold religious beliefs, since those beliefs are not based on the process of science.

18. In the simple version of the scientific method, which of the following is the correct order of actions that scientists take in their studies?
    a. theory, observation, hypothesis, test, results
    b. hypothesis, test, results, retest, observation, potential theory
    c. observation, hypothesis, test, results, retest, results, potential theory
    d. theory, test, results, hypothesis, revised theory
    e. observation, theory, hypothesis, test, results, retest, results

19. True/False: Religious accounts provide supernatural as opposed to natural explanations of the history of life on Earth.

20. True/False: Understanding the natural world through the application of scientific knowledge can give us a better understanding of our place in the Universe.

To the Point: Qualitative and Discussion Questions

21. Scientific knowledge is said to be simultaneously "tentative and reliable." Describe how this paradox is a strength of scientific knowledge, not a weakness.

22. The progress of modern science often is not linear, as may be suggested by the scientific method. What aspects of modern science do not fit neatly into the scientific method?

23. Pseudoscience is often intended to sound and look a lot like science. In what ways do pseudoscientific understandings differ from scientific ones?

24. A recent study by the National Science Foundation found that fewer than 25 percent of Americans understand the true nature of scientific knowledge. How would you describe the process of acquiring scientific knowledge to someone in the other 75 percent?

25. What are the criteria for a successful scientific theory?

26. What are some of the ways science has changed the world? Do you see these as benefits or challenges?

27. Name several of the toughest challenges we face today as a result of technological advancement.

28. What types of questions ultimately cannot be answered through scientific investigation?

29. How do astronomy and astrology differ?

30. What are some practical applications for early astronomical knowledge of the cycles of the Sun, Moon, and constellations?

31. At Newgrange, the Sun lights up a passageway exactly on the winter solstice. Judging by what you already know from experience, what is a solstice?

32. Recall what you have heard or read about climate change. Consider the source. What purpose or agenda might the speaker or writer have had? Can you cite examples that you judge to be based on science and others based on pseudoscience?

33. Humankind has been fascinated by the mysteries of the night sky since its earliest beginnings. Given our modern understanding of the Universe and our position within it, what thoughts, ideas, and questions come to mind when you look up at a starry sky?

34. Most world religions provide an account of the creation of Earth and the cosmos. Is a belief in such stories necessarily incompatible with an acceptance of the findings of astronomers and other scientists? Why or why not?

35. Do you agree or disagree with Martin Rees's statement that humanity has only 50/50 odds of surviving into the next century? Why or why not?

Going Further: Quantitative Questions

36. Humanity has existed in its culturally and anatomically modern form for approximately 50,000 years. Write this number in scientific notation.

37. How many orders of magnitude are there between the cost of the candy bar and the luxury-car rental in the example given in the text?

38. Using the values given in the text, estimate how many times larger Jupiter is than Earth.

39. How many orders of magnitude are there between the sizes of a dust particle and a proton?

40. How many years would 1,500 generations represent, if each generation was 25 years? Give your answer in scientific notation.

41. Which is larger, the number of powers of 10 between the sizes of a planetary system and a galaxy, or those between an atom and a proton?

42. What proportion of the time since *Homo sapiens* appeared on Earth have we been culturally and anatomically modern? Give your answer as a ratio.

43. Using the birth dates of you, your parents, and your grandparents (or those of a friend, if the information about your family is not available), what is the average span of a generation in your family? How many generations of that length would take you back to the first culturally and anatomically modern humans?

44. Write a million billion billion in scientific notation.

45. As of December 31, 2015, how many seconds (will) have passed since the end of the year that Copernicus (see Chapter 3) published his Sun-centered model of the Universe (1543)? (Hint: There are 3.16×10^7 seconds in a year.)

 If your instructor assigns homework in **smartwork5**, access it at the Digital Landing Page for *At Play in the Cosmos*: **digital.wwnorton.com/cosmos**

A UNIVERSE MADE,
A UNIVERSE DISCOVERED

The Night Sky and the Dawn of Astronomy

In this chapter you will learn about the basic features and motions visible on the sky and humanity's early attempts to understand them. After reading through each section, you should be able to do the following:

2.1 An Old Obsession
Explain how today's astronomical study evolved from the millennia-long history of humanity's fascination with the cosmos.

2.2 Dance of Night and Day: Basic Motions of the Sky
Describe the observed daily and yearly motions of the Sun and stars and explain why we see what we do.

2.3 Monthly Changes of the Moon
Describe the Moon's phases and their relation to the Moon's orbital position.

2.4 Celestial Wanderers: The Motion of Planets
Explain why the planets look and appear to move the way they do.

2.5 Stone and Myth: Astronomy before History
Explain the relationship between science and myth in humanity's pursuit of knowledge of our Universe and how it laid the groundwork for future discoveries.

2.6 The Greek Invention of Science
Summarize the Greeks' role in establishing rational explanations based on mathematical models for physical phenomena.

This photo, taken by leaving the camera shutter open for several hours, shows how Earth's rotation creates the appearance of stars revolving around the north celestial pole, which is located near the star Polaris (the "North Star").

2

2.1
An Old Obsession

CAROL LATTA never took any astronomy graduate courses. Instead, Latta got an MBA and spent most of her career working for Xerox, then the largest maker of copy machines in the world. "I was a product launch manager, among other things," she says. "It was a rewarding career," she says, "but it wasn't my passion."

Now Latta knows exactly where her passion lies—in the stars. On any given Friday night, you can find her, the current president of the Rochester Academy of Science, Astronomy Section, out under the dark starlit skies of upstate New York with her beloved telescopes. "When I found astronomy, I was in my mid-50s," she says. "It was like a lightbulb went on, and I knew, 'This is it for me.' I've found that I like to do observing with binoculars or a telescope." (Many are available at the academy's privately owned 17-acre observatory site.) "But what I *love* is doing outreach, showing others how to find the spiral arms of a spectacular galaxy like M64, showing them the bright jewel of the globular cluster M13 or perhaps the glow of a star-forming nebula like [M42] in Orion [**Figure 2.1**]. Besides observing and

outreach, my other favorite activity is using my imagination to envision some of the cosmic events I read about. I can't think of anything else that comes close to this," says Latta, describing her love affair with astronomy. "Nothing else I have found has the same richness in it."

No one pays amateur astronomers like Latta to spend hours building telescope domes, giving presentations to grade school kids, or holding public "star parties" where everyone gets a chance to look through a telescope. These people do what they do because they have become enchanted by the sky and its mystery. They are men and women who are just the latest in a long line of people obsessed with the stars. It's a line that stretches back 50,000 years or more into the dim origins of human culture.

In this chapter we will explore the basic motions of Sun, Moon, stars, and planets visible to anyone who takes the time to look. We will also unpack some of humanity's earliest attempts to understand the Universe and our place within it. Looking back is an essential step in understanding the modern story of astronomy and the development of the enormously powerful tool we call science. In these first attempts to make sense of the world, we can see the seeds of our own efforts today.

FIGURE 2.1
Deep-Sky Objects
Favorite targets for amateur astronomers' telescopes include galaxies, star clusters, and nebulas such as these, known as (A) M64, (B) M13, and (C) M42, respectively. They are called deep-sky objects because they lie far beyond our Solar System.

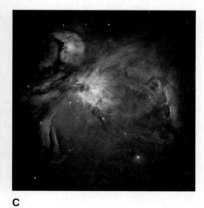

A

B

C

2.2
Dance of Night and Day: Basic Motions of the Sky

It is a fact of modern life that most of us have never seen the night sky in its full splendor: the radical invention of artificial illumination more than 100 years ago erased the darkness of night for most of the world's inhabitants. If we encounter a dark sky, it's usually for just a few nights on a camping trip (**Figure 2.2**), and even then we generally don't have enough time to become familiar with the night sky and its motions. For 99.9 percent of human evolution, however, our ancestors lived in a world without artificial light and couldn't help but notice what happened on the sky.

"Every day the Sun looks like it's moving from east to west," says Carol Latta. "Every night the stars look like they are doing the same. The daily motion of the Sun and stars is the most fundamental astronomical fact and the most basic illusion." We now know, of course, that it's Earth that moves, not the sky. "It certainly seems like Earth is stationary," says Latta. "It doesn't feel like we are moving, so you can understand

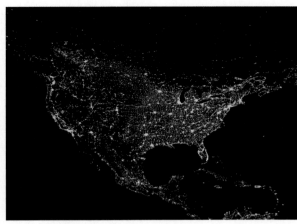

FIGURE 2.2 Light Pollution
(A) Light pollution (seen clearly in this satellite image) makes it impossible to see the skies as our ancestors saw them, but sometimes (B) a camping trip to a remote location can provide a sense of the seemingly countless stars in the night sky.

why the ancients thought it was the stars and the Sun that were in motion. But, of course, what's happening is simply that Earth is rotating." It was not until around the 1600s that the **rotation** of Earth—the planet's spin on its axis—was universally accepted. The story related here, the one we've now learned to be true, was hard won over thousands of years and reflects a triumph of science and human intelligence.

Daily Motions and the Celestial Sphere

Each morning, at a time that varies as the seasons progress, the Sun appears to rise above the eastern **horizon** (the boundary between land and sky from an observer's perspective). As the day winds on, the Sun

climbs until it reaches its highest **altitude**, or height above the horizon, at noon. It then descends toward the west, eventually setting by dropping below the western horizon. As the sky darkens, the stars appear as patterns of twinkling points of light whose positions relative to one another do not appear to change. As night progresses, stars also make a trek across the sky similar to the Sun's during the day, rising above the eastern horizon and setting into the western horizon (**Figure 2.3**).

THE CELESTIAL SPHERE. The exact path that the Sun and stars take through the sky relative to the horizon in their east-to-west daily motion depends on where on Earth you're standing and what season it is. More specifically, the Sun's and stars' paths depends on your **latitude**—your distance north or south as measured from the equator. (Your **longitude** on Earth, which measures your position from east to west, referenced from a line running through Greenwich, England, does not affect the path of the Sun and stars on the sky.) Since Earth is a sphere, we can measure distances from one place to another on it in terms of an angle or **angular size**, given in degrees, minutes, and seconds of arc. Here are a few examples of latitudes for specific locations on the planet: The North

rotation The spin of an object around an internal axis.

horizon The horizontal line defining the lower edge of the sky.

altitude The angular distance from a point on the sky to a point on the horizon directly below it.

latitude A measure of location specifying north-south position on Earth running from 0° at the equator to 90° north (North Pole) and 90° south (South Pole).

longitude A measure of location specifying east-west position on Earth.

angular size Angular distance measured on a sphere (such as Earth or the celestial sphere) in degrees, arcminutes, and arcseconds.

interactive
Movements of the Sky

A Daily path of the Sun

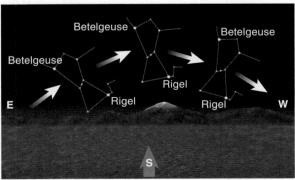

B Nightly path of the stars

FIGURE 2.3
Daily Path of Sun and Stars
(A) From northern latitudes, the Sun appears to rise in the east, reach its highest point in the south, and set in the west. (B) Stars also rise in the east, move across the sky, and set in the west. Here, we see the constellation Orion, which includes the stars Betelgeuse and Rigel, following that path over the course of a night.

Pole has a latitude of 90 degrees (90°) north; the South Pole has a latitude of 90° south; New York City's latitude is 40 degrees 43 minutes (40° 43′) north; and Sydney, Australia, has a latitude of 30° 51′ south.

Your latitude and the Sun's motion on the sky are related because of the *line of sight*, the line connecting your eyes to the Sun. Suppose you're standing somewhere on the equator (such as Quito, the capital of Ecuador) at noon. For you, at Quito, the Sun will take a path that brings it much higher on the sky at noon than it would be if you were someplace farther north, such as Seattle, Washington (**Figure 2.4**). Because Seattle is well north of Quito, the line of sight to the Sun in Seattle will always point more to the south than it will in Quito, and therefore the Sun will appear lower on the sky and closer to the southern horizon.

Before astronomers could understand the reason for the difference in the Sun's position at different latitudes, they needed a way to map the positions of objects like the Sun and stars on the sky. To do so, they invented the concept of the celestial sphere. The **celestial sphere** is like an imaginary transparent globe that surrounds Earth with the positions of stars and planets painted on the inside. Earth's equator, poles, and lines of latitude and longitude are extended out to this imaginary globe and act as reference points. "Take Earth's equator and extend it out into space. We call that line the **celestial equator**," explains Latta. "Take the points that mark Earth's North and South poles and extend them out into space. We call those points the celestial poles." Every day Earth rotates once on its axis. That means every day the Sun and stars appear to turn once around the celestial poles. "The star **Polaris** is very close to the **north celestial pole**," explains Latta, "so it looks like all the stars visible in the Northern Hemisphere turn around Polaris. That's why it's called the North Star. There is no star near the **south celestial pole**, so if you were in the Southern Hemisphere, you would see all the stars spinning around a relatively empty patch of sky at the southern celestial pole each night" (**Figure 2.5**).

"No matter where you are on Earth, only one-half of the celestial sphere is available to you at any given moment," explains Latta. Your location in latitude on Earth determines which part of the celestial sphere

celestial sphere An imaginary transparent sphere surrounding Earth, on which positions of stars, planets, and other celestial bodies are projected from extensions of Earth's coordinate system.

celestial equator The extension of Earth's equator onto the celestial sphere.

Polaris The star that is found very near Earth's celestial north pole (hence called the North Star) and around which the sky of the Northern Hemisphere appears to turn.

north celestial pole The extension of Earth's North Pole onto the celestial sphere.

south celestial pole The extension of Earth's South Pole onto the celestial sphere.

FIGURE 2.4
The Sun's Position at Different Latitudes
For most of the year, the Sun's position at noon, as seen from (A) Quito, on the equator, at latitude 0°, will be much higher from the observer's horizon than the position of the Sun seen at the same time from (B) Seattle, at latitude 47° north.

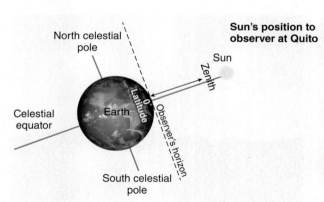

A Quito

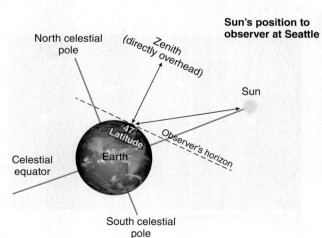

B Seattle

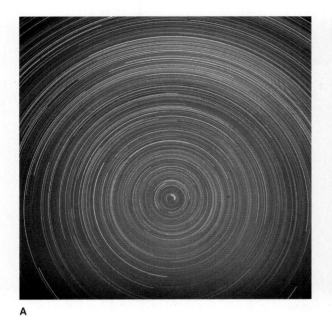

A

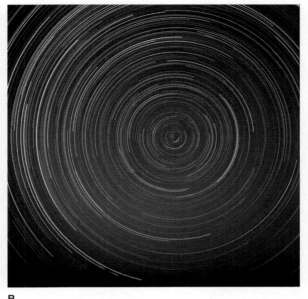

B

you can see. "Imagine extending the horizon around you out to infinity. You can only see above where the horizon line intersects the celestial sphere, and that determines which half of the sky you can see day or night [**Figure 2.6**]. If you were standing exactly at the North Pole, you would see only the northern portion of the celestial sphere. If you were standing at the South Pole, you would only see the southern half of the celestial sphere."

From where you're standing at any moment, the **zenith** is the point on the celestial sphere directly above your head. The **meridian** is an imaginary line that runs from the southern horizon, through the zenith, to the northern horizon. Thus, if you were standing at Earth's North Pole, your zenith would be the north celestial pole. If you were standing in a field outside of Chicago, your zenith and the north celestial pole would no longer overlap. How far apart would they be? Since Chicago is located at a latitude of about 42° north and the North Pole is at 90° north, the difference in angle between your zenith and the north celestial pole would be 90° − 42° = 48°. Another way to say this is that if you started at your northern horizon and then moved your gaze 42° upward, your line of sight would hit the north celestial pole (and the North Star).

What about Quito on the equator? Since Quito lies at 0°, the difference between the zenith and the north celestial pole is 90° − 0° = 90°. The north celestial pole will lie just at the northern horizon. The north-south position of an object on the celestial sphere is referred to as its **declination,** and it is measured similarly to latitude on Earth.

The part of the celestial sphere that you can see at any moment also depends on the time of day, as Earth swings around in its daily rotation. At noon you can see the half of the celestial sphere with the Sun at its highest position. Twelve hours later, at midnight, Earth has swung around 180°, halfway through its daily rotation. Now you are facing directly away from the Sun and seeing the opposite half of the celestial sphere.

Note that we can define a "day" in two ways using the celestial sphere. First is the **sidereal day**, the time Earth takes to rotate on its axis relative to the stars—meaning the time it takes for a particular point on the celestial sphere (a star's location, for instance) to cross

zenith The point on the sky directly overhead from an observer.

meridian A circular arc crossing the celestial sphere and passing through the local zenith and celestial pole.

declination The position of an object on the celestial sphere, measured similarly to latitude on Earth.

sidereal day The time taken by Earth to rotate on its axis relative to the stars.

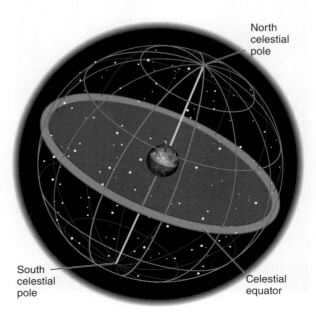

North celestial pole

South celestial pole

Celestial equator

FIGURE 2.6
The Celestial Sphere
The portion of the celestial sphere visible depends on the observer's latitude and includes one-half of the sky.

the same celestial meridian. Then there's the **solar day**, which is the time between successive crossings of the same celestial meridian on the sky by the Sun. Because the Sun appears to move against the fixed background of stars across the celestial sphere, these two measurements of the day will be slightly different. The solar day is 24 "solar" hours (by definition), but the sidereal day is only 23 hours and 56 minutes long.

APPARENT SIZE AND THE SMALL-ANGLE FORMULA.

Because the sky *appears* to us as an upturned bowl, a hemisphere, the distance between objects on the sky can also be measured in terms of angular size. The hemisphere of the sky extends through 180° of arc. The Moon, as it appears from Earth, has a width of about 0.5° of arc, and your pinky finger held at arm's length also takes up about 0.5° (**Figure 2.7**). (For a more familiar, down-to-earth example of angular measure, the monitor on a typical computer spans about 30° of arc in your field of view.)

While we're thinking about angles and the night sky, it's worth taking a moment to consider the relationship between how large something appears on the sky (which would be measured as an angle) and how large it actually is if you measure it with a ruler (sometimes called its linear size, measured in meters or kilometers). The relation between the two for astronomical applications is neatly expressed in something called the **small-angle formula**, discussed in **Going Further 2.1**.

solar day The time between successive crossings of the same meridian on the sky by the Sun.

small-angle formula A mathematical relation between the angular size of an object as measured on the sky and the object's distance and physical size.

arcminute A measure of angular size. There are 60 arcminutes in a degree.

Yearly Motions of the Sun

Imagine it's a hot summer day in a city in the northern United States. The Sun has been up since early morning, and it hangs high and oppressive on the sky.

Now imagine it's the middle of winter in the same city. The day is cold, short, and forlorn. People hustle across the windy street trying to fight off the chill. It's noon and the Sun hangs low in the southern half of the sky. Its slanting rays seem weak and ineffective.

These two paragraphs describe something you've probably noticed: measured at the same time of day, the Sun is higher on the sky during summer days than it is in winter. Technically, "higher" means closer to the zenith. This difference shows that along with the daily motion of the Sun and stars, there are yearly motions, which affect the lengths of days (the amount of daylight) and seasons.

FIGURE 2.7 Angular Size
Angular size is measured as the number of degrees of arc that an object spans in the field of view of the observer. (A) The Sun varies a little in apparent size because of its orbital motion but generally appears to span 0.5°, about the width of a pinky finger held at arm's length. (B) A computer monitor typically measures 30° of arc to the user seated in front of it.

A

30°

B

CHANGES IN THE SUN'S POSITION THROUGH THE YEAR. Imagine that each day at noon you put a thumbtack on a large map of the sky marking the Sun's position that day. After an entire year you would find that the thumbtacks marked out a thin figure eight on the sky (**Figure 2.8**). For someone living in the Northern Hemisphere, the thumbtack for December 21 would be very close to the lowest point of the figure eight, meaning that the Sun was closer to the southern horizon at noon on that day than on any other day of the year. The thumbtack placed 6 months later, on June 21, would be very close to the highest point on the figure eight. This was the day when the noontime Sun was closest to the zenith. On March 21 and September 21, the thumbtacks would be close to the middle of the figure eight. This cyclical motion of the Sun around the

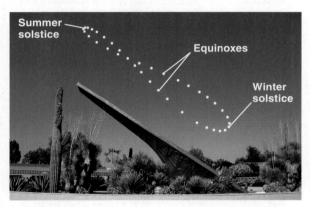

A

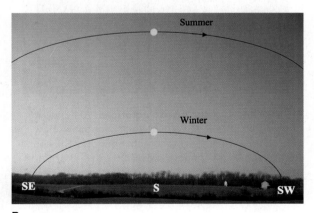

B

FIGURE 2.8 The Analemma
(A) If you were to take a picture of the Sun's position every day at the same time, the Sun's location would trace out a figure 8, called the analemma. Marking key events of the astronomical year on the analemma shows its relation to the seasons and to Earth's motion around the Sun. (B) Note also the seasonal altitude of the Sun apparent in this image. The Sun's altitude on the sky is higher in summer, lower in winter.

GOING FURTHER 2.1 THE LANGUAGE OF THE COSMOS
The Small-Angle Formula

You were probably a kid when you first noticed that big things look small when they're far away, and small things look big if you hold them right up to your eye. The ancient Greeks noticed this too, and they used geometry to discover how an object's apparent size, true size, and distance are related. When things are far enough away, the relationship can be determined by what is called the small-angle formula, which can be written as

$$a = 206{,}265'' \times d/D$$

In this formula, d refers to the true size of an object, and D refers to the object's distance from us. The small-angle formula requires the distance D and the physical size d to be expressed in the same units of length (meters or kilometers, for example). The term a refers to the object's apparent size—that is, how much of our field of view it appears to take up. We can measure this as an angle (see Figure 2.7) in units of degrees (°). For smaller apparent sizes, we subdivide each degree into 60 **arcminutes** (abbreviated 60′) and divide each arcminute into 60 *arcseconds* (abbreviated 60″), so that there are 3,600 arcseconds (60 × 60) in a degree. The small-angle formula is usually expressed in terms of arcseconds (″). The value 206,265 comes from converting radians (an angular size unit) into arcseconds.

Now let's apply this formula to some real astronomy. The Sun has a diameter or linear size of 1.39×10^6 kilometers (km). That would be its true size, d. We orbit the Sun at a distance of 1.49×10^8 km. That would be the distance, D. How large on the sky does the Sun appear to us? Using the small-angle formula, we would calculate as follows:

$$a = 206{,}265'' \times (1.39 \times 10^6 \text{ km})/(1.49 \times 10^8 \text{ km}) = 1{,}924''$$

We would then convert this number into degrees:

$$a = 1{,}924'' \times 1°/3{,}600'' = 0.5°$$

The Sun takes up about one-half of a degree on the sky, so its apparent size is about the same as your pinky finger held up at arm's length.

How large would the Sun appear to us if we were on the planet Neptune, increasing our distance from the Sun by 30 times, to $D = 4.5 \times 10^9$ km? Again, we would calculate

$$a = 206{,}265'' \times (1.39 \times 10^6 \text{ km})/(4.5 \times 10^9 \text{ km}) = 64''$$

or

$$a = 64'' \times 1°/3{,}600'' = 0.02°$$

From the distance of Neptune, the Sun would look much smaller.

Notice that you can rearrange the formula: If you know the apparent size a and the distance D, you can calculate the true size d. Alternatively, if you know the apparent size a and the true size d you can find the distance D.

figure eight (called an **analemma**) each year was recognized as far back as two millennia ago and was used as the basis for creating sundials to track time.

The analemma is just one way to see how the Sun moves around the sky over the course of a year. You can also track the solar motion by noticing where the Sun rises and sets each day. The path mapped out by the analemma—the Sun's noontime positions over the course of a year—reflects changes in where on the horizon the Sun rises and sets (**Figure 2.9**). If you were awake before sunrise each day (and your farmer ancestors certainly were), in the Northern Hemisphere you would notice that the position of the Sun as it peeks over the horizon moves progressively northward from December to June. On about June 21 (the **summer solstice**, the day of the year with the most sunlight), it stops its northward progression. From that point until about December 21 (the **winter solstice**, the day with the least sunlight), the Sun rises progressively southward. The autumn and spring **equinoxes** (September 21 and March 21) mark the midpoint of this yearly journey. On these two days the amount of daylight and darkness are roughly equal.

"The length of the day (meaning the amount of daylight) changes. That's the one thing everyone notices about the seasons, even those who live in places where the temperature is fairly constant all year," says Latta. "The change in the length of the day and the changes in the path of the Sun through the sky are both a result of the yearly orbital journey of Earth around the Sun and the tilt of Earth's axis of rotation relative to that orbit."

EXPLAINING THE SUN'S YEARLY MOTIONS AND THE SEASONS. Earth's **orbit**, its journey around the Sun, defines the year. "It's a 365-and-¼-day trip," explains Latta. Recall from Chapter 1 that the average distance of Earth from the Sun is about 1.49×10^8 km, a distance astronomers call an astronomical unit (AU). The shape of the orbit is not perfectly circular but takes the form of a slightly squashed circle called an **ellipse**. An elliptical orbit means that Earth swings a little closer and then a little farther from the Sun as it completes its yearly orbit. "One of the most common misconceptions people have is that summers are hotter than winters because Earth is closer to the Sun in summer," says Latta. "Not only is that wrong, because Earth is actually closer to the Sun when the northern latitudes are in winter; it doesn't even explain why the length of the day changes or why the Sun is higher in the sky in summer. The real culprit is Earth's tilt."

"If you love the seasons," she continues, "then you have to be thankful for one of those lucky coincidences

analemma The figure-eight pattern formed by charting the daily position of the Sun at noon over the course of the year.

summer solstice The day (approximately June 21 in the Northern Hemisphere) with the most hours of sunlight, when the Sun appears highest in the sky.

winter solstice The day (approximately December 21 in the Northern Hemisphere) with the fewest hours of daylight, when the Sun appears lowest in the sky.

equinox A day when the hours of daylight and darkness are equal. The autumn equinox occurs on approximately September 21; the spring equinox, approximately March 21.

orbit The cyclical path in space that one object makes around another object.

ellipse A geometric form in the shape of an oval or circle, in which the sum of the distances from two points (foci) to every point on the ellipse is constant.

obliquity or **axial tilt** The angle between a planet's spin axis and a line perpendicular to the planet's orbital plane.

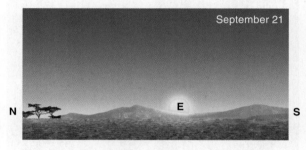

FIGURE 2.9 The Sun's Yearly Motions The direction of sunrise on the solstices and equinoxes shows how the Sun's daily path across the sky changes over the course of the year. These changes are all determined by Earth's axial tilt relative to the plane of its orbit.

that our planet happens to rotate at an angle of 23.5° to a line that's perpendicular to the plane of its orbit" (**Figure 2.10**). In other words, Earth's rotation axis is *inclined* with respect to its orbit, and the angle between the perpendicular to the orbital plane and the spin axis is called its **obliquity** (this is the same as the angle between the celestial equator and the orbital plane). The inclination and its direction with respect to the fixed stars do not change as the year progresses; the North Star is the

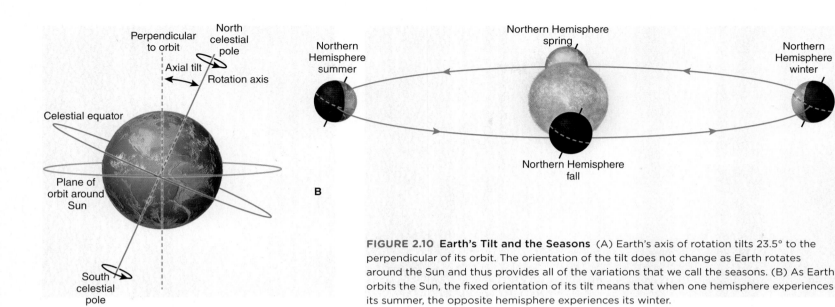

A

B

FIGURE 2.10 Earth's Tilt and the Seasons (A) Earth's axis of rotation tilts 23.5° to the perpendicular of its orbit. The orientation of the tilt does not change as Earth rotates around the Sun and thus provides all of the variations that we call the seasons. (B) As Earth orbits the Sun, the fixed orientation of its tilt means that when one hemisphere experiences its summer, the opposite hemisphere experiences its winter.

North Star in both winter and summer. "Since Earth's tilt always has the same orientation," says Latta, "the Sun appears to change its position on the sky as the year progresses. It will be low on the sky in northern winter when the Northern Hemisphere tilts away from the Sun. Six months later, the Sun will appear high in the summer sky (in the north) because the Northern Hemisphere is now tilted toward the Sun."

Earth's spin axis wobbles slightly, however, moving a little more than 1° per century. This phenomenon is known as **precession**. While the motion is small, a little more than 1 degree per century, the north celestial pole has moved enough that the Egyptians would not have recognized Polaris as the North Star.

The tilt of Earth's rotation axis, relative to its orbit around the Sun, is all you need to explain both the Sun's yearly motion through the sky and the seasons. For example, winter occurs in the Northern Hemisphere when the axis of rotation tilts away from the Sun, so the Sun's rays hit Earth at a shallow angle, causing their energy to be spread over a larger area (**Figure 2.11**). In summer, when Earth is tilted toward the Sun, the opposite effect takes place, and the Sun's rays strike the surface at a steeper angle, concentrating more of its energy in a smaller surface area.

Earth's tilt also explains the changing length of the days. There is less daylight during winter because the tilt of the rotation axis means that the Sun rises later and sets earlier, making the winter day shorter. In summer, the Sun rises earlier and sets later, making the summer day longer than the winter day. A longer day,

A Winter

B Summer

FIGURE 2.11
Angle of the Sun's Rays
The Sun's rays hit the ground more obliquely in winter and more directly in summer. Thus, the same amount of solar energy is spread over a larger area in winter, so the Sun's rays provide less heat. (A) At noon in winter, the angle between a tall pole and the Sun's rays is large, so the pole casts a long shadow. (B) At noon in summer, the Sun is almost overhead and its rays hit the ground at an angle closer to 90°, so the pole casts a short shadow.

precession The rotation of a planet's spin axis, similar to that of a wobbling top.

A

B

FIGURE 2.12
Seasonal Constellations
As Earth moves from one side of its orbit in the winter months to the other side in summer months, different constellations are visible on Earth's "night side." Some other constellations appear in the sky primarily in the summer months while winter constellations appear primarily in the winter months. In the Northern Hemisphere, (A) Sagittarius and Scorpius adorn the summer night sky, (B) while the constellations centered around Orion dominate the winter night sky.

constellation A group of stars that form a pattern, and the designated region of the sky surrounding them.

circumpolar constellation A constellation that, from the viewer's perspective, never rises or sets.

ecliptic The path that the Sun appears to follow against the background stars, as defined by Earth's orbit around the Sun.

zodiac The 13 constellations that lie close to the plane of the ecliptic, which the Sun appears to pass through over the course of the year.

with sunlight striking the ground at a steeper angle, leads to warm summers. A shorter day, with sunlight striking the ground at a shallower angle, leads to cold winters. Near the equator, however, Earth's rotational tilt has less effect. The Sun always sits high on the sky at noon, the amount of daylight does not change dramatically, and therefore seasonal changes are minimal.

Earth's tilt and its relation to seasonal changes in the position of the Sun on the sky pose a conceptual challenge for some people. Their difficulty often comes in translating an Earth-based view of the sky into a view of the Earth-Sun system as a whole.

SECTION SUMMARY
Yearly Motions of the Sun

- The Sun's daily position is higher (closer to the zenith) on the sky in summer than in winter.
- There are two ways to track the changes in the Sun's position over its annual cycle: by noting the position of sunrises and sunsets on the horizon, and by noting the highest altitude achieved by the Sun each day.
- The change in seasons is due to the tilt of Earth's axis in relation to the line perpendicular to its orbit (23.5°). This affects the number of hours of sunlight and how directly that light hits Earth.

Yearly Motions of the Stars

Like the Sun, the stars also change their positions on the sky throughout the year. To our ancestors who lived their whole lives without artificial light, the drama of the nighttime sky was the only show in town. One of the most obvious changes over the course of the year is the shifting positions of constellations as

the year progresses. **Constellations** are groups of stars that form familiar patterns. In modern astronomy, the sky is divided into 88 zones based on constellations, and astronomers refer to objects by their position within one of these constellation zones ("the star-forming clouds in Taurus," for example).

For our ancestors, however, what mattered was that not all constellations were visible all year long. Some constellations could be seen only in the winter or only in the summer. The stories that early peoples imagined about each constellation often reflected the season when it appeared (**Figure 2.12**).

There are also the **circumpolar constellations**, which are visible all year long and do not rise and set each night, because they are located close to the celestial pole. Ursa Minor (the "Little Bear") is a constellation that includes the North Star, so it is one circumpolar constellation in the Northern Hemisphere.

Of particular importance were the constellations that appeared along the path that the Sun charted on the sky. This may sound strange because, of course, no stars are visible when the Sun is up. But people could make note of the position of the Sun on the sky at a certain time each day (noon, for example). By doing so, they found that the Sun made a line across the background of the stars on the celestial sphere as the year progressed. That line is called the **ecliptic**, and the constellations of the **zodiac** lie in a band about 9° wide on either side of the ecliptic (**Figure 2.13**). As the year progresses, the Sun appears to march eastward along the ecliptic, and the constellations of the zodiac progressively become visible at night. (For those who follow astrology, a person's date of birth indicates an astrological sign that corresponds to one of these 12 constellations of the zodiac. Because of the way astronomers define constellations in terms of regions of the sky, there are actually 13 constellations in the Zodiac. Ophiuchus is the 13th zodiacal constellation.)

If we think of the year beginning in January (which is arbitrary), you can imagine Earth beginning on one side of its orbital track around the Sun. At night, the view of the sky faces the direction opposite the line linking the Earth and the Sun. That view of the celestial sphere determines which stars (and which constellations) are visible. As Earth moves around the Sun in its orbit and the year progresses, the portion of the celestial sphere we face at night continually shifts. "It's this changing view of the celestial sphere as Earth swings around in its orbit that changes the constellations you can see," explains Latta. "Amateur astronomers like me get excited as the year unfolds because it means our favorite objects lying in different constellations start appearing in the night sky for us to observe."

For example, Taurus is a constellation that appears at night in the northern sky in winter. That means the night side of Earth faces in the direction of Taurus in January. In other words, the half of the celestial sphere that is visible at night in January includes Taurus. Sagittarius is a constellation that is visible during the half of the year stretching from spring through summer and into fall, so the night side of Earth faces in the direction of Sagittarius in June. In other words, the half of the celestial sphere that is visible at night in June includes Sagittarius. Why, then, is Sagittarius a winter astrological sign? The Sun was thought to be in the "house" of Sagittarius during specific winter months, because the position of the Sun on the celestial sphere in the winter crosses through the constellation Sagittarius.

Since Earth moves around the Sun, and not the other way around, it is actually Earth's motion that defines the plane of the ecliptic on the celestial sphere. The Sun appears to chart a course against the background of fixed stars because the line of sight from Earth to the Sun points to different background stars as the year rolls on and Earth spins around in its orbit.

SECTION SUMMARY
Yearly Motions of the Stars
- Because of Earth's orbital motion, the apparent position of the Sun against the background stars changes over the course of the year.
- Some constellations are visible only during certain seasons.
- The Sun's annual path against the background of the stars is called the ecliptic.

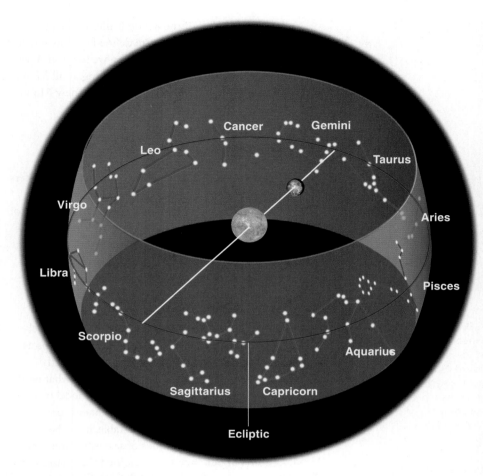

FIGURE 2.13 The Ecliptic and Zodiac
The ecliptic is the path that the Sun appears to follow through the sky over the course of the year, and the zodiac represents the constellations through which it passes. Astronomers also include a 13th constellation, Ophiuchus, although it does not appear in most traditional versions of the zodiac.

2.3
Monthly Changes of the Moon

The change of the Moon's phases poses the same challenge of moving from what we see from the ground to the view of the Earth-Moon-Sun system from space. The monthly round of lunar phases made such an enormous impression on early cultures that the Moon became a principal actor in the mythologies of almost every culture on the planet (**Figure 2.14**). In later ages, it became a primary driver of astronomical science.

The Moon rises and sets each day just as the Sun and stars do. Unlike the Sun, however, the Moon can be "up" during either the day or the night. This fact is directly related to changes in the Moon's appearance

**FIGURE 2.14
Ever-Changing View of the Moon**
Orbiting Earth, the Moon presents an ever-changing appearance to viewers on Earth, although these changes repeat in cycles. Here, we see a full Moon rising above the horizon.

called **phases**. Half of the Moon is illuminated by the Sun at all times. The phases are defined by how much of the illuminated half of the Moon we see from Earth. We see the entire illuminated half during a full Moon, and none of the illuminated half during a new Moon.

A Scientific Model for the Moon's Phases

The explanation for the Moon's motion and cycles offers a wonderful example of a **scientific model**. In any science, including astronomy, a model is an idea or set of ideas that enables people to create testable explanations of how they think a particular aspect of the world works. The world presents us with phenomena such as the monthly cycle of the Moon and its phases. In response, we try to imagine a changing configuration of bodies such as Earth, Moon, and Sun that enables us to "recover" what we observe—meaning that the model accounts for our observations. A model often helps us understand how something we cannot see directly (Earth, Moon, Sun, and their orbits) creates what we do see (the phases of the Moon). A testable model makes predictions that we can verify by experiments or observations. When a model is testable, it can be incorporated into the methods of scientific investigation.

Some of our ancestors imagined the phases of the Moon as cycles of victory and defeat of two warring gods, but we now know that the Moon is a spherical body orbiting counterclockwise (as viewed looking down at the North Pole) around a counterclockwise-spinning Earth on an approximately 30-day orbit and that the entire Earth-Moon system orbits the Sun once a year. We know this is true because we have tested and confirmed this model with everything from telescopes to space probes. Earlier in history, it was just a set of ideas that had to be tested. Let's see how this model of the Earth-Moon-Sun system explains what we see from the ground.

The Moon's phases occur as a direct result of its relative positions between the Sun and viewers on Earth. The Moon shines only by reflecting sunlight. That means we can see only the side of the Moon that is directly exposed to sunlight. The half of the Moon that lies in shadow will appear dark to us. To understand how this works, go into a dark room and place a flashlight on a table to represent the Sun. Face the flashlight and hold a tennis ball out to your side. The tennis ball is the Moon in this exercise. You should see that one-half of the ball is in shadow and the other half is illuminated (note also that you're seeing only

phase (of the Moon) The appearance of the illuminated portion of the Moon as seen by an observer on Earth, which changes cyclically as the Moon orbits Earth.

scientific model An idea or set of ideas used to create testable explanations.

one half of the ball, the side facing you). This configuration is exactly what happens with the Moon in its "quarter" phases, as we'll see shortly. Now hold the ball directly in front of you between your head and the flashlight. The illuminated side of the ball is now facing away from, and you can only see the shadowed side directly. This is exactly what happens with the Moon in its "new" phase, except that the dark side of the Moon cannot be seen at all when Earth, Moon, and Sun have the same orientation as your head, the ball, and the flashlight. Keep this model in mind as we now track through the Moon's phases in detail.

> **SECTION SUMMARY**
> **A Scientific Model for the Moon's Phases**
> ● A testable scientific model describes how a phenomenon works and why it appears to us as it does; the model makes predictions that we can compare with observations.

The Moon's Phases: What We See and Why

A new Moon occurs when the Moon is between Earth and the Sun—the point in its orbit when the Moon is closest to the Sun on the sky. The new Moon is easy to explain because it occurs when the Moon's orbital position is close to the line connecting Earth and the Sun (**Figure 2.15**). During the new Moon, the lunar face illuminated by the Sun points away from Earth and us and thus is essentially invisible to us.

As the Moon moves around Earth on its orbit, it appears progressively eastward of the Sun. This means that a few days after a new Moon, you might notice the Moon appearing in the late-afternoon sky as the Sun is sinking in the west. This phase is called waxing crescent.

The waxing crescent phase culminates in what is called the first quarter Moon. The first quarter phase, during which the Moon is at its highest altitude at sunset, occurs approximately 7 days after the new Moon. The illuminated part of the Moon always curves toward the Sun (since that's where the illumination is coming from). So from new Moon to first quarter Moon, the "horns" of the crescent point away from the Sun and toward the east.

The Moon continues its journey eastward relative to the Sun, rising later and later in the day. This is called the waxing gibbous phase and culminates in the full Moon, which rises in the east just as the Sun is setting in the west. Occurring about 7 days after the first quarter Moon, the full Moon will be at its highest point in the night sky at midnight (when the Sun is at its highest point on the other side of the world). The full Moon completes the first half of the lunar cycle, when the Moon has traveled halfway around its orbit and is on the opposite side of Earth from the Sun. The Moon appears full because the illuminated half of the Moon, the half facing the Sun, is now fully visible to viewers on Earth.

As the month continues, the Moon now begins to appear over the eastern horizon closer to midnight. This is called the waning gibbous phase and culminates in the third quarter Moon 7 days after the full Moon. The third quarter Moon rises at midnight and will appear overhead when the Sun is rising. At this point, the Moon has completed three-quarters of its orbit and is on the other side of Earth from its position at first quarter.

The final quarter of the Moon's cycle occurs as the Moon moves back toward the line connecting Earth and the Sun. During this approximately 7-day period, called the waning crescent phase, more and more of the Moon's illuminated face becomes hidden from viewers on Earth. In the waning crescent phase (on the way to the next new Moon), the horns of the crescent once again point away from the Sun. When you see a sliver of the Moon on the sky as you walk to class in the morning, you're seeing it as a waning crescent.

Finally, it's worth noting that the Moon takes about 29.5 days to cycle through its phases. This period is called the **synodic month**, and it is about 2 days longer than the time it takes for the Moon to make one orbit about Earth with respect to the background of distant stars—a period called the **sidereal month**. The difference between these two definitions of the month comes about because while the Moon is completing one cycle of its orbit around Earth, the entire Earth-Moon system is moving through its own orbit about the Sun (**Figure 2.16**). After one synodic month, Earth has moved approximately one-twelfth of the way around the Sun. That means the Sun is located in a different part of the celestial sphere than it was at the beginning of the month. The difference between the synodic month and the shorter sidereal month is the time it takes for the Moon to cross that extra portion of sky and catch up to the Sun.

SECTION SUMMARY

The Moon's Phases: What We See and Why

- As the Moon makes its monthly orbit around Earth, its phase changes because of its shift in position relative to both Earth and Sun.

- The Moon phases cycle as follows: new Moon, waxing crescent, first quarter, waxing gibbous, full Moon, waning gibbous, third quarter, waning crescent.

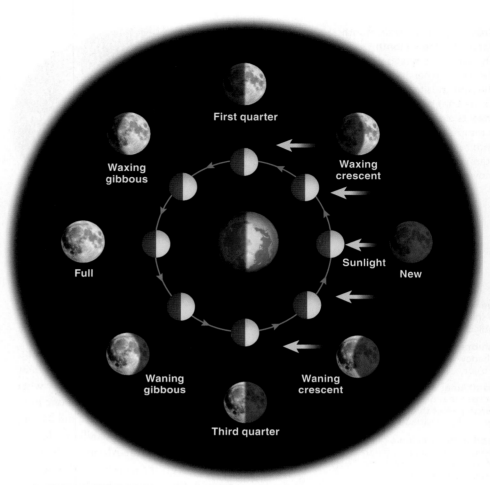

FIGURE 2.15 Phases of the Moon
The Moon's phase indicates its location relative to Earth and the Sun. One-half of the Moon's surface is always illuminated by the Sun. As the Moon orbits Earth, the proportion of the illuminated surface that can be seen from Earth changes from 0 percent (new Moon) to 100 percent (full Moon). The inner circle shows the Moon's position relative to Earth and the Sun. The outer circle shows the Moon's phase, as seen from Earth, at that point in its orbit.

Going Dark: Lunar and Solar Eclipses

interactive
Moon Phases

It is perhaps one of the most remarkable accidents in cosmic history, at least as far as human beings are concerned, that the diameter of the Sun is 400 times larger than the diameter of the Moon. Yet as seen from Earth, the disks of the Moon and the Sun have just about the same size on the sky (in other words, their angular diameters are about the same). That coincidence occurs because the Sun is not only farther away than the Moon and so appears smaller on the sky, but it is exactly far enough away that their angular diameters nearly match. From that accident of size and distance arises one of the most stunning forms of astronomical events: the solar **eclipse**. In this section we will explore the causes of solar eclipses and the related eclipses of the Moon (the lunar eclipses).

synodic month The amount of time it takes the Moon to cycle through its phases.

sidereal month The amount of time it takes the Moon to make one complete orbit in relation to the background stars.

eclipse The passing of one celestial body through the shadow of another.

FIGURE 2.16 Synodic Month versus Sidereal Month
The synodic orbital period of the Moon (when it returns to the same position with regard to the Sun) is 2.2 days longer than its sidereal period (when it returns to the same position with regard to the distant stars) because Earth has moved in its orbit as well.

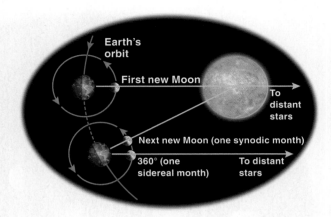

penumbra The inner region of a shadow cast by an extended object.

umbra The outer region of a shadow cast by an extended object.

total lunar eclipse An eclipse that occurs when the Moon passes through Earth's penumbral shadow.

partial lunar eclipse An eclipse that occurs when the Moon passes through Earth's penumbral shadow.

total solar eclipse An eclipse that occurs when a region of Earth's surface passes under the Moon's umbral shadow and the Sun's disk is fully blocked.

Eclipses are all about shadows. When the light from an extended object like the Sun hits the Moon or Earth, the shadow it casts has two parts: an extended **penumbra**, shaped like open-ended cone; and a smaller **umbra**, shaped like the closing tip of a cone.

A **total lunar eclipse** occurs when the Moon passes through the center of Earth's shadow, meaning that it passes first through Earth's penumbra and then through its umbra. Because this can happen only when the Moon, Earth, and the Sun lie along the same line, total lunar eclipses can occur only during a full Moon. Since the Moon "shines" via reflected sunlight, during a total lunar eclipse the face of the Moon will first darken as it passes through the penumbra. When the Moon passes through the darker umbra, however, it does not disappear. Sunlight passing through Earth's atmosphere is redirected into the umbra and onto the face of the Moon. Since blue light is more easily

scattered by the atmosphere than red light, during the umbral phase of a total lunar eclipse the Moon will take on a dusky reddish glow. The extended interval of the reddish hue will depend on the conditions in Earth's atmosphere at the time of the eclipse (**Figure 2.17**). A typical total lunar eclipse lasts a few hours as the Moon traverses the entirety of Earth's shadow.

You may wonder why we don't see a lunar eclipse at every full Moon. The answer is that the Moon is not usually directly behind Earth. The plane of the Moon's orbit has a 5° tilt relative to the plane of Earth's orbit around the Sun (the plane of the ecliptic). That means that during most full Moons, Earth's satellite is either above or below the planet's shadow. Sometimes the Moon passes only through the penumbra or only partially into the umbra, and in those cases we see a **partial lunar eclipse**.

A solar eclipse occurs when the Moon's shadow falls across a region of Earth. As **Figure 2.18** shows, even the penumbral lunar shadow is too small to cover all of Earth, and the umbra can only cover an even smaller area at any moment. Thus, unlike a lunar eclipse, which can be seen by everyone on the night side of Earth, a solar eclipse can be seen only by those living in the regions of Earth where the Moon's shadow falls.

Solar eclipses occur when the disk of the Moon blocks all or some of the disk of the Sun. This can happen only during a new Moon, when the Moon lies directly along the line joining Earth and the Sun. **Total solar eclipses** occur only where the Moon's umbra falls on Earth, creating a spot approximately 270 km

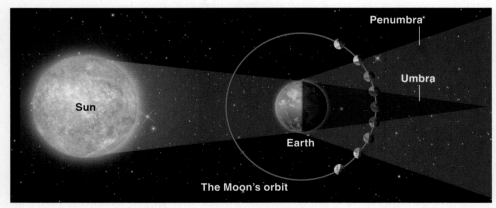

A

B

FIGURE 2.17 Total Lunar Eclipse
(A) Lunar eclipses occur when the Moon passes into Earth's shadow. As it passes into the penumbra, the Moon darkens, and as it passes into the umbra, Earth's shadow intensifies and grows across the face of the Moon until totality. As the Moon exits Earth's shadow, the steps reverse. (B) This time-lapse series of a lunar eclipse shows Earth's shadow falling across the face of the Moon. The reddening that occurs at totality is caused by refracted sunlight passing through Earth's atmosphere and reflecting off the Moon.

in diameter. As Earth turns under this spot and the Moon moves in its orbit, this umbral spot moves across Earth in what is called the *path of totality*.

Because they are remarkable events for anyone who experiences them, total solar eclipses must have had a profound effect on early cultures. As the Moon begins to cross in front of the Sun, it first appears as if an arc-shaped "bite" is being taken out of the solar disk. As **totality** is reached and the Moon completely obscures the Sun, the sky suddenly slips into darkness. Stars come out. Animals switch to nighttime behaviors. Streetlights switch on. And as if this were not enough, behind the disk of the Moon, the Sun's usually obscured outer atmosphere, called the solar corona, becomes visible. Lower regions of the Sun's atmosphere, such as the chromosphere and eruptive prominences (discussed further in Chapter 10), may also be seen (**Figure 2.19**). The duration of totality during a total solar eclipse is relatively short, lasting less than 8 minutes.

The Moon's orbit is not circular. Therefore, when it is at **apogee** (its farthest distance from Earth), it will appear smaller on the sky than when it is at **perigee**, the point closest to Earth. Earth's orbit is also not circular (its farthest and closest points from the Sun are called, respectively, **aphelion** and **perihelion**). That means the disk of the Sun will also sometimes appear larger or smaller (**Figure 2.20**). So sometimes, even in the umbral shadow, the disk of the Moon will not entirely block the disk of the Sun. When that happens, only an **annular eclipse** will occur, in which a thin ring of Sun appears to surround the darkened disk of the Moon (**Figure 2.21**).

Partial solar eclipses occur in those regions of Earth that fall under the Moon's penumbra. Usually regions within a few thousand miles of the umbra will see a scalloped portion of the Sun blocked by the lunar disk.

Just as we don't see a lunar eclipse at every full Moon, solar eclipses don't occur at every new Moon either. Once again, the reason is the 5° tilt of the Moon's orbital plane relative to Earth's orbit around the Sun. Solar eclipses can occur only when the Moon's tilted orbit carries it directly between Earth and the Sun. But when does this alignment occur? Just as the spinning Earth keeps its spin axis pointed in the same direction in space as it orbits the Sun (creating the seasons), the tilt of the Moon's orbit maintains a fixed direction in space as Earth goes around the Sun. That means the location *where the Moon crosses the Earth–Sun orbital plane* during its monthly revolution changes throughout the year.

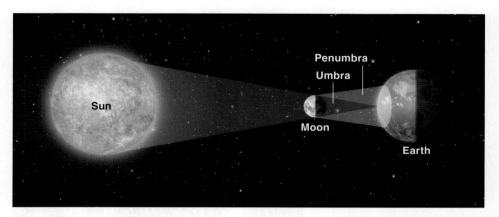

FIGURE 2.18 Total Solar Eclipse
Solar eclipses occur when the Moon's shadow falls across a region of Earth. Totality, lasting only a few minutes, is visible for observers in the umbral area, which is about 270 km in diameter, while partial eclipses are visible to those within the penumbral region.

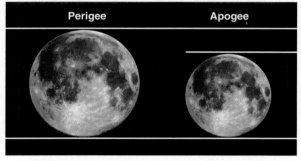

A

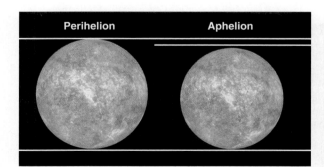

B

FIGURE 2.20 Relative Sizes of Moon and Sun on the Sky
Both the Moon and the Sun vary in their apparent angular size as seen from Earth, because of small eccentricities in the orbits of the Moon and Earth. (A) The terms *perigee* and *apogee* refer to the extremes of the distance between the Moon (or any object) and Earth. (B) The terms *perihelion* and *aphelion* represent the extremes of the distance between Earth (or any object) and the Sun.

FIGURE 2.19
The Sun's Atmosphere during an Eclipse
Because the brightness of the solar disk is blocked by the Moon during a solar eclipse, the less bright solar atmosphere, called the corona, becomes visible around the darkened Moon. The red features are solar prominences erupting from the Sun's surface.

totality The duration of total obscuration of the Sun or Moon during an eclipse.

apogee The distance of farthest approach for an object orbiting Earth.

perigee The distance of closest approach for an object orbiting Earth.

aphelion (pl. aphelia) The distance of farthest approach for an object orbiting the Sun.

perihelion (pl. perihelia) The distance of closest approach for an object orbiting the Sun.

annular eclipse A solar eclipse that occurs when the positions of Moon and Sun are such that the lunar disk is not able to fully block the solar disk.

partial solar eclipse An eclipse that occurs when a region of Earth's surface passes under the Moon's penumbral shadow.

FIGURE 2.21 Annular Eclipse
An annular eclipse appears when the angular size of the Moon is not as large as that of the Sun, because of the eccentricities of the orbits of Moon and Earth.

Astronomers call the line connecting the locations of such plane crossings the Moon's **line of nodes** (**Figure 2.22**). The Moon will touch the line of nodes twice in its monthly orbit, once when it crosses from above to below the plane of the Earth-Sun orbit, and once when it crosses from below to above. Only when the line of nodes aligns with either the full Moon or the new Moon can we see a lunar or solar eclipse.

line of nodes The line defined by the intersection of the Moon's orbital plane and Earth's orbital plane around the Sun.

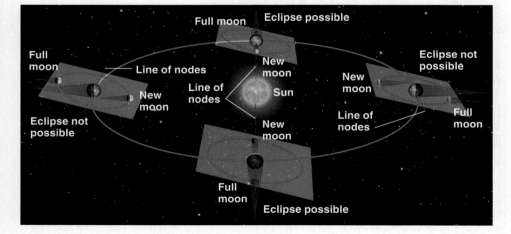

FIGURE 2.22
Line of Nodes
The Moon's orbit is tilted relative to the plane of Earth's orbit. Where the two orbits cross is called the line of nodes. The position of the line of nodes relative to the line connecting Earth and the Sun rotates as Earth completes its orbit. Eclipses are possible only when full or new Moons align with the line of nodes.

All of these factors make the occurrence of eclipses periodic, or cyclical—something ancient astronomers knew about 2,800 years ago.

Note that you should never, ever look directly at the Sun, even during a solar eclipse. Numerous guides available in print or on the Internet describe how to safely view a solar eclipse.

SECTION SUMMARY
Going Dark: Lunar and Solar Eclipses
- By coincidence, the Sun and the Moon appear the same size (or nearly so) on our sky.
- Lunar eclipses occur when the Moon passes through Earth's shadow.
- Solar eclipses occur when a region of Earth's surface passes under the Moon's shadow. Both lunar and solar eclipses can be partial or total.
- The Moon's orbit is tilted by 5° with respect to Earth's orbit around the Sun, so eclipses do not occur at every full or new Moon, but only when Moon, Earth, and Sun are all in alignment.

2.4
Celestial Wanderers: The Motion of Planets

The fourth celestial motion to consider, the motion of the other planets in our Solar System, played a critical role in the history of astronomy and human thinking about our place in the Universe. It is also something about which most citizens of the modern world know little. We are all familiar with the Sun, the Moon, and the stars, but few people can identify any of the planets on the sky, and fewer still can find and name them all.

There are five planets (besides the one beneath our feet) that can be seen with the naked eye: Mercury, Venus, Mars, Jupiter, and Saturn (**Figure 2.23**). In principle, it's easy to tell the difference between a star and a planet. "Planets look different," explains Carol Latta. "They don't twinkle. They also follow the ecliptic, just as the Sun and Moon do."

Stars are so far away from Earth that they appear as nothing more than points of light on the sky. Turbulence in Earth's atmosphere bounces the light from the point-like stars, scattering it from one moment to the next (see Chapter 4). Planets are so much closer to Earth than stars are that, in general, they appear as tiny disks instead of points on the sky. While individual rays of light from a planet are scattered just as much as those from a star as they pass through the atmosphere, scattering for a point-like star moves the point, whereas scattering a larger disk leaves you with a disk.

Basic Planetary Motion

"*Planet* means 'wanderer' in ancient Greek," says Latta, "because that's exactly what they do. The planets move relative to the background stars. Stars don't change their position relative to each other on the sky—at least they don't do so in a way that can be seen with the naked eye. Planets do." To understand the motion of planets across the sky, says Latta, "Imagine you went out every night at the same time and took a picture of the sky. If you compared the pictures over time, you would see the planets moving slowly eastward relative to the stars" (**Figure 2.24**). The eastward motion of the planets against the background stars was something our ancestors did notice. They did not have artificial illumination to drown out the night sky. They did not have televisions, computers, or smart phones to distract them at night. Instead, they noticed what changed on the sky. And so, night after night, month after month, they watched the planets' movement against the fixed stars.

As the planets marched across the sky, they would brighten and dim at different points in their motion. More important, they always appeared close to the ecliptic. However, some planets could be seen anywhere along the ecliptic relative to the Sun, while other planets, such as Mercury and Venus, could be seen only when they were close to the Sun, right before sunrise or right after sunset. Some planets

A

B

FIGURE 2.23 Planets Visible to the Naked Eye
The five planets that are visible without a telescope are known to all observers of the night sky. Their locations, brightness, and lack of twinkling make them easily identifiable as planets, not stars. (A) Since the orbits of all the planets and the Moon lie approximately in the same plane, these worlds are always found close to the ecliptic (the orbital plane of Earth). Venus and Jupiter outshine any stars at their brightest, and their light is steady. (B) Here, Jupiter shines more brightly than nearby stars in the twilight sky.

moved quickly across the starry field. Some planets crawled. Mars, for example, would take just a few months to cross a constellation. Jupiter would take years to cover the same distance.

SECTION SUMMARY
Basic Planetary Motion
- Planets "wander" along the ecliptic relative to the fixed stars, moving eastward nightly but with differing apparent speeds.
- Planets can be distinguished from stars because planets don't appear to twinkle.

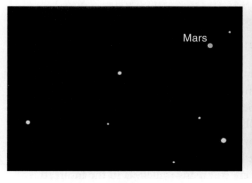

Time 1

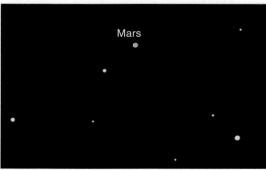

Time 2

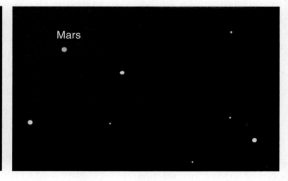

Time 3

FIGURE 2.24 Direct Planetary Motion
As the months pass, the planets move eastward against the background of the "fixed" stars, as illustrated here with Mars.

Carol Latta

It's hard not to catch Carol Latta's enthusiasm. After a career working at the upper levels of a major international corporation and raising two children, Latta found herself ready to take on something new. Returning to a local university, she rediscovered her enjoyment in learning science and then, unexpectedly, found a passion for astronomy. "It's not like I lived a life as a frustrated astrophysicist," Latta says. "I was happy with what I was doing in my career. But it seems like my education was entirely circular. I started out with math and science, then went to linguistics, and ended up working in business. It was only after I took my early retirement that I realized how much I missed math and science."

"At one point I took a course in astronomy at the Museum & Science Center in Rochester," she recalls. "Then my husband gave me a chance to go to Astronomy Camp in Arizona as a birthday present." That trip sealed Latta's fate. She returned to upstate New York determined to find a way to become actively involved in astronomy education. Joining the local amateur astronomy club, Latta became an expert at using a telescope. Night after night she learned how to find objects like galaxies and nebulas on the sky, and her love for astronomy grew.

These days Latta divides her time among her 13 grandchildren (and counting), acting as a teaching assistant for a local university's Astronomy 101 classes, and directing the affairs of the Rochester Academy of Science, Astronomy Section as its president. Latta's excitement and passion for astronomy remains as keen as that of any professional researcher. "I found that astronomy is something broad enough and deep enough and has so many different aspects that I could learn, be productive, and have so much fun all at the same time. In the end, becoming serious about astronomy was an obvious choice."

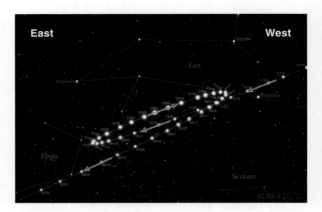

FIGURE 2.25 Retrograde Motion
The retrograde motion of a planet occurs when it moves westward on the sky relative to the distant stars. The retrograde motion of Mars is shown here. Retrograde planetary motion baffled astronomers for centuries and became a major issue in developing models for the Solar System.

The previous sections introduced each basic celestial motion and then offered the modern explanation for that motion. In the case of planetary motion, we must wait a bit and let the story play itself out. Hidden within the mystery of planetary motion is a key to understanding the essential nature of science and how it came to play such a pivotal role in human culture.

SECTION SUMMARY
Retrograde Motion
● Retrograde motion, in which planets reverse direction, remained a particularly challenging mystery for ancient astronomers.

2.5
Stone and Myth: Astronomy before History

The motions on the sky constitute the raw material of our astronomical heritage. Across the many thousands of years before written records (which begin around 2500 BCE), humans were just as filled with a sense of wonder by the Sun, Moon, stars, and planets as people are today. While the processes and methodology of science were still many millennia in the future, the basic need to make sense of the night sky led our ancestors to create explanations in the form of stories.

We often think of myths as false stories. But for the cultures that lived before history, **mythologies**

interactive
Retrograde Motion

Retrograde Motion

As human cultures became more adept at charting the behavior of the planets, one facet of their motion became a puzzle that took more than 3,000 years to solve: a phenomenon called **retrograde motion**. "Let's say you are following Mars's motion," says Latta. "Night after night you will see it moving eastward against the background of the fixed stars. Then, all of a sudden, the eastward motion stops, and the planet reverses course to move westward. Then the westward motion stops, and the planet returns to an eastward path. It makes a kind of loop, and that loop is called retrograde motion." These loops are easy to miss if you're not paying attention. "Mars's retrograde loop only covers a few degrees on the sky," says Latta. "That is only the width of a couple of Moons. You really have to look carefully over a long time to see it" (**Figure 2.25**).

The appearance of the retrograde loops was an enigma for early astronomers. Why did the planets move like that? What model could explain the planetary motions and their strange retrograde loops?

retrograde motion Apparent motion of a planet in a direction opposite to its normal motion.

mythology A collection of stories used to explain phenomena whose physical basis is not understood.

were interconnected collections of stories that took the basic "data" of what was seen on the sky and transformed those observations into explanations (**Figure 2.26**). The stories of myth not only told of how the cosmos came to be and why the sky looked the way it did; they also explained humanity's place in the Universe.

Astronomical Myths and Astronomical Science

"The relationship between astronomy as a science and ancient mythologies is really fundamental," says Professor Marcelo Gleiser of Dartmouth University. "At root they are both efforts to understand the Universe." Gleiser is a cosmologist who has studied everything from the physics of the early Universe to the origins of the first biologically active molecules. Gleiser has written widely on cosmology and its relation to myth. As he and others have shown, it is critical not to lose sight of the continuity between ancient mythology-based attempts to understand the world and the science practiced today by astronomers.

Myths often gave the sky special meaning and power. "As human cultures evolved," explains Gleiser, "the sky itself became a stage for stories that related directly to human activities. Constellations were attributed supernatural powers related to gods that ruled different aspects of life on Earth. There were gods of war in the sky and gods of love. There was the bull that represented fertility and the lion that represented power."

"Of course, the big difference between science and myth," says Gleiser, "is that myth focuses on using supernatural means of explanation. Science focuses on natural explanations using rationally based cause and effect. Still, the questions they both address and the motivations for asking those questions can have many similarities."

> ## SECTION SUMMARY
> **Astronomical Myths and Astronomical Science**
> - While myth explains observations on the sky in terms of supernatural powers of various gods, science looks for natural explanations.

Megaliths and Early Astronomy

Along with telling mythological stories, early cultures were watching the sky and translating their understanding into concrete forms as well. Artifacts from the Americas, Egypt, the Middle East, India, China, and other regions provide evidence that many ancient peoples systematically observed astronomical

A

B

FIGURE 2.26
Astronomical Myths
Myth was one way that ancient cultures accounted for the phenomena they observed on the sky. (A) This vase displays a myth based on agriculture: the Greek goddess Demeter (right) withholds the harvest during the winter months in retribution for the time each year that her daughter Persephone (left) must reside in the underworld. (B) In ancient Egyptian mythology, the falcon god, Horus, was said to be the sky. His right eye was the Sun and his left eye the Moon, and they traversed the sky when he flew across it.

megalith A large stone used as a monument or part of a monument.

phenomena. Across the world, there exist hundreds of giant stone monuments called **megaliths**, some of which date back 5,000 years or more (**Figure 2.27**). These megaliths are imposing structures that, in some cases, may have taken generations and enormous effort to construct. Most important for our understanding of astronomy's history, at least some of these prehistoric monuments align with celestial cycles.

Newgrange, the massive stone monument in Ireland (see Chapter 1), aligns with the rising Sun during the winter solstice. The most famous megalith site, however, is the enigmatic Stonehenge in England (**Figure 2.27B**), which consists of a series of circular rings. The outer circle is constructed from stone slabs weighing up to 50 metric tons each that had to be dragged from 40 km away. These stones are not set at just any location along the ring. Instead, their placement shows alignments with the motion of the Sun

A

B

FIGURE 2.27 Early "Observatories"
Many cultures left behind megaliths, such as these on the Isle of Lewis (A), that were symbols of homage in some cases and served astronomical functions in others. (B) Stonehenge, in the United Kingdom, served religious purposes but also marked the motions of the sky, which are vital to agrarian cultures that must accurately gauge when to plant and harvest.

Marcelo Gleiser

The tragic loss of his mother when he was just a child left Marcelo Gleiser searching for answers even as a boy. During his early years, growing up in Rio de Janeiro, Gleiser found those answers in tales of the supernatural—stories of ghosts and spirits. Later, in his teens, he discovered science.

"I gawked at the TV when Neil Armstrong and Buzz Aldrin planted the US flag on the Moon," he says. "And the destructive power of the H-bomb really terrified me. I was amazed that for the first time in history, we could obliterate civilization at the push of a few buttons." Stanley Kubrick's famous movie *2001: A Space Odyssey* left Gleiser with questions about life in the Universe. "I was really fascinated with the possibilities that there might be greater intelligences living somewhere in the vastness of space," he says. "Could *they* have been our creators?"

These questions propelled Gleiser toward a career in science. After earning his bachelor's and master's degrees in Brazil, Gleiser traveled to England and took his PhD from King's College London. Spending a few years as a researcher in theoretical particle physics and cosmology at both Fermilab and the Kavli Institute for Theoretical Physics, he eventually joined the faculty of Dartmouth College. In 1999 he was awarded the Appleton Professorship of Natural Philosophy.

One of the most popular writers on science in Brazil, Gleiser has also written two English-language books on the intersection between science and human culture: *The Prophet and the Astronomer* and *The Dancing Universe*. "Science is more than a way to make sense of the material world," Gleiser says. "It is a way of living, a very human creation and the product of the same curiosity that has moved our collective imagination for thousands of years. It is an expression of our deeply ingrained desire to make sense of the world and to know our place in the big scheme of things."

Paleolithic period The period extending from 2.6 million years ago (characterized by the earliest known use of stone tools) until 10,000 years ago.

Neolithic period The period extending from about 10,000 BCE to between 4500 and 2000 BCE, when farming was introduced.

FIGURE 2.28
Cave Drawings at Chauvet
The Chauvet-Pont-d'Arc Cave in southern France contains cave paintings dating back 30,000–32,000 years ago. The caves also harbor other evidence of life in the Paleolithic period. The paintings depict 13 species of animals, including horses, mammoths, bears, and cave lions.

and the Moon. The main axis of the monument, for example, faces the horizon where the Sun rises on the summer solstice, the longest day of the year.

Stonehenge had a practical role to play for its society. "Stonehenge was built by people who were part of an agrarian society," says Gleiser. "By knowing how the skies were moving, you actually knew when to plant, you knew when to harvest, when the rains would come, when the cold would come. The builders of Stonehenge wanted to create the clock that would

reproduce the motions of the sky." But to think of Stonehenge as only a clock is probably a mistake. "The location of burial mounds around the Stonehenge site also makes it clear that the site was used for religious purposes," says Gleiser. "By mimicking the motions of the skies, which were the realm of the gods, Stonehenge's creators were bringing that realm down to Earth."

SECTION SUMMARY
Megaliths and Early Astronomy
- Megalithic monuments like Stonehenge reflect both astronomical and religious purposes of ancient peoples.

Early Human Evolution and Astronomy

Although the evolution of human beings can be traced back at least 6 million years, only in the relatively recent past did human cultures begin to develop the technologies and structures that led to our modern societies. Sometime between 70,000 and 50,000 years ago, a flowering of consciousness occurred, in which toolmaking and the production of art exploded (**Figure 2.28**).

This earliest period of human culture occurred at the end of what is called the **Paleolithic period**, when Earth's climate was locked in the deep freeze of an ice age. But around 10,000 BCE, at the beginning of what is called the **Neolithic period**, conditions began to shift toward the warmer, wetter world we inhabit today. Since the dawn of their existence, humans had been nomadic hunter-gatherers. But as the climate changed, they began to put down roots and practice agriculture.

Farming changed everything. Agriculture allows surpluses to develop. If the harvest is good, a culture ends up with extra grain, as well as the need to store and protect the surpluses. As a result, larger permanent settlements tend to grow and diversify. The villages of the Neolithic period thus gave way to the first true cities, such as the cities of Çatal Hüyük in eastern Turkey (7000 BCE) and Sumer in southern Iraq (5000 BCE).

Life in these cities depended on their inhabitants developing specialized skills as leatherworkers or metalsmiths. A priestly class also arose whose job was attending to the supernatural aspects of life, which often meant attending to the night sky. Priests became the first true astronomers as they began keeping long-term records of lunar phases, lunar and solar eclipses, and planetary motions. The longevity of city-building civilizations also meant that these records could be passed from one generation to the next—a critical step in the development of astronomy. The planet Saturn, for example, completes its motion against the fixed background of stars over the course of almost 30 years. Without long-term records, observers of the sky could not recognize these patterns.

The most important city-building civilization for astronomy was the Babylonian Empire, centered in what is today southern Iraq. Babylon became the hub of a great civilization in 1700 BCE that lasted for at least 1,000 years. It was the Babylonians who first began to maintain detailed records of celestial motions on small slabs of stone that were called cuneiform tablets (**Figure 2.29**). Archaeologists have hundreds of years of these astronomical observations recorded in a series of tablets called the Enûma Anu Enlil. By cataloging both stars and constellations, the Babylonians developed schemes for predicting the risings and settings of the Sun and planets, as well as the length of the day. The so-called Venus tablet of Ammisaduqa from the first millennium BCE lists risings of Venus over a 21-year period and is the earliest evidence that planetary phenomena were recognized as periodic.

Most of the tablets, however, appear to be concerned more with astrology than with astronomy. But even though the observations were for astrological purposes, the Babylonians took their application in new directions, including the use of mathematics to calculate variations in the length of day across the year. "The Babylonians were the first to have a preoccupation with mapping the motions of planets in the sky," says Gleiser. "The skies were a kind of open book to them that only certain people would be able to interpret. Those people were priests who were, in a sense, the precursors of scientists and astronomers today."

Centuries later and 1,500 miles northwest of Babylon, philosophers in the city-states of Greece would explain the skies in the rational language of mathematics and natural cause and effect rather than divine intervention.

SECTION SUMMARY
Early Human Evolution and Astronomy
- With the development of farming and the ability to store food also came the creation of specialization within societies.
- Priests, who specialized in interpreting the supernatural, often had the task of tracking the motions of celestial objects.
- The Babylonians were the first to create long-term records of the motions of planets.

2.6
The Greek Invention of Science

"It all started with the Greeks," says Marcelo Gleiser. "They were the ones who discovered a new way of approaching nature. For the first time, rational arguments were brought to bear, rather than mythic or supernatural explanations. It was something very different and very new that began around 650 BCE."

Advances in Science

In cities dotting the Greek peninsula and the islands of the Aegean, a radical new culture arose during the last millennium BCE. Thales of Miletus (624–547 BCE), a Greek citizen who lived in Turkey, was a forerunner in rejecting supernatural explanations for phenomena and claimed that the world could be understood through reason alone. In the centuries that followed, Greek philosophers took up Thales's approach and began constructing rational models to explain the natural world, including the astronomical motions of the Sun, Moon, and planets.

One of the most influential Greek mathematician-philosophers was Pythagoras (570–500 BCE), who developed a mathematical way of understanding music and harmony. Pythagoras and his followers (called the Pythagorean brotherhood) extended this approach and saw mathematics as a language for describing the underlying patterns they saw on Earth and on the sky.

Plato (428–347 BCE), another influential Greek philosopher, continued the Pythagorean enchantment with numbers and mathematics. Above his famous school in Athens (called the Academy), he had inscribed the words "Let None Who Is Ignorant of Geometry Enter." Plato argued that behind the appearances of the world lay a more perfect mathematical world that acted as a kind of blueprint for all we see. The world we experience is just a corrupted version of this ideal world of mathematical forms. As part of this doctrine of ideals, Plato held that all the celestial motions must be manifestations of a perfect underlying mathematical order. He encouraged his students to find the mathematical forms that could recover, or "save," the observed celestial motions, such as the planetary retrograde loops. (Plato famously asked his students to "save the appearances," meaning to develop a mathematical model for the motions of the planets that could account for what appeared on the sky from the perspective of humans on Earth.)

FIGURE 2.29
Cuneiform Tablets
The Babylonians recorded planetary and stellar motions on cuneiform tablets such as this one, showing that they recognized the periodic nature of astronomical events. They also used their observations to make astrological predictions for the king.

geocentric model A model stating that Earth is the body around which all other Solar System objects orbit.

heliocentric model A model stating that the Sun is the body around which all other Solar System objects orbit.

parallax A displacement or difference in the apparent position of an object viewed along two different lines of sight

SECTION SUMMARY
Advances in Science
- The Greeks approached the world from the perspective of reason and pursued mathematics as the key to understanding the sky's patterns.
- Plato believed that while Earth was imperfect, celestial objects and motion were perfect and their motions should be based on perfect mathematical order.

Aristotle and the Geocentric Model

Plato's most famous student was Aristotle (384–322 BCE), whose ideas would dominate Western thinking about nature for more than 1,500 years. Aristotle was not a scientist in the modern sense of the word, because he never sought to validate his theories through rigorous experimentation. His mode of reasoning was to begin with a set of core beliefs that he thought were self-evident (clearly true) and combine these with select observations. In this way, Aristotle attempted to deduce how the world worked.

In the arenas of physics and astronomy, Aristotle argued that the cosmos was divided into two domains. First there was Earth, which was spherical and the center of all the cosmos. Surrounding Earth was the celestial domain of the planets and stars. To account for celestial motions, Aristotle built on ideas proposed by Eudoxus (409–356 BCE), another of Plato's students, arguing that 55 nested crystalline spheres composed the heavens. Each sphere was centered on Earth and rotated with a different speed (**Figure 2.30**). The Moon, Sun, planets,

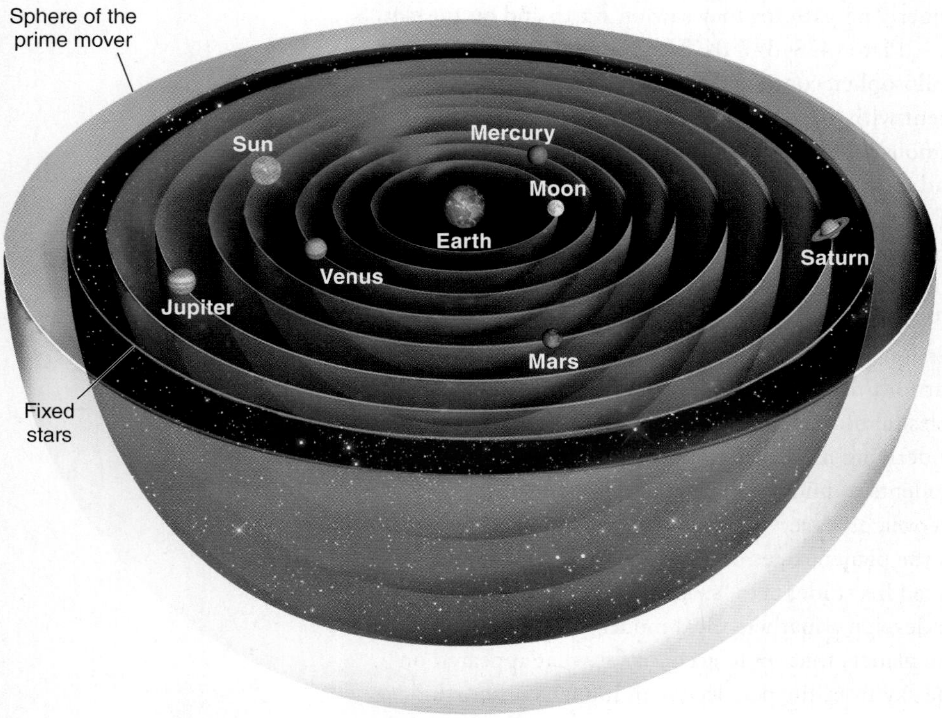

FIGURE 2.30
Aristotle's Model of the Universe
Aristotle's model placed Earth at the center of the Universe. All other heavenly bodies were attached to crystalline concentric spheres rotating at different uniform speeds around Earth. A simplified rendering of his model is shown here.

Sphere of the prime mover

Fixed stars

Sun, Mercury, Moon, Earth, Venus, Jupiter, Mars, Saturn

and stars were each attached to one of the spheres. The rotation of the crystalline spheres created the appearance of celestial motion across the sky on Earth.

In Aristotle's model, the terrestrial, or "sublunar," domain was the realm of change and decay. The realm of the crystalline spheres, in contrast, was unchanging and perfect. Each of the crystalline spheres guided the planets through mathematically perfect circular motions at a perfectly constant, or uniform, velocity. This emphasis on the perfection of the celestial realm, including the constant circular motion of the planets on their spheres, would play an important role in the evolution of astronomical thinking.

Many but not all Greek astronomical thinkers accepted this kind of **geocentric model**, in which the Sun and planets orbited Earth. It is a tribute to the imagination, creativity, and ingenuity of Greek thinkers that Aristarchus (310–230 BCE), who lived after Aristotle, proposed a **heliocentric model**, in which Earth was just another planet orbiting the Sun (the center of the Universe). Aristarchus's model was not widely accepted, however. His opponents argued that if Earth really moved through space, then there should be a great wind due to the motion.

Another argument against the heliocentric model was that if Earth orbited the Sun, then the changing view of the stars from one side of the orbit to the other would lead to changes in stellar position. This effect, called **parallax**, can be precisely defined as the apparent displacement of an observed object due to a change in the position of the observer. You can most easily see this effect by holding your finger up to your nose and closing only the left eye and then closing only the right eye. When you switch between views, notice that the positions of distant objects relative to your finger change depending on which eye you're looking from. The Greeks reasoned that if Earth was orbiting the Sun, the position of nearby stars relative to the farther ones should shift as Earth moved from one side of its orbit to the other. Since they did not observe any parallax, they believed that Earth must be stationary (**Figure 2.31**). We now recognize that the stars are too distant for us to observe the parallax without a telescope, but the true distances to the stars were too large for the Greeks to imagine.

SECTION SUMMARY
Aristotle and the Geocentric Model
- Most of the Greek philosophers believed that Earth was the center of the cosmos and subscribed to Aristotle's geocentric model of the heavens, based on the existence of crystalline spheres rotating around Earth.

Mathematical Advances

The reliance on mathematics and geometry enabled the Greeks to estimate the sizes of Earth, Moon, Sun, and Solar System. Sometimes their estimates were wildly wrong; for example, the Greeks believed that the distance to the last crystalline sphere, which held the stars, was smaller than what we now know to be the size of the Solar System. Still, in some cases their reasoning was remarkably sophisticated. Eratosthenes (276–195 BCE) was a philosopher, an astronomer, and the head of the great library of the city of Alexandria. He used the length of midday shadows observed at two different cities some 500 miles apart to infer the size of Earth. Using only the record of shadow lengths and geometry, he deduced a radius of Earth that was off by only 14 percent of the actual value.

Greek astronomers built on and improved the observations of the Babylonians. One of the most famous of all the Greek astronomers, Hipparchus (190–120 BCE), is credited with the development of trigonometry, which he may have used to construct the first reliable method for predicting solar eclipses. Hipparchus's most important contribution, however, was the development of the first comprehensive star catalog in the Western world. The catalog included a scheme for describing stellar brightness called the magnitude system, which remains in use today. Hipparchus's ideas and influence in astronomy were paramount in the ancient world and lasted for more than 300 years.

> ### SECTION SUMMARY
> **Mathematical Advances**
> - Eratosthenes used geometry to measure Earth's radius with only a small degree of error.
> - Hipparchus is credited with inventing trigonometry and with creating the first star catalog and a magnitude system for describing brightness.

Egocentrism Triumphant: The Ptolemaic Universe

Hipparchus's influence was eclipsed only by the appearance of the great Claudius Ptolemy (90–168 CE), whose efforts gave astronomers a new functional explanation for planetary motions (including retrograde loops). Ptolemy can rightly be considered the greatest of the Greek astronomers. His most important legacy was the fulfillment of Plato's demand to create a geometrically based model of the Universe using perfect, uniform circular motion to recover the behavior seen in the night sky.

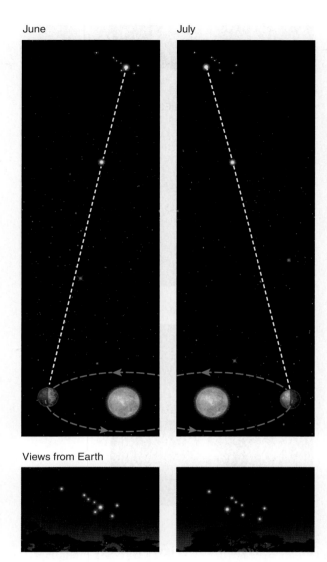

June   July

Views from Earth

FIGURE 2.31
Stellar Parallax
Stellar parallax occurs when the apparent position of a star changes as Earth moves from one side of the Sun to the other (this effect is judged against "background" stars, whose positions do not noticeably change relative to one another). Since this effect could not be observed with the technology of the time, the ancient Greeks reasoned that Earth did not move but was stationary at the center of the Universe.

Ptolemy's model was geocentric: it placed Earth at the center of the Solar System. How can a model so wrong be considered a triumph? Ptolemy's model worked. It did a good job of predicting (and prediction is a crucial feature of any model) all the regular celestial motions. So powerful and useful was Ptolemy's model that his book *Mathematical Syntaxis* (written in 150 CE) would remain the standard textbook in astronomy for nearly 1,400 years. The Arabic astronomers who carried science forward while western Europe was plunged into the Dark Ages called Ptolemy's book *Almagest*, or "The Greatest."

How did Ptolemy manage such accuracy in predicting planetary and other celestial motions with a model that was completely wrong? Building off the work of Eudoxus, Aristotle, and others, Ptolemy imagined the Sun, Moon, and planets all to be

FIGURE 2.32

Ptolemy's Geocentric Model of the Universe

Ptolemy's model of the Universe was based on the incorrect assumption that the Universe was geocentric. It included elaborate constructs such as (A) epicycles, small circular paths planets moved on as the the epicycles followed the larger orbit, and (B) equants, a point displaced from the center of Earth's orbit around which the other planets displayed uniform motion. Nonetheless, it did an admirable job of predicting the observed motions of the Sun and planets.

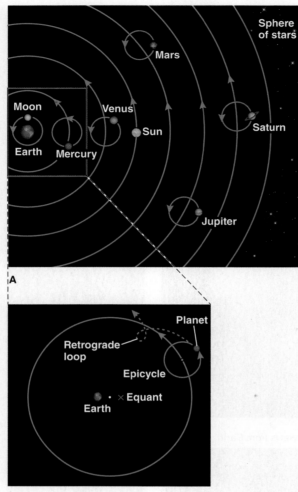

epicycle A secondary orbit whose center point orbits Earth; devised by Ptolemy to explain retrograde motion.

equant A point displaced from the center of a planet's orbit around which the planet displays uniform motion.

Occam's razor A principle stating that among competing hypotheses, the simplest one should be selected.

orbiting around Earth (**Figure 2.32**). To account for retrograde motion and the variable speed that planets manifest, Ptolemy's model included special mathematical devices. For example, the planets did not move directly in circular orbits about Earth. Instead, each planet moved with uniform speed on a smaller circle called an **epicycle**, which created the appearance of retrograde motions. It was the *center of the epicycle* that then moved with uniform speed around Earth. When the planet's motion on the epicycle was in the same direction as the orbital motion, it would appear to an Earth-based observer to move eastward on the sky. But when the planet looped around to the other side of the epicycle, it would appear to move westward on the sky. The combination of motions—the planet on its epicycle and the epicycle orbiting Earth—allowed Ptolemy's model to recover

retrograde motion with uniform circular motion and to predict very accurately how planets moved through the sky.

Ptolemy needed additional devices other than single epicycles to get his model to match observations. In some cases, smaller epicycles were required in addition to the main epicycle. Getting the variable speed of the planets in the night sky correct also required that Ptolemy assume Earth was slightly off center from the planetary orbits. Every planet's uniform motion was then adjusted relative to this new point, called the **equant**. In some later versions of Ptolemy's geocentric model, more than 100 epicycles and other devices were required to increase the accuracy of predictions.

In modern science there is a principle called **Occam's razor**, which states that explanations for natural phenomena should be as simple as possible when compared with other models. By today's understanding, Ptolemy's model, with its epicycles and the equant, clearly violates Occam's razor. It is important to recall, however, that astronomers at the time (and for the next 14 centuries) were under the sway of philosophical principles like the demand for the heavens to be "perfect," including perfect circular motion. It took great intellectual courage (and some personal risk) to finally reject the geocentric model.

Ptolemy can be celebrated as a culmination of the Greek genius, with its demand to know the world through reason and mathematics and its rejection of the supernatural. The Greek era changed all of human history and led directly to the flourishing science of our age. What will people say of our era 1,000 years from now? What will we leave behind that will be so lasting and of such consequence?

SECTION SUMMARY

Egocentrism Triumphant: The Ptolemaic Universe

● Ptolemy's geocentric model, though wrong, was quite successful at replicating the sky's motions and dominated astronomy for more than 1,400 years.

● Ptolemy's model required a complicated system of epicycles and equants to match the motions of the planets, including the explanation of retrograde motion.

● Ptolemy's work epitomizes the Greek foundation of rational thought that laid the basis for modern science.

chapter summary

2.1 An Old Obsession

For thousands of years, human beings have looked to the sky and tried to explain the phenomena they witnessed there.

2.2 Dance of Night and Day: Basic Motions of the Sky

The Sun, Moon, stars, and planets appear to move from east to west each day (and night), with variations based on the observer's latitude on Earth. It is Earth's rotation that creates this apparent motion, though it took people a long time to understand and accept that fact. The celestial sphere is an imaginary transparent globe surrounding Earth, on which all the stars and planets appear to rest. Its equator and poles are extensions of Earth's equator and poles. An observer at any position on Earth can see only half of the celestial sphere. For Northern Hemisphere observers, the celestial sphere appears to turn around the star Polaris each day, and Polaris appears at an altitude equal to the observer's latitude. Over the course of the year, the Sun's observed path changes, reaching higher altitudes in summer (with its highest point on the summer solstice) than in winter (with its lowest point on the winter solstice). In correlation with these changes, its rising and setting points move southward from the summer solstice to the winter solstice, and then head northward again from the winter to the summer solstice. Seasons are the result of Earth's tilt, which makes for longer days and more direct radiation from the Sun in summer. Although Earth's orbit is slightly elliptical, its varying distance from the Sun is not the cause of the seasons. A constellation is a recognizable group of stars and the area encompassing them. Which constellations can be observed depends on the observer's location on Earth and Earth's position relative to the Sun. Only constellations on Earth's night side can be observed; the others are lost in the Sun's glare. Circumpolar constellations—those located close to one of the poles—are visible all year. The ecliptic is the Sun's apparent annual path through the sky.

2.3 Monthly Changes of the Moon

The Moon shows phases as a result of its 29.5-day orbit around Earth. It reflects light from the Sun, and the phase depends on how much of its illuminated half we can see. The full Moon occurs when the Moon is opposite the Sun, so we see its entire face illuminated by sunlight; the new Moon occurs when the Moon is between Earth and the Sun. The order of the phases is new Moon, waxing crescent, first quarter, waxing gibbous, full Moon, waning gibbous, third quarter, and waning crescent. From one night to the next, the Moon moves eastward against the background stars.

2.4 Celestial Wanderers: The Motion of Planets

Like the Moon, planets move eastward against the stars. However, because the inner planets move faster than the outer planets in their orbits, they sometimes appear to be moving in the opposite direction—a phenomenon called retrograde motion.

2.5 Stone and Myth: Astronomy before History

Both myth and science are ways in which humans have tried to understand the Universe. Before science provided testable models, people developed elaborate myths about the supernatural origins of celestial phenomena, but both approaches seek to make sense of what can be observed. Early astronomers created megalithic monuments such as Stonehenge to record and use motions of sky objects, and those monuments may have served ceremonial purposes. Knowing when to plant was an important practical use.

2.6 The Greek Invention of Science

The study of astronomy grew from the birth of civilizations, especially in Babylon. Record keeping enabled knowledge of events spanning many years, such as the orbits of outer planets. The Greeks contributed to astronomy in many ways, such as the practice of mathematical modeling. Some of the most noteworthy developments were Pythagoras's geometry, Aristotle's geocentric model of the Universe involving crystalline spheres, Eratosthenes' measurement of Earth's diameter, and Hipparchus's invention of trigonometry and a star magnitude chart. Ptolemy built on their work to develop a detailed geocentric model of the Universe that was the accepted standard for 1,400 years, despite being fundamentally wrong in placing Earth at the center. Ptolemy's model went a long way toward incorporating the retrograde motion of planets and was an important example of the concept of interpreting the world through reason.

questions and problems

Narrow It Down: Multiple-Choice Questions

1. How does the altitude of the Sun at noon on the same day in the Northern Hemisphere's summer compare for two observers at latitudes 12° north and 54° north, respectively?
   a. It is the same for both observers because they are in the same hemisphere, experiencing summer.
   b. It cannot be determined without knowing their longitudes.
   c. It is 42° higher for the observer at 12° north because of the difference in latitude.
   d. It is 42° higher for the observer at 54° north because of the difference in latitude.
   e. The relative altitude of the Sun cannot be determined for the two locations without knowing the exact date.

2. Which observers on Earth can see Polaris on a clear night?
   a. all observers on Earth
   b. only observers above the Arctic Circle
   c. only observers in the Western Hemisphere
   d. only observers in the Southern Hemisphere
   e. only observers in the Northern Hemisphere

3. You are observing distant object A of 30 arcseconds in diameter. From your understanding of the small-angle formula, and compared with your observation of A, which of these statements is always true?
   a. A larger object at the same distance will appear the same size.
   b. A smaller object at the same distance will appear the same size.
   c. A smaller object at a greater distance will appear smaller.
   d. A smaller, less distant object will appear larger.
   e. A larger, less distant object will appear smaller.

4. The Sun is highest on the sky at noon on
   a. the winter solstice.
   b. the spring equinox.
   c. the summer solstice.
   d. the autumn equinox.
   e. any day, because the Sun reaches the same altitude daily regardless of season.

5. Which statement about constellations is true?
   a. Any group of stars can be called a constellation.
   b. A constellation includes a group of stars within specific boundaries in the same region of the sky.
   c. The stars that form a constellation must be in a configuration resembling an animal or human.
   d. The stars within a constellation are all located at the same distance from Earth.
   e. The stars within a constellation are all about the same brightness.

6. From which location are the same constellations above the horizon at any time of year?
   a. the North Pole
   b. the equator
   c. Rio de Janeiro, Brazil
   d. New York City
   e. No such location exists.

7. Warmer summertime temperatures in the Northern Hemisphere are due partly to
   a. longer days.
   b. a lower angle of the Sun's rays.
   c. Earth's being closer to the Sun in summer.
   d. the Sun radiating more energy in summer.
   e. the tilt of the Northern Hemisphere away from the Sun.

8. Which statement about Moon phases is true?
   a. In waxing phases, the lit portion of the Moon faces the eastern horizon.
   b. The new Moon has its whole face illuminated as seen from Earth.
   c. The Moon rises at sunset every day.
   d. The waning gibbous phase follows the full Moon.
   e. In waning phases, the lit portion of the Moon faces the western horizon.

9. You observe the full Moon just rising in the east. What time of day is it?
   a. sunrise (about 6:00 A.M.)
   b. noon (about 12:00 P.M.)
   c. sunset (about 6:00 P.M.)
   d. midnight (about 12:00 A.M.)
   e. midafternoon (about 3:00 P.M.)

10. Synodic and sidereal months differ because of
    a. the Moon's orbit.
    b. Earth's orbit.
    c. the Sun's orbit.
    d. the fact that Earth's year is not exactly 365 days.
    e. the different number of days in each calendar month.

11. The direct (and most typical) seasonal motion of the planets as observed from Earth is
    a. west to east with respect to the background stars.
    b. east to west with respect to the background stars.
    c. east to west at the same rate as the background stars.
    d. north to south with respect to the background stars.
    e. south to north with respect to the background stars.

12. Using only Stonehenge to calibrate astronomical motions, early people would *not* have been able to tell which of the following? Choose all that apply.
    a. when to plant
    b. when the longest day had come
    c. when Mars would appear
    d. when the Moon would be full
    e. when winter would begin

13. Which Greek philosopher is most closely associated with first rejecting supernatural explanations and arguing that reason alone could explain phenomena?
    a. Thales
    b. Aristotle
    c. Socrates
    d. Hipparchus
    e. Pythagoras

14. For what significant contribution is Eratosthenes most famous?
    a. inventing trigonometry
    b. measuring the circumference of Earth
    c. constructing the geocentric model of the Universe
    d. creating the first catalog of bright stars
    e. defining the four basic elements

15. Which of the following statements about parallax is true?
    a. Our two eyes enable us to use parallax to determine distances to objects.
    b. Earth's orbit provides astronomers an opportunity to use parallax.
    c. Most stars do not appear to shift position, because they are too far away for parallax to be observed.
    d. If all stars were on the surface of a celestial sphere and located at the same distance, we would observe no stellar parallax.
    e. all of the above

16. Which of the following was/were elements of Ptolemy's geocentric model? Choose all that apply.
    a. It provided a true explanation for why we don't feel a constant strong wind on Earth.
    b. It included epicycles.
    c. It accounted for retrograde motion.
    d. It supposed that Mercury and Venus orbit the Sun.
    e. It assumed that all planetary orbits were ellipses.

17. A lunar eclipse can occur at which Moon phase(s)? Choose all that apply.
    a. new Moon
    b. first quarter
    c. full Moon
    d. third quarter
    e. all of the above

18. Which of the following statements about solar eclipses is/are correct? Choose all that apply.
    a. A total eclipse is possible because the Sun and Moon sometimes appear to be identical in size.
    b. Solar eclipses can only occur at full Moon.
    c. Not all solar eclipses achieve totality.
    d. Prediction of solar eclipses became possible only with the advent of computers.
    e. A solar eclipse is visible to everyone on Earth equally.

19. True/False: Total eclipses can occur only when both the Moon and the Sun simultaneously pass through the line of nodes.

20. Eclipses are possible only when both Sun and Moon are at specific positions relative to Earth. How many times each month does this alignment occur?
    a. one
    b. two
    c. three
    d. four
    e. It varies widely.

To the Point: Qualitative and Discussion Questions

21. What are some factors that led to the advancement of human civilization and culture?

22. What likely uses did the ancients have for megaliths?

23. Name at least one important contribution associated with each of the following Greek thinkers: Thales of Miletus, Pythagoras, Plato, Aristotle, Eudoxus, Aristarchus, Eratosthenes, Hipparchus.

24. If the Moon crosses the meridian at midnight, what phase must the Moon be in?

25. Suppose that a month ago you saw the star Betelgeuse in the constellation Orion just rising at the eastern horizon at 8:00 P.M. Describe its position at the same time today.

26. Define retrograde motion and explain how Ptolemy's model represented it.

27. Define the celestial sphere. How is it a useful (if imaginary) tool?

28. From what location on Earth can you see every part of the celestial sphere over the course of the year?

29. How does the Sun's path across the sky differ in summer versus winter?

30. What is an analemma, and what gives it its characteristic shape?

31. If Earth's axis had no tilt relative to the plane of its orbit, how would the seasons differ from those we experience today?

32. If Earth's axis had a more significant tilt relative to the plane of its orbit, how would the seasons differ from those we experience today?

33. Explain the difference between sidereal and synodic months.

34. Describe and compare the models of the Universe defined by Aristotle and Ptolemy.

35. How did Aristotle use the lack of measurable parallax to disprove the heliocentric model championed by Aristarchus? Comment on the flaw in Aristotle's logic.

Going Further: Quantitative Questions

36. How many arcseconds are there in 4°?

37. How many degrees are there between the horizon and the zenith?

38. From your location, the Sun is at an altitude of 80° as it crosses the meridian on the summer solstice. Describe its altitude as it crosses the meridian one month later.

39. A star is at the zenith for an observer at latitude 44° north. What is its declination on the celestial sphere? (Note that astronomers use a "+" before the number for north declination and a "−" before the number for south declination.)

40. You observe the Moon's position on the sky at the same time on two consecutive days. Across how many degrees of sky has its position moved?

41. How many days are there between new Moon and full Moon?

42. You observe Mars with an angular diameter of 18″. What is its distance from Earth in kilometers? (Hint: The diameter of Mars is 6,792 km.)

43. A globular star cluster has an angular diameter of 20′. It is 25,000 light-years away. What is its diameter in light-years?

44. An object at a distance of 200 meters is 0.5 meter wide. What is its corresponding angular width in arcseconds? in arcminutes?

45. Comet Hale-Bopp has a core diameter of 40 km. At its closest approach to Earth, it was about 137 million km away. How large in arcseconds did its core appear to observers at that distance?

 If your instructor assigns homework in **smartwork5**, access it at the Digital Landing Page for *At Play in the Cosmos*: **digital.wwnorton.com/cosmos**

A UNIVERSE OF UNIVERSAL LAWS

From the Copernican Revolution to Newton's Gravity

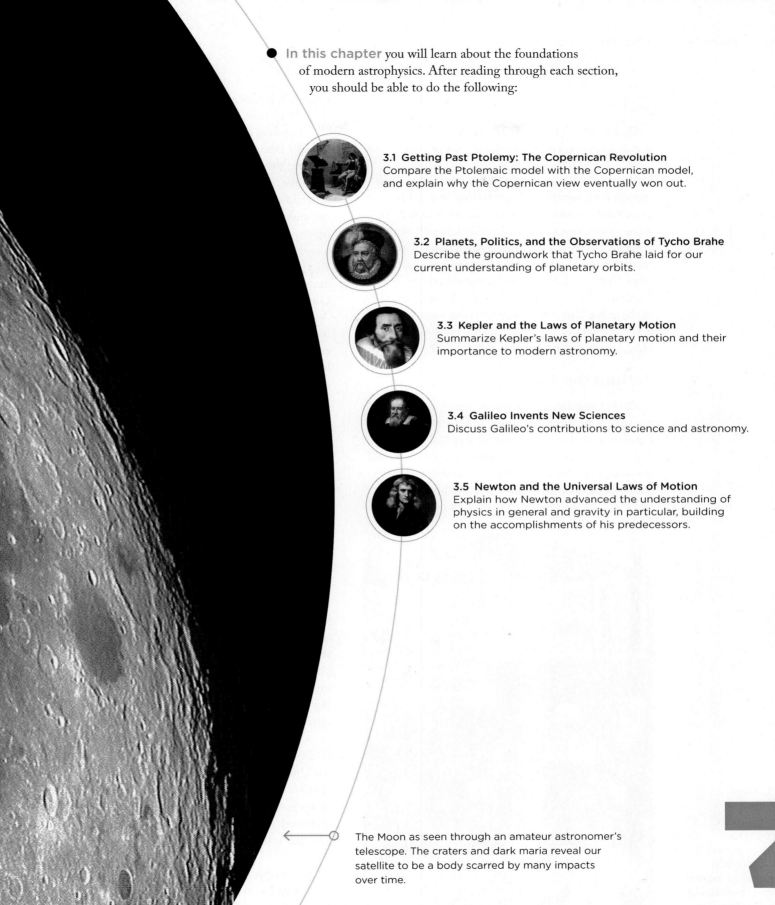

● **In this chapter** you will learn about the foundations of modern astrophysics. After reading through each section, you should be able to do the following:

3.1 Getting Past Ptolemy: The Copernican Revolution
Compare the Ptolemaic model with the Copernican model, and explain why the Copernican view eventually won out.

3.2 Planets, Politics, and the Observations of Tycho Brahe
Describe the groundwork that Tycho Brahe laid for our current understanding of planetary orbits.

3.3 Kepler and the Laws of Planetary Motion
Summarize Kepler's laws of planetary motion and their importance to modern astronomy.

3.4 Galileo Invents New Sciences
Discuss Galileo's contributions to science and astronomy.

3.5 Newton and the Universal Laws of Motion
Explain how Newton advanced the understanding of physics in general and gravity in particular, building on the accomplishments of his predecessors.

The Moon as seen through an amateur astronomer's telescope. The craters and dark maria reveal our satellite to be a body scarred by many impacts over time.

3

3.1

Getting Past Ptolemy: The Copernican Revolution

FOURTEEN HUNDRED YEARS is a long time—long enough for more than 50 generations (that's 50 grandmas) to come and go. But for more than 1,300 years—from about 200 CE to 1600 CE—Ptolemy's geocentric vision of the Solar System reigned supreme, though it was continually tweaked by Arabic astronomers. Until the 16th century, *everyone* believed that Earth was the center not only of the Solar System but of the entire Universe. Today everyone knows the Sun is at the center of the Solar System. How did such a radical change in perspective come about?

Setting the Stage

In the 14 centuries after Ptolemy, the Western world changed dramatically. The might of the Roman Empire, so vital during Ptolemy's day, had splintered and crumbled. In western Europe, the Dark Ages had descended and learning slowed to a crawl, as most people lived as peasants under the thumb of feudal lords. Eastern Europe fared better, but by 700 CE the torch of scholarship and learning had been passed to the vibrant young empires of the Middle East.

In places like Persia (modern Iran), scientific learning continued. From 500 CE onward through the next five centuries, astronomers were particularly energetic. They established new and better predictions for eclipses, made accurate measurements of Earth's tilt, and produced new, highly detailed star maps. The legacy of this dynamism can be found in the Arabic names we still use for many of the brightest stars, like Rigel, Aldebaran, and Deneb (**Figure 3.1**). But even during this explosion of new learning, the ancient Greek Ptolemy and his geocentric model held sway.

Ptolemy's work earned such deep respect because it explained so much of the night sky's behavior. Ptolemy's astronomy made accurate predictions that were in accord with an increasing body of observations, all made with the naked eye. Eventually, the power and vibrancy of the Islamic empires waned. Meanwhile, western Europe slowly awoke from the Dark Ages. Scholars from Spain to Poland rediscovered Ptolemaic astronomy, as well as classical Greek thinkers like Plato and Aristotle. This rediscovery was made possible by Arab scholars who translated and preserved the Greek works, which were later translated into Latin. By 1400, Ptolemy's view was firmly established as the dominant astronomical model across a reinvigorated Europe.

Ptolemy's work became popular in Europe not only because of its predictive power, but also because of the worldview it supported. The Catholic Church had been the dominant political power for centuries. Catholic scholars fused interpretations of the Bible with the Earth-centered astronomy of Ptolemy and the Earth-centered physics of Aristotle. Aristotle's physics—with its distinction between the lower, imperfect "sublunar" (below the Moon) realm and the higher, perfect celestial realm—also fit neatly into the church's vision of a perfect God in a perfect Heaven.

The combination of Ptolemy, Aristotle, and the Catholic Church meant that arguing against a geocentric Universe could lead to charges of heresy. Given the absolute authority of the church and its harsh punishments for heretics (burning at the stake was a common sentence), challenging Ptolemaic astronomy was not something to take on lightly.

FIGURE 3.1
Early Middle Eastern Astronomers
Middle Eastern scientists, such as those depicted here working in Turkey, kept the flame of astronomical knowledge burning during the Dark Ages. Part of their legacy remains in the names of stars such as Aldebaran, Deneb, and Rigel.

SECTION SUMMARY
Setting the Stage
● Ptolemy's geocentric Solar System was the dominant model of the Universe for 1,400 years and found a powerful supporter in the Catholic Church.

A Simpler Plan: Copernicus's Heliocentric Universe

"It was so much against what seemed to be common sense," says Owen Gingerich, one of the foremost historians of astronomy and an acknowledged expert on the Copernican revolution. Gingerich is speaking about the famous book *De revolutionibus orbium coelestium* (*On the Revolutions of the Celestial Spheres*) by Nicolaus Copernicus, who published it just before his death in 1543. The "revolutions" Copernicus was writing about were the orbits of planets, but by changing humanity's conception of the Universe, the book was also revolutionary in a political sense.

Copernicus was born in Poland in 1473, during the rise of the remarkable era of learning and upheaval known as the **Renaissance** (**Figure 3.2A**). He was broadly educated, studying law, medicine, and astronomy and spending a significant amount of his time as a scholar in Italy. By the time he reached his late thirties, he was already exploring alternatives to Ptolemy's model.

Copernicus disagreed with two features of the Ptolemaic model (**Figure 3.2B**). First, Copernicus was sure that the Sun, not Earth, held the central position in the Solar System. Second, Copernicus firmly believed in uniform circular motion. Recall that to account for the varying speeds of planets across the sky, Ptolemy was forced to make planetary motion constant relative to an arbitrary point off center from the orbit (the equant). This approach was aesthetically and philosophically offensive to Copernicus, so he set out to develop a full mathematical account of a heliocentric Solar System built on circles and uniform circular motion.

Thus the greatest scientific revolution in the history of humanity was driven more by aesthetics than by questions of data and predictions. Copernicus set himself the task of reordering the Solar System and developing a working model—based on planets moving on circular paths—that could make reasonable predictions for planetary motions. As it turned out, Copernicus's new model did not predict the motion of planets much better than Ptolemy's had, but it did tell a completely different story about the Solar System's architecture.

Copernicus's model had three critical features. The first and most obvious is that it placed the Sun at the center of the "planetary system." All planets including Earth (Copernicus left the Moon in orbit about Earth) moved around the Sun. It is noteworthy that, despite his desire for simplicity, Copernicus was forced to shift the center point of the orbits away from the Sun and add epicycles—though, as we will see, he did not need these in order to make the model reproduce the all-important retrograde motions.

The second critical feature of Copernicus's model was the daily rotation of Earth. "Most people could not envision Earth spinning around, because it seems quite solid to us," explains Gingerich. "People naturally asked, 'If it's spinning, why don't we fly off?'" Since Copernicus and his followers did not yet have a fully developed theory of gravity, it took some intellectual daring to imagine how Earth could spin each day on its axis without our

Renaissance A cultural movement spanning the 14th to 17th centuries that encompassed a flowering of education and art, much of which was based on principles established by the ancient Greeks.

A

FIGURE 3.2
Nicolaus Copernicus
(A) Nicolaus Copernicus (1473–1543) developed a heliocentric model for the motions of the Sun and planets in the Solar System. His controversial work was published just before his death. (B) Copernicus's heliocentric model contrasted sharply with Ptolemy's geocentric model.

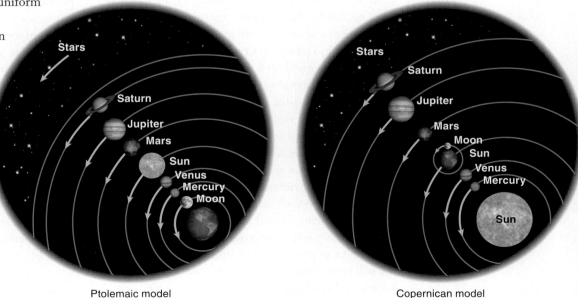

B

Ptolemaic model

Copernican model

feeling its effects. (Of course, the Copernicans could argue that we see effects of Earth's rotation every 24 hours, as daylight turns to night and back again.)

SECTION SUMMARY
A Simpler Plan: Copernicus's Heliocentric Universe
● Copernicus published his heliocentric model in 1543, taking great care to validate his controversial theory over long years of observation.
● Copernicus's model correctly depicted the relative positions of Sun and planets and the rotation of Earth.

The Copernican Model and Retrograde Motion

interactive
Retrograde Motion

The third and most important aspect of Copernicus's model was its ability to cleanly recover the retrograde motion of planets that had vexed astronomers for so long. Copernicus came to this explanation by assuming that the periods of planetary revolution—the time they took to complete one orbit—increased with further distance from the Sun. Thus Mercury, which is closest to the Sun, completes an orbit faster than Venus does. Venus completes its orbit faster than Earth. Earth completes its orbit faster than Mars. And so on.

Because of these different speeds, all retrograde motion is simply a catch-and-pass effect. Faster-moving inner planets "lap" the outer ones. As one planet passes another, its motion—projected against the background of fixed stars—appears to stop and change direction not once, but twice. This feature of Copernicus's model, along with orbits centered on the Sun, allowed retrograde motion to be explained naturally and simply without resorting to epicycles like those needed for Ptolemy's geocentric model.

To get a feel for how retrograde motion works, imagine you're a passenger in a car driving around a big circular racetrack with well-marked lanes. The track has an outer wall plastered with ads. Your car stays in its lane and keeps its speed steady. Imagine for a moment how the ads on the wall look to you as you speed around the track. Now imagine that your friends are driving in another car, in a lane farther from the center of the track than yours is. They're also moving more slowly than you are. Because of the differences in speed and the distance from the center (which determines the difference in each circle's circumference), you will complete one lap faster than they will. Now imagine that you keep your eye on them as you complete a lap. Don't forget that you will be watching their car as it moves against the background of ads on the wall. When you're approaching their car, you will see them moving smoothly against the ads. But then something interesting happens.

As you get close and prepare to lap their car, the line of sight connecting your eye, their car, and a point on the wall swings around. It goes from pointing toward your direction of motion to pointing backward, away from your direction of motion. This is what "catch and pass" means. If you made a film looking out your window, you would see your friends' car appear to slow, stop, move backward (relative to the background), slow, stop, and then move forward again. Note that it's the *projection* of their motion against the fixed background of ads on the wall that matters. Although their speed against the background appears to change, both you and your friends have really been moving at constant (though different) speeds the whole time.

Copernicus showed how each planet's motion against the background of fixed stars could be explained by this catch-and-pass effect (**Figure 3.3**). The geometry was complex, but the underlying idea was still much simpler than Ptolemy's.

SECTION SUMMARY
The Copernican Model and Retrograde Motion
● Copernicus's model solved the mystery of retrograde motion of the planets by suggesting that it was merely a visual effect caused by faster-moving inner planets lapping slower-moving outer planets.

The Size of the Universe

There was one more benefit to Copernicus's system. In a geocentric model, the size and speed of each planet's orbit was arbitrary. You could assume larger orbits with faster-moving planets or smaller orbits with slower planets. In Copernicus's model, the need to get the retrograde motions correct determined the sizes and speeds of the orbits. Thus, Copernicus was able to use his model to get an approximate sense of the size of the Solar System based solely on its motions.

The difference in size between the Ptolemaic and Copernican models was startling. The volume of the Copernican Universe, in terms of the distance to the "starry frame" surrounding the Solar System, was at least 400,000 times greater than that of the Ptolemaic cosmos. This increase was the first of many times that scientific astronomy would enlarge its estimate of the size of the Universe, making humanity's place ever smaller and less significant.

Copernicus did not publish his book until the end of his life, and there has been considerable debate about why he held back. Some scholars assume that Copernicus was afraid of persecution by the Catholic Church. According to Gingerich, it was probably the ridicule of other professional astronomers that Copernicus feared.

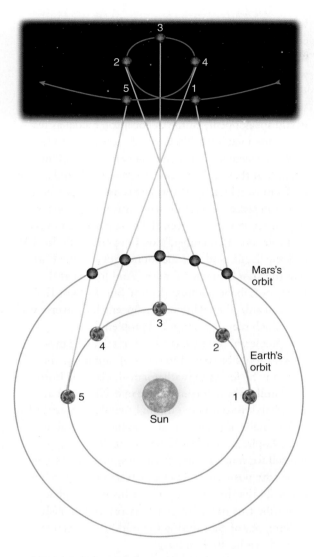

FIGURE 3.3 Retrograde Motion
The orbit of Mars appears to execute a "loop" against the fixed background of stars when viewed from the perspective of the more rapidly orbiting Earth. Thus, in the Copernican heliocentric model, retrograde motion occurs via a catch-and-pass effect due to inner planets orbiting faster than outer planets.

"Copernicus was reluctant to publish for fear that he would be 'hissed off the stage,'" Gingerich concludes. "He wanted to get things right." But while Copernicus may have been more concerned with his professional reputation than with charges of heresy, politics would soon intrude in astronomy, as Copernicus's idea began to make its way across Europe.

SECTION SUMMARY
The Size of the Universe

- The Copernican model provided an estimate of the Universe's size that was 400,000 times larger than in the Ptolemaic model.

3.2
Planets, Politics, and the Observations of Tycho Brahe

The northern nation of Denmark would host the next stage of the Copernican revolution under the guidance of Tycho Brahe, a brilliant and ambitious astronomer (**Figure 3.4**). The son of a Danish nobleman, Tycho was a contentious fighter who, as a youth, lost a considerable chunk of his nose in a sword duel while defending his reputation as a skillful mathematician. For the rest of his life Tycho wore a bronze nose "bridge" as a memento of the fight.

Tycho was a born scientist who developed many powerful methods of observation using only the unaided eye. On November 11, 1572, while looking at the constellation of Cassiopeia, Tycho observed the appearance of a tremendously bright "new star," which he called a *nova*. (He actually observed what is now known to be a supernova, the titanic explosion of a massive star.) The dramatic brightening of a star was such a rare event that it was deeply disturbing for people of the time, who saw these changing stars as ill omens cast in the otherwise constant sky.

Tycho carried out meticulous measurements of the new star before it faded after a few months. With his new methods of observation, Tycho demonstrated that the nova showed no parallax (see Chapter 2). Therefore, the new star could not be in the sublunar or atmospheric realm. It was as distant as the other stars—a conclusion that directly challenged Aristotle's ideas about the Universe. Remember that according to Aristotle, the heavens were perfect and unchanging. By showing that the new star was situated beyond the Moon in the celestial realm, Tycho put the first nail in the coffin of Aristotle's cosmos.

FIGURE 3.4
Tycho Brahe
The accurate naked-eye observations of Tycho Brahe (1546–1601) became the basis for a true understanding of planetary motion.

SECTION SUMMARY
Planets, Politics, and the Observations of Tycho Brahe

- Danish astronomer Tycho Brahe designed and used a highly accurate system for capturing naked-eye measurements of planets' motions and made meticulous records of his observations.

- Tycho also observed a supernova, a changing star, which contradicted ancient Greek ideas about unchanging celestial perfection.

Owen Gingerich

"I'm a lapsed astrophysicist," says Owen Gingerich, a professor emeritus at Harvard University. Recognized as one of the world's foremost experts on the Copernican revolution, Gingerich never expected to become a historian. Astronomy was his first love, and it was a romance that began early.

As far back as his childhood in Iowa, Gingerich knew with certainty that he wanted to be an astronomer. "I remember one very hot summer night," he recalls. "My mother took cots out in the backyard for us to sleep." Looking up into the sky, Gingerich asked, "Mommy, what's that?" "Those are the stars," she answered. "You've often seen them." It was a moment when the world opened up for the young boy. "I didn't know they stayed up all night," he recalls, laughing. His future path was set.

Gingerich went on to study astronomy at Harvard, where he eventually became a professor of astrophysics. "I was trained in the pioneering days of high-speed computing," he says, "and worked on studies of stellar atmospheres."

Gingerich's own turn to history, when it came, was more accident than intention. "There were all these wonderful anniversaries coming up," says Gingerich, referring to the 400th anniversary of Kepler's birth in 1971 and the 500th anniversary of Copernicus's in 1973. "That is what got me started working on the history of science as a way to understand how science works. I've been at it ever since."

3.3
Kepler and the Laws of Planetary Motion

FIGURE 3.5
Johannes Kepler
Johannes Kepler (1571–1630) used Tycho's meticulous observations of planet positions to develop his three laws of planetary motion.

Religious politics brought together Tycho Brahe and the next figure in our story, Johannes Kepler (**Figure 3.5**). "Johannes Kepler was a brilliant young high school teacher when the Counter-Reformation swept in," says Owen Gingerich. The Counter-Reformation was the often violent Catholic response to the rise of Protestantism. "Suddenly Kepler, along with all the other Protestant teachers, found himself out of a job," says Gingerich. Forced to look for work elsewhere, Kepler took up an offer to become Tycho's assistant in Prague. "Tycho always needed mathematically inclined assistants, and he quickly saw that Kepler was a brilliant mathematician." In Prague, Kepler took his first steps in formulating a more accurate vision of heliocentric planetary orbits.

Returning to the Copernican Model

While Copernicus's model of the Solar System could be described using geometry and was easier to work

with than the Ptolemaic system, the mathematics could not be expressed in simple form. Kepler was sure that Copernicus's Sun-centered model of the Universe was correct, but he was equally sure that nature's plan should be simple, elegant, and easy to express.

Kepler was one of the few astronomers who would speak openly of the heliocentric model as reality. "Copernicus's work was widely accepted as a recipe book for calculating planetary positions," says Gingerich, "but there was no way astronomers would accept it as a physical reality. It was just so much opposed to common sense." Common sense was not the only reason for resistance, however. Religion played an important role too. "Remember," says Gingerich, "Psalm 104 says the 'Lord God laid the foundations of the Earth that they not be moved forever.' And Joshua at the Battle of Gibeon commanded the Sun and not Earth to stand still. The reality of a heliocentric Universe was very much seen as a scriptural problem."

Kepler, however, was a kind of scientific mystic who could be moved to states of rapture by the harmony he found in mathematical relations. Born of a family whose fortunes were in decline, he had completed his schooling through scholarships awarded on the merit of his remarkable mathematical abilities. Despite his talents, Kepler struggled to support himself for much of his life and was unable to secure a university position. It was while teaching high school at age 24 that he claimed to have his first epiphany about the true nature of planetary orbits, famously walking out of class during the middle of a lecture to write down his flash of insight.

In Kepler's first mathematical model of the Solar System, planetary orbits were represented by nested three-dimensional geometric figures called Platonic solids. Kepler quickly wrote his theory into a book called *Cosmic Mysteries*. Although he ultimately found that Platonic solids did not match observations, Kepler's book earned him the attention of influential European astronomers, including Tycho.

During his time in Prague, Kepler gained access to a small portion of Tycho's data for the motion of the planet Mars. Kepler knew these were the most precise observations in existence, and with Tycho's data he was sure he could discern the true nature of planetary orbits—the laws governing their shape and their changes in speed. He begged Tycho to give him the rest of the Mars data. "Tycho was extremely wary about sharing too much data with somebody who was not yet a tried-and-true disciple," explains Gingerich. "It was not until Tycho died and Kepler got access to all the observing books that he could really make great progress."

Most important for the history of science, Kepler's final results could be described in terms of three general laws. The simplicity, elegance (mathematically speaking), and compactness of Kepler's three laws became a model for scientific descriptions of nature and ushered in a new era of scientific investigation.

> ### SECTION SUMMARY
> **Returning to the Copernican Model**
> - Tycho's death provided Kepler with access to previously withheld data on Mars.
> - Using only observational data, and without even understanding the underlying forces at work, Kepler defined three laws of planetary motion.

Kepler's First Law: Planets Move on Elliptical Orbits

Kepler recognized early on that uniform circular motion would never account for Tycho's detailed observations of planetary motion. He resolved to abandon the 2,000-year-old Greek prejudice for circles and to search for the true form of the orbits. After extensive experimentation, Kepler found that Tycho's precise data for Mars could be explained using an orbit based on the geometric form called an *ellipse*.

The ellipse had been known since the time of Pythagoras. Ellipses look like squashed circles (a circle is actually just one form of ellipse). You can draw a circle by pinning a length of string to a sheet of paper, keeping the string taut, and tracing a pencil around the pinned center point. An ellipse can be constructed in

a similar way, except ellipses have two points (called **foci**) that must be used in their construction. Imagine you have a piece of string and two pins. Place two pins at the points that will define the foci of the ellipse and tie each end of the string to the pins. Now pull the string taut around the pencil and trace out a closed line. The shape you trace out will be an ellipse. Notice that if the two pins are closer to each other, the shape sketched out becomes more circular. If the pins are farther apart, the shape looks more like a cigar.

Kepler found that Mars traces an elliptical path through space with the Sun at one focus (the other focus plays no role in the planet's motion). In time, he determined that all planets move on elliptical orbits. (Ultimately, the most dramatic example was provided by the highly elliptical orbit of Pluto, though the dwarf planet was not discovered until three centuries after Kepler's death.) This **first law of planetary motion** was an enormous advance. For the first time, a single shape could account for the path of the planets without need for epicycles or other geometric devices.

A few quantities associated with ellipses are important for describing planetary motion. The **semimajor axis** is half the length of the longest axis that can be drawn across the ellipse. The **semiminor axis** is half the length of the shortest axis that can be drawn across the ellipse (**Figure 3.6**). Another important quantity is the **eccentricity** (*e*) of the ellipse, which is the ratio of the distance from the ellipse's center to its foci, divided by the length of the semimajor axis. An eccentricity of 0 gives a perfectly circular orbit, while an eccentricity approaching 1 gives an orbit that is essentially an oscillation back and forth along a straight line.

interactive
Ellipses and
Kepler's First Law

focus (pl. foci) Either of two points interior to an ellipse that are used to define its shape. The Sun is always at one focus of a planet's elliptical orbit.

first law of planetary motion The principle, advanced by Johannes Kepler, stating that planets move on elliptical orbits with the Sun at one focus.

semimajor axis Half of the long axis of an ellipse.

semiminor axis Half of the short axis of an ellipse.

eccentricity A measure of the roundness of an ellipse, calculated as the ratio of the distance from the ellipse's center to its foci, divided by the length of the semimajor axis.

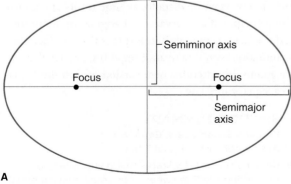

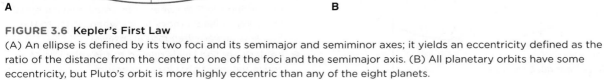

A

B

FIGURE 3.6 Kepler's First Law
(A) An ellipse is defined by its two foci and its semimajor and semiminor axes; it yields an eccentricity defined as the ratio of the distance from the center to one of the foci and the semimajor axis. (B) All planetary orbits have some eccentricity, but Pluto's orbit is more highly eccentric than any of the eight planets.

SECTION SUMMARY
Kepler's First Law: Planets Move on Elliptical Orbits

● Kepler's first law states that planets move on elliptical orbits (not on perfectly circular orbits, as had long been assumed).

Kepler's Second Law: Equal Areas Are Swept Out in Equal Time

Kepler's first law tells us the shape of a planetary orbit, but what about the planet's velocity? Kepler still had to account for planets' variable speeds as they trace out their orbits. By close analysis of Tycho's data, Kepler specified a simple rule for how a planet's speed changes as it completes an elliptical circuit. In doing so, he was able to abandon arbitrary mathematical contraptions like Ptolemy's equant.

Kepler's **second law of planetary motion** states that the line connecting a planet to the Sun will sweep out equal areas on its elliptical orbit in equal times. To understand what this means, imagine that you map out the orbit of Mars on a piece of graph paper with each square equal to one unit of area. Now imagine drawing a line connecting Mars to the Sun, which lies at one of the orbit's foci. Through intense

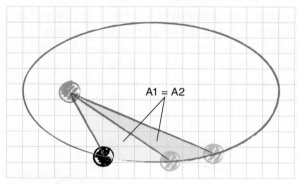

A Time interval: 1 month

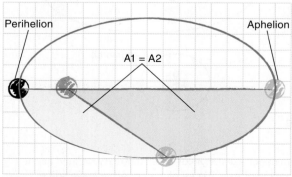

B Time interval: 5 months

study of Tycho's data, Kepler discovered that, every week, the line connecting Mars and the Sun sweeps out the same area—the same number of little boxes on the graph paper (**Figure 3.7**). The same relation held true if Kepler changed his unit of time from a week to a day or a month. In fact, it didn't matter what unit of time Kepler used, as long as it was held constant. Measure the area a planet sweeps out in its orbit in any two months, two weeks, or two seconds, and the area swept out in the two (equal) time intervals will be equal.

The importance of this law becomes apparent when you look at how a planet must move in order to fulfill the "equal area, equal time" rule. Kepler's second law tells us how the speed of motion changes as a planet traverses its orbit. Kepler found that planets speed up as they near the point closest to the Sun (*perihelion*) and slow down as they near the point farthest from the Sun (called *aphelion*).

Kepler's second law thus tells us more than where the planet's motion is fast and where it is slow. It tells us how to find exactly where the speed changes at every point in the orbit. That is its beauty and elegance.

Kepler's second law is based on an important principle called **conservation of angular momentum**, which relates speed and distance for rotating (or orbiting) objects. Conservation of angular momentum tells us that when an object is spinning around a central point, its speed will increase if its distance from the center decreases. This is the same physical principle that makes a spinning skater speed up when she pulls her arms inward and makes her rotate more slowly if she extends her arms (**Figure 3.8**). Similarly, according to Kepler's second law, a planet orbits the Sun quickly when its distance from the Sun is small (that is, at its perihelion) and orbits the Sun slowly when its distance is large (at its aphelion). The principle of conservation of angular momentum had not been formulated in Kepler's time; in fact, its formulation would take another 200 years. Kepler, in his genius, invented his own version of it in the "equal area, equal time" law.

SECTION SUMMARY
Kepler's Second Law: Equal Areas Are Swept Out in Equal Time

● Kepler's second law states that a line joining a planet and the Sun sweeps out equal areas in equal time; thus, planets move faster when closer to the Sun and slower when farther away.

● Kepler's second law is related to the principle of conservation of angular momentum.

Kepler's Third Law: The Period-Radius Rule

From his careful study of Tycho's data, Kepler developed a third powerful law that neatly expresses the relation between a planet's orbital period around the Sun and its average distance from the Sun. The **orbital period** (*P*) of a planet is the time the planet takes to complete one cycle around the Sun. Earth's period is a year. Mars has a period of 1.88 years.

A planet's distance from the Sun would never change if its orbit were perfectly circular (that is, if its eccentricity *e* = 0), because the radius of the planet's orbit would be constant. If, however, the planet moved on an elliptical orbit, its distance from the Sun (located at one focus of the ellipse) would be constantly changing. In that case, astronomers often consider the average radius of the planet's orbit (*R*; the sum of the closest and farthest approaches, divided by 2).

Kepler's **third law of planetary motion** is a relation between *P* and *R*. When the period is expressed in years and the average radius is expressed in astronomical units (AU, the average radius of Earth's orbit), Kepler's third law looks like this:

$$P^2 = R^3$$

In words, the third law tells us that a planet's period raised to the second power (squared) equals its average radius raised to the third power (cubed). This is a very powerful relationship because it is universal. Kepler found that every planet orbiting the Sun obeys this rule (**Figure 3.9**). The third law enables astronomers to convert the observation of a planet's period into knowledge of its average distance from the Sun, and vice versa. (**Going Further 3.1** on p. 60 gives more details on these calculations.) Compare the simplicity of Kepler's third law to the description of Ptolemy's model, with its equants and epicycles, and you will see what a dramatic change had occurred in astronomy in the 100 years separating pre-Copernican astronomy and Kepler's revolutionary discoveries.

With his three simple laws, Kepler was sure he had unveiled the true Copernican Universe. It is worth noting, though, that Kepler was never able to explain why his three laws existed. He imagined they might be the result of magnetic forces holding the planets in orbit around the Sun. The full story of gravity, orbits, and the reasons behind the three laws would have to wait another 50 years after Kepler's death for the genius of Isaac Newton. In the meantime, however, another revolution had to occur in the form of a different genius, who also believed in a heliocentric Universe. Far to the south of Kepler, another giant of the age was opening a new window on the night sky.

FIGURE 3.8
Conservation of Angular Momentum
Conservation of angular momentum makes a skater spin faster when his arms are pulled in closer to his body.

┌─ **SECTION SUMMARY**
Kepler's Third Law: The Period-Radius Rule

● Kepler's third law provides a formula for the period-radius relationship: the square of a planet's period in years equals the cube of its average orbital radius in astronomical units.

● Kepler's work shows that he fully supported the Copernican view of the Solar System.

interactive
Kepler's Third Law

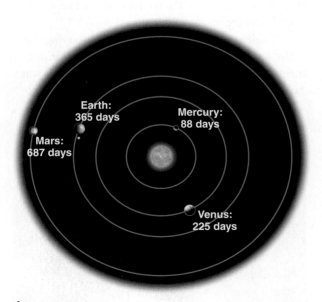

A

B

P^2 (yr)

R^3 (AU) ——→

Planet	Average orbital radius (AU)	Period (years)	Period (Earth days)
Mercury	0.387	0.24	88
Venus	0.723	0.61	225
Earth	1.000	1.00	365
Mars	1.524	1.88	687

FIGURE 3.9
Kepler's Third Law
(A) Kepler found the relationship between the period (in Earth years) and radius (in AU) of a planet's orbit to be $P^2 = R^3$. (B) This graph plots P^2 versus R^3 for the planets Mercury through Mars. All the points lie on a straight line, confirming the validity of Kepler's third law.

GOING FURTHER 3.1   THE LANGUAGE OF THE COSMOS
Working with Kepler's Third Law

Imagine that a new asteroid is discovered orbiting the Sun. When the asteroid is closest to the Sun (at perihelion), its orbit has a radius R_p = 3 AU. When the asteroid is farthest from the Sun (at aphelion), it has a radius R_a = 5 AU. How would we use Kepler's laws to find the asteroid's period?

The first thing we have to do in this problem is get the average radius of the asteroid's orbit. A simple way to find the average would be to add the distance from the planet to the Sun at perihelion (R_p) and at aphelion (R_a) and divide the result by 2:

$$R_{Avg} = \frac{1}{2}(R_p + R_a)$$

We know that the radii for the asteroid's aphelion and perihelion are 3 and 5, respectively, so we have

$$R_{Avg} = \frac{1}{2}(R_p + R_a) = \frac{1}{2}(3\ AU + 5\ AU) = \frac{1}{2}(8\ AU) = 4\ AU$$

Next we rearrange Kepler's third law and solve for the period. Remember that Kepler's third law states

$$P^2 = R^3$$

To solve for the period, we take the square root of both sides (which is the same as raising both sides to the power of 1/2), which gives us

$$P = R^{3/2}$$

We know that R = 4 AU, so we get

$$P = (4)^{3/2} = (2)^3 = 8\ \text{years}$$

Notice that 2 is the square root of 4. In equation form, this relation is written as $(4)^{1/2} = 2$. So when you see something like $(4)^{3/2}$, you can first take the square root and then raise the result to the power 3: $(4)^{3/2} = (2)^3 = 8$.

3.4
Galileo Invents New Sciences

Galileo Galilei was born in 1564 of a highly cultured family in Pisa, Italy (**Figure 3.10**). His father was a well-known musician and music theorist whose books on the harmonies of musical scales had affected Kepler's thinking about the structure of the Solar System. After briefly considering a life in the church, Galileo began medical studies at the University of Pisa. Mathematics and the sciences captured his interests more than cadavers, however. His talent was obvious to his teachers, and soon the young man was rising through the academic ranks on his way to becoming one of the most famous astronomers in Italy and across Europe.

Galileo and the Telescope

Much of Galileo's early success grew from his quick adoption of the newly developed telescope as a tool for astronomy. Galileo did not invent the telescope. Instead it was a Dutchman, Hans Lippershey, who first recognized that two appropriately sized lenses, placed at either end of a tube, could magnify the image of distant objects. Lippershey created his first telescopes in 1608, but most people saw them only as military tools. Galileo was the first to recognize the new invention's enormous potential for astronomy. In 1609 Galileo built his own telescope. Its magnifying power was dismal by modern standards, increasing the size of an image by a factor of only 8 (called 8× magnification). Later versions of the instruments Galileo used would have higher magnifying powers (30×).

Galileo set to work systematically observing the night sky with a series of ever-more-powerful instruments. In 1610 he published a short report of his work called *Sidereus nuncius* (*The Starry Messenger*). The

FIGURE 3.10
Galileo Galilei
(A) Galileo (1564–1642) was one of the greatest scientists in history and played an essential role in establishing the modern practices of science, such as the use of experiments and direct observations. (B) Galileo's telescopes were low-power refractors, much less powerful than those available to amateur astronomers today, but he made discoveries with them that revolutionized astronomy.

A

B

book's publication rocketed Galileo to fame and social status, turning him into a kind of Renaissance rock star. Combining his newfound fame and considerable confidence in his own abilities with his firm belief in heliocentric astronomy, Galileo was sure he could single-handedly win acceptance for Copernicanism.

Galileo's studies established a substantial list of key observations supporting the heliocentric worldview. His method of using direct observations to probe new ideas helped establish the scientific process discussed in Chapter 1. Such observations led to a number of important discoveries.

MOUNTAINS ON THE MOON. Although even a naked-eye observer can see that the face of the Moon looks mottled, Renaissance astronomers held that this "imperfection" was just an optical illusion. Most astronomers accepted Aristotle's claims (and those of the Catholic Church's scholars) that the Moon had to be a perfect, unblemished orb (like a heavenly Ping-Pong ball). When Galileo trained his telescope at Earth's nearest neighbor, however, he saw anything but a smooth sphere. To his astonishment, Galileo saw the lunar surface pockmarked with craters, ridges, and mountains. When he trained his telescope at the Moon's terminator—the line between its sunlit and shadowed sides—he quickly understood that the patterns he saw there were patterns of light and shadow (**Figure 3.11**). The features he observed on the Moon had to be high enough to cast long shadows across the lunar surface. Using his knowledge of geometry, Galileo was able to calculate shadow lengths and mountain heights. His result showed him that the mountains rose as high as 5 miles above the lunar surface. Through his

observations, Galileo demonstrated that the surface of the Moon was much like that of Earth, and in so doing, he dealt Aristotle's theory of differing sublunar and celestial realms its heaviest blow.

THE PHASES OF VENUS. Venus always appears near the Sun on the sky, so it is visible only close to sunrise or sunset (and therefore Venus is called the "morning star" or "evening star"). With his telescope, Galileo was able to resolve the disks of Venus and other planets. This in itself was a stunning advance. Previously, the planets had been mysterious, wandering bright points of light, but Galileo was able to see the face of the planets for the first time. By training his telescope on Venus throughout its 224-day orbit, Galileo saw the planet move through phases, as the Moon does (**Figure 3.12**).

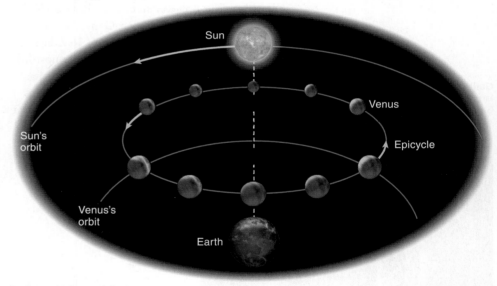

A Geocentric model

FIGURE 3.11 Lunar Mountains
Galileo used observations of shadows and geometry to compute the heights of mountains on the lunar surface.

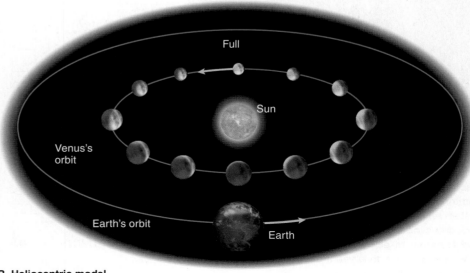

B Heliocentric model

FIGURE 3.12
The Phases of Venus
Galileo observed the phases of Venus with his telescope. He showed that the phases of Venus required by the (A) geocentric model were not the phases he observed with his telescope. His observations thus provided an additional piece of evidence for the validity of the (B) Copernican heliocentric model.

Recognition of the phases of Venus presented Galileo with a strong argument against geocentric models. In Ptolemy's astronomy, both Venus and the Sun orbited Earth. That meant the Sun and Venus would always remain at fixed distances from and orientations to Earth. It was therefore impossible to create the perspective from which Venus would ever appear as nearly full. A heliocentric model could naturally predict the correct phases for Venus, however, so Galileo's discovery served as evidence for the Copernican system.

THE MOONS OF JUPITER. Training his telescope on the planet Jupiter, Galileo made his most remarkable discovery. Not only did he make out the face of the planet but just a few months of observing showed him that the planet included its own family of orbiting satellites.

There are four **Galilean moons** (as these satellites came to be called): Io, Europa, Ganymede, and Callisto. Their orbital periods range from 16.7 days (for outermost Callisto) to 1.8 days (for innermost Io). These periods are short enough that even in less than a month of observation, Galileo could see the moons swing from one side of Jupiter to the other. It was clear that the inner moons orbited faster than the outer ones, just as Copernicus had predicted for the planets in his Sun-centered Solar System. While the discovery of Jupiter's moons did not prove that Copernicus was right, it was a blow for followers of Ptolemy because it showed conclusively that not every object in the Solar System orbited Earth. In essence, Galileo discovered that Jupiter formed its own mini-Copernican system (**Figure 3.13**).

SUNSPOTS. Training his telescope on the Sun, Galileo was able to make out dark spots appearing on the Sun's face. Like the mountains on the Moon, the existence of these solar "blemishes" stood in stark contradiction to the church-sanctioned "perfect realm" of the heavens that was the backbone of the Aristotelian cosmos.

THE MILKY WAY. Through the magnifying power of the telescope, Galileo was able to resolve the structure of the Milky Way (**Figure 3.14**). Rather than

Galilean moons The four largest moons of Jupiter: Io, Ganymede, Callisto, and Europa.

A

B

FIGURE 3.13 Galilean Moons of Jupiter Observing "stars" close to Jupiter executing periodic motions around the planet, Galileo reasoned that he had discovered moons orbiting the planet. The fact that Jupiter and its moons represented a mini-Copernican system represented another point weakening the geocentric model of the Universe. (A) Images of Jupiter and its moons taken by visiting spacecraft. (B) Galileo's notes showing the observed movement of the moons.

FIGURE 3.14 The Milky Way Galileo was able to resolve some of the individual stars that make up the Milky Way, showing that it was not just nebulosity, as had been believed.

being simply a band of clouds, or **nebulosity** (a word used by astronomers for clouds in space), as scholars of the day believed, Galileo saw that the Milky Way was composed of innumerable stars packed too close together to resolve with the naked eye.

With his newfound fame, Galileo felt sure he could convince the Catholic Church at least to remain open to shifting away from its adherence to Ptolemaic astronomy. But while Galileo was a scientific genius, his ability to discern the dark paths of politics appears to have been lacking. With a few famous false steps, he found himself dragged before the church's most feared judicial body—the Inquisition—on charges of heresy.

SECTION SUMMARY
Galileo and the Telescope
- Galileo pointed the newly invented telescope at the sky and made remarkable discoveries, including mountains on the Moon, the phases of Venus, the moons of Jupiter, sunspots, and the true nature of the Milky Way as a band of countless stars.

Galileo's Trial

For all his abilities, Galileo appears to have been arrogant and vain. As his writings and arguments in favor of the Copernican system accumulated, he acquired many enemies within the church. Galileo was sure enough of his own stature that he took no notice of the storm growing around him. "He was always trying to convince the hierarchy that the Copernican system was at least a reasonable alternative hypothesis," explains Owen Gingerich. In 1616, Galileo traveled to Rome and was given an order: he could think about the Copernican theory all he wanted, but he had to stop teaching or writing about it. Galileo agreed to this condition but still maintained that his mechanical philosophy described the natural world better than any alternative explanation. He was sure that, in time, he would win the church over.

In 1623 a cardinal named Maffeo Barberini became Pope Urban VIII. The new pope favored the geocentric model (as did many in the church) but had been willing to entertain Galileo's arguments in favor of Copernicus. With the accession of Urban VIII to the papacy, Galileo believed the time was right to deal his enemies, and the Ptolemaic system, a killing blow. He published a book called *Dialogue on the Two Chief World Systems*. The *Dialogue* featured a discussion among three gentlemen—Salviati, Sagredo, and Simplicio—on the merits of the geocentric and heliocentric worldviews. In a stunning act of poor

judgment, Galileo made Simplicio the proponent of geocentrism. Thus, the pope's favorite arguments for Ptolemaic astronomy were given to a character whose name meant "dumb" or "naïve." With the publication of the *Dialogue*, Galileo came into the sights of the Inquisition. "The clutch," says Gingerich, "was released from the wheels of persecution."

"When the Inquisition came to try Galileo, heliocentrism was essentially not discussed," explains Gingerich. "Galileo didn't have a chance to give any kind of arguments in its favor. Instead the trial entirely turned on whether or not he disobeyed papal orders in publicizing the heliocentric system."

In the end, Galileo was condemned for publicly advocating the Copernican model (**Figure 3.15**). To save himself, he was forced to recant his views on astronomy and claim that they were only intellectual arguments and not the physical truth. In return for his recantation, Galileo received a relatively modest sentence of permanent house arrest. His eyesight was failing him and, in despair, he wrote to his beloved eldest daughter, "This universe that I have extended a thousand times has now shrunk to the confines of my body."

SECTION SUMMARY
Galileo's Trial
- Galileo erroneously believed he had enough support within the church to change its views on Copernicus and the heliocentric model of the Solar System.
- After being tried by the Inquisition, Galileo was forced to recant his beliefs publicly and was placed under house arrest for the remainder of his life.

FIGURE 3.15
Galileo's Trial
The Inquisition accused Galileo of heresy for professing his discoveries, but the greater issue was his disobedience to the church. In particular, Galileo was found guilty of heresy for publishing the *Dialogue on the Two Chief World Systems* (on the geocentric and heliocentric theories) in violation of a 1616 decree making heliocentrism formally heretical.

nebulosity Clouds in space.

The Creation of a New Physics

Galileo did not spend his last years mourning his loss of freedom. Just as he had challenged the worldview of Ptolemaic astronomy earlier in his life, in his last years Galileo overturned Aristotelian physics. He accomplished this second feat not with a telescope but through a series of ingenious experiments using wooden wedges, balls, and water clocks.

Galileo's greatest achievement during these last years was the development of a true experimental method for physics. For centuries, scholars had relied on the authority of classical authors like Aristotle in their discussion of scientific topics. They addressed issues such as the relation between **force** and motion by arguing about what Aristotle had said in his philosophical works. Galileo instead decided to ask nature directly. His experiments tested different ideas rather than debating them. In this way he made a number of discoveries that would prove crucial for the development of Isaac Newton's later ground-breaking theories.

The most important outcome of Galileo's investigation was his determination of how velocity, acceleration, and **mass** are related in falling bodies. **Velocity** is the distance a moving object covers in a certain amount of time (in miles per hour or kilometers per second [km/s], for example). The words **speed** and *velocity* are sometimes used to mean the same thing. Both measure an object's change in distance with time. Physicists, however, use the term *velocity* to mean both the change in distance with time and the direction of that change. The direction turns out to be a key aspect in the definition of velocity. **Acceleration** is the change in velocity per time (measured in units such as kilometers per second per second [km/s^2]) (**Figure 3.16**). According to Aristotle's physics, a falling object moved downward with an unchanging (constant) velocity. Using experiments, Galileo proved this was not the case. Falling objects accelerate as they drop downward. Galileo measured this acceleration due to gravity, and we know its value to be 9.8 meters per second squared (m/s^2) in modern units. In other words, for every second an object falls, its velocity will increase by 9.8 meters per second (m/s).

Galileo also found that the acceleration of falling objects does not depend on their mass. This observation seems counterintuitive; common sense seems to demand that a hammer should fall faster than a feather. Our common sense often is fooled, however, by the force of air resistance that will slow the

force An interaction between two bodies that, if unbalanced, can change their state of motion.

mass A property of matter that is related to its resistance to changes in motion or inertia. Mass is measured in units of kilograms.

velocity The change of distance and the direction in which it is changing per unit of time.

speed The change in distance per unit of time.

acceleration The change of velocity per unit time.

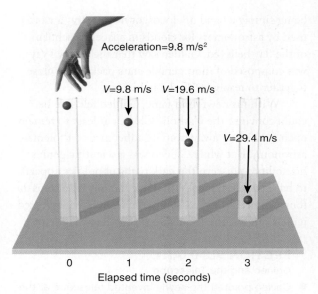

FIGURE 3.16 Acceleration Due to Gravity
At a constant acceleration, velocity is always changing, but in a uniform way. Here, a ball is dropped into a vacuum cylinder at constant acceleration due to gravity. Every second, the velocity of the falling ball increases by 9.8 meters per second.

feather's fall. Galileo's genius was the ability to idealize a situation—to ignore a particular aspect of a physical situation and thus more clearly understand it. Galileo was able to idealize away the effect of air resistance and conclude that the acceleration due to gravity occurs at the same rate for all objects. This conclusion contradicted Aristotle and opened the way for Newton to develop his universal law of gravity, discussed in Section 3.5.

Galileo wrote his final book, *Two New Sciences*, during the last years of his home imprisonment. It is in many ways his most important and insightful work, establishing many of the key features of what we now call the scientific method. While Galileo is sometimes outshined by the sheer creative power of his successor Isaac Newton, the world owes a great debt to the tireless and courageous Italian. It was Galileo's genius that first shed light on the Universe we have now come to take for granted.

SECTION SUMMARY
The Creation of a New Physics
- Galileo spent his final years pursuing an understanding of physics through experimentation.
- Galileo's most important achievement was defining the relationship of velocity, acceleration, and mass in falling bodies.

3.5

Newton and the Universal Laws of Motion

"We scientists are always looking for simplicity," says Liliya Williams, reviewing the progression from Copernicus through Kepler to Newton. Williams, a theoretical astrophysicist at the University of Minnesota, has spent a great deal of time thinking about the elegance of nature's laws. "When you find a description of nature that is simple, like an ellipse in the case of Kepler and planetary orbits, that means there is some deeper principle to which you are being led. You just have to walk one step further toward the underlying physics which makes the world behave so elegantly." In the search for the underlying physics of the Copernican model, that one further step was taken by Isaac Newton.

It is not an exaggeration to claim that Isaac Newton (**Figure 3.17**) was the greatest physicist, and perhaps the greatest scientist, who ever lived. No one else, except Charles Darwin or Albert Einstein, had so deep a vision of the structure of the world. As the poet Alexander Pope wrote, "Nature and Nature's laws lay hid in night / God said, 'Let Newton be!' and all was light."

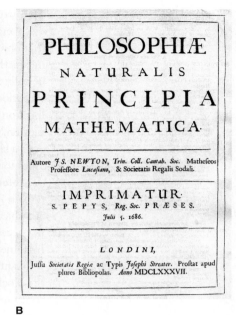

A   **B**

FIGURE 3.17 Isaac Newton
(A) Sir Isaac Newton (1642–1727) was one of the greatest natural scientists of all time. Basing his own explorations on the labors of Galileo, Kepler, and Tycho, he formulated a new physics in his famous work (B) *Principia*.

Newton's *Principia*

The irony of Newton's great scientific genius is how little time, over the course of his life, he actually spent doing science. He was born in 1642, the same year that Galileo passed away, and just 3 months after the death of his own father. Raised in farm country, the young Newton hated farming. Luckily, his gift for mathematics was recognized by a local teacher, who arranged for a scholarship to Cambridge (Newton's family at this time had sunk to the edges of poverty). At Cambridge, Newton's talent blossomed. Spurning studies of Aristotle, which composed much of the curriculum, the young Newton focused on the "new" astronomical writings of Copernicus, Kepler, and Galileo. In his third year at college, fate and the bubonic plague intervened, setting Newton on the road to immortality.

When the plague broke out in August 1665, the university closed and Newton returned to his rural home. There, in the space of just a few months, he began developing an entirely original vision of the relationship between force and motion—one that could encompass the physics of Earth and the heavens together. To make his new theory work,

Newton had to invent an entirely new branch of mathematics called calculus. He began all this before the age of 26 (though he did not publish much of it until later).

In 1687, Newton published his new physics in a book now known as the *Principia*. Forming the bedrock of the new physics were three universal laws of motion. "The cool thing about Newton's three laws," says Williams, "is that they are applicable to very different situations. Each law is a principle that can be applied to any object anywhere in the Universe at any time."

The three laws are usually listed as follows:

1. The **inertial law**. Objects in motion at constant velocity along a straight line stay that way unless acted on by a net force. Objects at rest stay at rest unless acted on by a net force.

2. The **force law**. The change in an object's acceleration due to an applied net force is in the same direction as the force and directly proportional to it, but inversely proportional to the object's mass.

3. The **reaction law**. For every applied force, there is an equal and opposite force.

Let's unpack each of Newton's laws to understand their meaning and importance.

inertial law The principle, advanced by Isaac Newton, stating that objects in motion at constant velocity along a straight line continue in that way unless acted on by a net force.

force law The principle, advanced by Isaac Newton, stating that the change in an object's acceleration due to an applied net force is in the same direction as the force and directly proportional to it, but inversely proportional to the object's mass.

reaction law The principle, advanced by Isaac Newton, stating that for every applied force, there is an equal and opposite force.

THE INERTIAL LAW. "The first law was actually discovered by Galileo," says Williams. "It recognizes that objects have this property we call inertia, which means they won't change their state of motion unless literally forced to do so. A wooden block sitting on a table doesn't spontaneously jump up and begin moving. You have to apply force, which sets the block into motion." If you have ever had to push a broken car even a short distance, you know how much force it takes to move a massive object.

Momentum, the tendency of a moving object to remain in motion, is a concept closely tied to inertia. An object that is moving at constant velocity will not accelerate or decelerate unless a force is applied. Imagine a wooden block moving through space at a constant speed. Unless some rockets are attached to the block to make it speed up or slow down, it will continue moving at a fixed rate forever. An object's momentum is the product of its mass multiplied by its velocity. We've already encountered angular momentum (the momentum associated with orbits and spin) in our discussion of Kepler's second law.

The first law also broadens the meaning of velocity to include direction. Thus, if an object has uniform velocity, it travels at a constant rate (meaning a constant speed) and in a constant direction. To change an object's direction (even if the speed does not change), a force is also needed (**Figure 3.18**). "The first law tells you that if no net force is acting on an object," says Williams, "then the object will either remain stationary or will continue at the same velocity (including direction) forever and ever."

THE FORCE LAW. "The second law," says Williams, "shows you what happens when an object does encounter a force." Whereas the first law states that changes in velocity demand a force, the second tells us explicitly what that relationship looks like. Put mathematically, the second law says that for a force F acting on a mass m, producing an acceleration a, the relation of the three quantities will always be

$$F = ma$$

or, in words, force equals mass times acceleration. The best way to understand this formula is to rearrange it:

$$a = F/m$$

This formula shows that when you apply a force to an object, the resulting amount of acceleration you see will depend on its mass. Notice that mass appears in the denominator (that is, the acceleration depends *inversely* on the mass). This means that a massive object (large m) will accelerate less (a will be small) than a low-mass object (small m) with the same force acting on it (a will be large). If you strap a weak rocket motor to a mountain-sized asteroid, it won't accelerate much. Strap the same rocket onto a Volkswagen (in space), and it will take off (that is, it will experience high acceleration; **Figure 3.19**).

"The amazing thing about Newton's second law," says Williams, "is it never specifies what kind of force we are talking about. The second law holds for any kind of force: gravity, magnetism, a horse pulling a cart. It doesn't matter. The second law is general and universal."

THE REACTION LAW. Newton's third law tells us that forces always come in equal and opposite pairs. When a skateboarder kicks off against a wall to get started, the wall pushes back. Perhaps the best example of the third law is a rocket engine operating in space. The fuel ignites in the rocket nozzle, creating a force that accelerates the gas outward, while an oppositely directed force accelerates the entire rocket in the other direction (**Figure 3.20**).

Together, Newton's three laws were revolutionary because, for the first time, people could understand, predict, and even control the behavior of moving

momentum The tendency of a moving object to continue in motion. Only the application of a force can change an object's momentum.

FIGURE 3.18 Newton's First Law
Newton's first law is the inertial law: Objects in motion at constant velocity along a straight line stay that way unless acted on by a net force. (A) Astronauts floating motionless in deep space (with no massive objects nearby) remain motionless. (B) Astronauts moving at constant speed remain moving at constant speed unless acted on by a force. (C) Astronauts wishing to change their velocity (accelerate, decelerate, or change direction) must exert a force such as using their rocket backpacks.

objects and the forces that affect them in situations as different as the motion of planets or the design of bridges or the trajectory of cannonballs.

SECTION SUMMARY
Newton's *Principia*

- Newton formulated three laws that revolutionized science: the inertial law, the force law, and the reaction law.
- Newton's laws are universal, in that they apply to all forces and all objects; they were an important step in science, linking forces and motion.

Newton's Universal Law of Gravity

In the century following Copernicus's publication of *De revolutionibus*, the dominance of Aristotle's concept of distinct sublunar and celestial realms had been slowly eroding. But it was Newton's brilliant conception of a universal law of gravity that united the heavens and Earth under one common physics.

Recall that Newton's second law ($F = ma$) never specifies what force is being discussed. Instead, it is a general law that tells us how an object (of mass m) will respond, via an acceleration a, to a force F.

Liliya Williams

Liliya Williams knew she was headed toward the abstract side of science early on. "I was born with an interest that steered me towards physics," she says, "preferably something that has very little to do with our everyday experience. I knew I wasn't going to do tabletop experiments."

Telescopes never worked for Williams. "I grew up in Moscow," she says. "When I was 9 or 10, somebody gave me a book on planets with colored pictures in it. My dad noticed my interest in astronomy and got me a telescope. I didn't take to it. I didn't enjoy it at all. But later I got hold of another book that was more mathematically inclined. That one I liked a lot."

Now a professor at the University of Minnesota, Williams is one of the world's experts on gravitational lensing—the use of Einstein's theory of relativity to predict how light will be bent and focused as it travels past large, invisible quantities of matter.

Williams still loves the beauty of mathematics and its applications in astrophysics after all these years. "It's when something just clicks that you feel great," she explains. "When one of those moments of realization comes and you understand something, oh boy! Even if it's something that has been understood for decades but you understand it for yourself the first time—that is a really wonderful feeling. Of course, it's even better when what you realized is something no one else has ever explored before, when it's something new in your own research."

Rocket carrying small car

A

Rocket carrying asteroid

Larger mass $\longrightarrow$ Smaller acceleration

B

FIGURE 3.19 Newton's Second Law
Newton's second law is the force law: The same force when applied to a large mass will yield less acceleration than when applied to a smaller mass. If the same rocket (same force) was attached to the smaller mass of a Volkswagen Beetle (A) and the larger mass of an asteroid (B), the Beetle would experience much higher acceleration than the asteroid.

FIGURE 3.20 Newton's Third Law
A rocket taking off from a launchpad is an excellent example of Newton's third law, the reaction law: For every applied force, an equal and opposite force arises. As hot gas flows out of the booster rockets, an impulse equal to the momentum of the escaping gas causes the rocket to move in the opposite direction.

GOING FURTHER 3.2   THE LANGUAGE OF THE COSMOS
Working with the Law of Gravity

So far, the mathematical computations in this book have focused on one example at a time. Another way to approach quantitative topics, however, is to think in terms of comparisons. Taking Newton's law of gravity as an example, suppose astronomers discover two planets, named (imaginatively) A and B. They are the same size (meaning they have the same radius). Their masses, however, are different. Planet A is 4 times more massive than planet B.

How much more force does a bowling ball on the surface of planet A feel compared with a bowling ball on planet B? In the comparative approach, the force equations for the two planets would look like this:

$$F_A = GM_A M_{ball}/R_A{}^2$$

$$F_B = GM_B M_{ball}/R_B{}^2$$

Planet A's equation could then be divided by planet B's equation:

$$F_A/F_B = (G/G)(M_A/M_B)(M_{ball}/M_{ball})(R_B/R_A)^2$$

All the quantities that are the same within each set of parentheses can be canceled:

$$F_A/F_B = (M_A/M_B)(R_B/R_A)^2$$

The planets are the same size ($R_A = R_B$), but planet A is 4 times more massive than planet B ($M_A = 4M_B$). Substituting these relations for M_A and R_A into the previous equation gives

$$F_A/F_B = (M_A/M_B)(R_B/R_A)^2$$
$$= (4M_B/M_B)(R_B/R_B)^2$$
$$= 4$$

Therefore, the force on the bowling ball on planet A would be 4 times larger than the force on planet B.

Newton then went further to provide an explicit relation for the strength of the particular force he called **gravity** between two masses (call them m_1 and m_2) that were separated by distance r between their centers of mass. For this *gravitational force*, Newton's second law took the form

$$F_g = Gm_1 m_2/R^2$$

Notice the label on the force, F_g. The subscript g is there to make sure we know that, in this case, we are talking about one specific kind of force: gravity.

The G in the equation is a number that doesn't change (it is the same for all masses and all distances), which Newton derived directly from observations. "G is an example of what physicists call a **constant of nature**," says Williams. "The size of G sets the strength of gravity. It's simply part of nature's architecture." In the course of this book, we will come across other constants of nature associated with other forces and other processes that shape our world.

Newton's law of gravity tells us that force depends on the mass of each of the two objects multiplied together: mass 1 times mass 2. If you increase the mass of one object by a factor of 2, then the force of gravity

between them will also increase by a factor of 2 (see **Going Further 3.2**).

Newton's law of gravity also tells us that the force between the two masses will decrease as the square of an increase in distance. In other words, gravity weakens with the *inverse square of the distance*. (Expressed mathematically, $F \propto 1/R^2$, where the symbol $\propto$ means "proportional to.") Thus Newton's gravitational force law is sometimes called the **inverse square law**: If you double the distance between two objects, the force between them drops by $2^2 = 4$. If you increase the distance between the objects by a factor of 3, the force drops off by a factor of $3^2 = 9$ (**Figure 3.21**). Newton's law does not explain *why* this is true; Newton simply discovered that the force of gravity depends on mass and distance.

"When you combine the inverse square law for gravity," says Williams, "with [Newton's] second law of motion, $F = ma$, then you can find how an object accelerates in response to gravity." By putting the inverse square law and Newton's second law together, we can see why an object's acceleration due to gravity does not depend on its mass. Recall what Galileo found in his experiments. "Galileo showed that if you

gravity   An attractive force between any two massive bodies that depends on the product of the bodies' masses and the inverse square of the distance between them.

constant of nature   A term that does not change from one situation or time to another. The value of a constant of nature must be measured to be determined.

inverse square law   A relationship whereby a quantity (such as gravity) decreases in proportion to the square of a variable (such as distance) as the latter increases.

GOING FURTHER 3.3 THE LANGUAGE OF THE COSMOS
Acceleration Due to Gravity

Let's say mass 1 is Earth (call it M_E) and mass 2 is an iron ball (call it M_i) and the distance between them (from Earth's center of mass, which is approximately its geometric center, to the iron ball) is R. If we want to know the acceleration of an iron ball when it drops, then we use Newton's second law as follows:

$$M_i a_i = F_g$$

Now we use Newton's law of gravity:

$$M_i a_i = GM_E M_i / R^2$$

Notice that something interesting happens in this equation. The mass of the iron ball (M_i) appears on both sides of the equation. That means we can divide by it and cancel it out of the equation:

$$a_i = GM_E / R^2$$

So the acceleration of the iron ball does not depend on its mass. If we used a second, less massive iron ball ($M_{i2} \ll M_i$), the same cancellation would occur and we would get

$$a_{i2} = GM_E / R^2 = a_i$$

Thus, the acceleration due to gravity for the massive iron ball and the smaller, less massive iron ball is the same. They fall with the same acceleration.

dropped a heavy object and a light object from the tower of Pisa, they would hit the ground at the same time," says Williams: "But he never showed the physics behind it. That is where Newton's law of gravity comes in. The acceleration due to gravity is the same for all objects because of the explicit form the law of gravity takes" (see **Going Further 3.3**).

Newton's law of gravity shows the important distinction between mass and weight. Mass is an intrinsic property of an object. Your mass does not change if you move from one planet to another. **Weight** is the downward force on the object produced by gravity. If you traveled to a different planet whose mass or radius differed from Earth's, your weight would change.

Using his law of gravity, Newton was also able to show that all three of Kepler's laws are simply a consequence of the inverse square law. Speaking of Kepler's first law of elliptical orbits, Newton once said, "Kepler guessed it was an ellipse, but I have proved it." As another example, with a little mathematics Kepler's third law, which states that a planet's period squared equals its average radius cubed, just pops out of Newton's rules of motion and his law of gravity. Even more important, Newton's version of Kepler's third law includes the mass of orbiting objects and thus allows astronomers to use their observations to measure the mass of distant planets, stars, and other celestial bodies (see **Going Further 3.4**, p. 70).

weight The force on an object that is produced by gravity.

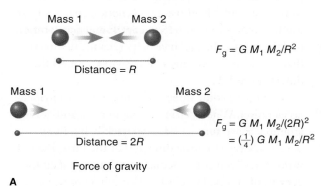

$$F_g = G\, M_1\, M_2 / R^2$$

$$F_g = G\, M_1\, M_2 / (2R)^2$$
$$= \left(\tfrac{1}{4}\right) G\, M_1\, M_2 / R^2$$

Force of gravity

A

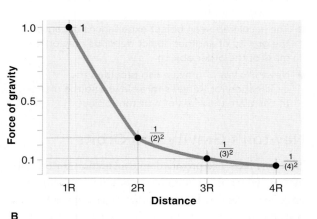

B

FIGURE 3.21
The Inverse Square Law
(A) The inverse square law says that the strength of gravity between two masses varies according to the inverse square of the distance between them, thus decreasing as the distance increases. (B) If we increase the distance between two objects by a factor of 2, the force of gravity between them decreases by a factor of 4.

✳ GOING FURTHER 3.4 THE LANGUAGE OF THE COSMOS
How to Weigh a Planet (Determine Its Mass)

In providing a mathematical basis for Kepler's third law, Newton's law of gravity was an enormous advance for astronomers, giving them, among other things, a way of measuring the mass of distant objects. Here's how it works.

Newton showed that for a planet with mass M_p orbiting the Sun, with mass M_{Sun}, the true form of Kepler's third law looks like this:

$$P^2 = [4\pi^2/G(M_{Sun} + M_p)]R^3$$

According to this equation, every planet should have a slightly different number in front of R^3, since the planet's mass (M_p) appears in the denominator. Why didn't Kepler have to adjust his third law a little for each planet?

As it turns out, the mass of the Sun, M_{Sun}, is so much greater than the mass of any planet, M_p, that you can ignore the planet's mass in the equation above. So, to a fair approximation, Newton's form of Kepler's third law can be written like this:

$$P^2 = (4\pi^2/GM_{Sun}) \times R^3$$

This equation can be used for every planet orbiting the Sun, confirming Kepler's discoveries. But how do we know the mass of the Sun? You can't just put the Sun on a scale. By rearranging Newton's form of Kepler's third law, however, we can "measure" the masses of astronomical objects. If we know a planet's period and its distance from the Sun, we can solve the equation for M_{Sun}, the mass of the Sun. The rearrangement looks like this:

$$M_{Sun} = 4\pi^2R^3/(GP^2)$$

To this day, astronomers use this relation to "weigh" an astronomical object by watching something else in orbit around it. In this way the mass of everything from asteroids to moons, planets, stars, and galaxies has been calculated. (In the back of this book you will find a table listing the masses of Solar System objects, including the mass of the Sun.)

Newton's law of gravity holds for objects on Earth and in the heavens. Aristotle's distinction between the two realms had finally been proved false.

SECTION SUMMARY
Newton's Universal Law of Gravity

● Newton determined that the force of gravity between two masses is proportional to their masses multiplied together and inversely proportional to the square of the distance between them. By experimentation, Newton defined the value of G, a naturally occurring constant that represents the intrinsic strength of gravity.

● The acceleration an object experiences because of the gravity of another object depends only on the mass of the other object.

● Newton's law of gravity also provided the explanation for Kepler's three laws and the means to calculate the mass of a distant object.

Newton's Gravity and Orbits

Newton's laws also provided an explanation for orbital motion, which is the foundation of much of astronomy. Newton's first law states that objects in motion will stay in uniform motion unless acted on by a net force. Remember that "uniform motion" means constant velocity, and velocity is both speed and direction. If you want to change an object's direction, you need to apply a force in that direction. Even if a planet orbits the Sun with a constant speed, its direction will always be changing. So for a planet to continue moving in orbit around the Sun, there must be a force that constantly changes its direction of motion. In general any force that keeps an object moving in a circle is called a *central force* because it points toward the center of the circle. For orbiting bodies that central force is gravity. Next, if forces produce accelerations, what kind of acceleration is at work when only the direction of motion changes in an orbit? Physicists use the term **centripetal acceleration** to describe the change in velocity (change in direction) that occurs as an orbiting planet swings around the Sun (**Figure 3.22**).

You have some direct experience with centripetal acceleration. When you're driving in a car and going around a sharp turn, you feel yourself being flung outward, away from the direction of the turn. The outward "force" you feel is actually an illusion called the **centrifugal force**. Here is what's really happening.

centripetal acceleration
Acceleration involving a force that keeps an object moving on a curved path. The acceleration is directed toward the center of curvature of the path.

centrifugal force The illusion of outward force when an object is moving on a curved path.

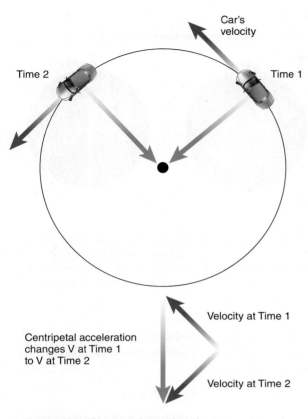

interactive
Circular Orbits

Before the car starts turning, you're moving in a straight line. As the car starts its turn, your body wants to keep moving in that straight line (consistent with Newton's first law). Since you are connected to the car by only your seat belt, at first you feel your body slide outward, along the seat, away from the turn. It feels as if there's a force, a centrifugal force, pushing you away from the direction of the turn. Only when the seat belt grabs hold of you is the real *centripetal force* applied to your body, forcing you to change direction along with the car. Of course, it was friction of the car wheels on the road that applied the centripetal force to the car itself.

Newton's laws also provide two relationships for understanding the properties of simple orbits. A basic question an astronomer often needs to answer is, "What's the velocity of an object in orbit?" The question is relevant for planets orbiting the Sun, Moons orbiting a planet, or a space probe orbiting Mars. Newton's laws provide an equation for the simplest kind of orbit: a perfect circle. Here is the expression for the **orbital velocity** of a satellite in a circular orbit V_c at a distance R from its parent body that has mass M:

$$V_c = \sqrt{GM/R}$$

Notice that the orbital velocity drops as the satellite moves farther away from the parent body. You might recall that this is exactly what Copernicus hypothesized when he introduced his heliocentric model: the inner planets orbit faster than the outer planets. Newton used his simple and elegant laws to show astronomers why Copernicus had to be right (**Figure 3.23**).

orbital velocity The velocity required to keep a body in orbit around another body.

FIGURE 3.22 Centripetal Acceleration
Physicists use the term *centripetal acceleration* to describe the acceleration required to change an object's velocity (change in direction) when it moves in a curved path, such as when a car rounds a corner or a planet travels in its orbit. For the car, the force of friction between the tires and the road produces the centripetal acceleration. In this illustration, the velocity of the car changes direction (red arrows) because of the inward-directed force (blue arrow), which yields the centripetal acceleration.

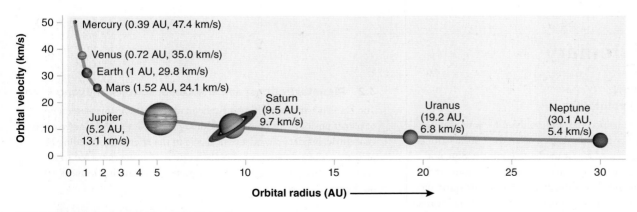

FIGURE 3.23 Orbital Radius and Velocity
A planet's orbital velocity decreases as the radius of the planet's orbit increases, just as the formula for circular velocity derived from Newton's laws predicts. (Keep in mind that the planets actually move on slightly elliptical orbits.)

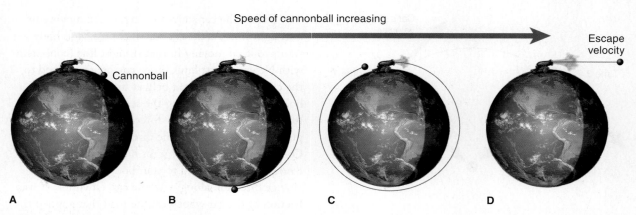

FIGURE 3.24 Cannonballs and Orbits
How fast does something have to be moving to be launched from the surface of a parent body into space? Newton thought about this problem in terms of a cannonball being launched from a powerful cannon: (A and B) Too little gunpowder and the cannonball will not reach escape velocity and will fall back to Earth. (C) Just the right amount will give the cannonball escape velocity and launch it into orbit. (D) Even more gunpowder, and the cannonball will be launched with much more than the escape velocity, sending it into deep space.

interactive
Escape Velocity

Another basic question for astronomers (and rocket engineers) is how fast something needs to be moving to be launched from the surface of a parent body into space. Newton thought about this problem quite a bit as he considered what an orbit really entails.

Think about launching a cannonball from a powerful cannon. If you don't use much gunpowder, the cannonball will arc through the air and drop a short distance away (**Figure 3.24**). If you use more gunpowder, the cannonball will make it farther from its launch position. Earth is essentially a sphere, so the farther the cannonball travels, the more it will be falling toward a surface that is, at the same time, curving away from it.

escape velocity The velocity required to escape the pull of gravity from a body.

Extending the idea a little further, Newton was able to find a relation for the speed that you would have to give the cannonball in order for it to escape the pull of Earth's gravity entirely. This **escape velocity** has a simple relation too:

$$V_e = \sqrt{2GM/R}$$

These two equations—for orbital velocity and escape velocity—are enormously powerful, and we will revisit them many times in the story that follows.

SECTION SUMMARY
Newton's Gravity and Orbits

● Newton's laws provided the basis for understanding orbits: the gravity from the parent body changes the direction of the orbiting object's velocity (centripetal acceleration) to keep it in orbit.

● Newton's laws also provided the mathematical relationships for orbital velocity and escape velocity.

●→ **chapter summary**

**3.1 Getting Past Ptolemy:
The Copernican Revolution**

Ptolemy's geocentric model remained the gold standard for interpreting the observed motions of the sky until the 1500s, when the Polish astronomer Nicolaus Copernicus developed a new model that put the Sun, not Earth, at the Solar System's center. Copernicus erroneously embraced the notion of uniform circular motion but correctly understood the spinning of Earth and the catch-and-pass effect of inner planets moving faster than outer ones that created the illusion of retrograde motion. His model also dramatically increased the projected size of the Universe.

3.2 Planets, Politics, and the Observations of Tycho Brahe

The Danish astronomer Tycho Brahe made meticulous measurements of planet positions and motions. Although he believed the Universe was geocentric, his data played a critical role in the later understanding of planetary orbits. In addition, his observation of a supernova led to the conclusion that the ancient notion of perfect, unchanging heavens could not be true.

3.3 Kepler and the Laws of Planetary Motion

The German mathematician Johannes Kepler believed in the Copernican model. He worked as an assistant to Tycho, and on Tycho's death he gained access to particularly valuable orbital parameters for Mars. Kepler used them for intense study that produced his three laws of planetary motion: (1) Planets move on elliptical orbits. (2) A line connecting a planet to the Sun sweeps out equal areas in equal time. (3) The period of a planet's orbit in years, squared, equals the average radius of the orbit in astronomical units, cubed.

3.4 Galileo Invents New Sciences

The Italian Galileo Galilei was the first astronomer to use the telescope for astronomical observations. What he saw would drastically change our view of Earth's place in the Universe. His belief in the Copernican model was bolstered by seeing sunspots, mountains on the Moon, the phases of Venus, and moons orbiting another world (Jupiter), and by identifying individual stars in the Milky Way. The Inquisition of the Catholic Church prevented Galileo from openly expressing his discoveries and views, and sentenced him to house arrest for the last years of his life, but he continued to advance science by using actual experiments as the basis for understanding physics.

3.5 Newton and the Universal Laws of Motion

Sir Isaac Newton revolutionized physics by discovering relationships between force and motion that hold true throughout the Universe. Building on the advances of Copernicus, Kepler, and Galileo, he conceived his three laws of motion: (1) The inertial law states that objects in motion at constant velocity along a straight line, as well as those at rest, maintain their motion (or lack thereof) unless acted on by a net force. (2) The force law states that the change of acceleration due to an applied net force is in the same direction as the force and directly proportional to it, but inversely proportional to the object's mass. (3) The reaction law states that for every applied force, an equal and opposite force arises. Equally important, Newton discovered the universal law of gravity, or inverse square law. He further deduced that the acceleration due to gravity on the surface of a world depends only on the mass and radius of the world and is thus the same for objects of different masses. Newton's work provided an understanding of orbits, including concepts such as orbital velocity and escape velocity.

●→ questions and problems

Narrow It Down: Multiple-Choice Questions

1. Ptolemy's model of the Universe was accepted as valid for over a thousand years. Which of these is *not* one of the reasons for its acceptance?
   a. It predicted the ever-changing positions of planets fairly well.
   b. An Earth-centered Universe was consistent with Catholic Church doctrine.
   c. Greek philosophy permeated many aspects of life (politics, medicine, physics).
   d. The lack of observable stellar parallax allowed for a stationary Earth.
   e. It correctly put the Sun at the center.

2. On which of the following point(s) was the Copernican model correct, according to our current knowledge? Choose all that apply.
   a. placement of the Sun at the center
   b. use of epicycles
   c. rotation of Earth
   d. shape of planetary orbits
   e. relative orbital speeds of the planets

3. Tycho Brahe is noted for all of the following achievements *except*
   a. observation of a nova.
   b. meticulous recording of observations.
   c. capturing accurate orbital data for Mars.
   d. innovative use of the telescope.
   e. developing a hybrid Sun- and Earth-centered model.

4. Which of the following statements about ellipses is true?
   a. A circle is a form of ellipse.
   b. An ellipse is a form of circle.
   c. All ellipses have four foci.
   d. Orbits are always ellipses with eccentricity equal to 0.
   e. Semimajor and semiminor axes are never equal.

5. Two ellipses have the same semimajor axes but different eccentricities. Which of the following statements is true? Choose all that apply.
   a. The ellipse with greater eccentricity has a longer semiminor axis.
   b. The area inside the ellipse with the smaller eccentricity is smaller.
   c. The semiminor axes of the two ellipses may be equal.
   d. One ellipse may be a circle.
   e. Both ellipses may be circles.

6. Which of the following statements is consistent with Kepler's second law?
   a. All planets move at the same speed when nearest the Sun.
   b. Orbits are actually areas, not paths.
   c. The orbital speed of a planet decreases when the planet is nearest the Sun.
   d. "Equal time, equal area" implies that planets move at constant speed in orbit.
   e. Planets always move faster when they are close to the Sun than when they are farther away.

7. A spinning figure skater spins faster as he brings his outstretched arms inward. This is an illustration of the physical principle underlying
   a. Kepler's first law.
   b. Kepler's second law.
   c. Kepler's third law.
   d. the law of gravity.
   e. none of the above

8. Which of the following statements is true regarding Kepler's third law?
   a. Periods and orbital radii are correlated only for planets in our Solar System.
   b. It says that the period and size of an orbit are correlated.
   c. It implies that orbital velocity increases as orbit size increases.
   d. It works only for orbits with eccentricity equal to zero.
   e. The relation changes to $P^2 = R^4$ for the outer planets.

9. Which of Galileo's observations disputed the perfect nature of the heavens above the sublunar region? Choose all that apply.
   a. mountains on the Moon
   b. phases of Venus
   c. individual stars in the Milky Way
   d. sunspots
   e. Galilean moons

10. Which of the following does *not* describe a contribution that Galileo made to science?
    a. He laid the groundwork for the practices of modern science, including observation and experimentation as the best processes for understanding nature and the Universe.
    b. With his discovery of the moons of Jupiter, he demonstrated that there was more than one center of motion in the Universe.
    c. He argued that objects of unequal mass will fall at the same rate.
    d. He theorized that there was a mutually attractive force proportional to distance to explain how and why planets orbit the Sun.
    e. He used his observations of the phases of Venus to argue that this planet clearly orbits the Sun, not Earth.

11. An astronaut is traveling in deep space far from any star or planet. To change his course, he must fire his rockets. Which principle does this example most clearly illustrate?
    a. Newton's law of inertia
    b. Newton's law of force
    c. Newton's law of reaction
    d. Newton's law of gravity
    e. none of the above

12. A gun recoils when it fires. Which principle does this example most clearly illustrate?
    a. Newton's law of inertia
    b. Newton's law of force
    c. Newton's law of reaction
    d. Newton's law of gravity
    e. none of the above

13. In the law of gravity, G is
    a. the gravitational force specific to Earth.
    b. a constant related to the strength of the gravitational force.
    c. the gravitational force specific to the Solar System.
    d. a value that differs depending on the objects that are nearby.
    e. immeasurable.

14. Given two objects, if the mass of one increases while the distance between them increases, how does the magnitude of the gravitational force change?
    a. It always increases.
    b. It always decreases.
    c. It always remains the same.
    d. It depends on the specific values of the mass and distance changes.
    e. It depends on the local value of G.

15. What is the effect on the force of gravity between two objects if the mass of one object remains unchanged while the distance to the second object and the second object's mass are both doubled?
    a. It always increases.
    b. It always decreases.
    c. It depends on the specific values of the two quantities.
    d. It depends on the local value of G.
    e. It cannot be determined.

16. Which of the following values is/are the same whether one is standing on Earth or on the Moon? Choose all that apply.
    a. one's mass
    b. one's weight
    c. one's weight-to-mass ratio
    d. the acceleration due to gravity
    e. the value of G

17. Two moons orbit a much larger planet. Moon α orbits at radius X. Moon β orbits at radius 4X. How does the planet's pull of gravity on each of the moons compare?
    a. The difference is so small that it can be considered zero.
    b. It is 2 times weaker for moon α.
    c. It is 4 times weaker for moon α.
    d. It is 4 times weaker for moon β.
    e. It is 16 times weaker for moon β.

18. Which of the following statements is/are true about orbital velocity versus escape velocity when considering a planet of radius R and a spaceship that will either orbit just above the planet's surface or attempt to escape the planet? Choose all that apply.
    a. They may be the same, depending on the radius of the central body.
    b. Orbital velocity is always greater.
    c. Escape velocity is always greater.
    d. They differ by a factor of 2.
    e. They differ by a factor of $\sqrt{2}$.

19. Which of the following describes Kepler's approach and contributions to astronomy?

    a. He constrained his models of planetary motion with the positional data from Tycho Brahe.

    b. He maintained the use of epicycles in planetary motion.

    c. He correctly showed that planetary orbits were based on geometric shapes called platonic solids.

    d. He ignored all the work of important astronomers who preceded him.

    e. He understood that planetary motions resulted from mutually attractive gravitational force.

20. Which of the following characteristics was/were included in Ptolemy's model of the Universe? Choose all that apply.

    a. Earth rotates on its own axis.

    b. The celestial sphere is perfect and unchanging.

    c. Planets move in circles on top of circles called epicycles.

    d. All the planets, as well as both the Sun and the Moon, revolve around Earth.

    e. The nested set of crystalline, celestial spheres is maintained as proposed by Aristotle.

To the Point: Qualitative and Discussion Questions

21. In what ways did the Greek philosophers contribute to the advancement of science? In what ways did they hinder it?

22. What factors contributed to the decline of the Catholic Church's influence in matters of science during the Renaissance?

23. What does the current use of star names such as Aldebaran and Alnitak tell us about some of the early astronomers?

24. Compare the features that Ptolemy's model required to explain retrograde motion with those in Copernicus's model.

25. Describe what one observes when a planet displays retrograde motion, and explain its cause.

26. Newton is often described as one of the greatest contributors to our modern understanding of the natural world. Summarize his most important contributions.

27. Kepler's second law implies that a planet increases its orbital speed as it gets closer to the Sun. What is the underlying principle of physics that explains this effect?

28. What strongly held beliefs of Ptolemy and Copernicus were later proved incorrect?

29. Describe the relationship between centripetal acceleration and a central force. Comment on the relationship between the direction of a central force and the direction of the velocity of an object executing circular motion.

30. What do we mean when we say that Newton's universal law of gravity is an "inverse square" law?

31. How would the value of the acceleration due to gravity for Earth, which is 9.8 m/s², differ if Earth's radius were larger but its mass were the same? Why?

32. The equation for Newton's universal law of gravity gives the magnitude of the gravitational force between two gravitating objects and includes the masses of both objects. The equations for both escape velocity and orbital velocity depend on the mass of only one of the objects. Explain why the escape velocity and orbital velocity are independent of the mass of the object whose velocity is being calculated.

33. How do some of today's major religions deal with scientific theories and discoveries (such as evolution, the age of Earth, birth of the Universe, advances in medicine)?

34. What scientists or visionaries in your lifetime would you consider comparable to those described in this chapter, in terms of their long-term impact?

35. The stories of early great astronomers and their contributions were all shaped by the intellectual climates that existed during their lives. How does today's intellectual climate encourage or discourage new discoveries in science? How might it differ for future generations?

Going Further: Quantitative Questions

36. What would be the semimajor axis, in astronomical units, of an elliptical orbit for a planet whose perihelion was at 4.5 AU and aphelion at 6.8 AU?

37. What would be the average distance from the Sun, in astronomical units, of a comet with a period of 47 years?

38. What would be the period, in years, of a planet in our Solar System whose average orbital radius was 26 AU?

39. Asteroid A has a mass of 2×10^{20} kilograms (kg), and asteroid B has a mass of 4×10^{18} kg. Assuming that the same force was applied to both (a shock wave from a supernova, for example), what would be the ratio of A's acceleration to B's acceleration?

40. How many times greater is the force of gravity on a 3-kg object lying on the surface of a moon than on a 3-kg object orbiting at a distance of three moon radii above the surface?

41. By what factor would the gravitational force of the Earth-Moon system change if the Moon were 3 times as far away and 3 times as massive?

42. What is the ratio of the orbital velocities of two satellites, each in circular orbit around Earth, given that satellite A orbits 1.7 times as far from Earth's center of gravity as satellite B? Give your answer in terms of B's velocity to A's velocity.

43. NASA is designing a spaceship that will land and launch from two moons of Jupiter. To plan correctly for the amount of fuel the spaceship will carry, the escape velocities of the moons need to be calculated. Moon A has a radius twice that of moon B, and a mass 5 times that of moon B. If moon B has an escape velocity of 1.2 km/s, what is the escape velocity of moon A?

44. What is the mass, in M_{Sun}, of a star observed to have a planet with an orbital radius of 4 AU and a period of 15 years?

45. A star is orbited by a planet at an orbital radius of 2.2 AU, and with a period of 1.6 years. How does its mass compare with that of the Sun?

 If your instructor assigns homework in **smartwork5**, access it at the Digital Landing Page for *At Play in the Cosmos*: **digital.wwnorton.com/cosmos**

A UNIVERSE OF UNIVERSAL LAWS

How Light, Matter, and Heat Shape the Cosmos

In this chapter you will learn about light and its interaction with matter, which is the principal way astronomers learn about the cosmos. After reading through each section, you should be able to do the following:

4.1 Light: The Cosmic Messenger
Describe the nature of visible light and where it falls on the electromagnetic spectrum.

4.2 Astrophysical Spectra
Describe the nature of blackbody radiation, and explain how the different forms of spectra are used.

4.3 Spectra and the Strange World of the Atom
Define the structure of an atom and how it accounts for the interactions of light and matter that we observe.

4.4 Telescopes
Identify the types of telescopes and compare their advantages and disadvantages.

4.5 Atmospheres and Their Problems
Explain how Earth's atmosphere affects our ability to observe at different wavelengths of light, and give examples of ways in which astrophysicists have addressed these issues.

The Rosette Nebula, a star forming cloud, glows in the red light of the H-alpha emission line of hydrogen atoms in the cloud.

4

4.1
Light: The Cosmic Messenger

FIVE MONTHS in Puerto Rico was not what Alyssa Goodman was expecting when she signed up for graduate studies in astrophysics at Harvard. But that was exactly where her graduate adviser sent her. "We had written this proposal to the Arecibo Observatory," she explains. "That's a giant radio telescope that lives in a collapsed limestone bowl down in Puerto Rico. It's 1,000 feet across, deep in the jungle, and very dramatic" (**Figure 4.1**).

Goodman's proposal was to use the extremely sensitive telescope to detect radio waves from vast clouds of interstellar gas and dust where new stars were forming. "We hoped to detect magnetic fields in the gas clouds using properties of the radio waves," says Goodman. "Everyone expected the signal to be so faint that we had to ask for a lot of telescope time. I mean a lot of time, like a half a year."

For a grad student living through wet winters in Boston, this project was not necessarily a bad thing. "I got a really good tan, okay, and yes, there was a pool, and yes, I lived in one of those little huts on the cliff just like Jodie Foster in *Contact*. But there was no Matthew McConaughey who came to the hut to visit me."

Somewhere in the middle of the project, however, Goodman found herself with time on her hands and decided to begin digging into the data she had already accumulated. "So I added up all the data from the previous 20 days of observing and, to my astonishment,

there it was! The signal was already there staring back at me. We had already actually detected the magnetic field!" The signals were coming from a star-forming cloud designated B1.

"I knew right then that this was going to be a really big deal," Goodman recalls, smiling at the memory. "I called up my adviser and said, 'What if I told you that I detected a magnetic field in B1?'" Goodman's thesis supervisor paused for a minute to take in the news and then replied, "I would say, 'Would you like to graduate tomorrow?'"

Goodman, now a professor of astrophysics at Harvard and a world expert on observations of star-forming clouds, looks back fondly on that moment. Her work is a prime example of how light captured in telescopes can be used to unlock the secrets of distant astronomical objects. By linking light to **matter** (anything made of subatomic particles such as quarks or electrons), astronomers can determine the strength of a magnetic field, the mass of an interstellar cloud, and the temperature of a star.

Astronomy Becomes Astrophysics

Newton's law of gravity opened the floodgates to a new understanding of the heavens. Using his inverse square law for gravitational force (see Chapter 3), astronomers were able to turn telescopic observations of celestial motions into detailed descriptions of forces between massive objects, even when those objects were invisible.

But gravity was only part of the story of astronomy from the 1700s to 1800s. Advances in the design and construction of telescopes during this period enabled astronomers to see deeper and with greater resolution. Optics, the study of light and its properties, was the science driving the advance in telescopes (**Figure 4.2**). With these powerful instruments, astronomers mapped the Solar System, determined the distances to the stars, and began a census of the Milky Way.

Telescopes could catch light, but they could not, by themselves, tell astronomers anything about the origin of the light. For that, astronomers needed the help of physicists, who had made strides toward understanding the nature of light and its relationship to matter. Most important, physicists had developed a new instrument called a **spectrograph**. "Spectrographs turned astronomy into astrophysics," says Goodman. "They let astronomers interrogate the light they catch in their telescopes and use it to figure out conditions in the stars."

matter Any substance made up of particles with mass that occupies space.

spectrograph A device for separating light from a source into its component wavelengths and measuring the energy at each wavelength.

FIGURE 4.1
Arecibo Observatory
Arecibo Observatory, located in Puerto Rico, is the world's largest single-dish radio telescope and largest single-aperture telescope, with a diameter of 305 meters.

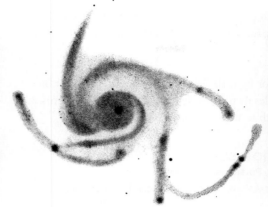

A **B** **C**

FIGURE 4.2 The Leviathan of Parsonstown
William Parsons, third Earl of Rosse and cofounder of the Royal Astronomical Society, carried out decades of astronomical observations, which were published in the leading astronomical textbook of the early 19th century. (A) His instrument was this 72-inch home-built telescope (nicknamed the "Leviathan of Parsonstown") at Birr Castle in Ireland. The telescope was the largest in the world from 1845 until the early 20th century and required two stone walls to support its great weight. (B) In the 1840s, Parsons studied and sketched "spiral nebulas" such as M51, the Whirlpool Galaxy, long before the existence of galaxies was generally accepted. (C) Although his instruments were primitive by today's standards, his sketch is remarkably accurate, as this photo of M51 shows.

---SECTION SUMMARY
Astronomy Becomes Astrophysics
● Advances in optics and the invention of spectrographs provided the observational data to draw new conclusions about light and matter.

Waves, Particles, and Light

What is light? As far back as Newton, scientists split into two camps on this question. Some, like Newton, argued that light was made of particles, like little bullets. Others, like the Dutch physicist and astronomer Christiaan Huygens, claimed that light was made of waves, like ripples on a pond. The differences between particles and waves are significant: a particle can be in only one place at a time; a single wave can spread over an extended region of space. By the early 1800s, experiments had shown light clearly demonstrating wavelike behavior, and scientists built strong foundations for understanding light in terms of wave properties.

There are many examples of waves in nature: wind-blown ripples on a pond, waves in the ocean, sound waves moving through the air. But any wave, including a light wave, can be described by three basic characteristics: wavelength, wave frequency, and wave amplitude. Any particular wave type (whether sound waves, ocean waves, or waves on a taut string) can appear in a range of wavelengths, frequencies, and amplitudes. **Wavelength** is the distance between

successive peaks or troughs of a wave (**Figure 4.3**). A wave's **frequency**, measured in hertz, is the number of peaks (or troughs) that pass a point in space in a single second. The **amplitude** of a wave is a measure of its strength.

The last property of waves that we need to consider is their speed. Wave speed is the distance an individual peak travels in one unit of time. For example, sound waves in air travel at approximately 343 meters per second (m/s). All light waves travel at the same speed when moving through a vacuum. Physicists call this speed c, and measurements yield $c = 2.99 \times 10^8$ meters per second (m/s). That's about a billion kilometers per hour. The tremendous velocity of light is why "live" television feeds from distant events like the Olympics on the other side of the world are possible. The TV signals are beamed around the planet via radio waves, making the trip in milliseconds.

wavelength The distance between successive peaks or troughs of a wave.

frequency The number of peaks (or troughs) of a wave that pass a point in space in a single second.

amplitude A measure of a wave's strength.

interactive
Waves

Electromagnetic Waves ($v = c = f\lambda$)

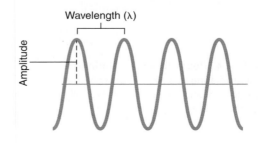

FIGURE 4.3
Wave Measurements
Wavelength and frequency are related properties in waves (like those of electromagnetic radiation—that is, light). As wavelength increases, frequency decreases when the wave speed is constant, as it is for light waves in a vacuum.

For light waves, the frequency (f), wavelength (λ), and speed (c) are related in this simple way:

$$c = f\lambda$$

Because for light the speed c is constant, this relationship enables us to convert easily from measurements of wavelength to measurements of frequency.

electromagnetic radiation
Oscillating electric and magnetic fields traveling through space.

visible light The form of light to which the human eye responds ranging from 400 nm to 700 nm.

electromagnetic spectrum The full range of electromagnetic radiation at different wavelengths running from gamma rays (very short) to radio waves (very long).

radio wave An electromagnetic wave with a wavelength greater than 0.1 meter.

microwave The shortest-wavelength radio wave, with a wavelength of 1 millimeter (1×10^{-3} meter) to about 0.1 meter.

infrared wave An electromagnetic wave with a wavelength of 700 nanometers (1×10^{-9} meter) to 1 millimeter, just longer than those of visible light.

ultraviolet light An electromagnetic wave with a wavelength of 10–400 nanometers, just shorter than those of visible light.

X-ray An electromagnetic wave with a wavelength between 10^{-11} and 10^{-8} meter.

gamma ray A high-energy electromagnetic wave with a wavelength less than 10^{-11} meter.

SECTION SUMMARY
Waves, Particles, and Light

- Light is a wave, whose characteristics can be described in terms of wavelength, frequency, and amplitude.

- In a vacuum, light travels at a constant speed referred to as c; frequency (f), wavelength (λ), and c are related by the equation $c = f\lambda$.

The Electromagnetic Spectrum

Finally, what are light waves made of? Water waves are water molecules being sloshed back and forth. Sound waves are traveling oscillations of air molecules being compressed and then relaxed. What about light waves?

It turns out that light is made of a different type of wave than water or sound. In the 1860s, physicist James Clerk Maxwell derived a set of equations predicting that light was actually composed of waves of **electromagnetic radiation**: oscillating electric and magnetic fields traveling through space (**Figure 4.4**). In an electromagnetic wave, the electric and magnetic fields are lined up perpendicular to each other, and both are perpendicular to the direction the wave is traveling. The laws Maxwell derived dictate that the electric-field oscillation induces changes in the magnetic field, which then in turn induce changes in the electric field. Thus, unlike sound waves or waves on the ocean, electromagnetic waves do not require a medium and can travel through the virtual vacuum of space.

According to Maxwell's theory, **visible light**—the kind to which your eye responds—is just one

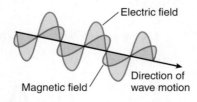

Electric field

Direction of wave motion

Magnetic field

FIGURE 4.4 Electromagnetic Radiation
An electromagnetic wave has two components, a magnetic field and an electric field, oscillating at right angles to each other.

form of electromagnetic radiation, with a particular range of wavelengths and frequencies. Maxwell predicted the existence of an entire **electromagnetic spectrum** of waves with a range of wavelengths running from very short to very long. In 1888, radio waves were first generated in an electromagnetism laboratory, proving Maxwell's theoretical predictions. Since that time, we have come to use the electromagnetic spectrum as the basis for everything from cell phones to computer chip manufacturing. The electromagnetic spectrum is made of waves with every possible wavelength, and visible light comprises only one narrow band of possible wavelengths. For that reason, in this book all forms of electromagnetic radiation are referred to as *light*.

Red light is the longest-wavelength visible light, representing electromagnetic waves with wavelengths of about 700 nanometers (1 nm = 10^{-9} meter). Blue light is the shortest-wavelength visible light, with wavelengths of about 450 nm. Our eyes are particularly sensitive to electromagnetic waves in the middle of this visible range, corresponding to yellow-green light at approximately 550 nm.

Radio waves are at the long-wavelength end of the electromagnetic spectrum, with wavelengths greater than 0.1 meter (in the FM part of the radio spectrum). **Microwaves**, like those produced by microwave ovens, fall on the short-wavelength end of radio waves, having wavelengths of 1 millimeter (1 mm = 1×10^{-3} meter) to about 0.1 meter. **Infrared waves** have wavelengths just longer than visible light, ranging from wavelengths of 700 nm to 1 mm, and **ultraviolet light** has wavelengths just shorter than visible light (10–450 nm). On the shorter end are **X-rays** (with wavelengths between 0.01 and 10 nm), which we have direct experience with via medical imaging. **Gamma rays**, with wavelengths of less than 0.01 nm, or 10^{-11} meters, are the shortest-wavelength electromagnetic radiation known. "From radio to X-rays, we experience a good chunk of the electromagnetic spectrum quite a bit in our everyday lives," says Goodman. "We use low-frequency radio waves for communications, we use X-rays for medical imaging, we use ultraviolet for bug zappers" (**Figure 4.5**).

SECTION SUMMARY
The Electromagnetic Spectrum

- Light is electromagnetic radiation that exists in a broad range of wavelengths and corresponding frequencies.

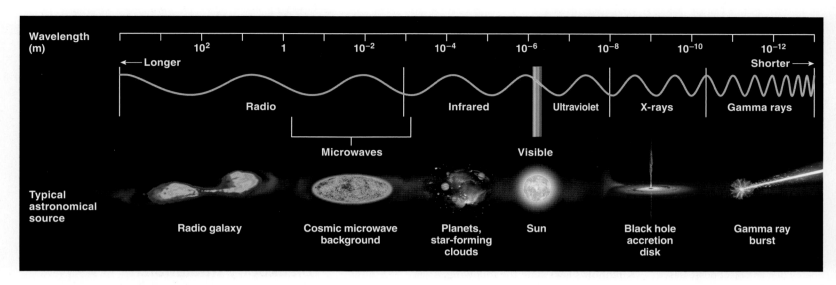

FIGURE 4.5 The Electromagnetic Spectrum
All parts of the electromagnetic spectrum are "light," but our eyes make us most attuned to visible or optical light. The wavelength of light varies across the continuum, as this diagram shows. While most astronomical objects radiate across a wide region of the electromagnetic spectrum, many emit a significant fraction of their light in a more limited wavelength region, as shown.

4.2
Astrophysical Spectra

Light from the Sun is called white light because it is composed of electromagnetic waves of many different wavelengths (which, in the visible part of the spectrum, correspond to different colors). When sunlight passes through a glass prism, the bending of the different wavelengths at different angles breaks up the white light into its constituent colors. This bending of light as it passes through any transparent medium is called **refraction**. Separating wavelengths like this is what astronomers call "taking a spectrum." A **spectrum** tells us how much energy (also referred to as intensity) the source has put into each wavelength or frequency range of light. For example, a blue star will emit more energy in the blue portion of the spectrum than a red star will. For example, for two stars with the same luminosity, a blue star will emit more energy in the blue portion of the spectrum than a red star will.

Kirchhoff's Three Laws

The invention of the spectrograph enabled physicists to take sensitive readings of the amount of energy emitted in each wavelength of light (**Figure 4.6**). They immediately began pointing their spectrographs at different luminous sources—wood fires, a chunk of burning sulfur, the Sun, and so on. Recording spectra from these objects, they were astonished to find that each gave off a distinct mix of colors. For some objects, a smooth, continuous blend of wavelengths would be present in the spectrum. This blend was called a continuous spectrum. For others, only a few wavelengths would appear in the spectrum. For still others, dark lines superimposed on a continuous spectrum would appear, as if all the energy had been taken out of the continuous spectrum at just a few wavelengths. Only after

interactive
Kirchoff's Laws

refraction The bending of light as it passes between two media, such as water and air.

spectrum (pl. spectra) The distribution of light intensity versus wavelength (or frequency).

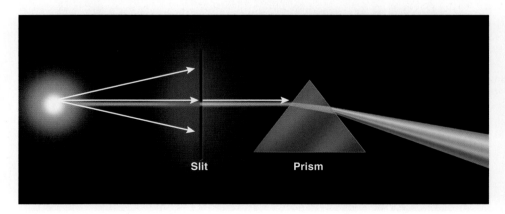

FIGURE 4.6 Spectrographs
Spectrographs separate light from a source into its component wavelengths. Modern spectrographs have become highly sophisticated, but even early versions enabled scientists to investigate the interactions of matter and light by separating the light into its component wavelengths.

emission line A bright line at a particular wavelength in a spectrum. Specific patterns of emission lines indicate the presence of a specific element in the source.

absorption line A dark line in a continuous spectrum. Specific patterns of absorption lines indicate the presence of a specific element in the source.

energy The ability to perform work. Energy comes in many forms, such as kinetic, thermal, gravitational, and electromagnetic.

kinetic energy Energy that is due to an object's bulk motion.

gravitational potential energy Energy that an object possesses because of its position in a gravitational field.

thermal energy or **heat energy** Energy that is due to random motions of atoms.

interactive
Blackbody Radiation

considerable effort were physicists able to group the different kinds of spectra and link them to specific physical conditions from different kinds of sources.

In unraveling the mystery of spectra, physicists had to take a crucial step that is common to the development of any science. Science always begins with a mass of observations that must be sifted through; the goal is to categorize these observations and identify a pattern.

This is exactly what happened with all the early spectra that scientists acquired. By the 1860s, building on the contributions of many other scientists, the German physicist Gustav Kirchhoff found the patterns hiding in the spectral data. Now known as Kirchhoff's laws, these patterns define the kinds of spectra that can be expected from different kinds of sources. In short, Kirchhoff's laws state the following:

1. An opaque body emits a continuous spectrum (**Figure 4.7A**).

   Example: the hot filament of an illuminated lightbulb.

2. Hotter, rarefied gas emits a sequence of bright lines, called **emission lines** (**Figure 4.7B**).

   Example: a neon sign.

3. Colder, rarefied gas *in front of a hotter, opaque body* produces a continuous spectrum with dark lines superimposed, called **absorption lines** (**Figure 4.7C**).

   Example: light from the solar surface passing through cooler gas in its atmosphere.

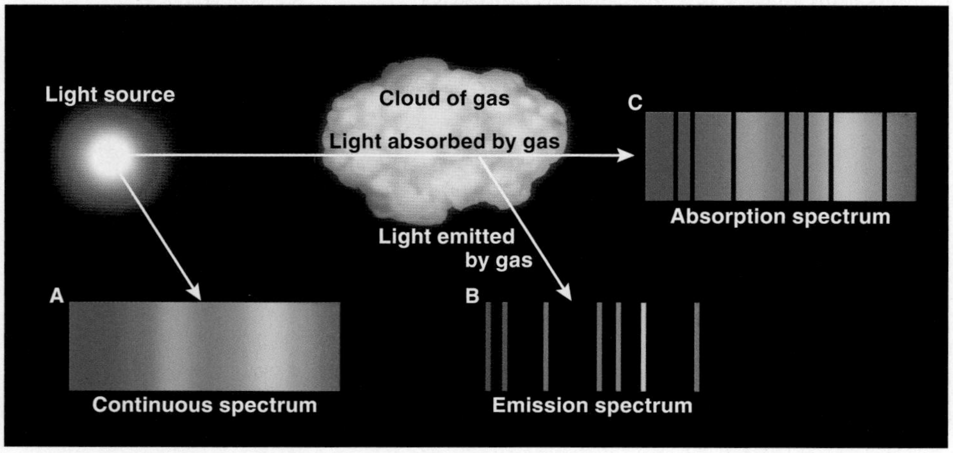

Kirchhoff's laws established a key link between light and matter, which astronomers needed in order to understand the distant objects they observed in their telescopes. The next step was to explain why the patterns of continuous spectra, emission lines, and absorption lines existed.

SECTION SUMMARY
Kirchhoff's Three Laws
- Light is emitted and received in a distribution of wavelengths and intensities called a spectrum.
- Gustav Kirchhoff observed patterns in spectra and summarized them into his three laws, which showed the link between light and matter.

Blackbody Radiation: Turning Heat into Light

Kirchhoff's first law focuses on continuous spectra. *Continuous* means that the distribution of energy across wavelengths is smooth. Light at all wavelengths is, to some degree, present in the spectrum. There is, however, more energy in some wavelength bands than in others. Matter can produce a continuous spectrum in a variety of ways, but the most important for physics and astrophysics is the link between heat and light leading to blackbody radiation.

HEAT AND ENERGY. *Heat* is another way of saying "thermal energy." Physicists define **energy** as the ability to do work (such as lifting a weight). Energy can take many forms—motion, gravitation, electromagnetism— all of which can be precisely defined and described mathematically. By the early 1800s, scientists had figured out how to express the energy locked up in an object's motion, called **kinetic energy**. A freight train moving at 80 kilometers per hour (km/h) has much more kinetic energy than a bicycle rolling along at 10 km/h. Energy is measured in units called joules (J). A baseball pitched at 90 miles an hour (40 m/s) has about 117 J of energy. Scientists had also come to understand **gravitational potential energy**—the energy an object has in a gravitational field. A book on a shelf 3 meters above the floor has more gravitational potential energy than the same book placed on a shelf 1 meter above the floor.

Thermal energy is nothing more than kinetic energy locked up in the random motions of atoms (**Figure 4.8**). The kinetic energy in a freight train is associated with its bulk motion, the whole train moving in one direction. The thermal energy of a box of hot gas (like an oven) is also associated with motion—the motion of gas atoms randomly colliding with one

FIGURE 4.7 Kirchhoff's Three Laws
Kirchhoff's laws linked the observed differences among the three kinds of spectra—(A) continuous, (B) emission, and (C) absorption—to the physical situation required to generate each one.

another and the walls of the oven. Faster random atomic motion means more heat. Scientists measure temperature in units called kelvins (K); in this scale, 0 means absolutely no motion (a state that is not actually possible). For comparison, water freezes at 32 degrees Fahrenheit (32°F), 0 degrees Celsius (0°C), and 273 K (the term *degrees* is not used in the Kelvin scale).

"Everything has a temperature," says Alyssa Goodman, "and temperature is a measure of how fast atoms are moving inside an object. If the object is very, very cold, the atoms or the molecules aren't moving very much. If something is very hot, they're moving a lot." Even in a solid, where atoms are locked in place, heat translates into vibrations of the atoms back and forth. Goodman is quick to point out the crucial relationship for an astronomer between heat and light. "The motion of all of these particles," she says, "generates lots of radiation."

THERMAL ENERGY AND BLACKBODY RADIATION. The radiation produced by the motions of atoms takes the form of electromagnetic waves, which are generated when charged particles accelerate or decelerate. The random motions in a warm, dense collection of atoms will naturally produce accelerations and decelerations as particles jiggle around. The adjective *dense* is important because it means that if enough atoms are packed into a small space, their collective random jiggling motion (acceleration and deceleration) will produce a characteristic pattern of light waves. This pattern is a special kind of continuous spectrum called a blackbody spectrum. Formally, **blackbody radiation** is the electromagnetic spectrum produced by the thermal motions of atoms in a dense object.

"Max Planck was the one who figured out the relationship between an object's temperature and how much radiation it emitted at every different wavelength," says Goodman. To make his breakthrough in understanding blackbody radiation, Planck first had to realize that light came in tiny bundles, or "quanta," of energy. He then figured out a relationship between the energy (E) and frequency (f) for these light quanta, which came to be called **photons**:

$$E = hf$$

In this equation, h was a new constant of nature called, appropriately, Planck's constant, which is equal to 6.626×10^{-34} m^2 kg s^{-1}. Planck's constant would come to define the world of the very small, including atoms and their building blocks. Using this relationship

between energy and frequency, Planck then made a remarkable theoretical leap. He was able to show how any object that could be considered a blackbody would always emit a *characteristic spectrum*, and that spectrum would depend only on temperature.

"It's really amazing," says Goodman. "Temperature is the only property of the source that sets the shape of its blackbody spectrum. A gold ball at $T = 300$ K (80°F) will produce the same spectrum as a plastic ball at the same temperature." Remember from Kirchhoff's first law that dense objects produce continuous (that is, blackbody) spectra. "If that definition seems really general," says Goodman, "that's because it needs to be. We all are radiating blackbody radiation right now."

Max Planck showed that all blackbody radiation has the same characteristic spectrum: a wide range of emitted wavelengths with a strong peak at a single wavelength. That peak represents the wavelengths (frequencies) around which most of the energy is emitted (**Figure 4.9**). A simple formula called **Wien's law** relates an object's temperature T with the wavelength λ_{max} where the maximum emission occurs:

$$\lambda_{max} = 0.0029 \text{ m K}/T$$

Note that the unit meter kelvins (m K) is needed so that dividing by T, measured in kelvins, gives the wavelength in meters.

"Wien's law explains a lot of things," says Goodman. "You heat up an iron poker and first it turns red, and then brighter red, then bright orange, until it finally glows white. The different colors reflect the change in temperature of the iron and where the wavelength of the spectrum's peak is." In Goodman's example, the temperature of the "red hot" iron poker puts the peak wavelength λ_{max} in the infrared, but in the visible portion of the spectrum it is dominated by

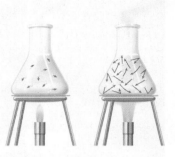

FIGURE 4.8
Thermal Energy
Thermal energy represents the random motions of the atoms in matter. Thermal radiation is the emission of photons resulting from the rapid motions of atoms in a dense object.

blackbody radiation
Electromagnetic radiation from a dense, opaque body, whose spectrum depends only on the temperature of the body.

photon A particle of light.

Wien's law The mathematical relation between a blackbody's temperature and the wavelength at which it radiates with peak intensity.

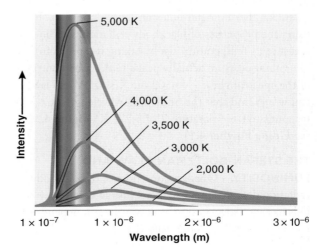

FIGURE 4.9
Wien's Law and Blackbody Radiation
A blackbody (an object emitting blackbody radiation) has a spectrum in which the wavelength of peak radiation depends only on the object's temperature.

A

B

FIGURE 4.10 Blackbody Spectra
Any warm, dense object—including the human body—produces a blackbody spectrum. An iron rod, heated in a blacksmith's oven, glows in the red region of the spectrum (A). When it cools down again, it will radiate most strongly in the infrared, as does this kissing couple (B).

FIGURE 4.11 Stars Radiate as Blackbodies
Astronomers can tell the approximate temperatures of stars just by seeing their color. The Sagittarius Star Cloud presents a wide range of temperatures, from hot, massive blue stars to cooler red ones.

red light. As the temperature rises higher, the peak shifts to shorter wavelengths and includes enough colors for the poker to appear "white hot." Wien's law also tells us that the iron poker is emitting radiation even before you stick it in the fire. The peak wavelength is just too long (in the infrared region) for your eyes to detect. This is, therefore, also the spectral region where human beings emit the peak of their blackbody radiation (**Figure 4.10**).

Wien's law is remarkably powerful for astronomy. Stars are big, dense balls of hot gas, so they fulfill the necessary conditions to be blackbodies. Therefore, just a quick glance at a star's color can tell you something about the conditions on its surface, even if the star is billions and billions of kilometers away. "We can't just go stick a thermometer in a star," says Goodman, "but using the properties of blackbody radiation, we can measure its temperature just by taking its spectrum." The Sun's spectrum actually peaks in the green portion of the spectrum at λ_{max} = 502 nm. Using Wien's law, you would find that the Sun's peak wavelength corresponds to a temperature T of 5,777 K (**Figure 4.11**; see **Going Further 4.1**).

THE STEFAN-BOLTZMANN LAW AND LUMINOSITY. One last consequence of blackbody radiation deserves mention here. Just as Wien's law tells us that a blackbody's peak wavelength depends on its temperature, the **luminosity** (energy output divided by time) also depends on temperature, as well as on the total surface area of the blackbody. The exact relation looks like this:

$$L = A\sigma T^4$$

where A is the surface area of the blackbody and σ is a constant of nature (like Newton's gravitational constant G and Planck's constant h). This formula is called the **Stefan-Boltzmann law**, and σ is a constant called the Stefan-Boltzmann constant. Note how strongly the luminosity depends on temperature in the Stefan-Boltzmann law ($L \propto T^4$) (equal to 5.7×10^{-8} watts m^{-2} K^{-4}). Thus, the law tells us that if two blackbodies have the same surface area but temperatures that differ by a factor of only 2, their energy output will differ by a factor of $(2)^4$ = 16.

SECTION SUMMARY

Blackbody Radiation: Turning Heat into Light

- Blackbody radiation is produced by the thermal motions of a dense object—that is, by the oscillating or "jiggling" motions of its atoms.

- In a blackbody spectrum, the wavelength of peak intensity depends only on temperature, so observing the spectrum of an object tells astronomers its temperature.

- The Stefan-Boltzmann law shows that the luminosity (energy output divided by time) produced by a blackbody depends on its total surface area and its temperature raised to the fourth power.

luminosity The total energy output per time of a light source.

Stefan-Boltzmann law The mathematical relationship between a blackbody's luminosity and its surface area and temperature.

nebula An interstellar cloud of dust, hydrogen, helium, and other ionized gases.

Emission and Absorption Spectra

By connecting random, thermal motions of atoms in dense collections of matter to the radiation they emitted, Planck explained Kirchhoff's first law. Explaining Kirchhoff's next two laws would prove to be more complex. Linking absorption and emission spectra to the structure of matter would demand redefining the very meaning of matter, light, and physics itself.

Physicists discovered that pure samples of elements produce their own characteristic emission spectra. A burning sample of hydrogen gas produces a very different set of bright emission lines than does a burning sample of helium. When mercury vapor is heated until it glows, the light it emits shows a few prominent bands of color in the visible part of the spectrum. Sodium, in contrast, produces two very bright lines in the yellow region of the spectrum (**Figure 4.12**). Careful study showed astronomers that each element had its own emission line "fingerprint."

"The discovery of elemental spectral fingerprints meant that astronomers could use all those emission lines to figure out what stuff in space is made of," says Goodman. "Now that is really remarkable." Recall that Kirchhoff's second law tells us that relatively hot, tenuous gas is the source of emission lines (*tenuous* means the gas will not be opaque to the radiation we're interested in; see Figure 4.7C). Space is full of hot, tenuous gases in the form of diffuse clouds. Using emission lines produced by these astronomical **nebulas** (**Figure 4.13**), astronomers began an inventory of the gas clouds' chemical composition. Says Goodman, "Astronomers started taking a census of the elements in the Universe."

Absorption line spectra proved to be just as useful for astronomers as emission line spectra were. Examining starlight with spectrographs, astronomers found the continuous blackbody spectra expected from a hot, dense gas. Superimposed on top of the stellar blackbody emission, however, was a series of dark bands. These absorption lines turned out to occur at exactly the same wavelengths as the emission lines that scientists had mapped out when studying heated samples of pure elements. The elemental fingerprints were showing up again, this time as absorption lines. But where was the gas that was doing the absorbing in those stellar spectra?

Recall that Kirchhoff's third law says that absorption lines occur when a relatively cold, tenuous gas is placed in front of a source of continuous radiation (see Figure 4.7B). In the case of stars, blackbody radiation comes from their dense surfaces. Above the surface is a rarefied atmosphere. As stellar blackbody radiation passes through the colder, more tenuous atmospheric

GOING FURTHER 4.1 THE LANGUAGE OF THE COSMOS
Wien's Law

Wien's law enables us to find the peak wavelength λ_{max} of emission coming from a blackbody at temperature T. At what peak wavelength would a star with surface temperature $T = 3{,}000$ K radiate? Let's begin with the Wien's law equation and plug in 3,000 K for T:

$$\lambda_{max} = 0.0029 \text{ m K}/3{,}000 \text{ K}$$
$$= 9.7 \times 10^{-7} \text{ m}$$

Now we have to convert from meters to microns (μm), which is the unit astronomers often use for infrared. If

$$1 \text{ μm} = 1 \times 10^{-6} \text{ m}$$

we can convert the result above as follows:

$$\lambda_{max} = (9.7 \times 10^{-7} \text{ m})/(1 \times 10^{-6})$$
$$= 0.97 \text{ μm}$$

Thus, most of the energy from the star will be radiated at a wavelength of 0.97 μm, which is in the infrared portion of the spectrum.

We can do the same for the Sun's spectrum, which peaks in the green portion of the spectrum, at $\lambda_{max} = 502$ nm. First we need to convert nanometers to meters:

$$502 \text{ nm} \times (1 \times 10^{-9} \text{ m/nm}) = 5.02 \times 10^{-7} \text{ m}$$

Wien's law then tells us the Sun's temperature is

$$T = 0.0029 \text{ m K}/\lambda_{max}$$
$$= 0.0029 \text{ m K}/(5.02 \times 10^{-7} \text{ m})$$
$$= 5{,}777 \text{ K}$$

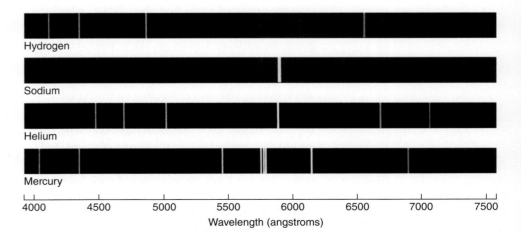

FIGURE 4.12 Emission Spectra of Elements
The emission spectra of hydrogen, sodium, helium, and mercury reflect the wide variation in energy levels, which are unique to each element.

Alyssa Goodman

Not every astronomer is featured on a public-television game show for kids. But then again, Alyssa Goodman is not your everyday astronomer, and it's more than her appearance on *Fetch! with Ruff Ruffman* that gives her those credentials.

Goodman's interest in astronomy has taken her to telescopes around the world, though the Boston area has remained her home base. After completing her undergraduate degree in physics at MIT, she earned a PhD in physics from Harvard. She won a President's Fellowship at UC Berkeley but returned to Harvard in 1992 as a professor of astronomy.

Goodman loves the process of observing best. "It's amazing to be up at 14,000 feet working all night under the stars," she says. "You get tired, but working with your collaborators while the data comes in is just one of the most exhilarating experiences you can have."

Goodman's interest in imaging—making pictures from her radio astronomy data—has led her to seek a deeper understanding of how data are represented and visualized in other fields. "We would like to see what the clouds we study look like in 3-D," she explains, "but of course we can't do that in astronomy. So we need a way of putting our images together into something that looks more like a 3-D picture." Searching for the tools she needed led her into an unusual collaboration with Harvard Medical School. "People in other fields, like health science, figured out how to do what we needed with our radio telescope data. We call the project astronomical medicine; it gives people 3-D views of what these very large regions of space look like." Goodman's interest in the image extends to her work as an artist. The beauty of astronomy is just one location where she has found a passion for visualization.

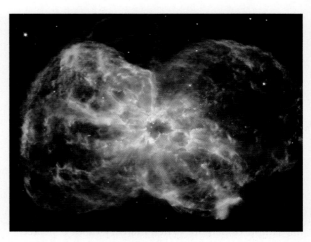

FIGURE 4.13 Nebular Spectra
Once the links between matter and the emission of light were fully understood, astronomers could examine the spectra of nebulas like this one (NGC 2440) and determine the elements that compose them.

FIGURE 4.14 Stellar Spectra
Superimposed onto the continuous spectra of the stars are the absorption lines created when the star's light (a blackbody spectrum) passes through its own cooler atmosphere. This image shows the spectra from a number of different stars (each spectrum is the rainbow band running blue to red). Superimposed onto the continuous spectra of the stars are the absorption lines created when the star's light (a blackbody spectrum) passes through its own cooler atmosphere.

gas, light is absorbed, producing the dark lines that astronomers see with their spectrographs (**Figure 4.14**). In this way, absorption lines enable astronomers to directly determine the chemical composition of stellar atmospheres and, by inference, the composition of the stars themselves.

None of the discoveries related to Kirchhoff's laws, however, explained why emission and absorption lines are produced in the first place. "It was very hard for physicists and astronomers to imagine what was going on," explains Goodman. "Why was there lots of light at some particular wavelength, or no emission at some other wavelength? And why did they get absorption or emission depending on the different circumstances in the gas? It was all kind of mystifying." Answers to these questions came only when new insight was developed into the nature of atoms, and scientists found that the "nanoworld" of atoms behaves very differently from the macroworld we inhabit.

4.3
Spectra and the Strange World of the Atom

The first important step to explaining Kirchhoff's laws came when physicists recognized the dual nature of light. For 100 years scientists had believed that light was a wave. At the beginning of the 20th century, they realized light could also act like a particle and that light energy comes in discrete bundles. Physicists call light particles photons, and their discovery was an early

SECTION SUMMARY
Emission and Absorption Spectra
● Interpreting emission and absorption spectra required understanding that each element can emit or absorb only certain, specific wavelengths of light.

○ interactive
Atomic Structure:
Absorption and Emission

example of a quantum phenomenon in physics (*quantum* is the German word for "package"). How can light be both a particle that exists in only one place at one time and also a wave that is spread out through space? Physicists and astronomers needed a way to reconcile the wave-particle duality of the photon—its ability to act as either a wave or a particle. Ultimately, the study of atoms and light opened up a new and fascinating domain of behavior that physicists called **quantum mechanics** or **quantum physics**. Quantum mechanics is the science of the very small—the world of molecules, atoms, and smaller particles.

Developing a Model for the Atom

Atomic spectra are astronomer Pat Hartigan's bread and butter. Like Alyssa Goodman, Hartigan studies star formation. He also uses spectral lines as a tool for understanding conditions in star-forming clouds hundreds of light-years away. "Every spectral line is associated with a particular kind of atom," says Hartigan. "So the questions scientists had to answer were, What's happening inside atoms to create spectral lines? Why do individual atoms emit light at only certain wavelengths?"

Atoms are composed of three basic types of particles. At the center of every atom is a **nucleus** composed of **neutrons** with no electric charge and **protons** with positive electric charge. Surrounding the nucleus are **electrons**, which carry negative electric charge. Left on their own, atoms are electrically neutral, meaning that there will be the same number of electrons orbiting the nucleus as there are protons inside the nucleus, making an overall electric charge of zero.

"In the early 20th century, physicists thought atoms might be like little solar systems," says Hartigan. "The electrons were like planets, and the nucleus played the role of the Sun." The hope was to use this solar-system model of the atom to explain line spectra as electrons shifting in their orbits around the nucleus, but the model could not explain why light was emitted at only specific wavelengths. Not until the young Danish physicist Niels Bohr made an imaginative leap was the problem of emission and absorption spectra solved.

Bohr imagined a very different kind of behavior, *quantum behavior*, for electrons in an atom. The **Bohr atom** (**Figure 4.15**), as his model was called, still had electrons orbiting a central nucleus, but the young physicist's radical change was to restrict the location of the orbits. In the Bohr atom, the electrons can "live" on only a specific set of well-defined orbits

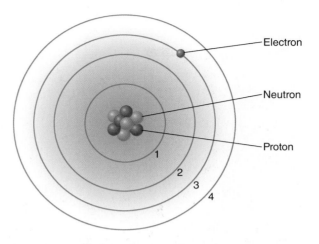

FIGURE 4.15 The Bohr Atom
In the classic model of the Bohr atom, neutrons (with no charge) and protons (with positive charge) make up the nucleus of the atom, while negatively charged electrons constantly move around the nucleus on specific orbits. This model served only as a starting point for understanding the quantum mechanical nature of atoms. Figure 4.16 reflects the current understanding of atomic structure.

with well-defined distances from the nucleus. The innermost orbit has a fixed radius r_0, referred to as the **ground state**. The electron in the Bohr atom can never get closer to the nucleus than the ground-state orbit. The next orbit, referred to as the first **excited state**, has a larger radius than the ground state: $r_1 > r_0$. The second excited state is an orbit with an even larger radius, and so on, up to a certain fixed number of orbits for each element. In this way, orbits in the Bohr atom form a series of concentric circles of ever-greater radius, all of which are centered on the nucleus.

> ### SECTION SUMMARY
> #### Developing a Model for the Atom
> ● Niels Bohr hypothesized that the electrons of an atom can occupy only specific ground or excited energy states.

Spectra and the Bohr Atom

By "quantizing" the orbits (allowing only certain orbits to exist), Bohr was able to offer a simple explanation for atomic spectra. In Bohr's model, both emission and absorption spectra occur when electrons jump from one orbit to another. Jumps from higher to lower orbits lead to the **emission** of a photon of a specific wavelength. Jumps from lower to higher orbits require the **absorption** of a photon of specific wavelength. In later years,

quantum mechanics or **quantum physics** A branch of physics dealing with physical phenomena at the level of molecular, atomic, and subatomic scales.

nucleus (pl. nuclei) The dense region at the center of an atom consisting of protons and neutrons.

neutron A subatomic particle with no net electric charge within an atomic nucleus.

proton A subatomic particle with a positive electric charge.

electron A subatomic particle with a negative electric charge.

Bohr atom A model of the atom, devised by Niels Bohr, in which electrons in discrete orbits surround a positively charged nucleus. Electron "jumps" between the discrete orbits determine the emission or absorption of light.

ground state The lowest energy state within an atom.

excited state Any energy state within an atom that is higher than the ground state.

emission The release of a photon of a specific wavelength when an electron jumps from a higher to a lower energy level.

absorption The capture of a photon of a specific wavelength, causing an electron to jump from a lower to a higher energy level.

FIGURE 4.16
Electron Orbits
Quantum mechanics implies that electrons spread out around the nucleus in a cloud. As electrons are excited into different orbits, their "probability cloud" changes shape. (A) This drawing shows hydrogen's first excited state. (B) A "photo" of a hydrogen atom in the third excited state. The photo was produced by combining many images of electron positioned around the nucleus.

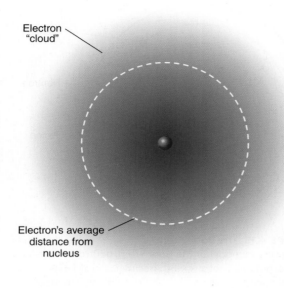

Electron "cloud"

Electron's average distance from nucleus

A

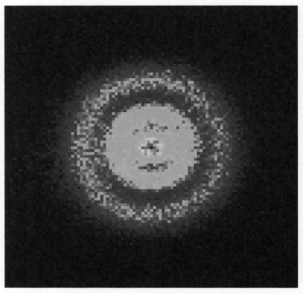

B

interactive
Spectrum Analyzer

physicists learned that the electrons are actually smeared out and that the orbits are best imagined as clouds of "probability." Still, the idea of jumps from one orbit to another persisted (**Figure 4.16**).

"The secret to the Bohr atom is the energy difference between the orbits," says Hartigan. Each orbit has a specific energy associated with its distance from the nucleus, so physicists call the orbits **energy levels**. Bohr's crucial step came in recognizing the relation between the energy levels and the wavelengths of the photons. In the Bohr atom, the energy difference between two orbits (E_{diff}) corresponds exactly to the photon energy (E_p) associated with an electron jumping between those orbits ($E_{diff} = E_p$). If the electron jumps down from a higher orbit to a lower one, it loses energy, E_{diff}, and gives up a photon with E_p. "The Bohr atom doesn't allow light with some random amount of energy to be emitted," says Hartigan. "Only photons with energy equal to jumps between orbits are going to appear."

The link between energy and frequency (or wavelength) comes from the equations $c = f\lambda$ and $E = hf$, introduced earlier in the chapter. From the second equation, the frequency f of an emitted photon depends on E_p. So an electron jumping down between energy levels can emit only a photon with frequency $f = E_{diff}/h$ (or wavelength $\lambda = c[h/E_{diff}]$ from the first equation, where h is Planck's constant). This is the origin of emission spectra. Only wavelengths

corresponding to quantum jumps between energy levels will be part of the spectrum.

Absorption is just the opposite process. For an electron to go from a lower energy level to a higher one, it must absorb a photon. The energy of the absorbed photon must equal the energy difference between the two orbits. "Bathe an atom in light representing a continuous spectrum, and it will see photons with all possible wavelengths," says Hartigan, "but only photons that can kick an electron between specific energy levels can be absorbed." Thus, absorption spectra show a continuous distribution of light with only certain wavelengths missing. Those are the wavelengths of photons that can be absorbed and bump electrons upward between energy levels (**Figure 4.17**).

Goodman uses a common analogy to understand the process of line emission and absorption in the Bohr model. "Think of it like a staircase," says Goodman. "It takes a certain amount of energy to go up a step." In the Bohr atom, however, the "steps" would not all have the same height. The old solar-system model for the atom, which allowed any orbit and any energy, was more like a ramp than a staircase. "You can be anywhere you want on the ramp," says Goodman. "But you are only allowed to be on one step or on another. You can't just be floating in between them."

The Bohr model explained Kirchhoff's laws in terms of atomic structure. According to Kirchhoff's second law, a hot, rarefied gas will produce

energy level Any one of the certain discrete values of orbital energy possible for an atom.

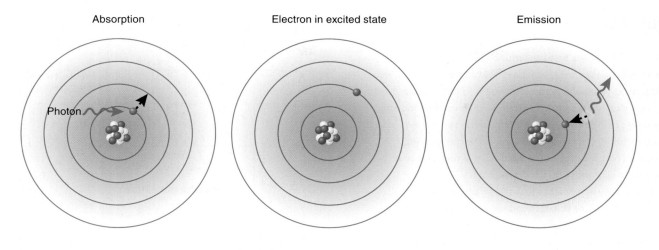

Absorption   Electron in excited state   Emission

Photon

FIGURE 4.17
Energy Levels of an Atom
Niels Bohr imagined—correctly—that electrons can occupy only certain discrete energy levels around the nucleus. The energy levels of the atom, each of which is higher than the preceding one, can be envisioned as successively larger orbits of the electron(s) around the nucleus. Moving between energy levels requires the absorption (for moving to a higher level, left) or emission (for moving to a lower level, right) of a specific wavelength of photon.

an emission spectrum. *Hot* is a critical part of the description for the Bohr atom explanation. In a hot gas, atoms move at high speed, careening off each other in wild collisions. Every time two atoms smash together, an electron can be bumped from the ground state to an upper orbit. That's why a lot of atoms in a hot gas are in excited states with electrons in upper orbits that could emit a photon and jump back down. Hot gases therefore produce emission spectra. The same logic in reverse applies to absorption spectra. Electrons in the atoms of a cold gas are in the ground state. In front of a light source generating a continuous spectrum, the electrons of these atoms are ready to absorb those photons. But they will absorb only those photons whose energy matches upward jumps to higher orbits.

In the Bohr model, each element has a different number of protons and electrons, so atoms of different elements have different sets of quantized orbits. In other words, the energy levels of helium (with two protons and two electrons) are different from the energy levels of hydrogen. Each element has its own energy staircase and its own spectral fingerprint of photons that it can absorb or emit. Bohr's model not only predicted why absorption and emission spectra exist; it also explained why each element has a unique spectral fingerprint.

The quantum mechanical description of the atom that Bohr provided enabled astronomers to forge a direct link between light from a distant object and the physical properties of that object. By taking spectra, astronomers could tell an object's chemical composition, its temperature, its density (how much matter there is in a cubic centimeter), even the presence of magnetic fields. "We would be lost without spectral

lines in astronomy," says Hartigan. "Those large, beautiful regions like the Orion Nebula and the Ring Nebula have all been explored in detail because we can use emission and absorption spectra to understand conditions in the gas."

SECTION SUMMARY
Spectra and the Bohr Atom
- For an electron to move up from one energy state to another, it must absorb a specific photon; to move down, it must emit a photon of specific energy.
- The absorption and emission of photons result in distinctive spectral lines that can be used to identify the elements present.

The Hydrogen Atom

"The hydrogen atom has only the one proton in the nucleus, so its energy levels are the simplest of them all," says Hartigan. Hydrogen's simplicity is good news for astronomers because about 75 percent of all the matter in the cosmos is hydrogen. The spectrum of hydrogen is therefore the key to understanding the properties of many astrophysical environments. Its spectrum also explains why so many gas clouds appear red.

The first two steps in hydrogen's energy level diagram are the most important for astronomers, because that's where electrons spend most of their time. Transitions into and out of the first two levels form two series of emission and absorption lines that are often very prominent.

"All of the jumps that start or end in the ground state produce photons in the **Lyman series**," says Hartigan. As discussed in later chapters, Lyman-series photons, which have wavelengths lying outside the visible range (in the ultraviolet), are critical to

Lyman series Emission or absorption lines of the hydrogen atom as an electron moves between the ground state and excited states.

Pat Hartigan

Pat Hartigan got his first subscription to *Sky and Telescope* in 1969 when he was just 9 years old. He has never lost his focus since. At 10 years of age he received his first telescope. "I went outside, swatted mosquitoes (this was Minnesota), and spent the whole night trying to find things in the sky," he explains. "Since then I just basically continued on doing the same thing (minus the mosquitoes)." Hartigan pauses for a moment to correct himself. "Well, I did want to be a rock star or a professional baseball player, but neither one of those things panned out."

Hartigan, a professor at Rice University in Houston, Texas, is an observational astronomer whose work has focused on studies of star-forming regions. He has traveled to some of the world's most remote observatories for his work. "I remember being on top of a mountain in Chile, the top of the Andes," he says. "It was spectacularly clear up there, and the Sun had just gone down. Up in orbit, the shuttle was about to rendezvous with the space station, and you could see them both in the sky. They were about a degree apart and brilliantly bright. I watched them tracking across the night sky in a perfectly smooth motion. I really had a sense of what the ancients were imagining with their celestial spheres and the perfect kinds of motions they wanted for the heavens."

"Another time, back when I was a graduate student, I was working with the 4-meter telescope in Chile. I was in the 'cage' at the top of the telescope. I remember looking down into the mirror—this huge, 4-meter-diameter block of glass. There, reflecting back at me, were all the brilliant stars of the Southern Hemisphere. It was like looking down into this ocean of starlight. That was really quite something."

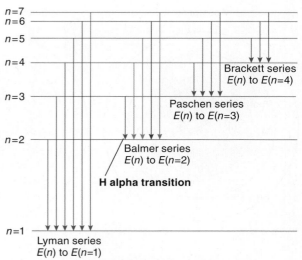

A Electron transitions for the hydrogen atom

400 nm   700 nm

H alpha line
656 nm
Transition *n*=3 to *n*=2

B IC 434, an emission nebula

FIGURE 4.18 Lyman and Balmer Series
(A) The simple hydrogen atom has many energy levels and therefore many transitions. Of these, the H alpha transition is responsible for the bright-red color of nebulas. (B) IC 434, an emission nebula that surrounds the dark Horsehead Nebula, produces a spectrum whose H alpha line is prominent.

understanding the evolution of matter in the Universe (**Figure 4.18**).

"All of the jumps that start or end in the first excited state involve photons in what is called the **Balmer series**," Hartigan continues. Astronomers often use the letter H (for "hydrogen") to refer to lines in the Balmer series. The jump from the first excited state to the second excited state is called the Balmer alpha (or H alpha) transition and requires a Balmer alpha photon. "Unlike the Lyman series, the brightest line of the Balmer series shows up in the visible part of the spectrum," says Hartigan. In fact, H alpha photons have a wavelength of 656.3 nm. That's in the red part of the spectrum. "There is lots of hydrogen gas in these clouds," says Hartigan. "That can mean a lot of red H alpha photons for our telescopes to capture."

SECTION SUMMARY
The Hydrogen Atom

● Two specific energy levels in hydrogen atoms result in emission and absorption lines called the Lyman series and the Balmer series.

● One particular line in the Balmer series produces visible, red photons; the abundance of hydrogen in the Universe explains why many gas clouds (nebulas) in space appear red.

Balmer series Emission or absorption lines of the hydrogen atom as an electron moves between the first excited state and higher energy states.

Spectra and the Doppler Shift

Something else light can tell us is the velocity of an object toward or away from Earth. Astronomers can use spectral lines as a kind of cosmic speedgun and thus can map out the motions of the entire Universe.

To do so, they take advantage of a phenomenon called the **Doppler shift**, which is the change in frequency and wavelength of waves brought about by an object's motion.

A Doppler shift occurs if either the receiver or the source of the waves is in motion. Consider for a moment sound waves, such as those produced by a moving police car and its siren. When the police car is moving toward you, sound waves pile up as the ambulance catches up a little with each wave that has been emitted. That means the sound's wavelength decreases, and the frequency increases (the siren has a higher pitch). When the police car is moving away from you, the sound waves are stretched out as the vehicle gets a little farther away from each emitted wave. The wavelength increases, and the frequency decreases (the siren has a lower pitch; **Figure 4.19**).

The same basic process occurs for light emitted by an astronomical object. Both the wavelength and the frequency of light change when the source is in motion. To make an accurate measurement of the object's motion, however, you have to know the light's original, unshifted wavelength. That's where emission and absorption lines become so powerful. For example, astronomers have measured the precise "at rest" wavelengths for all hydrogen spectral lines in Earth-based laboratories. Using these known wavelengths as a measuring stick, astronomers can compare them with the wavelengths they observe from an astronomical object. Any difference between the laboratory measurement of the spectral-line wavelengths and the astronomically observed wavelengths means that the object must be moving. "The whole spectrum with all the lines gets shifted," says Hartigan, "so astronomers can use the whole pattern of lines to determine velocity."

If the wavelength of spectral lines from the astronomical object is longer than the rest wavelength, the object is moving away from Earth, and the spectrum

interactive
Doppler Shift

Doppler shift A change in wavelength that results because either a wave source or an observer moves relative to the other.

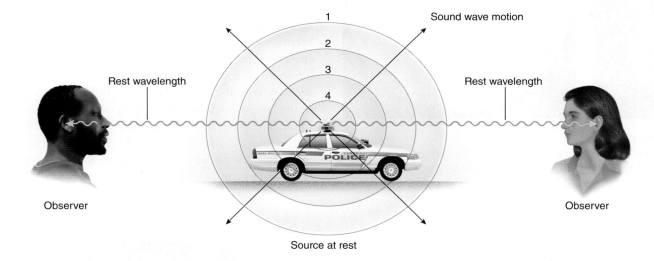

Sound wave motion

Rest wavelength

Rest wavelength

Observer

Observer

Source at rest

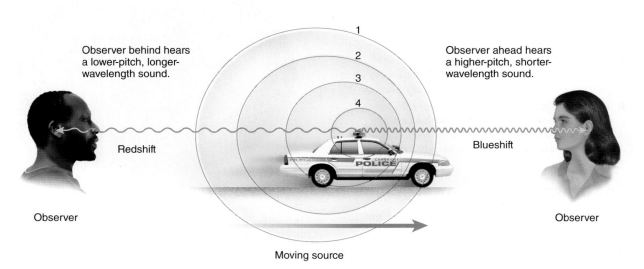

Observer behind hears a lower-pitch, longer-wavelength sound.

Observer ahead hears a higher-pitch, shorter-wavelength sound.

Redshift

Blueshift

Observer

Observer

Moving source

FIGURE 4.19
Doppler Shift
The Doppler shift is the reason that the sound of a police car coming toward us increases in pitch but the sound of a police car moving away decreases in pitch. The pitch remains constant when the police car is not moving.

FIGURE 4.20
Redshift and Blueshift
The pattern of spectral lines emitted by an object is redshifted when objects move away from us, and blueshifted when they move toward us.

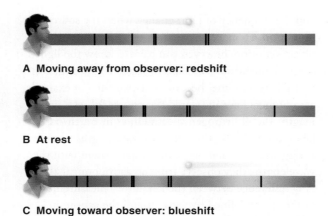

A **Moving away from observer: redshift**

B **At rest**

C **Moving toward observer: blueshift**

FIGURE 4.21 Apparent Size of the Sun and Moon
The Sun and Moon have approximately the same apparent size in our sky, but that's because of the coincidence that the Sun is 400 times larger while the Moon is 400 times closer. (Moon image from Luc Viatour/www.Lucnix.be)

interactive
Inverse Square
Law of Light

is said to be **redshifted**. If the wavelength from the astronomical object is shorter than the rest wavelength, the object is moving toward Earth, and the spectrum is said to be **blueshifted** (**Figure 4.20**).

SECTION SUMMARY
Spectra and the Doppler Shift
- The Doppler shift is caused by the apparent stretching of wavelengths as an object moves away, or their compression as the object approaches.
- Redshifted spectral lines (at wavelengths longer than expected) tell astronomers how fast an object is moving away, while blueshifted lines enable measurement of velocities toward the observer.

The Inverse Square Law for Light

The detailed links between matter and light were worked out as quantum physics and astrophysics matured. But there is one property of astronomical

redshift An increase in the wavelength of light that results when the source of the light moves away from the observer or the observer moves away from the source.

blueshift A decrease in the wavelength of light that results when the source of the light moves toward the observer or the observer moves toward the source.

objects that quantum physics cannot directly help determine: distance. There is no way for astronomers to lay a tape measure across the vastness of space, yet knowing the distance to an object can be critical to understanding its most basic properties. "Look at the Sun and Moon," says Hartigan. "They both have the same angular size in the sky. Are they really the same size? Of course not! The Moon is just closer than the Sun, but you wouldn't know that if you did not know their distance" (**Figure 4.21**).

The distances to stars, nebulas, galaxies, and other astronomical objects may be the most poorly known of all their properties. "The distance problem is magnified when you study something like a star," says Hartigan, "because all you can see is a point of light. You can always measure how bright it is and measure its spectrum and so forth, but what you can't tell, at least very easily, is how far away it is." To determine the distance to stars and other astronomical phenomena, astronomers have to be very clever and very lucky.

One property of light—brightness—can be used for determining distance. This makes sense intuitively, because as light radiates away from its source, its energy will be spread across a greater area the farther it travels. The apparent brightness of a light source decreases with the square of distance. This inverse square law for brightness looks just like the inverse square law for the force of gravity. A detector placed 2 meters from a lightbulb will record 4 times less energy than a detector placed 1 meter away (**Figure 4.22**). In equation form, this relationship can be written as

$$B = L/4\pi D^2$$

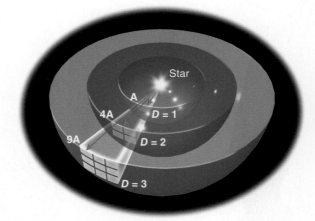

FIGURE 4.22 The Inverse Square Law for Brightness
The geometry of a sphere helps illustrate why brightness decreases with the square of the distance. With each doubling of radius, for example, the light must spread out over 4 times as much area.

where D is distance from the source, B is the brightness you would see at that distance, and L is the object's luminosity, which is equal to its intrinsic brightness. The "100 watts" of a 100-watt lightbulb is a measure of luminosity.

This equation holds the key for astronomers desperate to determine astronomical distances. For a class of objects whose luminosity L is already known, they can compare that luminosity to the brightness B measured in telescopes and solve the equation for distance (see **Going Further 4.2**).

A **standard candle** is an astronomical light source whose luminosity L is known. "The idea of a standard candle is to find objects of known luminosity," says Hartigan. "Then you simply look at how bright they appear, and that tells you how far away they are. You convert a measure of brightness into a measure of distance. It's very useful!"

More than 100 years ago, astronomers discovered that certain kinds of stars behave as standard candles, meaning that they have properties that enable astronomers to determine the stellar luminosity. Now there are certain types of galaxies, supernovas, and other objects that astronomers have also learned to use as standard candles. We will meet these different objects in later chapters. For now there are two important points to remember. First, the behavior of brightness as indicated by the inverse square law is what makes standard candles possible. Second, standard candles are rare, and astronomers are always on the hunt for new ones so that they can map out the true structure of the cosmos.

SECTION SUMMARY
The Inverse Square Law for Light

- The inverse square relationship of brightness to distance means that astronomers can calculate the distance to objects by comparing their apparent brightness with their luminosity.
- Standard candles are objects whose luminosity is known.

4.4
Telescopes

Understanding how light encodes information about the matter emitting it is only the first step for astronomers. The next step is to use telescopes to catch the light and accurately analyze the resulting data.

GOING FURTHER 4.2 THE LANGUAGE OF THE COSMOS
Brightness and Distance

Imagine that astronomers discover two stars (let's call them star X and star Y) having the same luminosity. In the language of math, this means

$$L_X = L_Y$$

Now imagine that star Y *appears* 25 times brighter than star X when viewed from Earth. In mathematical language, this means

$$B_Y = 25B_X$$

For each star we can rearrange the equation $B = L/4\pi D^2$ to solve for distance and find out which star is farther away:

$$D_X = (L_X/4\pi B_X)^{1/2}$$
$$D_Y = (L_Y/4\pi B_Y)^{1/2}$$

Now let's divide the first equation by the second and notice that the factors of 4π cancel out:

$$D_X/D_Y = [(L_X/L_Y)(B_Y/B_X)]^{1/2}$$

We can now use the fact that $L_X = L_Y$ to cancel luminosity from the result, and then we can substitute $B_Y = 25B_X$ into this equation to find

$$D_X/D_Y = (25B_X/B_X)^{1/2} = (25)^{1/2}$$
$$= 5$$

Now we can see that

$$D_X = 5D_Y$$

Star X is 5 times farther away than star Y.

Light Buckets and Resolution

"A telescope basically does two things for you," explains Pat Hartigan. "First, it collects a lot of light and concentrates it into one spot." Astronomers like Hartigan often think of telescopes as "light buckets." The bigger the bucket, the more light you can collect. For a telescope, the "bucket size" is measured by the size of its principal light collector (such as a lens, mirror, or radio dish), referred to as the telescope's **aperture**. The more collecting area the telescope has, the more light it can gather.

If the collector is circular, astronomers measure the aperture in terms of its diameter D (with radius

standard candle An object with known luminosity that can be used to determine distance.

aperture The diameter of a telescope's main light-collecting lens or mirror.

$R = D/2$). The light-collecting power of a telescope is directly proportional to the area of the collector. For a circular collector, the area is

$$A = \pi R^2 = \pi(D/2)^2 = \pi(D^2/4)$$

Thus, the light-collecting power of a telescope increases with the square of its diameter. Build a telescope whose aperture is 10 times larger in diameter than the one your neighbor's telescope has, and you will be able to catch 100 times more light—which, in turn, means being able to see objects 100 times fainter.

"The other thing that a telescope allows you to do," says Hartigan, "is to separate two objects that appear very close together on the sky so you can see them independently." Hartigan is describing the ability of telescopes to increase **angular resolution**. The resolution of an optical device (like your eye or a telescope) tells you the smallest angle on the sky that it can cleanly "see." If a telescope has a resolution of 100 arcseconds (100″), then any details of an astronomical object that extend across an angle less than 100″ will appear as a single blur and be "unresolved" (**Figure 4.23**). It's important to distinguish resolution from magnification. Many people think a telescope's job is **magnification**—to make something look bigger—but it's the resolution that matters most.

Resolution depends not only on the telescope's aperture, but also on the wavelength of light the telescope is collecting. Wavelength matters because as waves enter the telescope, they bend and spread out in a process called diffraction, which degrades resolution. The best angular resolution a telescope can achieve can be written as

$$a = 2.5 \times 10^5 \lambda/D$$

where a is the smallest angle (measured in arcseconds) that can be resolved, using a telescope with aperture D (measured in meters), when looking with light of wavelength λ (also measured in meters). The human eye has an aperture of about 6 millimeters (mm), which means that its resolving power in visible light (600 nm) is approximately 30″, or 1/120 of a degree. The Hubble Space Telescope (HST) has a 2.4-meter aperture. The angular-resolution equation then tells us that the HST has a resolution of 0.06″ at the same wavelength—500 times better than the human eye.

The angular-resolution equation shows that if a telescope is used to observe the same object with a shorter wavelength, its resolving power will be even higher. Observations in blue light will be better resolved than observations in red light. (Recall that most objects emit at many wavelengths.)

SECTION SUMMARY
Light Buckets and Resolution
- The aperture of a telescope determines its light-gathering power.
- Resolving ability depends on both the telescope's aperture and the wavelength of light being gathered.

Telescope Designs: Refractors and Reflectors

The first telescopes, called **refractors** (**Figure 4.24A**), used a large glass lens (called the **objective**) at one end of a long tube as a light collector. Refraction occurs when light passes through a glass prism. The large glass objective lens in a refracting telescope collects light by focusing all the waves that pass through it on a spot at the other end of the tube. An image of the source emitting the light forms there, and a smaller lens called the **eyepiece**, positioned at the back of the tube, magnifies that image.

The problem with making large refracting telescopes is the weight of the objective lens. Attempts to build large-aperture refractors led to glass lenses so heavy that they could not keep their shape. Another problem with glass lenses is that they bend blue light more strongly than red light. This problem is called **chromatic aberration**—different colors are refracted at different angles—and can be corrected by modifying the shape of the lens, but it's impossible to fix

angular resolution A measure of the ability of a telescope to separate or distinguish features in a distant object.

magnification The ability of a telescope to enlarge appearances in an image.

refractor An optical telescope that uses a lens to collect light and form an image.

objective In a refractor telescope, the large lens that gathers light from the object being observed and focuses the light rays to form an image.

eyepiece The small lens of a telescope through which the observer looks.

chromatic aberration The failure of a lens to focus all colors to the same point.

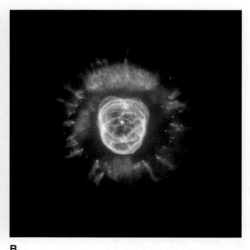

A   **B**

FIGURE 4.23 Telescope Resolution
Astronomers require the best resolution possible in order to gain the most information about an object. For a given wavelength, the larger the aperture of the telescope astronomers use, the finer the resolution. Here, planetary nebula NGC 2392 is seen with both (A) low resolution and (B) much higher resolution.

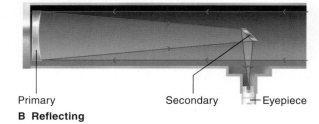

A Refracting

B Reflecting

FIGURE 4.24 Refractors and Reflectors
(A) Refractors use lenses to gather light. (B) Reflectors
use mirrors.

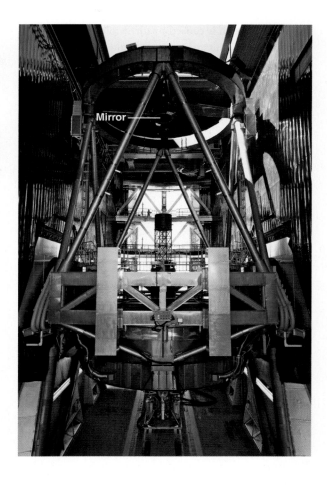

FIGURE 4.25
The Subaru Telescope
The Subaru Telescope is an
example of a modern large
research telescope; it has a
primary mirror more than
8 meters in diameter (and
observes in optical and
infrared light). The telescope
is shown here in its housing
on top of Mona Kai volcano
in Hawaii.

completely. These restrictions have limited the utility
of refractor telescopes.

Most telescopes used for research today are **reflectors** (**Figure 4.24B**). Instead of using a lens at the front
of the telescope to collect light, reflectors use a large
mirror (called a primary) at the back end. The curved
primary reflects and focuses the light to a second smaller
mirror (called the secondary) positioned above it, which
gathers and redirects the light. "The secondary redirects
the light to different locations depending on the design,"
says Hartigan. In reflectors with a Newtonian focus, the
secondary redirects the light to the side of the telescope,
where an eyepiece or an instrument can be placed.
Reflectors with a Cassegrain focus redirect the light back
down to an opening at the center of the primary mirror.

Reflecting telescopes can be made much shorter
than refractors because a mirror can redirect light far
more effectively than a lens can. They can also be made
much larger because the mirror can be supported from
below, helping it to retain its shape. The typical aperture size on a modern research telescope is 10 meters
across (**Figure 4.25**), more than 5 times larger than
the height of a typical human adult. Because reflectors
provide a larger aperture for much less cost, amateur
astronomers usually choose reflectors over refractors.
Currently, 20- and 30-meter reflecting telescopes are
under construction.

To avoid sagging and distortion from the weight
of the mirrors, astronomers build large reflectors from
mirrored segments (usually hexagonal) that can be
fitted together. To maintain the mirror's proper shape,
small positioning motors are mounted behind each
segment. As the telescope moves, the motors shift the
mirror segments to properly focus the light.

SECTION SUMMARY
Telescope Designs: Refractors and Reflectors
- Refractors, which use lenses at each end of a tube,
  have the disadvantages of chromatic aberration, as
  well as distortion due to the excessive weight of the
  lenses as aperture diameter is increased.
- Reflectors use mirrors instead of lenses to gather
  and focus the light, and they can be built with
  much larger apertures. Segmented mirrors and
  computer-controlled adjustments enable even
  larger apertures.

4.5
Atmospheres and
Their Problems

Earth's atmosphere keeps most incoming electromagnetic radiation at bay. The relatively thin layer of air
(containing nitrogen, oxygen, carbon dioxide, and
other gases) acts like a blanket, absorbing most
wavelengths of light and keeping them from reaching
Earth's surface. This is, in general, a good thing. For

reflector An optical telescope
that uses curved mirrors to
collect light and form an image.

example, sunburn is caused by ultraviolet (UV) radiation, which just manages to make it through the atmospheric blanket. If more-damaging wavelengths like X-rays penetrated to the ground, our planet would be sterilized and unable to sustain life.

Where the Light Shines Through

The fact that we can see the Sun and the stars means that our atmosphere allows the **transmission** of light at visible wavelengths; that is, it allows visible light to pass through without being absorbed. Astronomers use the term *transmission window* to describe the ability of light in a band of wavelengths to penetrate the atmosphere. "Each transmission window means you can do astronomy from the ground in that wavelength," says Pat Hartigan. Radiation will not be observed if the atmosphere is opaque at that wavelength (**opacity** is the ability of material to absorb radiation, thereby preventing light from passing through).

Figure 4.26 shows the opacity of the atmosphere at different wavelengths. "You can see the 'band pass' windows," says Hartigan, "where Earth's atmosphere is transparent. But there are other areas where it's completely opaque, because of carbon dioxide or water vapor or other types of absorbers in the atmosphere." The atmosphere's opacity drops strongly in the visible part of the spectrum. It also drops strongly in the radio part of the spectrum, making radio astronomy a fruitful endeavor. Notice that the atmosphere also allows some wavelengths of infrared radiation to penetrate the atmosphere. The atmospheric opacity in other parts of the infrared spectrum is low enough that radiation may not make it all the way to the ground but can penetrate far enough that telescopes mounted on high-flying airplanes can pick it up.

TWINKLING STARS AND THEIR CHALLENGES. Even if an astronomer wants to observe the sky in a transmission window, the atmosphere causes another kind of problem. Anyone who has flown on an airplane is likely to have had the uncomfortable experience of being bounced around by **turbulence**, the random jostling motions of moving air. A related form of turbulence is also what jostles light on its way down through the atmosphere, causing stars to appear to twinkle. Light rays from distant sources are bounced around as they head toward the ground, leading to blurry images. For astronomers, atmospheric turbulence poses a major problem. Even the most powerful telescope with the largest aperture will produce blurry, low-resolution images if the light it catches has passed through a turbulent atmosphere.

Astronomers have developed various strategies for dealing with the problems created by the atmosphere. The first is simply to place the telescope as high above sea level as possible. Modern visible-light and infrared telescopes are placed on mountaintops because the density of Earth's atmosphere drops quickly with height. Many state-of-the-art telescopes have been built on the extinct volcano Mauna Kea in Hawaii, which stands at 13,803 feet (4,207 meters) above sea level. At that height, its peak lies above 40 percent of the atmosphere and above 90 percent of atmospheric water vapor. Atmospheric opacity and turbulence are thus less of a problem at Mauna Kea's summit. When pursuing some kinds of longer-wavelength studies, such as infrared astronomy, it pays to put observatories in very dry climates, because water vapor is the primary cause of opacity at these wavelengths. The Atacama Desert in Chile is one of the highest, driest locations on Earth, and a number of telescopes are currently in operation there.

QUICK THINKING: ADAPTIVE OPTICS. Another strategy for dealing with atmospheric turbulence is a technique called **adaptive optics**. "Adaptive optics allows you to change the shape of the mirror depending on what the atmosphere above you is doing," explains Hartigan. The same motors that

transmission The passing of light through matter without being absorbed.

opacity The degree to which light is absorbed when passing through a material.

turbulence The random jostling motions of moving air.

adaptive optics A telescope technology in which mirror segments are rapidly adjusted to account for atmospheric turbulence and produce higher-resolution images.

FIGURE 4.26
Transmission Windows and Opacity
Not all wavelengths of electromagnetic radiation can be observed from Earth's surface. The atmosphere is transparent to visible and radio light but not to gamma-ray, X-ray, and most UV light.

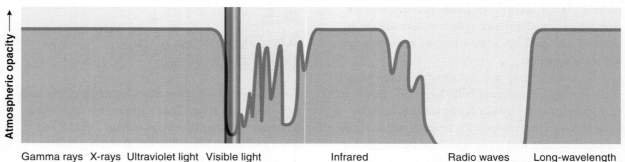

Atmospheric opacity →

Gamma rays   X-rays   Ultraviolet light   Visible light   Infrared   Radio waves   Long-wavelength radio waves

Wavelength →

keep the sections of the mirror in place can also make rapid, small-scale changes to its shape to sharpen an image that has been blurred by turbulence. "You can either look at a guide star or shine a laser out of your telescope," says Hartigan. "The reflected laser light becomes a guide star. Then you look to see how the image is shimmering around and adjust the mirror rapidly to compensate for the atmosphere's motion." Computers controlling an adaptive-optics system drive the motors to change the positions of the mirror segments many times in a single second. Using adaptive optics, astronomers can achieve very sharp vision in at least some wavelength bands, including infrared.

SECTION SUMMARY
Where the Light Shines Through

- Earth's atmosphere effectively blocks some wavelengths of light, so astronomical studies in certain wavelengths cannot be made from Earth's surface.
- Ground-based observations are subject to blurring caused by atmospheric turbulence.
- Strategies to reduce atmospheric effects on observing include placing instruments as high as possible in the atmosphere and using adaptive optics.

More with Less: Radio Telescopes and Interferometry

For most of human history, astronomy meant optical astronomy. Once the entire electromagnetic spectrum had been discovered, scientists recognized that studying the sky in other wavelengths would lead to fundamental new discoveries. Beginning in the 1950s, radio astronomy quickly became a major contributor to our understanding of the cosmos.

Most radio telescopes operate like reflectors. They have a curved dish, which acts as a collector, bouncing light to a smaller detector at the collector's focus. Because radio waves from distant cosmic sources are so faint and their wavelengths are so long, the telescopes must have large collecting areas.

The large wavelengths of radio waves can be quite useful for astronomers. Electromagnetic radiation tends to be absorbed by solid bodies that have a size similar to the radiation's wavelength. Dust particles often have a radius that is close to or larger than the wavelength of visible light, so absorption causes dusty regions of space to appear dark in optical images. Since the galaxy is full of dusty clouds, studies at visible wavelengths cannot see across galactic distances. Radio waves, with their long

wavelengths, can penetrate the dusty clouds, allowing radio astronomers to see from one end of the galaxy to the other.

The long wavelengths of radio light do, however, mean that radio telescopes have more restrictive limits on resolution. Recall that the minimum angular resolution of a telescope depends directly on the wavelength of light being used (see the angular-resolution equation on page 94). In other words, a single radio telescope, even a large one, will have low angular resolution compared with an optical telescope. But, in one of the major innovations of modern science, astronomers have found a way around this limitation.

"If you took a large telescope," explains Hartigan, "and blocked out a spot in the middle, you would find the brightness of your image would go down, but your resolving power, meaning the ability to split apart two closely spaced image features, would be just the same." Hartigan is describing the principles behind **interferometry**, a technique that combines the signal from an array of smaller telescopes to create the resolving power of a much larger one. "What matters for resolution," says Hartigan, "is the distance from one side of the aperture to the other. But you don't need a continuous mirror or collector to make an aperture." By carefully combining signals from multiple telescopes that can be widely separated, astronomers can achieve images with extremely high resolution.

One of the best-known radio interferometers is the Atacama Large Millimeter Array in Chile. ALMA consists of 66 giant radio telescopes that can be placed in different configurations. Combining the signal from so many individual telescopes, ALMA astronomers can create a single instrument with an aperture of 16 km (**Figure 4.27**).

interferometry The technique of combining the signal from many smaller telescopes to achieve the resolving power of a larger one.

A Atacama Large Millimeter Array

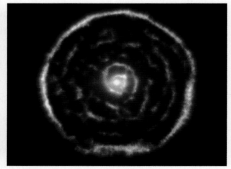

B ALMA image of dying star

FIGURE 4.27 Atacama Large Millimeter Array
(A) With its 66 radio telescopes, the Atacama Large Millimeter Array in Chile is the largest astronomical project in the world. Using interferometry, it is capable of acting as a single telescope with a 16-km aperture. (B) Shells of matter driven off a dying star are captured by ALMA.

In recent years, interferometry has been applied in other wavelength bands as well. Infrared telescopes have been combined to increase their resolution, and optical interferometry has been explored, but with limited success. "It's partly an instrumental problem and partly a problem with Earth's atmosphere and turbulence, which need to be taken into account when combining signals from different telescopes," says Hartigan. "It's just more difficult to do when the wavelengths are shorter."

SECTION SUMMARY
More with Less: Radio Telescopes and Interferometry
- Radio waves have proved to be very useful to astronomers because their long wavelengths allow them to penetrate interstellar clouds of dust.
- Because collecting light in the radio part of the spectrum means dealing with faint radio emissions and long wavelengths, resolving radio sources requires a very large aperture on the telescope.
- Interferometry, which allows separate, distant radio telescopes to function as one large telescope, is used primarily for radio astronomy.

Telescopes in Space

Getting a telescope completely above Earth's atmosphere is the best option for eliminating problems of turbulence and absorption. And for X-rays and gamma rays, there is no other option because our atmosphere has no transmission windows at these wavelengths. Putting a telescope in space is an expensive proposition, however, because designing precision equipment to work in the hostile environment of space requires extensive testing.

INFRARED. Infrared light extends from 700 nm to 350,000 nm (but astronomers use microns, μm, for measuring infrared light, where 1,000 nm = 1.0 μm). The atmospheric window for infrared light allows ground-based studies only at the shorter wavelengths called the near infrared (0.7–5 μm). Observations in the mid infrared (5–40 μm) and far infrared (40–350 μm) can be carried out only by a telescope that is positioned above most or all of the atmosphere. Being above the atmosphere means that the entire infrared spectrum is available to astronomers, so they can use the longer wavelengths of infrared light to see deep into objects like star-forming clouds and planet-forming disks.

The most famous infrared satellite has been the Spitzer Space Telescope (**Figure 4.28**). Spitzer was

A

B

FIGURE 4.28 Spitzer Space Telescope
(A) The Spitzer Space Telescope (shown here in an artist's rendering) gathers infrared light, which is particularly useful to astronomers for its ability to penetrate dust. (B) This Spitzer image of galaxy NGC 1097 shows its coiled structure, with a central ring of stars at its center.

the last of NASA's Great Observatories, which were designed to investigate the sky across a wide range of wavelengths. Spitzer was launched in 2003 with a mirror 85 centimeters (cm) in diameter and a 5-year supply of liquid helium to keep the spacecraft cool (at a temperature of about 5.5 K) and reduce heat radiation from its own "self-emission." Spitzer was launched into a special orbit that keeps it far from Earth and

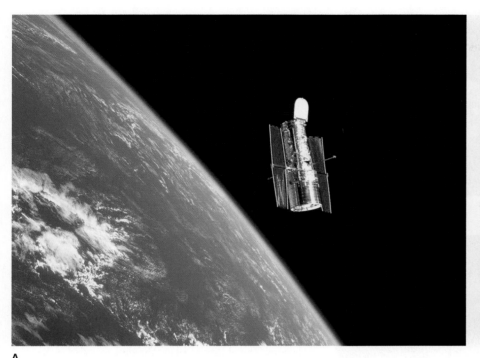

A

B

FIGURE 4.29 Hubble Space Telescope
(A) The HST is the most successful scientific instrument in history, with imaging capacity mainly in optical light (as well as some capacity in ultraviolet and infrared). (B) This Hubble Extreme Deep Field image, covering a speck of the sky only a small fraction of the diameter of the full Moon, shows 5,500 galaxies, some of which are in the early stages of evolution 13 billion years ago.

our planet's infrared heat radiation. (As of 2014, it was still active, though reduced to a smaller range of wavelengths than it started out being able to handle.) The James Webb Space Telescope, which is NASA's next large-scale space telescope platform, will be designed to work mostly at infrared wavelengths as well.

OPTICAL AND UV.   By eliminating atmospheric turbulence, space telescopes that operate in the visible range offer better resolving power than do similar-sized ground-based telescopes. The Hubble Space Telescope (HST) was the first of NASA's Great Observatories, launched in 1990 (**Figure 4.29**). It has been the most successful optical space mission to date.

The HST also has the capacity (although limited) to carry out UV observations, which cannot be made from Earth because of atmospheric absorption. Ultraviolet photons are produced in stellar atmospheres, at the inner edges of disks surrounding black holes, and in other environments, making UV studies critical to astronomy. An early and important UV telescope was the International Ultraviolet Explorer (IUE). Launched in 1978 with a 45-cm mirror, it carried out observations in wavelengths between 115 and 320

nm. The Extreme Ultraviolet Explorer (EUVE) and the Far Ultraviolet Spectroscopic Explorer (FUSE) pushed farther out in the UV spectrum, observing wavelengths between 10 and 120 nm.

X-RAYS.   The X-ray portion of the spectrum extends from 0.01 to 10 nm. Using Planck's energy-frequency relation, $E = hf$, you can see that such short wavelengths (high frequencies) mean high energy. Thus, X-rays come from extremely energetic astrophysical regions and events like supernovas, neutron stars, and galaxy clusters.

The only way to carry out a robust observational study of the sky in X-rays is to launch telescopes into space. However, because X-rays are so energetic, they tend to pass right through mirrors meant to collect and focus them onto a detector. To get around this problem, engineers build grazing-incidence mirrors for large X-ray telescopes, which gently deflect incoming X-rays many times to focus them on a detector.

The first X-ray telescopes rode on balloons and high-altitude rockets, but X-ray astronomy began in earnest with the launch of the Uhuru satellite in 1970. The current workhorse of X-ray studies is the Chandra

FIGURE 4.30

Chandra X-ray Observatory (A) CXO (shown here in an artist's rendering) presents astronomers with startling new details visible only in X-ray light. (B) The secrets of galaxy Centaurus A— including jets powered by the supermassive black hole—spring into view. This image shows a combination of wavelengths. X-rays are shown in blue.

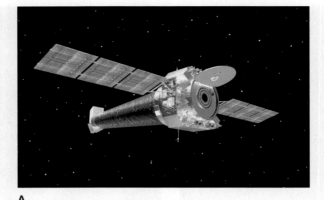

A

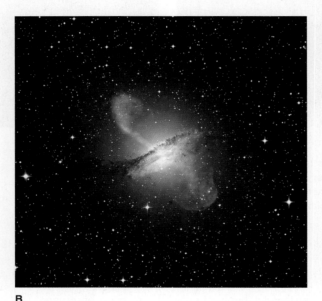

B

launched in 2008, observed the "gamma ray sky" with better resolution capabilities and flew with a suite of powerful new instruments.

Since gamma rays are so energetic, they are produced only in the most energetic celestial objects, such as exploding stars, supernovas, active galaxies with black holes at their centers, and neutron stars. The extreme high energy of gamma rays also means that they are not produced in large numbers. A typical gamma-ray observation may catch only a few photons per hour. However, a special class of objects, called gamma-ray bursts (GRBs), has puzzled astronomers for decades. Appearing as brief flashes of gamma rays that do not repeat, their source remains highly debated. Much of gamma-ray astronomy has focused on the nature of GRBs.

Space-based astronomy will continue to play a critical role in the future as detectors become more sophisticated and astronomers become more ambitious in their designs. The ability to place the detector above the atmosphere and its complications makes space an attractive place to carry forward astronomical studies. "Once you get above Earth's atmosphere, images are steadier and you can see things a lot more sharply," says Hartigan, who has spent years using the HST to study regions of star formation. "That really makes a huge difference for getting to the science."

SECTION SUMMARY
Telescopes in Space
- Putting telescopes in space enables astronomers to observe in all kinds of light (including X-rays and gamma rays) and to avoid turbulence, but it is expensive and limits the size of the telescopes.
- NASA has launched many space-based observatories to study light in infrared, optical, UV, X-ray, and gamma-ray wavelengths.

Telling Light's Stories

From Newton's law of gravity to blackbody radiation, we have covered a lot of ground in these first four chapters. There is still more physics that goes into astrophysics (you will meet Einstein and his theory of relativity in Chapter 14), but from here on you can start using all that you've learned about the basic structure of the physical world to begin gaining a perspective on the cosmos as a whole. In the remaining chapters of this book, we will explore the grand understanding of the Universe that astronomers have painstakingly built using the tools (both conceptual and technological) we've just learned about.

X-ray Observatory (CXO), launched in 1999 (**Figure 4.30**). Chandra is also a Great Observatories mission, and it continues to operate today.

GAMMA RAYS. Gamma rays are the light with the shortest wavelength, the highest frequencies, and therefore the highest energies of all electromagnetic radiation. Earth's atmosphere is completely opaque to gamma rays, so gamma-ray astronomy can be done only from space. Unlike X-rays, however, there is no efficient way to focus gamma rays. Therefore, gamma-ray telescopes are really gamma-ray detectors that provide very low angular resolution. A gamma-ray telescope can provide very little detail about the sources it "sees" compared even with X-ray telescopes. One of the most famous gamma-ray telescopes was the Compton Gamma Ray Observatory. It was the second of NASA's Great Observatories to be launched (in 1991) and the first to map the entire sky in gamma rays. More recently, the Fermi Gamma Ray Telescope,

●→ chapter summary

4.1 Light: The Cosmic Messenger

The ways that light, heat, and matter are linked provide scientists with information about distant objects. Advances in optics have provided the needed instruments, including telescopes and spectrographs. Light can act as both a wave (with qualities of wavelength, frequency, and amplitude) and a particle (called a photon). The electromagnetic spectrum represents all the forms of light with wavelengths running from very small to very large (gamma-ray, X-ray, ultraviolet, visible, infrared, microwave, radio).

4.2 Astrophysical Spectra

Spectra show the distribution of intensity (related to energy emitted) versus wavelengths in a light source. The kind of spectrum observed gives information about the physical characteristics of the source. Kirchhoff's laws describe the various conditions that produce each basic type of spectrum (continuous, absorption, or emission). Blackbody radiation comes from dense objects whose random atomic motions from heat create a characteristic spectral shape. The temperature of an object emitting a blackbody spectrum can be determined by the wavelength where the spectrum peaks via Wien's law. The luminosity of a blackbody also depends on temperature (and the object's surface area) and is described by the Stefan-Boltzmann law. The energy of light is determined by its frequency: the higher the frequency, the higher the energy. Unique spectral lines in emission and absorption spectra show which elements are present.

4.3 Spectra and the Strange World of the Atom

Quantum mechanics is the behavior of matter at the molecular, atomic, and subatomic levels. Niels Bohr developed a model of the atom in which protons and neutrons in the nucleus are surrounded by electrons that can exist only on a discrete series of orbits. The orbit closest to the nucleus is the ground state, and the successively higher orbits are the first excited state, the second excited state, and so on. Electrons jump directly between

levels, emitting energy in the form of photons when moving to lower states and absorbing energy when moving to higher states. The wavelengths of the photons are unique to the energy transitions available in the atom. Thus, every different kind of atom (each element) has a different absorption or emission spectrum. The Doppler shift is the apparent stretching or compression of light's wavelength based on the motion of an object away from or toward the observer; it can thus be used to derive the velocity and direction (toward or away) of a source. The inverse square law relates distance to brightness of an observed object. Standard candles are objects whose luminosity is known. By comparing the known luminosity of a standard candle to its observed brightness astronomers can determine its distance from Earth.

4.4 Telescopes

Telescopes gather and focus light. Larger apertures gather more light and increase the angular resolution of distant objects. Resolution also depends on the wavelength of the light. Refractors are telescopes using lenses as their light-gathering objectives, while reflectors use mirrors.

4.5 Atmospheres and Their Problems

Atmospheres such as Earth's limit the transmission of light, being transparent to some kinds of light through transmission windows while being opaque to others. Optical telescopes are often built at higher altitudes to limit the effects of atmospheric disturbance, and space telescopes must operate from space to effectively observe some kinds of infrared light, ultraviolet light, X-rays, and gamma rays. The Hubble Space Telescope observes in optical light, and the quality of its images is greatly enhanced by its location above the atmosphere. Radio light can be captured from Earth's surface. Radio observation has benefited from the development of interferometry, which allows multiple telescopes to function together as if they were one very large telescope.

●→ questions and problems

Narrow It Down: Multiple-Choice Questions

1. Which of the following is *not* a characteristic of waves?
   a. density
   b. wavelength
   c. frequency
   d. amplitude
   e. velocity

2. Which of the following equations correctly gives the speed of light c in a vacuum?
   a. $c = \lambda \times f$
   b. $c = \lambda/f$
   c. $c = f/\lambda$
   d. $c = \lambda^2/f$
   e. $c = f^2/\lambda$

3. Which of the following are *not* found in the electromagnetic spectrum? Choose all that apply.
   a. gamma rays
   b. cosmic rays
   c. sound waves
   d. ultraviolet waves
   e. infrared waves

4. Arrange the following regions of the electromagnetic spectrum from lowest to highest frequency: infrared, gamma-ray, optical, X-ray, ultraviolet, radio.
   a. X-ray, gamma-ray, ultraviolet, infrared, radio, optical
   b. gamma-ray, X-ray, ultraviolet, optical, infrared, radio
   c. gamma-ray, infrared, optical, radio, ultraviolet, X-ray
   d. radio, infrared, optical, ultraviolet, X-ray, gamma-ray
   e. optical, radio, infrared, ultraviolet, gamma-ray, X-ray

5. A spectrum typically displays which two characteristics of light?
    a. frequency and wavelength
    b. velocity and frequency
    c. intensity and wavelength
    d. intensity and amplitude
    e. velocity and amplitude

6. The presence of dark lines at specific wavelengths in an otherwise continuous spectrum can be explained most easily by which of the following?
    a. The original source is not radiating light at all wavelengths.
    b. The original source is radiating light most intensely at the wavelengths corresponding to the missing photons.
    c. A relatively hot gas between observer and source is radiating light at the wavelengths corresponding to the missing photons.
    d. A relatively cool gas between observer and source is absorbing light at the wavelengths corresponding to the missing photons.
    e. The spectrograph is not sensitive at the wavelengths corresponding to the dark lines.

7. A spectrum of a distant object reveals a sequence of known absorption lines that are all shifted to shorter wavelengths. What can be concluded about the object?
    a. It must be very massive.
    b. It must be highly magnetized.
    c. It must be emitting cosmic rays.
    d. It must be moving away from us.
    e. It must be moving toward us

8. Compare the spectra of blackbodies with different temperatures. Which of the following is true?
    a. The peak of emission in the spectrum of the hotter object is at a longer wavelength.
    b. The cooler object shows greater emission at infrared wavelengths than the hotter object does.
    c. The peak of the emission in the spectrum of the cooler object is at a longer wavelength.
    d. The total energy emitted per area by the cooler object is greater than that emitted by the hotter one.
    e. The emission from the hotter object is smaller at ultraviolet wavelengths than that of the cooler object.

9. In observing an oxygen emission line from a warm gas cloud that is expected to appear at 500 nm, the line is seen at 525 nm instead (5% longer). Which of the following is true?
    a. The object is moving away from the observer at 5% of the speed of light.
    b. The object is moving toward the observer at 5% of the speed of light.
    c. The observed emission line indicates that a chemical change has occurred.
    d. The velocity of the light is greater than expected.
    e. The observer's equipment is inaccurate, since emission lines do not change.

10. The formation of absorption lines in a spectrum emitted by a blackbody indicates which of the following? Choose all that apply.
    a. the existence of an intervening cloud of material cooler than the emitting source
    b. the specific elements of the atoms present in an intervening cloud
    c. the existence of an intervening cloud of material hotter than the emitting source
    d. a violation of the conservation of energy
    e. the interaction of matter and light

11. Which statement related to the energy states of electrons is true?
    a. Electrons can occupy any energy state that is bound to the atom.
    b. Electrons within an atom require the absorption or emission of a photon of any wavelength to change to any new energy state.
    c. The wavelengths of the photons emitted or absorbed are unique to a specific element.
    d. The energy state farthest from the nucleus is the first excited state.
    e. Electrons can be found traveling between the energy states.

12. Which of the following statements concerning the Doppler effect is *not* correct?
    a. The pitch of the whistle on an approaching train is higher than that of a stationary train.
    b. The pitch of the whistle on a receding train is lower than that of a stationary train.
    c. The light from a very distant, receding galaxy appears redshifted with respect to a nearby dwarf galaxy whose motion is observed to be negligible.
    d. The light from the approaching Andromeda Galaxy appears redshifted with respect to a nearby dwarf galaxy whose motion is observed to be negligible.
    e. The pitch of the whistle of a train is unchanged to a passenger on the train.

13. Stars A and B can be considered blackbodies. The peak wavelength λ_{max} of one star (A) is longer than that of another (B). What conclusion can you draw from this information?
    a. A is cooler than B.
    b. The blackbody curves of the two are identical, since both are stars.
    c. Star B must be closer to the observer.
    d. The colors of the two stars differ.
    e. Emission lines seen in the spectrum of A will be of greater intensity than those seen in B.

14. A typical amateur telescope is a 10-inch reflector. How does the light-gathering power of the Hooker 100-inch reflector on Mount Wilson compare with that of the amateur telescope?
    a. It is the same.
    b. It is 10× greater.
    c. It is 100× greater.
    d. It is 1,000× greater.
    e. It is 10,000× greater.

15. Which of the following is a disadvantage of large refractors?
    a. weight
    b. size
    c. chromatic aberration
    d. sagging lens
    e. all of the above

16. True/False: Optical telescopes are now routinely linked together through interferometry to produce the equivalent of a single telescope with a much larger aperture telescope.

17. True/False: Light cannot behave as both a particle and a wave, since those two objects are mutually exclusive.

18. Which of these NASA space-based observatories was designed primarily to study wavelengths of light to which Earth's atmosphere is transparent?
    a. Hubble
    b. Compton
    c. Chandra
    d. James Webb
    e. Spitzer

19. Why do stars twinkle?
    a. The stellar surface where the light is emitted is very turbulent.
    b. Stars are constantly changing their brightness.
    c. The distribution of absorption lines in their atmospheres is changing.
    d. The motion of Earth's atmosphere creates the appearance of twinkling.
    e. Our eyes are constantly moving.

20. Star I is observed to be brighter than Star II. Which of the following is/are true? Choose all that apply.
    a. Star II cannot be more luminous than Star I.
    b. Star II may be bigger than Star I.
    c. If their luminosity is the same, Star I must be closer.
    d. If they are the same size and distance from the observer, Star I is hotter.
    e. Star I must have a different chemical composition from Star II.

To the Point: Qualitative and Discussion Questions

21. Name and describe three parameters that we use to describe an electromagnetic wave.

22. Describe, in terms of frequency, wavelength, and energy, where visible light falls on the electromagnetic spectrum with respect to other regions of the spectrum.

23. What are the most important characteristics of objects that emit blackbody spectra?

24. Earth's atmosphere is mostly opaque to which regions of the electromagnetic spectrum? To which regions is it mostly transparent?

25. What characteristics of stars can astronomers learn by observing their starlight with a spectrograph?

26. Using Kirchhoff's laws, describe how two observers viewing the same gas cloud can observe different types of spectra—one seeing an emission spectrum, while the other sees an absorption spectrum.

27. In your own words, describe the relationship established by Wien's law.

28. The interaction of light with matter deals directly with the absorption and emission of photons. What happens to electrons in an atom when a photon is emitted? What happens when a photon is absorbed?

29. How do refracting telescopes differ from reflectors?

30. Under what circumstances can an element be observed to emit light with an observed wavelength different from those it normally emits?

31. Why do stars appear to twinkle when viewed from the ground?

32. What is the advantage of locating telescopes on mountaintops?

33. Describe how adaptive-optics systems improve observations made from ground-based telescopes.

34. Name the NASA Great Observatory that was specifically designed to observe in each of the following regions of the electromagnetic spectrum: infrared, optical/UV, X-ray, gamma-ray.

35. How might life on Earth differ if all wavelengths of light could pass through the atmosphere?

Going Further: Quantitative Questions

36. An electromagnetic wave has a frequency 5 times higher than that of another electromagnetic wave. How do their wavelengths compare?

37. How does the frequency of one light wave compare with the frequency of another, if the ratio of their wavelengths is 1:3?

38. The spectrum of a blackbody has a peak wavelength of 7.7×10^{-7} meters. What is its temperature, in kelvins?

39. The blackbody spectrum of a star with a surface temperature of 8,000 K will peak at what wavelength? Give your answer in meters.

40. Two stars are known to have the same luminosity, but one appears one-sixteenth (1/16) as bright as the other. How many more times distant is the dimmer star?

41. Star A is 9 times as far away as star B. Both appear to have the same brightness. What is the ratio of the luminosity of star A to that of star B?

42. You have a reflecting telescope with a 6-inch aperture. Your sister has one with a 10-inch aperture. What is the ratio of the light-gathering power of her telescope to that of yours?

43. Telescope A has a diameter of 2 meters, while telescope B's diameter is 5 meters. What is the ratio of the angular resolution of B to A when both are observing the same wavelength of light?

44. How would the angular resolution of each of the two telescopes in the previous question change if the wavelength being observed was 3 times as long?

45. If the HST had a diameter twice as large as it does, how would its angular resolution compare with its current resolution?

 If your instructor assigns homework in **smartwork5**, access it at the Digital Landing Page for *At Play in the Cosmos*: **digital.wwnorton.com/cosmos**

THE
ARCHITECTURE
AND BIRTH OF
PLANETARY SYSTEMS

In this chapter you will learn about the structure and origin of our Solar System. After reading through each section, you should be able to do the following:

5.1 The Rest of the Solar System
Describe the structures and objects that, along with the Sun and planets, compose our Solar System.

5.2 Just the Facts: A Solar System Census
Distinguish among the three classes of planets in terms of their composition and location in the Solar System.

5.3 Construction Debris: Asteroids, Comets, and Meteoroids
Identify the differences among asteroids, comets, and meteoroids and the importance of each type of body to Solar System studies.

5.4 Us versus Them: Our Solar System and Others
Explain how observations of the components, shape, and dynamics of our Solar System and others contribute to different theories about the formation of planetary systems.

5.5 Developing a Theory of Planetary System Formation
Describe in your own words the main components of the condensation theory for planetary system formation.

The Tarantula Nebula, seen in all its intricate beauty within the Large Magellanic Cloud, is the most active starburst region in the Local Group of galaxies. It is also one of the largest, with a diameter of 200 parsecs. Astronomers study such regions to learn about the fundamental processes by which stars and planets are born.

5.1
The Rest of the Solar System

DAVID JEWITT, a planetary astronomer at UCLA, remembers the moment the rest of the Solar System was discovered. "It was 1992 and my former student Jane Luu and I were sitting in the observing room of the University of Hawaii's 2.2-meter telescope at the top of Mauna Kea," he recalls. "It's a kind of dingy room 4,000 meters (14,000 feet) above sea level." At the time, Jewitt had already spent 5 years of his career in a search for the elusive **Kuiper Belt objects (KBOs)**—debris left over from the Solar System's construction almost 5 billion years ago. The Kuiper Belt was supposed to exist at the edges of the Solar System beyond Neptune's orbit, and *supposed to exist* is the key phrase. Up to that time, the Kuiper Belt was just an idea originally proposed by Gerard Kuiper in 1951. No one really knew if the Kuiper Belt was out there or if it was simply a theory gone wrong. Jewitt had spent a lot of years staring at fuzzy images trying to answer that question. "We had to compare pictures taken just 20 minutes or so apart to see if anything moved." Stars would not move across the sky that fast, but an object orbiting the Sun, even out beyond Neptune's orbit, would.

"In the time between just two pictures we took that night, we could tell we had a discovery," says Jewitt, speaking of the object they called 1992 QB1 (**Figure 5.1**), a modest name for a discovery that would rewrite textbooks. "We took more and more images of the thing as the night progressed. We were able to follow its motion, which was exactly what we expected for a KBO: the right direction, the right speed, everything. After spending night after night after night after night not getting detections, finally nailing a KBO was pretty exciting."

The 250-kilometer (km) chunk of icy rock that Jewitt and Luu discovered rocked the astronomy world and opened a new window on the structure and origin of the Solar System. As Jewitt puts it, "Most of the real estate in the Solar System occupied by significant bodies has only been discovered in the last 20 years." The discovery of 1992 QB1 began a renaissance in the study of our Solar System and was also a crucial first step in a chain of events that would eventually kick Pluto out of the club of planets. "What we really discovered that night," says Jewitt, "is that the Solar System is not a 'known' place."

Jewitt's groundbreaking discovery followed several decades of revolutionary space exploration. After the dawn of the space age in the late 1950s, humanity began sending robotic probes to the planets of our Solar System. Almost all the worlds orbiting our Sun have now been visited at least once. Asteroids and comets have also been targeted by space missions, returning valuable information (and actual samples of material in a few cases). These missions, along with intense study from ground-based telescopes, have provided an entirely new understanding of our Solar System, allowing astronomers to piece together the story of its origin and evolution.

Planets and the 2,000-Year-Old Question

It was a cold day in February 1600 as the hooded figure was led out to the plaza. The crowd, which had gathered to watch the spectacle of an execution, was hushed as Giordano Bruno—an ex-Dominican monk and philosopher who had been declared a heretic—was tied to the stake. The torches were readied, the straw at his feet was set aflame, and Giordano Bruno became a martyr for science (**Figure 5.2**).

The charges of heresy against Bruno focused on details of Catholic Church theology, but undeniably, it was Bruno's astronomy that had led him to this fateful moment. Bruno was a confirmed Copernican who had argued vehemently that the Sun—and not Earth—was

Kuiper Belt object (KBO)
A small body made of ice and rock that orbits beyond Neptune's orbit, in the Kuiper Belt.

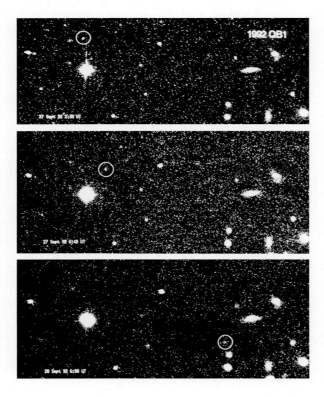

FIGURE 5.1
Kuiper Belt Object 1992 QB1
The identification of Kuiper Belt object 1992 QB1, followed by the discovery of other similar objects, validated the proposal that the Kuiper Belt existed beyond Neptune's orbit. This series of images provided evidence of the orbiting KBO, seen as a bright, moving dot (circled) in an otherwise unchanging field of stars.

the center of the Solar System. But Bruno had gone further still. The Sun was not alone in hosting a family of planets, he claimed. Other stars had their own worlds in orbit, and some of these planets were inhabited just like Earth. It was a radical argument made at a time when radicalism was not tolerated.

Bruno was not the first to ask whether other planets existed beyond the confines of our own Solar System. Even in the age of classical Greece, more than 2,000 years ago, philosophers wondered whether other worlds, orbiting other stars like our own, might exist. Only in our own time has this most ancient of astronomical puzzles been answered once and for all.

In 1995, two Swiss astronomers announced the discovery of a Jupiter-sized planet orbiting the Sun-like star 51 Pegasi. Within a decade, more than 100 other **extrasolar planets**, or **exoplanets**, had been discovered in orbit around other nearby stars. Today, the number of known exoplanets can be counted in the thousands. After millennia of having only one example of a planetary system to study, astronomers can finally ask how our Solar System compares with others. "We have found planetary systems everywhere," says Jewitt, "and most of them do not look like ours." Knowing just that fact alone is an important step toward answering many other questions.

Exoplanet system studies, along with the study of our own Solar System, have enabled astronomers to take the first strong steps toward a complete theory of how planets form. With these steps, they have also moved closer to understanding the grandest question of all: Where might life form elsewhere in the Universe?

This chapter describes the basic architecture of our own planetary system and that of others. It begins with the eight planets and assorted debris orbiting our own star and then proceeds to examine the diversity of worlds orbiting other stars. With this galaxy-wide view in place, the chapter then explores how planetary systems form and what can be expected as astronomers probe ever deeper into the Universe in search of life-harboring worlds.

SECTION SUMMARY

Planets and the 2,000-Year-Old Question

- Although astronomers have amassed a wealth of information about our Solar System and its components, much exploration remains to be done.
- Planetary systems are now known to be common in the Universe, and recently discovered exoplanetary systems have provided astronomers with the long-awaited answers to questions about the nature of our planet and our own Solar System.

5.2
Just the Facts: A Solar System Census

The process of science always begins with the simple act of collecting data. And before we can ask how planetary systems form, we have to know what they look like. Exploration of the planets orbiting our own Sun reveals a few basic facts that will set the stage for an *explanation* of how this system ended up as it did. The first step is to take a simple census of the Solar System (cataloging what's out there) and describe its architecture (how things are arranged). But those data still provide only a partial view. As David Jewitt points out, "We are still discovering basic facts about the Solar System as a whole, and we're still pretty ignorant about what's out there, far beyond the known planets."

The Inhabitants of Our Solar System

Planets are, of course, the most obvious inhabitants of the Solar System. A planet is a body that orbits a star and is massive enough for self-gravity to have pulled it into a spherical shape and to have cleared smaller objects from the neighborhood of its orbit. There are eight planets in our Solar System. In order of increasing distance from the Sun, they are Mercury, Venus, Earth, Mars, Jupiter, Saturn, Uranus, and Neptune, and their distances from the Sun range from 0.39 astronomical unit (AU) (for Mercury) to 30.10 AU (for Neptune). Recall that 1 AU = 1.5×10^8 km. At least 168 known satellites, or **moons**, orbit these planets. Most of these moons can be found orbiting the giant planets, and their count comes in as follows: Jupiter has at least 67 confirmed moons, Saturn has at least 62, Uranus has at least 27, and Neptune has at least 14.

Although Neptune is the outermost planet in the Solar System, there are many more objects at greater distances from the Sun. Pluto used to be considered a planet, but astronomers now recognize that its small size, and the fact that its gravity is not sufficient to clear its neighborhood of debris make it a member of a family of objects called **dwarf planets**. One dwarf planet occupies the region between Mars and Jupiter called the **asteroid belt**, but astronomers suspect that most reside in the region beyond Neptune's orbit called the **Kuiper Belt** (the region David Jewitt was exploring), which may be home to as many as 100,000 objects with diameters of more

FIGURE 5.2
Giordano Bruno
During his life, Giordano Bruno (1548–1600) advocated radical astronomical ideas, such as the existence of planets around other stars. His heretical ideas eventually led to his being burned at the stake by the Inquisition.

---

extrasolar planet or **exoplanet** A planet orbiting any star other than the Sun.

moon A natural satellite of a planet.

dwarf planet A body that orbits a star and has enough mass to become spherical by self-gravity but that has not cleared its neighboring region of small objects.

asteroid belt The disk-shaped region from 2.1 to 3.5 AU (as measured from the Sun) where many asteroids orbit.

Kuiper Belt The disk-shaped Solar System zone outside Neptune's orbit, from 30 to 50 AU, containing small bodies made of ice and rock.

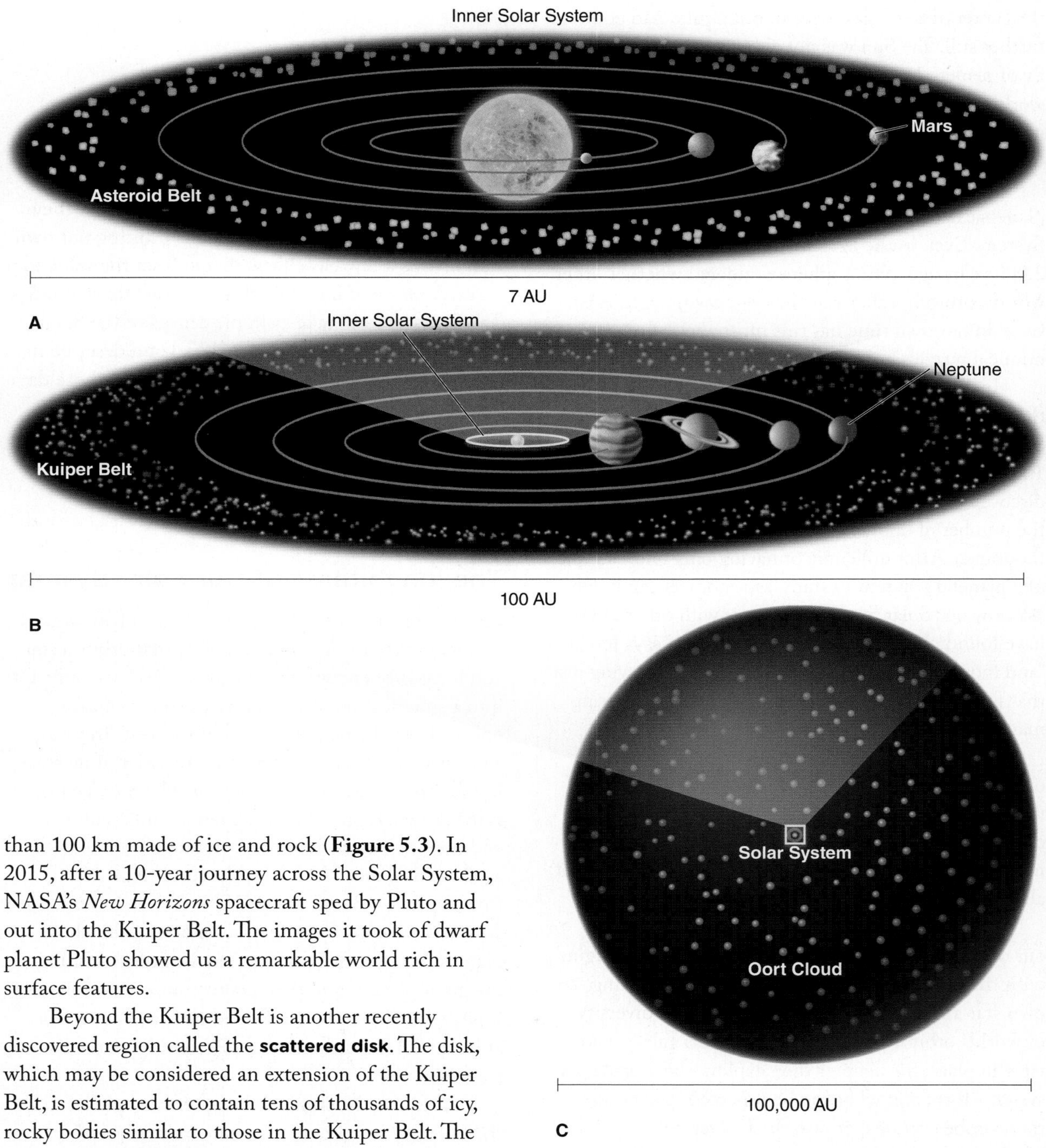

FIGURE 5.3
The Scale of the Solar System
The Solar System requires views on different scales to display all its major components: Sun, planets, asteroid belt, Kuiper Belt, Oort Cloud. Note that Planets and Sun are not shown to scale. (A) The inner Solar System, including the terrestrial planets and the asteroid belt. (B) The gas giant/ice giant region and the Kuiper belt. (C) The scale increases dramatically in the outer Solar System, where the Oort Cloud stretches across more than 100,000 AU.

scattered disk The disk-shaped region beyond the Kuiper Belt that contains many small bodies made of ice and rock. Their highly elliptical orbits may reach perihelion as low as 30 AU and aphelion as high as 100 AU.

asteroid A rocky object that orbits the Sun and has a diameter from hundreds of meters to 1,000 km.

comet A body of rock and ice ranging in size from hundreds of meters to 1,000 km and typically orbiting the Sun on a highly elliptical orbit.

than 100 km made of ice and rock (**Figure 5.3**). In 2015, after a 10-year journey across the Solar System, NASA's *New Horizons* spacecraft sped by Pluto and out into the Kuiper Belt. The images it took of dwarf planet Pluto showed us a remarkable world rich in surface features.

Beyond the Kuiper Belt is another recently discovered region called the **scattered disk**. The disk, which may be considered an extension of the Kuiper Belt, is estimated to contain tens of thousands of icy, rocky bodies similar to those in the Kuiper Belt. The discovery of the Kuiper Belt and the recognition of dwarf planets as a separate class of objects are recent developments in astronomy (occurring in just the past decade or two), and both are important parts of the story of planetary system formation.

Along with planets, moons, and dwarf planets, the Solar System is home to many smaller objects, such as asteroids, comets, and meteoroids. **Asteroids** are mostly rocky objects that range in diameter from hundreds of meters to 1,000 km. There are as many as several million asteroids in the Solar System with

diameters greater than 1 km. Most reside in the asteroid belt, but some orbit the Sun in the company of Jupiter (preceding and following the giant planet in the same orbit), and a handful of others move through the inner Solar System in orbits that can take them close to Earth.

Comets are made up of rock and ice and, like asteroids, range in size from hundreds of meters to

1,000 km. Most comets move on highly elliptical orbits that take them close to the Sun and then far out into space. There are about 4,000 known comets, but astronomers currently estimate that more than a trillion comets populate the Solar System, most residing in the **Oort Cloud**, a vast spherical region that has a radius of about 50,000 AU and surrounds the Solar System, as shown in Figure 5.3.

Finally, **meteoroids** are smaller chunks of rock ranging in size from a grain of sand to a boulder. Because meteoroids are so small—and so numerous—astronomers cannot accurately estimate their number.

SECTION SUMMARY
The Inhabitants of Our Solar System

- Astronomers have mapped out the large-scale components of our Solar System, but they are still discovering and enumerating individual objects.

- The primary categories of objects in the Solar System are planets and dwarf planets, moons, comets, asteroids, and meteoroids.

- The Kuiper Belt, scattered disk, and Oort Cloud are regions discovered to exist beyond the region of the planets.

Three Flavors of Planets

"There are three different kinds of planets," says Jewitt, "and each one represents a separate domain of the Solar System." Mercury, Venus, Earth, and Mars are relatively small, rocky worlds with orbits relatively close to the Sun. Jupiter and Saturn are massive worlds composed mainly of gas with relatively small, rocky cores. Uranus and Neptune are also large worlds (although smaller than Jupiter and Saturn) with rocky cores, but they are composed mostly of ices. **Table 5.1** displays the properties of the eight planets along with a few other representative Solar System objects. Looking over the table, do you see any patterns that strike you as important in terms of offering clues to the history of planetary systems? Where, for example, are the massive planets located compared with the smaller ones?

Beginning with the inner Solar System, we find a class of **terrestrial planets** (*terrestrial* comes from the Latin word *terra*, which means "land"). Mercury, Venus, Earth, and Mars are all terrestrial planets because they are composed mostly of solid material. While all of the terrestrial planets have some degree of

Oort Cloud A spherical region at the outer limits of the Solar System, with a radius of about 50,000 AU, that is believed to contain potentially trillions of comets.

meteoroid A rock ranging in size from a grain of sand to a boulder and orbiting the Sun.

terrestrial planet A planet consisting primarily of solid material.

TABLE 5.1 ••• **Properties of Solar System Objects**							
Object	Class	Mass (Earth Masses)	Radius (km)	Density (kg/m³)	Average Orbital Distance (AU)	Orbital Period (yr)	Number of Known Satellites (as of 2015)
Mercury	Planet	0.055	2,440	5,430	0.39	0.24	0
Venus	Planet	0.82	6,052	5,240	0.72	0.62	0
Earth	Planet	1.00	6,378	5,510	1.00	1.00	1
Mars	Planet	0.11	3,397	3,930	1.52	1.88	2
Jupiter	Planet	317.82	71,492	1,330	5.20	11.86	67
Saturn	Planet	95.16	60,268	690	9.52	29.42	62
Uranus	Planet	14.54	25,559	1,270	19.19	83.75	27
Neptune	Planet	17.15	24,764	1,640	30.10	163.72	14
Pluto	Dwarf planet	2.2×10^{-3}	1,185	2,030	39.48	248.02	5
Eris	Dwarf planet	2.8×10^{-3}	1,163	2,520	68.05	561.40	1
Ceres	Dwarf planet	1.5×10^{-4}	476	2,090	2.77	4.60	0
1992 QB1	Kuiper Belt object	Unknown	80	Unknown	43.9	291.00	0
Vesta	Asteroid	4.47×10^{-5}	251	3,456	2.36	3.63	0
Earth's Moon	Moon	0.012	1,738	3,340	2.6×10^{-3} (3.84×10^{5} km from Earth)	27.32 days	0
Sun	Star	3.33×10^{5} (1.989×10^{30} kg)	109 (6.96×10^{5} km)	1,409	Within the Sun	N/A	Trillions

FIGURE 5.4
The Planets
Each of our eight planets is unique, but they can be grouped into terrestrial, gas giant, and ice giant worlds, shown here with proper size scale (distance, however, is not to scale).

gas giant A planet consisting mostly of hydrogen and helium that has a mass many times that of Earth.

Earth mass The mass of Earth, 5.97×10^{24} kg, used as a unit of measure for other objects.

ice giant A planet consisting mostly of ices (water, methane, carbon dioxide, and other compounds) that has a mass many times that of Earth.

uncompressed density The density an object would have if the effect of gravity were excluded.

atmosphere, these gases represent only a tiny fraction of each planet's mass (**Figure 5.4**).

Beyond the terrestrial planets is the domain of the giant planets, which come in two forms. "Jupiter and Saturn are **gas giants**," says Jewitt. "They're primarily made of hydrogen and helium. They are similar, but not equal, to the Sun in their composition, having an extra component of heavier elements." Jupiter's mass is 318 times the mass of Earth, and Saturn contains about 95 **Earth masses**. "They're giant because they're much more massive than Earth," explains Jewitt. Farther out are Uranus and Neptune, considered the **ice giants**. Uranus has about 15 Earth masses, and Neptune has 17 Earth masses. "The thing about these two planets," says Jewitt, "is that most of their mass is not in the form of gas. Of the 15 or 17 Earth masses in Uranus and Neptune, only one or two of those are hydrogen and helium. The rest consist of molecules like water, methane, carbon dioxide, and lots of other compounds frozen into ices."

The differences between the terrestrial and giant planets are not confined to location, size, mass, and composition. Other properties distinguish the different types of worlds and provide clues for understanding their origin and evolution. The terrestrial planets have relatively slow rotation rates, while the giants tend to rotate quickly (Jupiter's "day" is just 10 hours). Terrestrial planets have few or no satellites. The giants all have 13 or more moons. Terrestrial planets have weak magnetic fields, while the giants have strong magnetic fields. No terrestrial planet has rings, but all the giants are surrounded by ring systems. Even the orbital separation between the two types of planets differs: The inner terrestrial worlds are relatively close to each other, each orbit separated by less than 1 AU. In

contrast, each giant planet's orbit lies many astronomical units distant from the next.

The strong distinctions between the properties of giant and terrestrial planets and between the gas and ice giants must be due to the physics of planetary systems. "The Solar System clearly has these three planetary domains of rock, gas, and ice," says Jewitt. "It's probably divided up this way because of the processes by which the planets were built." Finding a theory of planetary system formation that naturally explains these different domains and their different properties is a fundamental problem that astronomers have been seeking to solve for decades.

A critical clue for the development of that theory comes from the different densities of the planets. *Density* is defined as the *mass per volume* of a material and is measured in kilograms per cubic meter (kg/m^3). A gaseous atmosphere is obviously less dense (has fewer kilograms per cubic meter) than a chunk of rock. To peel apart the story of the Solar System, however, astronomers have to do more than just look at density. They must also take gravity out of the picture. The **uncompressed density** of a planet is the density it would have if gravity were not squeezing matter in addition to the usual interatomic forces that hold matter together. Once gravity is factored out, astronomers can see the original density of the material making up each planet. "The uncompressed density is what you would get if you measured a sample of the stuff in the lab," says Jewitt.

Looking just at the terrestrial planets, astronomers found a clear pattern—a smooth decrease in uncompressed density of the planets as one moves farther out in orbital radius (**Table 5.2**). The fact that the

TABLE 5.2 ••• Compressed and Uncompressed Density of the Inner Planets		
Planet	Average Compressed Density (kg/m³)	Uncompressed Density (kg/m³)
Mercury	5,430	5,300
Venus	5,240	4,400
Earth	5,510	4,400
Mars	3,930	3,800

Note: The uncompressed densities of the terrestrial worlds show a clear pattern: a decrease as orbital radius increases.

uncompressed densities of terrestrial planets drop with distance shows astronomers something important: the process of planet formation must somehow be sensitive to the distance from the Sun. We will return to this critical clue in Section 5.5 when we begin exploring theories of planetary system formation.

SECTION SUMMARY
Three Flavors of Planets

- Our Solar System contains three types of planets: small, dense, rocky terrestrial planets orbiting relatively close to one another nearest the Sun; gas giants made of hydrogen and helium that lie on larger orbits; and ice giants composed of both gas and heavy compounds mostly frozen into ices and orbiting farthest from the Sun.

- Differences in the uncompressed density of planets, measured by factoring out the effect of gravity, provide insights into the planets' composition and formation.

5.3
Construction Debris: Asteroids, Comets, and Meteoroids

The Sun contains 99.86 percent of the mass of the Solar System. As Table 5.1 shows, the planets, with Jupiter leading the pack, make up the bulk of the remaining mass. But the other occupants of interplanetary space, even though they account for only a small percentage of the Solar System's mass, are also important. Between and beyond the planets lie smaller orbiting bodies, including asteroids, comets, and meteoroids, which are essentially debris left over from the construction of the Solar System. By studying them, astronomers have gained essential insights into the distant past, when the worlds of our Solar System were being assembled.

Asteroids

Asteroids represent one class of Solar System debris. Asteroids are small objects (relative to planets and moons) composed mostly of rock and metal. The best-known asteroids reside in the asteroid belt, lying between Mars (at 1.52 AU) and Jupiter (at 5.20 AU). Most of the main-belt asteroids have orbits with average radii ranging between 2.1 AU and about 3.5 AU from the Sun (**Figure 5.5**).

Science fiction movies tend to portray the asteroid belt as a dense swarm of tumbling, house-sized rocks. The truth is a little less cinematic. The real asteroid belt is mostly empty space, and the average distance between asteroids is about a million kilometers. This distance is so large that it would take a passenger jet months to cross. But given the force of gravity and enough time, collisions between asteroids do occur, and astronomers have benefited from this process. "We actually know quite a bit about these main-belt asteroids," says David Jewitt, "because we have pieces of them here on Earth in the form of meteorites left over from collisions."

The four largest objects in the asteroid belt make up half the mass of the entire belt system. Ceres is the largest object in the belt, with a diameter of 940 km and a mass of 9.4×10^{20} kg. Ceres is so large, in fact, that its gravity has pulled it into a quasi-spherical shape, and astronomers consider it to be a member of the dwarf planet family of objects along with Pluto. After Ceres, the largest known object in the asteroid belt is asteroid 4 Vesta, with a diameter of 540 km

Mathilde Gaspra Ida

FIGURE 5.5
Asteroids
Images of asteroids Mathilde, Gaspra, and Ida clearly illustrate their rocky, irregular nature. They lack sufficient mass for gravity to pull them into spheres.

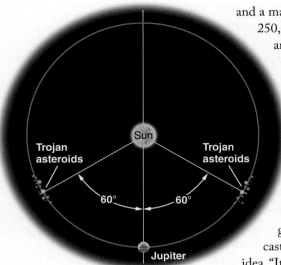

FIGURE 5.6
Trojan Asteroids
The Trojan asteroids lie along Jupiter's orbit at 60° preceding and following the planet. These asteroids lie at stable points where the gravity of Jupiter and the Sun and the motion of Jupiter in its orbit all balance out.

planetesimal An irregular rocky object, typical of those from which planets are believed to have formed by accretion.

Trojan asteroid An asteroid co-orbiting with Jupiter in one of two groups: one group preceding Jupiter at a specific distance and the other group following Jupiter at the same distance.

Earth-crossing asteroid (ECA) An asteroid whose highly elliptical orbit crosses Earth's orbit and that is therefore a potential risk for collision with Earth.

and a mass of 2.6×10^{20} kg. Most of the 250,000 known main-belt asteroids are far smaller and less massive than Ceres or 4 Vesta, and only a few reach a diameter of 100 km.

For many years, astronomers debated the origin of the asteroid belt. Some scientists saw it as the remains of a planet that had been broken apart by either a collision or the force of Jupiter's gravity. However, various facts cast doubt on this initially popular idea. "It's very difficult to smash a planet once you've made it," explains Jewitt. "You'd have to hit a fully formed planet with something just as big and have it moving very fast. There's no evidence that anything like this kind of collision happened." In addition, if the "broken-planet" theory were accurate, all asteroids in the belt should have a similar composition. Instead, asteroids have a range of compositions, suggesting that they did not originate from a single large body.

A second theory, which most astronomers now have more confidence in, is that the belt represents a desert of **planetesimals**, rocky objects that might have been the building blocks of planets. According to this second theory, planetesimals in the region between Mars and Jupiter could not form into a planet because the gas giant's gravity kept the planetesimals from coagulating into a larger body. "It looks a lot more like the asteroids came from a number of objects," explains Jewitt, "with different initial formation histories. Those origins get reflected in their properties, like the different ratios of rock to iron we see in different asteroids."

Beyond the collection of asteroids in the asteroid belt, there are at least two other important collections of asteroids in the Solar System. The so-called **Trojan asteroids** are two groups of objects that move around the Sun along the same orbit as Jupiter, with one group always preceding and the other group always following the gas giant (**Figure 5.6**). There may be a million or more of these Trojan asteroids with sizes greater than 1 km.

A third collection of asteroids, the **Earth-crossing asteroids (ECAs)**, is of enormous importance for human civilization. Most asteroids move along orbits

with fairly low eccentricity ($e < 0.3$). Recall from Chapter 3 that low eccentricity (e closer to 0) means circular orbits, while high eccentricity (e closer to 1) means highly elliptical orbits. An asteroid on a low-eccentricity orbit will not change its distance from the Sun very much. This means it will not intersect the orbits of any planets. There is, however, a family of asteroids that have high eccentricities and periodically cross Earth's orbit. "These Earth crossers are mostly coming from the asteroid belt," explains Jewitt. "They were kicked onto elliptical orbits by Jupiter, and now they cut across the paths of the terrestrial planets." As Jewitt explains, ECAs last only so long. "Either they're going to hit one of the planets," he says, "or they're going to experience a slingshot effect from a planet's gravity into the Sun. They might even get shot out of the Solar System entirely."

As their name suggests, Earth-crossing asteroids may sometimes become Earth-striking asteroids. When this happens, the consequences can be apocalyptic. "We know the surfaces of the terrestrial planets have been hit by asteroids, because they're all covered with craters," says Jewitt. "And we know this is true for Earth as well, even though craters here tend to get weathered away over time." Geologists have compiled a catalog of more than 180 impact craters on Earth, some of which must have been caused by asteroid-sized objects. "Some of them are frighteningly big," says Jewitt. "They can stretch across hundreds of kilometers." The most famous impact happened 65 or 66 million years ago and is associated with the extinction of the dinosaurs. "It was probably a 10-km-sized body that hit us back then," says Jewitt. "It was a global extinction event that killed almost every surface-dwelling species." A more recent collision occurred when a much smaller body exploded over Tunguska, Siberia, in 1908, flattening forests for thousands of square kilometers (see Figure 5.7B).

In 2002, a 100-meter-wide asteroid called 2002 MN passed a mere 120,000 km from Earth, well inside the Moon's orbit. An object this size has the energy of a 50-megaton nuclear weapon and could do catastrophic damage if it made impact near a populated area. Although the odds of a collision in any given year are quite small, the consequences are so far-reaching that in the late 1990s astronomers began a systematic effort to find all ECAs. Since then, thousands of ECAs have been discovered, and hundreds of those have been listed as "potential hazards." This dubious distinction is given to ECAs with sizes

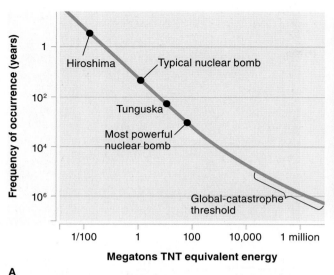

A

B

FIGURE 5.7 Asteroid Impacts
(A) The explosive power of asteroid impacts represents an ever-present danger to our planet and its life-forms. Impacts with energies greater than global-catastrophe threshold will create mass extinctions. The asteroid whose impact resulted in the extinction of the dinosaurs was only about 10 km in diameter. That asteroid had nearly 10,000 times the power of the nuclear bombs that leveled Hiroshima and Nagasaki during World War II. (B) This image shows the aftermath of the Tunguska event of 1908.

of 150 meters or more and with orbits that bring them within 7,500,000 km of Earth. **Figure 5.7A** shows the frequency of impacts as a function of the impact energy.

Why don't astronomers already know where all the ECAs are? For that matter, why can they only estimate the number of asteroids in the Solar System? The trouble is, finding asteroids is an extremely tricky business. Most asteroids are small and faint, reflecting only a small amount of sunlight, so determining both the total number of asteroids and their orbits requires long hours of effort with as-yet-incomplete results.

Determining asteroid properties has also been a difficult task. Identifying properties such as composition, size, or shape requires indirect methods, such as interpreting reflected sunlight. From these studies, astronomers have determined that a number of distinct classes of asteroids exist, two of which account for most of the asteroids in the Solar System. Asteroids of the first main class are called carbonaceous (rich in carbon), or **C-type asteroids**. These are very dark and reflect little sunlight. Asteroids of the second main class are called **S-type asteroids**. They contain more **silicates** (rocks made with silicon and oxygen) and are more reflective. A full 75 percent of the Solar System's asteroids are S-type, 15 percent are C-type, and the rest fall into other classes.

The distinction between the mineralogy of the two major classes of asteroids is important because it highlights different early evolutionary phases of the Solar System. "From the ages of meteorites, we know that all the asteroids were formed around 4.5 billion years ago," says Jewitt. "But the C-type asteroids were never strongly differentiated," he explains. **Differentiation** refers to the process by which gravity within forming planetesimals forces heavier material (like iron) to sink to the center. "It appears that many C-type asteroids escaped differentiation. They represent more primitive material." The S-type asteroids are probably chunks of planetesimals that were on their way toward forming larger bodies, so they were subjected to heat, compression, and differentiation (all from gravity) when collisions smashed them apart.

Although telescopes and meteorites tell us a lot about asteroids, to learn more we have to visit them. This is exactly what a flotilla of spacecraft has done since 1990. A number of missions from several regions, including the United States, the European Union, and Japan, have made the journey to at least eight asteroids. In 2005, for example, the Japanese *Hayabusa* space probe reached asteroid 25143 Itokawa with the intention of bringing samples of the rocky body back to Earth.

From these missions, astronomers have gained important insights into the nature, structure, and

C-type asteroid An asteroid that is carbonaceous (rich in carbon) and is typically dark.

S-type asteroid An asteroid that contains mostly silicate rock and is typically more reflective than C-type asteroids are.

silicate A mineral compound containing silicon and oxygen.

differentiation The process in planet formation whereby, during a planet's molten stage, the denser materials sink to the center while less dense materials rise to the surface, forming distinct layers.

origins of asteroids. Two of the most important asteroid encounters came in 1997 and 2000, when the Near Earth Asteroid Rendezvous (NEAR) mission visited the C-type main-belt asteroid Mathilde and the S-type near-Earth asteroid Eros. Images showed a crater so large on Mathilde that scientists wondered how the object could still be intact. "The answer," says Jewitt, "came when we looked at Mathilde's density and found it to be so low that it wasn't really a solid object." Instead of being a solid rock, Mathilde must be a rubble pile—a loosely porous collection of smaller rock fragments and dust held together by gravity. Images and measurements of Eros showed that, unlike Mathilde, it was solid and heavily cratered from millions of years of collisions with other smaller Solar System objects. Recently both Vesta and Ceres have been visited by the *Dawn* spacecraft. An additional recent discovery has been the detection of water vapor enveloping Ceres, which must be continually outgassed from the body.

SECTION SUMMARY
Asteroids
- Hundreds of thousands of asteroids are known to exist, primarily in the asteroid belt beyond the orbit of Mars.
- Two special groups of asteroids are the Trojan asteroids and the Earth-crossing asteroids.
- Asteroids are primarily either carbonaceous (C-type) or silicate (S-type).

Comets

Like asteroids, comets are a form of debris left over from the formation of the Solar System, but comets are very different from asteroids in composition. "While asteroids are mostly rock and metal," says Jewitt, "comets are maybe 50 percent ice with some dirt thrown in." Thus, astronomers think of comets as similar to "dirty snowballs" (although "icy mudballs" may be a more accurate term). In their properties, locations, and orbital trajectories, comets offer clues to the formation of the Solar System and, perhaps, the history of our own planet.

The difference between asteroids and comets already offers a clue about the processes at work in the early Solar System. Rocky asteroids formed within a critical distance from the Sun that is demarcated by the **snow line**. "Inside the snow line, solar radiation is strong enough that water vaporizes," Jewitt explains. "But outside the snow line, water will be frozen and can remain as ice." Comets must have formed beyond

snow line The distance from the Sun or other star beyond which it is cold enough for water to exist as ice.

A

B

FIGURE 5.8 Comet Hale-Bopp and the Bayeux Tapestry
(A) The magnificent Comet Hale-Bopp, discovered in 1995, was visible to the naked eye for 18 months. (B) A comet—believed to be Halley's Comet—appears on this medieval tapestry, which tells the story of the Norman Conquest of England. Comets were viewed in that era as ominous events.

the snow line because they incorporate so much frozen water and other icy compounds along with rocky material.

Comets often make dramatic celestial appearances (**Figure 5.8A**). Taking just a few weeks (or, at most, months) to move across the fixed background of stars, comets first appear as bright, glowing objects, and they appear to grow larger over time to reveal luminous tails

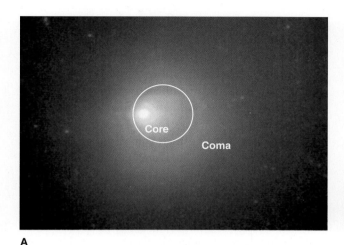

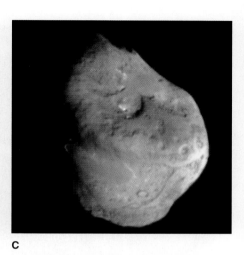

A   **B**   **C**

FIGURE 5.9 Comet Structure (A) The wide, diffuse comas of comets have been observed for thousands of years, but recent up-close studies of comets reveal that the outer surfaces of their nuclei are composed of crusts left over as ice is burned away. (The core is the nucleus and the inner region of expelled plasma.) (B) The crusted nucleus, coma, and jets of Halley's Comet are clearly seen in this image, taken by the European Space Agency's *Giotto* spacecraft from 600 km. (C) This image of Comet Tempel 1, taken far from the radiation and solar wind of the Sun, shows only the comet's nucleus; the coma is completely absent.

that can extend many degrees of arc before the comet eventually fades and disappears. In premodern times, the appearance of a comet was often interpreted as a sign (usually a bad one) and, as such, was a source of considerable anxiety (**Figure 5.8B**). Astronomers now understand the physics of a comet's fleeting appearance.

A comet's glowing appearance comes from changes in its **nucleus** as it approaches the Sun. A comet's nucleus can range in size from 100 meters to 50 km. The nucleus is a mix of rock, dust, water ice, and icy frozen gases like carbon monoxide, carbon dioxide, ammonia, and methane. Although these ices make up the bulk of the comet's mass, space missions have shown that comets' nuclei have rocky surface crusts that are remarkably dark. Astronomers use the term **albedo** to describe how much light an object reflects. Comet surfaces have very low albedos of just 2–4 percent, about half that of the black tarmac that covers most roads. The comparison with road tar is appropriate: some scientists believe comet crusts are dark because surface ice, burned away by sunlight, leaves behind a sticky goo of rock and dust the color of asphalt. But if the rocky, dusty, tarlike surface of a comet nucleus reflects so little light, what is it that glows brightly when a comet makes its appearance on the sky (**Figure 5.9**)?

Most comets move in highly elliptical orbits, taking them from the cold, outer Solar System into the domain of the terrestrial planets and eventually

very close to the Sun (some even end up plunging into the Sun). As the comet nucleus enters the inner Solar System, sunlight warms the nucleus, vaporizing the ices below its crust. **Jets** of high-pressure water, carbon dioxide, methane, and ammonia vapor break through the surface. The jets of vapor shoot into space and are ionized by the Sun's radiation. In this way the nucleus of the comet becomes surrounded by a glowing cloud of gas and dust called the cometary **coma**. Light reflected off dust blown into the coma and direct emission lines from its ionized gases are what make the coma visible to observers on Earth. Recall that emission lines form when electrons in atoms or ions are driven into excited orbits; the electrons then emit light when they cascade back down (see Chapter 4).

The tails of comets, the most spectacular part of their appearance, are also linked to material blown off the nucleus as it warms up on its journey through the inner Solar System. Actually two tails form as the comet moves along its orbit. The **ion tail** is made of ionized gas blown off the coma. The solar wind sweeps up these ions and drives them outward away from the direction of the Sun. The **dust tail** is composed of small grains of solid matter blown off the nucleus. The dust particles are blown away from the comet by the pressure of sunlight, but after leaving the vicinity of the coma, they move on their own orbits, and hence the dust tail tends to be curved. Note that both tails always point away from the Sun, even when the

nucleus (pl. nuclei) The main body of a comet, composed of ice and rock.

albedo The reflecting power of an object, measured as the ratio of light reflected to light received.

jet A stream of a liquid, gas, or small solid particles shot outward from a comet nucleus or other object in a focused beam.

coma The nebulous envelope around the nucleus of a comet, formed when the comet passes close to the Sun.

ion tail The tail of a comet that is composed of gas and reacts to the solar wind and thus always points directly away from the Sun.

dust tail The tail of a comet that is composed of dust driven off the coma by solar radiation pressure; its particles then move on their own orbits. It always points away from the Sun.

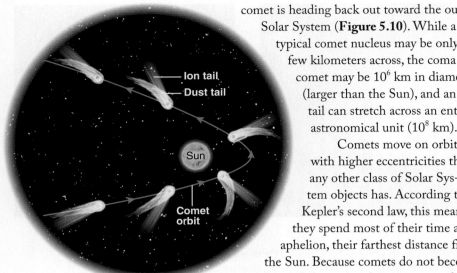

FIGURE 5.10
Ion Tail and Dust Tail
Comets have two tails. The ion tail is pushed by the solar wind in a direction opposite to the Sun. The dust tail is made up of particles initially driven away from the comet by the pressure of sunlight, but the tail is then shaped by the comet's own orbital motion around the Sun. The tails lengthen as the comet approaches the Sun and essentially disappear as the comet approaches aphelion.

short-period comet A comet that originated in the Kuiper Belt or scattered disk and has an orbital period of 200 years or less.

long-period comet A comet that originated in the Oort Cloud and has a highly eccentric orbit and an orbital period from 200 years to millions of years.

comet is heading back out toward the outer Solar System (**Figure 5.10**). While a typical comet nucleus may be only a few kilometers across, the coma of a comet may be 10^6 km in diameter (larger than the Sun), and an ion tail can stretch across an entire astronomical unit (10^8 km).

Comets move on orbits with higher eccentricities than any other class of Solar System objects has. According to Kepler's second law, this means they spend most of their time at aphelion, their farthest distance from the Sun. Because comets do not become bright until they enter the inner part of the Solar System, astronomers must reconstruct comets' orbital paths using observations and detailed calculations. These studies show that comets originate from two distinct locations in the Solar System and therefore fall into two distinct classes, which differ in orbital period and orbital inclination. (Inclination is a measure of the tilt between an object's orbit and the ecliptic—the plane of Earth's orbit.)

The first class, **short-period comets**, have orbits that take 200 years or less to complete. These orbits have low inclination, meaning they are usually aligned with the ecliptic plane. In addition, their direction of orbital motion is the same as that of all the planets in the Solar System. Halley's Comet, with a 76-year orbit, is an example of a short-period comet. Its aphelion lies just beyond Neptune.

Long-period comets, in contrast, can have periods of thousands of years or even a million years. In general, the orbits of long-period comets are not confined to the ecliptic, and their aphelia take them far beyond the domain of the outer planets. "The long-period comets have very large eccentricities," says Jewitt, "and they come from all directions with equal probability."

The majority (80 percent) of known comets belong to the long-period family, and short-period comets make up just 19 percent of known comets. Comets whose orbits cannot be determined or that might leave the Solar System altogether make up the rest. From orbital analysis, astronomers can tell where the different families of comets originate. "The Kuiper Belt is the likely source of the short-period comets," says Jewitt, explaining that occasional gravitational tugs from Neptune get these bodies started on new

orbits. "The short-period comets have relatively small eccentricities and inclinations," Jewitt explains. "Their orbits are not circular, but they're not very eccentric either. These relatively circular, relatively uninclined orbits are all consistent with an origin in the Kuiper Belt, although some may have originated in the scattered disk as well."

The long-period comets come from much deeper in space. In the 1920s, the Dutch physicist Jan Oort was the first to recognize that the Solar System must be surrounded by a spherical "cloud" of comet nuclei at distances of 50,000 AU or more from the Sun. This Oort Cloud is now recognized to be the source of the long-period comets. In the same way that the gravitational influence of an outer planet can knock a Kuiper Belt comet nucleus onto a higher-eccentricity orbit, a passing star may sometimes come close enough to disturb comets in the Oort Cloud and knock them onto Sun-bound elliptical orbits.

Comets on highly elliptical orbits must, by definition, cross the orbits of some or all of the planets. Considerable evidence suggests that encounters with gas giants alter the orbits of comets over time. But gravitational interactions with planets can do more than simply adjust cometary orbits. Collisions between planets and comets had long been hypothesized, but in 1994 the comet Shoemaker-Levy 9 was captured by Jupiter as it passed the giant planet on its way toward the inner Solar System. Astronomers around the world watched with delight as the comet was torn apart by Jupiter's gravity and the fragments eventually plunged into the planet's atmosphere (**Figure 5.11**).

According to one theory, collisions between Earth and comets in the early history of the Solar System may have been responsible for bringing our young world the bulk of its water. "Earth formed hot," says Jewitt. "If you overbake a cake in the oven, it's going to come out as something dry and crisp. The early Earth was probably a lot like that overbaked cake. So its water and other volatile materials had to be added later, after the surface cooled down. Since comets are big, icy things, they are the natural candidates for water delivery systems."

Comparisons between the composition of Earth's oceans and the ice composition of comets, however, place some doubt on this hypothesis. Some scientists, Jewitt among them, now believe that asteroids—not comets—may have delivered the early supply of water to Earth. "There is a new class of main-belt 'comets,'

A

B

FIGURE 5.11

Comet Impact with Jupiter

In 1994, comet Shoemaker-Levy 9 broke apart because of Jupiter's tidal forces. (A) The comet's 21 fragments were observed stretching across more than 1 million km prior to the collision. (B) The fragments struck Jupiter with spectacular results. Dark regions in this ultraviolet-light image of Jupiter's atmosphere, captured by the Hubble Space Telescope, represent explosive energy depositions as comet fragments plunged into the Jovian atmosphere. The black dot above Jupiter's equator is its moon Io.

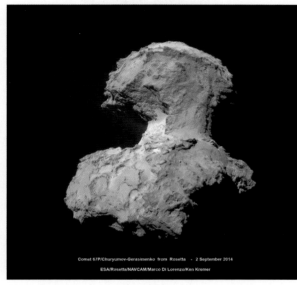

A

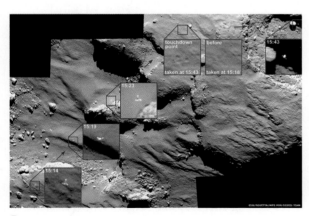

B

FIGURE 5.12 Comet 67P/Churyumov-Gerasimenko
(A) This image was taken by *Rosetta*'s probe at a distance of 56 km from Comet 67P/Churyumov-Gerasimenko. (B) This mosaic and time sequence of images shows the *Philae* lander on its descent to Comet 67P/Churyumov-Gerasimenko. The image marked "touchdown point" shows an imprint of the lander on the comet; the image marked simply "15:23" shows the lander sailing over the comet's edge before dropping back again.

some of which are basically water-rich asteroids," says Jewitt. "There are also asteroids which, even though they don't have ice, have minerals in them which include water as part of the molecular structure." If there are significant quantities of these **hydrated** minerals (minerals that contain water molecules) in asteroids, then an impact with Earth would break the minerals apart when the asteroids were destroyed. The water molecules would then be freed and might have become part of Earth's oceans. "These kinds of asteroids," says Jewitt, "tell us that a lot of water must have been present in the Solar System in the past in a variety of forms, so there are different ways Earth's oceans may have formed."

In 2014 the European Space Agency's *Rosetta* spacecraft rendezvoused with Comet 67P/Churyumov-Gerasimenko. What *Rosetta* found was an unusually shaped comet nucleus approximately 4.3 × 4.1 km across with a narrow "neck" connecting two wider regions (**Figure 5.12A**). *Rosetta* released a lander called *Philae* designed to attach itself to the comet (given its weak gravity) via harpoons and screws embedded in the lander's legs. Unfortunately, these systems failed because of a remarkably hard ice layer just below the surface of the comet. Although *Philae* "bounced" at least once, it eventually came to rest on the comet's surface (**Figure 5.12B**).

SECTION SUMMARY

Comets

- Comets are composed of ice and rock. As they approach the Sun, they have a bright coma, which is formed from gases and dust blown off the nucleus by the solar wind, and two tails—the ion tail and the dust tail—both of which always point away from the Sun.

- Short-period comets originate in the Kuiper Belt or scattered disk. Long-period comets come from the Oort Cloud and have highly elliptical orbits.

- Comets or asteroids may be the source of Earth's water.

hydrated Containing water.

Meteoroids

Meteoroids are the smallest class of debris in the Solar System, ranging in size from a grain of sand to a boulder. The bright trails that **meteors** leave on the sky as they enter Earth's atmosphere are among the most striking events that people get a chance to see on the sky.

Meteoroids become meteors when they are captured by Earth's gravity and plunge into the atmosphere at speeds as high as 20 kilometers per second (km/s), or more than 70,000 kilometers per hour (km/h). Meteoroids act like supersonic jets as they plow through the air. Extremely high temperatures build up in front of the fast-moving meteoroid as air molecules pile up. The glowing trail of gas is what astronomers formally call a meteor (**Figure 5.13**). Millions of meteors appear in the atmosphere each day (although they can usually be seen only during the night); most of them occur at heights between 50 and 95 km above Earth's surface.

A particularly remarkable example of a meteoroid occurred on February 15, 2013, over the city of Chelyabinsk, Russia. Thousands of people on their way to work watched as a fiery ball streaked across the sky and flared brightly (**Figure 5.14**). Soon afterward, a shock wave swept across the region, shattering windows and damaging thousands of buildings. The event was caused by an object 20 meters in diameter

meteor A meteoroid that has entered Earth's atmosphere. Heat from friction causes radiation that may be seen as a very brief streak of light or as a fireball.

meteorite A rock that was a meteoroid, then a meteor, and has reached Earth's surface.

meteor shower The appearance of frequent meteors, over a period of a few hours or several days. The meteors seem to emanate from a common location on the sky, which is caused by Earth's passing through the debris deposited along the orbit of a comet.

FIGURE 5.13
Meteors
This streak of light was created by a meteor, a small piece of solid matter being heated by friction with Earth's atmosphere.

FIGURE 5.14
Chelyabinsk Fireball
The fireball created by the Chelyabinsk meteoroid as it exploded 29.7 km aboveground. Even at that height, the blast wave from the airburst damaged thousands of buildings.

A

B

C

FIGURE 5.15 Meteorites
Meteors that make it intact to the ground are called meteorites. (A) An iron meteorite on Mars, spotted by the rover *Opportunity*. (B) A fragment of the crust torn off the asteroid Vesta by a collision. (C) A fragment of asteroid TC3, which exploded over Sudan in 2008.

with a mass of more than 12 million kilograms (kg). It was the largest known natural object to enter Earth's atmosphere since the Tunguska event.

Most meteor-causing meteoroids are the size of small rocks, and these usually burn up (evaporate) in the atmosphere. Larger meteoroids can, however, make it to the ground, at which point they are called **meteorites** (**Figure 5.15**). These lumps of space rock are treasured finds for astronomers and planetary geologists, offering a rare glimpse of the raw material of Solar System formation. "It's rare that someone tracks a meteor all the way to the ground and finds a meteorite," says Jewitt. "The majority of meteorites are discovered by accident long after they fell." Many meteorites come from the South Pole. "They fall onto the ice, get buried, and then ice flow under the surface brings them to the surface years later," says Jewitt. "A meteorite is pretty obvious on an ice field, so a lot of scientists travel to the South Pole to hunt for new finds."

Studies of meteorite composition and analysis of meteor tracks on the sky show that most meteorites are chunks of asteroids from the main belt. But not all meteors originate in asteroids. During **meteor showers**, unusually high numbers of meteors occur. Just as important, the meteors in a shower all appear to radiate from a single point on the sky. Meteor showers occur at fixed times each year. For example, the Leonid meteor shower, in which meteors appear to radiate from a point in the constellation Leo, occurs in November.

The source of these showers has been tracked to material blown off comets. Over time, dust particles and pebbles ejected from a comet become distributed along its orbit. If Earth crosses the comet's orbit as the planet moves around the Sun, it will pass through this trail of debris. The resulting meteors will, from the perspective of someone on the ground, all appear to radiate from the same location on the sky. That point on the sky will be the position of the comet's debris trail (**Figure 5.16**).

Remarkably, some meteorites appear to be chunks of other planets. Meteorites that are chemically similar to the Moon or even Mars have been found. These finds provide strong evidence that asteroid and comet impacts with planets and moons can eject chunks of these bodies into space. Eventually, some of these planet fragments are captured by Earth's gravity and fall to the ground as meteorites.

Ample evidence shows that Earth is routinely struck by large meteoroids capable of considerable (though not apocalyptic) damage. The Barringer Meteor Crater in Arizona stretches 1.2 km across and is 170 meters deep (**Figure 5.17**). It is the result of an impact 50,000 years ago by a 50-meter-wide object that weighed some 200,000 tons (note that 50 meters is too small to be considered an asteroid or comet). Atmospheric and geologic forces tend to erase the evidence of meteor craters on Earth, but space-based imaging has helped astronomers find other examples. A more recent large impact on Earth was not really an impact at all. "A good example of a recent meteor strike is the Tunguska impact in Siberia in 1908," explains Jewitt. "It was a rather small body, maybe only 50 meters across, that exploded in midair, but it delivered a big enough punch to knock over 2,000 square miles of pine forest." The Tunguska blast had the force of a 5- to 10-megaton nuclear weapon. Had it occurred over a modern metropolitan area, the casualties would have been in the millions.

If they do not wipe out cities, meteorites can be a great boon to astronomers, giving them a chance to examine pristine material left over from the birth of the Solar System. "Meteorites give us access to a huge number of different source bodies in the asteroid belt," says Jewitt. "They put all our theories into a context." The clues that meteorites offer, along with clues from

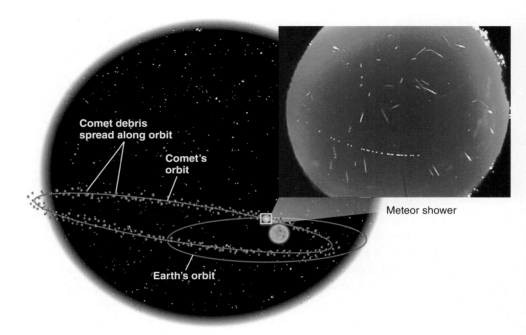

Meteor shower

FIGURE 5.16 Meteor Showers
Meteor showers occur when Earth's orbit takes it through clouds of small particles typically left by comets. Meteor showers are named for the constellation from which they appear to originate. Other than that, there is no connection between the meteoroids and the stars that make up the constellation.

FIGURE 5.17
Barringer Meteor Crater
Most Solar System debris that reaches Earth's surface is in the form of small meteorites, which provide information about their sources and the Solar System's history. Larger objects have enormous destructive power, however. The meteorite that struck in Arizona and created this crater 50,000 years ago had the approximate energy of 10 megatons of TNT.

other sources, provide a strong foundation for building a story of Solar System formation.

SECTION SUMMARY
Meteoroids
- Meteoroids are chunks of rock and metal ranging in size from a grain of sand to a boulder. When on the ground, they are called meteorites.
- Meteor showers result when Earth encounters residual comet dust along its orbit.
- Meteorites are rich sources of information about the early history of the Solar System.

David Jewitt

The skies of London may not seem like the best place for a young boy to find astronomical inspiration, but in the 1960s the hazy skies of England's urban capital were dark enough to let David Jewitt see his future.

"I grew up in an industrial part of London, where the city lights were pretty bright," he explains. "One night I was riding my bike home from the park. I looked up and saw a bunch of meteors crossing the sky. I didn't really know what they were, but I must have seen 10 or 15. The main thing that got me was they all came from the same direction." With this celestial vision in his mind's eye, Jewitt pedaled his bike straight home and asked his mother the simplest of questions: "What are those things?"

"She told me they were shooting stars," recalls Jewitt. With a stubborn curiosity that is the hallmark of scientists, budding or otherwise, Jewitt replied, "How can a star shoot?"

The meteor shower Jewitt saw that night set the course for his life. "The thing I really remember was I saw something that my parents didn't understand. Nobody could tell me what it was. That really impressed me." Eventually, Jewitt's grandparents bought him a small telescope. "It was about as big as Galileo's," he recalls. "I watched the phases of the Moon. I watched sunrise and sunset on various lunar craters. I thought it was all amazing. I hadn't imagined that a person in North London and of my age and lack of knowledge could see things like that."

The years of backyard observing paid off for Jewitt. After attending the University of London, Jewitt went on to earn a PhD at the California Institute of Technology, ending up as a professor in the Earth and Space Sciences Department at UCLA. Jewitt's discovery of the first Kuiper Belt objects in 1992 marked a turning point in studies of our Solar System. "There are some things we do know," he says, "and there are lots of things we don't know. And then there are things we thought we knew, but we're learning we have to go back and rethink them. It's that last aspect of science that keeps it all interesting."

5.4
Us versus Them:
Our Solar System and Others

Now that we have a complete census of the Solar System, we can begin to ask what insights its structure (shape) and dynamics (motion) offer us as we try to build a theory of its origin. Our next step will be to compare the information we have about our own Solar System with what we know about planets orbiting other stars.

Evidence for the Origin of Our Own Solar System

interstellar gas Ionic, atomic, or molecular gas found in diffuse clouds in the regions between the stars.

Looking down on the Solar System from "above"—that is, looking down at the Sun's north pole—reveals the first clue about the origin of our Solar System. All the planets in the Solar System orbit the Sun in the same counterclockwise direction. The majority of asteroids and most short-period comets also have the same orbital direction. But it is not only the orbits that have the same rotational direction: "Most of the planets are spinning in the same direction," says David Jewitt, "and their sense of spin is the same as the Sun's spin. So, if you look down from the top of the Solar System almost all the planets are spinning counterclockwise, and all of them are orbiting counterclockwise as well. That's important."

Venus is the only planet with a different spin direction (clockwise). Most astronomers, however, believe that Venus began with the same direction of spin as all the other planets and the Sun. Its current reverse rotation, they suggest, resulted from a powerful collision with another large body early in the Solar System's evolution.

The fact that almost everything is spinning and orbiting in the same direction points to a common origin for the material that formed the planets and the Sun. Having spins and orbits aligned any other way would be very unlikely. If planets were randomly captured bodies, for example, formed in other locations in the galaxy, it would be unlikely that every one of them would settle into an orbit with the same rotational sense.

"The inference," says Jewitt, "is that the rotation and spins of the Sun and the planets have all been picked up from whatever formed them together." The common direction of spin and rotation tells astronomers that both the Sun and the planets must have formed out of a large object, such as a cloud of **interstellar gas** that was itself rotating. The parent cloud must have imparted its sense of rotation to all the spinning/orbiting material in the newly formed Solar System.

The next important clue about the Solar System comes not from its motions but from its shape. If we could look down along the Sun's spin axis, we would see the planets orbiting at different distances. But if we looked at the Solar System along the Sun's equator, we would find that all the planets move on orbits that are roughly aligned in the same plane (**Figure 5.18**). "Basically, all the planets in our Solar System have small inclinations," says Jewitt. The largest inclinations are about 6° or 7° for the planets and the smallest is about 3.5°. "The planetary orbits are essentially aligned into a disk," Jewitt continues. "And that means they probably formed in a disk."

There are other indications that the primitive material that formed the planets must have

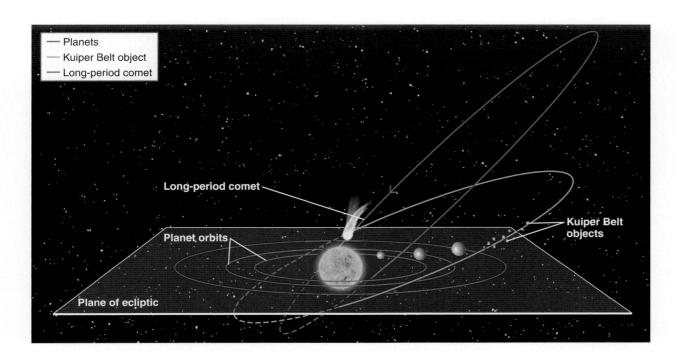

FIGURE 5.18
Inclination of Orbits
Planets and asteroids orbit very close to the plane of the Sun's equator, while the orbits of Kuiper Belt objects are typically somewhat inclined. By comparison, long-period comets from the Oort Cloud have orbits with far greater inclination.

had a disk shape. The asteroid belt looks like a disk, or annulus, of orbits; and the Kuiper Belt, which extends beyond Neptune, has the shape of a fat disk or doughnut of orbits. The exceptions to this rule, however, are the long-period comets. These originate in the spherical Oort Cloud and come diving into the Solar System on highly inclined orbits. "The material in the Oort Cloud didn't form out there," explains Jewitt. "The comets probably formed in the disk and then were gravitationally scattered by close encounters with the giant planets."

Thus, the shape and motions of material in the Solar System tell us something elemental about how it must have formed. The rotations and spins of the planets tell us that the Solar System must have originated from a single spinning object. The orientation of planetary orbits tells us that they must have formed from some kind of disklike configuration of source material.

Later in the chapter we will revisit these details about our own Solar System—including the shape, the motions, and even the composition of the planets—to see how they have contributed to the modern story of Solar System formation. But for now, the next step in thinking about *a general theory of planetary system formation* is to see what has been learned from other systems.

"Until we started finding other planets beyond those orbiting the Sun," says Jewitt, "we did not know whether our Solar System was a representative example. Are we kind of average, or are we some kind of freak? Now we know the answer to that question, and it turned out to be pretty surprising."

SECTION SUMMARY
Evidence for the Origin of Our Own Solar System

- All planets in the Solar System orbit the Sun in the same direction the Sun rotates, and most planets rotate in that same direction too, providing evidence that the planets and the Sun all formed from the same similarly rotating object.
- The alignment of planetary orbits in a disk strengthens the evidence for their formation in a disk.

Discovering Exoplanets

Answering the 2,000-year-old question—Do planets exist outside our Solar System?—was bound to be vexing. "The problem with finding planets in systems beyond our own is very simple to state and very difficult to solve," says Heather Knutson. Knutson is an astronomer at Caltech in Pasadena who studies exoplanets. "These planets are very far away," she explains. "They are very faint, and they are parked right smack next to very, very bright objects: their parent stars."

Knutson provides an analogy that underscores the difficulty of exoplanet searches. "Imagine that someone in California sets up a large spotlight, like the kind they have at those Hollywood movie openings," she begins. "Now imagine they point the spotlight at New York City. Next to the spotlight they let some fireflies out of a can. The fireflies flutter around a few feet away from the spotlight. Your job in New York is to take a picture that can separate the fireflies from the spotlight." As Knutson emphasizes, that is going to be a pretty difficult task. But as she says, "It's the same-size problem astronomers have to deal with in finding planets around other stars."

It took a lot of ingenuity and hard work before scientists could be sure that they had discovered conclusive evidence of exoplanets. As far back as the 1880s, some astronomers claimed to have detected a planet orbiting the binary stars 70 Ophiuchi (the term *binary stars* refers to two stars in orbit around each other). In the 1950s and 1960s, attention turned to Barnard's Star (a low-mass star just 0.14 times the mass of the Sun and 6 light-years from Earth) and claims for an entire family of planets residing there. As late as 1990, two radio astronomers claimed to have found a planet orbiting a neutron star, the dead cinder of a once massive sun. In each instance, the excitement grew and then faded as detailed examination of the evidence revealed the "planet" to be no more than an illusion. (In 1992, however, planets were confirmed to be orbiting a type of dead star called a *pulsar*; see Chapter 13.) How do we know now that planets orbit other stars? Technological innovations in the past 30 years have brought about two important methods of accurately determining the existence and properties of extrasolar planets.

The first method is based not on finding the planet directly but on finding its gravitational influence on the parent star. Although we may say that one object orbits another—such as a planet orbiting a star—in actuality both objects rotate around a common **center of mass** that can be thought of as the average location of all total matter in the system. If two equal-mass objects orbit each other, then the center of mass lies at the midpoint of their separation. That will be the point around which both objects orbit. If one body is much more massive than the other, as in the case of the Sun and Earth (the Sun is more than 300,000 times as massive as Earth), then the center of mass will lie almost at the center of the more massive object. But *almost at the center* is not the same as *at the center*. If Earth and the Sun were all there was to the Solar System, the Sun would still complete a small orbit around the Earth-Sun center of mass, which is 449 km from the Sun's center. Earth, of course, would also orbit this point, but at a distance of 1 AU. Because the Sun has a radius of 696,000 km, the gravitational influence of Earth, if it were alone, would induce a tiny (449 km) wobble in the Sun's position at a speed of 0.09 meters per second (**Figure 5.19**). What astronomers needed was a way of detecting these tiny back-and-forth motions of the star created by the gravity of its orbiting planets.

In the mid-1990s, new high-precision instruments were developed that enabled astronomers to track Doppler shifts from stars moving with velocities as small as a few meters per second. With this precision, they could finally look for planet-induced motions of stars, called **reflex motion**. This approach is called the **radial velocity method** (**Figure 5.20A**).

Once the new instruments were tested, the search was on. To claim that they had detected a planet, astronomers knew they would have to watch the star (and hence the planet) move back and forth for a full orbit. "As the star moves toward you (and away from you), you see this little Doppler shift back and forth in wavelength," says Knutson. "By measuring the period

center of mass The point around which two orbiting bodies move. It lies closer to the more massive of the two objects.

reflex motion The movement of a star in response to the gravity of an orbiting planet.

radial velocity method An exoplanet detection method that measures changes in a star's velocity as it orbits a common center of mass with one or more planets.

FIGURE 5.19
Center of Mass
A planet does not orbit its star; rather, the star-planet pair orbits its common center of mass (indicated by the X in the figure). Regular changes that can be observed in the star's velocity—moving toward and then away from us—indicate the presence of the otherwise unseen planet.

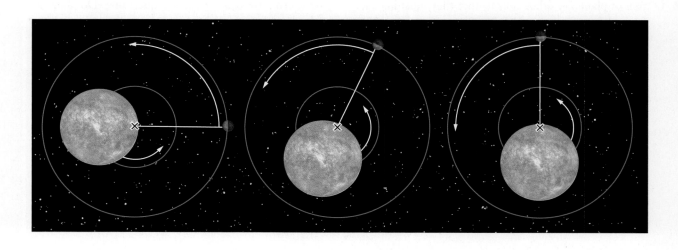

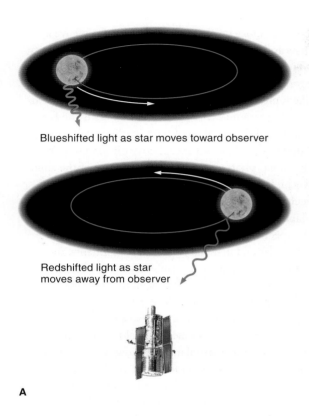

Blueshifted light as star moves toward observer

Redshifted light as star moves away from observer

A

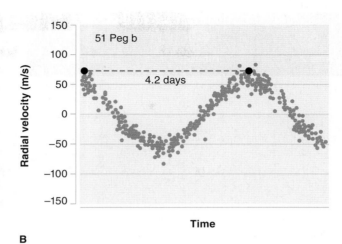

B

FIGURE 5.20 Radial Velocity Method
(A) Doppler shift of light allows astronomers to determine the motion of a star toward and away from Earth. (B) This method successfully identified the presence of 51 Peg b and astounded astronomers by the very short orbital period of the planet, indicating its close proximity to the star.

of that shift you get the planet's orbital period, and by measuring the amplitude of that shift you get the mass of the planet. Massive planets mean bigger wobbles." This means that gas giants will exert a greater gravitational pull and will make their stars wobble back and forth more than lower-mass terrestrial-type planets do. There is a complication, however. The radial velocity method provides no way of knowing the inclination of the orbit, so the masses it provides are always lower limits—they are the minimum mass the orbiting planet might have.

When the radial velocity method was first put into practice, astronomers' expectations of what a planetary system should look like were based on the one example they had—our own Solar System. Because the gas and ice giants in our system have very large orbits that take years to complete, astronomers expected they would have to wait years to find the signature of an orbiting gas giant. To their surprise, the first radial velocity detection showed a planet with an orbit much shorter than those of our gas giants. In 1995, astronomers announced the discovery of a planet orbiting the star 51 Pegasi, which they named 51 Peg b. From the gravitational reflex motion, astronomers derived its minimum mass to be about half that of Jupiter. But the new planet's orbit was the real surprise: it lasted just

4.2 days (**Figure 5.20B**). An orbit of such short period suggests that this giant planet circles its star at a distance of 0.05 AU, well within the orbit of Mercury in our system. The very first detection of an extrasolar planet found a planetary system with a gas giant hugging its own star. Clearly, there are other planetary systems that look nothing like our own.

Since the discovery of 51 Peg b, astronomers have found more than 400 extrasolar planets by the radial velocity method. Up to now, it has been one of the most successful means of detecting the presence of planets beyond our Sun. By careful examination of the radial velocity curves of a star, astronomers can even see evidence for multiple planets orbiting other stars. To date, astronomers have found evidence for hundreds of planetary systems—stars with more than one planet orbiting them. Of course, the more distant a planet is from its star, the longer its period will be. This means that astronomers have to wait longer to detect "outer planets" in a system.

Although the radial velocity method has been extremely successful, it does not tell astronomers what kind of planet is being discovered. What, for example, is the planet's physical size? For this kind of information, astronomers turn to the **transit method**, in which they look for small decreases in starlight as

interactive
Exoplanet:
Radial Velocity

transit method An exoplanet detection method that measures changes in the brightness of a star as a planet passes in front of it.

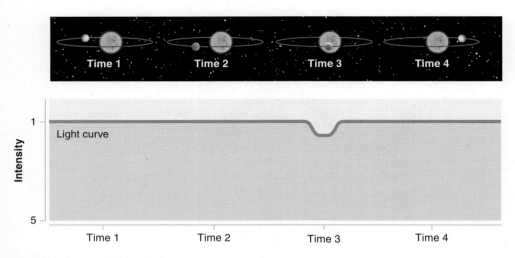

FIGURE 5.21 Transit Method
The transit method of exoplanet detection uses the periodic decrease in starlight received as an orbiting planet passes in front of the star. By accurately tracking the amount of starlight blocked, astronomers can infer both the presence of the planet and many of its properties.

interactive
Exoplanet: Transits

planet's radius," explains Knutson. "So if you have a mass and you have a radius, you can calculate a density, which means you are suddenly in the business of categorizing planets. You can say, 'Is your planet a gas giant like Jupiter? Is it a small, rocky planet more like Earth?'"

Even more exciting for astronomers is the fact that for a brief window of time, starlight will pass through the planet's atmosphere (if it has one). As the stellar photons glance along the edge of the alien world, some will be absorbed by the atoms and molecules in the atmosphere. Absorption lines from the planetary atmosphere will become superimposed on the spectrum of the star. Once astronomers separate the stellar and planetary spectral components, the constitution of the alien atmosphere can be derived.

the planet passes in front, or transits, the distant star (**Figure 5.21**). When combined with the radial velocity method, the transit method is enormously powerful. "If you know the planet passes in front of the star, then you know the orbital inclination," says Knutson, "and you know the true mass of the planet from your Doppler measurements." But the transit method can provide even more information.

"By measuring the amount of light the planet blocks when it goes in front of its star, you learn that

The transit method has been the most successful for smaller exoplanets. In particular the Kepler mission, which used a space telescope designed to find transits, revolutionized the field by discovering thousands of new exoplanet candidates. "To detect something Earth-sized moving at 1 AU from its star, we would have to be able to see Doppler shifts of a few centimeters per second," says Knutson. "The best we can do now is a meter per second, so we are not ready to detect Earth-sized planets with Earth-like orbits using the radial velocity method yet."

Of course, what astronomers want most of all is a direct image of an extrasolar world (**Figure 5.22**). "You can just go straight for the really hard thing," says Knutson, "which is taking an image of the planet

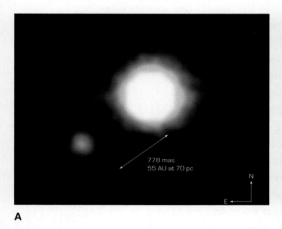

A

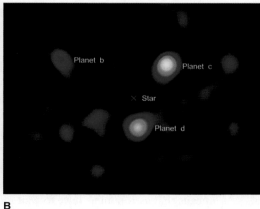

B

C

FIGURE 5.22 Direct Imaging of Exoplanets
(A) In what may be the first direct image captured of an exoplanet, the brown dwarf 2M1207 and a giant planet candidate are shown here in this 2004 near-infrared image. (B) Three gas giants are seen orbiting the star HR8799 in this visible-light image taken in 2010 using a 1.5-meter portion of the Palomar Observatory's Hale Telescope. (C) Astronomers first captured this visible-light image of what appeared to be a gas giant orbiting a Sun-like star at a distance of more than 300 AU in 2008. They confirmed in 2010 that this interpretation of the pair is correct.

where you can see the star in one part of the image and the planet separate from the star in the other part. We're starting to be able to do that. We have images where we see planets separate from stars, but as you would imagine, they are mostly very large planets, very far from their stars—many times the distance of Jupiter. So that technique doesn't get us to things like Earth or things at Earth's distance." A number of astronomical teams have directly imaged an exoplanet. In a few cases, these astronomers have relied on the planet's own infrared glow to capture the image. In general, however, direct detections remain rare.

Only time will tell whether recent imaging claims are true. Astronomers are also currently hard at work developing methods to screen out the light of the star, and a number of future space-based missions are planned whose sole objective is the direct detection of extrasolar worlds.

Perhaps the most remarkable aspect of exoplanet searches is that astronomers are now able to make educated predictions about how many planets orbiting other stars they can expect to find. "There is still a lot of debate, but most astronomers would accept that somewhere between 30 and 50 percent of all stars like the Sun have low-mass planets orbiting around them," says Knutson. "When you consider that there are hundreds of billions of stars like the Sun, you see that there are a lot of planets. When we talk about the possibility for life elsewhere in the galaxy and the Universe, I think it starts to sound a little bit more plausible when you consider that planets must be so easy to make."

SECTION SUMMARY
Discovering Exoplanets
- Exoplanets are difficult to find because they are so far away and so close to a bright star, but scientists believe that one-third to one-half of Sun-like stars may have low-mass Earth-like planets.
- Three primary methods have been used to identify exoplanets: (1) the radial velocity method, (2) the transit method, and (3) direct imaging of exoplanets.

The Planets We Don't Have and the Ones We Do: Super-Earths

If all we knew was the Solar System, we would think there were basically two classes of planets: rocky terrestrial worlds and giants (the massive gaseous kind and the less massive icy kind). In our Solar System, Earth is the most massive rocky world. The next-most-massive planet after Earth is the ice giant Uranus, with about 14.5 times the mass of Earth (14.5 M_E). Thus, in our Solar System, simply no planets exist in the intermediate range of 1–14 M_E.

This absence is, however, not the rule. One of the most exciting developments in the field of exoplanets has been the discovery of planets more massive than ours but still too small to be true giants. These worlds are called super-Earths, and their structure remains one of the most intriguing mysteries in modern astronomy. Formally, a **super-Earth** is a planet with a mass larger than Earth's but substantially smaller than Uranus's. For worlds with masses below the ice giants' but close to the 14-M_E limit, the term **mini-Neptune** is sometimes applied.

The first super-Earth to be found orbiting a main-sequence star was a 7.5-M_E world called Gliese 876d (it orbited Gliese 876, which was already known to have two orbiting gas giants: Gliese 876b and Gliese 876c). Since then, hundreds of super-Earths have been discovered (transit studies have been particularly useful), including several worlds, Gliese 832c among them, that orbit in their host star's **habitable zone**, where liquid water (and perhaps life) could exist on the surface (**Figure 5.23**). But surfaces are the problem when it comes to super-Earths. What astronomers really want to know about the worlds in this new class is whether they have surfaces like ours and, if so, what those surfaces might be like.

super-Earth A planet with a mass greater than Earth's but substantially less than 15 Earth masses.

mini-Neptune A planet with a mass less than 15 Earth masses but substantially greater than Earth's mass.

habitable zone The region around a star where liquid water can exist on the surface of a planet.

FIGURE 5.23 Super-Earth Gliese 832c
An artist's depiction of Gliese 832c, a 5.2-M_E "super-Earth" that orbits its host star in the "habitable zone," where liquid water (and hence potentially life) can exist on the surface.

Iron

Silicate

Carbon

Water

Carbon monoxide

Hydrogen

|—————|
15,000 km

FIGURE 5.24 Predicted Sizes of Planets with Same Mass but Different Composition
The size of a planet is determined not just by its mass but also by its composition. Astronomers must use detailed models to determine the structure, and hence radius, of a planet based on composition (among other things). This figure shows six different versions of a planet, all of which have the same mass but different elemental and chemical abundances.

In many cases, it is not yet possible to determine what a super-Earth should look like, given its mass and radius (which then yields a density). For some worlds with mass greater than Earth's, a range of configurations are possible from the data. The super-Earth might have a rocky core with a thick gaseous atmosphere, or it might be a "water world" composed mostly of H_2O molecules in both liquid and gaseous form. There is a great deal of overlap in what models predict for these different kinds of worlds so it is not yet possible for observations to distinguish between them (**Figure 5.24**).

It is noteworthy that astronomers have also begun finding Earth-sized planets using transit methods. In 2014 the first Earth-sized exoplanet in the habitable zone of its star was found. The planet, called Kepler-186f, orbits a star half the mass of our Sun, a so-called *M-dwarf star*, and is one of at least four other planets. M-dwarf stars are likely to play an important role in the search for life in the Universe because they are the most numerous kind of star in the galaxy.

SECTION SUMMARY
The Planets We Don't Have and the Ones We Do: Super-Earths

- Super-Earths, planets with masses between that of Earth and that of Uranus, have been found orbiting in the habitable zones of stars.

- Current technology does not allow us to determine the composition of super-Earths or whether they may have surfaces suitable for supporting life.

hot Jupiter A Jupiter-sized planet orbiting very close to its star, usually at a fraction of an astronomical unit.

planet migration A change in a planet's orbit after the planet forms.

Learning from Exoplanets

With 20 years of exploration and thousands of exoplanets discovered, what have astronomers learned about planetary systems that they could not have learned from the study of our own? More important,

what new insights has the discovery of exoplanets provided into how planetary systems are assembled in the first place?

"I think the biggest thing that we've learned," says Knutson, "is that we have a lot to learn. We see that most planetary systems don't look anything like ours."

The first big surprise for astronomers was the location of exoplanets. In our system, there is a clean division between the terrestrial planets that define an inner Solar System, and the giant planets (of the gas and ice varieties) that define an outer planetary domain. But the first exoplanets discovered showed that this tidy distinction was not universal. "When we started looking for planets around other stars," says Knutson, "the very first thing that we found was a population of Jupiter-like planets (in terms of mass) that orbited right next to their host stars."

The initial discoveries were all **hot Jupiters**—a new class of planet in which a massive Jupiter-sized body moves on an orbit of extremely small radius (**Figure 5.25**; **Going Further 5.1**). Astronomers use the term *hot* to refer to any planet orbiting 0.1 AU or less from its star. One particularly important hot Jupiter was HD 209458 b, an exoplanet orbiting a star similar to our Sun (in mass, radius, and temperature) in the constellation Pegasus, 150 light-years from Earth. HD 209458 b has a semimajor axis of just 0.045 AU and a period of just 3.5 days. What made HD 209458 b special is that it was the first hot Jupiter that astronomers detected using the transit method, so

FIGURE 5.25 Artist's Rendering of Hot Jupiters
Hot Jupiters are giant gas exoplanets orbiting much nearer to their stars than our Jupiter orbits the Sun. The gas giants in these systems probably did not form so close to the star. Their orbits decreased over time through a process called migration.

its mass could be determined with fairly high precision at 0.69 times that of Jupiter. With such a large mass, the exoplanet had to be a gas giant. Transit observations confirmed that it was, showing the atmosphere of HD 209458 b to be rich in hydrogen while also containing carbon, oxygen, and water vapor.

Detailed calculations of hot Jupiters like HD 209458 b show that exposure to intense stellar radiation heats and puffs up their atmospheres, extending their outer boundaries and lowering the planet's overall density. Transit observations have confirmed this result, adding yet another dimension to the depth with which this new class of planets has been explored.

"The hot Jupiters were a huge surprise," explains Knutson, "because we know such a massive planet can't form so close to its star." The gases and ices that make up the bulk of a gas giant cannot exist so close to a star during the period when its planets are forming. Thus, the massive planets found on "hot" orbits must have formed at larger distances from the star and then migrated inward. "You must form these planets out beyond the snow line," says Knutson, "and then move them in closer to the star." The recognition that **planet migration**—a change in a planet's orbit after it forms—is a common occurrence in planetary systems is one major conclusion to come from exoplanet studies (**Figure 5.26**). "Somehow you must take a planet from the outer part of your system and move it in close," says Knutson. "But how do you get it close without shoving it all the way into the star?"

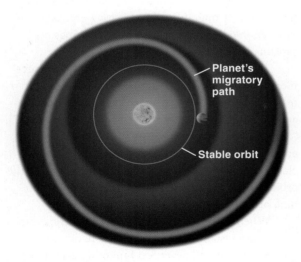

FIGURE 5.26 Planetary Migration
Gravitational interactions between a planet and both disk gas and planetesimals can cause dramatic changes to a planet's orbit. Here, the planet slowly spirals inward and ultimately reaches a stable orbit closer to the star.

GOING FURTHER 5.1   THE LANGUAGE OF THE COSMOS
Orbit, Speed, and Velocity

Back in Chapter 3 you learned an expression for the velocity of objects in circular orbits. Now we can use that formula to get a better idea of how orbital properties change depending on the architecture of a planetary system or a disk of gas orbiting a young star in which new planets will be born.

Let's start with the formula for circular velocity:

$$V_c = \sqrt{\frac{GM}{R}}$$

Using this formula we could, for example, calculate that Earth orbits the Sun with a speed of about 30 km/s (you can try it yourself after this example is done). Let's use this formula to compute some numbers for a kind of orbit very different from those we see in our Solar System.

Imagine we're exploring the space around a star similar to the Sun but with slightly higher mass [$M = 1.5$ solar masses = $1.5\ M_{Sun} \times (1.99 \times 10^{30}$ kg)]. As we've seen, astronomers have been finding planets orbiting other stars in so-called hot orbits very close to the star. Let's imagine we find a planet in a hot orbit around this star and want to calculate its velocity if it orbits at a distance of 0.05 AU.

The circular-velocity formula tells us the orbital speed:

$$V_c = \sqrt{\frac{GM}{R}} = \sqrt{\frac{(6.67 \times 10^{-11}\ \mathrm{m^3 kg^{-1} s^{-2}}) \times [1.5\ M_{Sun} \times (1.99 \times 10^{30}\ \mathrm{kg})]}{0.05\ \mathrm{AU} \times (1.5 \times 10^{11}\ \mathrm{m})}}$$

$$= 1.63 \times 10^5\ \mathrm{m/s} = 163\ \mathrm{km/s}$$

Now let's find the orbital period of this planet. We cannot just use Kepler's formula $P^2 = R^3$ because, as noted in Chapter 3, it does not account for the different mass of the star. Even if we do not recall "Newton's version" of Kepler's third law (which does include the mass of the star), we can reason out what the period should be by remembering how velocity, distance, and time are related.

The period is just the time it takes a planet to make one orbit around the star. The distance a planet covers in one orbit of average radius R will just be the circumference of a circle of that radius: $C = 2\pi R$. This means the planet's period in its orbit around a star will be

$$P = \frac{2\pi R}{V_c}$$

Using the formula for V_c,

$$P = \frac{2\pi R}{\sqrt{\frac{GM}{R}}} = \frac{2\pi \sqrt{R^3}}{\sqrt{GM}}$$

(Note that if you square this formula, you will get back Newton's version of Kepler's third law.) When we use this formula for our hot orbit, we find

$$P = \frac{2\pi \sqrt{[0.05\ \mathrm{AU} \times (1.5 \times 10^{11}\ \mathrm{m})]^3}}{\sqrt{\{(6.67 \times 10^{-11}\ \mathrm{m^3 kg^{-1} s^{-2}}) \times [1.5\ M_{Sun} \times (1.99 \times 10^{30}\ \mathrm{kg})]\}}}$$

$$= 289{,}226\ \mathrm{seconds}$$

$$= 3.35\ \mathrm{days}$$

Thus, a planet in a hot orbit has a "year" that lasts just a few days.

Heather Knutson

For Heather Knutson, a life in astronomy began with accidents of location. "I was an undergraduate at Johns Hopkins University," she explains. "I wasn't sure what I wanted to study, but right across the street from the physics department was the Space Telescope Science Institute." STScI, as it is known, is the operations center for the Hubble Space Telescope (HST). From inside the futuristic building, technicians and astronomers oversee the orbiting telescope, changing where it points for different observations and maintaining its overall health.

"That turned out to be a wonderful opportunity for me," says Knutson. "I was able to work there and got to meet real astronomers. I got my hands on real data from the HST and had a chance to see how science really works. From that experience I saw how I could take the physics I was learning in class and apply it to the unsolved astronomical problems. After a while I could tell those problems were far more interesting to me than some of the other paths that I could go down within physics."

Knutson's time at STScI showed her another aspect of life in astronomy that piqued her interest. "I noticed these conference flyers everywhere for astronomical meetings. Looking at their locations, I noticed a theme: they were all in really interesting places." Knutson saw meetings on galaxy formation being held in Hawaii, meetings on cosmology held in Italy, and meetings on exoplanets held in Brazil. She was hooked. "Looking back, one of the things I have thoroughly enjoyed about doing astronomy, in addition to the science, has been the opportunity to travel."

When she sat for this interview, Knutson was a postdoc at UC Berkeley. (A postdoc is a 3-year position taken after a student finishes her PhD.) After that, Knutson had to look for a permanent position, something not easy to find in astronomy. "It is kind of chancy, but I'm definitely glad that I did this," she said. "I always felt like even if I never find a permanent job, the time I spent studying extrasolar planets while in grad school was so enjoyable that it was really worth it to me." Knutson is now a professor at Caltech.

scattering The rearrangement of the orbits of planets due to mutual gravitational interactions.

This question is one of the new challenges facing astronomers in their quest to build a complete theory of planetary system formation.

Even as astronomers debate the mechanisms driving migration, they recognize that its consequences are dramatic. Migration means that the architecture of a planetary system when it forms may not be what it ends up with. Many fall into the "hot" category, and a sizable fraction of these are hot Jupiters. Astronomers have also found planets with a range of masses from Neptune-sized bodies to super-Earths in "hot" orbits, so it is clear that migration can be an almost universal phenomenon.

Hot Jupiters, mini-Neptunes, and super-Earths were a dramatic discovery, but the bulk of massive planets are not on hot orbits. Most planets discovered to date have orbits with radii that keep them farther from their parent stars. It is noteworthy, however,

that most massive exoplanets are on orbits that are smaller than what we see in our own system. Many exo-Jupiters and exo-Neptunes, for example, have orbital radii of a few astronomical units or less, which in our Solar System is the domain of the terrestrial planets (and asteroid belt). Extremely large orbits also occur in exoplanetary systems. In some cases, there is evidence for gas giant–sized or ice giant–sized worlds with orbits of 100 AU or more, which is much larger than anything seen in our Solar System.

Location, however, was just one surprise waiting for astronomers in their exploration of exoplanetary systems. The shapes of many exoplanet orbits, and the histories they imply, have also forced astronomers to revise their thinking about planetary systems and their formation. "When we started looking at extra-solar planets," explains Knutson, "not only did we see very big planets very close to their stars; we also saw lots and lots of planets with large orbital eccentricities. That means these planets move on very elliptical orbits instead of the more circular orbits we see in our Solar System."

The planet with the highest-eccentricity orbit in our Solar System is Mercury, with $e = 0.2$, and six of the eight planets have eccentricities of 0.05 or less. In contrast, more than 80 percent of all exoplanets discovered to date have eccentricities above $e = 0.1$, and nearly one-fifth have eccentricities of 0.5 or greater. How do planets end up with such highly eccentric orbits? Like migration, gain in eccentricity may be an essential part of early planetary system evolution. "If two young planets have a close encounter, perhaps because one is migrating, then a strong gravitational interaction occurs that can really shake things up," explains Knutson.

The process is called **scattering**—the rearrangement of orbits due to gravitational interaction. Scattering can take a planet that was on a circular orbit and fling it onto a highly eccentric orbit (**Figure 5.27**). "If you have a bunch of planets that begin at different distances from the star," explains Knutson, "and close encounters lead to scattering, then in time all the orbits can be moved around, and they can all attain high orbital eccentricities."

When planets end up on highly eccentric orbits, a planetary system can become far more unstable than our own. Venus will never feel Jupiter's gravitational influence up close because both planets are on fairly circular orbits with very different radii. But a planet on a highly eccentric orbit will have many chances to cross paths and interact with other orbiting bodies.

"It seems like we might have been lucky in our Solar System," says Knutson. "All our planets are on fairly circular orbits, so everything is pretty stable. There probably haven't been any catastrophic encounters between planets except for a long time in the past, and what we are left with has become very safe."

Evidence for scattering interactions in exoplanetary systems also comes in the form of the orbital inclinations of planets relative to the host star's spin. While almost all the planets in our system have low inclination angles—that is, they rotate in a plane that is roughly perpendicular to the Sun's spin axis— exoplanets show a wide range of inclinations. Again, scattering interactions between exoplanets are the likely culprit, tossing orbiting bodies in different directions, including out of the plane of the original disk from which the system formed. "It's clear from exoplanet studies that planetary systems may be far more dynamic than we thought," says Knutson. "You form things with one configuration, but the final configuration that you see may be very different."

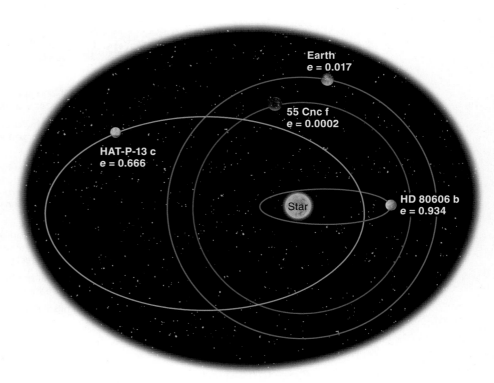

FIGURE 5.27 Eccentricity of Exoplanet Orbits
Some exoplanets exhibit the nearly circular orbits typical of the planets in our Solar System, but many do not. Highly eccentric orbits are evidence of gravitational interactions between planets, called scatterings. In this diagram, three exoplanet orbits appear in red, purple, and green, and Earth's orbit is shown in blue.

SECTION SUMMARY
Learning from Exoplanets

- Among the new types of planets discovered are gas giants very close to their stars, called hot Jupiters.
- Many exoplanets have large orbital eccentricities and orbits inclined to the plane of the stars' spins.
- Some of these differences provide evidence for migration and scattering interactions between sibling planets.

5.5
Developing a Theory of Planetary System Formation

After touring planetary systems near and far, we are now in a good position to take all the observed facts and distill them into a set of critical points that must be explained by a theory of planetary formation.

Let's begin with four key points we know about our own Solar System. (These facts are summarized in **Table 5.3**.)

1. Planets in a system all tend to orbit in the same direction, which is also the same direction as the parent star's spin.

2. Planets tend to orbit with low inclination; that is, all planetary orbits tend to be closely aligned, and they are arranged in a disklike configuration.

3. Planets in our system are highly differentiated in terms of composition and location. The inner planets are rocky and dense. The outer planets have lower density and contain significant amounts of gas or ice.

4. Significant debris from the era of planet construction remains in our Solar System, taking the form of asteroids, comets, and Kuiper Belt objects.

Now let's add two further points from studies of exoplanets:

5. Planets can migrate. They may change their orbital position from the location of their formation. Close encounters between planets or planetesimals are likely to occur during the migration process and can lead to catastrophic gravitational interactions in the form of scattering.

6. Planets can exist on highly eccentric orbits. These eccentric orbits may be the result of scattering events, and they may lead, in turn, to further scattering events. In some planetary systems the planets have high-inclination orbits.

TABLE 5.3 ••• Key Facts for Planet Formation Models

Feature	Effect	Cause
Orbit and rotation	Planets generally orbit and spin in the same direction as their star's spin.	Conservation of angular momentum. Planets formed from a single rotating cloud.
Inclination	Planets (in our Solar System) orbit with low inclination and are arranged in a disklike configuration.	Conservation of angular momentum forms disk, causing most planets to orbit with low inclinations.
Composition and location	Planets in our Solar System are highly differentiated in terms of composition and location: inner planets are rocky and dense; outer planets are less dense and contain mostly gas or ice.	Planets formed from disk with a temperature that decreased with distance from the star.
Debris	Debris from planet construction remains, represented in our system by asteroids, comets, and Kuiper Belt objects.	Planets formed from collisions between smaller objects.
Migration (extrasolar planets)	Planets can change orbital position from the location of their formation.	Interactions between planets and disk or other planets were common.
Eccentric orbits (extrasolar planets)	Planets can exist on highly eccentric orbits.	Interactions between planets were common.

Condensation Theory and Conservation of Angular Momentum

interactive
Conservation of
Angular Momentum

The goal of any successful theory of planetary system formation would be to explain all six facts in one simple, all-encompassing model. Scientists believe they have a candidate in what they call the **condensation theory**. Remarkably, elements of the condensation theory have been around for almost 400 years, finding their beginnings with philosophers such as René Descartes (1596–1650) and astronomer Pierre-Simon Laplace (1749–1827).

"The basic idea of the condensation theory," says Knutson, "is that planetary systems are born when a spinning cloud of interstellar gas and dust collapses under its own weight." That spinning cloud represents condensation theory's starting point. From there, the theory goes on to tell a detailed story of gravity, gas flows, radiation, and chemistry—combining them all to explain how a system of worlds is constructed.

Dense interstellar dust clouds (**Figure 5.28**) are well known to astronomers, and they will be explored in some detail in association with star formation in Chapter 12. For now, imagine one of these clouds floating in space. Astronomers also assume that the entire cloud is slowly rotating with some initial spin that is a remnant of the cloud's formation. "The inward pull of gravity in the cloud will be balanced by the outward push of its internal pressure," explains Knutson, noting that the pressure comes from internal heat in the cloud.

But this balance between internal pressure and gravity is *unstable*: "Any bump or nudge (perhaps from another cloud) will let gravity win over pressure, and the cloud will start collapsing in on itself."

At this point, a fundamental principle of physics called *conservation of angular momentum* takes over, shaping the cloud's collapse and setting the stage for planet formation. Recall from Chapter 3 the example of the spinning figure skater. She begins the spin with her arms extended. As her arms pull inward, her rate of rotation increases until she is spinning so fast as to be just a blur. "These skaters are using the principle called conservation of angular momentum to their best advantage," says Knutson, who explains how conservation of angular momentum *relates changes in an object's rate of rotation to changes in its size.*

In particular, conservation of angular momentum states that when a spinning object decreases in size, it must increase its rate of rotation (the opposite holds true if a spinning object increases in size). When applied to a collapsing cloud, conservation of angular momentum tells us that, like the skater, the spin rate of the cloud around its axis of rotation must increase as the cloud gets smaller (**Going Further 5.2**). But while the skater decreases her size by only a factor of 2 (pulling her arms in), the cloud decreases its size by a factor of 1,000.

Any gas that begins on the cloud's axis of rotation is not spinning at all. That material will drop straight to the center and start forming a star. But parcels of gas that do not lie on the axis of rotation have a

condensation theory A theory of planetary formation in which planets are created from a disk of gas and dust around a newly formed star.

FIGURE 5.28 Interstellar Dust Clouds
Stars (and hence planets) form from dense clouds of interstellar gas such as this one imaged by the Hubble Space Telescope. (The black rectangle at the top left corner of the image is an artifact of the space telescope.)

GOING FURTHER 5.2   THE LANGUAGE OF THE COSMOS
Conservation of Angular Momentum

Physicists are always looking for "conserved" quantities. In physics, *conservation* means "no change." When trying to understand the evolution of complicated systems such as a gravitationally collapsing cloud of interstellar gas, knowing what does not change (what is conserved) and what does change can often be the key to calculation and understanding.

In its simplest terms, conservation of angular momentum relates changes in an object's size (R) to changes in its velocity (V). Imagine we're following a blob of gas of mass M in a collapsing, rotating interstellar cloud. The blob begins rotating around the cloud's center at a distance R_1 from the center and with a velocity V_1. Conservation of angular momentum relates those initial values of radius and velocity to the blob's radius and velocity sometime later in the collapse. The relationship looks like this:

$$MR_1V_1 = MR_2V_2$$

Because it is assumed that the mass of the blob, M, does not change, we can cancel it out and then solve for the velocity at the later time:

$$V_2 = V_1(R_1/R_2)$$

A typical interstellar cloud has a size of $R_1 = 1$ parsec (pc) $= 3 \times 10^{16}$ meters. The typical size of a protostellar disk is only $R_2 = 100$ AU $= 1.5 \times 10^{13}$ meters. Thus, the conservation of angular momentum tells us that the rotational velocity of the gas will increase 2,000-fold as it spirals inward. Once the velocity of rotation reaches the orbital velocity (see Chapter 3), the blob will stop collapsing and go into orbit around the new star.

different fate. As this gas falls inward, its rotation forces it to trace out a spiral path. The closer it gets to the center, the more its rate of rotation increases because of conservation of angular momentum. Eventually, rotation can balance gravity and stop the gas from getting any closer to the center. But there is more to the story. If we imagine that the cloud starts as a sphere, then some of the gas begins its spiral above or below the sphere's equator. These gas parcels will eventually collide at the midplane as they spiral inward, creating a swirling disk of gas surrounding the newly formed star now referred to as a **protoplanetary disk**. In this way, what began as a spherical rotating cloud ends up as a rotating disk surrounding a new star (**Figure 5.29**).

"Conservation of angular momentum takes you a long way to explaining planetary systems," says Knutson. By beginning with a collapsing, rotating gas cloud, we can already explain why all the planets in our Solar System orbit in a plane and in the same direction as the Sun's spin. Conservation of angular

momentum also explains why all but one of the planets spin in the same direction. When material in the disk eventually creates planets, conservation of angular momentum tells us that the direction of a planet's spin will also have the same orientation as the overall direction of rotation for the original cloud. "That is a lot of bang," says Knutson, "for just a few assumptions, the force of gravity, and this principle of conservation of angular momentum."

SECTION SUMMARY
Condensation Theory and Conservation of Angular Momentum

- The condensation theory, which states that planetary systems form from a spinning, collapsing interstellar cloud of gas and dust, accounts for many of the observed characteristics of our Solar System and others.
- Conservation of angular momentum explains the physics behind many of these results.

protoplanetary disk The disk that surrounds a young star, from which planets may form.

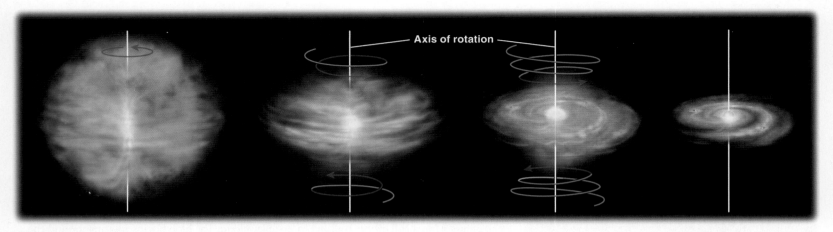

Axis of rotation

FIGURE 5.29
Planet Formation
Planets form out of a spinning disk of gas that surrounds a newborn star. The disk and star both originate in a rotating, collapsing cloud of interstellar material. Material along the cloud's axis of rotation falls straight in to create the initial protostar. Material from other regions of the cloud spirals inward and forms a disk along the cloud's midplane.

Condensation Theory, Dust Grains, and Planet Formation

The rest of the story, including the existence of different kinds of planets, requires more science than just conservation of angular momentum. This is where condensation comes into play in condensation theory. To understand the formation of planets, astronomers must focus on the physics occurring within the disk. In particular, they focus on the disk's density, its temperature, and the kind of dust grains that can form at different locations.

The cloud that forms the star and disk is already full of **interstellar dust grains**. These are tiny bits (1 micron or 0.001 mm) of solid matter that form in the winds of dying stars and are ejected into space. What matters for the story of planet formation is the fate of these grains at different locations in the disk. The temperature in the dusty disk close to the star (where it absorbs the most light) is very high, and it drops at larger radii. Calculations show that the inner edge of the disk (at about 0.1 AU) can have temperatures of 1,500 K, while the temperature at 10 AU (where Saturn is today) would be only 100 K. "Interstellar dust grains can't survive at the high temperatures close to the star," says Knutson. "They get turned into vapor." But the grains are less affected in the outer parts of the disk. As the entire disk evolves, it slowly cools by radiating energy out into space. Except for regions very close to the star, the cooling of the disk allows new dust grains to condense out of the gas and alters those that already exist.

"Dust condensation is a bit like the way raindrops form in a cloud," says David Jewitt, "where the drops condense out of water vapor." But the new grains that form in the disk are not like the old interstellar grains. While the original gas cloud contained a fairly uniform mix of grain sizes and compositions, the new grains that condense out of the cooling disk are very sensitive to the location of their birth.

Figure 5.30 shows how temperature in the disk falls with distance from the star. The figure also shows the temperatures at which different materials condense into grains from the background gas. Close to the star, everything is in the vapor state, but farther out the gas cools and grains begin to condense. In the higher-temperature regions, comparable to the region around the planet Mercury, the only kinds of material that can condense from the vapor to the solid state (grains) are metals. Farther out, about where Venus, Earth, and Mars are now, the gas temperatures are even lower. At these distances and temperatures, silicates (mixtures of silicon and oxygen) and other rocky materials can also begin to form dust grains. Eventually, beyond the snow line, the temperature gets low enough for water ice to form, and at even larger radii, ices of other compounds such as ammonia and methane can condense.

"The spatial sequence of condensation temperature and radius—first metals, then rocky material, then simple ices, then more complex ices—does a wonderful job of explaining the sequence of planets," says Knutson. One point to notice is that the farther out one gets in the Solar System, the more the composition of the planets begins to look like that of the rest of the Universe. Recall that the most abundant elements in the cosmos are hydrogen and helium, and everything else represents only a tiny fraction of the total. Temperatures in the inner disk, however, keep the most abundant elements (hydrogen and helium) from condensing into grains at those distances, so these elements do not become part of the inner planets. Beyond the snow line, when large enough rocky cores form, they can rapidly accrete surrounding gas, building up their mass into the gas and ice giant range.

interstellar dust grains Tiny bits of solid matter found in diffuse clouds in the regions between the stars.

SECTION SUMMARY
Condensation Theory, Dust Grains, and Planet Formation
- The compositions of planets in our Solar System at different distances from the Sun are consistent with the condensation theory.
- Dust grains condense at different distances from the central star depending on their melting points. This condensation creates different abundances of materials in the disk at different distances from the star.

Accretion, Fragmentation, and Planet Building

The existence of asteroids and comets in our Solar System also fits the general picture of condensation theory and adds more detail to the story. These smaller bodies have provided crucial information about the next steps after condensation. For example, how do young planetary systems go from making grains to making rocky planets (or the rocky cores of giant planets)?

"The answer to that question," explains Jewitt, "is a process called **binary accretion**, where collisions between pairs of objects let larger and larger structures get put together." Collisions between grains, which are small and sticky, quickly lead to the construction of pebbles. Collisions between pebbles lead to rocks. Collisions between rocks lead to boulders. Collisions between boulders lead to planetesimals, rocky bodies the size of asteroids. If this accretion process happens far enough from the star, significant quantities of ices will be included in the planetesimals. This is the likely origin of comets. Eventually, the objects are large enough for gravity to begin compressing and heating the planetesimal interiors. As planetesimals grow even larger, gravity pulls them into a spherical shape, and the heaviest elements differentiate, sinking to the center of the body. Iron and nickel will, in this way, form the dense metallic cores of the young planets. Eventually, full-sized terrestrial planets, the rocky cores of gas giants, and moons are formed through the accretion process.

On the journey from microscopic grains to full-sized planets (or moons), the competing process of **fragmentation** can occur alongside accretion. Shattering collisions break asteroid-sized planetesimals apart into smaller bodies. Fragmentation is still at work in the asteroid belt and is the source of both stony and iron meteorites. Note that iron meteorites come from fragmented cores of differentiated planetesimals.

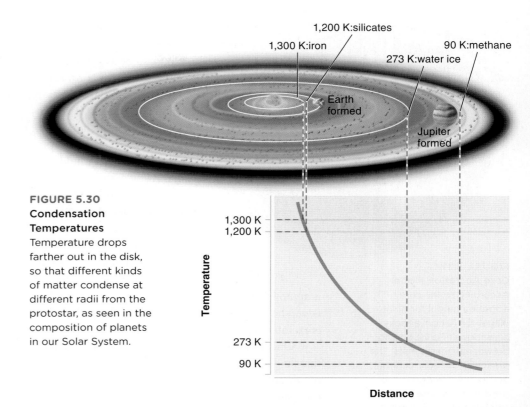

FIGURE 5.30
Condensation Temperatures
Temperature drops farther out in the disk, so that different kinds of matter condense at different radii from the protostar, as seen in the composition of planets in our Solar System.

The timescale for the accretion process depends on the density of gas and dust in the protoplanetary disk because higher densities mean more frequent collisions. In the inner part of our pre–Solar System disk, the density was high, and it took only 100 million years to build the terrestrial planets. The formation of the outer planets is a more complicated story, and scientists are still not sure how the gas and ice giants formed.

"There are basically two models for the formation of giant planets like Jupiter," says Jewitt. In the first model, an icy terrestrial planet grows by binary accretion up to a critical mass of about 5–10 Earth masses. "When that small core planet reaches the critical mass," Jewitt explains, "it has enough gravity to start pulling in gas from the surrounding disk. You get a very rapid flow of gas onto this core, taking the planet all the way up to Jupiter's or Saturn's mass." This idea is called the **core accretion model** (**Figure 5.31**). According to Jewitt, among most Solar System astronomers the core accretion model is the preferred idea for the formation of Jupiter and Saturn.

The problem with core accretion is that it may take too long to build up the rocky center. If the disk loses its gas before the core has a chance to reach its critical mass, the idea cannot work.

"That is why there is this other model," says Jewitt. "It starts with an enormous disk of gas that

binary accretion The process by which the collision of two objects results in the formation of one larger object.

fragmentation The shattering of solid bodies, such as planetesimals, caused by collisions.

core accretion model A model of giant gas planet formation in which an icy terrestrial planet grows by binary accretion up to a critical point and then rapidly pulls gas in from the surrounding disk.

FIGURE 5.31

Core Accretion Model

According to the core accretion model for giant planet formation, a rocky core forms first via collisions between planetesimals. The gravity of the core can then pull in gas. Over time, the gravity of the core pulls in and accretes more and more gas, eventually creating a gas giant. This drawing shows the assembly of a giant planet via core accretion, with the flow of time moving counterclockwise around the circle.

FIGURE 5.32

Hydrodynamic Instability Model

According to the hydrodynamic instability model, small regions of denser gas in the disk begin attracting material around them in a runaway process that quickly forms a giant planet.

would be unstable to its own gravity." Just as star and disk began as a gravitationally unstable cloud, some astronomers claim that planets may form from the gravitational collapse of gas within the disk itself. "Parts of the disk would just contract under their own gravity," says Jewitt. "The planets would form directly without needing a core." This second idea is called the **hydrodynamic instability model** (**Figure 5.32**). "It tends to be more popular for people who study extrasolar planetary systems," says Jewitt. "The fact that we don't know which model is appropriate, and we still talk about this very basic issue, shows our state of ignorance on the subject. That is exciting because it shows we have lots to learn."

One thing astronomers are certain of, however, is that far enough out in our Solar System, low disk densities meant that the process of turning planetesimals into planets never finished. This explains the presence of the Kuiper Belt and objects like Pluto. The Kuiper Belt represents the outer regions of the Solar System where planet formation began but could never be completed.

While everything about the condensation theory seems to fit our Solar System, what have we learned from exoplanets? Do the profound differences in planetary systems other than our own contradict the theory? The answer is that the basic ideas of condensation theory still hold, but studies of exoplanets show us the overwhelming importance of gravitational interactions among planets, planetesimals, and disk material.

"As early as 1980, well before exoplanet discoveries, we already had evidence that Jupiter must have formed slightly farther out in the Solar System than

its current location," says Jewitt. By gravitationally interacting with planetesimals early in the history of the Solar System, Jupiter must have scattered smaller objects. "All those scattering events flung the planetesimals out of the Solar System," says Jewitt, "and they let Jupiter move slowly inward."

In many exoplanet systems, this process may be much more dramatic than it appears to have been in ours. Early in the history of an exoplanetary system, young Jupiter-sized planets can have direct gravitational interaction with nearby gas in the disk, thus causing migration. A giant planet can scatter material in the disk and migrate all the way from the outer system where it formed to the inner system, where it can become a hot Jupiter. "We still aren't entirely sure what halts inward migration," says Heather Knutson. "Some planets may migrate all the way into their stars."

Once the planets have formed and settled into relatively stable orbits, the gas in the disk is disturbed by winds and radiation from the star. In time, all that is left of the former gaseous disk is whatever stable planets managed to form, accompanied by a swarm of rocky planetesimals and icy planetesimals farther out. There are enough of these asteroid-sized planetesimals early on that their interactions with planets can still cause migration. It's worth noting that migration does not always have to take a planet inward to an orbit closer to the star. "If conditions are right, the planet can migrate outward," says Jewitt. This process most likely moved Saturn, Uranus, and Neptune out to larger radii than those where they were born and may be responsible for the very large orbits that some exoplanets exhibit.

The gravitational interactions between planets and planetesimals produce not only migration but also scattering, kicking small bodies into highly eccentric and highly inclined orbits. In our own Solar System, the spherical Oort Cloud of comets at large distances from

the Sun was probably the result of scattering between icy planetesimals and giant planets. In exoplanetary systems, the predominance of observed high-eccentricity orbits also may be a result of scattering between planets.

Thus, the condensation theory provides a useful framework for understanding the formation of both our own Solar System and the exoplanetary systems we continue to discover. But it is a framework only. Many questions remain unanswered, such as how giant planets form. Even more important, we should expect many more surprises to be in the works as astronomers continue their observations and discoveries.

SECTION SUMMARY
Accretion, Fragmentation, and Planet Building

- By means of binary accretion, pairs of objects collide and stick together, so that grains become pebbles, then boulders, then planetesimals, then planets.
- Differentiation explains the formation of cores of planetesimals and planets. Fragmentation explains how asteroids formed from shattered planetesimals.
- The formation of gas and ice giants is still not fully understood, but two theories have been posited. According to the core accretion model, a core forms and then its gravity pulls in gas. According to the hydrodynamic instability model, gases within the disk collapse.

hydrodynamic instability model A model of giant gas planet formation in which small regions collapse to form planets within a gravitationally unstable disk.

●→ chapter summary

5.1 The Rest of the Solar System

Much of the Solar System is still being explored, but it is known to include several regions far beyond the realm of the eight planets, as well as smaller objects too numerous to count. For much of human history, we have not known if ours was the only planetary system, but recent discoveries have shown definitely that it is not.

5.2 Just the Facts: A Solar System Census

The primary regions of the Solar System include the region of the planets, the Kuiper Belt, the scattered disk, and the Oort Cloud. The eight planets fall into three classes differentiated primarily by location, composition, and size. The four terrestrial planets are small, rocky worlds with solid surfaces that orbit nearest the Sun. Gas giants Jupiter and Saturn, composed mostly of hydrogen and helium, orbit out beyond the terrestrial planets and are much more massive. Farthest from the Sun, the midsized masses of ice giants Uranus and Neptune contain ices of molecules. Other objects in the Solar System include dwarf planets, icy-rocky Kuiper Belt objects, asteroids, comets, and meteoroids.

5.3 Construction Debris: Asteroids, Comets, and Meteoroids

Some of the material in the Solar System is found in smaller objects "left over" from the construction of larger bodies. Asteroids are irregular rocky or metallic objects, typically either carbonaceous or silicate, the largest of which are less than 1,000 km in diameter. Collections of asteroids are found in the asteroid belt between the orbits of Mars and Jupiter, within Jupiter's orbit (the Trojan asteroids), and on Earth-crossing orbits. Comets, composed of ice and rock, originate in the Kuiper Belt or Oort Cloud and generally follow long, elliptical, highly inclined orbits. As a comet approaches the Sun, material is blown off its nucleus, forming jets and a cloud of gas called a coma; two long tails then form—an ion tail and a dust tail—driven by solar emissions and gravitational effects. Rocky meteoroids are the smallest kind of Solar System debris, ranging in size from a grain of sand to a boulder, and may have originated from asteroids, comets, or planets. When they encounter Earth's atmosphere, they are seen as glowing trails of gas and are called meteors. Meteor showers occur when Earth's orbit intersects dust left behind by an earlier orbit of a comet. Meteoroids large enough to survive their plunge through the atmosphere are called meteorites.

5.4 Us versus Them: Our Solar System and Others

In our Solar System, all planets, most asteroids, and many comets orbit in the same direction, and most planets also rotate that way. The planets' orbits are aligned in a disk, as are the asteroid belt, Kuiper Belt, and scattered disk. Successful methods for finding exoplanets include the radial velocity method, which looks for changes in the star's velocity as it orbits a common center of mass with one or more planets; the transit method, which looks for dips in a star's brightness as a planet passes in front of it; and direct imaging. More than 1,000 exoplanets have been detected to date, and astronomers believe that as many as half of all stars have planets. Surprising differences in exoplanets compared with planets in our Solar System provide insights into how planetary systems originate. Hot Jupiters—gas giants very close to their stars—reflect planet migration. The highly eccentric, highly inclined, and exceptionally large orbits of some exoplanets suggest that scattering by gravitational interaction has occurred.

5.5 Developing a Theory of Planetary System Formation

Astronomers have sought a model of planetary system formation that fits all of the facts they have amassed by studying our Solar System and others. The prevailing model is the condensation theory, which states that planetary systems form from a spinning cloud of interstellar gas and dust that collapses into a disk following the principle of conservation of angular momentum. Dust grains condense in the disk at different distances from the central star depending on their melting points: first metals, then silicates, then water ice, then ices of heavier compounds. Grains then grow by binary accretion into larger and larger objects orbiting in the disk, the composition of each determined by its distance from the star. Differentiation, collision, and fragmentation change the sizes, shapes, and structures of the objects. The largest bodies, planets, become spheres by self-gravity, and gravitational interactions among them may cause their orbits to move and change. The exact mechanism of giant gas planet formation is still unknown, but several theories exist, and investigation continues.

●→ questions and problems

Narrow It Down: Multiple-Choice Questions

1. What is the defining characteristic of the Trojan asteroids?
   a. They are the largest known asteroids.
   b. They are rocky.
   c. They orbit ahead of and behind Jupiter.
   d. They orbit within Saturn's rings.
   e. They are the remnants of an inner terrestrial planet.

2. What determines the path of the ion tail of a comet?
   a. the Sun's gravity
   b. the direction of the solar wind
   c. Earth's gravity
   d. conservation of angular momentum
   e. the solar magnetic cycle

3. What makes a typical Earth-crossing asteroid difficult to detect and track?
   a. its density
   b. its albedo
   c. its lack of an atmosphere
   d. the low eccentricity of its orbit
   e. its low velocity relative to background stars

4. A rocky object with a diameter of 10 cm is found lying atop otherwise untouched Antarctic ice. It is determined to have originated from space. Therefore, the rock is
   a. a meteorite.   d. an asteroid.
   b. a meteoroid.   e. a planetesimal.
   c. a meteor.

5. Detailed observations of a long-period comet, near aphelion, would most likely *not* include which of the following? Choose all that apply.
   a. jets   b. nucleus   c. coma   d. dust tail   e. ion tail

6. Which of the following characteristics of our Solar System does *not* directly support the condensation theory of planetary system formation?
   a. the planets' orbital direction   d. Venus's rotational direction
   b. Saturn's rotational direction   e. the positions of the gas giants
   c. the positions of the ice planets

7. Using technology similar to that available on Earth today, a distant alien astronomer searching for evidence of planets around the Sun is likely to be most successful with which of the following?
   a. direct imaging of the terrestrial planets in visible light
   b. detection of Mercury via the transit method
   c. detection of Pluto via the transit method
   d. detection of Mercury via the radial velocity method
   e. detection of Jupiter via the radial velocity method

8. Observations of the earliest exoplanets and exoplanetary systems differ from observations of our Solar System in a number of ways. In which way(s), however, are they similar to observations of our Solar System? Choose all that apply.
   a. They show planets with high orbital eccentricity.
   b. They show planets orbiting in the same direction around the host star.
   c. They show gas giants orbiting very close to the host star.
   d. They show planets with masses between 1 and 15 M_{E}.
   e. They show planets with atmospheres.

9. You observe a single meteor during 1 hour of observing the night sky. This meteor was most likely
   a. from a comet.
   b. part of a meteor shower.
   c. composed of ice.
   d. from an asteroid.
   e. a Kuiper Belt object.

10. Pluto was reclassified from planet to dwarf planet in 2006. Which of the following is an official criterion for being called a planet that is *not* met by Pluto?
    a. It has enough mass that its self-gravity has caused it to assume a nearly round shape.
    b. It orbits the Sun.
    c. It has at least one moon.
    d. It has gravitationally cleared the neighborhood around its orbit.
    e. Its name has a Greek origin.

11. Place the following items in sequence from smallest to largest: 1 light-year (ly), 10 AU, Oort Cloud (radius), Kuiper Belt (radius), orbit of Trojan asteroids (radius).
    a. Trojans, 10 AU, Kuiper Belt, Oort Cloud, 1 ly
    b. 10 AU, Trojans, Oort Cloud, Kuiper Belt, 1 ly
    c. Kuiper Belt, Trojans, 10 AU, 1 ly, Oort Cloud
    d. 1 ly, 10 AU, Trojans, Oort Cloud, Kuiper Belt
    e. Trojans, Kuiper Belt, 10 AU, Oort Cloud, 1 ly

12. Using your knowledge of orbital velocity and the orbital radii of the following objects, place them in sequence from highest to lowest *average orbital velocity*: 1992 QB1, Ceres, Halley's Comet, Jupiter, Earth.
    a. Ceres, Earth, 1992 QB1, Halley's Comet, Jupiter
    b. Earth, Ceres, Jupiter, Halley's Comet, 1992 QB1
    c. 1992 QB1, Halley's Comet, Jupiter, Ceres, Earth
    d. Halley's Comet, Earth, 1992 QB1, Ceres, Jupiter
    e. Earth, 1992 QB1, Ceres, Jupiter, Halley's Comet

13. Which of the following processes is *not* evident in the current structure and distribution of planets in our Solar System?
    a. ongoing planet scattering
    b. previous eras of planet differentiation
    c. previous eras of planet formation inside or outside a snow line
    d. previous eras of planet formation within a disk
    e. ongoing comet collisions with planets

14. The search for exoplanets is challenging because of a number of constraints. Which of the following is *not* a constraint?
    a. the orientation of many planetary orbits in relation to their star
    b. the proximity of planets to their star
    c. the distance to exoplanetary systems
    d. the existence of appropriate Sun-like stars
    e. the orbital periods of many exoplanets

15. Which of the following would you expect to differ significantly between a long-period comet and an asteroid? Choose all that apply.
    a. radius
    b. distance between location of formation and the Sun
    c. inclination of orbit
    d. albedo at perihelion
    e. distance at aphelion

16. In which of the following ways are the terrestrial and giant planets in our Solar System similar?
    a. average number of moons
    b. approximately spherical shape
    c. elemental composition
    d. average rotation speed
    e. orbital velocity

17. Which of the following is/are potential results of planet scattering? Choose all that apply.
    a. change in orbital inclination of a planet
    b. change in the proximity of a planet's orbit to its star
    c. presence of a large gas planet very close to its star
    d. presence of a terrestrial planet very far from its star
    e. highly eccentric orbit of a planet

18. Comparing the orbital inclination of Pluto to those of other planets, what conclusion can we draw?
    a. It is higher, implying that Pluto likely underwent gravitational interactions with other planets.
    b. It is higher, implying that Pluto likely was captured from deep space.
    c. It is lower, implying that Pluto was likely formed by a different mechanism than other Solar System bodies were.
    d. It is lower, implying that Pluto likely has never had gravitational interactions with other bodies.
    e. None of the above.

19. Observations yield radius and mass measurements for an exoplanet. Which of the following properties can be determined from these two quantities? Choose all that apply.
    a. the existence of a core in the planet
    b. the planet's status as a water world
    c. whether or not the planet is a gas giant
    d. the eccentricity of the planet's orbit
    e. the planet's density

20. True/False: An exoplanet at 0.5 AU from its Sun-like star, with an uncompressed density of 980 kg/m^3, is likely to have migrated to its current position after formation.

To the Point: Qualitative and Discussion Questions

21. How is the discovery of the Kuiper Belt connected to Pluto's reclassification as a dwarf planet?

22. How can Solar System objects be identified among background stars?

23. What views, later confirmed correct, were part of the basis for Giordano Bruno's persecution?

24. How do the orbits of the Trojan asteroids, Ceres, and 2002 MN differ?

25. How would the spectrum of a comet's nucleus compare with that of its coma?

26. What category of comets has the greatest potential to be hazardous to Earth?

27. You spot 20 meteors within an hour. What kind of object is their likely source?

28. How does the wavelength of a star's light change as a result of Doppler shift when it is being pulled by an orbiting planet away from us on our line of sight? How about when it is being pulled toward us on our line of sight?

29. The melting and freezing temperatures of material that makes up planets correlate strongly with each planet's distance from the Sun. What does that correlation suggest about the formation of planetary systems?

30. How are the characteristics of our Solar System explained by the condensation theory?

31. List several everyday occurrences that demonstrate conservation of angular momentum.

32. What is your reaction to the evidence that exoplanets exist in abundance?

33. What is your definition of a day? How does a day on Earth compare with a day on Jupiter?

34. What actions should Earth's inhabitants take with regard to Earth-crossing asteroids? How might the danger these asteroids represent affect Earth, both positively and negatively?

35. Astronomers have found super-Earths among exoplanets. What characteristics would such a planet need for life to be possible on it?

Going Further: Quantitative Questions

36. What is the ratio of the orbital velocity of a terrestrial planet orbiting at 3 AU from its star to that of a giant planet orbiting at 12 AU?

37. How many times greater is the escape velocity from a giant planet with a mass of 298.5 Earth masses (M_E) and a radius of 10 Earth radii (R_E) to that of a terrestrial planet with a mass of 2.4 M_E and a radius of 2 R_E?

38. What is the ratio of Halley's Comet's orbital velocity at perihelion (0.6 AU) to that at aphelion (35.1 AU)?

39. Calculate the orbital velocity, in kilometers per second, of a planet orbiting 7.8 AU from a star of 4 M_{Sun}.

40. What is the ratio of the period of an asteroid orbiting at 1.8 AU to that of another asteroid orbiting at 3.3 AU?

41. How many times greater is the period of a planet orbiting its star compared to the period of a sibling planet orbiting at half the distance?

42. A spinning molecular cloud has decreased in radius by a factor of 23. By what factor has its velocity increased?

43. A portion of dust in a protoplanetary disk is rotating at 450 m/s. If the disk decreases in size by a factor of 400, what will the rotational velocity of the dust be?

44. How much greater is the velocity of a blob of gas rotating in an interstellar cloud at a radius of 0.74 pc to another blob of gas at a radius of 1.7 pc?

45. What would be the period, in years, of a Kuiper Belt object with an average orbital radius of 52 AU? (See Going Further 3.1.)

 If your instructor assigns homework in **smartwork5**, access it at the Digital Landing Page for *At Play in the Cosmos*: **digital.wwnorton.com/cosmos**

HOME BASE

Earth and the Moon

● **In this chapter** you will learn about our home (planet Earth) and its satellite (the Moon). After reading through each section, you should be able to do the following:

6.1 Discovering Change: Crocodiles in the Arctic
Summarize the features and processes that characterize Earth and the Moon.

6.2 Earth Inside and Out
Describe Earth's internal structure and the method by which it was discovered.

6.3 Earth's Near-Space Environment
Explain the nature and origins of Earth's atmosphere and magnetic fields.

6.4 The Closest Desolation: Earth's Moon
Describe the internal structure, external features, and probable origin of the Moon, as well as its gravitational interactions with Earth.

Earth and the Moon as seen as by the *Galileo* spacecraft on its way to Jupiter.

6.1

Discovering Change: Crocodiles in the Arctic

THERE ARE NO hotels or convenience stores deep in the Arctic. So when research takes John Tarduno, a geophysicist at the University of Rochester, to desolate ice-bound islands, it's all tents, cookstoves, and very warm sleeping bags. Tarduno studies the history of Earth's magnetic field—a discipline called *paleomagnetism*. "We are looking for rocks with a lot of iron in them," says Tarduno. "When these rocks formed millions of years ago, the iron particles got oriented in the direction Earth's magnetic field had back then. Since Earth's magnetism is constantly changing, we go up there looking for what are, essentially, fossil compasses."

It was on one of his expeditions to Ellesmere Island, just 500 miles from the North Pole, that Tarduno and his students found fossils of another kind altogether. "We found a champsosaur," explains Tarduno, "which is a kind of prehistoric crocodile-like creature" (**Figure 6.1**). Tarduno's champsosaur meant that millions of years ago, the Arctic we now experience as frozen tundra was once a warm swamp. "That was a big 'Wow!' moment for us," says Tarduno, "one of the most exciting I have ever had."

Tarduno's trips to the Arctic have shown him not only how much Earth has changed since the distant past, but also how much the planet remains in flux. "Up at the poles you get to see climate change firsthand," explains Tarduno. "The first time I was in

comparative planetology The study of planets' characteristics based on comparison with other known planets.

radiometric dating A technique used to date a sample of material by comparing relative amounts of radioactive atoms with the "daughter" atoms they decay into.

isotopes Forms of an element with different numbers of neutrons in the nucleus.

decay The process by which some isotopes of an element transform over time into other elements, emitting subatomic particles and energy in the process.

half-life The period of time over which the total number of nuclei of a radioactive isotope will drop by one-half.

FIGURE 6.2 Earth and Its Moon
This image, called *Earthrise*, was captured during the voyage of *Apollo 8* as it orbited the Moon. For many people this iconic image offered a new vision of the unity and fragility of Earth.

the Arctic was in 1990. Back then you never thought about walking around without a parka, even in the summertime. That's how cold it was all the time. Now I take students up there and they walk around in T-shirts. We have seen a lot of dramatic changes over just that short a timescale."

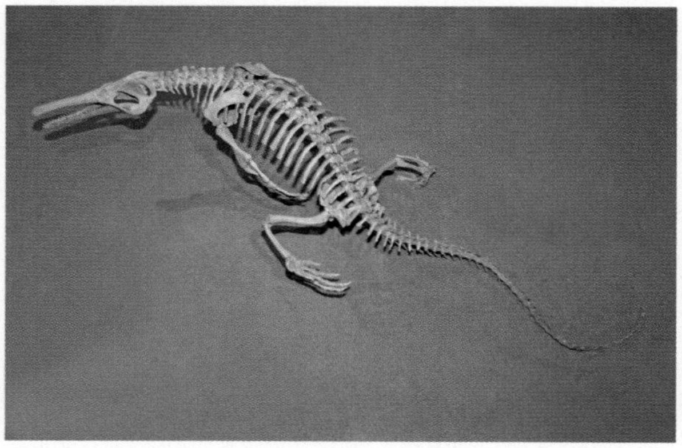

The history of a planet like Earth is a history of change driven by powerful forces and enormous energies. The research of atmospheric physicists, ocean scientists, and geophysicists shows us how Earth and the Moon have evolved to create a unique home for life in the Solar System (**Figure 6.2**). Through the intense study of our home world, scientists have also laid the groundwork for understanding all the worlds in the Solar System. Thus we begin our study of **comparative planetology**—comparing the different worlds in the Solar System with Earth and the Moon.

Earth and Moon: A Unique System

Earth and the Moon do not resemble any other planet-satellite system in the Solar System. Earth, the largest terrestrial planet, is about 4.5 billion years old (see **Going Further 6.1**). Unlike the other planets, Earth's surface is covered in liquid water. It has supported a complex biosphere for billions of years. These features are obvious. But some of the differences between Earth and other planets are less immediately obvious. For example, the Moon is so large (27 percent

FIGURE 6.1 Champsosaur Skeleton
John Tarduno's Arctic expedition found the unexpected: a champsosaur, or prehistoric crocodile, providing evidence of swamp conditions where now there is extreme cold and very low humidity.

GOING FURTHER 6.1 THE LANGUAGE OF THE COSMOS
Measuring Eternity: Radioactive Dating

Although human beings were not around for the formation of ancient meteorites or the first era of Earth's geologic activity, we have found a kind of "stopwatch" that reveals the age of rocks and meteorites. With a process called **radiometric dating**, scientists are able to use the relative amounts of certain types of atoms to render remarkably precise ages and recount the story of Earth's evolution.

As discussed in Chapter 4, an atom of a given element can have only a fixed number of protons. Carbon, for example, always has exactly six protons. The number of neutrons that make up an element is not fixed, however. Versions of an element with different numbers of neutrons are called **isotopes**. Some isotopes of an element are stable, but others are radioactive, meaning that they will destabilize and transform over time into other elements, emitting particles and energy in the process.

FIGURE 6.3 **Radioactivity and Half-Lives**
The predictable average decay times of radioactive elements with known half-lives into their daughter elements provides an accurate means of determining the age of Earth and the Moon.

This process of **radioactive decay** is the clock hiding in atomic nuclei. Even though scientists can never predict when a single unstable nucleus will decay (it's an inherently random quantum process), they can predict with great certainty how large collections of unstable nuclei will behave. Each unstable isotope has a distinctive timescale called a **half-life**, T_h. The total number of nuclei of that isotope will drop by one-half over the course of a half-life (**Figure 6.3**).

To get a better understanding of how radioactivity works, let's look at the formula for how the fraction F of an unstable isotope with half-life T_h changes over time. We use lowercase t to denote the specific time period in the problem we're considering.

$$F = \left(\frac{1}{2}\right)^{t/T}$$

From this equation we can see that after one half-life, or $t = T_h$, the fraction of the original unstable nuclei left is 1/2. What is the fraction left of 5 half-lives, or $t = 5T_h$?

$$F = \left(\frac{1}{2}\right)^{\frac{5T}{T}} = \left(\frac{1}{2}\right)^{5} = \frac{1}{32} = 0.031$$

$$0.031 \times 100 = 3.1\%$$

After 5 half-lives, only 3.1 percent of the original nuclei remain.

Many unstable isotopes are found in nature, and they all have different half-lives. Uranium-238 (92 protons + 146 neutrons), for example, has a half-life of 4.5 billion years. Using the equations above, if we began with 1,000 uranium nuclei, after 4.5 billion years only 500 would be left. Since uranium-238 decays into "daughter" nuclei of lead-206 (82 protons + 124 neutrons), you would also expect that if you had no lead-206 to begin with, then after 4.5 billion years 500 lead-206 nuclei would have appeared. If scientists know the ratio of the isotopes present when a meteorite or other object formed, they can then use the ratios they measure today to determine the age of the meteorite.

of Earth's radius) that an alien visiting our system might consider Earth and the Moon to be a binary planet rather than a planet and its satellite. "The Moon is about 1 percent the mass of Earth," says Jonathan Lunine, a planetary geologist at Cornell University who has studied many Solar System bodies. "Although that sounds small," Lunine continues, "it's actually a

far larger ratio compared to the moons of Jupiter, Saturn, Mars, Uranus, and Neptune" (**Figure 6.4**). According to Lunine, only Pluto's moon Charon is comparable in relative size, but Pluto is considered a dwarf planet, not a planet.

Earth's unique role as a life-bearing world and the fact that it has such a large moon are probably

interactive
Radiometric Dating

FIGURE 6.4

Comparison of Planet and Moon Sizes

In our Solar System, only the dwarf planet Pluto and its moon Charon are closer to each other in relative size than Earth and the Moon are. Jupiter's largest moon, Ganymede, whose radius is 27 times smaller than the gas giant's, is more typical of planet-moon systems in our Solar System.

Jupiter

Ganymede

Callisto

Io

Europa

Earth

Moon

Pluto

Charon

core Earth's innermost layer, composed primarily of iron. The inner core is solid; the outer core is in a fluid state.

mantle The layer between Earth's core and crust, composed of silicate rocks. High pressures found in the mantle cause material to deform and flow.

crust Earth's solid outer layer.

basalt A type of volcanic rock that forms at the ocean floor.

lithosphere Earth's solid outer layers, consisting of the crust and the uppermost regions of the mantle.

related. Because of its large mass, the Moon has had an important effect on Earth's evolution by creating tides and lengthening the day. Just as important, the Moon's gravitational force has kept the spinning Earth from wobbling too much, and this effect has kept Earth's climate relatively steady. No other moon in the Solar System has played such an important role in its parent planet's life.

Yet for all their uniqueness, studying Earth and the Moon reveals much about the evolution of the other worlds in our Solar System. We can apply what we learn from a detailed study of Earth and the Moon to other planets, and we can apply what we learn from other planets to understanding our own world. What we always find is a story of evolution: planets forming and differentiating; surfaces changing from forces both within and without; and in many cases, atmospheres growing to blanket their worlds, profoundly changing the surface histories of those planets. And for Earth, at least, we must add the all-important evolution of life, which changed land, sea, and sky.

SECTION SUMMARY

Earth and Moon: A Unique System

● The Moon's large size relative to Earth has played an important role in making Earth suitable as a life-bearing world.

6.2

Earth Inside and Out

Earth's internal structure is dominated by a metallic **core**, made mostly of iron, with a radius that is about one half the radius of the entire planet (**Figure 6.5**). The inner part of the core is solid; the outer core appears to be fluid molten iron. Surrounding the core is an extended **mantle** that is composed of silicate rocks (made of silicon, oxygen, and other elements). Lying on top of the mantle is the thin layer of **crust**, also composed mainly of silica rocks, which includes both the continents and the ocean floor (**Table 6.1**). However, the two types of crust differ somewhat in their composition: the ocean floor is composed of **basalts**, a type of volcanic rock, and the continents are composed of granites. Fully 71 percent of Earth's crust is covered in oceans, which can reach depths of about 11 km. On the continents, the highest mountains extend almost 9 km above sea level. Together, the crust and the solid upper portion of the mantle are referred to as Earth's **lithosphere**.

How, exactly, have geophysicists learned that Earth's internal structure has this series of layers? "Many people think scientists explore Earth's interior by drilling," says John Tarduno, "just like they do

TABLE 6.1 ••• **Earth's Layers**				
Layer	**Depth from Surface**	**Primary Material**	**Temperature**	**Major Processes**
Continental crust	30–50 km Greatest height: 9 km above sea level	Silicate rocks	Average: 286 K Range: 184 K 450 K–650 K at boundary of mantle	• Earthquakes • Continental drift • Volcano formation
Oceanic crust	5–10 km	Silicate rocks	450 K–650 K at boundary of mantle	• Earthquakes • Continental drift • Volcano formation
Oceans	Greatest depth: 11 km	Water	309 K in Persian Gulf to 271 K near North Pole	
Mantle	From crust to 2,890 km	Silicate rocks	750 K–1,150 K at boundary of crust About 4,000 K at boundary of core	Convection that drives continental drift
Outer core	2,890–5,150 km	Molten iron-nickel-sulfur-oxygen mix	4,000 K–5,000 K (estimated)	Dynamo effect creating geomagnetic field
Inner core	5,150–6,370 km	Solid iron-nickel alloy	5,000 K–6,000 K (estimated)	

looking for oil. But that's not how it works." The deepest well drilled to date extends a mere 12 km downward, which is only a fraction of the crust. The distance to the center of Earth is another 6,000 km below that (see Table 6.1).

Shake, Rattle, and Roll: Seismic Waves

Although scientists cannot drill below the crust, they can "see" into the interior using a kind of planetary vibration called **seismic waves**, generated during strong earthquakes. "Strong earthquakes are like a hammer hitting a bell," says Tarduno. "In a sense, the whole planet will ring in response to a really strong quake." The seismic vibrations created by earthquakes do not remain in the crust. They radiate away from the quake, traveling downward and taking different paths depending on the material they encounter as they travel.

Since Earth is a sphere, the waves will eventually make it back to the surface again hundreds or thousands of miles away. "A strong earthquake sends out so much energy," explains Tarduno, "that the waves it generates can make it all the way through the planet."

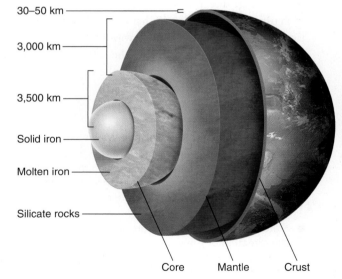

30–50 km

3,000 km

3,500 km

Solid iron

Molten iron

Silicate rocks

Core   Mantle   Crust

FIGURE 6.5
Earth's Interior Structure
Earth has four major layers: the thin, rocky crust; the mantle, where convection occurs; and a two-part core. The outer core remains molten, while the inner core is solid because of the extreme pressure.

Scientists pick up the vibrations from large earthquakes using sensitive instruments called seismographs, which can detect even tiny surface vibrations. By mapping the global distribution of vibrations from different earthquakes at different locations on the planet, scientists have built up a detailed picture of Earth's interior.

seismic wave A propagating wave in Earth's interior caused by an earthquake or a volcano.

Two kinds of seismic waves are used for mapping out the internal structure of the planet. First are **P-waves**, where *P* stands for *pressure*. "In a seismic P-wave," says Tarduno, "you are actually seeing compression and expansion of different materials at different conditions deep within Earth." P-waves travel through the material in Earth's interior much the way sound waves travel through air. As a P-wave passes, atoms and molecules oscillate back and forth in the direction of the wave's propagation. The other kind of seismic waves are called **S-waves**, where *S* stands for *shear*. In an S-wave, material is distorted in a direction perpendicular to the wave's motion. "If you take a rope and yank it sideways," says Tarduno, "you can make an S-wave."

What makes these two kinds of waves so important for studies of Earth's interior is a key difference in the way they propagate. "P-waves will transmit through solids and liquids," says Tarduno, "but S-waves only go through solids." Thus, S-waves traveling into a region of fluid will be stopped entirely (**Figure 6.6**). In addition, both types of waves can change their direction of propagation as they encounter materials of different density. "The process is called refraction," says Tarduno, "and it's the same thing that makes a ruler half-submerged in a glass of water appear bent. The waves change direction as they pass through regions with different density" (see Chapter 4). By tracking the time it takes for P-waves and S-waves from a strong quake to reach various points on the surface, scientists can calculate the range of densities the waves must encounter as they move through Earth.

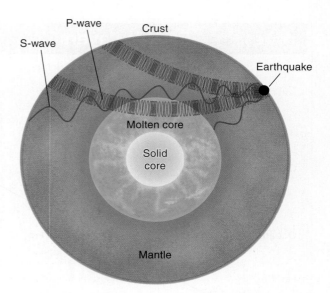

FIGURE 6.6 P-waves and S-waves
Two kinds of seismic waves, P-waves and S-waves, result from earthquakes and have different abilities to pass through liquid and solid matter. Tracking them enables scientists to probe the properties of Earth's deep interior. Waves change direction (refract) as they propagate through material of different densities.

SECTION SUMMARY
Shake, Rattle, and Roll: Seismic Waves

- Geologists study Earth's interior using seismic waves created by earthquakes, which make it all the way through the interior of the earth and can be detected around the planet with instruments called seismographs

- Material in a P-wave (pressure wave) oscillates in the direction of the wave's propagation. Material in an S-wave (shear wave) oscillates perpendicularly to the direction of the wave.

- P-waves pass through both solids and liquids while S-waves pass through solids only. Because of this difference, scientists can use seismic wave propagation to model Earth's internal structure.

P-wave or **pressure wave** A seismic wave in which atoms and molecules oscillate back and forth in the same direction as the wave's propagation.

S-wave or **shear wave** A seismic wave in which atoms and molecules oscillate perpendicular to the direction of the wave's propagation.

Earth's Interior

From studying P-waves and S-waves, scientists understand that the part of Earth we experience, the crust, is really just a thin layer of rock that literally *floats* on top of denser rock below. Under the continents, the crust reaches to depths of 60 km, while under the oceans it may extend downward for only 5–10 km.

Seismic studies also show that below the crust is a remarkable layer of flowing rock—the mantle—whose density increases with depth. "The material in the mantle is not melted," says Tarduno, "but the high pressure in there allows it to slowly move." How slow is the mantle rock's motion? "The mantle moves fastest at the upper layers," says Tarduno. "There you get top speeds of around 10 centimeters a year." The idea of rocks that move may seem unusual, but the tarmac that makes up most roads behaves the same way. Under a hammer blow, tarmac will appear solid, either shattering or bouncing the hammer back. Under high pressures, however, such as the pressure of a million-kilogram weight, the tarmac will ooze and deform like warm taffy.

Although the mantle is deep below the surface, scientists can get a good look at its makeup during volcanic eruptions. The magma that spews out in an eruption is not, however, the slowly oozing kind of mantle rock. Instead, it's mantle material that, in local regions where temperature and pressure conditions are high enough, has completely melted so that it flows like a liquid. A volcano is a region where melted mantle material (molten rock) manages to break through the crust. These mantle **plumes** are the source of one kind of volcanic activity, called **shield volcanoes**. The Hawaiian Islands are shield volcanoes that formed where a plume from deep in the mantle rose through the crust. Plumes also provide evidence that the material in the mantle is in a state of constant movement—a slow, steady series of rising and falling motions called **convection**.

Convection occurs when material deep in a planet (or a star) is heated. The heat makes the material expand, becoming less dense than its surroundings. Like a balloon held underwater, the lower-density rock becomes buoyant and begins to rise. As the material rises, however, it cools and eventually begins to sink back downward, where it gains heat again and restarts the whole process (**Figure 6.7**). The presence of mantle convection also shows that Earth is hotter in the center than it is in the outer layers. "When radioactive elements in the bottom layer of the mantle decay, they release heat," says Tarduno. "That is what sets the mantle convection in motion."

Seismic waves have also provided information about Earth's core, which lies below the mantle. Scientists have seen that S-waves from earthquakes never reach seismic stations located directly on the other side of Earth. That means S-waves attempting to pass directly through, or close to, the center of the planet are absorbed. In contrast, most P-waves can make it from one side of Earth to the other. This difference in the motions of S-waves and P-waves tells scientists that Earth must have at least a partially liquid core. "The core is actually made of two components," says Tarduno. "We have the solid inner core, and we have the molten outer core. Both the inner and outer cores

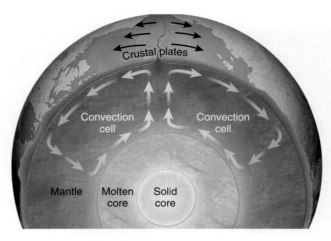

FIGURE 6.7
Convection in the Mantle
Convection in Earth's mantle is the engine that drives the motion of the crustal plates. Convection is shown here in a schematic of the convection patterns and their effect on the spreading ocean floor.

are mostly iron." The presence of the iron core and the rocky mantle means that early in Earth's history the planet must have *differentiated*. The heavy elements that would form the core sank to the planet's center.

SECTION SUMMARY
Earth's Interior
- The crust is a thin layer of silicate rock.
- Beneath the crust, the solid, denser mantle is under such pressure that it deforms and flows on timescales of centuries.
- The iron core has a molten outer layer and an inner solid layer.
- Earth's layers differentiated because of gravity, as the heavier components sank inward.

Earth's Living Crust

The planet's crustal surface is in a state of constant change driven by a variety of forces, including volcanic eruptions and "weathering" (also called **erosion**) via the flow of wind and water (**Figure 6.8**). The most

plume Molten rock from a planet's mantle that breaks through the crust.

shield volcano A volcano formed from a mantle plume that tends to have shallow, sloping sides.

convection The transfer of heat from one place to another by fluid motion.

erosion or **weathering** The process by which surface features are worn away by the action of wind and water (including ice).

A

B

FIGURE 6.8
Earth's Surface
Earth has been resurfaced throughout its evolution by processes originating both within its interior, such as (A) earthquake-created tsunamis and (B) volcanic eruptions, among other forces.

A

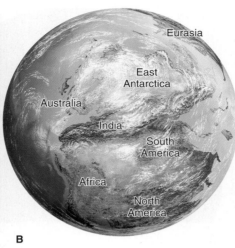

B

FIGURE 6.9
Continental Drift
(A) Alfred Wegener (1880–1930), shown here during a 1930 expedition to Greenland that cost him his life, developed the theory of continental drift in 1912, but his theory was not generally accepted until the 1960s. He based his theory partly on how the continents seemed to fit together like puzzle pieces. (B) This image shows how today's continents can be imagined as a single landmass by joining their boundaries.

dramatic changes in the planet's surface, however, occur so slowly that they are virtually invisible in human lifetimes. Over long periods, they have reshaped the planet's surface again and again.

The idea that Earth's continents move over time—a phenomenon known as continental drift—was first proposed by Alfred Wegener in 1912 (**Figure 6.9**). Noticing that fossils on one side of the

Atlantic were identical to those on the other side, Wegener began to hypothesize that the continents had once been part of a single landmass and had slowly drifted apart in time. Wegener was literally laughed out of scientific meetings when he proposed his theory. Not only was his evidence too thin, but he could not explain why the continents would be moving. It was not until seismographic studies revealed the properties of Earth's interior and its slowly convecting mantle that the force driving continental drift, called **plate tectonics**, was recognized. "Plate tectonics was a complete paradigm shift," says Tarduno. "It was not accepted until the 1960s." Wegener died in 1930 and never got to see his radical idea become mainstream.

Scientists now understand that Earth's crust is made of large blocks, or **plates**, that float on the mantle. "There are 12 plates that make up Earth's surface," says Tarduno, "and they are always moving relative to one another." As the mantle material reaches the top of its convective circulating motions, it pushes the surface plates along. Places where crustal plates meet are locations of powerful and often violent geologic activity. **Figure 6.10** shows where the majority of volcanoes

plate tectonics The movement of Earth's crustal plates due to underlying mantle flow.

plate A large segment of Earth's crust that floats on the mantle.

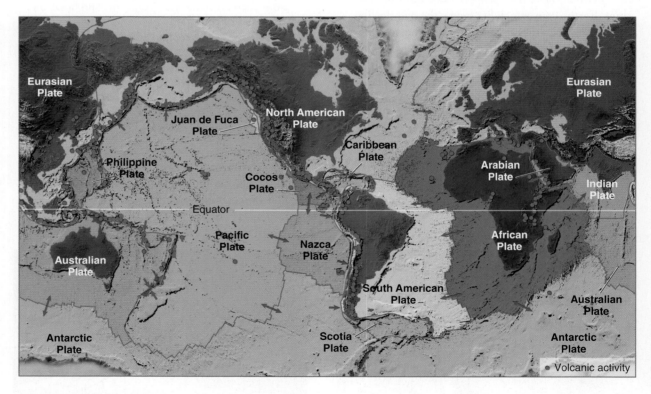

FIGURE 6.10 Continental Plates
Earth's crustal plates have been mapped in detail. Here, we see different plates coded with different colors. The darker shades within each plate represent continents. Earthquakes occur with greatest frequency and force at the junctures of the crustal plates, as pent-up energy is released in the sudden plate movement. The red arrows indicate regions of seafloor spreading where new crust is being created.

and earthquakes take place on Earth. Overlaid on this map is the location of the various crustal plates. Wherever two plates meet, there is a strong likelihood of violent geologic forces being unleashed. Volcanic eruptions and earthquakes occur because the plates are sliding past each other, colliding, or spreading apart. In all cases, tremendous energies locked up in moving trillions of tons of rock from one part of the planet to another are released at these plate boundaries, often with devastating consequences.

In regions where plates are sliding past each other, the movement is not continuous. Pressure builds up at the plate edges until it is released suddenly as an earthquake, and the plates move a few fractions of a centimeter past each other. The San Andreas Fault in California, responsible for many of the region's large and small tremors, comes from shearing (sideways motions) between the Pacific Plate on one side and the North American Plate on the other.

When plates collide, crustal material may be thrust upward to create mountain ranges. This process of "mountain building" is occurring right now in the Himalayas, where the Indian Plate is currently moving into the Eurasian Plate. In other cases, the collision between two plates will force one to dive downward in a process called **subduction**. In subduction zones, the sinking crustal plate will melt and eventually be reabsorbed into the mantle.

Collisions between plates can also cause volcanic activity. Where one tectonic plate slips under another, melting crustal material rises to the surface, creating what is called a composite (or explosive) volcano. This form of volcanism is usually a violent event as gases build up, only to vent in a tremendous explosion. Composite volcanoes tend to rise sharply and have steep slopes because of their rapid formation. In

1980, the powerful eruption of Mount Saint Helens in Washington State, a composite volcano, killed 57 people and flattened trees for 230 square miles. Other well-known composite volcanoes are Mount Shasta in California and Mount Fuji in Japan.

Plate tectonics can also affect shield volcanoes, the type that forms above mantle plumes. The Hawaiian Islands formed over a mantle plume, but because the Pacific Plate (the seafloor) is moving, the plume creates one island after another. The northwesternmost island, Kure Atoll, is older than the southeasternmost island, Hawaii, by about 28 million years because the plate continues to move while the plume stays in the same location (**Figure 6.11**).

At the center of the Atlantic Ocean is a region, called the Mid-Atlantic Ridge, where two plates are moving away from each other. In the space between them, molten magma, raw material from the mantle, wells up between two plates, filling in the gap formed as the plates move apart. As the magma cools, it becomes new ocean floor crust and is carried along with the plates (**Figure 6.12**). "You can see that the oceanic crust has been constantly re-forming," says Tarduno. "The oldest oceanic crust is 200 million years old. The continents, on the other hand, are billions of years old."

Tectonic plates move at a rate of about 10 centimeters (cm) per year, the same speed at which the mantle flows below them (this is just a bit faster than your fingernails grow). Although it takes hundreds of millions of years to appreciably alter the shape of the continents, scientists have been able to use highly accurate GPS measurements to track the shifting plates.

Combining current data with fossil records and geologic studies, scientists have reconstructed plate tectonic movement over the entire history of the

subduction A geologic process in which one edge of a crustal plate is forced below the edge of another.

FIGURE 6.11
Mantle Plumes
Volcanic island chains, such as the Hawaiian Islands, form because a crustal plate moves over a mantle plume, forming island after island over millions of years.

FIGURE 6.12
Formation of the Ocean Floor
Ocean crust forms at oceanic ridges, where two continental plates are moving apart, allowing magma to well up and form new crust at the bottom of the ocean.

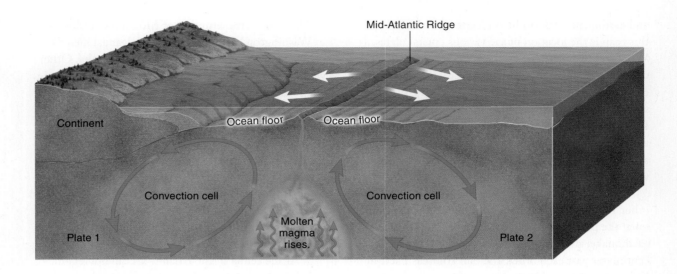

FIGURE 6.12
Formation of the Ocean Floor
Ocean crust forms at oceanic ridges, where two continental plates are moving apart, allowing magma to well up and form new crust at the bottom of the ocean.

planet. With this long view they can see, for example, that 250 million years ago all of Earth's continents formed a single giant landmass called Pangaea. Before the era of Pangaea, the continents were separate, and afterward they separated again. "If you visited Earth 250 million years in the future," says Tarduno, "it would be a very different-looking place. The maps we have now wouldn't make any sense."

SECTION SUMMARY
The Geosphere: Earth's Living Crust

- Plate tectonics is the slow movement of Earth's landmasses, as large blocks of crust called plates are moved by convection of the mantle.

- When plates collide or slide past each other, the enormous energy generated can cause earthquakes and volcanoes; individual plates may be either subducted or raised to form mountain ranges.

- Island chains occur when the crust moves over a mantle plume.

6.3
Earth's Near-Space Environment

interactive
Keeping an Atmosphere

troposphere The lowest level of Earth's atmosphere, reaching a maximum altitude of 17 km; the location of most of Earth's weather and clouds.

When astronauts orbiting Earth look out over the planet's curved horizon, they can see a thin layer of blue-white haze blanketing the oceans and continents below. Earth's atmosphere constitutes a minute fraction of the planet's mass (1/1,200,000), but it has been a principal driver of Earth's history as a life-bearing world. "Sure, the atmosphere has affected the evolution of life," says John Tarduno, "but it has worked the other way around too. Life has actually had a significant effect on the evolution of the atmosphere."

A Thin Life-Giving Veil: Earth's Atmosphere

The primary gases that compose the atmosphere (in order of volume) are nitrogen (78.09 percent), oxygen (20.95 percent), argon (0.93 percent), and carbon dioxide (0.039 percent). Small amounts of other gases, including methane (CH_4), helium, and ozone (O_3), also exist. In some cases these trace gases play an important role in regulating Earth's climate or the kinds of solar radiation that penetrate the atmosphere to reach the ground. Water vapor is also an important component of the atmosphere, but its content varies from one location to the next. On average, water in the gaseous state makes up about 1 percent of the atmosphere's composition.

The density of the atmosphere drops continuously from the surface upward. This decrease in density is so swift that by the time you reached the summit of Denali, the tallest mountain in the United States and only about one-tenth the height of the atmosphere, more than 50 percent of the atmosphere's mass would lie below you. The temperature of the atmosphere does not, however, decrease evenly with height, and this variation creates fairly well defined atmospheric layers (**Figure 6.13**).

The lowest layer of the atmosphere, called the **troposphere**, contains 80 percent of its mass. The troposphere extends just 17 km above ground level at Earth's equator, and just 7 km above at the poles. Sunlight warming the ground, which is then reradiated upward, is the principal mechanism warming the troposphere. Because this heating starts at the ground and because the atmosphere's density decreases with height, the tropospheric temperature also decreases with height. The troposphere is where the majority of Earth's weather is generated, and most forms of clouds occur within this single layer. Convection, driven by the absorbed sunlight, also occurs in the troposphere. The circulation of warm air upward and cold air downward makes the layer relatively turbulent, meaning the air does not flow smoothly in sheets but shows circulating ("bumpy") motions. Commercial jetliners cannot always avoid this turbulence, because they never reach altitudes of much more than 12 km, which is well within the troposphere.

The **stratosphere** is the atmosphere's next layer, extending up to heights of 50 km. Unlike the troposphere, the stratosphere's temperature *increases* with height. This shift is called a **thermal inversion**, and it occurs because the stratosphere contains a layer relatively rich in molecules of **ozone**, composed of three oxygen atoms (O_3). Ozone is extremely effective at absorbing ultraviolet light from the Sun, which leads to an increase in atmospheric temperature at these heights. Ultraviolet light is also harmful to biological compounds. "There is a thin region of the stratosphere where ozone can form and remain stable," says Tarduno. "It's really critical for life on Earth, as we've learned over the past few decades."

The emission of certain chemicals, called chlorofluorocarbons (CFCs), into the atmosphere has been shown to destroy stratospheric ozone. In response, every year giant **ozone holes**—substantial, localized decreases in ozone abundance—have begun appearing above the Antarctic and the Arctic, because the large seasonal changes in sunlight there affect the ozone chemistry. Worldwide bans of the chemicals have helped stabilize the decrease in stratospheric ozone, but many scientists remain concerned about its long-term health (and hence our own).

Above the stratosphere lie the **mesosphere** and **ionosphere**. Stretching from about 50 to 85 km, the mesosphere is the layer in which most meteoroids burn up. Its temperature drops with altitude. The ionosphere (also called the thermosphere), extending up to 1,000 km from the surface, gets its name because most atoms in this layer are ionized, meaning that they've lost one or more of their electrons. During solar storms, the Sun emits high-energy particles, which collide with atmospheric ions and drive them into excited states. The interactions with solar radiation heat the low-density outer regions of the ionosphere. As in the stratosphere, temperatures in this outer layer of the atmosphere increase with height until they plateau at about 300 km (**Figure 6.14**).

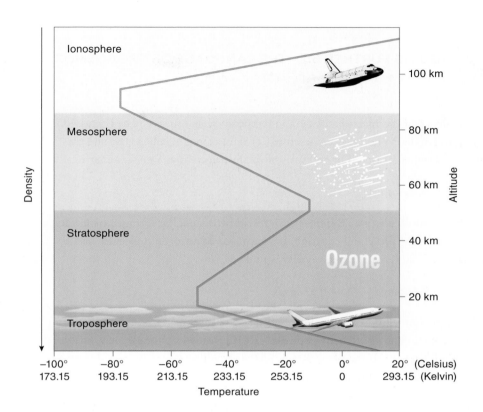

FIGURE 6.13 Earth's Atmospheric Layers
Earth's atmosphere decreases in density as altitude increases, but temperature (the orange line) actually increases in the stratosphere and again in the ionosphere, in what is called thermal inversion. Note that clouds (weather) and airplane travel are confined to the troposphere. Meteors typically burn up within the mesosphere, and the region we normally think of as "space," where the Space Shuttle flew, is within the ionosphere.

FIGURE 6.14 Our Atmosphere
In 1999, astronauts aboard the space shuttle *Discovery* took this image of Earth and the denser portions of the planet's atmosphere while orbiting within the ionosphere (about 300–500 km above Earth).

stratosphere The second level of Earth's atmosphere, extending from about 20 km to about 50 km.

thermal inversion An increase in temperature that occurs with increasing altitude (whereas temperature usually decreases with increasing altitude).

ozone A molecule composed of three oxygen atoms: O_3.

ozone hole An atmospheric region near either of Earth's poles and within the stratosphere that shows a substantial decrease in ozone abundance.

mesosphere The third level of Earth's atmosphere, extending up to 85 km.

ionosphere or **thermosphere** The fourth (and highest) level of Earth's atmosphere, extending up to 1,000 km, where most of the atoms are ionized.

GOING FURTHER 6.2 THE LANGUAGE OF THE COSMOS
How to Keep an Atmosphere

The dependence of thermal velocity (V_t) on temperature and mass can be expressed in a simple formula. When the temperature (T) is measured in kelvins (K) and the particle mass (m) is measured in multiples of the hydrogen atom's mass, we have

$$V_t = 0.16 \text{ km/s} \sqrt{\frac{T}{m}}$$

Nitrogen molecules are 28 times heavier than hydrogen. For nitrogen molecules, then, $m = 28$ in the preceding equation. If a gas made of nitrogen molecules had a temperature of 350 K, its thermal velocity would be

$$V_t = 0.16 \text{ km/s} \sqrt{\frac{350}{28}} = 0.57 \text{ km/s}$$

This equation helps explain why planets retain or lose their atmospheres (or at least some constituents of their atmospheres). Recall the concept of escape velocity, discussed in Chapter 3. If the planet's mass (M) and radius (R) are expressed in terms of multiples of Earth's values, we have

$$V_e = 11.2 \text{ km/s} \sqrt{\frac{M}{R}}$$

If an atmospheric particle has a thermal velocity that is much higher than the planet's escape velocity V_e, that particle will eventually experience a collision that knocks it upward, and then it will escape into space. Since different types of atmospheric particles have different masses, the lightest constituents of an atmosphere could have $V_t > V_e$, while heavier constituents have $V_t < V_e$. Therefore the lighter particles will eventually escape from the atmosphere, leaving only the heavier elements.

Imagine there is a planet with a mass one-tenth that of Earth ($M = 0.1$) but a radius twice as large as our planet's ($R = 2$). The escape velocity for this world is

$$V_e = 11.2 \text{ km/s} \sqrt{\frac{0.1}{2}} = 2.5 \text{ km/s}$$

Let's also assume that the planet has a temperature of 350 K. Nitrogen molecules with $V_t = 0.57$ km/s will not escape from the atmosphere of this world. What about hydrogen with $M = 1$? Let's use the thermal-velocity equation again:

$$V_t = 0.16 \text{ km/s} \sqrt{\frac{350}{1}} = 3.00 \text{ km/s}$$

Since $V_t > V_e$, we would expect this world to end up eventually with a nitrogen-rich and hydrogen-depleted atmosphere.

An important point here is that the thermal velocity of a type of particle does not just have to be greater than the escape velocity in order for a planet to slowly lose that species. The thermal velocity is just an average of all the velocities that a particular species will have at the atmosphere's temperature. Some particles will be moving slower, and some will be moving faster. The particles moving faster than the escape velocity will eventually be lost.

Although scientists consider the ionosphere part of the atmosphere, its upper reaches are what most people consider "space" in the sense that space shuttles and the International Space Station both orbit within the bounds of the ionosphere.

SECTION SUMMARY
A Thin Life-Giving Veil: Earth's Atmosphere
- The layers of atmosphere from lowest to highest are the troposphere, stratosphere, mesosphere, and ionosphere.
- The density of the layers continuously decreases with altitude, but temperature varies with height because of other factors.

The Creation of Earth's Atmosphere

The history of life on our planet is intimately connected with the history of the atmosphere, and life has played a strong role in determining the atmosphere's form. In fact, the layer of gas surrounding the planet today is not its original atmosphere. Earth's first atmosphere was probably very similar to the composition of the solar nebula and the gas giant planets—containing mostly hydrogen, helium, and other light gases. Early in the planet's history, however, these light elements escaped into space.

The ability of a planet to retain different gases directly affects what kind of atmosphere it will have. The composition of a planet's atmosphere depends on the balance between its escape velocity and the **thermal velocity** of atmospheric particles (atoms and molecules). The thermal velocity is the speed that a particle obtains from random collisions as it careens around. "Like the name implies, the thermal velocity depends on temperature," says Tarduno. "In a hot atmosphere, atoms and molecules move at high speeds. In a cold atmosphere, they move more slowly." If an atmospheric particle's thermal velocity is higher than the planetary escape velocity, it will eventually escape into space. But the thermal velocity also depends inversely on a particle's mass (that is, the mass appears in the denominator). Therefore, hot atmospheres are prone to lose their lightweight particles before they lose their heavy ones (see **Going Further 6.2**).

Earth's escape velocity is low enough and the temperature of its first atmosphere was high enough that all the light elements eventually escaped into space. "The formation of the Moon, which was likely a pretty violent event, may also have contributed to losing the first atmosphere," says Tarduno. What replaced the primal atmosphere were "outgassed" compounds, meaning that they were released by geologic processes such as volcanoes. from the crust and from volcanoes. Much of the second atmosphere may also have been delivered to Earth via comets and other planetesimals that were rich in heavier **volatile** (easy to vaporize) materials, such as methane and ammonia.

While there remains intense debate about the early stages of the atmosphere's evolution, scientists agree that the rapid increase in oxygen was one of its most important transitions. As Tarduno puts it, "Clearly one of the most fundamental events in Earth's history is something called the Great Oxygenation Event." Roughly 2.3 billion years ago, the oxygen content of the atmosphere increased dramatically. The cause of this oxygenation was photosynthesis by simple single-celled organisms, called cyanobacteria, in which oxygen was produced as a by-product (they "breathed in" CO_2 and "breathed out" O_2). While this photosynthesis was occurring, most of the oxygen was being absorbed chemically by minerals. "All those red rocks you see out West are basically from this period," says Tarduno. "Those rocks had iron in them, which absorbed the oxygen, making the minerals in the rocks rust colored." When the minerals became oxygen saturated, the biologically produced oxygen began to accumulate in the atmosphere.

SECTION SUMMARY
The Creation of Earth's Atmosphere

- The retention of gases in a planet's atmosphere depends on their thermal velocity in relation to the planet's escape velocity.

- Earth's original atmosphere, composed of light gases, was probably lost to space, while the current atmosphere was likely created from a combination of incoming materials, outgassing, and the activity of life.

Taking the Heat: The Greenhouse Effect

Each second, the Sun-facing side of Earth encounters 1.366 kilowatts per square meter (kW/m^2) of solar radiation. Thus, more energy falls on the hemisphere facing the Sun in 1 second than is generated by all the power plants on Earth in a year. What happens to this energy once it reaches the planet?

For the most part, the Sun can be considered a blackbody. Given its temperature, 5,778 K, Wien's law tells us that most of the Sun's energy is emitted in the visible parts of the electromagnetic spectrum. Earth's atmosphere is transparent to those wavelengths, so most of the solar energy falling on the upper layers of the atmosphere makes it down to the planet's surface. The surface, including the oceans, is quite effective at absorbing the incoming solar energy, and in doing so it heats up. If there were no atmosphere to consider, it would be easy to predict what would happen next: the temperature of Earth would increase until the energy it radiated back into space (as its own blackbody) would balance the energy it was receiving from the Sun. The temperature at which that balance would be achieved can be calculated to be about 250 K (−23°C, or −10°F), well below the freezing point of water (note that at this temperature, Earth's outgoing radiation is mostly in the infrared). But our planet is not a giant frozen snowball, and since the incoming energy rate from the Sun is fairly constant over short times, the extra warming must be due to the atmosphere. Somehow the atmosphere must trap some of the infrared energy that Earth would otherwise radiate back into space, holding it in and providing extra warmth for the planet (**Figure 6.15**).

interactive
The Greenhouse Effect

thermal velocity The average speed of particles in a gas, which depends on the temperature of the gas.

volatile Easily vaporized (transformed into a gaseous state) at relatively low temperatures; or, a volatile substance.

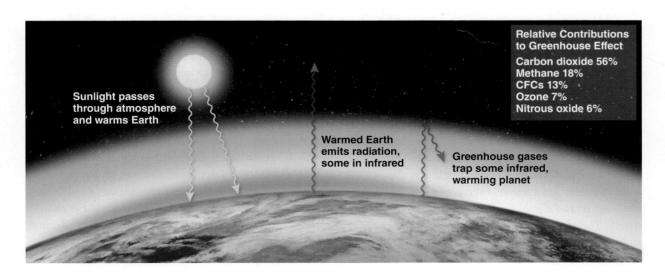

Sunlight passes through atmosphere and warms Earth

Warmed Earth emits radiation, some in infrared

Greenhouse gases trap some infrared, warming planet

Relative Contributions to Greenhouse Effect
Carbon dioxide 56%
Methane 18%
CFCs 13%
Ozone 7%
Nitrous oxide 6%

FIGURE 6.15
The Greenhouse Effect
Sunlight in primarily visible wavelengths warms Earth's surface (yellow wavy lines), which then reradiates energy back into space in the infrared (red wavy lines). Greenhouse gases like CO_2, water vapor, and methane absorb the infrared radiation, thus increasing the planet's overall temperature.

John Tarduno

John Tarduno's research as a geophysicist has taken him to some of the remotest corners of Earth, from the frozen wastelands of the Arctic to the mountains of New Zealand. But his globetrotting for science was not something he could have predicted when he was young.

"Geology had always been a hobby when I was a kid, but I was mainly interested in physics," says Tarduno. As an undergraduate, he saw a job opening advertised in a campus physics lab, and a merger of hobby and profession became possible. "The guy I was working for ran a solid-state physics lab, meaning they studied properties of solids like rocks. They were looking for ways to extract measurements of Earth's magnetic field when the rocks had first formed. That was my first experience with paleomagnetism and how I learned about the whole field of geophysics. It was just a natural fit."

Over the years Tarduno's work in the field looking for new paleomagnetic samples has taken him to some extraordinary places and, a few times, gotten him into some dangerous situations. "One time working in the Arctic, there was this narrow, steep-walled ravine that I wanted to explore," Tarduno recalls. "So I had the chopper put me down at one end and told him to pick me up on the other at the end of the day. But I had no way to know if I could make it to the other side, and if I didn't find a path I would be stuck in there because there would be no way to land a helicopter in the ravine. I was lucky because I found a natural terrace that I could move along, but it was 500 feet down and if things hadn't worked out I would have been in trouble." The hard-won samples Tarduno collected that day turned out to be key to recovering the record of a particularly turbulent part of Earth's magnetic history. "I guess it was worth it," Tarduno says, smiling.

---

greenhouse gas An atmospheric gas that absorbs infrared radiation and keeps it from being reemitted into space.

greenhouse effect The phenomenon by which the fraction of incoming solar energy that would have been radiated back into space is trapped by greenhouse gases, leading to an increase in the average temperature of the planet.

greenhouse forcing An increase in greenhouse gas concentrations that leads to a change in a planet's climatic state.

The agents raising Earth's temperature above freezing are called **greenhouse gases**. The Sun-warmed surface of Earth reemits most of its energy in the infrared. Certain molecules in the atmosphere, such as CO_2, water vapor, and methane, are extremely effective at absorbing the infrared radiation emitted by Earth's surface and keeping it from escaping into space, although their relative effects differ and some are more effective than others. By trapping a fraction of the energy radiated by Earth, gases like CO_2 and water vapor create an atmospheric **greenhouse effect**. "Without the greenhouse effect, the planet would be a terribly cold place," says Tarduno. "Just as the windowpanes in a greenhouse allow solar energy in but do not allow Sun-warmed air (or infrared radiation) out, greenhouse gases raise the temperature of a planet." In essence, the greenhouse effect increases the fraction of incoming solar energy that stays within Earth's surface and atmospheric system. The greenhouse effect is an essential part of atmospheric dynamics for many of the planets in the Solar System, including Venus and Mars.

"We have tremendous evidence that Earth has been warmer and much cooler over the geological record," explains Tarduno. "All the evidence points to these changes coming from natural variations in the concentration of greenhouse gases from something we call **greenhouse forcing**." *Forcing* in this case means changing the physical and chemical conditions, which can lead to climate change. Variations in the climate have been driven not only by CO_2, but also by natural variations in methane. "Methane gets stored in many places in Earth, including permafrost," says Tarduno. "Melting the permafrost can release methane, a more potent greenhouse gas than CO_2." Drops in methane may also have strongly reduced the greenhouse effect in the past and led to snowball Earth phases, super ice ages when the planet was almost entirely covered in snow and ice.

Thus, for Earth the greenhouse effect is crucial to the planet's ability to maintain a comfortable environment for life. Unfortunately, defining "comfortable" may become problematic for human civilization, as the greenhouse effect has taken on new importance for us.

Carbon dioxide is a by-product of combustion—that is, burning things. The industrial revolution, which began in the 1800s and led directly to the society we have today, was based on the combustion of fossil fuels, such as coal and oil. Before the industrial revolution, the combustion of coal, oil, and natural gas was negligible. Now, however, fossil fuel combustion occurs at a rate that produces about 21.3 billion tons of CO_2 per year. Over the last century, the CO_2 levels in the atmosphere have increased by 20 percent, and they are expected to continue rising at a level of 4 percent each decade (**Figure 6.16**). After exhaustive study, scientists are certain that all this CO_2 added to the atmosphere will lead to an increased greenhouse effect. As more of the incoming solar radiation is trapped, more energy is added to Earth's atmospheric and ocean systems. The result will be some form of climate change. *Climate* means long-term behavior of weather systems. While Earth as a whole will get warmer, what humanity can expect are changes in long-term trends in regional rainfall, snowfall, ice coverage, the number and intensity of storms, and so on.

It is currently impossible to predict with certainty how severe the change in climate will be, but the possibilities range from the improbable end of human life to the equally improbable "no big deal." Because climate affects everything from agricultural production to transportation, human civilization is fairly sensitive to climate change. Efforts to both halt continued climate change and deal with effects we cannot change are likely to play an important part in the history of the next century.

SECTION SUMMARY
Taking the Heat: The Greenhouse Effect

- A planet is a blackbody, absorbing and reradiating energy from the Sun. Gases in Earth's atmosphere create a greenhouse effect, trapping some of the reradiated energy.

- Increased combustion of fossil fuels has increased greenhouse gases, including CO_2, in the atmosphere which are causing climate change as Earth heats up. A warming Earth will also release more methane, another strong greenhouse gas.

Our Magnetic Blanket: The Magnetosphere

"Earth has a pretty strong magnetic field," says Tarduno, "just like a number of other planets in the Solar System." The field extends from deep in Earth's interior far out into space. The region of space around a planet where the planet's magnetic field exerts a strong influence is called the **magnetosphere**. Only recently have scientists come to understand how Earth's magnetic field is generated and the important role it plays in the life of the planet.

Magnetic fields are created by the motion of charged particles. A wire carrying charged particles (an electric current) will generate a magnetic field (there is a magnetic field surrounding every wire with current running through it in your home). The influence works both ways, however, as magnetic fields exert forces on charged particles. For example, the trajectory of an electron moving through a region with a magnetic field will be deflected by the field. Strong enough fields deflect the electron's motion so completely that the particle ends up trapped by the magnetic field. Particles move in a helix, rotating *around* the "field lines" while also traveling *along* the field lines. Understanding these basic features of magnetic phenomena provides the building blocks for understanding all planetary magnetic fields, including Earth's.

Overall, Earth's magnetic field has the same shape as a bar magnet—what scientists call a **dipole field**. There is a **north magnetic pole** and a **south magnetic pole**, with magnetic-field lines bowing outward to connect them. Earth's field also shows considerable variation from one point to another beyond what would be expected for a dipole field, and there are

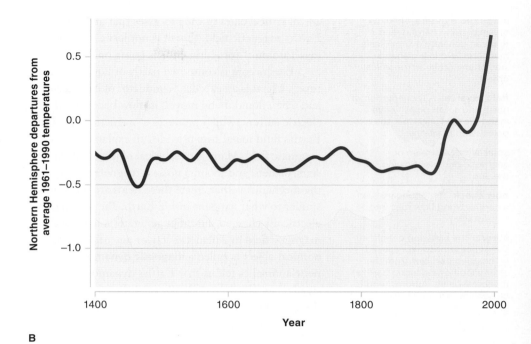

B

FIGURE 6.16 Atmospheric Temperature
Variations in atmospheric CO_2 concentrations have occurred since the industrial revolution, which ushered in processes that released much higher levels of CO_2. Since then, the global average temperature of the planet has risen. This graph reconstructs temperatures for the last 600 years. The light blue band shows the range of uncertainty in the temperature reconstructions.

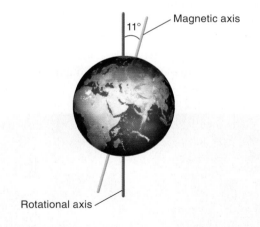

FIGURE 6.17
Earth's Magnetic Axis
Earth's rotational and magnetic axes are closely but not perfectly aligned with each other. The two currently differ by 11°.

places on the planet's surface called magnetic anomalies, where the field may be stronger or weaker than average by a few percent.

In addition to the magnetic anomalies, the north and south poles of Earth's magnetic field are not aligned with the axis of Earth's rotation. The spin and magnetic axis are misaligned by 11° (**Figure 6.17**). Even more important, the magnetic poles move. During the 20th century the north magnetic pole wandered 1,100 km. The origin of these features, the anomalies, and the movement of the magnetic poles

magnetosphere A region of space around a planet where the planet's magnetic field exerts a strong influence.

dipole field The magnetic-field configuration consisting of a north and south pole and magnetic-field lines connecting them.

north magnetic pole The point on Earth's Northern Hemisphere at which the planet's magnetic field points vertically downward.

south magnetic pole The point on Earth's Southern Hemisphere at which the planet's magnetic field points vertically upward.

magnetic dynamo The flow of electrically charged fluid that can generate a large-scale magnetic field in a planet, a star, or any other astronomical object.

field reversal A reorientation of north and south poles within a magnetic field such that the two poles swap positions.

solar wind A stream of energetic particles flowing off the Sun.

bow shock The boundary where a strong flow runs into an obstacle.

aurora The emission of light in the upper atmosphere caused by charged particles from the solar wind flowing along Earth's magnetosphere. Auroras tend to be seen at upper latitudes.

FIGURE 6.18
Earth's Magnetic Field
Earth's magnetic field extends far into space and interacts with the solar wind at a region called the magnetosphere. The magnetosphere shields Earth by deflecting charged, high-energy solar particles away from the planet.

are all critical clues to the processes that generate Earth's magnetic field. "Earth is not just a bar magnet," says Tarduno. "Something else is going on."

Earth's core is composed partly of liquid metal (iron). Metal is an excellent conductor of electricity, and when liquid metal moves, its flow becomes an electric current, which produces a magnetic field. Thus, Earth's fluid metal core provides an explanation for its magnetic field. Many astrophysical objects (galaxies, stars, planets, and moons) possess magnetic fields, and the mechanisms that create these fields are probably similar to what happens inside Earth. The moving, electrically charged fluid that generates a large-scale magnetic field in a planet, a star, or any other astronomical object is called a **magnetic dynamo**. "Magnetic anomalies tell us that Earth's dynamo is not steady," says Tarduno. "The flows at the core generate the fields by changing their patterns over time."

The most dramatic evidence for change in the dynamo and Earth's field consists of what are called **field reversals**. On timescales of many hundreds of thousands of years, Earth's field loses its large-scale dipole pattern. The north and south magnetic poles disappear, only to reappear with their orientation reversed. A record of the field-flipping is preserved in the rocks composing the spreading seafloor in places like the Mid-Atlantic Ridge. "You can see the field direction changing back and forth in the orientation of iron filaments in the seafloor rocks," explains Tarduno.

Although Earth's terrestrial magnetic field was discovered centuries ago, scientists did not know that

the magnetic field extended far out into space until the dawn of the space age, when the satellite Explorer 1, launched in 1958, discovered strong currents and magnetic fields as it orbited the planet. The magnetosphere is the result of Earth's magnetic field interacting with the **solar wind**—the stream of energetic particles flowing off the Sun. Because of this interplay between solar wind and terrestrial field, the magnetosphere is not symmetric. The solar wind exerts a pressure on the magnetic field, driving it into a teardrop shape. The magnetosphere extends only about 70,000 km on the side facing the Sun, but it can extend 20 times farther on the side facing away from the Sun (**Figure 6.18**).

The distortion of the magnetic field by the solar wind shows us that Earth's field acts as a planetary buffer from the powerful and deadly stream of solar particles. As charged particles from the Sun flow past Earth, its magnetic field deflects some of them via a **bow shock**. "The bow shock is a lot like the bow wave that forms in front of a fast-moving boat," says Tarduno. Particles that are not deflected away from Earth at the bow shock will be trapped in Earth's field. They spiral around the field lines and simultaneously move toward the poles. Collisions with these charged particles drive the atoms and molecules in the atmosphere at the poles into excited states. As they drop back to lower levels, they emit photons, which produce the glowing **auroras** visible in the polar regions (**Figure 6.19**).

Sometimes powerful solar storms occur when blasts of matter and magnetic fields from the Sun erupt into space. If these eruptions hit Earth, they can trigger powerful "space storms" in the magnetosphere, leading not only to bright auroras but also to dangerous torrents of energetic charged particles. These high-energy particles can destroy satellite electronics, pose a serious radiation threat to orbiting astronauts, and, on rare occasions, lead to so much current flowing down to Earth that power grids can be quickly overloaded. In 1989 a space storm sent enough current surging into the atmosphere that it took only 90 seconds to bring down Quebec's electric grid, leaving more than a million people without electricity.

Earth's magnetosphere may have played an important role in the early history of the life on our planet. If energetic solar wind particles routinely made it to the planet's surface, the damage they would have caused to biological molecules might have made long-term evolution difficult, increasing mutations and wreaking havoc on living systems. Thus, we may all have the magnetosphere to thank for our current existence on the planet.

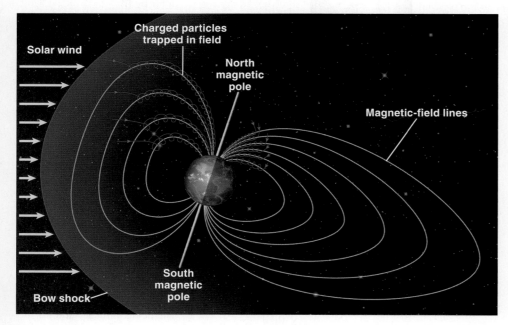

Solar wind

Charged particles trapped in field

North magnetic pole

Magnetic-field lines

South magnetic pole

Bow shock

FIGURE 6.19 Auroras
An aurora is created in Earth's atmosphere when charged particles from the Sun become trapped in the magnetic field. They spiral down along the field and are funneled toward the poles. Eventually they collide with atmospheric molecules and atoms, which then emit light, giving rise to auroras.

SECTION SUMMARY
Our Magnetic Blanket: The Magnetosphere
- Earth has a dipole magnetic field with magnetic field lines connecting the north and south magnetic poles.
- The magnetic field is generated by Earth's dynamo—its rotating fluid iron core. Earth's field changes structure continually over long timescales.
- Most of the solar wind is deflected by Earth's magnetic field, but some particles enter Earth's atmosphere, causing auroras and potential problems for human technological systems.

6.4
The Closest Desolation: Earth's Moon

The magnetosphere is not the only entity in space close to Earth that affects conditions on the planet's surface. Just as important is the Moon, a mere 385,000 km away, which exerts an enormous influence on Earth and its daily life.

Tides and the Earth-Moon Connection

Anyone can see the periodic rise and fall of Earth's oceans by spending a day at the beach. Walk along the shoreline in the morning, and the waves will come only so far up the beach. Take the same walk a few hours later, and the waves will be rolling in much farther. Carry out the same experiment a few weeks later, and the timing of the risings and fallings will have shifted. This is the phenomenon of **tides**—a twice-daily change in the local height of the sea level—with regular modulations over periods of months to a year.

In the middle of the ocean, the typical variation in sea level will be about a meter. The impact of this change on a particular coastline—how far it moves any waterline inland—depends on the slope of the coast. If the topology of the land is such that water is funneled through channels, the tidal variation can be extreme. Tides in the Bay of Fundy in Nova Scotia, Canada, where the seawater must flow through a narrow causeway, reach daily variations of 20 meters—large enough for ships to be completely grounded during low tide (**Figure 6.20**).

tide The effect of differential gravity on an extended object. Tides lead to distortion of spherical mass distributions, among other effects.

A

B

FIGURE 6.20
High and Low Tides
The Bay of Fundy, on the Atlantic Coast border of the United States and Canada, shows an extreme example of the difference between (A) low tide and (B) high tide.

Jonathan Lunine

For Jonathan Lunine, a career in astronomy and planetary sciences was an accident of geography: his family lived in New York City just a few blocks from the Hayden Planetarium. "The planetarium was definitely part of it. My parents took me there all the time." Tragedy, however, almost steered Lunine in a different direction. "I was 14 or 15 when my father died, and I began to get this guilt complex that I needed to help save the world. I decided I was not going to become an astronomer but instead would become a medical doctor."

In his senior year of high school, Lunine asked his doctor for some medical school catalogs to see what courses he would have to take. "He talked me out of it," Lunine recalls. Instead, the doctor offered him a ticket to go see astronomer Frank Drake talk about the giant radio astronomy dish in Arecibo, Puerto Rico. That was all the push toward his childhood love that Lunine needed. "After Frank Drake was done lecturing, I remember coming back down in the elevator and thinking, 'I really want to be an astronomer, not a doctor.'"

Drake wasn't the only scientist to make an impression on the young Lunine. "I read Carl Sagan's *The Cosmic Connection* at the age of 14," he explains. "I was so enchanted by it that my mother encouraged me to write to him. I did, and then he wrote back a two-page letter on Cornell stationery. Having a famous scientist like Sagan actually write a letter to this little junior high school student made a big impression on me, so now if a kid writes to me, I try to respond because I remember how much Sagan's letter changed my life."

differential force of gravity
The different strengths of the gravitational force from a massive body felt on the different parts of an extended object because the force decreases as the inverse square of the distance.

tidal bulge A distortion from spherical symmetry that is caused by tidal force.

spring tide A tide that is greater than average because it is reinforced by the linear alignment of Sun, Earth, and Moon at the new and full Moon phases.

neap tide A tide that is less than average because the Sun's gravity partially cancels out that of the Moon at the Moon's first and third quarter phases.

The tides have been known since prehistory (the first sea voyages were braved in canoes some 20,000 years ago). Understanding the cause of the tides, however, had to wait for Isaac Newton and his theory of gravity. Recall that the strength of the gravitational force is inversely proportional to the distance between two objects. If the objects are extended, meaning they are not simply tiny points of mass, then astronomers must also consider the differences in force from one side of the object to the other. The origin of Earth's tides lies in this concept of the **differential force of gravity** due to the Moon. Jonathan Lunine explains how tides work. "A small particle that's located on the surface of Earth, facing the Moon, is going to feel a stronger gravitational force from the Moon than a particle that's at the center of Earth or a particle that's all the way on the other side of Earth. It's the differential force of gravity that matters in making tides."

The water on the side of Earth closest to the Moon feels a stronger gravitational pull than the water on the side of Earth farthest away from the Moon. Thus the Moon creates a **tidal bulge** of water on Earth. There are actually two bulges, one pointing toward the Moon and one pointing away. The second bulge may seem counterintuitive, given the preceding arguments about differential force. But just as Earth's water on

the Moonward side is pulled toward the Moon, Earth itself is pulled a little toward the Moon, relative to the ocean on the opposite side of the planet. The differential force of gravity means that the oceans on the side of Earth facing away from the Moon are "left behind," creating a second tidal bulge. As Earth rotates once each day, the tidal bulges maintain their relative positions, pointing toward *and* away from the Moon. Thus, Earth rotates under the tidal bulges, which, from the perspective of the land, become giant waves flowing around the planet twice a day (**Figure 6.21**).

The Moon is not the only astronomical body controlling the tides. Although the Sun is about 400 times farther away from Earth than the Moon is, it is 27 million times more massive. Since the force of gravity depends directly on an object's mass and inversely on the square of its distance, the Sun can still exert a strong differential force on Earth's oceans. "Because the Sun is so massive," Lunine explains, "we now have a very complex situation where we have actually two bodies raising the tides." Thus, variations in the height of the daily tides will depend on the position of Earth, Sun, and Moon. When the Moon is either new or full, the Sun and Moon line up relative to Earth, and their combined gravitational effects produce large tidal variations called **spring tides** (**Figure 6.22A**). When the Sun and Moon are at right angles to each other (when the Moon is in its first or third quarter phase) the tidal effects of Sun and Moon tend to partially cancel each other out, creating weak tidal variations called **neap tides** (**Figure 6.22B**).

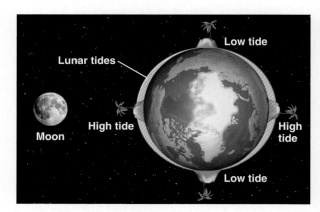

FIGURE 6.21 Tidal Bulges
The tidal force that the Moon produces on Earth leads to two tidal bulges in Earth's oceans. The first occurs on the Moon side, caused by the oceans pulling toward the Moon; the second occurs on the far side as Earth pulls away from the ocean. Earth's rotation leads to two high tides each day (shown schematically here for an island). The high tides occur as the island passes under the Moon and as it passes through the point opposite the Moon.

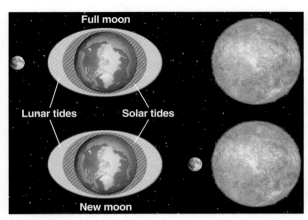

A Spring tides

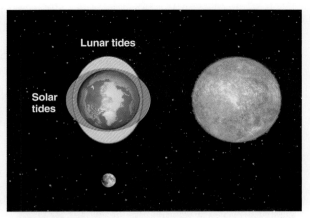

B Neap tides

FIGURE 6.22
Spring and Neap Tides
Spring tides (A), when the tidal effects of Sun and Moon align, are greater than neap tides (B), when the two tidal effects are at right angles to each other.

THE LONGEST DAY? "**Tidal forces** shape a lot of the behavior we see in the Solar System," says Lunine. A local example of this importance can be found in the direct, long-term consequences that the Moon and its tidal forces have had on the evolution of Earth's day.

Earth rotates on its axis once every 24 hours. This was not, however, the length of the day 20 million years ago. Nor will it be the length of the day 20 million years in the future. Earth's rotation, and hence the length of its day, have been gradually slowing down since its formation. Fossil records of animals, like coral, whose daily growth cycles create patterns that can be examined today, show that rather than 365 daily cycles in a year, there used to be as many as 400 day-night cycles per year. The culprit slowing Earth's day down is once again the tidal effect of the Moon (and to a lesser degree the Sun).

Although the reasons for this deceleration of Earth's spin are complicated, one way to understand them is to think about tidal bulges in Earth itself. Differential gravity not only raises a bulge in the oceans, but also deforms the crust, the mantle, and the entire Earth. More important, these tidal bulges do not line up directly with the Moon. Instead, Earth's rotation keeps the tidal bulge slightly "ahead" of the line connecting Earth's center with the Moon's center. That means there is always a small but important component of gravitational force between the Moon and Earth's tidal bulge. This force pulls back, *against* the direction of Earth's spin, decreasing its rate of rotation and slowing the length of the day (**Figure 6.23**). The effect is small, only 0.0017 second every century, but over time it adds up. In another 500 million years, the expression "there are only 24 hours in the day" will have to be changed. In that far-flung future, people

will have more time to work, play, and sleep, as each day will last about an extra two and a half hours.

TIDAL LOCKING. Another example of how one body tidally affects the rotation of another can be found in the dramatic effect Earth has had on the Moon's rotation. According to Newton's third law, for every force, there is an equal and opposite force. Thus, while the Moon raises tides on Earth, Earth is also raising tides on the Moon. Since Earth is far more massive than the Moon, it has a greater effect in distorting the Moon's shape and creating tidal drag. The effect of Earth's differential gravity on the Moon's spin is so strong that our satellite's rotation has been permanently altered. In an effect known as **tidal locking**, the Moon now rotates only once for every orbit it completes around Earth (**Figure 6.24**). "The perfect match between the Moon's rotation and its orbit is entirely a tidal effect," says Lunine. "It's Earth's gravity

tidal force Force acting on an object that is due to differential gravity from a second object.

tidal locking The one-to-one relationship between the rotation and revolution periods of an orbiting body that is created by tidal effects, such that the orbiting object always shows the same side to the object being orbited.

FIGURE 6.23
The Moon's Effect on Earth's Rotation
The Moon's tidal effect on Earth has caused Earth's rotation speed to slow over long periods of time through tidal friction. The Moon pulls on the tidal bulge, slowing Earth's rotation and lengthening the day.

FIGURE 6.24
Tidal Locking
Tidal locking of the Moon means that its rotation and orbital periods are equal, so it shows the same side to Earth at all times.

Moon's rotation

crater An approximately circular depression in the surface of an object formed by a high-velocity impact of a smaller body.

terracing The steplike appearance of steep walls in an impact crater created by slumping of the walls due to gravity.

FIGURE 6.25
Cratering
The energy and impact angle (among other properties) of an incoming asteroid or meteoroid determines the size and features of the resulting crater on the Moon. Shown here are examples of a peak (A) in Keeler crater, of terracing (B) in Copernicus crater, and of rays (C) extending from Tycho crater.

acting on bulges in the Moon, braking the Moon's rotation." Tidal locking, a process by which larger bodies exert tidal effects on smaller orbiting ones, occurs throughout our Solar System.

SECTION SUMMARY
Tides and the Earth-Moon Connection

- The Moon causes two tidal bulges on Earth. Because of Earth's rotation, these manifest in the oceans as traveling waves that raise and lower local sea levels twice a day.

- The Sun also has tidal effects on the rotating Earth, so the alignment of the two bodies with Earth magnifies tides (spring tides) or the misalignment reduces them (neap tides).

- Tides have slowed down Earth's rotation time and thus increased the length of the day.

- Earth's gravity has tidally locked the Moon so that its periods of rotation and orbit are equal.

The Moon's Face: Craters and the Lunar Surface

Tidal locking of the Moon means that only one lunar hemisphere is visible to us from Earth. Until the space age, humans were unable to see the "back" side of the Moon, sometimes called its "dark side." This term, however, is a misnomer, since the side of the Moon we cannot see faces directly toward the Sun during the new Moon phase of its orbit, when the portion of the lunar globe that faces us is completely in shadow. Not until the 1960s, when space probes orbited the Moon, did we have a chance to see its other side.

With a complete view of the lunar orb, scientists could finally begin to piece together the story of planetary evolution written on the Moon's surface. "I suppose you could say the most important feature of the Moon's surface is that it is heavily cratered," says Lunine. **Craters** are surface features caused by impacts with fast-moving Solar System debris—everything from small meteoroids to large asteroids or comets. When an object impacts a body like the Moon, the energy of its motion is converted into tremendous heat as the object rapidly decelerates. Both the object and the impacted lunar material are vaporized. A circular crater is formed as ejecta, material from the impact, is blown outward. As the gouged-out crater material settles back, its walls will slump, creating an effect called **terracing**. The lunar material may also rebound, creating a *peak* at the center of the crater. Debris blown outward during the impact may also fall back to the Moon, sometimes creating *rays* that extend radially away from the crater (**Figure 6.25**).

Few impact craters are visible on Earth, because our planet is still geologically active. "Tectonic activity and erosion are all very effective processes at

Peak

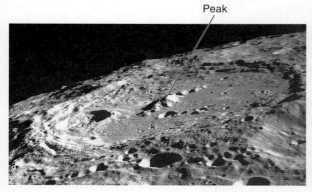

A

Terracing

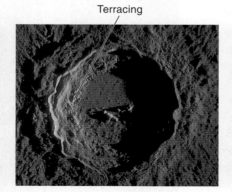

B

Rays

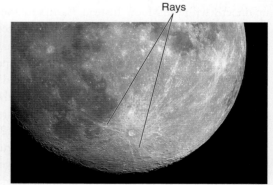

C

obscuring craters," explains Lunine. "Since Earth's surface is always being modified, turned over, and rebuilt, impact craters have not been well preserved except in a few places."

Because the Moon stopped being geologically active billions of years ago, it has preserved an excellent record of nearly all its impacts over the history of the Solar System. *Nearly* is the key word here, because even a quick glance at the side of the Moon that we can see from Earth shows us a second obvious feature of the Moon's surface besides cratering. "The other dramatic features we see on the Moon," explains Lunine, "are the dark, smooth regions called the **maria**." *Maria* (Latin for "seas") were so named because early observers thought these dark regions might be oceans. Instead, the maria date back to one of the Moon's last periods of geologic activity. Even these vast, smooth regions, which are volcanic in origin, may be tied to impacts. At the time the maria formed, the Moon's interior was still relatively hot. "The maria may have been formed by impacts that were powerful enough to heat the Moon's lower crust, creating vast flows of molten lava," says Lunine (**Figure 6.26**).

Scientists also find some evidence for volcanic activity on the Moon at the time, such as a few small domes pushed up by lava and winding channels, called sinuous rills, through which lava may have flowed. However, there probably was never the kind of large-scale volcanic activity on the Moon that we see on other worlds.

Geologists have been able to use the maria and the heavily cratered regions of the Moon, called the **lunar highlands**, to deduce the history of both the Moon and the Solar System. "On parts of the Moon like the lunar highlands, we see a record of the time when the rate of bombardment by debris left over from planet formation was very large," says Lunine. Laboratory dating of lunar samples has enabled scientists to determine that many of these large craters on the Moon were formed about 4.1–3.8 billion years ago. This epoch is called the **Late Heavy Bombardment**, and it affected all the inner planets. After this era, lunar impacts declined as the debris left over from the formation of the planets was cleared out of the Solar System.

It's worth noting that the other side of the Moon shows only heavy cratering of the kind seen on the lunar highlands, and no maria. "It may be that the impacts that formed the maria caused an asymmetry inside the Moon itself," says Lunine, "which then reoriented itself

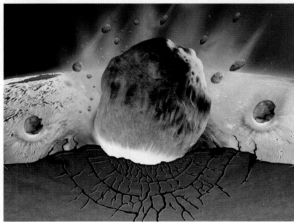

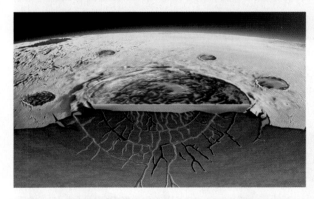

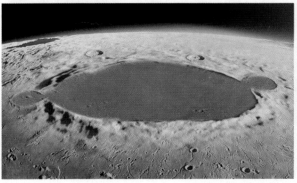

FIGURE 6.26 The Formation of Maria
Maria formed on the Moon when the crust fractured from impacts and molten rock filled in former craters.

maria (sing. mare) Large, dark plains on the Moon, formed by ancient volcanic eruptions.

lunar highlands Older, lighter-colored, heavily cratered portions of the Moon's surface.

Late Heavy Bombardment A hypothetical epoch, from 4.1 to 3.8 billion years ago, during which a large number of asteroids collided with bodies in the inner Solar System.

so that the maria now face the Earth" (**Figure 6.27**). Some researchers attribute the differences to a slight offset of the Moon's core from its geometric center, but the asymmetry between the near and far sides of the Moon remains an open research question.

Although the Moon, having no atmosphere, is certainly a lifeless world, evidence has indicated that some craters may be home to significant quantities of water

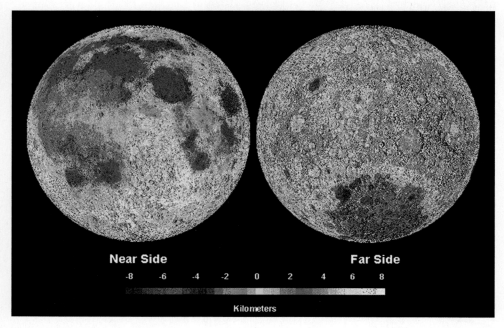

Near Side   **Far Side**

-8   -6   -4   -2   0   2   4   6   8

Kilometers

FIGURE 6.27 Lunar Terrain
In 1994, the *Clementine* lunar mapping mission provided these maps of the Moon's near and far sides, showing altitude differences. In these topographic maps, the lower maria appear blue, in sharp contrast to the lunar highlands, shown in red.

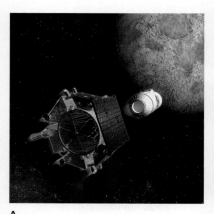

A   **B**

FIGURE 6.28 *LCROSS* Mission
(A) To search for water on the Moon in preparation for future exploration and the potential establishment of a permanent station on the Moon, NASA launched the *LCROSS* mission, which was purposely driven to impact in a shadowed crater. Analysis of the debris plume by observers on Earth provided proof that there is indeed water on the Moon. (B) The collision site appears here as a cloudy spot seen 20 seconds after impact.

hidden in shadows or underneath the lunar soil. In 2009 the *LCROSS* probe fired a high-speed projectile into the southern crater Cabeus and analyzed light from the debris cloud that the impact created. The results provided strong evidence for water in the debris (**Figure 6.28**). The origin of the water is probably linked to the comet or asteroid that created the crater in the first place, though this point is still hotly debated among scientists. Whatever its origin, the presence of water on the Moon may be a boon to humans who eventually hope to build permanent lunar bases. More recent probes such as NASA's moon mineralogy instrument on India's Chandrayaan-1 mission have found further evidence for small amounts of water molecules in lunar soil.

SECTION SUMMARY
The Moon's Face: Craters and the Lunar Surface

- Asteroids and meteoroids that impacted the Moon billions of years ago created craters as well as maria—giant depressions that became filled with lava, thus obscuring smaller craters within them.
- The number of craters in the lunar highlands provides evidence of the Late Heavy Bombardment period, about 4 billion years ago.
- There is strong evidence of water delivered via impacts in craters on the Moon.

Inside the Moon

"Of course, we know less about the Moon's interior than we do about Earth's," says Lunine. "Much of what we know comes from the *Apollo* missions. The astronauts left seismic detectors on the lunar surface (**Figure 6.29**). Then, NASA generated artificial earthquakes by sending spent rocket stages to crash into the Moon's surface." Using the S-wave and P-wave data returned from the lunar seismic stations, scientists have been able to build a fairly complete picture of the Moon's interior.

The structure of the Moon is, in many ways, similar to that of Earth. Both bodies have a metal core that is mostly iron, with smaller amounts of elements such as oxygen and sulfur. The Moon may have a two-layer core like Earth's, with fluid iron in the outer zone and solid iron in the inner zone, but the Moon's magnetic field is weak, indicating that no dynamo process is ongoing in the core (**Figure 6.30**).

"The big difference between Earth and the Moon," says Lunine, "is the size of the core." The Moon's core constitutes only 20 percent of its radius. Earth's core constitutes a full 45 percent of its radius. "That fact immediately tells us something important,"

FIGURE 6.29 Moonquakes
Apollo 11 astronaut Buzz Aldrin is shown here beside a seismic detector placed on the Moon to record potential moonquake activity or impacts from asteroids on the Moon's surface.

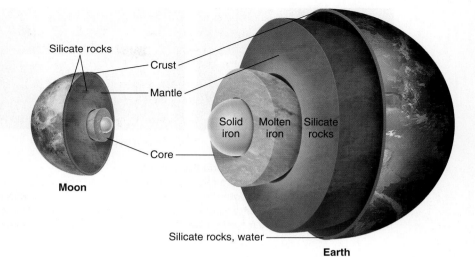

FIGURE 6.30

Comparison of Earth and Moon Interiors
Earth and the Moon have strikingly similar interiors (although the Moon's iron core is smaller proportionally and its outer core is only partially melted), suggesting that they originated from the same material and providing clues to the Moon's formation.

says Lunine. "The Moon is greatly impoverished in iron relative to Earth. It's a clue we can use to understand the Moon's history."

The Moon's mantle layer, like Earth's, is composed of rocks under high enough pressure for the mantle to slowly flow, but there are no longer large-scale circulation patterns in the Moon's mantle. Regular, though weak, moonquakes have been detected and they are believed to originate deep within the mantle. The quakes occur every month or so, indicating that the tidal stresses associated with Earth continue to affect the Moon's interior. The mantles of both the Moon and Earth are low in iron relative to what is found in primitive meteorites, suggesting that both Earth and the Moon differentiated as they formed.

Above the lunar mantle lies a rigid layer of crust. Unlike Earth's crust, where the ocean floors and continents are composed of different materials, the Moon's crust shows no such difference. "On the Moon," says Lunine, "there's just a single type of crust, which probably has not changed much since its original formation." The Moon's crust is quite thick relative to its size, making up 3 percent of its total radius. Earth's crust, in contrast, makes up just 0.6 percent of the planet's radius. In addition, Earth's crust has been shaped by plate tectonics and, on the surface, by weathering, while the Moon's crust has been shaped primarily by volcanism and impacts.

SECTION SUMMARY
Inside the Moon

● The Moon has a two-layer metal core, a mantle, and a crust, but it has no atmosphere.

● Compared with Earth's core, the Moon's core accounts for a much smaller percentage of its volume.

● The Moon's crust is much thicker relative to its size than Earth's, but it has no tectonic plates.

Making Earth's Moon

"People have been arguing about the origin of the Moon for centuries," says Lunine. "It was only after the *Apollo* Moon landings, though, that the story finally became clear." Rock samples returned by both American and Russian missions could be chemically analyzed, giving scientists the hard data they needed to track the Moon's history all the way back to its formation.

One of the earliest models for the Moon's formation was the so-called **capture hypothesis**, according to which the Moon formed somewhere else in the Solar System. "The idea was that, somehow, the Moon made a close approach to Earth and was gravitationally captured," explains Lunine. A second idea, called the **fission hypothesis**, states that a very young and still-molten Earth suffered a close encounter with another body that pulled material into orbit, which formed the Moon. "People imagined pulling out a piece of the molten Earth by tidal forces the way you would pull on taffy," says Lunine. "Eventually a molten blob split off from Earth and solidified to form the Moon." Other ideas, such as the Moon condensing from material orbiting around the young Earth, have also been proposed. Unfortunately, though each of these ideas

capture hypothesis
A hypothesis of the Moon's origin stating that the Moon formed elsewhere and was captured intact by Earth's gravity.

fission hypothesis
A hypothesis of the Moon's origin stating that the Moon formed from material pulled into orbit from the still-molten Earth by a close encounter with another body.

 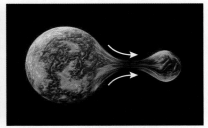

A Capture hypothesis   B Fission hypothesis

FIGURE 6.31 Hypotheses for Moon Formation
Two early explanations for how Earth acquired the Moon were the capture hypothesis and the fission hypothesis. (A) According to the capture hypothesis, Earth's gravity snared a passing Moon-sized object. (B) According to the fission hypothesis, rotation pulled material away from a molten Earth to form the Moon.

impact hypothesis The currently accepted hypothesis of the Moon's origin, which states that a large, Mars-sized body striking Earth ejected mantle material into orbit, which gravitationally coalesced into the Moon.

FIGURE 6.32
The Impact Hypothesis
The generally accepted explanation for the formation of the Moon is the impact hypothesis: a collision of a Mars-sized body with Earth caused a segment of Earth (crust, mantle, and a small amount of core) to be torn away. This material then formed a ring around Earth, which eventually coalesced into the orbiting Moon.

explains some aspects of the Moon, none of them successfully account for all the data (**Figure 6.31**).

One persistent problem for the capture hypothesis is that it's very hard for two bodies to be arranged in such a way that makes gravitational capture possible. "You can't just have the Moon wander in and go into orbit," explains Lunine. "Even trying to get a spacecraft to drop into orbit around another planet requires a lot of fuel to fire rocket engines in just the right way. In general, for an object the Moon's size, a passing body either collides or just keeps going." The fission hypothesis is also inconsistent with scientists' understanding of Earth's history, as our planet was probably never fully molten. "Early on there may have been what could be called a magma ocean covering the surface," says Lunine. "But Earth became essentially solid a short time after its formation."

Despite the difficulties, each theory on its own provides a piece of the puzzle that the modern theory of lunar origin has embraced. "It turns out that each of these simpler models got a part of it right," explains Lunine. "They were all just too oversimplified."

The model that is generally accepted today arose from two striking clues provided by the lunar missions. The first clue was that the rocks on the Moon contained very little iron compared with what is found on Earth. The second clue came from lunar seismic studies showing that the Moon's core accounts for a much smaller fraction of total volume than does Earth's core. Together these data pointed to a Moon that was seriously depleted in iron. Lunar rocks were also low in elements, such as potassium, that are often found in Earth minerals. The lack of lunar potassium was an important discovery because potassium's low melting temperature means it can be considered a volatile material that can be easily vaporized. The lunar samples did, however, show some chemical similarities to Earth rocks. The relative abundance of oxygen isotopes—forms of oxygen with more or fewer neutrons in the nucleus—was the same in Moon rocks as in Earth rocks.

After several years of analysis of the *Apollo* samples, scientists were able to find a coherent story that fit these data. "If you struck Earth with a big object," explains Lunine, "you can imagine gouging out a piece of the mantle after Earth's core had already differentiated. That meant the impact would miss most of the iron core." Mantle material blown off the planet by the impact would be very hot, so the volatile elements like potassium would have quickly evaporated away. "The debris that you're left with would then have the right chemistry," explains Lunine. "It would be poor in iron, poor in volatile elements, but unaltered in terms of oxygen isotope ratios." In the final chapter of the story, some of the mantle material released by the giant impact entered orbit around Earth and gravitationally coalesced into a young Moon. This basic idea, called the **impact hypothesis** (**Figure 6.32**), accounts for much of what is known about lunar structure and chemistry.

One critical problem with the impact hypothesis is to figure out what kinds of impacts could make the idea work. Almost all impacts either cause the material to be ejected, leaving the orbit of Earth, or cause most of the material to fall back on Earth. "It has not been trivial to find out what would and would not work," says Lunine. "But a number of people using very elaborate computer models have proved that the right kind of glancing blow of a body about the size of Mars would get enough material into Earth's orbit to form the Moon."

The acceptance of the impact hypothesis represented an important turning point in not only lunar but all planetary science. "By seeing the Moon created as the result of an impact," explains Lunine, "scientists also recognized that impacts were a crucial and even a dominant process occurring across the entire early Solar System."

SECTION SUMMARY
Making Earth's Moon

- Several hypotheses have been proposed for how Earth acquired the Moon: the capture hypothesis, the fission hypothesis, and the impact hypothesis.
- The Moon's small iron core and lack of iron in its crust support the impact hypothesis.

●→ chapter summary

6.1 Discovering Change: Crocodiles in the Arctic

The relative sizes of Earth and the Moon are unusual for our Solar System, since most planets have much smaller satellites. The Moon's influence on Earth has played a large role in the evolution of life.

6.2 Earth Inside and Out

Earth's layers include the two-part core, the mantle, and the crust. All of Earth's layers below the crust (which is 60 km at its deepest) have been identified through the study of two kinds of seismic waves that propagate through Earth: pressure (P) waves and shear (S) waves. P-waves cause the expansion and compression of material in the direction of the wave and can move through solids and liquids. S-waves force material to oscillate from side to side and move only through solids. The mantle's rock is so hot and under such pressure that it flows at a rate of centimeters per year. Earth's crust floats on the mantle, and its movements, known as plate tectonics, are driven by convection in the mantle. The various plates continue to move and be reshaped, causing the formation of mountains and islands, volcanic activity, and earthquakes. Earth's layered structure is evidence that our planet differentiated early in its history.

6.3 Earth's Near-Space Environment

Earth's atmosphere is primarily nitrogen and oxygen. Its layers, from the planet's surface upward, are the troposphere, stratosphere, mesosphere, and ionosphere. Their density decreases with altitude, but temperature variations are more complex. Earth's current atmospheric composition originated with contributions from incoming comets and planetesimals. Its high oxygen content originated with the earliest life-forms. Much of the Sun's energy directed toward Earth passes through the atmosphere. Some is reemitted outward, but some is absorbed by greenhouse gases, notably carbon dioxide, water vapor, and methane. This greenhouse effect causes Earth's temperature to be higher than it would be without an atmosphere. Scientists conclude that the increase of greenhouse gases since the industrial revolution is heating Earth's average temperature and driving climate change in ways that are certain to affect Earth's population. Earth has a magnetic field created by the combination of its rotation and its liquid core, as well as a magnetosphere that is shaped by the solar wind and protects the planet from much of the harmful radiation and high-energy particles. Particles that do reach Earth create auroras and can have harmful effects such as power grid and satellite failures.

6.4 The Closest Desolation: Earth's Moon

Tidal forces are caused by the difference in gravity from a distant mass experienced from one side of an extended body to the other. On Earth, tidal forces from the Moon create a tidal bulge that is observed primarily as daily cycles of ocean level changes, most visible at coastlines. The Sun also exerts tidal effects on Earth and either magnifies or diminishes tide levels based on the relative positions of Moon and Sun. Tidal effects also have lengthened Earth's rotational period and have locked the Moon's rotation period equal to its orbital period. The Moon's surface differs from Earth's because it lacks Earth's weathering and tectonic plate effects, so that craters, many of them created during the Late Heavy Bombardment period, remain highly visible. The near side of the Moon has smooth areas called maria (formed by lava flows), while the far side is more uniformly cratered. There is evidence that the Moon contains water delivered via comet and asteroid impacts. The Moon has the same general pattern of internal structure as Earth, and their relative sizes and compositions provide evidence of the Moon's formation. The currently accepted hypothesis is that a large planetesimal collided with Earth in a glancing blow, ejecting crust and some mantle but little or no core. The orbiting debris then coalesced into the Moon.

●→ questions and problems

Narrow It Down: Multiple-Choice Questions

1. Which of the following does *not* describe an effect that the Moon has had on Earth?
   a. limiting the wobble in Earth's spin
   b. lengthening Earth's day
   c. creating the tides
   d. stabilizing Earth's climate
   e. decreasing Earth's orbital speed

2. Which of the following items describe(s) features of Earth believed to be unique among the eight planets in our Solar System? Choose all that apply.
   a. presence of life
   b. liquid water on surface
   c. solid surface
   d. presence of one or more moons
   e. presence of an atmosphere

3. Where is solid material found in the layers of Earth's structure?
   a. crust only
   b. continents only
   c. crust and inner core
   d. crust and outer core
   e. mantle and core

4. In which of the following ways are Earth and the Moon similar? Choose all that apply.
   a. length of the orbital period around the Sun
   b. relative sizes of structural layers
   c. length of day
   d. geologic activity
   e. kinds of material that make up each world

5. The distance from New York City to Tokyo, Japan, is 10,787 km. If you traveled downward through Earth that distance, where would you find yourself?
   a. in the crust
   b. in the mantle
   c. in the liquid core
   d. in the solid core
   e. outside Earth

6. A forward surge in a crowd is an example of what kind of wave?
   a. S-wave
   b. P-wave
   c. both S- and P-waves
   d. neither S- nor P-wave
   e. It is not a wave at all.

7. When volcanoes erupt, the spewed material comes from which layer(s) of Earth? Choose all that apply.
   a. crust
   b. mantle
   c. molten core
   d. solid core
   e. ocean floor

8. Which of the following is/are considered evidence of tectonic plates? Choose all that apply.
   a. matching fossils in separated landmasses
   b. the shapes of the continents
   c. convection in the mantle
   d. observed motion of the continents
   e. distribution of hurricanes

9. Arrange the layers of Earth's atmosphere in the correct order of increasing altitude.
   a. troposphere, stratosphere, mesosphere, ionosphere
   b. ionosphere, mesosphere, stratosphere, troposphere
   c. stratosphere, troposphere, mesosphere, ionosphere
   d. troposphere, ionosphere, mesosphere, stratosphere
   e. stratosphere, mesosphere, troposphere, ionosphere

10. Earth's first atmosphere escaped because of
    a. the planet's low density.
    b. the planet's high temperature.
    c. lack of O_2.
    d. the high velocity of Earth's spin.
    e. violent winds.

11. The temperature at Earth's surface directly depends on which of the following? Choose all that apply.
    a. the amount of energy received from the Sun
    b. the presence of radio waves in the Sun's spectrum
    c. the concentration of CO_2
    d. the concentration of CH_4
    e. the presence of cosmic rays at the surface

12. The Moon's surface shows more cratering than Earth's because
    a. it was hit more than Earth because of its location.
    b. it formed earlier than Earth, when more debris was present in the Solar System.
    c. its surface is softer and thus less resistant to cratering.
    d. the forces that resurface Earth do not exist on the Moon.
    e. Earth deflected Solar System debris, which then bombarded the Moon.

13. Which of the following features may occur when an object impacts a world, causing a crater? Choose all that apply.
    a. rays
    b. a peak
    c. terracing
    d. a surface fracture
    e. dust in the atmosphere

14. Which of the following statements regarding magnetic fields is/are true? Choose all that apply.
    a. Earth's North Pole has always been the magnetic north pole.
    b. The presence of a magnetic field implies the presence of a solid iron core.
    c. Any large, rotating body will have a magnetic field.
    d. Earth's magnetic field protects it from the solar wind.
    e. Earth's magnetosphere is symmetric.

15. Which of the following statements about the Moon is true?
    a. The side of the Moon not visible from Earth is always dark.
    b. The two hemispheres of the Moon (near and far) are symmetric in structure and similar in appearance.
    c. Maria appear only on the near side of the Moon.
    d. Most of the Moon's craters have disappeared because of weathering effects.
    e. An impact crater always takes the exact shape of the object that created it by striking the Moon.

16. The strongest tides occur at what Moon phase(s)? Choose all that apply.
    a. first quarter
    b. waxing crescent
    c. full Moon
    d. third quarter
    e. new Moon

17. If the rate of change in the length of Earth's day remains constant at its current level, what will its day length likely be 100 million years from now?
    a. 24 hours
    b. a little longer than 24 hours
    c. a little shorter than 24 hours
    d. many times longer than 24 hours
    e. The value cannot be determined from the available data.

18. If Earth were tidally locked with the Sun but the Moon's orbital period remained as it is now, what would be the frequency of high lunar tides experienced on Earth?
    a. once per year
    b. twice per year
    c. once per lunar orbit
    d. twice per lunar orbit
    e. The frequency cannot be predicted from the information given.

19. True/False: If the Sun were cooler but Earth's temperature was the same, on average, as it is today, a weaker greenhouse effect could be the cause.

20. True/False: The greenhouse effect has been detrimental to life on Earth throughout its evolution.

To the Point: Qualitative and Discussion Questions

21. Describe the two kinds of seismic waves.

22. How do scientists use seismic waves to "see" inside worlds?

23. What evidence indicates that continental drift continues today?

24. If there were no convection in the mantle, how might Earth's geologic history have differed?

25. What are the possible outcomes when two tectonic plates interact?

26. How does Earth's magnetic field protect life on the planet?

27. How does Earth's Moon differ from the moons around other planets?

28. If Earth were the size of a basketball (roughly 75 cm, or 30 inches, in circumference), what objects might approximate the size of the Moon?

29. How long is a day on the Moon? In other words, how long does one cycle of day and night last?

30. Compare the layers of Earth and the Moon. How are they similar, and how do they differ?

31. What is required for a world to have a magnetosphere? Describe the shape of Earth's magnetosphere.

32. How does an aurora occur?

33. What does the degree of cratering tell astronomers about the timing of events that shaped the Moon's surface?

34. What are the three hypotheses regarding how Earth acquired the Moon? Describe the model that is generally accepted today as the valid one.

35. What are some of the arguments presented by those who believe climate change cannot be created or influenced by humanity?

Going Further: Quantitative Questions

36. Gravity decreases with the square of the distance between objects. How much greater is the force of gravity by the Moon on the side of Earth closer to it versus the opposite side of Earth? Use an average distance of 384,400 km from the Moon's center to Earth's center, and a diameter for Earth of 12,756 km. Give your answer as a percentage.

37. The escape velocity for Earth's Moon is 2.38 km/s. Calculate the escape velocity, in kilometers per second, for a moon with radius 2,173 km and mass 1.1×10^{23} kg.

38. The moon of a distant planet has a mass 2 times that of Earth's Moon and a radius 3 times that of Earth's Moon. What is its escape velocity, in kilometers per second?

39. By what factor would the gravitational force between Earth and the Moon be greater if the mass of each body were twice as great and the distance were half as great as they are today?

40. Assume that a rock sample initially contained 5,000 U-238 nuclei 6.75 billion years ago. How many lead-206 nuclei would it contain today?

41. A rock sample is determined to have a concentration of uranium-238 (U-238) that is consistent with 0.5 half-life having passed. What percentage of U-238 remains, and how old is the rock?

42. An analysis of a rock sample indicates that 12.5% of the expected concentration of U-238 remains. How many half-lives have passed, and how old is the rock?

43. A planet has a mass twice that of Earth and a radius 1.5 times Earth's. What is its escape velocity? Is this planet likely to have retained hydrogen in its atmosphere if the temperature is 350 K?

44. Would oxygen be retained in the atmosphere of a planet whose surface temperature was 250 K and whose escape velocity was 4.6 km/s? (Molecular oxygen is 32 times heavier than a hydrogen atom.)

45. What is the thermal velocity of nitrogen in the atmosphere of a planet whose surface temperature is 380 K? Would nitrogen be retained if the planet had 3 times the mass of Earth and 4 times Earth's radius?

 If your instructor assigns homework in **smartwork5**, access it at the Digital Landing Page for *At Play in the Cosmos*: **digital.wwnorton.com/cosmos**

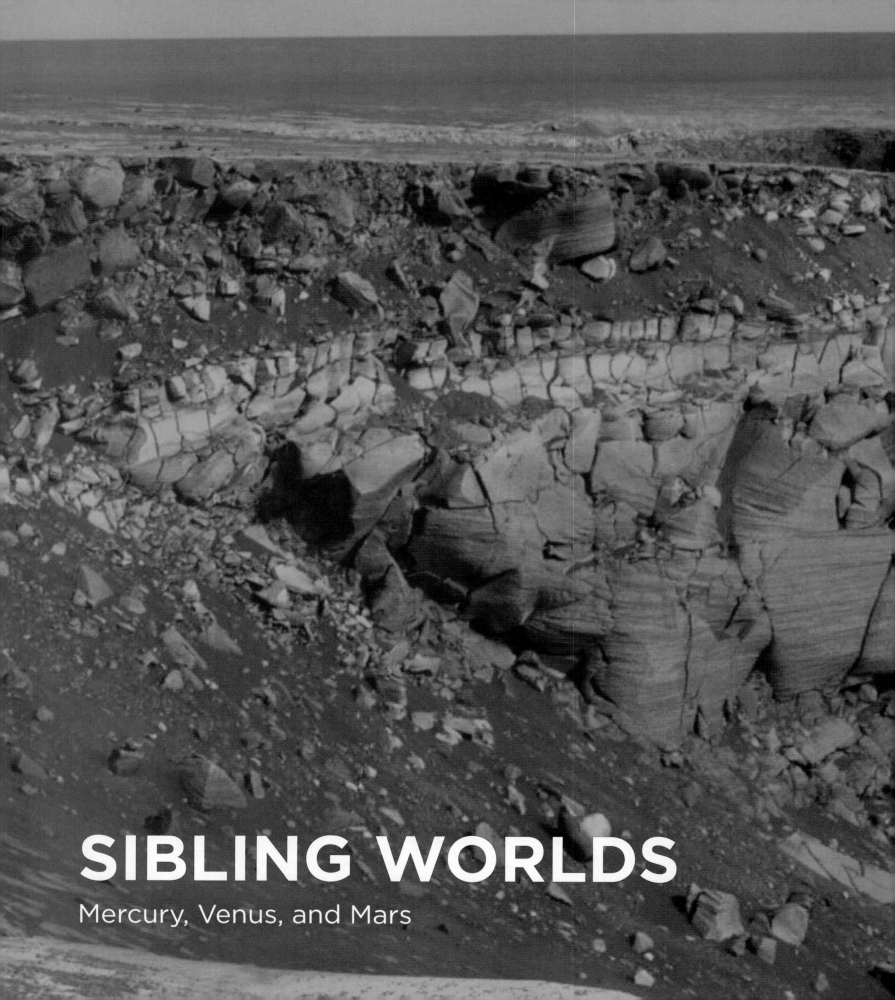

SIBLING WORLDS

Mercury, Venus, and Mars

In this chapter you will learn about the three terrestrial planets other than Earth and how they compare to our own planet. After reading through each section, you should be able to do the following:

7.1 Planet Stories
Explain how scientists use comparative planetology in their study of the terrestrial planets.

7.2 Mercury: Swift, Small, and Hot
Describe the characteristics and evolution of Mercury and how it has been shaped by its close proximity to the Sun.

7.3 Venus: Hothouse of the Planets
Describe the characteristics and evolution of Venus and how its atmosphere differs from those of Earth and Mars.

7.4 Mars: The Red Planet of Change
Describe the characteristics and evolution of Mars, and summarize the findings of the search for water and life on its surface.

Earth or Mars? As rovers have explored the surface of the Red Planet they have found landscapes that look remarkably like the deserts of our world. This image shows Cape Saint Vincent, one of the cliffs of the Victoria crater. This image was taken by the *Opportunity* rover.

7

7.1
Planet Stories

thermal history
The evolutionary pattern of heating and cooling for a planet.

FIGURE 7.1
Mars Rover
Approximately 6 years after the rovers *Spirit* and *Opportunity* landed on Mars, (A) a larger rover named *Curiosity* was safely landed and began exploring the planet. (B) A landscape image of the Martian surface captured by *Spirit*.

A

B

"ALL THAT WORK and it comes down to five minutes of hell," says Jim Bell of Arizona State University. Bell was one of the scientists who developed high-resolution stereo cameras for NASA's wildly successful *Spirit* and *Opportunity* Mars rovers. "The rover project started 10 years before the landings, and it was an enormous task. We knew whatever we built would have to survive the shock of launch, the shock of landing, and the harsh Martian environment." The camera development was just one part of building the complicated rovers.

On June 10, 2003, the Boeing Delta II rocket carrying *Spirit* lifted off the pad at Cape Canaveral. Seven months later, the truck-sized capsule that housed the rover dove into the Martian atmosphere. "We were all there in mission control as *Spirit* started its descent," recalls Bell. "It was crazy and exhilarating and terrifying all at the same time. You have a decade of your life wrapped up in this project. At that moment, you think, 'Oh my God, did I waste 10 years and it's all about to come to nothing?' And then a second later you're thinking, 'Oh, my God, I'm about to get a view of this world nobody's ever seen before!'"

Bell did not have the luxury of facing his fear and excitement alone in the control room. "You are going through all this and there's CNN, MSNBC, and all the other news crews with the cameras trained on you. The whole blasted world was watching over our shoulders live to see if we screwed up."

But they didn't screw up. Both rovers survived their landings (*Opportunity* arrived in January 2004) and almost immediately began sending images from Mars. The rover mission, which was planned to last a mere 3 months, turned into a years-long exploration that, as of this writing, still continues. In 2012, *Opportunity* was joined on the Martian surface by the rover *Curiosity*, which had its own harrowing high-speed 7-minute ride through the atmosphere to nail a perfect landing (**Figure 7.1**).

The rovers that Bell helped build represented another step in humanity's ongoing exploration of the Solar System over the last 50 years. After centuries of questions about the planets, we finally have enough answers to begin a true science of comparative planetology.

Comparative Planetology: Data, Distraction, and Focus

The overarching goal of the Mars rover program was simple: Follow the water. Earlier Mars missions had revealed tantalizing hints that Mars, a barren, frozen desert world today, might once have been a water world with streams, rivers, and perhaps oceans on the surface. Since life as we understand it requires the presence of liquid water, NASA has used the hunt for H_2O as the *guiding principle* in developing new Martian orbital satellites and the rovers.

To begin our own study of comparative planetology, we must also establish some guiding principles. Chapter 6 explored Earth and the Moon. This knowledge of our own world and its satellite provides a foundation to ask about the other planets in our Solar System: How are they like Earth? How are they different? If we're not careful, though, we will quickly become lost in an ocean of facts. Which of these facts are important, and how do they relate to one another? What questions and guiding principles will enable us to make the best sense of the data?

In light of the recent discovery of planets orbiting other stars, one of the first questions and guiding principles for organizing our exploration of the Solar System might be a focus on evolution: What do the planets in our Solar System tell us about general principles of the origins and evolution of planetary systems, including our Solar System? A second question and guiding principle might be how well each planet in our Solar System would fare as a home for life. Extending this inquiry into the past and into the future, we can ask whether, at any time in the past, conditions on a certain planet could have supported life. Similarly, we can look forward across the next few centuries and imagine which planets might be best suited for building permanent bases or colonies for human habitation. The presence of water in any accessible form is an obviously related question.

This chapter focuses on the inner "rocky" worlds of the Solar System: Mercury, Venus, and Mars. The discussion of each planet proceeds in four sections, starting at the center and moving outward:

1. *Interior structure.* What does the internal structure of the planet tell us about its origins and evolution? Does it have a core? Does it have a mantle?

2. *Surface evolution.* What surface features exist on the planet today, and what do they tell us about the planet's origin and evolution? What do they tell us about the planet's suitability for life? Is there volcanic activity? Is there plate tectonic activity? Is there evidence for water on or below the surface?

3. *Atmosphere.* Does the planet have an atmosphere? If so, how has it evolved over time, and how have the changes affected the surface? Was the atmosphere ever conducive to life?

4. *Near-space environment.* Does the planet have a magnetic field? Does the planet have any satellites? Is there anything about the planet, such as its location or the shape of its orbit, that has affected its evolution?

With these questions and our guiding principles in hand, we are ready to begin. But before we turn to any particular terrestrial planet, we can use what we learned from our study of Earth and the Moon to get a handle on the evolution of rocky planets in general.

─ **SECTION SUMMARY**
Comparative Planetology:
Data, Distraction, and Focus

● Guiding principles help to focus and direct research questions, as well as to gather and analyze data.

● In keeping with the guiding principles of this chapter—planets' origins, evolution, and suitability for life—we will consider each planet's internal structure, surface evolution, atmosphere, and near-space environment.

The Five Stages of Terrestrial Planetary Evolution

Since the 1970s, increasingly sophisticated data-gathering techniques have helped scientists understand that all the terrestrial planets must have shared certain key phases in their evolution. "The inner planets all condensed out of the hotter inner parts of the protosolar disk," says Jim Zimbelman, a planetary scientist in the Center for Earth and Planetary Studies at the Smithsonian National Air and Space Museum. "That means they ended up with silicates and metals rather than ices and gas. They were dense rocky worlds, and that shaped their future development in ways that led to very different evolution from that of planets further out."

The early years of a planet's life (much like those of a human being's life) have a huge impact on its subsequent history. "The first billion or so years of a planet's life set the stage for much of what is to come," says Zimbelman. "One thing we know from basic principles is that small planets lose their heat faster than big planets, and a planet's **thermal history** drives its whole geophysical and geochemical evolution" (see **Going Further 7.1**). According to scientists, the

GOING FURTHER 7.1   THE LANGUAGE OF THE COSMOS
The Big Chill: The Rate of Planetary Cooling

One of the most important factors governing a planet's evolution is size, which determines how rapidly the planet cools. Recall from Chapter 4 that all dense objects radiate heat into space as blackbodies. Thus, planets are continually losing their internal heat. Since this internal heat ultimately powers a planet's volcanism, its plate tectonics, and its dynamo-generated magnetic field, any significant cooling of a planetary interior will mean an end to a planet's further evolution.

How can we determine more accurately how fast one planet will cool relative to another? A planet's store of internal heat depends on, among other things, its volume. The more volume a planet has, the more total heat energy it holds. A planet's volume is the volume of a sphere:

$$V = \tfrac{4}{3}\pi R^3$$

However, the rate at which heat flows out of the planet depends not on its volume, but on its surface area, since the surface is the only place the heat can escape. The surface area of a sphere is

$$A = 4\pi R^2$$

This gives us a way of thinking about the cooling time for the planet, which is the characteristic time for it to lose most of its interior heat. The total heat (which depends on the volume) divided by the rate of heat escaping (which depends on surface area) tells us how long before the planet cools down. We can write this as

$$t_{cool} \propto \frac{\tfrac{4}{3}\pi R^3}{4\pi R^2} \propto R$$

So cooling time depends on the planet's radius, which makes sense. Smaller planets will cool faster than larger planets. Notice that t_{cool} is *proportional* to volume over area (that's what the $\propto$ symbol means). To write out the full expression for the cooling time, we would include factors related to the composition of the planet and its temperature. But for our purposes here, all we need to see is how cooling time, a planet's volume, and its surface area are related, because that relation alone shows us that cooling time is proportional to the planet's radius.

To see this explicitly, take the ratio of the cooling times for two planets, a large planet A and smaller planet B, with radii $R_A > R_B$:

$$\frac{t_{cool}(A)}{t_{cool}(B)} = \frac{R_A}{R_B} > 1$$

What we now see is that bigger planets (like planet A) will always have longer cooling times relative to smaller planets (like planet B). So we should expect that small planets cool down more quickly than big ones.

FIGURE 7.2
Evolution of a Terrestrial World
Astronomers have identified five distinct stages of evolution for the terrestrial planets. (A) During differentiation, denser material sinks to the planet's core. (B) During cooling, heat from the planet's interior escapes into space. (C) During cratering, impacts with asteroids and comets shape the surface. (D) During the flooding stage, magma from below and substances such as water fill in lowlands. (E) During weathering, surface features are eroded by winds and liquid flows.

Heavier material sinks.
Lighter material rises.

Proto-core forming

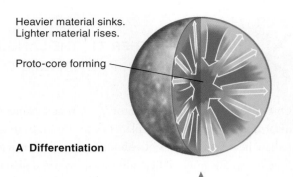

A Differentiation

Heat flows from core outward and escapes at surface.

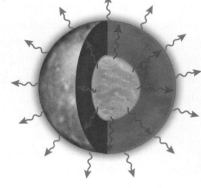

B Cooling

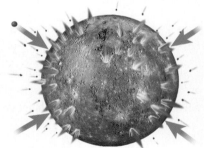

C Cratering

Magma, water, or other liquids fill in the lowlands.

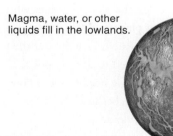

D Magma flooding

E Weathering

planetary differentiation The separation by gravity of a planet into different layers running from densest at the core to less dense in the outer regions.

planetary cooling The process by which a planet loses the heat acquired during its formation.

cratering The change in surface features of a planet due to impacts with comets, asteroids, and meteoroids.

evolution of the terrestrial worlds can be divided into approximately five stages (**Figure 7.2**).

As a planetary system forms, rocky planets are created by collisions between planetesimals. As these mountain-sized bodies collide to build up the planet's mass, the first key process, **planetary differentiation** (**Figure 7.2A**), occurs as material is melted first by the violence of collisions and then by the force of gravity squeezing material in the newborn protoplanet together. Heavy elements, such as iron, sink to the center of the newly formed world, and lighter materials, such as silicates, rise toward the surface. Says Zimbelman, "This process of differentiation is why all the terrestrial planets have a core-mantle-crust structure" (**Figure 7.3**).

The second evolutionary phase is **planetary cooling** (**Figure 7.2B**). "A planet's internal heat flows up to the surface and escapes into space," explains Zimbelman. The process is similar to what happens when freshly baked bread cools after being taken out of an oven. "Some of the heat a planet has will be left over from its original formation. Some of it may be from the decay of radioactive materials. The process of differentiation itself—making a core—actually releases enormous heat. It takes time for all of that heat to work its way out as the planet cools." Since the timescale for cooling depends on the planet's size (longer for large planets, shorter for small ones), the cooling phase may overlap with subsequent phases in the planet's evolution.

The battering of a planet's surface by comet and asteroid strikes marks the third evolutionary phase, called **cratering** (**Figure 7.2C**). "The cratering story happens throughout the evolution of the planets," says Zimbelman, "but early on it is really intense." In our Solar System, most of the planetesimals that had not yet become part of newly formed planets were cleared away during the Heavy Bombardment phase of Solar System evolution. During this epoch, about 4 billion years ago, the gravity of the young planets in the Solar System swept up many planetesimals, leading to intense cratering. During the earliest phases of the planets' lives, the bombardment may even have kept the surface partially molten.

The Heavy Bombardment phase represents the period of most dramatic cratering, but even later large impacts will remain a force that can reshape planetary surfaces. The effects of these later impacts will be most dramatic if the planet is still hot enough to allow for volcanism and the existence of partially melted rock close to the surface. "The later impact phases weren't catastrophic like the Heavy Bombardment," says

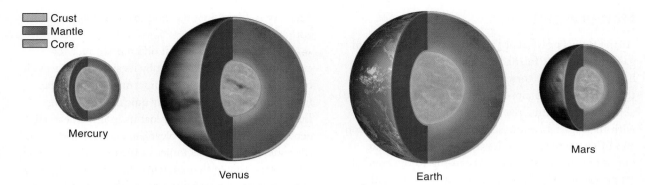

Crust
Mantle
Core

Mercury

Venus

Earth

Mars

FIGURE 7.3 Differentiation and Interior Structure
After differentiation, each terrestrial planet is composed of similar layers of increasing density from surface to center.
These are the core (which may have both molten and solid portions), mantle, and crust. The relative sizes (in radial extent)
of these layers differ for each planet, depending on the composition of material from which each planet was formed.

Zimbelman of our own Solar System's history, "but there was a period when collisions could form big basins that were later flooded by lava. On the Moon this led to the maria we see today, but it happened on other worlds too." During this **magma flooding** phase (**Figure 7.2D**), smooth lowlands can be created after asteroid collisions or through volcanism. And if significant amounts of water (or other liquids) exist on the surface, a second kind of flooding creates oceans.

The last evolutionary phase that occurs on planets with atmospheres (and perhaps oceans) is the slow process of weathering (**Figure 7.2E**). Over time, the flow of atmospheric gases, as well as the condensation and evaporation of those gases into ice or water, will alter surface features on the planet. Weathering can both erase older features, including craters, and create new ones, such as extended regions of sand dunes.

┌ **SECTION SUMMARY**
│ **The Five Stages of Terrestrial Planetary Evolution**
● Scientists believe that the terrestrial planets all experienced five basic stages in their history: planetary differentiation, planetary cooling, cratering, magma flooding, and weathering.

7.2
Mercury: Swift, Small, and Hot

The history of Mercury (**Figure 7.4**), the innermost planet in the Solar System, has been shaped by two important facts. First, its 88-day, 0.39-astronomical-

unit (0.39-AU) orbit means that the Sun has exerted a dominant influence on its evolution. Second, with a radius of 2,440 kilometers (km), Mercury is a small world. It is just 38 percent the size of Earth and 1.4 times the size of the Moon. In many ways, Mercury is like the Moon's larger cousin. Because of their relatively small sizes, the two worlds have experienced similar histories, and they are equally inhospitable to life.

interactive
Anatomy of Terrestrial
Planets

FIGURE 7.4 Mercury's Surface
This *Messenger* spacecraft image of Mercury shows a heavily cratered surface.

magma flooding The creation of smooth "lowlands" via lava flows that fill in craters.

Mercury's Interior

At the radius of Mercury's orbit, the original disk of gas and dust that formed the planets would have been very hot. The torrent of solar radiation tearing through the gas so close to the young star would have kept most elements in a vapor state. Only material with the highest condensation temperatures, like iron and other metals, would have been able to "freeze out" and form solid grains. According to the condensation theory of planet formation, Mercury would therefore be expected to have a high percentage of metals relative to planets more distant from the Sun. And in fact, measurements of Mercury's density yield 5,440 kilograms per cubic meter (kg/m^3), a value so high that the planet must have a large iron core. However, because Mercury's iron content is so high that even the condensation theory cannot account for it, scientists have theorized that Mercury must have collided with a metal-rich body early in its history. "It is certainly possible that Mercury was subject to a large impact," says Jim Zimbelman. "We've been finding evidence of really big whacks in other places, and we've realized that these large impacts are more prevalent and more important to the overall history of the terrestrial objects than was previously believed."

The delivery of the extra metals left Mercury, after differentiation, with an iron core that takes up a far larger fraction of its volume (42 percent) than the core of any other terrestrial world. Earth's iron core, for example, accounts for only 15 percent of its volume.

Mercury is, for all intents and purposes, a cannonball with a rocky crust.

After its formation and differentiation, Mercury's internal history was set by its size. Being a small object, Mercury radiated the internal heat from its formation into space relatively quickly (see Going Further 7.1). Once Mercury lost its internal heat, all geologic activity such as volcanism ended, and there is no evidence of plate tectonics in the planet's history. Scientists estimate that Mercury has been geologically dead for more than 4 billion years. The loss of heat so early in its life led to a contraction of the planet. As the inner zones shrank, the crustal material was compressed, squeezing shut fissures and other paths for molten material to make it to the surface. Thus, while the Moon and Mercury bear many surface similarities, the extensive lava flows that occurred on the Moon did not occur on Mercury.

SECTION SUMMARY
Mercury's Interior

- Mercury's large iron core is consistent with condensation theory but may suggest other factors at work, such as a collision with another iron-rich body.
- Mercury's small size meant early loss of heat, and the planet is believed to be geologically inactive.

Mercury's Surface

Having lost its heat and its volcanism early, Mercury retains a fairly pristine record on its surface of encounters with Solar System debris. The surface shows intense cratering, and in many ways the entire planet looks much like the jagged highlands of the Moon without its broad maria. One of Mercury's most prominent features is the Caloris Basin, a giant impact crater stretching 1,400 km (**Figure 7.5A**). The force of the impact that formed the Caloris Basin was so great that the planet's surface rippled like the surface of a pond after a heavy stone is thrown in. Rings of mountains 1,500 km in radius and as high as 3 km surround the impact crater (**Figure 7.5B**).

The one feature of Mercury's surface that is unique to this inner terrestrial world is the long lines of cliff systems that run for hundreds of kilometers. Discovery Rupes, for example, is an escarpment on Mercury approximately 650 km long (**Figure 7.6**). These steep cliffs run straight across craters, so they must have formed after the Heavy Bombardment period. Since Mercury has no plate tectonics, the best explanation for what caused the ridges to form is the cooling and shrinking of the

FIGURE 7.5
Caloris Basin
(A) Caloris Basin, at 1,500 km long, is one of the Solar System's largest impact basins. This *Messenger* image uses colors to enhance the chemical, mineralogical, and physical differences between the rocks on Mercury's surface. (B) The mountains surrounding Caloris Basin probably formed from the ejecta of the powerful impact that created the crater.

A Caloris Basin

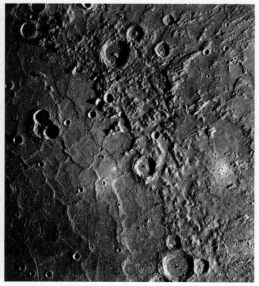

B Mountains

planet's interior. When Mercury's interior shrank, its skin (the crust) had to wrinkle, fold, and collapse to adjust, like the skin of a dried fruit. "The existence of the cliffs gets us back to the size of the core and the speed of Mercury's cooling," says Zimbelman. In going from a liquid to a solid, the volume of the core shrinks a lot when it freezes. The shrinkage forces so much 'crustal shortening,' or wrinkling of the crust, that the result is those scarps.

┌─ **SECTION SUMMARY**
Mercury's Surface
● Mercury's surface features include dense cratering, signs of very large impacts, and long ridges and cliffs caused by rapid cooling and shrinking.

Mercury's Atmosphere

Mercury has been visited just twice by spacecraft. In 1973, *Mariner 10* made three sweeps past Mercury, photographing its surface and using other instruments to probe conditions on the planet. In 2011, the *Messenger* spacecraft entered orbit around the innermost terrestrial world.

Mercury does not have an atmosphere. Although it does manage to hold on to a thin layer of hydrogen and helium, this layer is simply solar wind particles that Mercury's gravity can only temporarily trap. Eventually even this thin blanket of gas escapes back into space. The reason for Mercury's inability to retain an atmosphere can also be traced back to its small size and its proximity to the Sun. Going Further 6.2 showed how the ability to retain atmospheric particles depends on a planet's gravity and temperature, as well as on the mass of the particles. Mercury's low mass means that its escape velocity is also low (you could calculate it yourself right now). Its proximity to the Sun means that its surface temperature (on its daytime side) rises as high as 700 K. Such enormous temperatures drive even the heaviest atmospheric atoms to speeds high enough that they escape into space.

It is worth noting, however, that the temperature at Mercury's poles, where the sunlight glances off the planet's surface, can be 125 K, far below freezing (273 K). Therefore, craters near Mercury's poles could contain frozen water, just as deeply shadowed craters near the Moon's poles are known to retain large quantities of water ice.

┌─ **SECTION SUMMARY**
Mercury's Atmosphere
● Because of extreme heat and low gravity, no atmosphere was retained on Mercury, but the poles could contain frozen water.

Jim Zimbelman

Jim Zimbelman's interest in planets began early. "When I was in middle school, I was interested in all sorts of 'sciency' kinds of things," says Zimbelman. "But the turning point came when my dad got a small telescope. That's what really got me hooked specifically in astronomical things. But what I really loved best of all were the planets."

Zimbelman's childhood passion stayed with him from middle school through his undergraduate years. But by the time he started graduate work, he was in for a surprise. "When I got to UCLA to start a master's program in astronomy, I discovered that all of the classes I wanted were taught in a different department," Zimbelman recalls. In the late 1970s, when Zimbelman began his graduate education, astronomy and planets were parting ways. Astronomers studied stars and even more distant objects, like galaxies. If students like Zimbelman wanted to study something as nearby as the planets, they needed to take classes in geology. "I had to transfer my application from an astronomy PhD to an earth and space sciences PhD." Finishing his PhD in 1984, Zimbelman went on to play a significant role in the study of volcanism on all the terrestrial planets.

Zimbelman, a former chair of the Smithsonian Center for Earth and Planetary Studies, attributes his broad perspective on celestial science to his family background. "One thing I appreciated about my father was while he was a minister, he was open to listening to discussions of things." This open-mindedness has always been an encouragement to Zimbelman in his scientific studies of the sky.

Mercury's Near-Space Environment

Mercury supports a weak magnetic field. The strength of Mercury's magnetism is only one ten-thousandth that of Earth (in other words, the ratio of Mercury's to Earth's field strength is 10^{-4}). With such a large iron core, Mercury might be expected to host a strong magnetic field. Mercury's core, however, solidified 4 billion years ago, so no dynamo like Earth's is possible now. Mercury's fields are most likely remnants of **fossil magnetism** (magnetic fields imprinted in the planet's rocks and minerals) left over from the period when the iron core was still melted and generated its own field. Currently, Mercury shows a dipole magnetic field whose center is not located at the planet's center. Instead, the center of Mercury's magnetic field is located north of the center along the planet's rotational axis.

Mercury's proximity to the Sun has affected its evolution in ways even more profound than driving its temperatures to such extremes. Mercury is close enough to the Sun that its rotation period has been strongly modified by the Sun's gravity. If Mercury were in a circular orbit, the Sun's tidal effects would have locked its rotation to its 88-day orbital period; that is, Mercury would always present the same face to the Sun, as the Moon does

fossil magnetism Magnetic fields imprinted in a planet that were created by an earlier active dynamo phase.

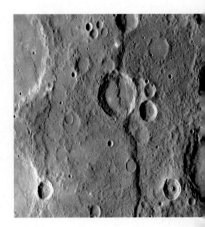

FIGURE 7.6
Discovery Rupes
Discovery Rupes is a dramatic line of cliffs produced by the shrinkage that occurred as Mercury cooled. In places, these cliffs run across craters, indicating that the cliffs must have formed after the impacts.

to Earth. The real situation is a bit more complicated. Mercury's orbit is fairly elliptical by our Solar System's standards, with an eccentricity of $e = 0.206$. Mercury's highly eccentric orbit means it has a larger variation in orbital velocity and distance as it swings in close to the Sun (perihelion, at 0.31 AU) and swings farther away from the Sun (aphelion, at 0.47 AU). These swings make simple tidal locking impossible. Instead, Mercury ends up in a state of **tidal synchronization**, with its spin modified to present the same face to the Sun every *other* orbit. Thus, Mercury's solar day (the time between two successive noons) lasts two full orbits, or 176 days. The time it takes to spin once around on its axis, however, is just two-thirds of an orbit, or 59 days (**Figure 7.7**).

With strong tidal influence from the Sun; an airless, heavily cratered surface; and long-extinct volcanism, Mercury does indeed appear to be a larger cousin to Earth's Moon.

tidal synchronization The modification of the spin rate of a satellite (such as a planet or a moon) by the gravity of the body it orbits such that the satellite's rotation and orbital periods are synchronized (for example, the satellite completes two rotations for every three of its orbits).

FIGURE 7.7
Mercury's Orbit and Rotation
Because Mercury's orbit is fairly eccentric, tidal forces from the Sun have not produced tidal locking. Thus, Mercury spins on its axis every 59.5 days, whereas its orbit is 88 days. This means there are three rotations for every two orbits—an example of tidal synchronization. As a consequence, it takes two orbits for the Sun to be overhead at the same spot on Mercury.

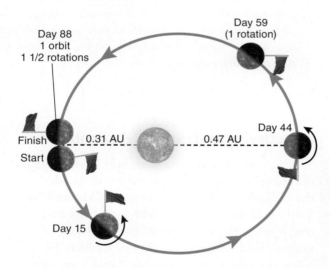

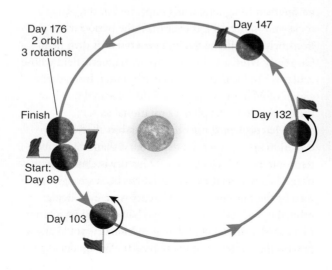

SECTION SUMMARY
Mercury's Near-Space Environment
- Mercury's small magnetic field is seen as a fossil from its early history, when the planet had a molten core supporting a magnetic field–generating dynamo.
- Mercury's elliptical orbit has caused its rotation period to be tidally synchronized, but not tidally locked, with the Sun.

7.3
Venus: Hothouse of the Planets

As we take our next step outward from the Sun and our next step deeper into comparative planetology, we encounter Venus, which should be a slightly less massive cousin of Earth. Exploration of Venus, however, has shown how two planets that began from very similar places have ended up leading very different lives.

In terms of the most basic planetary characteristics, Venus and Earth are near-twins. Venus (**Figure 7.8**) has a radius of about 6,050 km, 95 percent the size of Earth. Its mass is 4.87×10^{24} kg, which is about 82 percent that of Earth. Together these numbers give Venus an escape velocity of 10.3 kilometers

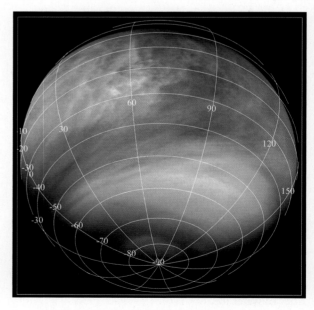

FIGURE 7.8 The Clouds of Venus
In this ultraviolet image of Venus, taken by the European Space Agency's *Venus Express* probe, the structure of the planet's clouds is visible. Latitude and longitude lines are imposed on the image to show location of poles and equator.

per second (km/s), 92 percent of Earth's. Venus's density is 5,240 kg/m³, and its uncompressed density is 4,400 kg/m³. Both these values are also very close to the values for Earth. Venus is closer to the Sun than Earth is, with an average orbital radius of 0.72 AU. That 28 percent difference in distance (Earth orbits at 1 AU) means that Venus receives almost twice as much solar energy than Earth gets, but this fact alone cannot account for the most stunning item on Venus's fact sheet. While Earth has an average temperature range somewhere around freezing (273 K), the temperature on Venus is 730 K. It is a world covered in a thick blanket of perpetual cloud and extreme heat, all a result of trapped solar energy that has run amok. "Venus," says Jim Zimbelman, "is the poster child for why we don't want the greenhouse effect to go too far."

If Earth is a Garden of Eden, Venus is as close to hell as you can find in a planet. How could two worlds turn out so differently? To answer that question, we have to explore how interior, surface, atmosphere, and space environments are linked.

Venus's Interior

Among the most remarkable facts about Venus are its orientation and rotation rate. While all the other planets spin essentially in the same direction (counterclockwise, if you were looking down on the Solar System from the north celestial pole), only Venus spins clockwise. Along with this retrograde rotation, Venus's spin is also very slow. A "day" on Venus (one rotation) lasts 243 Earth days (**Figure 7.9**). The slow spin and retrograde rotation pose a challenge to astronomers trying to explain the formation and evolution of Venus. Many astronomers now argue that Venus must have suffered a collision with another body that altered its rotation rate and direction. It also makes understanding the internal structure of Venus harder. Generally, scientists can use the rotation of a planet (or other object) to understand something about how mass is arranged inside it. But because Venus spins so slowly, this method does not work well for determining its internal composition. Thus, astronomers are left with Venus's density as a principal clue to its internal structure.

There is, however, another way to "see" the internal arrangement of a planet's mass. In the 1990s the *Magellan* probe mapped Venus using radar to penetrate the clouds and return topographic information. "By accurately mapping the shape of the surface

Length of day in Earth days

Earth
Mars
Moon
Mercury
Venus

FIGURE 7.9
Relative Day Lengths on the Terrestrial Worlds
In this graph depicting the relative length of one complete day-night cycle on each of the terrestrial worlds and the Moon, each Sun represents one Earth day (24 hours).

with radar and simultaneously following the precise motion of the spacecraft in orbit, we could figure out a lot about how Venus's mass is distributed," explains Zimbelman. "When we combined this kind of data with models of the planet's interior, it really helped us understand what is going on below the surface."

Because Venus's density is so similar to Earth's, scientists were able to use the same models they had constructed for our planet's structure and adjust them for a planet of Venus's size. From these models they concluded that the structure of Venus must look a lot like that of Earth. Both planets have an iron core of roughly the same size, and each planet has a mantle of silicate rocks overlying it, also of roughly the same size. The two planets might thus be expected to have cooled at about the same rate, so Venus should still retain a significant amount of heat from its formation, as well as the heat generated by radioactive decay.

SECTION SUMMARY
Venus's Interior
- Venus is Earth's twin in radius and mass, but it had a very different history that left it wildly unsuitable for life.
- Venus's density, core size, mantle, crust, and composition are similar to Earth's.
- Venus's slow retrograde rotation is a Solar System anomaly.

Venus's Surface

Given the similarities in core and mantle, one might expect Venus's crust to look similar to Earth's, but it turns out the high atmospheric temperatures have changed the properties of the crust. "Raising the surface temperature hundreds of degrees changes how heat escapes from the crust," explains Zimbelman. "The surface is already hot, so you don't have to go down very far before the crust melts. That means the

A

B

FIGURE 7.10 *Venera 9*
In 1975 the Soviet Union produced the first-ever images from the surface of another world via its *Venera* spacecraft sent to Venus, where atmospheric and surface conditions limited their functioning life span to minutes or hours. (A) *Venera 9*. (B) A surface image from one of the landers.

ejecta blanket A layer of pulverized material that is blown out of an impact site (a crater).

crust is thinner, and unlike the small number of crustal plates like you have on Earth, Venus may have thousands of plates. The thinner crust means that the kinds of plate tectonics we're used to here won't happen on Venus. There the thin crust gets scrunched up easier, leading to entirely different kinds of behavior that manifest in a very different kind of surface."

A

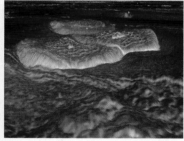

B

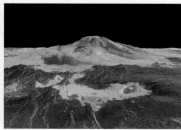

C

FIGURE 7.11 **Venus's Surface**
(A) Radar data obtained by the *Magellan* mission yielded this image of the surface of Venus. The image was processed to improve contrast and to emphasize small features, and color-coded to represent elevation. (B) Computer-enhanced images showing a three-dimensional perspective of surface features such as (B) the 25-km-diameter dome-like hills of the Alpha Regio region and (C) the volcano Maat Mons.

The first ground-view images of Venus were transmitted to Earth on October 22, 1975, when the Russian space probe *Venera 9* (**Figure 7.10**) parachuted through the planet's dense atmosphere and dropped heavily onto the surface. Where almost all other Venus probes had failed, *Venera 9* managed to function in that scorching, toxic environment for a full 53 minutes, beaming back images of Earth's sister world. The pictures were not encouraging. The surface of Venus looked flat, dry, and broken, a world smoldering under a sulfuric haze. None of the subsequent (all Russian) spacecraft that landed on Venus lasted very long, so we have few images at optical wavelengths of the planet's surface. That limitation has not, however, prevented the study of Venus's geology.

While the unyielding clouds covering Venus keep its surface invisible to us at optical wavelengths, radio waves can penetrate the clouds. *Magellan*'s radar mapping of Venus enabled scientists to develop detailed "images" of its surface down to scales of 100 meters. By bouncing radio waves from orbiting spacecraft down to the surface and timing the delay in the return signal, scientists can build up topological maps of the entire planet's surface (**Figure 7.11**).

"Everybody assumed we would find lots of impact craters on Venus," says Zimbelman, "just like on the Moon, Mercury, and the older parts of Mars. That didn't happen." Instead only 900 or so craters were discovered. "That's not many for an object as big as Venus," explains Zimbelman. "The implication is that the entire surface somehow resurfaced itself within the last billion years."

FIGURE 7.12 Danilova Crater
The impact crater Danilova on Venus and its broad, irregular ejecta blanket, as imaged by the *Magellan* spacecraft.

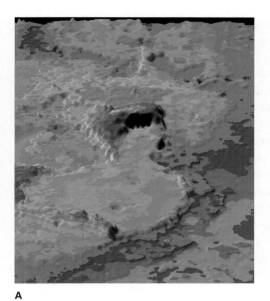

A

FIGURE 7.13 Venus's Terrain
(A) A perspective view of Ishtar Terra from *Pioneer* Venus radar imaging. Approximately equal in size to Australia, Ishtar is a large plateau rising 3.3 km above the surrounding lowlands, bounded by relatively steep slopes. At 11 km high, Maxwell Montes (in the upper middle part of the image) is the highest elevation on Venus, and thus the coolest, only 380°C. Color-coded altimetry shows elevations in 0.5- and 1-km intervals. Blue and green colors mark low elevations; yellow and red mark high elevations. (B) This map of Venus uses colors to show height above the surface: yellow and red indicate high elevation, and Ishtar Terra is at the top. Another high-elevation region called *Aphrodite Terra* can be seen just below the equator to the right.

The relative lack of craters is consistent with volcanism occurring on Venus. The meteor craters on Venus tend to be quite large. In many cases, irregularly shaped **ejecta blankets** (layers of ejected material) surround the craters, indicating that the meteors fragmented into irregular shapes in the dense atmosphere before they hit the ground in many pieces (**Figure 7.12**).

The rest of Venus's surface features have been strongly shaped by volcanism. Ishtar Terra is a region of raised highlands that extends across 5,600 km (larger than Australia). In its center is a plateau, most likely a great lava flow, called the Lakshmi Planum. An 11-km-high mountain called Maxwell Montes lies at one edge of Ishtar Terra (**Figure 7.13**). Jumbled mountainous terrain defines much of Ishtar Terra's boundaries.

Whereas Earth's continents are composed of low-density granites that allow them to "float" on the mantle below, the material composing Ishtar Terra and most of Venus's terrain appears to be the same basaltic material that makes up Earth's seafloors. These basalts well up from lava flows to form different kinds of structures. "Venus's thin crust," explains Zimbelman, "means that molten materials like those that generate lava flows on Earth or Mars or anywhere else had good access to the surface."

The volcanoes on Venus all seem to be shield volcanoes (**Figure 7.14**), which form when lava plumes rise from deep within the mantle, swelling up under crustal material. (Recall from Chapter 6 that the Hawaiian Islands are shield volcanoes.)

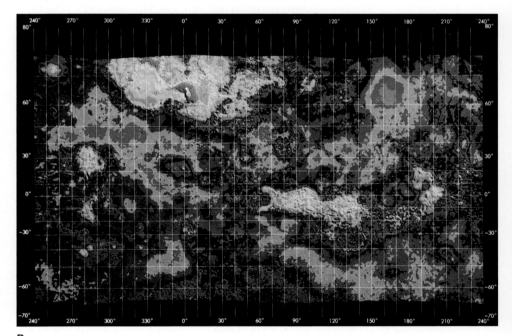

B

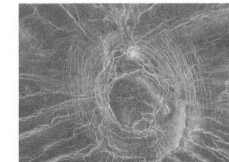

**FIGURE 7.14
Shield Volcanoes on Venus**
Coronas surround the largest shield volcanoes on Venus—evidence of extensive surface distortions and huge plumes of lava pushed upward.

These volcanoes tend to release their energy gradually and have sloping sides. "In the shield volcano case," explains Zimbelman, "the lava can flow fairly easily, and that's what makes the shield's gentle slopes."

As discussed in Chapter 6, composite (or explosive) volcanoes, caused by collisions between tectonic plates, are found in various places on Earth and are characterized by much more violent eruptions and steeper sides. "We don't see steep-sloped volcanoes like Mount Fuji on Venus, and that's consistent with the planet not having Earth-like plate tectonics," says Zimbelman.

The fact that only shield volcanoes are seen on the surface of Venus tells scientists about both the thinness of the crust and the flow of materials below the crust. Circular lava domes are seen in many places on Venus's surface, resulting from the upwelling of magma that distorts crustal material. Larger, broad, circular volcanoes are also seen with craterlike **calderas** at or near their centers where lava has welled up to the surface and then withdrawn (a caldera is the well at the top of a volcano). The largest shield volcanoes on Venus are surrounded by **coronas**, which are circular or oval uplifted regions. Coronas show evidence of extensive surface distortions and huge plumes of lava pushed upward. The lava plumes on Venus do not appear to be part of general convection systems in the mantle, but there is evidence that limited mantle flows may have forced crustal material to buckle into the mountainous regions of the highlands.

Any theory for the surface evolution of Venus has to remain grounded in one fact: scientists believe that all the surface features we see on the planet now are less than 1 billion years old. Something catastrophic appears to have happened to the planet in its relatively recent past that led to a complete resurfacing. "It may have been an era of supervolcanism," says Zimbelman, "or it may have been the crust breaking apart and overturning like ice floes in the Arctic (you might think of Venus's crust as made of 'rockbergs'). The jury's still out, though."

The dramatic and powerful event that happened on Venus, whatever its nature may have been, limits our ability to understand the deep history of the planet. For instance, Venus might not always have had its runaway greenhouse effect and its extreme temperatures. Perhaps very early in its history (more than 4 billion years ago), Venus was a water world like Earth. There is no way at present to know, and the mystery of Venus's surface only underscores the fact that planetary surfaces and atmospheres can change dramatically over the course of time.

caldera The depression at the top of a volcano.

corona A large oval structure surrounding volcanoes on some Solar System bodies, formed by upwelling of material.

SECTION SUMMARY
Venus's Surface

● Extreme heat and high pressures have made the study of Venus's surface by landers nearly impossible.

● Lava flows, large shield volcanoes, and jumbled mountainous terrain dominate the topography, but plate tectonics of the form seen on Earth has not been at play on Venus.

● Major resurfacing occurred a billion years ago.

Venus's Atmosphere

You run out of superlatives pretty fast when describing the atmosphere of Venus. Unlike any of the other terrestrial worlds, Venus hosts an extremely dense blanket of gases that have turned the world inhospitable to life (**Figure 7.15**). With a surface temperature above 730 K and sulfuric acid rain falling within the planet-enveloping layer of clouds that never parts (the rain never reaches the ground), it is a planet of extremes.

Perhaps the best place to start in understanding Venus's atmosphere is its composition. On Earth, carbon dioxide (CO_2) constitutes just 0.039 percent of the atmosphere. The composition of Venus's

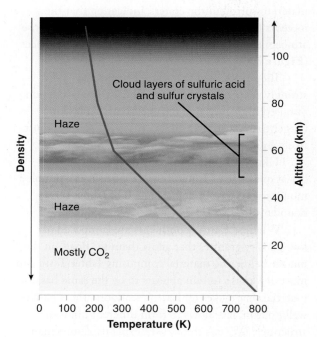

FIGURE 7.15 Venus's Atmosphere
Layers of Venus's atmosphere show differences in composition and density, as well as in temperature, which steadily decreases with altitude. Temperature and density both increase steadily with depth, so the atmosphere is hottest and densest near the surface. Although the atmosphere is more than 96.5 percent carbon dioxide by mass, several distinct cloud layers exist, consisting of sulfuric acid and sulfur crystals.

GOING FURTHER 7.2   THE LANGUAGE OF THE COSMOS
Planets without a Blanket: Temperature before Greenhouse Warming

What average surface temperature does a planet attain with no greenhouse effect? We can answer this question with a short calculation that shows what would happen on a planet containing no greenhouse gases in the atmosphere. Specifically, we want to know what surface temperature a planet with no atmosphere would attain purely through the surface absorption and reemission of solar energy.

Let's begin with the energy radiated into space by the Sun (its luminosity), which we will call L_{Sun}. We calculate this energy using the Stefan-Boltzmann formula, which relates L_{Sun} to the spherical Sun's radius R_{Sun} and temperature T_{Sun}:

$$L_{Sun} = 4\pi R_{Sun}{}^2 \sigma T_{Sun}{}^4$$

Now, how much of that solar radiation is captured by a planet of radius R_p orbiting at a distance D from the Sun? We call this L_{sp}:

$$L_{sp} = L_{Sun}\left(\frac{\pi R_p{}^2}{4\pi D^2}\right)$$

In this equation, the factor in the denominator tells us the surface area of the sphere of radius D across which all the solar radiation emitted by the Sun must pass. The term on the top is just the cross-sectional area of the planet. The ratio of the two tells us how much of the original solar radiation is captured by the planet.

Next we have to account for how much of the captured solar energy is simply reflected back into space. We call the albedo of the planet a, which tells us the fraction of incoming energy that is reflected, so the amount of energy that the planet absorbs can be represented by $L_{sp}(1 - a)$. Thus we have

$$L_{pa} = L_{sp}(1 - a)$$

Finally, we expect that the planet, bathed in sunlight, will eventually come to a stable temperature, T_p, at which it will radiate as a blackbody:

$$L_p = 4\pi R_p{}^2 \sigma T_p{}^4$$

We find the actual value of T_p by equating the solar energy the planet absorbs with the blackbody energy it radiates into space:

$$4\pi R_p{}^2 \sigma T_p{}^4 = (1 - a)\left(\frac{\pi R_p{}^2}{4\pi D^2}\right)4\pi R_{Sun}{}^2 \sigma T_{Sun}{}^4$$

Now we just have a couple of steps to solve for T_p. First let's divide both sides by the constants 4π and σ to cancel them out. That leaves us with

$$R_p{}^2 T_p{}^4 = (1 - a)\left(\frac{R_p{}^2}{4D^2}\right)R_{Sun}{}^2 T_{Sun}{}^4$$

Now we can divide both sides by the radius of the planet to cancel it out. Then we can take the fourth root of both sides (this is just like taking the square root twice). This gives us the final answer

$$T_p = (1 - a)^{1/4}\left(\frac{R_{Sun}}{2D}\right)^{1/2} T_{Sun}$$

Note that the planet's radius is not part of the final answer (because it canceled out). Thus, the planet's temperature depends only on its albedo, its distance from the Sun, and the properties of the Sun (its radius R_{Sun} and temperature T_{Sun}).

Let's use this formula to find the temperature that an airless Earth would have. Using $D = 1.5 \times 10^{11}$ meters, $T_{Sun} = 5{,}778$ K, $R_{Sun} = 6.69 \times 10^8$ meters, and an albedo for Earth of $a = 0.306$, we find

$$T_p = T_E = 249 \text{ K}$$

This is well below freezing. Thus, without the atmosphere and its greenhouse gases, our planet would be a dead, frozen rock.

atmosphere, however, is dominated by CO_2 (because of volcanism), which represents 96.5 percent of its mass. Carbon dioxide is a powerful greenhouse gas. The bonds between the carbon atom and two oxygen atoms are extremely effective at absorbing and reradiating heat energy (light in the infrared). Earth and Venus have the same total amount of planetary CO_2. However, much of the CO_2 on Earth is bound up in rocks and trapped under the oceans. Venus may have had an atmosphere similar to Earth's long ago; but when it lost its oceans, CO_2 was liberated into its atmosphere, causing Venus's surface temperatures to rise.

While Earth's small proportion of atmospheric CO_2 has allowed it to maintain mild temperatures (see **Going Further 7.2**), Venus's overwhelming abundance of the compound has driven the planet into its extreme state via a runaway greenhouse effect. Here the term *runaway* refers to evidence that Venus may not always have been as it is today. "A runaway greenhouse effect means a situation where once you pass a certain point in temperature, you can't go back," says Zimbelman. "The hotter it gets, the more the greenhouse effect increases, which makes it even hotter still." Scientists are not sure what would have triggered Venus's runaway greenhouse effect, but the culprit may have been the evaporation of the planet's water early in its history. "Water vapor is an effective greenhouse gas," says Zimbelman, "so if all the water went into the atmosphere, it would heat up quickly."

Venus's atmosphere is incredibly dense as well—about 90 times as dense as the air we breathe on Earth. The density and heat drive the formation of at least three permanent layers of clouds surrounding the planet. These clouds contain both sulfuric acid droplets and tiny sulfur grains.

SECTION SUMMARY
Venus's Atmosphere
- Venus's atmosphere is dominated by carbon dioxide, which produces a powerful greenhouse effect.
- Venus's atmosphere is extremely dense and covered by sulfuric acid clouds.

Venus's Near-Space Environment

Venus is a world without a substantial magnetic field, which, given its size, came as a surprise to scientists. It's not clear yet why Venus doesn't have a field, but it may have to do with the planet's slow rotation rate. "Without a decent rotation rate, you can't produce the circulating flows of molten iron in the core that create a magnetic field through the dynamo process," says Zimbelman.

The lack of a magnetic field has important implications for Venus's atmosphere. While the sheer mass of Venus's gaseous blanket protects it from disruption by the solar wind, the interaction with that wind may have changed the atmosphere's chemistry over time. "Because Venus lacks a strong magnetic field, the solar wind bumps straight into the upper atmosphere," says Zimbelman. "With no magnetic field acting as a buffer, the solar wind can strip away particles at the top of the atmosphere, and this is what may have happened to water on Venus." Ultraviolet radiation also played an important role in removing the water from Venus. Ultraviolet light dissociates the H_2O molecule. The liberated hydrogen atoms easily escape the gravitational hold of the planet. Once a planet loses its hydrogen atoms, it has lost its water for good. Thus, vaporizing water may have led to Venus's runaway greenhouse, and the planet now has very little water and is one of the driest environments in the Solar System.

It is notable that only Mercury and Venus, the two innermost worlds, are free of satellites. Every other planet from Earth out to Neptune has at least one moon. The lack of moons for these two inner worlds may be due to the nature of debris in their vicinity during their formation or to the kinds of collisions they endured. As we've seen, Venus's unusual rotation is thought to be the result of a dramatic collision with another large body. The energy of this early impact may have gone into reorienting Venus's rotation and not generating an orbiting moon (as scientists believe happened to an early Earth).

Surveying all we know about Venus, we find a world that might have been very Earth-like in its ability to harbor a planetary environment conducive to life. In fact, Venus may have been much more like Earth in its past. But planetary evolution worked out very differently for Venus than it did for Earth, and at some point 4 billion or so years in the past, the runaway greenhouse effect began on Venus and the planet ended up remarkably hot and dry. Venus shows us something very important in our exploration of planets: things change. This potential for transformation will be even more apparent as we move on to the final stop in our tour of terrestrial planets: the fabled world of Mars.

SECTION SUMMARY
Venus's Near-Space Environment
- Lacking a core dynamo, Venus has very weak fossil magnetism.
- No moons orbit Venus.
- Water was lost to space at the upper edges of the atmosphere, leading to a runaway greenhouse effect.

7.4
Mars: The Red Planet of Change

Late in the 19th century, astronomer Percival Lowell looked at Mars through a 24-inch refracting telescope and convinced himself he was seeing the handiwork of a dying civilization (**Figure 7.16**). Tracing what he thought was a network of straight lines across the planet, Lowell hypothesized that the Martians had built canals to transport water across a planet that had become an arid wasteland, desperately trying to save their civilization.

While it soon became clear that Lowell was mistaken, he was not alone in his fascination with life on Mars. H. G. Wells had already published his book *War of the Worlds* (1898), in which Martians undertake a nearly successful invasion of Earth. That story has since been reimagined once as a fake radio news program that inspired widespread panic (in 1938) and twice as a film (in 1953 and 2005).

The irony of this fascination with the Red Planet and life there is that Mars does, in fact, hold the best chance as another home for biology in the Solar System. But current conditions on the red planet suggest that we're in no danger of invasion anytime soon. Today Mars is a cold, dry, and empty world (**Figure 7.17**). Billions of years ago, however, it might have been a planet sporting a far bluer and perhaps even greener hue than the dust-swept planet we see now.

"Mars is a relatively small world by Solar System standards," says Jim Bell. "Its mass is just a tenth of Earth's, and it's only about half the size of our world." The smaller mass gives Mars a surface gravity that is just a bit more than one-third as strong as what we experience on Earth. Orbiting at 1.52 AU, it receives 56 percent less sunlight at its surface than Earth does. Its larger orbit also gives it a longer year, lasting almost 687 days. However, its day is quite similar to ours: Mars rotates once on its axis every 24 hours and 37 minutes.

Mars's Interior

In 1976 the first two landers to touch down on the Martian surface—*Viking 1* and *2*—carried seismographs with them. "Unfortunately, they both failed," says Bell. "The first seismograph never turned on, and the second was unable to filter out the effects of strong winds flowing past the lander." In the absence of seismographic data, planetary scientists used the same methods to infer the internal structure of Mars as they used for Mercury and Venus: density, rotation, volcanic activity, and the presence or absence of magnetic fields.

"Mars is pretty widely thought to have a core," says Bell, "but we don't know much about its size." The uncompressed density of Mars is 3,800 kg/m^3, the lowest of all the terrestrial planets, so Mars must have a lower metal content than the others. The presence of a magnetic field can also tell scientists whether Mars has

 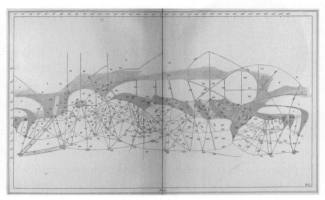

A   **B**

FIGURE 7.16 The "Canals" of Mars
(A) Percival Lowell (1855–1916) erroneously interpreted the markings he observed on Mars. (B) This drawing shows features that Lowell believed to be canals that transported water from the polar regions to more central latitudes. He concluded that intelligent life existed on Mars.

FIGURE 7.17 Martian Terrain
High-resolution images of Mars show a planet with dramatic features such as huge canyons and towering volcanoes. Many forces have shaped Mars across its history, from volcanism to weathering.

a planetary dynamo created by a molten iron core. A number of Mars probes have carried magnetometers to the red planet, but none have detected a strong global magnetic field. In 1997 the orbiting *Mars Global Surveyor* found evidence for fields as high as 0.1 percent of Earth's, but these appeared to be local anomalies related to differences in composition in the crust. "Mars probably used to have a magnetic field," says Bell. "That implies that, like Earth, Mars once had a partially

Jim Bell

"I was definitely one of those people who always knew they wanted to be an astronomer," says Jim Bell. "One of the great things about growing up in the early 1970s was being around to watch guys drive a car on the Moon. That was very, very cool, and it changed how I looked at the world and at space exploration."

Another astronomical landmark in Bell's early life also came to him from TV. "There was this program called *Cosmos* by the Cornell University astronomer Carl Sagan," Bell explains. "Kids today have no idea how influential it was. I try to describe to my own students how different things were back then for someone interested in science. I ask them to imagine a world where there are only three big networks on the television and pretty much none of them ever covered science." Sagan took viewers on a tour of everything from the theory of relativity to the birth of stars and on to the origins of the Universe. "Sagan was something no one had seen before—a charismatic scientist (in his signature turtleneck) who could translate science into plain English," recalls Bell. After Sagan's grand tour, Bell became a convert, determined to make a place for himself as a professional scientist.

After years of study, Bell's dream of helping to explore the Solar System became reality through his involvement in a series of major missions. Meanwhile, he became a professor at Cornell and then later at Arizona State University. For all his success, Bell has never forgotten the power of his early inspirations. "When eventually I got on the faculty at Cornell and was able to meet and talk with Carl Sagan, it was just a dream come true," he says. "In many ways it still is."

molten iron core." The present-day size of Mars's iron core remains a contentious issue among scientists studying the planet. "Altogether there's some uncertainty, but the data are consistent with the presence of a smaller core on Mars than we might have expected."

Even though the seismographs on the first landers failed, the presence of so many orbiters responding to Mars's gravitational field and landers giving precise measurements of the planet's spin rate has enabled extremely accurate determination of its rotation. That, in turn, gives scientists a measure of how the mass inside Mars is distributed. "Based on these kinds of measurements, it's likely there is a basic mantle and crust structure very similar to that on Earth and the other terrestrial planets," says Bell. In addition, the ability to directly probe the soil and rock on Mars has given scientists a unique view of Mars's composition. "The volcanic rocks on the surface of Mars are very similar to garden-variety volcanic rocks on Earth. It's a lot like the stuff coming out of the volcanoes on Hawaii, for example." These rocks are basalts, similar to those found on Earth's seafloor and on Venus. "Mars, Earth, Mercury, Venus are all basically made of the same stuff," says Bell, "which is not surprising since it's completely consistent with the standard condensation model of planet formation."

SECTION SUMMARY
Mars's Interior
- As a planet that may have been habitable in the past, Mars has long held a fascination for scientists.
- From direct measurements and calculations, scientists believe that Mars has a mantle and crust like Earth's, surrounding a small, dense core.

Mars's Surface

No terrestrial planet other than Earth has such a rich variety of surface features as Mars. Along with the volcanoes and meteor craters seen on other terrestrial worlds, Mars hosts continent-spanning canyons, polar caps of water ice and frozen CO_2, and a host of features that seem to point to a time when liquid water ran free on the planet's surface. And while the atmosphere of Mars is thin, its winds are strong enough to have shaped the long-term evolution of the surface terrain.

Most of the craters visible on Mars are large and indicate that small meteoroids usually burn up in the atmosphere and do not make it to the surface. In addition, wind erosion has smoothed away many craters, so some of the record of the Heavy Bombardment phase has been lost.

Some of the most dramatic features on Mars are the result of volcanism. "Most of the volcanoes on Mars appear to be of the shield, or effusive, type," says Bell. "The slopes are long and broad." Bell notes that there is some evidence for explosive volcanism on Mars. The planet also hosts the Solar System's largest volcano, Olympus Mons (**Figure 7.18**), a towering shield volcano that rises 24 km above the average surface level of the planet. The volcano is so high that the caldera at its peak has, essentially, risen above the bulk of the planet's atmosphere. The giant Olympus Mons is not an isolated feature but is part of an extended region called the Tharsis Bulge, which is home to a number of volcanoes. On average, the Tharsis region rises 10 km above the planet's radius.

In a sense, the entire Tharsis Bulge is a shield volcanic region. A continual process of magma upwelling from deep inside the Martian mantle pushed Tharsis and its volcanoes to such enormous heights. The fact that these volcanoes have grown so large offers compelling evidence that moving plates cannot exist on Mars. "If there were plate tectonics on Mars," says Bell, "you would have a chain of volcanoes like what you see

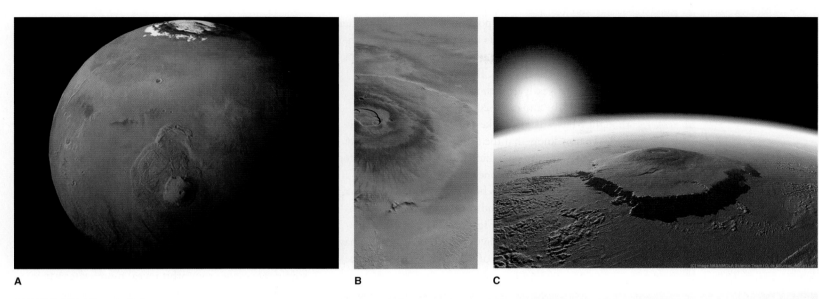

A   B   C

FIGURE 7.18 Tharsis Bulge
(A) The Tharsis Bulge is a large shield volcanic region, seen here in the middle of the image of Mars. (B) One of Mars's volcanoes, Olympus Mons, has the highest elevation of any mountain in the Solar System and covers an area as large as the state of Arizona. (C) Thousands of point elevation measurements from the laser altimeter on board the Mars Global Surveyor spacecraft were processed to create this computer-generated view of Olympus Mons. This volcano stands more than 30,000 meters above its base, and yet its slopes are only about 6 degrees, similar to that of its cousin volcano, Mauna Loa in Hawaii.

in Hawaii instead of the supervolcano that is Olympus Mons. Mars seems to be a one-plate planet." Although little is known about why any planet has, or does not have, plate tectonics, many scientists believe that the difference between Mars and Earth with respect to plate movements may simply be the presence of water. "Mars may have had more water in the past," says Bell, "but it did not have a gigantic ocean interacting with the crust like Earth does. All that water may be what's lubricating the whole process on our world."

Mars is also home to one of the largest canyon systems in the Solar System. Valles Marineris is a giant network of canyons that appears like a gash across the planet's face (**Figure 7.19**). Stretching more than 4,000 km, it's longer than the entire North American continent. At its widest, the canyon stretches 600 km from one side to the next. In some places, Valles Marineris drops 7 km to the canyon floor. (For comparison, the Grand Canyon is only 2 km deep.) The Valles Marineris system of canyons appears to be a direct result of crustal "faulting" associated with the nearby Tharsis Bulge. As the bulge rose, the surrounding surface was forced to respond, breaking into sections, some of which sank as others rose. The Valles region is relatively old, but along with the Tharsis region, it demonstrates that Mars was once a far more dynamic planet than it is today. "As Mars cooled, its surface activity slowed," says Bell. "But at one time it must have been really dramatic."

A

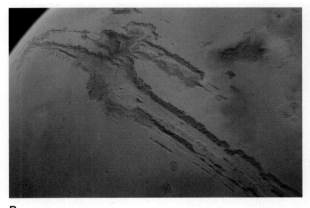

B

FIGURE 7.19 Valles Marineris
(A) Valles Marineris is a system of canyons resembling an enormous gash in the side of Mars; it extends more than 4,000 km in length, 600 km across, and 7 km deep at its extremes. (B) In this computer-generated image, data from Mars's orbiters have been used to create a three-dimensional representation of Valles Marineris.

Mars, like Earth, has polar ice caps, though neither Venus nor Mercury exhibits them. Even a small Earth-based telescope can see the northern and southern caps and their seasonal changes. The presence of the ice caps and their yearly growth and retreat were reasons that some 19th- and early-20th-century astronomers thought Mars was much like Earth and possibly inhabited. We now know that the caps are composed mainly of frozen CO_2—the same kind of dry ice that makes fog used in special effects at concerts.

Scientists distinguish between **seasonal ice caps**, which grow and retreat, and **residual ice caps**, which are present all year long (**Figure 7.20**). Earth's polar ice caps are currently residual, but global warming may lead to a seasonal melting of the Arctic (northern) ice sheets. Since Mars has an orbital tilt that is similar to Earth's, the planet experiences seasons in which the strength of sunlight varies as the planet orbits the Sun. Thus, in the northern winter, temperatures drop at the poles while the south experiences summer and rising temperatures. The northern and southern poles are not symmetric, however, because of Mars's elliptical orbit. The difference in size between the residual and seasonal caps in both the north and south is noteworthy. The southern seasonal cap covers 4,000 km, and the northern seasonal cap covers 3,000 km. The northern residual cap is 1,000 km across, while the southern residual cap barely crosses 300 km.

One important difference between the northern and southern caps is the role of water. Spectral measurements made from orbit reveal more water vapor in the atmosphere above the northern residual cap than above the southern cap, leading scientists to conclude that prodigious quantities of water may exist in the north. In 2008 the *Phoenix* lander touched down in the northern polar regions and immediately found evidence supporting this hypothesis when its mechanical scoop revealed water ice lying a few centimeters below the ground.

seasonal ice cap A polar region of frozen material such as CO_2 and water that increases and decreases in area and thickness with the seasons.

residual ice cap A polar region of frozen material such as CO_2 and water that stays constant in size with seasonal changes.

A

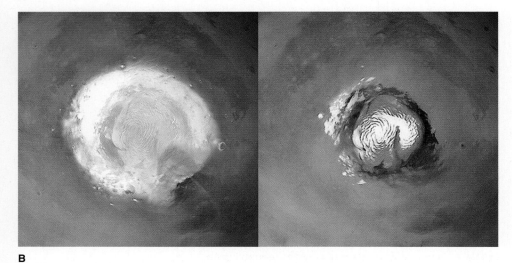

B

FIGURE 7.20 Polar Ice Caps on Mars
Seasons on Mars are reflected in changes in the polar ice caps. (A) North and south polar ice caps. (B) North polar ice cap in early spring (left) and late summer (right).

> ┌─ **SECTION SUMMARY**
> **Mars's Surface**
> - Surface features on Mars are dramatic: vast canyons, high mountains, and broad shield volcanoes.
> - Ice caps of CO_2 regularly grow and shrink on both poles, and the North Pole may hold large quantities of water.

Mars's Atmosphere

"As important as Mars's atmosphere is to the planet," says Bell, "there is actually not much of it to go around." Pressure at ground level, the base of the Martian atmosphere, is 150 times less than what is found on Earth. Another way of looking at the sparseness of Mars's atmospheric blanket is to note that the mass of gas surrounding Mars is 0.001 of that surrounding Earth and 0.0001 of that around Venus. Most of Mars's atmosphere is CO_2, which accounts for 95 percent of the gas by mass. Nitrogen makes up 1.6 percent, and oxygen makes up only 0.13 percent. Clearly, oxygen tanks will have to be included in any visit to Mars by humans (**Figure 7.21**).

While Mars's atmosphere may be thin, there is enough of it to give the planet weather. As Mars landers like *Viking*, *Pathfinder*, and *Phoenix* dropped to the ground, they measured the conditions in the atmosphere. The data they collected told scientists that the Martian atmosphere is layered like Earth's. In the Martian troposphere, CO_2-ice clouds form at high altitudes, and water-ice clouds form below them. The cycle of surface heating and cooling by sunlight drives Martian weather via convection. The thin atmosphere retains little heat, leading to almost 100-K changes between daytime and nighttime temperatures. While daytime temperatures on Mars can, at best, be almost comfortable for humans, nighttime temperatures would be deadly cold.

It never rains on Mars, but fog appears to form in canyons and craters as water ice evaporates in the morning with the rising Sun. Days on Mars are usually marked by, at most, light breezes. Mars, however, is subject to periodic strong dust storms that appear to begin in the southern highlands. Air flowing down from higher elevations picks up the light surface dust. The convection flow then carries the dust high into the troposphere. "From these humble beginnings you can end up with vast planet-engulfing dust storms," says Bell. "The storms can sweep across the planet, taking weeks before they subside." Wind speeds within the storms are extreme, sometimes reaching 400 km per hour (km/h). Because Mars's atmosphere is so thin, however, these storms are unlikely to pose much of a challenge to future astronauts. Bell says that the strong winds in the Martian storms "will feel only like being pummeled by feathers to a properly suited astronaut."

The difference between Earth's and Mars's atmospheres resides in the absence of life-forms on Mars and the difference in mass. The oxygen on our planet all comes from biological activity. Since Mars has no biological activity (that we know of), the oxygen in its atmosphere has become bound in rocks. The lower atmospheric pressure (low total mass) on Mars, however, is directly related to the planet's lower mass. The escape velocity of Mars is 5.0 km/s, compared with Earth's 11.2 km/s. Lighter elements such as nitrogen and argon, which are an important part of Earth's atmosphere, simply cannot be held by Mars's gravity, and over time they have escaped into space. The same is true of water vapor. Thus, Mars's atmosphere may have been far heavier and denser at early epochs in its evolution. The smaller mass of Mars also makes the atmosphere more fragile with respect to large-scale impacts. Some scientists have suggested that early asteroid impacts may have blasted significant fractions of the planet's gases into space.

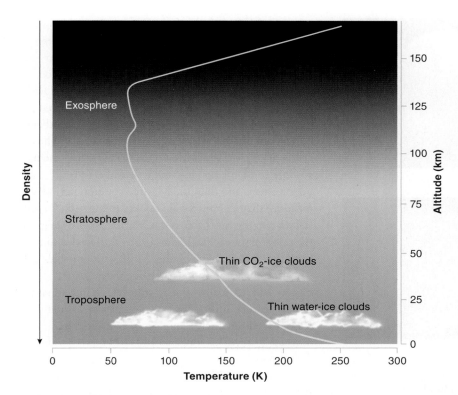

FIGURE 7.21 Mars's Atmosphere
The thin atmosphere of Mars exhibits layers of different densities and temperatures. The general pattern is a steady decrease in both temperature and density until reaching over 100 km, well into the exosphere, where the temperature rises again because of the absorption of solar ultraviolet radiation.

SECTION SUMMARY
Mars's Atmosphere

- Despite its very thin atmosphere, Mars experiences violent dust storms.
- Mars's low gravity has allowed heavier gases to escape its surface.
- CO_2 is the most plentiful gas in Mars's atmosphere; oxygen is only a trace element.

Mars's Near-Space Environment

As mentioned in the discussion of Mars's interior, the Martian magnetic field is many times weaker than that of Earth and does not envelop the entire planet. Without such a protective magnetic blanket, Mars's atmosphere has suffered erosion at high altitude through direct interaction with the solar wind, and this interaction is yet another reason for the current depleted state of the Martian atmosphere.

A

B

FIGURE 7.22
Deimos and Phobos
Mars has two small moons:
(A) Deimos and (B) Phobos.
Their origin remains a subject
of debate.

While Mars differs from Earth in its lack of a magnetic field, it is the only terrestrial planet other than Earth to host natural satellites. Mars has two moons: Phobos and Deimos (**Figure 7.22**). Phobos, with a diameter of just 22.2 km, and the smaller Deimos, with a diameter of 12.6 km, are tiny compared with Earth's Moon. Both moons are tidally locked and have orbits aligned with the planet's equator. Phobos orbits just 9,377 km above the Martian surface. Being so close to the planet, its orbital period is only 7.7 hours. This is less than the Martian rotation period, meaning that Phobos would appear to move backward (relative to the Sun) across the Martian sky twice in each day-night cycle. Deimos orbits farther out, at a distance of 23,760 km, giving it a 30.4-hour orbit.

Both Martian moons are irregularly shaped objects with dark, low-albedo surfaces. The dominant surface feature on Phobos is a giant, 9-km-diameter crater called Stickney. Phobos is also covered in unusual "grooves" as much as 100–200 meters wide and 20 km long. While many of them appear to radiate away from Stickney, analysis by the Mars orbiter *Express* showed that they are actually centered at the apex of Phobos's distorted oblong geometry, making their origin unclear. Deimos also shows craters, though the largest is only 2.3 km across. Deimos's surface holds other surprises. "It's very smooth," explains Bell. "Deimos looks like some powdery material is covering much of the surface. We haven't seen any asteroids that look like that." Scientists still do not understand what gives Deimos its unique appearance.

Given the moons' small sizes and their proximity to the asteroid belt, it would be tempting to think of Phobos and Deimos as captured asteroids. "That's the standard story," says Bell. "Both moons are very small, they're both irregular in shape, and they have spectra that look somewhat asteroidal." This theory, however, doesn't quite work out. Both the orbital dynamics of Mars's moons and some of their surface features point to a more complex history than one in which Mars simply captured a couple of asteroids.

"Phobos is extremely close to Mars," says Bell, "and it's slowly spiraling downward. In 10 million years or so (the blink of an eye in Solar System timescales), it will crash into Mars." The short life of Phobos appears as an oddity for scientists. "Has Phobos been there for billions of years and it's just a coincidence that we are seeing it near the end of its existence?" asks Bell. "That's an argument that doesn't carry a lot of weight."

The age of Phobos is not the only problem for astronomers trying to explain the origin of Mars's moons. "From theoretical studies we find it's impossible to capture an object the size of Phobos so close to Mars," says Bell. Calculations suggest that both moons were disrupted by tidal forces and torn apart long before they were captured into orbit. According to Bell, "People are starting to consider whether some unique situation enabled these objects to be captured, while others are exploring if a really giant impact might have launched material into orbit to form these moons."

Whatever the origin of tiny Phobos and Deimos, the lack of a large moon like Earth's has had a dramatic impact on Mars and its evolution in a way that may have affected its ability to maintain life. The difference has to do with Mars's obliquity, the angle between its orbital plane and its spin axis. While the direction of Earth's spin axis does slowly rotate like a wobbling top (a process called precession; see Chapter 2), its obliquity is remarkably stable, varying only slightly from its 23.5° tilt. Mars has not been so lucky. "The obliquity of Mars has changed by as much as 60° over timescales of just tens of millions of years," says Bell. "It must have been traumatic for the planet's climate. The spin axis can suddenly end up aligned almost with the Sun for a million years or more."

The cause of Mars's wild swings is the gravitational influence of Jupiter and the other giant planets. On Earth these forces are mediated by the gravitational influence of the large Moon. "Our Moon keeps Earth's spin stable," says Bell. "Mars has no such luck, and it has suffered for it."

SECTION SUMMARY
Mars's Near-Space Environment
- Mars has no global magnetic field, so the solar wind buffets the planet, changing its atmosphere significantly.
- Two tiny moons of uncertain origin orbit Mars.
- Without the stabilizing factor of a larger moon, Mars's obliquity has varied greatly over time.

Evidence for Water on Mars

NASA's Mars Exploration Program has been successful in its attempts to find evidence of water on the planet. From orbital images to orbital spectroscopic studies to direct exploration with landers and wandering rovers, the evidence that Mars once supported liquid water on its surface has grown, to the point of being almost incontrovertible. The evidence for *substantial* quantities of water existing today on Mars, in the form of subsurface ice, has also grown but remains less than conclusive.

High-resolution images of the Martian surface show ample evidence that liquid water once ran freely, sculpting features similar to those on Earth. Long, meandering features called **runoff channels** stretch for hundreds of kilometers on Mars. These features—found most often in the southern highlands—take the same form as river systems on Earth. They provide evidence that long ago, rivers flowed on Mars, transporting water from high elevations down to the lowlands. With water often comes flooding, and the Martian surface also shows evidence for dramatic, perhaps catastrophic, flows in the form of **outflow channels**, which created teardrop-shaped "islands" around craters and broad systems of plateaus and cliffs. Outflow channels appear to have been created by geologic sculpting in the form of vast amounts of water flowing around or clearing away obstacles. Images taken by *Mars Global Surveyor* even indicate the presence of vast river delta systems, where soils transported from upstream in the highlands were dumped as a river met a larger body of water.

Can astronomers conclude from these images that Mars was once a "blue" world of water lakes and water oceans? For Bell and some other scientists, the conclusion that vast expanses of water once existed on Mars is pushing it too far. "We just can't make a claim yet with great confidence," he says. "The pictures don't tell you whether it's water or beer that was flowing on Mars, but they do show that there was liquid, and even that is a big difference from what we see on Mars now."

The best evidence for water on Mars has come from studying minerals on the planet's surface. "The story of the rovers," says Bell, "is one of direct geologic investigation. Only by getting close can you really find the smoking gun." The rovers were equipped with different instruments to collect samples and analyze them using techniques like spectrometry. These instruments have enabled scientists to determine the composition and the mineralogy of the Martian surface. "Just like the periodic table of elements is universal," says Bell, "lots of minerals are universal too." With the rovers, scientists could get close to volcanic rocks or surface clays, collect samples, and see what they were made of. "What we discovered," says Bell, "was a variety of hydrated minerals. That means minerals requiring water for their formation."

One of the early water-related discoveries made by the rover *Opportunity* was the detection of tiny spherical, pebble-like structures found in the Martian soil in the Meridiani Planum region just south of the Martian equator. NASA scientists nicknamed these **Martian spherules** "blueberries" (**Figure 7.23**) because of their blue appearance in false-color images. Analysis of the blueberries showed they were composed of hematite, a mineral associated with the presence of water. The spherical shape led some scientists to believe that the blueberries crystallized out of a water-rich solution (spherical mineral deposits are the expected shape when formed in water). Other processes, such as volcanism, can create such spherical shapes. The discovery of blueberries both on the surface and mixed uniformly through deeper layers of Martian soil, however, has led *Opportunity* mission scientists to add them to their list of evidence for water on Mars. "As far as I am concerned," says Bell, "this is really unambiguous evidence that liquid water must have been present on Mars when these rocks formed."

The presence of small amounts of subsurface water ice today has been confirmed by the *Phoenix* lander at the Martian north pole (**Figure 7.24**). Evidence for large quantities of subsurface water in other parts of the planet is not as conclusive, however. In 2005, images taken from orbit by *Mars Global Surveyor* of a steep cliff system showed a streak-like feature running down the slope that had not been there 6 years

FIGURE 7.23
"Blueberries"
Tiny spherical rocks nicknamed "blueberries," discovered via the *Opportunity* Mars rover, have been identified as hematite, a mineral associated with the presence of water. This false-color image shows the spherical hematite deposits scattered on the Martian surface.

FIGURE 7.24 Water Ice on Mars
To the surprise and delight of specialists working on the *Phoenix* Mars lander mission, soon after touching down the lander found evidence for water ice just below the surface. Shown here are the white patches in the scooped regions.

runoff channel A long, meandering feature on Mars that resembles a river system on Earth.

outflow channel An extended region of scoured ground on Mars that includes features indicating high rates of ground fluid flow.

Martian spherule Any of the tiny pebble-like structures ("blueberries") found in Martian soil, which may have crystallized out of a water-rich solution.

FIGURE 7.25 Evidence for Subsurface Water on Mars?
Images of a canyon wall taken over time by Mars Global Surveyor reveal the appearance of dark streaks that may have been produced by the movement of subsurface liquid or melting ice.

before (**Figure 7.25**). Some scientists conjectured that subsurface water broke out, ran down the cliff, and then quickly evaporated into the Martian atmosphere. While these images are intriguing, scientists cannot yet say with certainty whether the streaks were created by running water or by something else.

SECTION SUMMARY
Evidence for Water on Mars

● Spectrographic measurements, surface features, and the presence of spherical rocks strongly suggest that Mars once had liquid water on its surface and harbors subsurface water today.

● Landers have found water ice just below the surface.

The History of Mars

The presence of liquid water in the past—and the possibility that it entails for the evolution of life—is what makes Mars such a fascinating planet. But it also reminds us that Mars has evolved and changed over time, just as Earth has. After intense study, scientists can even identify the key periods in Mars's history.

Scientists call the period from about 4 to 3.5 billion years ago in Mars's history its **Noachian period** (**Figure 7.26**). "Mars was much more active early in its history," explains Bell. "It had a magnetic field. It had a partially molten core. It had an active interior, active volcanoes on the surface. It's completely reasonable to assume that CO_2, sulfur dioxide, and water were coming out of those volcanoes, so the planet had a thicker atmosphere, providing a greenhouse effect

Noachian period The period of Mars's geologic history from about 4 to 3.5 billion years ago.

Hesperian period The evolutionary period of Mars from about 3.5 to 3 billion years ago.

Amazonian period The evolutionary period of Mars that began about 3 billion years ago.

that warmed the planet's surface above the melting point of ice. Put it all together, and Mars may have been a much more Earth-like place than what we see today." Bell is quick to caution that *Earth-like* does not mean just like Earth. "There may have been ponds, lakes—and they may have been intermittent—but that doesn't mean Mars was an ocean world. The evidence so far doesn't point in that direction." Thus, Mars may have had lots of water on its surface but perhaps not to the degree that oceans covered large parts of its surface.

Still, conditions in the Noachian period on Mars (such as a thicker atmosphere with more greenhouse warming) might have made it conducive for life to begin. "Early Mars was probably really interesting," says Bell. "By our usual definition, it would have been a habitable place."

Some 3.5 to 3 billion years ago, in an era called the **Hesperian period**, everything changed. "The core may have cooled and solidified," says Bell. "We are not sure exactly why, but the planet stopped being volcanically active." As volcanism slowly declined, fewer gases were driven into the atmosphere. Just as important, the decline in the magnetic field as the core solidified meant that the solar wind was able to erode the top of the atmosphere, and particles were lost to space because of Mars's low escape velocity. "Give these processes a billion years to work, and you can thin out Mars's atmosphere pretty well, reducing any greenhouse effect that was warming the surface. You end up with a pretty icy planet."

Once the surface froze, Mars no longer provided what scientists think of as habitable conditions for life. "It might be," speculates Bell, "that microbes, if they did exist, took refuge underground where there was a flux of heat coming from deeper regions."

Eventually the heat continued to drain away, and the planet entered its deep freeze in what scientists call the current **Amazonian period**. "Since the beginning of the Amazonian about 3 billion years ago," says Bell, "there has just been basically nothing going on except the wind moving across the surface, the occasional impact crater, or maybe an occasional burst of very, very late volcanism." Bell points out that the boundaries of the three periods are subjects of intense debate, and the story might still have significant holes in it. "We don't know if the upper layers of the planet's interior are active. We don't know if the core is partially or totally solidified. We don't know the rate of erosion of the upper atmosphere by the solar wind. We certainly don't know how much subsurface water exists and if

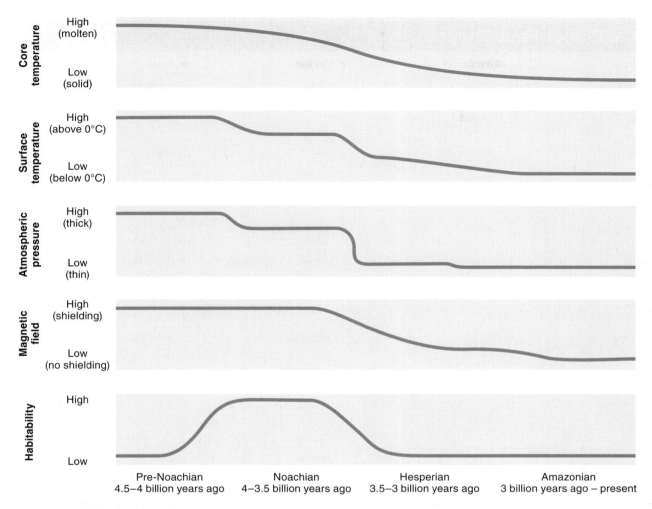

FIGURE 7.26 The History of Mars
The periods in the history of Mars are well defined, although uncertainty remains about the dates of transitions. Mars's evolution is characterized in general by decreases in temperature, atmospheric density, and the probability of the existence of microbial life over time.

any of it is liquid or not. These are all questions we will have to do our work to answer in the future."

One thing Mars does show us, however, is that planets that begin in the range of habitability can change with time. That is a conclusion of fundamental importance for us as inhabitants of the only planet in our Solar System still harboring conditions favorable for life.

The key features of the terrestrial planets are summarized in **Table 7.1**.

SECTION SUMMARY
The History of Mars
- Astronomers identify at least three distinct periods in Mars's history.
- Mars shows us that planets that begin in the range of habitability can change over time.

Outward Ho!

With our tour of the terrestrial planets complete, we are now ready to turn our gaze outward to the outer planetary domains of the Solar System. While our concern with the inner planets focused on surfaces and their evolution, our concern with the outer planets will have to shift. The outer planets are gas and ice giants and, as such, do not even have surfaces. What they do have, however, are environments unlike anything we experience on Earth. From a cornucopia of water-rich moons to gossamer ring systems, the outer Solar System is a frontier that has only recently been opened to human exploration. It is strange and awesome territory, and it has its own story to tell us about planetary evolution and the possibilities of life.

TABLE 7.1 ••• Terrestrial Planet Fact Sheet

	Mercury	Venus	Earth	Mars
Radius (km)	2,440	6,052	6,378	3,397
Radius (Earth radii)	0.38	0.95	1	0.53
Mass (kg)	3.30×10^{23}	4.87×10^{24}	5.97×10^{24}	6.42×10^{23}
Uncompressed Density (kg/m^3)	5,300	4,400	4,400	3,800
Magnetic Field	Fossil magnetism only	No	Significant	Very weak or none
Atmosphere	None	Primarily carbon dioxide	Nitrogen and oxygen	Primarily carbon dioxide
Atmospheric Density Compared to Earth	N/A	90 times as dense	1	1/150th as dense
Surface Temperature (K)	93–703	730	288–293	183–268
Polar Caps	No	No	Water ice	Carbon dioxide and water ice
Orbital Radius (AU)	0.39	0.72	1	1.52
Orbital Period (Earth days)	88	224	365	687
Rotation Period (Earth days)	59	243	1	1.03 (24 h 37 min)
Liquid Water	No	No	Yes	No?
Plate Tectonics	No	No	Yes	No
Age of Surface	4 billion yr	1 billion yr	Continually resurfaced	3–4 billion yr
Surface Features	Many craters, cliffs, high mountains	Few craters, shield volcanoes, high mountains	Oceans, continents, mountains, volcanoes, vegetation	Canyon network, shield volcanoes, craters
Tidal Effects	Tidal synchronization of rotation and orbit	No	Ocean tides due to Moon's orbit	No
Moons	No	No	One	Two: Phobos and Deimos
Other	Orbital eccentricity 0.206	Sulfuric acid rain; retrograde rotation	Millions of life-forms	Evidence for surface water in past

● → chapter summary

7.1 Planet Stories

Comparative planetology is the study of similarities and differences between planets in order to learn more about their natures and the underlying processes that shaped them. Comparative planetology is especially pertinent now, given the growing inventory of newly discovered exoplanets. The terrestrial planets in our Solar System all have core-mantle-crust configuration but show differences in interior structure, surface evolution, atmosphere, and near-space environment. The thermal history of these planets has driven their evolution through five distinct stages: differentiation, cooling, cratering, magma flooding, and weathering.

7.2 Mercury: Swift, Small, and Hot

Mercury, closest to the Sun, is hot and lifeless, and it orbits faster than any other planet. Its high density suggests a large iron core. It has been geologically dead for 4 billion years and bears many ancient craters. Because of its small size, it cooled quickly, leaving distinctive surface features such as cliffs. Any initial atmosphere has evaporated, and surface temperatures range widely, from 700 K in daylight to 125 K at the poles. Mercury has a weak dipole magnetic field offset from its center. Mercury's orbit and rotation periods have been influenced by tidal synchronization, creating a solar day lasting 176 Earth days.

7.3 Venus: Hothouse of the Planets

Venus, swathed in a dense atmosphere (mostly CO_2), is uniformly hot, with a surface temperature of 730 K because of a runaway greenhouse effect. The nearly Earth-sized planet has retrograde rotation that takes 224 Earth days, most likely the result of a collision early in its history. Because of cloud cover and hostile conditions, Venus's surface cannot be studied visually, but radar has revealed a flat, dry, broken surface with few craters and many shield volcanoes. These findings suggest that the thin crustal plates have allowed volcanism to reshape the surface relatively recently in the planet's history. Venus has no magnetic field.

7.4 Mars: The Red Planet of Change

Mars is the terrestrial planet that is likely to have harbored water on its surface earlier in its history. Mars is similar to Earth in its rotation period and seasonal changes, as well as in the presence of polar ice caps. Its surface has been studied by orbiters, landers, and rovers, which have produced evidence of subsurface water but no definitive evidence for the existence of current or past life. Volcanism on Mars has produced vast shield volcanoes, and the enormous Valles Marineris suggests violent phenomena. A thin CO_2 atmosphere creates strong winds and planet-enveloping dust storms. Two small, irregular moons, whose origin remains a mystery, orbit Mars. Astronomers posit three distinct evolutionary periods for Mars: the Noachian, the Hesperian, and the Amazonian.

● → questions and problems

Narrow It Down: Multiple-Choice Questions

1. Which of the following characteristics do the four terrestrial planets have in common? Choose all that apply.
   a. a significant magnetic field
   b. plate tectonics
   c. an atmosphere
   d. the presence of one or more moons
   e. a rocky crust

2. Which of the following statements accurately describe(s) all the terrestrial planets? Choose all that apply.
   a. They condensed from the part of the solar disk that had no solid hydrogen compounds (ices).
   b. They differentiated into core-mantle-crust.
   c. They maintain presence of water in some form.
   d. They have at least one moon.
   e. They display some evidence of cratering.

3. Which of the following is *not* a source of heat in the interiors of the terrestrial planets?
   a. friction between atmosphere and surface
   b. collisions from other bodies
   c. gravitational compression
   d. radioactive decay
   e. differentiation

4. Four planets that formed together have equal density and the following radii—planet W: 2,000 km; planet X: 2,500 km; planet Y: 2,700 km; planet Z: 5,000 km. Ignoring any other factors, which planet would be expected to cool fastest?
   a. planet W, with radius 2,000 km
   b. planet X, with radius 2,500 km
   c. planet Y, with radius 2,700 km
   d. planet Z, with radius 5,000 km
   e. There is no way to know.

5. A caldera is which of the following?
   a. the depression at the top of a volcano
   b. a long ridge formed by planet shrinkage
   c. a depression that filled with lava
   d. a crest formed by the meeting of two crustal plates
   e. a channel dug by the glancing blow of an asteroid

6. Which of the following bodies exhibit(s) rotation patterns clearly influenced by tidal effects? Choose all that apply.
   a. Moon
   b. Mercury
   c. Venus
   d. Deimos
   e. Mars

7. Which body has the longest "day"? (A day is defined as the interval between the time at which a particular point on the body faces the Sun and the time at which the same point faces the Sun again.)
   a. Mercury
   b. Venus
   c. Earth
   d. Mars
   e. Earth's Moon

8. Which of the following statements about Venus is/are *not* true? Choose all that apply.
   a. It rotates retrograde.
   b. Its volcanoes are all shield volcanoes.
   c. Its surface was mapped by Russian landers.
   d. It has crustal plates.
   e. On average, it has as many craters per square meter as Mercury does.

9. Which of the following geologic features is/are found on Mars? Choose all that apply.
   a. Ishtar Terra
   b. Valles Marineris
   c. Discovery Rupes
   d. Olympus Mons
   e. Stickney Crater

10. Rank the terrestrial planets by the ages of their surfaces, oldest to newest.
    a. Mercury, Venus, Earth, Mars
    b. Mercury, Mars, Venus, Earth
    c. Mars, Earth, Venus, Mercury
    d. Venus, Mercury, Mars, Earth
    e. Mars, Mercury, Earth, Venus

11. Which of the following statements about the atmosphere of Venus is *not* true?
    a. Its composition has changed significantly over time.
    b. It is 90 times less dense than Earth's atmosphere.
    c. It is mostly CO_2.
    d. It contains sulfuric acid clouds.
    e. Its heat and acidity quickly disable spacecraft that land on the surface.

12. Imagine a terrestrial planet, Minutia, with a mass m and radius r. A second planet, Garganzo, has mass of $1.4m$ and radius of $1.4r$. How does the surface gravity on Garganzo compare with that on Minutia?
    a. It is about the same.
    b. It is greater.
    c. It is less.
    d. It depends on the density of the two planets.
    e. It depends on whether there are atmospheres on either or both.

13. Which two planets are closest in rotation period?
    a. Mars and Venus
    b. Mars and Mercury
    c. Mercury and Earth
    d. Venus and Earth
    e. Earth and Mars

14. Which of the following describe(s) commonalities in the ice caps of Earth and Mars? Choose all that apply.
    a. Both planets have ice caps on both poles.
    b. Both planets' ice caps have residual components.
    c. The composition of the ice is the same on both planets.
    d. Seasonal changes occur in the ice caps of both planets.
    e. Each planet has roughly identical ice caps in its northern and southern hemispheres.

15. If Deimos were orbiting at its current orbital radius around Earth instead of around Mars, how would its orbital velocity differ?
    a. It would be the same.
    b. It would be slower because of Earth's greater radius.
    c. It would be slower because of Earth greater mass.
    d. It would be faster because of Earth's greater radius.
    e. It would be faster because of Earth's greater mass.

16. Which of the following affect(s) the climate of a planet? Choose all that apply.
    a. greenhouse gases
    b. obliquity
    c. orbital radius
    d. rotation period
    e. presence of a large moon

17. During which of the following periods of Martian evolution was it most likely that life would form?
    a. Noachian
    b. Hesperian
    c. Archian
    d. Amazonian
    e. Paleozoic

18. The radius of planet APC-11 is r, while the radius of planet APC-23 is $2.2r$. How do their cooling times compare?
    a. The question can't be answered without knowing the volumes of both planets.
    b. Both will take about the same time to cool.
    c. APC-11 will take about twice as long to cool as APC-23.
    d. APC-11 will take about half as long to cool as APC-23.
    e. APC-11 will cool about 4 times as fast.

19. You observe a planet orbiting at 2 AU from a distant 5-billion-year-old star. The planet has minimal cratering, a strong magnetic field, tall composite volcanoes, and a high degree of tilt, compared to Earth. From observations of our Solar System, which of the following is/are likely to be true of the planet? Choose all that apply.

    a. It has a large molten core.

    b. It has multiple moons.

    c. It has tectonic plates.

    d. It has been geologically dead for billions of years.

    e. It experiences no seasons.

20. Which of the following is the proper order of the stages of planetary evolution?

    a. weathering, magma flooding, differentiation, cratering, cooling

    b. cratering, magma flooding, differentiation, cooling, weathering

    c. differentiation, magma flooding, cratering, cooling, weathering

    d. differentiation, cooling, cratering, magma flooding, weathering

    e. cooling, differentiation, magma flooding, cratering, weathering

To the Point: Qualitative and Discussion Questions

21. Describe at least three surface features or characteristics of Mercury's surface, and indicate how each provides evidence for the planet's history.

22. Describe at least three surface features or characteristics of Venus's surface and indicate how each provides evidence for the planet's history.

23. Describe at least three surface features or characteristics of Mars's surface and indicate how each provides evidence for the planet's history.

24. What is meant by the term *planetary differentiation*, and what is the end result?

25. Rank the four terrestrial planets by their current degree of visible cratering.

26. What is the phenomenon of plate tectonics, and where in our Solar System is it known to exist?

27. Describe the different kinds of volcanoes found on each of the terrestrial planets.

28. What are the requirements for the formation of a planetary magnetic field?

29. What factors have changed on Mars that have made it less suitable (even hostile) for the evolution of advanced life-forms?

30. Describe the differences in the number, size, and shape of moons for the planets in the inner Solar System.

31. Views change frequently about the evidence for life on Mars. Do a search for recent scientific articles on this topic. How convincing do you find the evidence? What evidence might be more convincing?

32. Impacts by planetesimals, asteroids, and comets on young planets are a normal part of planetary evolution. For each of the terrestrial planets and moons, name at least one effect caused by such collisions.

33. The Martian day is slightly longer than Earth's day. Engineers and scientists working with Mars orbiters and rovers are sometimes required to adjust to Martian time for maximum mission efficiency. How might this adjustment affect rover mission personnel?

34. An enhanced greenhouse effect on Earth due to climate change could create more Venus-like surface and atmospheric conditions on Earth. Describe how such changes might affect life on Earth.

35. Manned spaceflight to Mars may represent a much-desired exploration to many people, but it is fraught with peril. What would be some of the greatest challenges to the physical and psychological well-being of the astronauts involved?

Going Further: Quantitative Questions

36. How many times greater would Venus's escape velocity be if it had the radius it does but mass equal to Earth's?

37. Calculate the escape velocity, in kilometers per second, for a planet with mass equal to Mars's and radius equal to Mercury's.

38. What is the orbital velocity, in kilometers per second, of a planet with 2 times the orbital radius and 4 times the mass of Mars? (See Going Further 5.1.)

39. The orbital velocity of Venus is 35.0 km/s. What would this value be, in kilometers per second, if its orbital radius were half as great?

40. The radius of an exoplanet is 3.0 times that of Earth. What is the ratio of Earth's cooling time to the exoplanet's cooling time?

41. A terrestrial planet with radius r cooled to 287 K over 1.4 billion years. How long would you expect another terrestrial planet with radius $2.7r$ to cool to 287 K?

42. Use Kepler's third law to calculate the orbital periods of two hypothetical planets that are 1.2 and 2.5 AU, respectively, from the Sun.

43. Calculate the stable temperature that Mars would have without an atmosphere. (Mars's albedo is 0.15.)

44. Find the stable temperature of an airless world with albedo 0.220 orbiting 7.8×10^{11} meters from a star with radius of 1.7×10^9 meters and surface temperature of 6,400 K.

45. A planet with no atmosphere and an albedo of 0.550 orbits a star with temperature of 8,700 K and radius of 4.5×10^{10} meters. The planet's orbital radius is 9.1×10^{12} meters. What is the planet's stable temperature?

 If your instructor assigns homework in **smartw⬤rk5**, access it at the Digital Landing Page for *At Play in the Cosmos*: **digital.wwnorton.com/cosmos**

GAS, ICE, AND STONE

The Outer Planets

● **In this chapter** you will learn about the giant planets, the large, ringed, moon-rich siblings of the terrestrial planets. After reading through each section, you should be able to do the following:

8.1 Giant Planets on a Roll
Explain how the vast distances between the Sun and the giant planets create conditions quite different from those on terrestrial planets.

8.2 The Giant Planets: Structures and Processes
Summarize the similarities and differences in the processes shaping the giant planets and the structures that have evolved as a result of those processes.

8.3 Jupiter: King of Planets
Describe the characteristics that make Jupiter and its satellites unique and important to the structure and evolution of the Solar System.

8.4 Saturn: Lord of the Rings
Describe the characteristics of Saturn and its satellites.

8.5 Uranus and Neptune: Ice Giants Discovered in Twilight
Distinguish the characteristics and properties of the ice giants from those of the gas giants.

Jupiter, the largest planet in our Solar System, is worthy of its regal name (Jupiter was the king of the Roman gods). The remarkable banded structure in its atmosphere results from the planet's rapid rotation and heat flowing from its interior. The prominent black spot (lower left) is the shadow of one of Jupiter's four largest moons.

8.1
Giant Planets on a Roll

LINDA SPILKER had always dreamed of seeing Saturn's rings up close. But she never imagined that the first detailed view would come on a spool of computer paper rolled out across 10 meters of hallway deep in the basement of NASA's Jet Propulsion Laboratory (JPL).

Spilker was fresh out of college when she joined JPL and the *Voyager* team. The two *Voyager* probes, launched in 1977, were designed to make a grand tour of the giant planets before heading out of the Solar System and into interstellar space. Spilker, one of the few women working with the *Voyager* science team at the time, had a ringside seat for humanity's early explorations of Saturn and its beautiful, but still mysterious, rings (**Figure 8.1**).

"We were studying the rings by watching how a background star dimmed and brightened as *Voyager 2* passed behind ring material," Spilker recalls. *Voyager 1* images had already shown a wealth of substructure in the rings that looked a lot like grooves on a phonograph record. "Those images were just incredible," says Spilker, "but with our **stellar occultation** method, where we used the blocking of starlight by ring material, we were able to detect structures much smaller: less than 100 meters."

This was 1979, however, long before the era of high-resolution desktop computer screens. "In those days you couldn't easily put a plot up on your screen," she recalls. "What we did was we ran the long sheet of paper through a printer and had a computer plot the points on it as the star went behind the rings. I

remember unrolling it in the hall and walking down its length. It felt like I was walking across Saturn's rings and looking through them to the star, just as *Voyager* was doing so far out there in the outer Solar System. It was an amazing feeling."

Beyond the Snow Line

The outer realms of the Solar System bear little resemblance to what is found among the terrestrial worlds. "It's a different kind of Solar System out there," says Amanda Hendrix, who was the deputy project scientist of the *Cassini* mission to Saturn from 2010 to 2012. Remarkably, you live among the first human generations to see any part of that dark domain up close. Only through the most advanced technologies, including the robot deep-space probes that Spilker and Hendrix oversee, have we been able to explore the outer Solar System.

To get a sense of the difference between the inner and outer domains of the Solar System, let's briefly consider two fundamentals: light and warmth. "In the outer Solar System the Sun is faint, it's small, and it does not provide as much heat," says Hendrix. The farther away you travel from a light source, the fainter it appears (**Figure 8.2**). Jupiter, the closest planet of the outer Solar System, orbits at 5 astronomical units (AU) from the Sun. At that distance the Sun appears much dimmer than it does from Earth, consistent with the inverse square law for brightness, which you learned in Chapter 4. "At this distance," Hendrix explains, "sunlight has dropped in its intensity by 25 times compared with what we get on Earth." Not only does the Sun appear dimmer; it appears smaller too. Rather than the 32-arcminute (32′) diameter of the solar disk that we

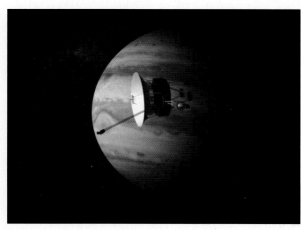

FIGURE 8.1
Voyager
(A) The twin *Voyager* spacecraft were launched in 1977. Here we see an artist's representation of one of the *Voyager* spacecraft passing Jupiter. (B) Both spacecraft sent back detailed images of their explorations, such as this close-up of Saturn and its previously mysterious rings.

A   **B**

Views of the Sun and its relative effect on each planet

	Earth	Jupiter	Saturn	Uranus	Neptune

Sun's angular size (arcminutes, or ′)	32′	6′	3′	2′	1′
Sun's relative brightness	1	0.037	0.01	0.003	0.001
Planet's equilibrium temperature (K) (no atmosphere)	255	124	95	59	59

FIGURE 8.2 The Sun and the Giant Planets
The Sun appears much smaller and less bright on the giant planets' skies and correspondingly provides less light, leading to much lower equilibrium temperatures.

see in our sky, for an observer on Jupiter the Sun would extend across just 6′—looking only one-fifth as large as it appears from Earth. Thus, Jupiter, Saturn, Uranus, and Neptune all exist in a world of half-light. "There is so little light," she explains, "that we have to adjust the cameras on probes like *Cassini* to get good images. The exposures have to be a lot longer than what you need around Earth or Mars, because of the Sun's distance."

In addition to illumination, sunlight brings heat. Without the brunt of the Sun's warmth that we enjoy, the outer realms of the Solar System are colder as well as darker. Going Further 7.2 showed how to calculate a planet's equilibrium temperature (if the planet had no atmosphere) if its distance from the Sun is known. At the location of Earth, a planet without an atmosphere would be chilly but bearable, with a temperature of about 255 K (or –0.67°F). For Jupiter, however, that temperature drops to 124 K (–236°F), and at the distance of Neptune's orbit a planet's equilibrium temperature drops to just 59 K (–353°F).

The decreasing temperatures farther from the Sun also had an enormous influence on the formation of the giant planets. Recall from Chapter 5 that all the giants lie outside the snow line. "The snow line is an imaginary boundary at somewhere around 3 AU from the Sun (depending on how it's calculated)," says Hendrix. "Inside the snow line, a newly forming planet accumulates lots of silicates and metals. Outside the snow line, water is stable as a solid. That means once

you get beyond the snow line, you should expect water ice and other frozen compounds to play a big role in both planets and moons."

Even beyond the snow line, sunlight does still play a role, however. "We see seasonal variations on the giant planets in response to how much sunlight they get, and we see some kinds of chemistry happening on both the giant planets and their moons that depend on energy from the Sun," says Hendrix. "So the Sun is still affecting these outer worlds, but it just doesn't play as important a role as it does for the inner planets."

As we begin to explore the outer planets, we must again develop a set of guiding questions, as we did for the terrestrial planets. This chapter, like Chapter 7, focuses on the evolution of planets and the possibilities for life. The outer Solar System, however, raises profound new questions about both issues. What kinds of processes shaped the outer, giant planets into configurations so different from the inner, rocky worlds? Do we need to think more broadly about where and how life might exist?

interactive
Equilibrium Temperature

SECTION SUMMARY
Beyond the Snow Line

- Because of their great distance from the Sun, the outer planets receive less light and warmth than the terrestrial planets do. Solar radiation is still important enough to affect the planets and produce seasons.

- The ice and gas giants lie beyond the snow line, where water and other compounds freeze.

8.2
The Giant Planets: Structures and Processes

Just as all the terrestrial planets share certain structural similarities, so do the giant planets, and these similarities anchor our understanding of their evolution. "A lot of what we see in the outer Solar System reflects the conditions out there beyond the snow line when the planets formed from the protosolar nebula," says Amanda Hendrix. "In the end, the questions come down to what kinds of physics and chemistry the temperatures in that initial disk of gas and dust allowed."

Raw Facts: Size, Mass, Density, and Composition

The most obvious common attribute of the giant planets is that they are huge in both size and mass, compared with their terrestrial cousins. Jupiter, the largest of the giants, has a radius 11 times larger than Earth's radius, and its mass is a whopping 318 times that of Earth (**Figure 8.3**). Jupiter's volume is so great that 1,321 copies of Earth could be hidden beneath Jupiter's clouds. Saturn, the second-largest planet, has a radius 17 percent smaller than Jupiter's, though its mass is only about one-third that of its larger sibling. Uranus and Neptune are smaller and are almost twins in terms of size and mass: both planets' radii are roughly 4 times that of Earth, and their masses are roughly 15 and 17 times Earth's, respectively. "The sheer volume and mass of these big planets determine a lot of the amazing behavior we see around them," says Hendrix.

As **Table 8.1** shows, the giant planets all have much lower densities than the terrestrial planets. These densities are found by using the planet's motion to determine its mass (via Newton's version of Kepler's third law; see Chapter 3) and then dividing that mass by the planet's volume, which can be found via measurements of its radius (the volume of a sphere = $\frac{4}{3}\pi R^3$.) "The low densities mean that the giants must be made mostly of material that is lighter than the silicate rocks and metallic cores composing the terrestrial worlds," says Hendrix. Jupiter and Saturn are composed mainly of the lightest elements, hydrogen and helium, in gaseous form, and they are therefore called *gas giants*. Uranus and Neptune, in contrast, contain larger amounts of heavier elements like carbon, oxygen, and nitrogen. These heavier elements are locked into chemical compounds with the abundant hydrogen, creating substances like methane (CH_4), ammonia (NH_3), and water (H_2O). At the distances of Uranus's and Neptune's orbits, temperatures are low enough for these compounds to have frozen in solid form (or at least into a slush). Thus, the outermost planets are called *ice giants*.

SECTION SUMMARY
Raw Facts: Size, Mass, Density, and Composition

- The giant planets are larger than the terrestrial planets in radius and mass; Jupiter is by far the largest with 11 times Earth's radius and 318 times Earth's mass.

- The icy nature of Uranus and Neptune reflects their composition (containing more heavy elements like carbon and nitrogen) and their formation in the colder reaches of the Solar System.

Jupiter   Saturn   Uranus   Neptune   Earth

FIGURE 8.3 Giant Planet Radii
The relative sizes of gas and ice giant planets, compared with Earth (which is the largest terrestrial planet). All of the giant planets have ring systems, though not all are visible in these images.

TABLE 8.1 • • • Giant Planet Fact Sheet				
	● Jupiter	● Saturn	● Uranus	● Neptune
Type	Gas giant	Gas giant	Ice giant	Ice giant
Mass (Earth masses)	318.26	95.14	14.54	17.09
Radius (km)	69,911	58,232	25,362	24,622
Volume (Earth volumes)	1,321	764	63	58
Density (kg/m^3)	1,330	690	1,300	1,760
Orbital radius (AU)	5.2	9.53	19.2	30.1
Orbital period (Earth years)	12	29	84	165
Rotation period	9 h 50 min	10 h 14 min	17 h 14 min	16 h 6 min
Number of known satellites (in 2013)	67	62	27	13
Primary elements/compounds	H, He	H, He	H, He, CH_4, NH_3, H_2O	H, He, CH_4, NH_3, H_2O

Internal Structure

The giant planets lack surfaces, so there is no distinct transition between their atmospheres and their interiors. If you could dive downward into a giant planet, you would find the density and pressure increasing continuously as you penetrated deeper into the atmosphere. Rather than eventually crashing into a hard surface, you would find a gradual transition between gaseous and liquid states as the pressure around you increased and you passed smoothly from the atmosphere into the interior. At their centers, all of the giant planets contain cores. "Certainly there's some kind of metal/rock core in each of these worlds," says Hendrix, "but the conditions will be extreme and there is a lot we have to learn about how materials behave under these kinds of conditions" (**Figure 8.4**).

These extreme conditions are, in part, the result of enormous pressure. The masses of the giant planets are so great that gravity squeezes down hard on their interiors, creating pressures high enough for unusual kinds of physics to kick in. The surface pressure on Earth is measured in a unit called a *bar* (the pressure at sea level is about 1 bar). At the center of Earth, underneath all that rock and metal, pressures can be

interactive
Anatomy of Gas/Ice
Giant Planets

FIGURE 8.4
Giant Planet Interiors
The relative densities of the giant planets led astronomers to conclude that the gas giants consist predominantly of hydrogen in increasingly dense forms, including a planet-covering "ocean" of molecular hydrogen atop a metallic liquid hydrogen layer surrounding a relatively small "rocky" core. The ice giants have larger rocky cores with overlying layers of water, ammonia, and methane, mainly in the form of ices, as well as liquid hydrogen oceans.

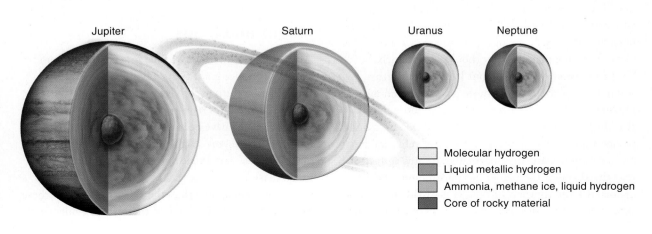

Jupiter   Saturn   Uranus   Neptune

☐ Molecular hydrogen
☐ Liquid metallic hydrogen
☐ Ammonia, methane ice, liquid hydrogen
☐ Core of rocky material

4 million times surface pressure, or 4 **megabars**. In the central regions of Jupiter, the pressures may reach 30–45 megabars. Greater mass thus creates greater pressure, which determines what kinds of structures exist between the core and atmosphere.

In addition, all the giant planets produce more heat than they receive from the Sun (although for Uranus, the excess is tiny). The source of this heat is thought to be gravitational contraction. Unlike the terrestrial planets, the giant planets are still slowly shrinking. As the same amount of mass is squeezed into a smaller space, heat is generated from atoms colliding more frequently and more energetically with one another. This heat eventually makes it to the surface, where it escapes as radiation.

┌─ **SECTION SUMMARY**
  Internal Structure
● The giant planets have no solid surfaces but show increasing pressure reaching extremes toward the cores.
● Changes in pressure with depth allow different structural layers to form inside giant planets.

Atmospheres and Rotation

The gas giants Jupiter and Saturn are widely considered the most beautiful objects in the Solar System, in large part because of the banded cloud structures in their atmospheres. Uranus and Neptune show atmospheric structures that are similar to but less dramatic than those of both Jupiter and Saturn. "The remarkable atmospheric patterns we observe in the gas giants, and to a lesser degree in the ice giants, originate in the development of clouds forming at different layers," explains Hendrix. Depending on the pressures and temperatures in each layer, different kinds of compounds will condense into clouds at different heights, giving each planet a distinctive appearance.

The inner and outer worlds' atmospheres share similarities that are helpful in understanding the behavior of both. "Lots of basic processes are the same from one planet to another," says Hendrix. "But then when you talk about Jupiter and Saturn, you're talking about much more gas and much more pressure variation. That makes for a lot more drama." Another key difference between terrestrial and giant planet atmospheres is the giant planets' remarkably rapid rotation. Saturn, in particular, rotates so fast that the shape of the planet has become noticeably distorted. "The rotation of the giant worlds makes the behavior in their atmospheres—including storms—a

lot more dramatic and more long-lived." Recall from Chapter 6 that rotation shapes Earth's atmospheric currents, leading to large-scale patterns of circulation as convection drives hotter gas to rise and cooler gas to fall. "The rapid rotation of the giant worlds sends this process into overdrive," says Hendrix.

┌─ **SECTION SUMMARY**
  Atmospheres and Rotation
● The atmospheres of the giant planets are much deeper and more layered than those of any terrestrial planet, with strong pressure variations that lead to multiple banded cloud layers.
● Rapid rotation contributes to conditions such as large, long-lived storms.

Magnetic Fields

Another structural similarity across the giant planets is the presence of strong, large-scale magnetic fields. "Everything's bigger in the giant planets, so it's not terribly surprising that they have these giant magnetic fields around them," says Hendrix. The giant planets' rapid rotation and the presence of circulating gas flows from lower to higher depths drive the fields. "What is surprising is how much diversity there is in the structure of the fields among the giants." Differences among the giant planets in the location of conducting layers, where a field creating a dynamo is active, lead to remarkable differences in the orientation and alignment of the giant planets' magnetic fields. In all cases, the fields extend well beyond the planets' atmospheres to form protective regions around their local systems of moons (**Figure 8.5**).

┌─ **SECTION SUMMARY**
  Magnetic Fields
● All four giant planets have large magnetic fields, although the sources vary according to their internal structure.

Moons and Rings

Although moons are rare in the inner Solar System, every gas and ice giant hosts a family of satellites. Jupiter and Saturn each have more than 60 natural moons, 27 satellites have been found to date orbiting Uranus, and 13 have been found around Neptune (**Table 8.2**). "There are a lot more moons orbiting the giant planets," says Hendrix, "because they have so much mass and, therefore, stronger gravity." What is even more remarkable than the sheer numbers of these moons is the fact that some of them show evidence of

megabar A unit of pressure equal to a million times Earth's surface pressure.

TABLE 8.2 ••• Selected Moons of the Giant Planets

A unique characteristic of the giant planets is the large number of moons orbiting each, as well as the variety of sizes and orbital patterns of those moons. Regular moons have prograde orbits that are aligned with the equatorial plane of the planet, while irregular moons may have retrograde orbits and tilted orbits.

Planet	Moon	Radius (km)	Mass (kg)	Orbital Period (days)	Unique Features
Jupiter	Io	1,821	8.93×10^{22}	1.8	Active volcanism; lava and sulfur plumes
	Europa	1,565	4.80×10^{22}	3.6	Iceberg surface; deep subsurface ocean
	Ganymede	2,634	1.48×10^{23}	7.2	Grooved surface; subsurface water
	Callisto	2,403	1.08×10^{23}	16.7	Subsurface water ocean but geologically dead
Saturn	Pan	14	4.95×10^{15}	0.58	Orbits in Encke Division
	Prometheus	43	1.60×10^{17}	0.61	Shepherd moon for F ring
	Pandora	40.5	1.37×10^{17}	0.63	Shepherd moon for F ring
	Mimas	198	3.74×10^{19}	0.94	Gravitation effects on Cassini Division
	Enceladus	252	1.08×10^{20}	1.37	Liquid water geysers
	Titan	2,575.5	1.35×10^{23}	15.95	Only Solar System moon with substantial atmosphere; liquid methane lakes
Uranus	Miranda	236	6.59×10^{19}	1.41	Faults, ridges, valleys; innermost moon
	Ariel	579	1.35×10^{21}	2.52	Long surface cracks
	Umbriel	584.5	1.17×10^{21}	4.14	Geologically dead
	Titania	788.5	3.53×10^{21}	8.71	Ancient craters
	Oberon	761.5	3.01×10^{21}	13.46	Ancient craters
Neptune	Proteus	210	5.04×10^{19}	1.12	Active geysers; orbits in rings
	Triton	1,352.5	2.14×10^{22}	5.88	Retrograde orbit; thin atmosphere
	Nereid	170	2.70×10^{19}	360.13	Very eccentric orbit

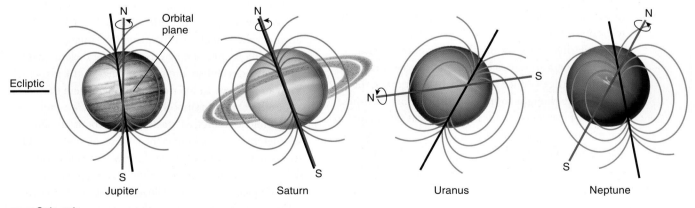

—— Spin axis
—— Magnetic axis

FIGURE 8.5 Giant Planet Magnetic Fields
Each of the four giant planets has a significant magnetic field, but the field's orientation varies greatly depending on the planet's spin axis and center. Here we see the four giant planets with their spin axes (red line) positioned relative to their planes of orbit around the Sun and the relative inclinations and orientations of their magnetic fields. The orientations of their magnetic fields are also shown (black line).

Amanda Hendrix

It can take decades for a project like *Cassini* to go from proposal to design to assembly to launch and then, finally, arrival. Amanda Hendrix was with the project from its launch in 1997. Seven years later, Hendrix's most cherished moment in an impressive career came when she viewed the *Cassini* data. "We were having a little image-viewing celebration as the data were coming back from *Cassini*," says Hendrix. "We were looking at images of Saturn's outer moon Rhea. There was so much detail I was stunned. I am a moon person, so this was really exciting to me. Finally I was seeing one of those moons up close!" The *Cassini* mission has been wildly successful, mapping out Saturn, its moons, and its spectacular ring system.

Hendrix is a planetary scientist whose own journey began even earlier than *Cassini*'s. "I was in second grade when I learned about the Solar System," recalls Hendrix. "I loved it." As she grew older, Hendrix added astronaut to her list of ambitions. "I thought being an astronaut and an astronomer would tie in pretty naturally."

The mix of Hendrix's interests in the science of planets and in the engineering aspects of being an astronaut led her to major in aerospace engineering as an undergraduate. The University of Colorado, where Hendrix did her PhD studies, also had a large planetary sciences program. "I was doing planetary science through Colorado's aerospace engineering department, and that worked out fine," she says. As an aerospace graduate student, Hendrix also analyzed data collected on the Moon and asteroids from the ultraviolet instrument on the *Galileo* Jupiter mission.

thick layers of ice, under which may lie kilometers-deep subsurface oceans. "There is a lot of water on these moons," says Hendrix, "and that makes us wonder if they might be places where life evolved."

Each of the giant worlds also hosts a system of rings. While Saturn's beautiful banded arcs are the most famous of the planetary rings, robotic probes and telescopic observations have revealed narrower, less dramatic rings surrounding the other outer worlds as well. The rings appear to be made of relatively small orbiting particles of ice and rock. In most cases, the particles are thought to have relatively short lifetimes. They don't stay in orbit for long, and that fact points to the moons as the origin of ring particles in many cases. Astronomers believe that small bits of rock and ice that were blasted off the orbiting moons may continually supply the rings with new material.

SECTION SUMMARY
Moons and Rings

- The giant planets' gravity has given each many diverse moons, some of which have thick layers of water ice or subsurface oceans.

- Each of the outer planets has a ring system, most likely supplied and maintained by its moons.

metallic hydrogen Hydrogen at a pressure and temperature high enough to have atoms closely spaced but with electrons able to move freely as in a metal.

8.3
Jupiter: King of Planets

Jupiter, named after the ruler of the Roman gods, is without a doubt the king of the planets. The gas giant accounts for 63 percent of all the mass in the Solar System other than the Sun, and its enormous gravity has a profound effect on much of the other 37 percent. In 1994, astronomers watched in amazement as a comet called Shoemaker-Levy 9, captured by Jupiter's gravity, was first pulled apart and then plunged in fragments into the planet's atmosphere. It was the first time a collision between a planet and a comet was seen, though the Solar System's meteor craters give ample evidence that such collisions occurred in the past. "It just so happens that we were lucky to catch this one," says Amanda Hendrix.

Astronomers gained a treasure trove of new insights watching Shoemaker-Levy 9 hit Jupiter. "We learned about Jupiter's atmospheric chemistry," says Hendrix. "We learned about its atmospheric structure. We even learned about Jupiter's rings because the passing comet fragments created a tilt or warp in the rings." Most important, however, astronomers saw direct evidence that giant planets like Jupiter may act as a kind of planetary vacuum cleaner, sweeping the Solar System of wandering debris that could pose a threat to life-bearing planets.

Jupiter's Interior

The composition of Jupiter is much like that of the Sun: by mass it contains about 75 percent hydrogen and 25 percent helium, along with traces of other heavier elements. In this sense, Jupiter's balance of elements reflects the composition of the interstellar cloud out of which the Sun and all the planets originated.

Using Jupiter's known mass, size, and elemental abundances, scientists can build models of the planet from atmosphere to interior. Beginning where the atmosphere merges smoothly into a liquid is a molecular hydrogen ocean that reaches down about 16,000 kilometers (km), making it the largest internal planetary structure in the Solar System. Below that depth, pressures get so high in the liquid hydrogen that atoms are tightly squeezed together, allowing electrons to become mobile, which in turn makes the material strongly conducting. Here a layer of **metallic hydrogen** begins.

Models of Jupiter also predict a 15-Earth-mass ($15\text{-}M_E$) core made from a mix of rocky and metallic materials. The weight of the other 303 M_E of H and He above the core drives temperatures at the center to 30,000 K—more than 5 times hotter than the surface of the Sun. Given these temperatures, it would be wrong to think of Jupiter as a gas ball with a terrestrial planet at its center. Instead, the central regions of Jupiter are best thought of as an Earth-sized spherical core of heavy elements at high temperatures and incredible pressures. "We still have lots of questions about the conditions at the Jovian core," says Hendrix. "There is a lot about the physics there that we don't understand."

Because it is still contracting, Jupiter gives off approximately 1.7 times more energy than it receives from the Sun. The contraction is slow enough that Jupiter has shrunk only 50 percent over the last 4.5 billion years. The energy released by the planet as it contracts keeps the interior temperatures high and drives the planet's remarkable atmospheric convection.

SECTION SUMMARY
Jupiter's Interior
● Jupiter's core of heavy elements is surrounded by a layer of metallic liquid hydrogen, and above that lies a molecular hydrogen ocean.

Jupiter's Atmosphere

In 1995 the *Galileo* spacecraft approaching Jupiter released a small probe that descended into the Jovian atmosphere. Falling first through a clear sky composed of nearly pure hydrogen, the probe then dropped into the first (or upper) layer of clouds, which is composed of ammonia droplets. A middle layer composed of ammonia sulfide lies below the first, and deeper still lies a layer of water clouds. "Each layer exists because temperature and pressure vary, and with them, the composition of droplets that can condense under those conditions," says Hendrix (**Figure 8.6**). Such **droplet condensation** also occurs in Earth's sky, when water vapor condenses from the gaseous state into the liquid state to form clouds.

The clouds in Earth's sky are white, but Jupiter's clouds show a remarkable range of colors. This is surprising because all of Jupiter's cloud constituents (ammonia, hydrogen sulfide, and water) should form pure-white crystals. Astronomers hypothesize that small amounts of complex compounds, which form when the Sun strikes the upper atmosphere, "pollute" the Jovian clouds, yielding the hallucinogenic mix of colors. "This is one of the places on Jupiter where sunlight really does matter," says Hendrix. "It may be the interaction with solar energy driving chemical reactions which gives the clouds their color."

The most striking feature of Jupiter's clouds is their horizontal banded structure. Astronomers call these bands **belts** and **zones** (**Figure 8.7**).

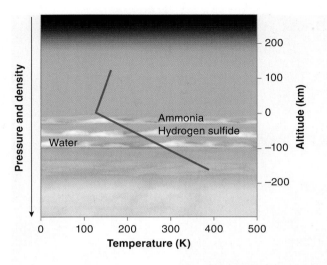

FIGURE 8.6
Jupiter's Atmosphere
Jupiter's atmosphere is characterized by three distinct cloud layers—consisting of ammonia, hydrogen sulfide, and water—each of which condenses at altitudes where the pressure and temperature allow droplets of that compound to form. The reference point marked by 0 on the altitude scale distinguishes the point where temperature stops decreasing with distance from the center and the hydrogen-dominated gas atmosphere layer begins.

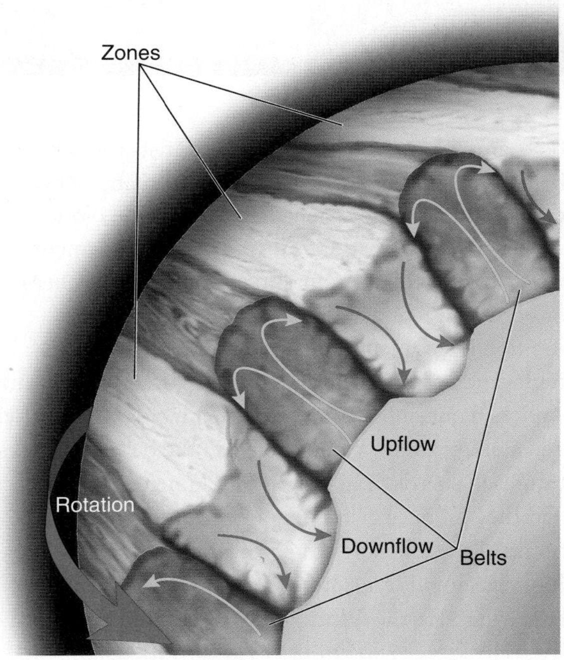

FIGURE 8.7 Zones and Belts
Zones are high-pressure areas that are rising and therefore form high in the atmosphere, where sunlight received is stronger. Belts are low-pressure areas that are falling lower in the atmosphere, where sunlight received is weaker.

Observations made by visiting space probes (particularly a flyby of the *Cassini* probe in 2000) show that the darker belts represent high-pressure material that has risen from deeper in the atmosphere. The belt material cools as it reaches the top of Jupiter's cloud deck. The zones, in turn, consist of low-pressure material sinking back down into the atmosphere, picking up heat as it goes. Thus, convection initiated by heat released from the interior is the engine driving the

droplet condensation The transformation of molecules from a gaseous to liquid state in the form of small droplets in an atmosphere. It generally occurs where temperature is decreasing.

belt A rising, high-pressure cloud region that forms a horizontal band in Jupiter's atmosphere because of convection and the planet's rapid rotation.

zone A sinking, low-pressure cloud region that forms a horizontal band in Jupiter's atmosphere because of convection and the planet's rapid rotation.

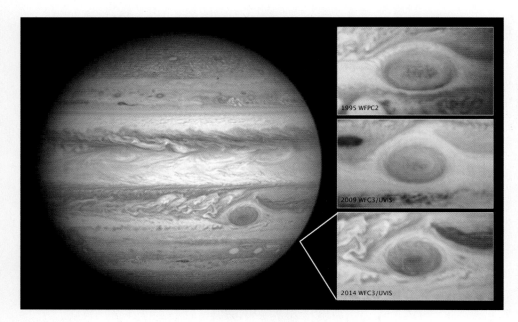

FIGURE 8.8
Jupiter's Great Red Spot
The Great Red Spot is a dynamic, evolving high-pressure system that has been observed for over 300 years. Recently, the Great Red Spot has appeared to be decreasing in size, as shown in Hubble Space Telescope images.

belts and zones. Before *Cassini*, scientists thought the belts were sinking and the zones were rising. This is a clear example of how new data can change old ideas.

On Earth, weather is driven by the movement of winds that circulate around in cyclonic patterns because of both convection and Earth's rotation. Jupiter's rotation is so much more rapid than Earth's that its high- and low-pressure systems have been permanently stretched out in the counterrotating bands of rising belts and sinking zones. Winds between the belts and zones can reach speeds of hundreds of kilometers per hour, and they produce circulating hurricane-like patterns. Some of these features can grow into vast storms like Jupiter's famous long-lived Great Red Spot, which is larger than Earth. Winds in the Great Red Spot have been clocked at 250 km per hour (km/h), making it, by Earth's standards, a category 5 hurricane that has persisted for at least 300 years, though it changes in size and shape over time (**Figure 8.8**).

SECTION SUMMARY
Jupiter's Atmosphere
- Jupiter has at least three distinct cloud layers composed mostly of ammonia and water.
- Convection causes pressure differences that, along with rapid rotation, create counterrotating belts and zones, as well as giant, long-lived storms like the Great Red Spot.

Jupiter's Magnetic Field

The 50,000-km-deep layer of metallic liquid hydrogen in Jupiter's interior is the source of its powerful magnetic field. Discovered in the 1950s via radio emission from the planet, Jupiter's magnetic field was first detected directly in the 1970s by the *Pioneer* and *Voyager* flybys. The *Galileo* mission in the 1990s then gave scientists a detailed laboratory for exploring Jupiter's magnetism and its effect on the nearby environment.

Jupiter's magnetic field, like Earth's, extends into space, creating an elongated magnetosphere around the planet. The strength of Jupiter's field, however, is so large that its magnetosphere extends more than 600 million km backward, reaching all the way to Saturn's orbit (**Figure 8.9**).

Jupiter's magnetosphere acts as a trap for charged particles. Some of these particles come from the solar wind, and some are ions ejected from Jupiter's highly volcanic moon Io. Streams of charged particles are channeled by the magnetic lines of force to the northern and southern Jovian poles, where they collide with atmospheric atoms to create vast auroras thousands of times larger and more energetic than those on Earth.

SECTION SUMMARY
Jupiter's Magnetic Field
- Jupiter's magnetic field creates a large magnetosphere and intense belts of high-energy particles.
- The flow of charged particles along the field lines produces vast auroras at Jupiter's poles.

Jupiter's Moons

Jupiter is known to host a family of more than 60 moons, though the exact number is bound to change, since many of these satellites are nothing more than captured asteroids. Four of Jupiter's moons stand out, however: Callisto, Ganymede, Europa, and Io are worlds large enough to be seen with a small telescope or even binoculars. In fact, it was Galileo who discovered these four rapidly orbiting moons with his first telescopes. These "Galilean" moons not only hold clues to the origin of Jupiter (and the Solar System) but also may be critical in understanding where and when life might form besides on Earth (**Figure 8.10**).

CALLISTO, DEAD WORLD FULL OF WATER.
The farthest Galilean moon from Jupiter, Callisto orbits at 1.9 million km from the planet's center with an orbital period of 16.7 days. "All the moons of Jupiter, including Callisto, are powerfully influenced by tidal forces from the giant planet," says Hendrix. "That means all their rotation rates have been locked to their

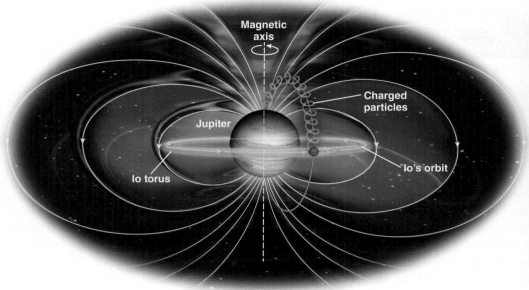

A

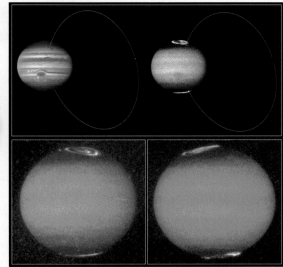

B

FIGURE 8.9
Jupiter's Magnetic Field
(A) Jupiter's enormous magnetic field is shown here, including the torus of material blown off of Io. (B) Auroras around Jupiter's poles, shown in these visible-light (top) and ultraviolet (bottom) images taken by the Hubble Space Telescope, provide further evidence of magnetic forces focused at the poles.

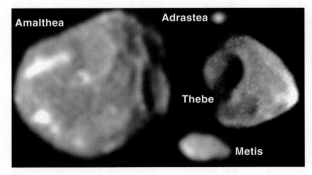

A

B

FIGURE 8.10 The Galilean Moons
(A) The four Galilean moons of Jupiter, so named because they were first observed by Galileo, have been the subject of intense study because of their varied surfaces and internal structures. (B) Some of the smaller moons of Jupiter have been imaged only from great distances, so fuzzy images exist to date.

orbit periods." Callisto, however, is far enough away from Jupiter that internal heating from Jupiter's tidal forces is relatively weak and the moon is, essentially, geologically inactive. The tidal forces arise because the one side of the moon feels Jupiter's gravity more strongly than the other (the force of gravity decreases as the inverse of the distance squared). This difference in force leads to both a squeezing and a stretching of the moon's interior, which increases internal temperature. The surface features visible today have not changed much since they were laid down billions of years ago when they first formed.

TABLE 8.3 ••• Densities of the Galilean Moons	
Moon ●	Density (kg/m³)
Io	3,530
Europa	3,010
Ganymede	1,940
Callisto	1,830

Similar to the way the planets formed from a disk orbiting a young Sun, the Galilean moons formed from a disk swirling around young Jupiter. Thus the range of densities in the moons shows an order similar to what we see in the planets orbiting the Sun (**Table 8.3**). Callisto is farthest out and has the lowest density: 1,830 kilograms per cubic meter (kg/m³). "At that distance," Hendrix explains, "temperatures in the disk would have been lowest." Thus, Callisto contains significant amounts of water ice mixed with heavier material such as silicate rocks. It does not appear to have differentiated to produce a core.

Callisto's surface is composed of dark ice shattered by craters from meteor impacts (**Figure 8.11**). Most of the giant planets' moons with icy surfaces have become darker with age as small meteorite impacts have vaporized the ice, leaving larger proportions of rock behind. The presence of significant amounts of water also changes the shape of larger impacts.

The *Galileo* space probe carried sensitive magnetometers on board as it orbited the gas giant and explored its moons. When Jupiter's powerful tilted magnetic field swept across Callisto, the space probe saw variations in magnetic energy, indicating the presence of liquid water beneath the moon's crust. "We think a layer of liquid water 10–100 km deep may be hiding beneath Callisto's ice-rock crust," says Hendrix. "That was a really exciting discovery, and Callisto is likely not alone in harboring such an underground ocean."

GANYMEDE, AN ALMOST ACTIVE MOON.

Unlike Callisto, Ganymede appears to have remained geologically active for some time after forming. With a higher density, 1,940 kg/m³, it may have differentiated into a rocky core lying below an ice-rock mantle. Ganymede's higher density is an indication that its location in the early moon-forming disk had a higher temperature, making the moon less able to hold on to volatiles. Its presence closer to Jupiter, however, has made its interior more susceptible to the big planet's tidal forces. "The extra internal heating left its mark on Ganymede's surface," says Hendrix.

Numerous craters appear on Ganymede's icy surface, just as they do on Callisto; however, there are vast "grooved" regions of the moon where resurfacing must have occurred. These grooves not only contain fewer craters, indicating that they are newer, but they cut right across older craters, erasing half of their structures. The shapes of these grooves, along with their spectra, indicate that water-rich flows of muddy material must have streamed across the surface before

A

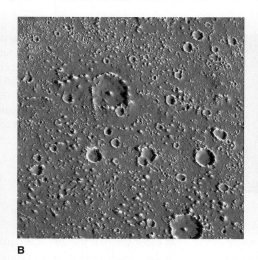

B

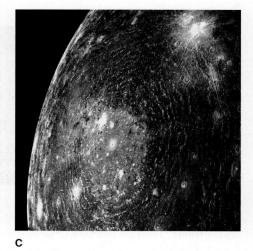

C

FIGURE 8.11 Callisto
Callisto's dark surface is marked by bright patches from impacts, as seen in this full-globe image (A) and this close-up (B) of craters, both captured by *Galileo*. (C) A multiringed basin is surrounded by ejecta and rings of mountains from the impact that created it.

turning into the hardened ice-rock structures seen today (**Figure 8.12**). Although the grooves are clearly newer than the cratered regions around them, it is not clear whether Ganymede remains active today. "If activity really has stopped," says Hendrix, "then the question is, why? What made this moon so active earlier in its history and then shut that activity down?"

One answer may lie in Ganymede's orbit. A perfectly circular moon on a perfectly circular orbit around a perfectly circular planet would still feel tidal forces, but those forces would not change over time. The moons of Jupiter, however, are not on perfectly circular orbits. The gravitational interactions between the moons, as well as with Jupiter, have nudged the moons into somewhat elliptical orbits. Thus, as a moon gets closer to Jupiter, it feels an increased tidal stretching in its interior, compared with when it is more distant. These changes in tidal stretching that occur throughout its orbit release energy. While Ganymede's orbit has a relatively low eccentricity now, in the past the orbit may have been more elliptical, which might explain the energy that drove the grooved flows and other younger features of the moon's surface. If Jupiter's massive gravitational field eventually pulled Ganymede into a more circular orbit, then the heating would have dropped off and the moon's surface activity would have diminished.

The *Galileo* probe's magnetometer found evidence for subsurface water on Ganymede, just as it did on Callisto. "The evidence points to a 5-km-thick layer of liquid water lying more than 150 km below Ganymede's surface," says Hendrix.

EUROPA, WATER WORLD. Europa, the second-closest Galilean moon to Jupiter, has become the focus of intense interest and debate over the last three decades. Europa has a radius of 1,561 km, making it slightly smaller than Earth's Moon. With a density of 3,010 kg/m³, Europa is clearly a rockier world than either Callisto or Ganymede, and scientists believe it has a differentiated metal core.

Europa's surface has an albedo of almost 0.7. That means almost 70 percent of the sunlight striking its surface is reflected back into space. Such a high albedo not only is indicative of water ice but also implies ice that is young and relatively pure. Unlike the ice seen on Jupiter's two outer moons, Europa's surface is young enough that meteor impacts have not vaporized the water away to leave a higher fraction of dark, "dirty" material (**Figure 8.13**). "Something remarkable is happening on Europa," says Hendrix.

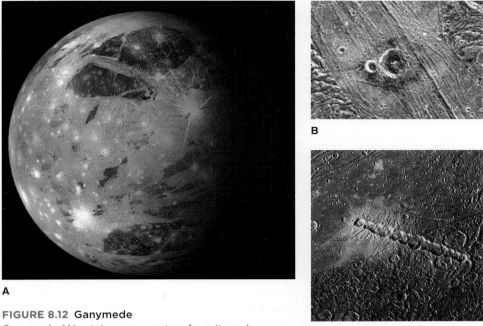

FIGURE 8.12 Ganymede
Ganymede (A) retains many craters from its early existence, but it also shows many long grooves (B) that formed because of resurfacing driven by tidal heating from Jupiter. (C) The 13-crater feature pictured here, called Enki Catena, is consistent with the idea that Ganymede once collided with a disintegrating comet that had been torn apart by Jupiter's gravity.

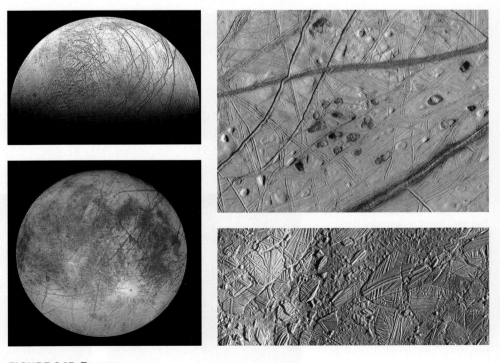

FIGURE 8.13 Europa
The bright, young surface of Europa is composed of ice and lies atop a water ocean 100 km deep. The planet is constantly resurfacing itself as "plates" move or break apart, allowing water from underneath to rise up and form new surface ice.

The apparent youth of Europa's surface is confirmed by its relative lack of craters. The brightest feature, called Pwyll (named for Pwyll Pen Annwn, a character from Welsh mythology), appears as a circular bull's-eye. "Pwyll looks more like concentric cracks on the surface of an icy pond than an impact crater on a rocky crust," says Hendrix. That pond analogy can be taken much further. A close inspection of Europa's surface indicates the absence of any rocky crust. Instead, Europa appears to be covered entirely by ice in constant motion. Jumbled, blocky features can be seen along with long systems of cracks, some of which extend more than 1,600 km. Taken together, these features show that Europa's surface is composed of moving "icebergs." When two icy plates move apart, water from below wells upward to freeze in the new space.

Using magnetic-field measurements, scientists have found strong evidence that below Europa's jumbled icy surface lies a liquid ocean potentially 100 km deep. Thus, despite its dense rocky interior, Europa is a water world: its entire rocky surface is covered by an ocean that, in turn, lies below kilometers of thick ice. The presence of so much water, especially water in liquid form, makes Europa a prime candidate for theories of life forming elsewhere in the Solar System.

The secret to Europa's oceans may rest, once again, with tidal forces. "Being closer to Jupiter than either Ganymede or Callisto is, Europa is constantly stretched and compressed," says Hendrix. "As that heat works its way toward the surface, it keeps the water ocean liquid and drives all the resurfacing of the icy crust."

IO, VOLCANIC CHILD. Tidal forces are also the key to understanding Jupiter's innermost Galilean moon, Io. Io's density, 3,530 kg/m³, makes it the densest of the Galilean moons. Io might be expected to be covered with impact craters, since all of the debris captured by Jupiter's gravity would pass by it. Instead, Io, even more than Europa, is a world where craters disappear almost as fast as they form. "Io's surface is not just young," says Hendrix, "it's newborn."

When *Voyager* passed Io in 1979, it saw a moon bathed in sulfurous reds and yellows. Instead of craters, it saw active flows of lava stretching across hundreds of kilometers. Turning back as it passed the small moon, the space probe's cameras caught a stunning plume of gas rising from one of Io's 150 visible active volcanoes. Sixteen years later, when *Galileo* pulled into orbit, a few of the older volcanoes were still visible, but entirely new features had appeared, including curtains of bright lava erupting from faults in the surface (**Figure 8.14**). "We could watch as the surface of Io was continuously being formed and re-formed," says Hendrix. "It was simply amazing."

The continuous volcanism on Io is a direct result of intense tidal forces working within the moon. Io moves on a fairly elliptical orbit because of gravitational interactions with the other Galilean moons. The changes in tidal stretching caused by the elliptical orbit continuously pump energy into Io's interior on every orbit, resulting in a permanent state of volcanism. That activity has shaped the moon's evolution and has prevented it from having a subsurface ocean, as all the other Galilean moons are believed to harbor in one form or another.

Plumes of sulfuric gas released by Io's volcanoes are blown hundreds of kilometers into space. While much of this material escapes the moon, some of it returns to the surface. Brighter patches on Io may actually be sulfur dioxide that is released as a gas in volcanic eruptions and then condenses and drifts

FIGURE 8.14

Io

(A) Io, the most actively volcanic world in the Solar System, owes its activity to its position closest to Jupiter, which subjects the moon to the strongest tidal forces. (B) Volcano Ra Patera is seen here spewing volcanic gases from Io's surface. (Copyright Calvin J. Hamilton; www.solarviews.com) (C) Here, Io is color coded blue to yellow to red in increasing brightness, and the brightest spot is the volcano Pillan Patera.

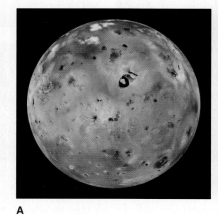

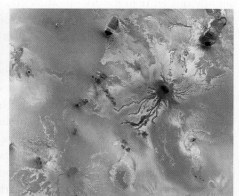

A

B

C

back to the surface as snow. Other elements, mostly sulfur, remain in the gaseous state, forming a thin atmosphere for Io.

Ionized gases escaping Io's meager gravity are swept up by Jupiter's powerful magnetic field, creating a vast **ion torus**, or doughnut-shaped region of emission, where ions blown off of Io in volcanic eruptions spiral around the magnetic field and emit radio photons.

SECTION SUMMARY
Jupiter's Moons
- Callisto harbors a subsurface water ocean but is geologically dead.
- Ganymede, the largest moon of Jupiter, has differentiated layers and subsurface water. Its surface shows evidence of earlier periods of geologic activity, which may or may not continue today.
- Europa harbors a deep water ocean under a young icy surface in constant motion, probably the result of tidal forces.
- Closest to Jupiter and its tidal forces, Io is extremely geologically active, spewing lava and sulfur gas from its many volcanoes.

Jupiter's Rings

Before *Voyager* passed Jupiter, scientists did not know that the giant planet was a ringed world. As the probe passed the massive planet, it turned its camera back toward Jupiter's dark side to reveal the faint outlines of a ring. These images made it clear that Jupiter, like Saturn, hosts a thin assembly of orbiting particles.

Detailed study by the *Galileo* probe demonstrates the presence of four separate rings around Jupiter. The innermost ring is a relatively thick torus of dust called the "halo ring." Beyond it comes the exceptionally thin "main ring," and two relatively thick but faint "gossamer" rings beyond this complete the set.

The outermost rings are named Amalthea and Thebe in honor of the two tiny moons that appear to be the principal source of their dust particles. The other two rings are also the result of dust ejected off small moons (in this case Metis and Adrastea), as these tiny worlds are bombarded by meteoroids. As with most of the rings in the Solar System, astronomers believe their constituent particles are not long-lived. "The dust won't be in the rings forever," says Hendrix. "Eventually the particles will spiral down and fall into Jupiter's atmosphere." While there are

still significant uncertainties, the lifetime of a dust particle in Jupiter's rings appears to be on the order of 100–1,000 years. "That means there has to be a continual resupply of ring material," says Hendrix. "Without the moons, the rings would not last."

SECTION SUMMARY
Jupiter's Rings
- Like Saturn's, Jupiter's ring system includes multiple rings, fed by several of its moons.

8.4
Saturn: Lord of the Rings

Galileo's first telescopic observations of Saturn in 1610 demonstrated that something was clearly different about the planet, which was then thought to be the Solar System's outermost world. Through the low resolution of a refracting telescope with a diameter of less than an inch, all Galileo could see were features that looked like "ears" on either side of Saturn. But Saturn's spin axis is tilted 27° to the plane of its orbit, and its rings possess the same tilt. Therefore, as Saturn moves through its orbit, its rings change their orientation relative to the Sun (**Figure 8.15**).

When the ring plane was aligned with the line of sight from Earth, Galileo was astonished to see Saturn's ears disappear completely. "Has Saturn eaten his children?" Galileo wrote. (Saturn was the Roman god of agriculture and is said to have devoured his own children to keep them from taking his place.) It was not until 1655 that Dutch astronomer Christiaan Huygens trained a larger telescope on Saturn and recognized that the planet was surrounded by rings, not ears. Today we rely on powerful cameras aboard space probes like *Voyager* and the 1997 Saturn probe *Cassini* to enhance our understanding of Saturn's rings (**Figure 8.16**).

Saturn's Interior

With a radius of 58,232 km, Saturn is the second-largest planet in our Solar System. "It would take about 764 Earths to fill up the volume of Saturn," says Linda Spilker.

Saturn's elemental composition is similar to Jupiter's. "You're looking at 80–90 percent hydrogen

ion torus (pl. tori)
A doughnut-shaped region of emission where ions spiral around the magnetic field.

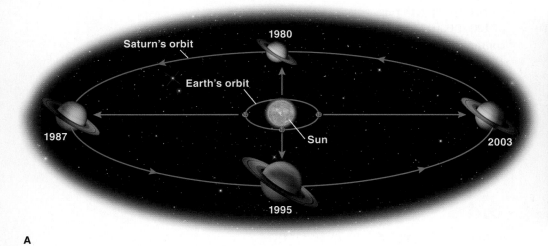

A

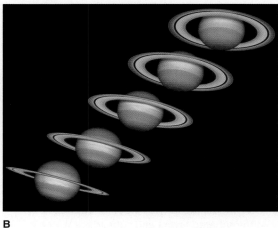

B

FIGURE 8.15 Saturn's Tilt
(A) Like Earth, Saturn maintains its tilt relative to the plane of its orbit throughout its trip around the Sun. (B) Thus, as seen in these images taken by the Hubble Space Telescope annually from 1996 (lower left) to 2000 (upper right), the view of the rings from Earth changes from practically invisible to broad in a continuous cycle.

A Galileo Galilei, 1616 **B Christiaan Huygens, 1659** **C *Voyager 2* spacecraft** **D *Galileo* spacecraft** **E *Cassini* spacecraft**

FIGURE 8.16 Saturn's Rings
(A–B) The earliest images of Saturn are those drawn in the 17th century by prominent astronomers, who saw the planet's rings without understanding their true nature or origin. (C–E) NASA space missions in the 20th and 21st centuries and the Hubble Space Telescope have provided stunning details about the nature and composition of the rings.

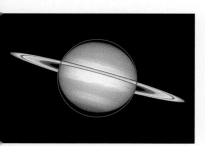

FIGURE 8.17
Saturn's Oblateness
The gas giants, Saturn in particular, exhibit a noticeable oblate shape because of their rapid rotation. The red circle drawn around the planet shows what it would look like if it were perfectly spherical.

in Saturn's atmosphere, with helium coming next in abundance," says Spilker. "Then there are also trace amounts of various other gases." Despite the compositional similarities, the interior of Saturn is not identical to that of its larger cousin. Like Jupiter, Saturn has a layer of molecular hydrogen over a layer of metallic hydrogen, all surrounding a rocky/metallic core of heavy elements, but Saturn is unique in that it has a very low average density: 690 kg/m³. It is the only world with an average density less than that of water.

The cause of Saturn's low density is likely tied to its lower mass combined with its rapid rotation (see **Going Further 8.1**). Saturn has less mass than Jupiter, so it has less gravity to pull the gas inward. Saturn also spins at a high rate, rotating once every 10.6 hours. Thus, Saturn's rapid rotation counteracts its lower gravity (relative to Jupiter's), reducing both

the density of gas in the equatorial regions and the average density of the whole planet. The rapid rotation also affects the planet's shape. "Saturn is the most **oblate** of all the planets," says Spilker. "Its radius at the equator is 10 percent higher than the radius at the poles" (**Figure 8.17**).

Like Jupiter, Saturn emits more energy than it receives from the Sun—about two and a half times more energy, in fact. This extra energy is more than astronomers would expect from their models and has left them with something of a mystery. "While some of the heat could be left over from Saturn's formation," says Spilker, "the numbers don't really add up."

For some scientists, the answer to this mystery might lie in a helium "rain" going on inside Saturn. Astronomers suspect that deep in Saturn's ocean of hydrogen, temperature and pressures may be just right

GOING FURTHER 8.1 THE LANGUAGE OF THE COSMOS
Rotation and Planetary Oblateness

Why does rotation change a planet's shape? The gas giants, Jupiter and Saturn, both spin fast enough to turn them from perfect spheres into fairly oblate objects that are larger at the equator than at the poles. How does this happen?

In general, planets all rotate at a much slower speed than that which a satellite would need to stay in orbit just above the planet's surface (the orbital speed; see Chapter 3). The planet spins as a single body because internal forces (like the force between atoms in rocks or pressure in gases) provide support against gravity and keep everything moving at the same rotational speed. But for very rapidly spinning planets, we have to begin thinking about the role of the *centripetal force*. Recall from Chapter 3 that this is the force needed to keep a freely orbiting object moving in a circle. Whenever a mass (let's call it m_2) moves in a circle with radius R and speed V, we can write the equation for the required inward-directed (centripetal) force as

$$F_c = m_2 V^2 / R$$

As discussed above, for slowly moving planets, it's not gravity but internal forces that keep everything in the planet moving at speed V. But in rapidly rotating planets, the demands of the centripetal force increase.

With this in mind, we can use the formula above to think about the shape of the gas giant planets. The trick is to think about a blob of gas in the planet's atmosphere. For a rapidly spinning spherical world, material at the equator, which, since it lies at the largest distance (R) from the spin axis, requires the highest centripetal acceleration to keep it moving with the rest of the rapidly rotating planet. Material at the poles is not spinning at all.

Now let's look at the ratio of the centripetal force to the force of gravity. This will give us an idea of how much the demands of rotation are beginning to affect the planet's dynamics. Recall that the force of gravity between the blob of mass m_2 and the planet of mass m_p can be written as

$$F_g = (G m_p m_2 / R^2)$$

Notice that we're using R as both the radius of the planet and the radius of the circular path of the blob. That's because we're imagining our gas blob to be sitting in the outer parts of the planet and we're only considering gas at the equator. The ratio of the two forces is

$$F_c / F_g = (m_2 V^2 / R) / (G m_p m_2 / R^2)$$

We can cancel all the identical terms on top and bottom to find that

$$F_c / F_g = (V^2 R) / (G m_p)$$

Now let's compare this ratio for three planets: Earth, Jupiter, and Saturn:

$$(F_c / F_g)_E = 0.0001$$

$$(F_c / F_g)_J = 0.0021$$

$$(F_c / F_g)_S = 0.040$$

These numbers tell the whole story. For Earth, the centripetal force at the equator is a tiny fraction of the gravitational force. That means the internal forces are what keep Earth rotating as a solid body, and they have no problem keeping Earth an (almost) perfect sphere. For Jupiter, that ratio is still small, but it is 20 times larger than what occurs on Earth. For Saturn, the ratio of forces is 20 times as high as on Jupiter.

Thus, we should expect that the internal forces alone cannot keep the gas giants rotating as a purely spherical body and that they will begin to be distorted the most at the equator, where the spin matters most. Equatorial gas will move slightly outward, making the centripetal force (with its $1/R$ dependence) lower. Gas at the equator will therefore bulge outward relative to the poles, making both planets fairly oblate. Since Saturn has such a high ratio of centripetal force to gravitational force, it is the most oblate of the planets, and this high oblateness contributes to its low density.

for the initially well-mixed helium atoms to condense into individual droplets. Just as water droplets condensing in Earth's atmosphere are denser than the air, the condensed helium is denser than hydrogen, so the droplets sink deeper into Saturn's interior, releasing gravitational energy and heating the planet's interior.

SECTION SUMMARY
Saturn's Interior

● Saturn has an oblate shape because of its rapid rotation, and is composed mainly of hydrogen and helium. It is the least dense of all the planets.

● Saturn's apparent excess of radiated energy may be due to internal helium rain.

oblate Roughly spherical, showing flattening in the equatorial plane.

Saturn's Atmosphere

Saturn's atmosphere does not offer the same visual spectacle as Jupiter's. "Saturn is smoother and has this overall golden haze," says Spilker. "We don't tend to see huge colorful storms on Saturn like the Great Red Spot of Jupiter." Astronomers believe the difference between Saturn and Jupiter may be simply that Saturn is farther from the Sun, so its atmosphere is colder. "The colder temperatures may allow a layer of haze to form at the top of the atmosphere," says Spilker, "and that masks cloud structure deeper down" (**Figure 8.18**). The lower temperatures found on Saturn also mean that its cloud layers form deeper within the atmosphere.

Using the *Cassini* space probe, scientists like Spilker have been able to view these cloud structures with imagers tuned to longer-than-visual wavelengths. "When we go deeper, we do start to see an atmosphere that's more dynamic and looks more like Jupiter's." But while Saturn shows the same belt and zone structure as Jupiter, there are important differences. The wind patterns on Saturn show fewer alternating bands than do those on Jupiter. Just as important, the wind velocities on Saturn are enormous, reaching speeds as high as 1,800 km/h (almost 5 times as high as those on Jupiter). "The high wind speeds are partly due to Saturn's fast rotation," explains Spilker.

A particular mystery associated with Saturn (and not seen on Jupiter) is the presence of strong and strangely shaped polar storms. "There's a hurricane-like structure at the south pole that we were able to image," says Spilker. "We actually saw the shadow of

regular satellite A satellite orbiting in the same direction as its planet's movement around the Sun and with a low inclination to the plane of the planet's orbit.

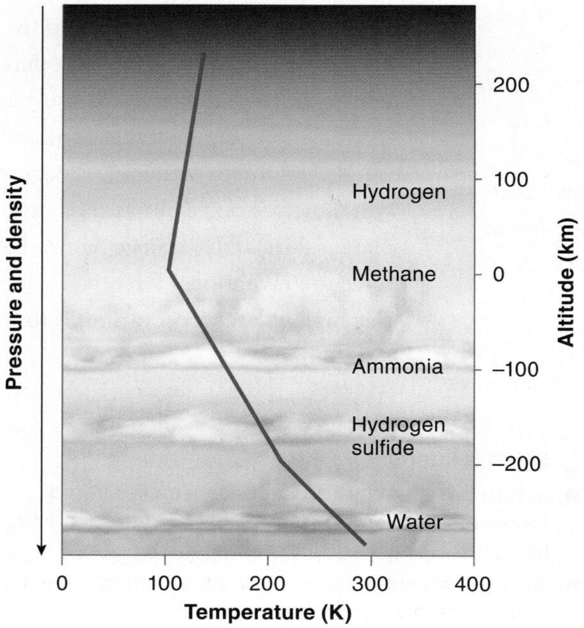

the huge wall of clouds that makes the hurricane's eye. That was amazing. The storm at the north pole is even more peculiar. It has a six-sided hexagon-shaped pattern." That storm was first seen with the *Voyager* probe and was still visible when *Cassini* arrived 25 years later (**Figure 8.19**). How such sharp corners can be maintained on a storm vortex remains a puzzle. "There has been lots of speculation about how you end up with two very different kinds of vortices at the different poles on Saturn," says Spilker, "but we still don't have any firm answers."

SECTION SUMMARY
Saturn's Atmosphere
- Haze obscures Saturn's cloud layers somewhat, but high winds and oddly shaped polar storms have been observed.

Saturn's Magnetic Field

Saturn's magnetic field is 20 times weaker than Jupiter's. The weaker field leads to fewer trapped, charged particles, so Saturn does not show the extended belts of radiation that are seen around Jupiter. Vibrant auroras are, however, observed around Saturn, making it clear that charged particles are being funneled to the planet's poles.

The near-perfect alignment of Saturn's magnetic field with its rotation axis is a huge puzzle for scientists. "Is it really so perfectly aligned or are processes in the atmosphere just making it appear symmetric?" asks Spilker.

The magnetic field on Saturn must be formed by swirling convective motions in the metallic hydrogen layer deep within the planet. But Saturn's lower mass and lower gravity suggest that the region where compression can squeeze hydrogen into metallic form is smaller than on Jupiter, accounting for Saturn's weaker field. Whether this difference in size of the metallic hydrogen layer causes such strong alignment between the magnetic and rotational poles is still unknown.

SECTION SUMMARY
Saturn's Magnetic Field
- Saturn's magnetic field, which is closely aligned with its axis of rotation, is many times weaker than Jupiter's.

Saturn's Moons

Scientists group Saturn's 62 moons into two types. There are the 24 **regular satellites,** which tend to move on prograde orbits (moving in the same direction as

FIGURE 8.18
Saturn's Atmosphere
The colder atmosphere of Saturn, compared with Jupiter's (see Figure 8.6), produces lower cloud layers and a haze that hides deeper layers.

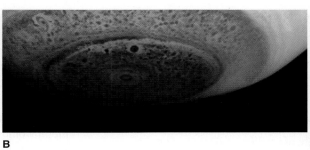

A

B

C

FIGURE 8.19 Storms at Saturn's Poles
(A) This *Cassini* image shows that Saturn's north pole is encircled by a hexagonal feature that is at least decades old. (B) This *Cassini* image of the south pole revealed a massive cyclone-like storm. (C) The Hubble Space Telescope captured an enormous storm at Saturn's equator in 1994.

Saturn's rotation) and stay close to the plane defined by Saturn's equator. "These moons probably formed out of the same disk of material which gave birth to Saturn," says Spilker. The **irregular satellites** tend to be farther away from Saturn and move on highly inclined orbits. "The irregular satellites can be retrograde or prograde," says Spilker, "and are most likely material Saturn captured at some point later in its history." Two of the regular moons that astronomers have found most fascinating are Titan and Enceladus.

TITAN. Titan, Saturn's largest moon, is larger than the planet Mercury. "If Titan had formed on its own," says Spilker, "it would be a planet in its own right." Titan shows an uncompressed density of just 1,500 kg/m³, meaning that it must contain significant amounts of ice. But what really sets Titan apart is that it is the only moon in our Solar System with a substantial atmosphere. In 2005, *Cassini* released the *Huygens* probe, which parachuted into Titan's atmosphere and returned remarkable and fundamentally new views of the icy moon (**Figure 8.20**).

"Titan's atmosphere is very dense," says Spilker, who explains that the pressure on that moon's surface is 50 percent higher than what we experience on the surface of Earth. The composition of the atmosphere is also remarkable. "Titan's atmosphere is composed mostly of nitrogen, very similar to Earth's atmosphere," says Spilker, "but there's pretty much no oxygen." Titan's atmosphere does, however, have a wealth of complex **hydrocarbons**: methane, ethane, propane, acetylene, and others. Together these compounds make for strange weather on Titan.

"It's very cold on Titan," says Spilker. "Daytime temperatures are only about 94 K." With so little heat

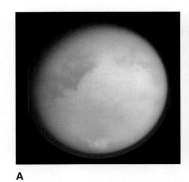

A

B

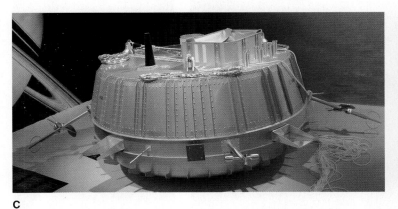

C

D

FIGURE 8.20 Titan
(A) Although visible light shows only the cloud tops of Titan's atmosphere, infrared light and false colors present these colorful views of Titan. (B) Using *Cassini*'s synthetic aperture, radar scientists made this mosaic of Titan's north polar region, revealing large lakes and tributary networks composed of liquid ethane and methane (shown in blue). Some regions were not mapped and show up as blank spots in the map. (C) The *Huygens* probe (shown here in replica) landed on Titan's surface. (D) It beamed back images from the surface for 90 minutes before ending operation.

irregular satellite A satellite orbiting in the opposite direction from its central object or on a highly inclined orbit in relation to the planet's orbital plane.

hydrocarbon An organic compound consisting entirely of hydrogen and carbon.

available, hydrocarbons in the atmosphere can condense into rain, and astronomers have observed lakes of liquid methane on Titan. "We can even see river channels where the methane rain collects and flows downhill to fill the lakes," says Spilker.

ENCELADUS.   A tiny moon only about 480 km in diameter, Enceladus has become the other celebrity in Saturn's family of satellites. Given its tiny size, Enceladus should be frozen solid and inactive. "That is not the case," explains Spilker. "We were all absolutely amazed and delighted to see active geysers of water vapor and ice crystals blasting out at Enceladus's south pole." (**Figure 8.21**).

The plumes of material erupting from Enceladus are important for a number of reasons. First, they appear to be the source of one of Saturn's outer rings (called the E ring). Second, the presence of liquid water on Enceladus raises the possibility, once again, that life can form on a moon of the outer Solar System. "When we flew near these plumes to investigate their composition," says Spilker, "we saw compounds with carbon, nitrogen, and oxygen in them, as well as salts—table salt, potassium salt, and so on." These

are the kinds of minerals scientists would expect if an underground ocean was providing the water for Enceladus's plumes.

Some scientists have argued that tidal forces from one of Saturn's other moons, Dione, might be the plumes' energy source, but others have questioned whether the amount of heat generated by another moon would be sufficient to keep water in liquid form on Enceladus. For now, the plumes remain a wonderful mystery for astronomers like Spilker to solve.

SECTION SUMMARY
Saturn's Moons

- Titan, the only Solar System moon with a substantial atmosphere, features methane lakes and rain on its surface; hydrocarbons exist in abundance.
- Tiny Enceladus spouts geysers of water, which feed one of Saturn's rings; salts found in the plumes suggest an underground ocean.

Saturn's Rings

Although Saturn's main rings extend 7,000–80,000 km from the planet's edge, they are a mere 10 meters thick vertically. From Earth, astronomers can see that Saturn's spectacular rings are actually a system of separate smaller ring systems. Earth-based telescopes resolve three main sections, called the A, B, and C rings. Once space probes like *Voyager* and *Cassini* arrived, however, scientists saw a number of new rings, as well as fascinating new features within the larger systems (**Figure 8.22**).

The composition of Saturn's rings is the first and most pressing question. "They are made primarily of water ice," explains Spilker. "The best data show the rings to be 99 percent water ice with a tiny amount of non-icy material contaminant." Contaminant mass likely comes from micrometeorite bombardment adding to the icy material over time. These ring particles range from dusty ice grains all the way up to objects meters in diameter. "Even though you have this wide range," explains Spilker, "the typical size of material seen in the large rings is about a centimeter" (**Figure 8.23**).

The next question is when and how the icy material got there. "When we got to Saturn with *Voyager* and saw these bright, pristine, icy rings," says Spilker, "we immediately concluded they must be made of material much younger than the age of the Solar System." Early estimates of the rings' age leaned toward the 100-million-year range. "But based on what we learned from *Cassini*," Spilker continues, "our studies

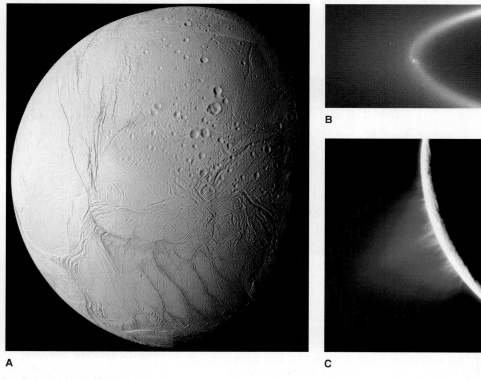

A **B** **C**

FIGURE 8.21 Enceladus
(A) Enceladus as imaged by the *Cassini* mission. Its highly reflective surface is believed to be composed of water ice. (B) Enceladus, which orbits in Saturn's E ring, as initially observed from Earth. (C) Scientists have discovered that geysers on Enceladus vent water mixed with salt and hydrocarbons from cracks in its surface.

FIGURE 8.22
Saturn's Ring Structure
Though Saturn's rings are made up of many small particles, they are remarkably stable, thanks in part to the moons that sweep out gaps and provide gravitational "shepherding." The individual rings and gaps, or divisions, have been observed for centuries, although discovery of the dimmer rings required images beamed back from visiting spacecraft.

show it may be more complicated." According to scientists, the massive B ring might have formed with Saturn many billions of years ago. The other rings, however, might have formed in other ways. "Maybe a comet got a little too close and was broken apart by tidal forces from Saturn," says Spilker. The smaller rings may be linked to material stripped off of moons, such as the E ring that is fed by Enceladus's plumes.

Moons play another important role in Saturn's rings: the gravitational force of the moons can keep ring particles in place. **Shepherd moons**, for example, maintain narrow rings, as can been seen for the thin F ring, which is contained by the moons Prometheus and Pandora orbiting inside and outside it (**Figure 8.24**). The combined gravitational force of the two moons focuses the motion of the ring particles that they act

on. Without the gravity of the shepherd moons, the material in the F ring would slowly wander from its current orbits, and the ring would eventually disappear.

Many substructures are seen within the larger rings, and these, too, probably originated in small satellites embedded within the rings. The two big gaps in the Saturn ring system, the **Cassini** and **Encke Divisions**, both have moons that help govern their orbits: material in the Cassini Division responds to gravitational tugs from Mimas (which lies farther out from Saturn), and the Encke Division is maintained by the moon Pan, which orbits inside it. The gravitational force from these moons clears the space around them, creating gaps devoid of material.

The phenomenon of shepherd moons in Saturn's rings highlights another important feature of Solar

FIGURE 8.23
Ring Particles
The particles that compose Saturn's rings range in size from grain-sized ice particles to boulder-sized objects. The average size of a ring particle is about 1 centimeter (0.01 meter).

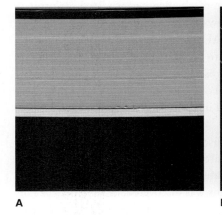
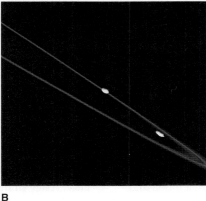
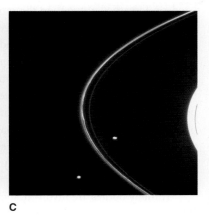

A   B   C

FIGURE 8.24 Shepherd Moons
Shepherd moons of Saturn seen in action by the *Cassini* orbiter. (A) The moon Daphnis creates ripples in the A ring. (B–C) Two views of Pandora and Prometheus guiding the F ring.

shepherd moon A planetary satellite whose gravity confines a ring to a narrow band.

Cassini Division A 4,800-kilometer-wide region between the A and B rings of Saturn's rings that has been shaped by Saturn's moon Mimas.

Encke Division A 325-kilometer-wide gap within Saturn's A ring that has been shaped by Saturn's moon Pan.

Linda Spilker

"Girls don't go into science," Linda Spilker's high school guidance counselor told her. It was the mid-1970s, and Spilker had already caught the science bug. "I grew up with the space race. I remember watching all the astronaut missions from *Mercury* to *Gemini* and then *Apollo*. I think that really fixed my attention on space." When Spilker was 9, her parents got her a telescope. "I was just so thrilled. I remember taking it out, looking at Jupiter and seeing the little moons as pinpoints of light. From that point I was hooked."

Despite her guidance counselor's discouraging words, Spilker's parents were supportive and knew their daughter could do anything she set her mind to. "My mom loved math," recalls Spilker. "She took algebra when she was a girl. She loved it but got teased so much she dropped the course. I know she always regretted that."

Completing her degree in physics at California State University, Fullerton, Spilker worked for JPL and eventually received her PhD from UCLA in geophysics and space physics. Now Spilker is the *Cassini* project scientist, holding responsibility for the science operations of the entire billion-dollar mission.

System dynamics: the role of **orbital resonances**. When two objects orbit the same body, such as two moons orbiting a giant planet or two planets orbiting a star, they will periodically pass each other, at which point the gravitational force between them is at its strongest. Usually, these tugs happen at different points in the orbits and the effects average out. But if the orbital periods are a whole-number ratio, then the close approaches and strong gravitational interactions will happen at the same place (or places) over and over again. These repeated synchronized accelerations can have strong effects on the bodies by either stabilizing or destabilizing their orbits, depending on the nature of the resonance. The less massive object is always destabilized, or perturbed, more than the more massive object.

Astronomers describe resonances using the ratio of the orbits, and the more perturbed object always appears first in the ratio. A two-to-one (2:1) resonance, for example, means that the more perturbed object orbits twice for every one time that the less perturbed object orbits. Thus, for instance, the dwarf planet Pluto is in a 2:3 orbital resonance with Neptune, meaning that Pluto is the more perturbed object, and it completes two orbits for every three orbits that Neptune completes.

If many bodies are involved, such as a moon and ring particles, then an orbital resonance that produces many close encounters will destabilize the ring particle orbits and clear material away, leading to a gap. For example, the Cassini Division represents a 2:1 orbital resonance with Mimas. In contrast, an orbital

orbital resonance The synchronized gravitational accelerations that arise when objects orbiting a body have periods that are whole number ratios of each other.

resonance that avoids close encounters allows the ring particles to stay in their orbits.

SECTION SUMMARY
Saturn's Rings
- Saturn's rings include many sections, with a range of sources and histories.
- Water ice and a small amount of contaminants make up the individual ring "particles"; these average a centimeter in diameter but span a wide range of sizes.
- The ring divisions and other features are formed and regulated by the gravity of Saturn's moons.
- Orbital resonances occur when objects orbiting a body have periods that are whole-number ratios of each other. Among other effects, orbital resonances can create gaps in rings.

8.5
Uranus and Neptune: Ice Giants Discovered in Twilight

The outermost planets in our Solar System are almost twins in terms of their size and composition. Like any human twins, however, they have acquired distinct properties through time and evolution (**Figure 8.25**).

Across most of human culture's 50,000 years, only five planets were visible from Earth: Mercury, Venus, Mars, Jupiter, and Saturn. Not until 1781 did William Herschel spot a fuzzy object that appeared too large to be a star. Watching carefully over many nights, Herschel found his mysterious object moving against the background stars but at too slow a speed to be a comet. He soon recognized it to be a seventh planet. Using Newtonian mechanics, he calculated the orbit and found the new world moving around the Sun at a distance of 19 AU and a period of 84 years. While initially inclined to call the new planet "George" in honor of the English king, Herschel was persuaded to retain the mythological tradition in planet naming, and he called it Uranus for the Roman god who fathered Jupiter.

While the discovery of Uranus came through serendipity, the discovery of the eighth planet, Neptune, was anticipated, and a triumph of Newtonian science. After Uranus was discovered, astronomers began charting its orbit in great detail. But a discrepancy was soon found in the new planet's motion that became more pronounced with time. All solutions to Newton's equations that described Uranus moving solely under the gravitational influence of the Sun and other known

A

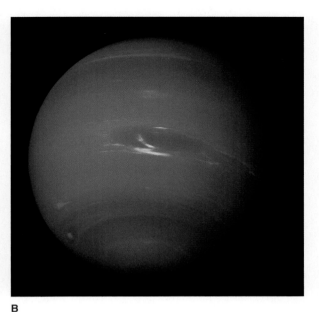

B

FIGURE 8.25
Uranus and Neptune
Ice giants (A) Uranus and
(B) Neptune are similar in size,
structure, appearance, and
composition.

planets failed to predict the motions astronomers saw on the sky. Only when astronomers added an as-yet unknown planet tugging on Uranus could they predict the observed motions.

Using Newtonian mechanics, astronomers predicted where this eighth planet should be and estimated its mass. In 1846, the planet was found almost exactly where predictions had put it, on an orbit with average radius of 30 AU from the Sun and a period of 165 years. The confirmation of a prediction as specific as the location of Neptune is one of the hallmarks that distinguishes science from other human endeavors.

Ice Giant Interiors

Given how similar in size, mass, and density Uranus and Neptune are, astronomers do not expect the internal structures of the two planets to differ significantly. The low densities of Uranus and Neptune indicate large reservoirs of light elements. They also indicate the presence of a rocky core at the center of each planet. Mathematical models of internal structure suggest that the core takes up about 20 percent of the planetary radius of Uranus, and somewhat more for Neptune. These cores would then be about the size of Earth but probably contain 10 times as much mass as our planet. Above each core lies a slush of light-element ices like methane and ammonia mixed in a bath of liquid hydrogen. Above the slush lies a large ocean of mostly liquid molecular hydrogen. Unlike the gas giants, neither Uranus nor Neptune possesses enough mass (and hence gravity) to squeeze the liquid hydrogen into metallic form.

SECTION SUMMARY
Ice Giant Interiors
- Both Uranus and Neptune have rocky cores beneath a layer of ices and liquid hydrogen, and both have top layers of liquid molecular hydrogen.

Ice Giant Atmospheres

The only way astronomers can get direct information on the composition of ice giant atmospheres is by using spectroscopy. From these measurements, both planets appear to have chemical makeups similar to those of the gas giants: about 80 percent hydrogen and 18 percent helium. Methane turns out to be the next-most-abundant component (less than 2 percent on each world). One key difference between the gas and ice giants is that Uranus and Neptune show much less ammonia in their atmospheres. The lower temperatures on these worlds allow ammonia to condense into crystals, which fall to the interior (**Figure 8.26**).

When *Voyager 2* visited the ice giants (the only time a probe has approached these worlds), it saw Uranus as a hazy greenish orb and Neptune appearing far more blue. These colors are easily explained by the relative abundance of methane, which is very effective at absorbing longer-wavelength (red) visible light. Neptune has more methane than Uranus, so more long-wavelength light is absorbed out of the sunlight striking Neptune, leading to a deeper blue color.

The atmospheres of Uranus and Neptune appear far less structured than those of the gas giants (**Figure 8.27**) in terms of bands of different colors. Because

FIGURE 8.26
Ice Giant Atmospheres
The atmospheres of the ice giants display far fewer features than those of the gas giants, with only a single cloud layer in each planet's atmosphere.

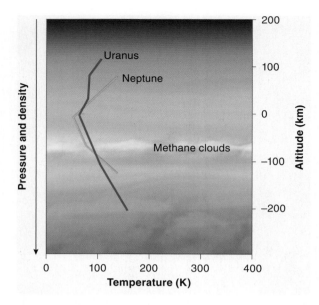

FIGURE 8.27 Weather on the Ice Giants
High winds and other complex features can be seen in the ice giant atmospheres, though to a lesser degree than in the gas giants. (A) This color-enhanced image taken by the Keck II telescope shows the brightest cloud feature ever observed on Uranus. (B) This HST image of Neptune shows the strong band structure, where winds may exceed 300 km per hour (670 miles per hour) in alternating directional bands.

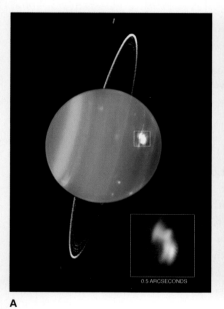

A

B

these outermost planets are colder, clouds form lower in the atmosphere (at higher temperatures) below the overlying haze. Storms and some banded cloud structures are seen on both planets, however. Infrared observations that can see deep into the planets' atmospheres have shown individual bands of clouds rotating around Uranus at 500 km/h. On Neptune,

Earth-sized storms, such as the Great Dark Spot seen during *Voyager 2*'s visit, have been observed to appear and vanish over intervals of a few years.

One significant difference between Uranus and Neptune is the inclination of their spins. Neptune rotates around an axis that is inclined 28° to the plane of the ecliptic (within the range of most other planets), but Uranus's axis of rotation is inclined a remarkable 98° from the ecliptic, such that its spin axis is almost in the plane of its orbit (**Figure 8.28**). As Uranus orbits the Sun, it experiences the most dramatic seasons possible. During its northern "summer," the north pole is pointed almost directly at the Sun, which therefore never sets over the entire northern hemisphere. Half a Uranian year later, the south pole points toward the Sun and the northern hemisphere is enshrouded in darkness. Astronomers are still at a loss to explain how Uranus ended up with such an extreme axial tilt. A collision with another large object is one possibility, but that far out in the Solar System, where the density of planetesimals would be lower, it is unclear how probable such an event would be.

SECTION SUMMARY
Ice Giant Atmospheres
- Uranus and Neptune are composed mostly of hydrogen and helium, with enough methane to produce their green and blue colors, respectively.
- Strong winds and large storms occur in the two planets' atmospheres.
- Neptune's spin axis is tilted about 28°, similar to that of Earth and most other planets, but Uranus's spin axis lies almost in the plane of its orbit, giving it extreme seasonal variations of sunlight.

Ice Giant Magnetic Fields

Without the oceans of metallic hydrogen like those of the gas giants, the ice giants might be expected to lack strong magnetic fields. However, both ice giants possess fields that are inherently 100 times stronger than Earth's.

The most peculiar aspect of these fields is not their existence but their inclination. Both Uranus's and Neptune's magnetic axes are significantly offset from the planets' spin axes—by 59° and 47°, respectively. The magnetic field on each of the ice giants is not centered in the planet but is shifted to a position at some distance from the planet's core (see Figure 8.5). No mechanism for creating the ice giants' off-center fields has been universally accepted by scientists, but potential explanations rest on currents of ammonia ions flowing in the slush layers deep within Uranus and Neptune.

● The ice giant magnetic fields are off center and greatly inclined to the planets' spin axes.

Ice Giant Moons

THE MOONS OF URANUS. At least 27 moons orbit Uranus. Most of these are less than 400 km in diameter, and even Titania, the largest, is smaller than Earth's Moon. Remarkably, all of Uranus's moons orbit in its wildly tilted equatorial plane. Thus the moons and Uranus itself experience the same extremes in seasons. All of Uranus's moons have low albedos, which may result from solar and cosmic radiation breaking down surface molecules. In addition, the less geologically active a moon is, the darker its surface should be, since any geologic events would bring up new, lighter material to the surface. All of Uranus's five large moons have densities less than 1,700 kg/m³, indicating a composition that includes significant amounts of water ice. This is to be expected for objects forming so far beyond the snow line.

The outermost large moons, Titania and Oberon, show surfaces pocked with ancient craters. Both these worlds appear geologically dead, experiencing no resurfacing since, perhaps, they were born with Uranus. Umbriel and Ariel are next in the order of decreasing orbital distance. Neither of these worlds shows evidence for ongoing geologic activity; however, Ariel has long cracks running across its surface, perhaps the result of tidal stresses experienced during the moon's evolution.

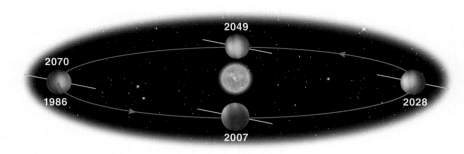

FIGURE 8.28 Uranus's Axis Tilt
Because its spin axis is nearly aligned with the plane of its orbit, the north and south portions of Uranus experience long periods of darkness during its 84-year orbit.

Of all Uranus's moons, Miranda, the innermost, is most intriguing. Miranda shows some cratering, but its face is gashed by enormous faults, ridges, and valleys (**Figure 8.29**). Instead of a geologically dead world, as some astronomers expected, Miranda turns out to be a buffet of planetary features. Some researchers have proposed that Miranda may have been torn apart by some unknown disaster and gravitationally reassembled a number of times.

THE MOONS OF NEPTUNE. Neptune has six regular moons and seven irregular ones—satellites that orbit either retrograde or nonequatorially (**Figure 8.30**). The largest of Neptune's moons, Triton, is the only satellite of its size in the Solar System to orbit in the retrograde sense. Some scientists speculate that Triton's retrograde orbit means that it did not form along with Neptune but was captured later. All the regular moons' orbits lie inside Triton's, and all of the other irregular moons' orbits lie outside Triton's.

Geologically, Triton is bright and appears to continually reshape its surface. When *Voyager 2* passed Triton, scientists were amazed to see geysers

FIGURE 8.29
Miranda
Uranus has a broad collection of moons, some related to its rings in properties and origins, some with highly irregular orbits. Its innermost moon, Miranda, presents a variety of dramatic surface features.

Triton   Proteus   Nereid

FIGURE 8.30
Moons of Neptune
Neptune's moons are dominated by Triton, its only spheroidal moon, which accounts for 99.5 percent of the mass of Neptune's satellites; its other major moons, Proteus and Nereid, are tiny by comparison and have irregular surfaces.

of nitrogen blown from the surface into space. These geysers appear to be the source of Triton's thin nitrogen atmosphere (it is 100,000 times less dense than Earth's). The remarkably low temperatures on Triton (just 37 K) allow some of the nitrogen to turn to frost and drift back down to the planet.

Along with the nitrogen geysers, the lack of craters on Triton indicates that it has been active fairly recently. Circular features resembling lakes suggest that liquid water from below the surface may periodically well up, flooding landforms before it freezes. Strangest of all is the "cantaloupe" terrain in the moon's southern regions, composed of numerous dimples, ridges, and valleys. The jumbled features in these regions indicate a surface that has experienced significant activity, probably the result of Neptune's tidal forces.

┌─ **SECTION SUMMARY**
  Ice Giant Moons

● All of Uranus's large moons have densities suggesting the presence of significant quantities of water ice, and all have orbits aligned with Uranus's tilted equator.

● Uranus's moon Miranda shows signs of a violent past; the other large moons display older, geologically dead surfaces.

● Neptune has six regular moons and seven irregular ones, many with highly eccentric orbits.

● Neptune's moon Triton has a bright, young surface; a thin atmosphere; nitrogen geysers; and cantaloupe-like terrain. Triton is on a retrograde orbit.

Roche limit The closest that a satellite can get to its central object before tidal forces from that object pull the satellite apart (because the satellite can no longer be held together by its own gravity).

Ice Giant Rings

Uranus and Neptune both host rings. In fact, Uranus's rings were discovered before either Jupiter's or Neptune's were. In 1977, astronomers were using the stellar occultation method to look for new moons around Uranus. As the planet's path took it close to a background star, the astronomers kept careful watch for unseen Uranus-orbiting objects that might block out the starlight. Instead of finding moons, however, the researchers saw the star wink on and off 40 minutes before Uranus occulted the star and 40 minutes after. The symmetry of the occultations suggested that they were caused by rings.

Uranus hosts at least 11 distinct rings with radii ranging from 37,000 to 51,000 km. Each ring is relatively narrow. The innermost ring, called 1986U2R, is the widest, at 2,500 km, but it is diffuse and difficult to see. The other rings have widths of 100 km or less. The rings are separated by large gaps, some of which are 1,000 km wide.

All of Uranus's rings are within its **Roche limit**, which is the closest distance an object can approach and not be torn apart by tidal forces. The rings are presumably composed of comet or asteroid material torn apart as it wandered too close to the planet. Uranus's rings are quite dark, which means they are unlikely to be composed of relatively pure ices like Saturn's large rings. Instead, the particles have likely been coated with material blown off of Uranus's moons. As in other ring systems, Uranus's moons play a shepherding role, their gravity keeping the rings sharp and preventing the rings' particles from drifting away (**Figure 8.31**).

All of Neptune's five rings lie within its Roche limit as well, so Neptune's rings are likely to be similar in origin to Uranus's. Neptune's rings are a mix of both broad and narrow particle systems, however. Of particular interest is the outer Adams ring, which is only 50 km wide and shows bright knots or clumps. In shorter exposures (when the camera is open for a brief time and collects less light), only arcs of material can be seen rather than a true ring. The origin of these clumps is not yet clear, though an unseen shepherd moon may be the culprit.

┌─ **SECTION SUMMARY**
  Ice Giant Rings

● The rings of Uranus are narrow, widely spaced, and dark with dust from small bodies within the Roche limit.

● The rings of Neptune are also within the Roche limit and vary in width.

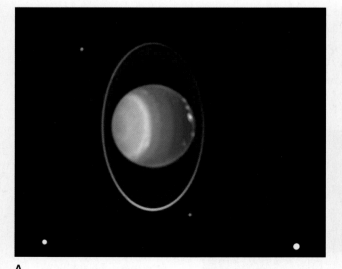

A

B

FIGURE 8.31 Ice Giant Rings
The Hubble Space Telescope captured these images of the thin, rocky rings of (A) Uranus and (B) Neptune.

chapter summary

8.1 Giant Planets on a Roll

The gas and ice giants formed outside the snow line from molecules and compounds that became solid (froze) at lower temperatures. Located at successively greater distances from the Sun, they receive relatively little solar radiation compared with the terrestrial planets.

8.2 The Giant Planets: Structures and Processes

The giant planets are larger in mass and radius than the terrestrial planets, but much less dense. All lack hard surfaces and show a continuum of increasing density and pressure from the outer edges of their atmospheres to their metallic/rocky cores. The enormous internal pressure due to the gravity of overlying material creates extreme internal conditions such as liquid hydrogen oceans. Each giant produces more energy through its slow contraction than it receives from the Sun. Their rapid rotation drives strong weather systems in their atmospheres, such as banded winds and giant storms. Each giant planet exhibits a strong magnetic field (driven by rapid rotation), a ring system, and many diverse moons.

8.3 Jupiter: King of Planets

Jupiter comprises two-thirds of the Solar System's mass (excluding the Sun) and has played a stabilizing and protective role for the Solar System by gravitationally sweeping up comets and asteroids. Its interior layers include a molecular hydrogen ocean, a layer of metallic hydrogen, and a rocky and/or metallic core at high temperature. The energy from Jupiter's continued contraction helps drive convection in the counter-rotating belts and zones of the atmosphere, where winds reach hundreds of kilometers per hour, and where the hurricane-like Great Red Spot has existed for at least 300 years. A vast, elongated magnetic field surrounds Jupiter, drawing ions from the tidally locked volcanic moon Io. The known moons of Jupiter are dominated by the four large, tidally locked Galilean moons. Callisto and Ganymede are geologically dead moons whose surfaces hide water oceans, though evidence suggests that Ganymede was once extremely active and remained so for some time after its formation. Europa's icy surface is characterized by the constant motion of its many plates of ice over a water ocean. Io is covered with active volcanoes, which derive their heat from Jupiter's strong tidal forces. Not visible from Earth, Jupiter's dark, segmented rings were discovered by the *Voyager* spacecraft.

8.4 Saturn: Lord of the Rings

Saturn shows the same structure as Jupiter but has a lower density because of its smaller mass and rapid rotation (it is the only planet with a density lower than water). The rapid rotation has also given Saturn an oblate shape. It drives high wind speeds within the belt/zone atmospheric structures, and it is responsible for large storm systems at the poles. The known moons of Saturn can be grouped into regular satellites (orbiting prograde in the plane of Saturn's equator) and irregular ones (with highly inclined and some retrograde orbits). The largest moon, Titan, has a dense atmosphere with complex hydrocarbons. Geysers of water have been observed emanating from the surface of the moon Enceladus. Saturn's rings have a complex structure, with a number of distinct ring systems and gaps that are shaped and maintained by shepherd moons.

8.5 Uranus and Neptune: Ice Giants Discovered in Twilight

The ice giants Neptune and Uranus are similar to each other in size, mass, density, and composition. Their structures include rocky cores; slushes of methane, ammonia, and hydrogen; and oceans of liquid molecular hydrogen. Their atmospheres contain hydrogen, helium, and methane and display storms, belts, and zones like those of the gas giants, although less prominently. A higher methane concentration in Neptune's atmosphere gives it a bluer color, compared with greenish Uranus. While Neptune shows a typical tilt of 28°, Uranus lies nearly on its side, thus experiencing dramatic seasons. The magnetic fields of both planets are offset significantly from their spin axes. Among Uranus's moons, most noteworthy is Miranda, with dramatic faults, ridges, and valleys. Neptune's moons are split roughly evenly between regular and irregular types. The largest, Triton, is the only large satellite in the Solar System to orbit retrograde, and it exhibits nitrogen geysers on its jumbled surface.

questions and problems

Narrow It Down: Multiple-Choice Questions

1. Which of the following planets formed within the snow line? Choose all that apply.
   a. Earth
   b. Mars
   c. Jupiter
   d. Saturn
   e. Uranus

2. In which of the following environments are ice and gas giant planets most likely to harbor life?
   a. atmosphere of the planet
   b. surface of the planet
   c. subsurface ocean on one of the planet's moons
   d. atmosphere of one of the planet's moons
   e. none, since these planets orbit where it is too cold for life to exist

3. The spectrum of an exoplanet reveals significant amounts of methane with smaller amounts of ammonia. Considering this information alone, to which planet in our Solar System might we expect the exoplanet to be similar?
   a. Earth
   b. Mars
   c. Jupiter
   d. Saturn
   e. Uranus

4. Metallic hydrogen layers in a giant planet are characterized by which of the following? Choose all that apply.
   a. presence of mobile electrons
   b. molecules of hydrogen
   c. shiny surface
   d. high pressure
   e. liquid state

5. Which of the following is the most likely source of the heat radiated by the ice and gas planets?
   a. solar radiation
   b. magnetic fields
   c. falling rain
   d. planet contraction
   e. nuclear fusion

6. Arrange the following planets by rotation period, longest to shortest:
   a. Mercury, Venus, Mars, Saturn, Uranus
   b. Venus, Mercury, Mars, Uranus, Saturn
   c. Saturn, Uranus, Mars, Mercury, Venus
   d. Mars, Mercury, Venus, Uranus, Saturn
   e. Uranus, Saturn, Mars, Venus, Mercury

7. A distant exoplanet is determined to have rings. From examples in our Solar System, what is the likelihood that it will also have moons?
   a. quite high
   b. quite low
   c. There is no way to know; rings and moons are not related.
   d. 100 percent, since rings and moons are always observed together
   e. zero

8. Which of the following does *not* describe characteristics or effects of Jupiter?
   a. It has benefited life on Earth by sweeping up Solar System debris.
   b. It has a ring system.
   c. It has the largest mass of any object in the Solar System.
   d. It has a strong tidal influence on its moons.
   e. It has distinct cloud layers varying in color.

9. Which of the following describes a commonality between Earth and Jupiter?
   a. The heaviest elements (atomic weights greater than iron) sank to the center of the planet during its formation.
   b. Three distinct cloud layers are present in the atmosphere.
   c. Each acquired a large moon by the same process.
   d. The planet is still shrinking.
   e. The mass ratio of planet to moon(s) is the same.

10. Earth does not show the patterns of belts and zones seen on Jupiter. What is the principal reason for this difference?
    a. Jupiter is farther from the Sun than Earth is.
    b. Jupiter has more mass and hence stronger gravity than Earth has.
    c. Jupiter's rotation rate is higher than Earth's.
    d. Jupiter has more moons than Earth has.
    e. Jupiter's magnetic field is stronger than Earth's.

11. What evidence tells us that the Galilean moons evolved together with the planet? Choose all that apply.
    a. The fact that all are tidally locked with Jupiter.
    b. Their relative densities.
    c. The orientation of their orbits.
    d. Their orbital periods.
    e. The quantity of hydrogen in their composition.

12. In which of the following ways is Io unique among the Galilean moons?
    a. It is strongly influenced by tidal forces from Jupiter.
    b. It has a very high concentration of volcanoes.
    c. Its orbital period is measured in (Earth) days rather than (Earth) months or years.
    d. It is bigger than Earth's Moon.
    e. It was first observed by Galileo.

13. Which of the Galilean moons is *not* believed to have subsurface water?
    a. Callisto
    b. Ganymede
    c. Europa
    d. Io
    e. All are believed to have subsurface water.

14. Why weren't the rings of Jupiter, Uranus, and Neptune discovered until the 20th century if Saturn's rings were discovered centuries earlier?
    a. The rings of the other three planets are made of dust rather than ice and thus reflect less sunlight.
    b. They have large magnetic fields.
    c. They have no shepherd moons.
    d. They are composed of highly energetic particles that don't emit light.
    e. The planets are all at greater distances than Saturn.

15. If Saturn's axis were tilted 5° instead of 27° to the plane of its orbit, which of the following would be true about the appearance of the rings?
    a. They would be easier to observe with a small telescope.
    b. From Earth we would be able to see the rings as if looking down from Saturn's north pole.
    c. The maximum apparent extension of rings above and below the equator of Saturn would be smaller.
    d. They would never disappear.
    e. They would never be visible.

16. Picture the following five Solar System planets placed on the surface of a vast, deep cosmic water ocean. Which would sink fastest?
    a. Saturn
    b. Mercury
    c. Mars
    d. Jupiter
    e. Uranus

17. Moon 1 orbiting a gas giant is found to have a 5:4 orbital resonance with a second moon, moon 2. Which of the following statements describes their orbits?
    a. Moon 1 completes 5 orbits for every 4 completed by moon 2, and feels the stronger perturbation.
    b. Moon 2 completes 5 orbits for every 4 completed by moon 1, and feels the stronger perturbation.
    c. Moon 1 completes 5 orbits for every 4 completed by moon 2, and feels the weaker perturbation.
    d. Moon 2 completes 5 orbits for every 4 completed by moon 1, and feels the stronger perturbation.
    e. There is no way to tell from the given information.

18. Which characteristic is *not* likely to be present in a moon that formed from the protoplanetary nebula with its planet?
    a. prograde orbit
    b. prograde rotation
    c. similar density
    d. highly inclined orbit
    e. similar composition

19. Which of the following features differs significantly between Uranus and Neptune? Choose all that apply.
    a. inclination of their axes to vertical
    b. presence of cloud bands
    c. strong inclination of the magnetic field to the spin axis
    d. degree of seasonal impacts
    e. density

20. True/False: Two planets spin at the same rate, but planet A is more massive than planet B; planet A is therefore less oblate than planet B.

To the Point: Qualitative and Discussion Questions

21. How is the temperature within versus beyond the snow line related to the kinds of materials found in each region?

22. How are seasons related to the tilt of a planet's axis?

23. Would the stellar occultation method work for discovering rings around exoplanets?

24. Describe the key differences between the gas giants and the ice giants.

25. How do the densities of the terrestrial worlds compare with those of the giant planets, and what does this difference tell us?

26. How do the rotation periods of the giant planets compare with those of the terrestrial planets? What conditions on the giant planets are determined by their rotation rates?

27. Moons orbit within and near the rings of the giants. In what ways do these moons and rings interact?

28. In what ways has the presence of Jupiter been favorable to conditions for life on Earth?

29. Name two important characteristics of each of the four Galilean moons of Jupiter. Which are candidates for microbial life?

30. How do tidal forces influence conditions on the moons of the giant planets? Name at least one dramatic example.

31. Describe at least five features of Saturn's complex rings.

32. Describe the ways in which Uranus and Neptune are close enough to be considered "twins." In what significant ways do the two planets differ?

33. Saturn is 10 times as far from the Sun as Earth is and rotates in about 10 hours. Describe how day and night, as experienced from an observer near the surface of Saturn, would differ from our experience on Earth.

34. Astronomers had a good opportunity to learn about Jupiter's atmosphere by studying the effects of the Shoemaker-Levy 9 comet collision with the giant planet. What are some other events or situations that have offered similar opportunities in astronomy or other areas of study?

35. From what you've learned about the eight planets of the Solar System, which seem the most compelling as places to explore? Why?

Going Further: Quantitative Questions

36. Calculate the equilibrium temperature, in kelvins, of an airless planet that orbits the Sun and has an albedo of 0.45 and an orbital radius of 7 AU. (See Going Further 7.2.)

37. Use Kepler's third law to calculate the orbital radius, in astronomical units, of an imaginary planet orbiting the Sun with an orbital period of 46 years. (See Going Further 3.1.)

38. Jupiter's mass is 318 times that of Earth. Given that the gravitational pull of any object extends forever and depends on the mass of the object and its distance, how many times greater would Jupiter's gravitational force be, compared with Earth's, on an object twice as far from Jupiter as it is from Earth? (See Going Further 3.2.)

39. Calculate Jupiter's imaginary mass, kilograms, if Io were observed orbiting at twice its distance from the planet with a period 2.5 times as long, compared with Io's current orbital parameters. (See Going Further 3.3.)

40. Calculate the mass, in kilograms, of a planet whose moon you observe orbiting with a period of 5 days and an orbital radius of 6.0×10^6 km. (See Going Further 3.3.)

41. The ratio of centripetal force to gravitational force for planet Z is 0.0013. If Z's rotational velocity tripled but its radius and mass remained the same, by what factor would this ratio change?

42. What would the ratio of centripetal force to gravitational force be for Saturn if its mass doubled while its rotational velocity and radius remained the same?

43. The ratio of centripetal force to gravitational force of a planet is 0.022. What would the ratio be if the planet's rotational velocity was 80 percent as fast, its mass increased by 50 percent, and its radius increased by 15 percent?

44. What is the ratio of the brightness of the Sun observed from Mars compared with the brightness observed from Neptune? (See Going Further 4.3.)

45. At perihelion, Uranus is 2.74×10^9 km from the Sun. If you were able to view Uranus from that distance, what angular size in arcseconds would it have? (See Going Further 2.1.)

 If your instructor assigns homework in **smartw✲rk5**, access it at the Digital Landing Page for *At Play in the Cosmos*: **digital.wwnorton.com/cosmos**

LIFE AND THE SEARCH FOR HABITABLE WORLDS

In this chapter you will learn about the science of life's origins on Earth and other planets. After reading through each section, you should be able to do the following:

9.1 The Origin of Life and the Bottom of the World
Apply recent findings about the origin and evolution of life on Earth to understand how the search for life elsewhere in the Universe is conducted.

9.2 What Is Life, and Where Can It Exist?
Describe the characteristics of life on Earth, as well as its molecular basis and the implications that these characteristics pose for where and how life may form elsewhere.

9.3 The Origins of Life
Summarize the major steps in the origin and evolution of life on Earth.

9.4 Searching for Other Life in the Universe
Give examples of ways in which astronomers have framed the search for life beyond Earth and how that search has been pursued.

An artist's conception of what one of the thousands of recently discovered exoplanets (planets orbiting other stars) might look like. Here, we see a relatively Earth-like world with oceans and continents.

9

9.1

The Origin of Life and the Bottom of the World

CHRIS MCKAY takes his science very seriously. In pursuit of his research, he has hiked through arid deserts, scrambled up dangerous cliffs, and chainsawed through 10 feet of solid Antarctic ice. Daunting as these undertakings are, the kind of determination they require may be the only way to understand questions as big as the origin of life on Earth and elsewhere in the Universe.

A recent stint while camped at the bottom of the world gives an example of McKay's workdays. "We went to the Antarctic to study these special microbial 'masses' living at the bottom of Lake Untersee in Antarctica," McKay explains. "But the lake is permanently frozen over, so we had to fight our way through the ice." Once they broke through, McKay and his team squeezed into wet suits for a dive into the lake's frigid waters. "Diving is the only way we can get down to the lake bottom and study the microbes." McKay and other scientists believe that the microbes in Lake Untersee may be cousins to some of Earth's first lifeforms (**Figure 9.1**). "If we can understand the biology of these critters, we might learn something about how life on this planet got started."

Camping out for a month in the Antarctic to scuba dive at the bottom of a frozen lake may seem like a strange way to reach back 4 billion years to the time when life began on Earth. It is, however, just one example of the way research in the field of **astrobiology** is carried out. Astrobiology is a relatively new science whose goal is to place life into its planetary (and cosmic) context. The study of astrobiology has recently become a diverse, worldwide endeavor, incorporating telescopic studies of newly discovered exoplanets, biochemical experiments on how molecules "learn" to self-replicate, and searches for signals from extraterrestrial civilizations. More important, it represents a turning point in the way humanity asks the most profound of questions: What is life, and where does it exist?

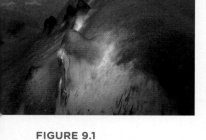

FIGURE 9.1
Life in the Antarctic
Chris McKay and his fellow scientists found microbial colonies in Lake Untersee in the Antarctic in the form of mats. These colonies may provide links to the first life on Earth.

astrobiology
The interdisciplinary study of life in its astronomical context.

Life, the Universe, and the Mother of All Questions

Whether planets exist outside our Solar System is one question human beings have asked for thousands of years (see Chapter 5). Just as ancient is the question of whether life exists on planets besides our own. As yet, no definitive evidence has been found for life—intelligent or otherwise—existing anywhere else in the Universe. But consider how much depends on the answer we eventually find to the question of extraterrestrial life. What, for example, would it mean if we never discovered life anywhere other than Earth? If we knew that our planet was the only world in the entire galaxy harboring life, how would our perspective be different? Would we see all life as sacred, or would we feel a profound sense of meaninglessness?

Or what if we found evidence for simple lifeforms (bacteria, grasses, maybe even some animals) on many planets but never found any evidence for intelligent life? Would we come to see ourselves as cosmically empowered—the eyes and ears of the cosmos—or would we see ourselves as a mistake and go mad from loneliness?

On the other side of the coin, it is not hard to imagine how finding (and perhaps meeting) other intelligent species could lead to dramatic consequences for our culture. Just being able to communicate with an intelligence that did not share our human perspectives about questions of life and death and existence would doubtless lead to a monumental shift in human culture.

But why explore the question of life in the cosmos just after finishing a study of planets? While some scientists have hypothesized that life might form in exotic places such as interstellar clouds, everything we *currently* understand suggests that planets are essential for life to begin and develop through evolution. This chapter therefore focuses on what can be inferred from the life we can study on Earth.

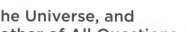

SECTION SUMMARY
Life, the Universe, and the Mother of All Questions
- There are profound implications to learning that extraterrestrial life does, or does not, exist.
- Scientists generally believe that a planetary environment is required for life.

9.2

What Is Life, and Where Can It Exist?

To study life in an astronomical context, we must first define what constitutes life. This is particularly important if we are interested in the origins of life, when nonliving (abiotic) material first made the transition into a living system.

From Chris McKay's perspective, however, defining life poses serious challenges. Life, according to McKay, is a *process*, not a thing. He makes an analogy between life and fire. "It's very hard to come up with a closed, concise definition for fire that captures everything from smoldering cigarette butts to the center of the Sun," he says. "But we do have a very good scientific (thermodynamic) understanding of it as a process." Thus, asking for a definition of what life is may be less useful than asking what life *does*.

The Machinery of Life

Let's start by summarizing the basic features of living things (**Figure 9.2**):

1. *Organic molecules*. In terms of chemical composition, living organisms are made up of organic molecules—that is, molecules that include the element carbon. In addition, many of these molecules have structural characteristics that allow them to form long chains or rings. Carbon is unique in that the structure of its six electron orbits enables a diverse array of bonds with other elements. While some other elements, like silicon, have similar properties, carbon is remarkably well suited for building life's chemistry.

2. *Metabolism*. The consumption of energy (often supplied in the form of chemicals) and, in the process, the production of waste is an essential feature of living organisms.

3. *Reproduction*. Through reproduction, living organisms create new versions of themselves by various means. These new versions are sometimes identical copies, but more often they are not.

4. *Mutation*. The ability to adapt and evolve, an essential aspect of living organisms, depends on mutation. Mutations are random changes in an organism occurring during reproduction that may or may not prove beneficial for survival.

5. *Sensitivity*. Living organisms also show sensitivity to their environments in that they notice and respond to changes in ways that enhance survival.

All five of these features are deeply connected with the processes of evolution. For McKay, this is the most important aspect of any understanding of what is and what is not life. "When you get down to it," McKay says, "the best definition of life is 'a material system that undergoes Darwinian evolution.'"

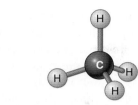

A Organic molecule (methane)

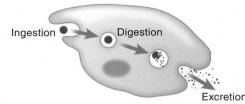

B Metabolism

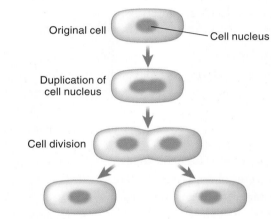

C Reproduction

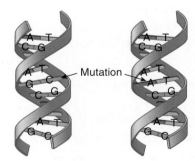

D Mutation

E Sensitivity

FIGURE 9.2
The Basic Characteristics of Living Things
Living things can be defined as having five characteristics: (A) composition based on organic molecules, (B) metabolism, (C) reproduction, (D) mutation, and (E) sensitivity.

It is worth noting that a few examples of living things do not share all these characteristics. Viruses, for example, lack the biosynthetic machinery needed for reproduction and require a host cell to replicate. In addition, some nonliving things show one or more of these traits. Crystals, for example, show growth that could be likened to some aspects of living systems. Despite these exceptions, the five criteria listed above at least provide a place to start a discussion of where, when, and how life might form on other planets.

SECTION SUMMARY
The Machinery of Life
- Scientists define living things as being composed of organic molecules and exhibiting metabolism, reproduction, mutation, and sensitivity.
- Exceptions exist, such as viruses, which meet some but not all of these criteria.

GOING FURTHER 9.1 THE LANGUAGE OF THE COSMOS
Calculating Habitable Zones

Recall from Going Further 7.2 the formula for a planet's temperature based on its distance from the star it orbits and the properties of that star:

$$T_p = T_* (1-a)^{1/4} \left(\frac{R_*}{2D} \right)^{1/2}$$

In this formula, T_* is the star's temperature, R_* is the star's radius, D is the distance from the star at which the planet orbits, a is the planet's albedo, and T_p is the temperature of the planet (without any greenhouse effect).

We can use a version of this formula to calculate the habitable zone around a star. To do so, we must solve this equation for D and then use the new expression to find the inner boundary of the habitable zone, where the temperature of the planet T_p will just equal the boiling point of water (100°C, or 373 K). We can then find the outer boundary of the habitable zone by setting T_p to the temperature at which water freezes (0°C, or 273 K).

Because we want to think about water under conditions where a non-greenhouse-gas atmosphere exists and we want to consider more realistic stars than mere blackbody radiators, we use a modified (and simpler) version of the equation than what would come from directly manipulating the equation above:

$$D = \left(\frac{T_o}{T_p} \right)^2 [(1-a)L_*]^{1/2}$$

where D is measured in astronomical units and L_* is the luminosity of the star measured in solar units (meaning our Sun has a luminosity of $L_* = 1$). The term T_o expresses the role of planetary rotation. Rotation rates must be accounted for because rapidly spinning planets will smooth out the absorption of solar radiation better than slowly spinning ones. For fast-rotating planets, $T_o = 279$ K; for slow-rotating planets $T_o = 394$ K.

Now let's use this second equation to find the inner boundary of the habitable zone for a slow-rotating planet that orbits a star that is 4 times as luminous as the Sun ($L_* = 4.0$). We can use $a = 0.3$ for an Earth-like planet:

$$D = \left(\frac{394}{373} \right)^2 [(1-0.3)4]^{1/2} = 1.87 \text{ AU}$$

This distance is beyond Earth's orbit. The outer boundary of the habitable zone for this star lies at

$$D = \left(\frac{394}{273} \right)^2 [(1-0.3)4]^{1/2} = 3.49 \text{ AU}$$

This is well beyond Mars's orbit. Thus, increasing the luminosity of a star changes the habitable zone considerably.

Note, however, that none of these calculations include the effect of greenhouse gases, which will warm the surface of a planet by raising the temperature at which it reradiates energy back into space. Thus, using this formula alone we would come to the surprising conclusion that Earth is not in the Sun's habitable zone! Without Earth's atmosphere, the development of life on our planet would have been impossible. More complete calculations take greenhouse warming into account and tend to shift the habitable zone outward from the star. The simple formula provided above for D is nevertheless useful because it gives us a simple way of understanding the basic physics involved and yields a first approximation of the size and location of a star's habitable zone.

Planets and Life: The Habitable Zone

Most scientists agree that for a planet to harbor life, it must have liquid water. The reasoning is that water is a highly effective solvent for biochemical reactions. That is, water readily acts as a medium in which compounds can be dissolved and made ready for further chemical combinations. Scientists therefore tend to define a planetary system's habitable zone, the region around a star where life might form, as the range of distances from the star where planets can retain liquid water on their surfaces. For example, if a planet is too close to the star, surface water will just boil away. If it is too far from the star, surface water will remain eternally frozen. Thus, the most basic definition of a habitable zone relies on a simple calculation of the range of distances from a star at which a planet can keep liquid water on its surface. The surface temperature (assuming that life will form and evolve on the surface) must lie in the range between water's freezing and melting temperatures. In other words, the planetary temperature T_p must range between 273 K and 373 K.

You have already learned how to calculate the surface temperature of a planet that lacks an atmosphere. The formula given in Going Further 7.2 used the parameters for the Sun (its temperature and radius) to find a planet's surface temperature, given its distance from our star. We can also manipulate this formula to give us the *range of distances* from the Sun that defines our Solar System's habitable zone (see **Going Further 9.1**). Using this formula, we find that between 0.47 and 0.88 astronomical units (AU) constitutes a first approximation of the Sun's habitable zone. The surprising result that Earth is not in the Sun's habitable zone appears only because we used the simplest form of the habitable-zone equation, which does not include the all-important effect of atmospheres in trapping solar radiation. Even though these simple equations get the answer wrong by a bit, they are an important first step in capturing the basic physics of the problem, as sunlight warms the surface of the planet.

When scientists carry out a more detailed calculation that includes the role of atmospheres, they find that the habitable zone for our Solar System today extends from 0.99 to 1.68 AU from the Sun. Notice that in these detailed calculations, Earth is just at the inner edge of the Sun's habitable zone (**Figure 9.3**).

Of course, not all stars are like the Sun: some are much hotter and brighter, and some are much cooler and dimmer. Therefore, the habitable zones around other stars may look very different from our own. Some of the coolest, lowest-luminosity stars have narrow habitable zones with small orbital distances. In the stellar classification system, these stars are referred to as type M. The habitable zones of very bright, very hot stars (called type O stars) will be at large orbital distances from the star (**Figure 9.4**).

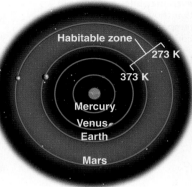

FIGURE 9.3
The Habitable Zone of Our Solar System
The habitable zone around the Sun includes bodies at a distance of about 1 astronomical unit (AU), making water-based life on Earth possible. Depending on the way the zone is defined, Venus may also lie within the habitable zone, but its runaway greenhouse atmosphere has rendered it too hot for liquid water.

interactive
Habitable Zone

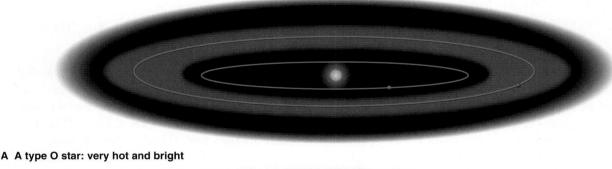

A A type O star: very hot and bright

B A type M star: very cool and dim

FIGURE 9.4
Habitable Zones for M and O Stars
Each star's habitable zone depends on the star's luminosity, which in turn depends on temperature and radius, so the habitable zones (shown here in green) of (A) large, hot O stars are broader and lie at greater distances from the star than the habitable zones of (B) smaller M stars.

extremophile An organism that thrives under extremes of temperature, pressure, and other conditions.

polymer or **macromolecule** A large molecule containing many atoms, often with a chain structure.

A

B

FIGURE 9.5

Tubeworms and Extremophiles in a Black Smoker
(A) Extremophiles such as these tube worms have been discovered flourishing deep in the oceans, far from light, in the hot, sulfurous environment around deep-ocean vents called black smokers. (B) The bacterium *Nautilia profundicola* is a microbe that survives near deep-sea hydrothermal vents.

Other factors must be considered as well. Hot, bright stars tend to have very short lifetimes—some as short as a few million years, which may be too short for life to form. Magnetic activity on a star in the form of giant flares might also affect the development of life. These kinds of issues add complexity to how a planetary system's habitable zone is determined.

SECTION SUMMARY
Planets and Life: The Habitable Zone
- Assuming that life on Earth is typical, scientists consider water to be essential for life.
- The habitable zone is the region around a star where the surface temperature allows liquid water to exist; in our Solar System today, Earth and Mars lie in this zone.
- The length of the star's life and the nature of the star's magnetic field are additional factors to be considered.

Expanding the Habitable Zone

As scientists have learned more about the limits of life on our own world, new discoveries have expanded the definition of habitability. Since the 1970s, scientists have found a bewildering array of life-forms thriving in conditions so harsh that a biologist in the early 20th century would not have dreamed they were possible. These newly revealed organisms, called **extremophiles**, live deep in the ocean and deep below Earth's surface. "No one really expected it," says McKay of the Earth-based discoveries shaking astrobiology for the last few decades. "We have found ecosystems completely independent of anything that goes on at the surface."

Among these news-making extremophiles are giant tube worms, crabs, and shrimp, which were first discovered in 1977, living a mile below the ocean surface. These colonies of extremophiles surround deep-sea geothermal vents. The vents, called "black smokers" for the plumes of hydrogen sulfide they belch into the ocean, presented scientists with an entirely new class of ecosystem that did not need direct sunlight to survive. The organisms' biochemistry relied on the heat and raw chemical material provided by the black smokers. Unlike anything on the surface, these deep-ocean ecosystems gave scientists a hint that life could form and thrive far from surface sunlight and surface water (**Figure 9.5**).

Far more compelling for McKay's thinking about habitable zones, however, is another class of extremophiles: bacteria found thriving deep underground. Scientists have discovered colonies of microbes living

5 miles down in the subterranean reaches of South African gold mines. "These creatures get their energy from sources we never imagined," says McKay. "The South African extremophile bacteria are powered by the radioactive decay of unstable molecules in the rocks. Radioactive decay! Sunlight and surface water play no role. Radioactivity sets the biochemistry in motion. It's amazing!"

Extremophiles that feed on nonsolar energy sources represent the first step in a process of discovery that has radically altered the concept of planetary habitable zones. The second step is the direct exploration of alien worlds. "When you take the discovery of liquid water below the surface of ice-covered moons like Europa and Enceladus," says McKay, "and put it together with our understanding of terrestrial extremophiles, you can see why the definition of 'habitable zone' had to change." Direct exploration has shown that liquid water exists in many places in our Solar System where scientists had not expected it. With the discovery of watery moons around ice and gas giants and the possibility of extremophile biology, scientists have learned to be more liberal in considering the habitats for life.

SECTION SUMMARY
Expanding the Habitable Zone
- The discovery of extremophiles, as well as the subsurface water found on some giant planets' moons, suggests that life may be able to flourish in a broader range of conditions than was previously believed.

Molecular Engines of Life: DNA, RNA, and Proteins

For all the diversity of life on our planet—in both the forms it takes and the conditions it thrives in—the molecular basis for that diversity is remarkably simple. McKay explains, "If we ask the question 'What is life made of?' the answer is really rather surprising. Life on Earth is not made up of a large number of different types of molecules. Instead, it's made up of a small set of molecules used over and over and over again."

Terrestrial life relies on a few classes of **polymers** (also called **macromolecules**). These are giant molecules that look like long chains, often folded over on themselves. The polymers are themselves made up of smaller molecular units that form the chain. "The diverse organisms we see on Earth come from the combinatorial complexity of those molecular units," says McKay, referring to the immense number of ways the building blocks can be put together and taken apart. "In a sense, life is like Legos. A Lego kit is a small number of standardized bricks, and you can make all sorts of diverse objects by the way you put these bricks together."

The most important classes of these biologically active molecules are DNA, RNA, and the very wide-ranging class of polymers known as proteins. **Deoxyribonucleic acid (DNA)** consists of two long polymer strands that are wound around each other in a characteristic double-helix shape (**Figure 9.6A**). The strands are called the "backbone" of the DNA molecule, and they are made of **nucleotides**, whose building blocks are sugar molecules and phosphate groups (molecules made of phosphorus and oxygen atoms, among other things). The backbone strands are linked together by molecules called **nucleobases**. In DNA there are just four possible nucleobases, denoted by their first letters: adenine (A), thymine (T), guanine (G), and cytosine (C). In locking the two backbone strands together, these bases can pair up only in certain ways. Adenine will link only to thymine (AT), and guanine will link only to cytosine (GC).

Ribonucleic acid (RNA) is also a polymer made of nucleotide building blocks (**Figure 9.6B**). In RNA, however, the base thymine is replaced by another nucleobase, called uracil (U), which links to adenine (UA). Unlike DNA, RNA has just a single strand. A strand of RNA can fold up to create structures in which different parts of the strand will be linked to each other through base pairing, just as the two separate strands of DNA are linked together.

Proteins, the workhorses of cell biology, are polymers that play many roles within a cell. Some proteins form the architecture of the cell—cell walls or the propeller-like structures that allow subcellular components to move around. Other proteins act as enzymes, helping certain chemical processes along. Still other proteins act as signalers that tell a cell when to begin or end different kinds of activities. All these proteins are made up of smaller molecules, called *amino acids*, that are strung together in the polymer chain. The sequence of amino acids in the long chain of a protein is what determines its function in a cell. To function properly, proteins must fold into very specific shapes.

The combinatorial complexity of the building blocks for these polymers (nucleobases for RNA and DNA, amino acids for proteins) is crucial. "There are just about 20 amino acids that terrestrial life uses," explains McKay. "By stacking them in different ways, you can make a bewildering array of proteins that do an enormous range of structural and functional things for life."

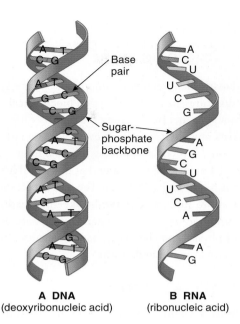

Base pair

Sugar-phosphate backbone

A DNA
(deoxyribonucleic acid)

B RNA
(ribonucleic acid)

FIGURE 9.6
DNA and RNA
(A) DNA and (B) RNA are related complex molecules that are necessary for life as it exists on Earth.

The relationships among DNA, RNA, and the proteins also make life possible. Because of the specific way that bases link up in the double helix (AT, GC), DNA acts as information storage for living systems. The sequence of bases on each strand of DNA becomes a kind of molecular language, or "code," that can store information about how to build the working parts of a cell: the proteins. To make protein building occur, the double helix of the DNA can be unzipped by RNA, which then transcribes the sequence of bases (reading the DNA code). Other forms of RNA then translate the code into sequences of amino acids that become the proteins that cells need to do their work.

DNA also can *self-replicate*, or produce copies of itself. When a cell grows and splits, the daughter cells take copies of the DNA molecules (and their instructions) with them. Thus, DNA molecules represent the self-replicating **genes** that are passed down from one generation of life to the next.

SECTION SUMMARY
Molecular Engines of Life: DNA, RNA, and Proteins

- DNA is the blueprint for life; it consists of two strands of nucleotides linked by four possible nucleobases: adenine (A), thymine (T), guanine (G), and cytosine (C). The sequence of the bases forms a code for how to build proteins. DNA is self-replicating.

- RNA is a single strand made up of nucleobases. RNA transcribes the code of DNA, placing amino acids in the correct sequence to create proteins.

- Proteins perform many functions within cells; the specific function is determined by a protein's sequence of amino acids.

deoxyribonucleic acid (DNA)
A molecule that encodes genetic information. It consists of two polymer strands wound around each other in a double-helix shape. DNA is the biological basis for life on Earth.

nucleotide A polymer built of sugar molecules and phosphate groups that link to compose the backbones of DNA and RNA.

nucleobase A polymer that links nucleotides in DNA and RNA.

ribonucleic acid (RNA)
Single-stranded polymer that transcribes the code of DNA, placing amino acids in the correct sequence to create proteins.

protein A polymer made of amino acids that assists in cellular functions.

gene A section of a DNA or RNA molecule that is the code for a particular trait or function.

9.3
The Origins of Life

The remarkable interactions among DNA, RNA, and proteins represent the fundamental biochemical machinery of life. But to study the origin of life, scientists must first explain an even more remarkable transition. How could nonliving material have combined in just the right way that made it able to transform into living material? The task for scientists who want to study **abiotic synthesis**—the creation of life from nonlife—is to understand, using only the laws of physics and chemistry, how molecules of DNA, RNA, and proteins could form.

Prebiotic Assembly

The study of abiotic synthesis began in earnest in the 1920s with the work of Russian scientist Alexander Oparin and British researcher J. B. S. Haldane. Both men proposed that the prebiotic molecules that act as the building blocks of life could form through natural processes in the early Earth. In the modern understanding of biochemistry, amino acids count as prebiotic molecules. They are necessary for life but are not living organisms themselves.

In 1953, Stanley Miller and Harold Urey, chemists at the University of Chicago, decided to test one version of this theory (**Figure 9.7**). The purpose of Miller and Urey's work was to create a simulated version of the early Earth in a test tube. "The essence of the Miller-Urey experiment (which has been repeated a thousand times)," explains Chris McKay, "is to take simple molecules like methane, hydrogen, and water; stick them in a jar; add energy; and see what happens. The results turn out to be astounding." Specifically, Miller and Urey created an "atmosphere" of hydrogen, ammonia, and methane in one flask and connected it to an "ocean" of liquid water in a separate flask. They also set up a high-voltage discharge in the atmospheric flask to act as lightning and a source of ultraviolet light.

After running the experiment for a week, Miller and Urey found that brown goo had formed in their test-tube ocean. When they chemically analyzed this goo, they saw that it had formed a rich soup of prebiotic molecules, such as glycine, lactic acid, and urea. "It turns out," says McKay, "that the Lego blocks life uses for building its macromolecules could be reproduced quite easily from nonbiological processes." What's noteworthy is that 10–15 percent of the carbon atoms in the original flask had been converted into molecules such as amino acids, sugars, lipids (the basis of fat), and a variety of other organic acids.

abiotic synthesis The creation of life from nonlife.

A

B

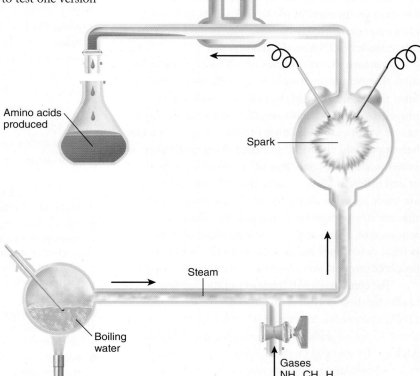

FIGURE 9.7
The Miller-Urey Experiments
In 1953, (A) Stanley Miller and (B) Harold Urey at the University of Chicago conducted (C) an experiment in an attempt to replicate the conditions and materials they believed to be present in Earth's early environment. The results showed that the molecules required for life could be produced in a short time frame.

C

The Miller-Urey experiment was hailed as an instant classic of biochemistry, and proof that, in principle, nonliving material, along with the natural processes of physics and chemistry, could create the molecular basis of life. "It was a major breakthrough and a major surprise," says McKay. "People thought that we were just moments away from solving the origin-of-life problem."

Although the Miller-Urey experiment showed that life's building blocks *could* be synthesized on conditions like those of the early Earth, with more research scientists also began to see how the story was more subtle and more complicated. Studies focusing on the atmosphere of the early Earth, for example, made it clear that there would not have been much hydrogen. In addition, volcanism would have produced much more CO_2 and N_2 in the atmosphere, yielding a very different set of chemical reactions than what Miller and Urey saw in their test tube. Modern studies have therefore focused on environments such as volcanic vents in the seafloor as the molecular factories where prebiotic molecules might be assembled. More important, as scientists came to understand the relationships and functions of proteins, DNA, and RNA, their perspectives changed.

"The Miller-Urey experiment happened back when the origin of life was viewed as a chemistry problem," explains McKay. "Now we know it's also an information-processing problem. It's a software problem as much as a hardware problem." In other words, scientists also need to understand how self-replicating molecules, which can store information and process it via the construction of new molecules, were created. The famous Miller-Urey experiment was a first step, showing one pathway for assembling life's building blocks, but researchers now look to other processes and perspectives for understanding the creation of the first "information-rich" self-replicating molecules on the young Earth.

SECTION SUMMARY
Prebiotic Assembly

- Although great progress has been made, the transformation of nonliving molecules into life, called abiotic synthesis, is not fully understood today.
- By simulating the expected state of Earth's early atmosphere and ocean, the Miller-Urey experiment produced many of the molecules required for life.
- Scientists later learned that early-Earth conditions differed from those in the experiment, suggesting that other natural paths may have initiated life.

Chris McKay

There was a time when Chris McKay liked to go winter camping for fun. Now, as an astrobiologist who spends months at a time living in a tent in the remotest parts of the world, McKay likes to stay at home in his off-hours. "After you've camped for a month in Antarctica, there is no attraction to go winter camping for fun anymore." Nevertheless, McKay loves his science. "To do the kind of research that interests me, you have to get pretty extreme. If you are going to go out and find extremophiles, it means you're not going to be sitting in an office somewhere wearing a white lab coat."

Science came naturally to McKay, who was always an adventurer. "I was always interested in science," he says. "Sometimes I'd skip school to ride my motorcycle, come back for a physics lecture, and then leave again. The next day my physics teacher would come up to me and say, 'You know, I am sure you were in class yesterday, but according to the roll you were absent.'"

Despite his less-than-stellar attendance record in high school, McKay went to Florida Atlantic University as an undergraduate and then did graduate work at the University of Colorado. The timing with one of astrobiology's founding events was fortuitous. "It was 1976," he says. "That was when the two *Viking* probes landed on Mars." Watching those missions propelled McKay into astrobiology, a field that had yet even to be named. "It was fate in a lot of ways that got me into this," says McKay. "There are a thousand interesting questions to ask about the world. I happened to be at the right place and the right time to end up spending my life asking astrobiological ones. But I am pretty happy with the way it worked out."

From Nonlife to Life, and the Heroism of Time

At the molecular level, the story of life's creation focuses on one aspect of our definition of life: self-replication. "That is where the software aspect of the problem comes in," says McKay. "If you make an amino acid in a prebiotic experiment, there's information in that molecule, but it's purely in the structure of the molecule. There is no algorithm for producing more amino acids coded within it. So going from a molecule where the information is the molecule itself, to a molecule where information about making new versions of itself is all coded within the molecule, that's quite amazing." McKay likes to use the analogy of a computer memory chip. "The shape of the chip does not depend on what is written on it. In the same way, you have to figure out how to build a stable molecule (in terms of its structure) in which information—like instructions for making copies of itself—can be coded into the molecule." In other words, the ability to store and pass on information is the most important part of DNA that scientists must understand. The molecule's shape is just a means to that end.

Thus, beginning with prebiotic molecules—like amino acids or nucleic acids and bases—scientists

GOING FURTHER 9.2   THE LANGUAGE OF THE COSMOS
Time, Chance, and Life

Building the first self-replicating molecules required time for chemistry and random chance to work their magic. Let's look more closely at how scientists figure out how long it will take for a molecule with certain kinds of properties to emerge from a series of random combinations.

Imagine that we want to build a self-replicating molecule out of nucleotides, which form the spines of DNA and RNA. There are just four distinct kinds of nucleotides (since each includes one kind of nucleobase). We wish to a form a molecule that is a chain of six nucleotides, with each of the six in the chain chosen randomly from the four nucleotides available. The number of ways that these four nucleotide types can be combined into a string of six is

$$N = 4 \times 4 \times 4 \times 4 \times 4 \times 4 = 4^6 = 4{,}096$$

One type of RNA-replicating molecule in our bodies (called a ribosome) has 192 nucleotides. If we wanted to randomly assemble such a 192-nucleotide molecule, the number of possible combinations would be $N = 4^{192} = 4 \times 10^{115}$. That is a huge number. How long would it take to create just the right molecule by going through all those combinations?

Chemical studies show that a single example of a 192-nucleotide molecule can be assembled in about 0.32 second. In that case, how long would it take to get a specific combination—one that has the property of self-replication? The answer comes from multiplying the number of combinations by the time each one takes to be assembled: it takes $4^{192} \times 0.32$ second $= 1.26 \times 10^{115}$ seconds to run through all the combinations.

Next we need to calculate the total age of the Universe, 13.7 billion years, in seconds:

$$(1.37 \times 10^{10} \text{ years})(3.15 \times 10^7 \text{ seconds/year}) = 4.32 \times 10^{17} \text{ seconds}$$

Even at a rate of one combination tried every 0.32 second, there would not be enough time in all the Universe to try every possibility that leads to a specific 192-nucleotide self-replicating molecule. Something else must have happened.

Scientists now believe that building a self-replicating molecule like a ribosome was a multistep process. Research indicates that the smallest molecule that can self-replicate is about 24 nucleotides long. A molecule of this size would not provide the information storage needed for even the simplest living organism, but it could have served as a subunit of a longer, more information-rich self-replicating molecule.

Studies show that a single example of a 24-nucleotide molecule can be assembled in just 0.04 second. How long would it take to build the right 24-nucleotide "subunit"? Using calculations similar to those above, we find

$$4^{24} \times 0.04 \text{ second} = 1.13 \times 10^{13} \text{ seconds} = 3.6 \times 10^5 \text{ years}$$

Therefore, it takes only 360,000 years to go through all the possibilities and get a specific self-replicating subunit. Further calculations show that it takes only another 5,000 years or so to assemble the 192-nucleotide molecule from the smaller subunits. Thus, it takes much less time to build up a large information-rich self-replicating molecule by constructing smaller parts and then putting those parts together.

must imagine how ongoing random chemical reactions led to more complex molecules that could store the information and enable self-replication. The key concept that scientists use is the idea that the early Earth was acting as a kind of giant chemistry experiment. Chemically active sites like pools of gloopy warm water or deep-ocean vents at the seafloor allowed for endless trial and error as new combinations came into being and then broke apart. If some of these new molecules had the ability to self-replicate even partially, that may have been enough to put the wheels of self-replication in motion. Once a self-replicating

molecule appears in a primordial soup of prebiotic chemicals, its numbers will increase in time relative to the other molecules. "You can take this kind of logic a step farther," says McKay. "If two kinds of self-replicating molecules appear, then whichever one reproduces itself faster will come to dominate in terms of sheer numbers." As an analogy, imagine that brown beetles in a population breed faster than green beetles; as time goes on, more and more beetles will be brown. The same is true of chemical reactions, where molecules that can self-replicate will become predominant among other molecules.

The question remains, however: How do you build a self-replicating molecule? If DNA contains the information needed to produce copies of itself, could a strand of DNA have appeared in the soup of random chemical reactions early in Earth's history? Most researchers think not. Studies of chemical reactions and their speeds suggest that the time it would take to assemble a working DNA molecule from random reactions is longer than the age of the Universe (10^{10} years). After years of study, researchers now believe that smaller segments of RNA (called **monomers**) formed first. These then combined into longer polymers of RNA, a few of which might have had the ability to self-replicate (see **Going Further 9.2**). Once self-replication begins, it leads to huge populations of the self-replicating molecule and enables further adaptations.

The key ingredient in this story is *time*. Human scientists can't wait more than a few decades to see whether their experiments pan out. Nature is not so impatient. By allowing the prebiotic soup to percolate for hundreds of thousands of years, nature sorted through countless possibilities until self-replication randomly occurred. When people question whether life could have formed from nonlife, they often fail to take into account the vast stretches of time that were available for the process to work.

SECTION SUMMARY
From Nonlife to Life, and the Heroism of Time
- Given sufficient time, nature can sort through enough chemical possibilities on a young, habitable planet to allow self-replicating molecules to appear.
- Scientists believe that small segments of RNA developed and then combined to form a few self-replicating molecules; once such a molecule exists, it proliferates.

Evolution 101

Life on Earth today exhibits a stunning array of forms and functions, ranging from tiny single-celled diatoms floating in the ocean to massive multicellular animals like elephants roaming the land. In between, we find everything from mosses to mushrooms to mollusks to muskrats. In addition, we see living beings that are astonishingly well tuned to their environments. Walking sticks, a type of insect, can mimic the trees on which they move, and eagles have feathers with special fringes that allow them to dive from high altitudes at tremendous speeds (**Figure 9.8**). Where did all this diversity and adaptation come from?

Scientists believe that once self-replicating molecules emerged, the stage was set for the processes of natural selection to create more complex and more

monomer A small molecular unit that can combine to form larger, more complex polymers.

A

B

C

D

FIGURE 9.8
Evolutionary Adaptations of Animals
(A) The bald eagle's keen eyesight, aerodynamics, and sharp beak and talons make it a formidable predator. (B) In the ocean depths, where food is scarce, the anglerfish has developed a dorsal fin with bioluminescence to lure other fish and an expandable body to swallow fish twice its size. (C) The anteater has an elongated snout and an extendable, thin tongue for extracting termites from their nests. (D) The Arctic hare uses its adaptation of snow-colored fur to avoid becoming prey.

FIGURE 9.9
Charles Darwin
Charles Robert Darwin (1809–1882) was an English naturalist who conducted meticulous studies of related species of plants and animals, establishing that all have descended over time from common ancestry, with variations occurring through natural selection. In 1859 he published his findings in his book *On the Origin of Species*.

genetic drift Changes in genes due to random events other than gene mutation.

diverse forms of life over time. For scientists, the term *evolution* is central to an understanding of the biological world. As McKay puts it, "Without doubt, evolution is the most powerful force in biology. It works because all life carries genetic information, which can be altered."

The mechanisms of evolution were first articulated by Charles Darwin in the 1800s (**Figure 9.9**). Darwin recognized that traits of living organisms—such as the color of a beetle's shell or the shape of a bird's beak—would be passed down from parents to offspring. In other words, traits were *hereditary*. The basis of evolution was the idea that living organisms contained structures called genes, and these genes controlled which traits an organism displayed. There were genes for shell color in beetles and genes for beak length in birds. According to Darwin, these genes were passed from parent to offspring. (We now know that these genes are, in fact, the DNA contained inside every cell.)

If the distribution of these genes changes from one generation to the next, or if entirely new genes are created, then new traits can emerge. Four basic processes in living systems—natural selection and three others (**Figure 9.10**)—lead to such changes in genes and their distribution within organisms:

1. *Mutation*. Mutations randomly change the genes for a certain trait within an individual. Certain kinds of radiation, like X-rays or cosmic rays (high-energy protons), can rearrange DNA. The altered DNA can cause a mutation in the organism. For example, green beetles might have mutated genes that lead to offspring with brown coloration. After the mutated genes appear, brown beetles may become more frequent in the population through reproduction.

2. *Migration*. In migration, individuals from one population with a certain trait move into new territories and breed with populations that have different traits. For example, brown beetles might move into a region populated by green beetles. If the two populations interbreed, in time the genes for brown beetles become more frequent in the green-beetle population.

3. *Genetic drift*. In **genetic drift**, certain traits (genes) become more common because of random events other than gene mutations. Imagine that in one generation, two brown beetles happened to have four offspring survive to reproduce. At the same time, several green beetles were killed when someone stepped on them before they had any offspring. The next generation of beetles would therefore have a higher ratio of brown to green

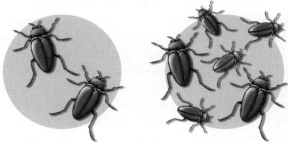

A Mutation

B Migration

C Genetic drift

FIGURE 9.10 Three Processes of Evolution
In addition to natural selection, the processes by which species evolve are (A) mutation (a mutated gene leads to a brown beetle among a population of green beetles); (B) migration (brown beetles move into an area populated by green beetles, and interbreeding changes the frequency with which genes for brown and green shells appear); and (C) genetic drift (random effects lead to changes in gene frequency and numbers of brown or green beetles).

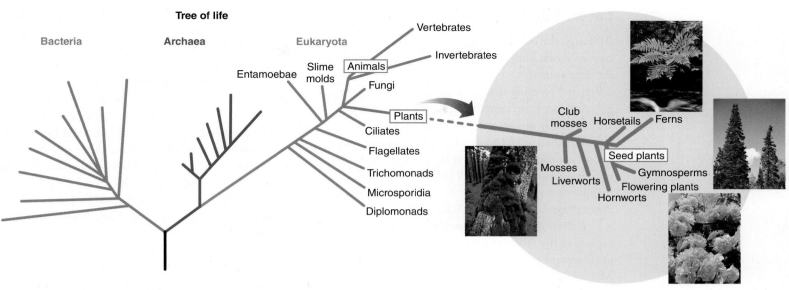

FIGURE 9.11 Tree of Life
The "tree of life" represents our current understanding of evolutionary relationships between species. Organisms that share more common ancestry are more closely related; for example, rosebushes and pine trees are more closely related than either is to a moss or fern.

beetles than the previous generation had—but just by chance.

4. *Natural selection.* **Natural selection** is the process by which some traits (determined by genes) may make one individual more likely than others of its species to survive in its environment. If green beetles are easier for birds to spot (and hence eat), for example, then brown and blue beetles are more likely to survive to produce offspring and pass on their genes for brown or blue coloration. In the next generation, brown and blue beetles will make up a greater proportion of the population than in the previous generation.

Natural selection is the most important among these four processes for tuning an organism to its environment. Natural selection means that, for example, mutations that create a trait favorable for survival in a particular environment give that organism a better chance to survive and produce more young, which also have the mutated gene. Environments, however, can change. Imagine that green beetles are well hidden in a forest environment. If a drought kills the forest, these beetles will find themselves suddenly exposed and vulnerable. But any mutations resulting in beetles of different colors that more closely match this new environment may lead to individuals that are better able to survive and pass these newly mutated genes on.

Natural selection, in conjunction with migration, mutation, and genetic drift, contributes to a "flow of genes" from one generation to the next. Given enough time, these changes in genetic inheritance may become so pronounced that a new species with significant new traits emerges. "It's these processes," says McKay, "that allowed evolution to create organisms that can be as complex as those we see today, and as adapted to their environment."

The raw material of evolutionary studies consists of the relationships among these species—relationships that can be represented as a "family tree" called a **phylogeny** (**Figure 9.11**). Species have evolved over time from other species, and an earlier species that links later ones is called a *common ancestor*. Thus, for example, mosses, ferns, pine trees, and roses are all different, but they share enough common traits that they must all have a common ancestor. Pine trees and roses, however, are more similar to each other than either is to a moss or a fern, so pine trees and roses must have had a common ancestor that evolved later than the common ancestor for all four plants.

SECTION SUMMARY
Evolution 101

- Charles Darwin's theory of evolution was based on the idea that traits of living things are hereditary—passed down via genes from one generation to the next.

- Genes and their frequency in living organisms change through physical migration of species, random mutations, genetic drift, and natural selection.

natural selection
The process by which an organism that is better adapted to its environment is preferentially selected to survive and reproduce over less adapted organisms.

phylogeny Relationships among species.

Lateral Gene Transfer

Although the notion of genetic inheritance being handed down from parents to offspring is appropriate for large animals and plants, microorganisms are different. "To think that a microorganism primarily gets its genes from parents is a mistake, because they are constantly swapping genes," says McKay. This process, called *lateral gene transfer*, occurs when one microorganism literally injects its genes into a neighbor. "It's a genetic free-for-all for these kinds of organisms compared to what we think of when we talk about plants and animals," McKay adds.

Lateral gene transfer also provides an explanation for one of the most potent examples of evolution: the development of antibiotic resistance among bacteria. McKay explains, "We begin with a population of microbes. Then we introduce a new element into the environment: antibiotics. The microorganisms respond in that most of them die. But some of them don't die; they develop mutations that allow them to survive to deal with the antibiotics. That mutation spreads through the population. We now have antibiotic-resistant strains, a variety of types."

The example of antibiotic resistance demonstrates the reality of evolution even on timescales that are small compared with a human life. On timescales much larger, we can see evolution's grand drama in the story of the entire planet.

SECTION SUMMARY
Lateral Gene Transfer

- Microbes can swap genes laterally, meaning that genes are exchanged within a single generation rather than passed from a previous generation to the next.

Mileposts in the History of Life

Try this thought experiment: Stand up and hold up both arms on either side of your body. Now imagine that your outstretched arms represent the entire history of life on Earth. The tips of your left fingers represent the origin of life on the planet some 3.7–4 billion years ago. Where do human beings appear on this time line? Somewhere close to your heart? Near your right armpit? Maybe close to your right wrist?

Not even close.

Conclusive evidence indicates that human beings appeared on the planet at the point represented by the tip of the fingernail of the longest finger on your

outstretched right hand. Humanity, with its remarkable intelligence and self-consciousness, is a very new addition to the tree of life (**Figure 9.12**). But if human beings and intelligence are new, what was happening in terms of evolution for those first few billion years? The remarkable answer is, not much.

Most scientists believe that single-celled microorganisms appeared soon after the first self-replicating molecules were synthesized. These single cells had genes (their DNA) floating freely within the cell walls. The cell nucleus had yet to evolve. Fossil evidence indicates that these microorganisms lived in giant colonies forming dense "mats" (thick, floating layers) in lakes or shallow-water coastal regions. "For 3 billion years, the story of life on Earth was pretty much nothing but single-celled organisms," says Woody Sullivan, an astrobiologist and astronomer at the University of

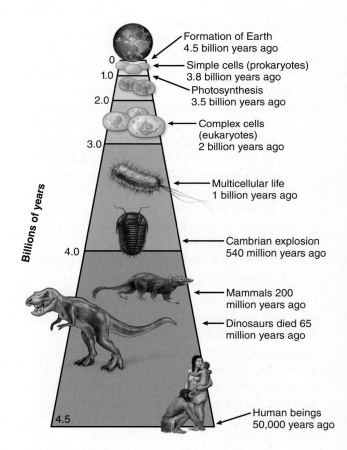

FIGURE 9.12 Time Line for Life on Earth
Life on Earth has a time line spanning most of the 4.5 billion years since the planet's formation. Simple cells developed very early, but it took over a billion years more for complex cells to appear. On this scale, human beings have only very recently appeared.

Washington who has pursued the question of extraterrestrial life for 40 years. "Then something remarkable happened."

One important fact about the first billion or so years of life on Earth is that there was little or no oxygen in the atmosphere. The "respiration" of early life-forms relied on "breathing" in CO_2, breaking it down in biochemical reactions, and then expelling oxygen (it was really photosynthesis in the cells that liberated the oxygen). For billions of years, these **anaerobic** (non-oxygen-breathing), single-celled organisms were the only kind of life on the planet (**Figure 9.13**), and the oxygen they created was absorbed by the sea and earth through chemical reactions. But approximately 2.5 billion years ago, these *sinks* (mechanisms or locations for absorption) of oxygen became saturated. The oceans could no longer soak up the oxygen expelled by a planet's worth of microorganisms. The atmosphere began to change, and oxygen built up rapidly in a short period called the **Great Oxygenation Event (GOE)**, which led to a dramatic change in the nature of life on the planet (**Figure 9.14**). (As evolution proceeded, there would be other periods when the oxygen content would rise even higher, eventually reaching the high level we experience today.)

Much of the existing microbial life could not survive in the new oxygenated atmosphere. This change affected mostly microbes that lacked cellular nuclei,

A Stromatolites

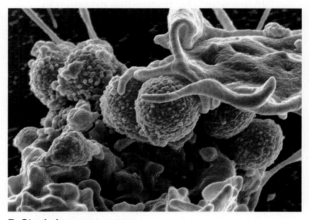

B *Staphylococcus aureus*

FIGURE 9.13
Examples of Anaerobic Bacteria
The earliest life-forms were anaerobic. Oxygen was not present in the atmosphere at this time, and these organisms did not use it in their metabolism. (A) Precambrian stromatolites in the Siyeh Formation, Glacier National Park. These 3.5-billion-year-old geologic formations contain fossilized anaerobic microbes, which may be the earliest known form of life on Earth. (B) *Staphylococcus aureus*, a common anaerobic bacterium that is found on human skin and mucous membranes and has become resistant to many modern antibiotics. It will carry out aerobic respiration if oxygen is present, but it is capable of switching to anaerobic respiration if oxygen is absent.

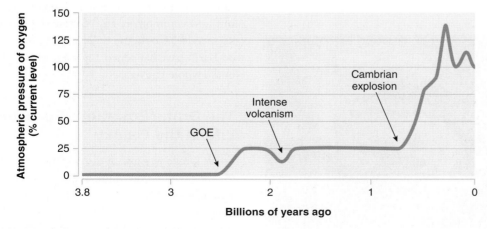

FIGURE 9.14 Oxygen Levels on Earth over Time
The dramatic increase in atmospheric oxygen that occurred with the Great Oxygenation Event (once land and sea had become saturated with oxygen) meant that the atmosphere could then support the growth in number and diversity of oxygen-breathing animals that would occur later in the evolution of life on Earth. This graph shows oxygen content in terms of percentage of current values (the current value is 100 percent). As is evident, later events also produced changes in the oxygen content of the atmosphere until it reached its present levels.

anaerobic Describing an organism that requires the absence of free oxygen to survive.

Great Oxygenation Event (GOE) The period in Earth's history during which oxygen built up relatively rapidly in the planet's atmosphere.

called *prokaryotic cells*. But other kinds of microbes changed and adapted. In particular, **eukaryotic cells**—those that stored their DNA in the central repository of a cellular nucleus—were better able to adapt. In fact, because oxygen allows for energy to be chemically stored and transferred better than CO_2 does, the organisms that adapted to breathe oxygen suddenly found themselves with fuel to burn. The increased activity drove the creation of entirely new forms of life in which many cells cooperated to form a single multicellular organism. Over time, life became more complex and varied. But the real turning point came even later, when evolution went through a remarkable period of creativity.

"That is when this tremendous explosion in evolution occurred," says Sullivan. "It was just 540 million years ago, during the Cambrian period. It was like evolution went crazy." During this brief period, called the **Cambrian explosion**, almost all the basic branches of the tree of life that we know today were formed. It was the Cambrian explosion that led to plants like mosses, ferns, and trees, as well as animals like sharks, dinosaurs, and mammals (**Figure 9.15**).

Ultimately, a species of mammal developed that learned to walk upright on two feet and use simple tools. Over time (though very little time, compared with the billions of years life had already existed), these creatures grew in complexity, progressing from burning twigs for cooking to burning rocket fuel for launching telescopes into space. And they became intelligent enough to wonder whether other intelligent life-forms might exist in the Universe.

Did the trajectory of life that we see occurring on Earth happen anywhere else in the galaxy or in the Universe as a whole? Now that we've learned about life on Earth in detail, it is time to ask some more detailed questions about life elsewhere.

eukaryotic cell A cell that stores its DNA in a cell nucleus.

Cambrian explosion The period 540 million years ago when the diversity of life expanded at an accelerated rate.

SECTION SUMMARY
Mileposts in the History of Life

- Once self-replicating molecules appeared, single-celled anaerobic microorganisms without cell nuclei evolved.

- Once a critical mass of oxygen was achieved, the Great Oxygenation Event changed the atmosphere, allowing new, oxygen-breathing organisms to evolve.

- Eukaryotic cells (microbes with nuclei) evolved, and their more efficient energy storage and transfer enabled multicellular life to evolve.

- The number and variety of species increased dramatically during the Cambrian explosion 540 million years ago.

9.4
Searching for Other Life in the Universe

Scientists now have detailed theories about the evolution of single-celled organisms, the Cambrian explosion, and the evolution of human intelligence. But how common are any of these steps in a cosmic perspective? Could the trajectory of life that occurred on Earth have happened anywhere else in the galaxy or in the whole Universe?

The Drake Equation: Guesstimating Life in the Galaxy

Frank Drake was one of the first modern astronomers to take the question of life elsewhere in the galaxy seriously, and his name is attached to the equation most often associated with that question. "The Drake equation," explains Woody Sullivan, "has been the standard way of understanding what we do and do not know about extraterrestrial life for five decades."

Drake was a radio astronomer by trade, and his early work focused on using astronomical observing

A A trilobite   **B** Dinosaur

FIGURE 9.15 Life from the Cambrian Explosion
Animals such as (A) trilobites and (B) dinosaurs were some of the highly diverse kinds of life that emerged from the Cambrian explosion.

A

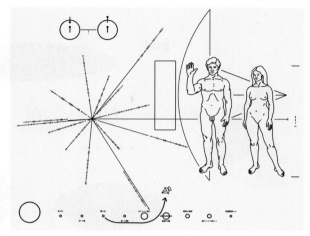

B

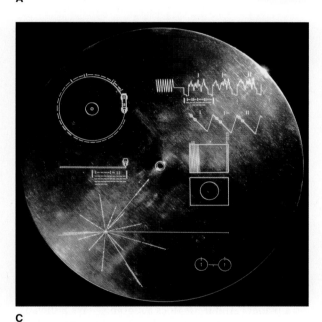

C

FIGURE 9.16

Frank Drake's Contributions to the Search for Extraterrestrial Intelligence
Frank Drake pioneered the search for extraterrestrial intelligence and developed the Drake equation. (A) He is shown here with a radio telescope, one of the primary tools in the search for extraterrestrial intelligence. He also helped create two physical messages sent into space: (B) the Pioneer plaque and (C) the Voyager Golden Record.

interactive
Drake Equation

technologies to search for evidence of intelligent life in other parts of the galaxy. Drake was asked by NASA to host a workshop on the problem in 1961 (**Figure 9.16**). "It was during the preparation for the meeting," says Sullivan, "that Drake came up with the idea for breaking down the problem of intelligent life on other planets into pieces that could be expressed in a single equation." The ultimate intention for the workshop was to determine the likelihood of finding evidence of an extraterrestrial intelligence by searching a sample of stars. But the answer to that question depended on just how many intelligent civilizations existed in the galaxy. "What Drake was after," says Sullivan, "was a way to formulate that second question."

Drake decided to organize much of the workshop around a discussion of that single issue—how many intelligent, technologically advanced civilizations exist in the galaxy (Drake called the number N_c). "He set up discussions of all the factors which would determine N_c," explains Sullivan. "There was a half-day discussion of how many planets might be out there and a half-day discussion of how many of these planets might have life on them and another half day of talking about which planets with life might have evolved intelligence and so on."

The most significant outcome of the workshop was the equation that summarized these discussions. Over the years there have been many versions of the Drake equation, but they all seek to express the same idea—how to estimate the number of technological civilizations in the galaxy using our best knowledge (or best guesses) for the factors that affect the probability of their existence. One often-used form of the Drake equation is this:

$$N_c = N_* \times N_p \times f_h \times f_l \times f_i \times f_c$$

FIGURE 9.17

The Drake Equation in Pictures

Several versions of the Drake equation have been used since it was first proposed, but all of them attempt to formulate the questions to be answered to estimate the number of technological civilizations in the galaxy (N_c). The terms are: number of stars in the galaxy (N_*), number of planets around each star (N_p), fraction of planets in the habitable zone (f_h), fraction of planets that evolve life (f_l), fraction of planets that evolve intelligence (f_i), and fraction of a star's lifetime in which a civilization exists (f_c).

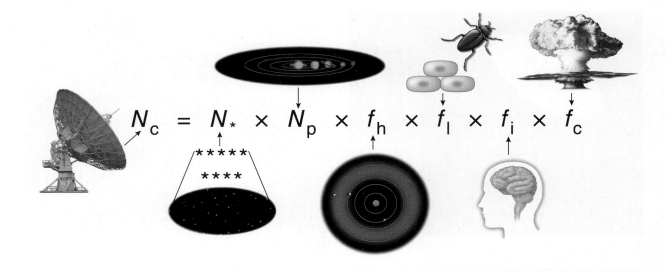

$$N_c = N_* \times N_p \times f_h \times f_l \times f_i \times f_c$$

Expressed in words, the equation says that (N_c) will be the product of the number of stars in the galaxy (N_*), the number of planets orbiting each star (N_p), the fraction of those planets in each star's habitable zone (f_h), the fraction of those planets on which life evolves (f_l), the fraction of those planets with life that develops intelligence (f_i), and the lifetime of an average intelligent civilization expressed as a fraction of the star's life (f_c) (**Figure 9.17**).

"The Drake equation does not allow you to actually predict how many technical civilizations there are in the galaxy," says Sullivan. "It's not an equation like $F = ma$; instead, it's a probabilistic statement. It lets us sum up our ignorance." The Drake equation has been useful in astrobiology because it helps scientists organize their thinking about life and intelligence in the galaxy. "People use it," says Sullivan, "because it forces them to think about the factors that affect life and intelligence. It lets us see how many of those factors we know well and how many we don't know well. It also helps us see whether there's any hope of determining some of the values over the near term." Thus, according to Sullivan, once the Drake equation was formulated, the next step was to figure out the values for each of the factors. As Sullivan puts it, "Moving from left to right in the equation is an exercise in going from knowledge to guesswork."

The number of stars in the galaxy is well known and equals about 400 billion. Until the 1990s, that was the only factor in the Drake equation that anyone was sure of. The recent revolution in exoplanet studies, however, has changed that. "The discovery of exoplanets has given astrobiology a tremendous boost in the last 15 years," says Sullivan. For example, the Kepler satellite mission, which relied on observations of planet transits, has provided thousands of new exoplanet candidates. "Basically, we now have some idea—based on real data—about how many ecologically suitable planets per star there are out there." Through the database from the Kepler mission and other observation programs, it is now possible to make reasonable estimates for both the second and third terms (N_p and f_h) in the Drake equation, where N_p is at least a few and f_h is approximately 0.3. The rest of the factors remain in the realm of conjecture.

"The other three terms are also going to require empirical survey—going out and looking," says Sullivan. Scientists simply don't know enough from just looking at Earth to say how easy it is to get life going on a planet. "And if life does get going," says Sullivan, "we can't say with any certainty if it will develop intelligence." Those two terms, f_l and f_i, are really questions of biology and evolution. Since we've seen these processes work on only one planet (ours), it's hard to know how they might work on others. "I call this the $N = 1$ problem," says Sullivan. "If you only have seen biology and evolution on a single world, how can you extrapolate to all possible planets?"

The $N = 1$ problem gets even tougher for the last factor of the Drake equation. If intelligence develops and leads to a technological civilization that can build radio telescopes, how long will that civilization last? "It's possible that intelligent technological civilizations might be popping up all over the place," says Sullivan. "But if each one only lasts a short time, then the odds of two civilizations getting in touch go way down." The final factor of the Drake equation has implications beyond the search for intelligent extraterrestrial

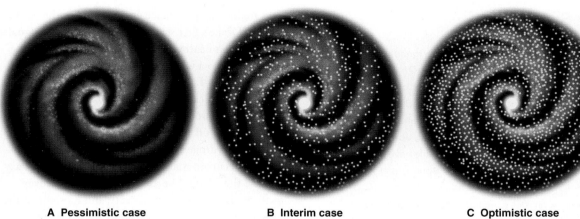

A Pessimistic case   **B Interim case**   **C Optimistic case**

FIGURE 9.18
Three Sets of Results for the Drake Equation
At this time, four of the seven terms in the Drake equation remain undetermined. Thus, we must make assumptions about their values to calculate the number of technological civilizations in the galaxy (N_c). Results range from the pessimistic case, that there is only one intelligent civilization in the galaxy and we are it (A), to the optimistic case, that there are many thousands or more (C).

life. Modern humans have a vested interest in knowing the lifetime of technological civilizations, since that is exactly what we are. "We have lasted for 100 years with radio technical capability," says Sullivan. "But how many more will it be? Another hundred? Another thousand? Another million?"

Because the last four factors of the Drake equation cannot be directly measured, scientists do their best to make logical arguments based on what we do know of molecular biology (f_l), evolutionary theory (f_i), and sociology (f_c). Even when reasonable scientists make their best estimates, the numbers for N_c span a huge range. "The optimistic estimates can put N_c in the tens of billions and make the galaxy appear to be teeming with intelligent life," says Sullivan. "Pessimistic estimates end up with N_c of just a few hundred. And really pessimistic ones can even lead to N_c less than one, which means that you would have to look at many galaxies before you found another technological culture. Those seem pretty depressing" (**Figure 9.18**).

SECTION SUMMARY

The Drake Equation: Guesstimating Life in the Galaxy

- The Drake equation is an attempt to evaluate the probability of extraterrestrial life and technological civilizations existing elsewhere in the galaxy.

- The factors in the Drake equation are the number of stars in our galaxy, the number of planets orbiting each star, the fraction of planets in the habitable zone, the fraction of those planets that develop life, the fraction of the planets with life that develop intelligence, and the lifetime of an average intelligent civilization.

- The optimistic case for extraterrestrial life could be tens of billions in our galaxy; the pessimistic case could be in the hundreds or fewer.

The Search for Other Minds

The Drake equation has proved enormously powerful for guiding our thinking about life in the galaxy (and hence the Universe), but ultimately, scientists simply have to collect more data. "There is no substitute for actually getting out to the telescope and looking," says Sullivan. And that is exactly what researchers involved in the Search for Extraterrestrial Intelligence (SETI) are doing.

SETI has been called everything from a wild-goose chase to the most important project in history. Sullivan himself is well aware of the hint of science fiction that comes with any consideration of extraterrestrial civilizations. "There have been many stories and movies based on the premise of aliens visiting Earth for one reason or another," he says. "But those of us involved with SETI feel that we shouldn't wait for them to visit, but rather actively search for possible distant extraterrestrial civilizations." Figuring out just what to do constitutes the science of SETI.

The fundamental premise of all SETI research is that extraterrestrial civilizations may be emitting signals that we can detect here on Earth. Those signals may be purposeful, such as a beacon beamed into space. They may also be unintentional, such as some kind of radiation leakage from other technologies that the civilization uses. Beginning from the premise that these signals exist, SETI researchers have to take the next step and ask detailed questions about their nature and the best strategy for finding them.

"One important question," says Sullivan, "is about what kind of radiation the ETs might be emitting." Because radio-band electromagnetic radiation has very long wavelengths, it can propagate over extremely long distances and not be absorbed. That makes radio waves

Woody Sullivan

A 9-pound Russian satellite was all it took to ignite Woody Sullivan's scientific ambitions. "I was in the eighth grade on October 4, 1957," says Sullivan, now a professor of astronomy at the University of Washington. "That was the day Sputnik was launched." Sputnik was the first artificial satellite to be lofted into orbit. While many Americans worried about the Russians dropping bombs from space, for Sullivan the whole enterprise was thrilling. "I thought, 'Wow, people have launched into space!'" he recalls. "I had been interested in science before that, but Sputnik really pushed me over the edge to astronomy."

Sullivan earned an undergraduate degree in physics from MIT, where he developed a passion for radio astronomy. But only while working on his PhD at the University of Maryland did he begin thinking about SETI. "I did my graduate thesis on water vapor in the interstellar medium," he explains. "It turns out water vapor has a spectral line with a wavelength of 1.35 cm (which is in the radio band). This was a new discovery at the time, and I ended up doing much of the early research on the subject. That's what led me to think about life in terms of water and radio signals."

Later, while Sullivan was reading about Project Cyclops, a NASA plan to build 1,000 large radio dishes to scan the sky for extraterrestrial signals, he got to thinking about something more than direct signals. "I started thinking about 'leakage,'" Sullivan says. Radio leakage means signals that a technological culture does not intend to send but could be detected anyway. "We are constantly leaking radio emission from our radar and other transmitting operations," Sullivan explains. "I started to think about what detection of leaked radio signals could reveal about a civilization." His work on the topic has become a classic SETI study.

again makes an appearance at this point in the SETI story. In Drake's first search for extraterrestrial signals, he focused on nearby stars similar to the Sun, tuning his telescope to a wavelength of 21 centimeters (cm), or a frequency of 1,420 megahertz (MHz), similar to what your microwave oven uses. Hydrogen, which is by far the most abundant element in the Universe, naturally gives off radiation at 21 cm.

Since hydrogen is so abundant, the **21-cm emission line** occurs in many places throughout the galaxy. "We use the 21-cm line all the time," says Sullivan, "to study the temperature, density, and motions of the gas between stars." Frank Drake reasoned that any technologically capable culture would also know about the 21-cm hydrogen line and would use it as a natural communication channel, since other scientific civilizations would think to use it as well. Since hydrogen is part of the all-important water molecule, many SETI scientists call the region of the 21-cm line the "cosmic watering hole" (**Figure 9.20**). "The idea is that different civilizations would naturally come to this frequency to communicate," says Sullivan, "just like a watering hole on Earth where people from different

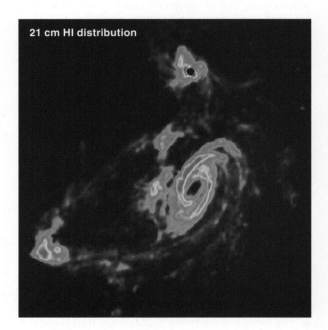

21 cm HI distribution

FIGURE 9.20 Cosmic Watering Hole
Hydrogen is part of the water molecule, so SETI scientists refer to the 21-cm line as the "cosmic watering hole," since advanced alien life would naturally come to this frequency to communicate. The term references the gathering effect of watering holes on Earth. As this 21-cm image of the galaxy M81 shows, emission in this particular wavelength is ubiquitous. (Red represents the brightest regions of 21-cm emission.)

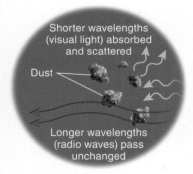

Shorter wavelengths (visual light) absorbed and scattered

Dust

Longer wavelengths (radio waves) pass unchanged

FIGURE 9.19
Radio Waves Can Travel Long Distances
Unlike light of shorter wavelengths, such as visible light, radio waves are not absorbed by interstellar dust and therefore propagate over long distances. This makes them a favored candidate for SETI programs.

a favorite of SETI scientists. "When we think about what kind of light ETs might use to beam a signal into space, we scratch our heads and say, 'Given our own technical capabilities and given our knowledge of physics, what is the best way to communicate over a hundred, a thousand, or ten thousand light-years?' Radio almost always comes out as the best candidate" (**Figure 9.19**). Recent advances in technology, however, have made researchers realize that powerful lasers could also be used to transmit information across interstellar distances. This idea has led a number of research teams to begin setting up searches in the optical or near-infrared portions of the spectrum.

Anyone who has flipped around the stations on a car radio knows that there are many frequencies (and hence wavelengths) to choose from in any of the electromagnetic wavelength bands. For example, if SETI scientists use radio telescopes to look at stars for evidence of intelligent life, to what frequencies should they tune their receivers?

"This is where SETI becomes an exercise in out-guessing the aliens," says Sullivan. Frank Drake once

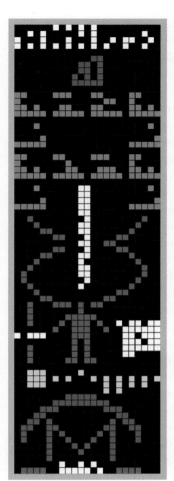

FIGURE 9.21
Frank Drake's Message to Extraterrestrials
This 1974 message, beamed toward star cluster M13 from the Arecibo radio telescope, used binary bits (ones and zeros) to form a symbolic picture when arrayed 23 × 73. Its content, developed by Frank Drake with collaborators such as Carl Sagan, included representations of the fundamental chemicals of life, the formula for DNA, a crude diagram of our Solar System, and simple pictures of a human being and a telescope.

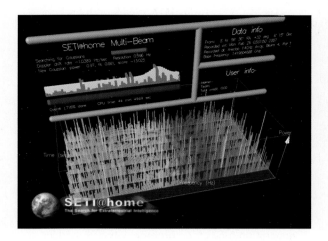

FIGURE 9.22
SETI@home Screen Saver
SETI has sought assistance from the public in analyzing the voluminous data it captures. Individuals can offer their home computers to download and search data for meaningful patterns via the SETI@home program.

21-cm emission line A wavelength of radio light emitted by hydrogen and considered by some to be a natural communication channel between Earth and other civilizations.

steady signals with information coded into them?" (**Figure 9.21**).

One way or another, SETI scientists are faced with the reality of sifting through vast quantities of data. They will have to look at many stars and carry out their observations at many frequencies. To sort through the mountains of data, scientists at the SETI Institute have even enlisted the help of the public, through the SETI@home project. Participants download the SETI@home software to their home computers. When the machines are not being used, the software takes over the central processor and uses it to analyze chunks of radio observations for possible extraterrestrial signals (**Figure 9.22**).

Despite the widespread fascination of the subject, SETI has always been controversial work. Most of the studies done so far have piggybacked on more traditional radio astronomy projects. In the 1980s and 1990s, NASA had plans to harness existing radio telescopes with powerful electronics to carry out ambitious searches for extraterrestrial signals. But pressure from a few vocal members of Congress killed the project before it was completed. "The project was axed because it was an easy target," says Sullivan. "People could make fun of it and say that searching for little green men was a waste of money. It's a pity. The project needed a relatively small amount of money to ask what is really the most important question we have in science: Are we alone?" (**Figure 9.23**).

SETI activity is still going on, thanks to the tireless work of some dedicated researchers (like astronomer Jill Tarter of the SETI Institute) and the generosity of private donors. The work, however, has barely started. To date, only a small percentage of the sky has been explored, even in the 21-cm line. We may have to wait a long time before we can make any

FIGURE 9.23
Area 51: Making a Joke of SETI
The concept of contact by aliens is so prevalent in our culture that it is often the subject of satire, as evidenced by this whimsical response to stories of aliens abducting cattle for experiments.

villages gather." (It's worth noting that microwave ovens are tuned to about the 21-cm wavelength because it makes water molecules in food vibrate rapidly enough to heat up the food.)

Drake based his search strategy on the 21-cm line, but other researchers have designed searches that scan as many frequencies as possible. "Some folks felt like too many eggs were being put in the 21-cm basket," says Sullivan, "but the problem with scanning lots of frequencies is the huge amount of data you generate. A search like that can easily overwhelm your processing and analysis resources."

Wavelength isn't the only domain in which SETI researchers have to make informed guesses about extraterrestrial behavior. The nature of the signal itself is a matter of debate. "The signal is going to have to be different enough from natural astrophysical radio emission that you can tell that it's due to an intelligent source," says Sullivan. "But what form will that take? Should we be listening for some kind of pulsing beacon like a 'beep beep beep,' or should we look for

Fermi paradox The conclusion that no intelligent life exists in our galaxy beyond Earth, since the billions of years following the origin of the galaxy would have allowed advanced civilizations to have spread across interstellar space and reached Earth.

A

B

FIGURE 9.24
Alien Civilization
What if other intelligent beings exist? What would they and their civilization be like? Theoretical physicist and cosmologist Stephen Hawking warns that aliens from an advanced civilization might not be as benevolent as (A) the lovable ET, but more like (B) the predatory, violent creature from the movie *Alien*. Until actual contact occurs, only the imagination of science fiction authors and artists provides models of what we might expect.

definitive statements about intelligent civilizations in space.

SECTION SUMMARY
The Search for Other Minds

- The Search for Extraterrestrial Intelligence, or SETI, is based on the premise that extraterrestrial civilizations are emitting signals, such as radio waves, that can be detected.

- Some searches scan many frequencies to increase their chances, and some focus on the 21-cm hydrogen line because hydrogen is so abundant and is a component of water.

- The scope of the investigation is an obstacle, as is generating funds for a program whose practical value is questioned by some.

Life and Its Implications

In the 1950s, the Nobel Prize–winning physicist Enrico Fermi was having lunch with his friends when the subject of life elsewhere in the galaxy came up. Fermi was known for his brilliance and the quickness with which he could dissect a problem. After thinking for a minute, Fermi asked his friends a penetrating question: "If ETs exist," he said, "why aren't they here already?"

What Fermi realized in that brief instant was that if even one civilization were to develop interstellar space travel and begin exploring other worlds, then they should have quickly spread across the entire galaxy. Even though it might take hundreds of years to travel between nearby stars, the galaxy is billions of years old. Assuming that a civilization was born relatively soon after the galaxy's formation, there should have been more than enough time for it already to have swept across our part of space.

One answer to the so-called **Fermi paradox** is simple: there are no other intelligent species out there in the galaxy. The implications of this cosmic isolation would be profound. What would it mean for us as a species if we were the only life-forms with self-consciousness in the galaxy—or the entire Universe (**Figure 9.24**)? Would humanity be destined to spend its entire history knowing only its own perspective of life and death, meaning and existence? For some thinkers, this situation is the species-level equivalent to an eternity of solitary confinement.

But there is another answer to Fermi's paradox that some people find even more disturbing. Perhaps we have not found other examples of intelligence because they are smart enough *not* to announce

themselves. Many animals in the wild are careful to cover their tracks to avoid predators, so perhaps we are being unwise in allowing our TV, radio, and radar signals to leak so freely into space. While SETI researchers often assume that extraterrestrials will be benign, some scientists, including Stephen Hawking, argue that caution is a better strategy. In Hawking's view, we should not purposely announce our presence by beaming signals into space until we have a better understanding of what's out there.

Whatever the answer to Fermi's paradox might be, if and when any evidence of extraterrestrial life is found, it will surely be one of the greatest discoveries in history. Even finding nothing more than microbes on another world would tell us that life was not a single accident on a single world. Knowing that the process of life's formation and evolution happened on more than one planet would force us to alter dramatically our understanding of the Universe.

What might happen, however, if we discovered not just life-forms, but intelligent life-forms, other than our own in the Universe? It is difficult to speculate, but many scientists and philosophers believe such knowledge would have the potential to change human culture. How would we understand our own history, for example, if we knew there were other histories out there? How would people understand their religion if they suddenly found out that other intelligent beings existed who were not mentioned in any scripture? Would knowing that at least one other civilization exists allow us to assume that technologically adept cultures can survive past our own point in evolution? While much would depend on the nature of what we learned about and from the new civilization, questions like these demonstrate how significant and powerful such a discovery would be.

As astrobiology matures through discoveries of exoplanets, explorations of our own Solar System, and new studies of life on Earth, we appear to be moving closer to an answer to these questions. Within your lifetime, astronomers may announce evidence, either direct or indirect, for life on another planet. And on that day our view of our place in the Universe will change forever.

SECTION SUMMARY
Life and Its Implications

- The Fermi paradox asks, "If extraterrestrials exist, why aren't they here already?"

- Generating signals from Earth to other civilizations is controversial because it may alert hostile civilizations to our presence.

●→ chapter summary

9.1 The Origin of Life and the Bottom of the World

A living organism can be defined as something that is composed of organic molecules and exhibits metabolism, reproduction, mutation, and sensitivity to its environment. Assuming life requires water, scientists define the habitable zone around a planet as the region where temperatures allow for liquid water on the planet's surface. In other planetary systems the boundaries of such a zone around a star depend primarily on the size and temperature of the star. Moons of exoplanets also provide potential environments for life. The discovery of extremophiles in deep-sea vents and rocks deep within Earth has broadened the known range of viable living environments and helps scientists understand how the earliest life may have evolved on Earth.

9.2 What Is Life, and Where Can It Exist?

It is difficult to come up with a comprehensive definition of life, but definitions often include the presence of organic molecules, metabolism, reproduction, mutation, and sensitivity. Every star has a habitable zone around it, where liquid water could exist on a planet's surface. Polymers (or macromolecules) including DNA, RNA, and proteins are the basis for life and its diversity. DNA is a two-strand helix, capable of self-replication, with a backbone of nucleotides linked by nucleobases. RNA has similar structure but in a single strand. Both carry genetic information. Proteins, made of amino acids, perform various functions within living cells.

9.3 The Origins of Life

An early study of abiotic synthesis—how life formed from nonlife—was the Miller-Urey experiment, which showed that prebiotic molecules could have formed in Earth's early environment. Complicated polymers can be built from smaller segments that are then linked. Producing the exact polymers molecules for self-replication is a matter of chance and would have required great stretches of time. Once a self-replicating molecule is built, however, it will proliferate. The DNA of a species can be altered by several means: spontaneous mutation, migration of populations with interbreeding, and genetic drift. Natural selection is the process by which any changes that enhance survival lead to further reproduction and increase in the frequency of the genes that code for those changes. In this way, life diversifies to thrive in various environments. Microbes, in contrast, can mutate asexually by lateral gene transfer.

9.4 Searching for Other Life in the Universe

Life on Earth evolved within the planet's first billion years, but the evolution of intelligent life is, relatively, a very recent phenomenon. The path of evolution started with one-celled organisms without nuclei, which evolved into organisms with cell nuclei. The atmosphere of early Earth had little oxygen, and the initial microbes did not require it for their metabolism, though they did produce it as a by-product. Thus, biological activity led to the buildup of oxygen in the atmosphere in a long-term phenomenon called the Great Oxygenation Event, after which oxygen-breathing organisms came to dominate evolution. The diversity of complex organisms multiplied exponentially during the Cambrian explosion. SETI (Search for Extraterrestrial Intelligence) is a scientific search for intelligent life on other worlds. The Drake equation specifies the factors believed to be important in assessing the likelihood of such life. Radio signals, particularly at the 21-cm hydrogen line, are considered by some to be the most likely means of detecting intelligent life beyond Earth. Recent discoveries about the prevalence of exoplanets have made the probability of extraterrestrial life much higher, but the question remains open.

●→ questions and problems

Narrow It Down: Multiple-Choice Questions

1. Which of the following is not a basic feature of all living things on Earth?
   a. consumption of energy
   b. sexual reproduction
   c. mutation
   d. sensitivity to environment
   e. organic molecules

2. The original definition of "habitable zone" included which of the following?
   a. surface temperature of a planet in the range of 273–373 K
   b. presence of surface water
   c. known existence of life-forms
   d. Sun-like central star
   e. availability of carbon

3. Why has the classic idea of a habitable zone based only on a planet's distance from the Sun and liquid water on its surface been reconsidered? Choose all that apply.
   a. Not all life on Earth is carbon based.
   b. Extremophiles have shown that life can exist in a broader range of conditions of temperature and pressure than was previously considered.
   c. Not all terrestrial life exists on Earth's surface.
   d. Water has been detected on giant planet moons.
   e. Life has been discovered in the dry regions of Mars.

4. You find a polymer that contains the nucleobases A, G, and C. What kind of molecule can it be?
   a. only DNA
   b. only RNA
   c. either DNA or RNA
   d. an amino acid
   e. DNA, RNA, or an amino acid

5. Which of the following is/are true of nucleotides? Choose all that apply.
   a. Each is a combination of a sugar molecule and a phosphate.
   b. They are components of DNA.
   c. They are components of RNA.
   d. They are polymers linked to other polymers by nucleobases.
   e. They are proteins.

6. The pattern of nucleobases in a strand of DNA is TGCAACG. When the strand splits for reproduction, which nucleobases will attach, in what sequence?
   a. TGCAACG
   b. CATGGTA
   c. CGTTGCA
   d. GCAACGT
   e. ACGTTGC

7. The sequence of bases in DNA directs which function?
   a. regulating oxygen synthesis
   b. building proteins
   c. facilitating mutations
   d. regulating reproduction
   e. regulating metabolism

8. The results of the Miller-Urey experiment supported which of the following conclusions about the early Earth?
   a. Life on Earth was inevitable.
   b. All available carbon would have been converted into organic molecules.
   c. The molecules produced in the experiment represented the specific molecules that formed life on Earth.
   d. Life was easy to produce.
   e. Amino acids and organic compounds would have formed readily.

9. Using Earth's evolutionary history as a model, what is the probability of finding intelligent life on a planet that has liquid water at or just below its surface and is orbiting in the habitable zone of a 1-billion-year-old Sun-like star?
   a. very likely, because of the presence of water
   b. very likely, because of the planet's location
   c. very likely, because of the type of star
   d. very unlikely, because of the age of the system
   e. There is not enough information to assess the probability.

10. Which of the following statements about natural selection is *not* true?
    a. Different traits are more favorable in different environments.
    b. Natural selection can lead to better adaptation of an organism to the environment.
    c. Traits that enhance the probability of reproduction tend to proliferate.
    d. Less adaptive traits die out in one generation.
    e. Natural selection may begin with a random genetic change.

11. Which of the following terms does *not* describe a process that can affect the frequency of a gene in a population?
    a. migration
    b. mutation
    c. genetic drift
    d. natural selection
    e. replication

12. Which of the following is/are examples of traits that can be tied to heredity? Choose all that apply.
    a. Men who shave their heads as cultural norm also shave their sons' heads.
    b. The tallest sibling in a family produces taller children than his siblings do.
    c. Pancreatic cancer runs in a family.
    d. A child, whose parents both lack a specific antigen in their blood, also lacks that antigen.
    e. A chameleon can change its skin color at will.

13. Which of the following statements about lateral gene transfer is/are *not* true? Choose all that apply.
    a. It occurs only in reptiles.
    b. It occurs only in microorganisms.
    c. It occurs only asexually.
    d. It represents a two-way sharing of genes.
    e. It occurs within one generation.

14. Which of the following characteristics accurately describe Earth's first life forms? Choose all that apply.
    a. self-replicating
    b. single-celled
    c. multicellular
    d. aerobic
    e. anaerobic

15. The Great Oxygenation Event is believed to have been caused by what? Choose all that apply.
    a. saturation of the ocean and land with CO_2
    b. reduction in CO_2 due to a reverse greenhouse effect
    c. saturation of the ocean and land with O_2
    d. proliferation of anaerobic organisms
    e. proliferation of aerobic organisms

16. Which of the following accurately describe(s) the Drake equation? Choose all that apply.
    a. It tells us exactly which kinds of stars can harbor planets with life.
    b. It calculates the true probability of extraterrestrial life.
    c. It names the factors that contribute to the total probability of life elsewhere.
    d. It provides a way to gauge how knowledge of the various factors is progressing.
    e. Its value has not changed since it was first conceived.

17. Recently acquired knowledge based on exoplanet discoveries has addressed which terms of the Drake equation?
    a. N_* (number of stars in the galaxy)
    b. N_p (number of planets orbiting each star)
    c. f_l (fraction of planets on which life evolves)
    d. f_i (fraction of planets with life that develop intelligence)
    e. f_c (lifetime of an average intelligent civilization expressed as a fraction of the star's life)

18. Why is the 21-cm line believed to be a prime candidate for the wavelength an extraterrestrial would send or recognize if sent? Choose all that apply.
    a. It is radiated by hydrogen, the most abundant element in the Universe.
    b. It falls in the radio part of the spectrum, whose long wavelengths are most likely to travel unimpeded over large distances.
    c. An advanced civilization would be aware of its relevance.
    d. It is definitive proof of water and therefore life.
    e. It is easier to detect than all other frequencies.

19. True/False: In calculating the habitable zone for the Sun, astronomers assume that life on Solar System planets requires water. If a planet orbiting at 1 AU from a Sun-like star had life based on a molecule with much higher melting and boiling points, the habitable zone for that star would be significantly farther from the star than the Sun's is.

20. True/False: All known living things on Earth contain carbon.

To the Point: Qualitative and Discussion Questions

21. What are the distinguishing characteristics of organic molecules?

22. Some moons of planets outside the habitable zone in our Solar System appear to have liquid water on or below their surfaces. What are the possible sources of the heating required to produce this environment?

23. Why are very hot stars not good candidates for highly evolved life?

24. What sources of energy other than sunlight have been discovered for extremophiles on Earth?

25. Describe how DNA differs from RNA in structure and function.

26. What additions to our understanding of the early Earth change the interpretation of the Miller-Urey experiment?

27. Discuss the importance of time in abiotic synthesis.

28. As the environment changes, how does natural selection help a particular species adapt?

29. What is meant by the term "Cambrian explosion"?

30. Why do astronomers believe that radio wavelengths are the most likely means of interstellar communication?

31. What is the Fermi paradox?

32. To date, what evidence is there for the existence of life beyond Earth?

33. Is the search for intelligent extraterrestrial life worth what it costs? Should SETI be expanded or eliminated?

34. Some people believe that evolution is only a theory and that there are other equally valid theories for the creation of different species on Earth. How would you respond to that claim?

35. How would it change your religious or philosophical views if the existence of intelligent extraterrestrial life were confirmed? if it were conclusively refuted?

Going Further: Quantitative Questions

36. Use the Drake equation to calculate N_c if N_* is 100 billion, N_p is 8, f_h is 2, f_l is 0.5, f_i is 1, and f_c is 2×10^{-5}. What does your answer mean for the likelihood of finding intelligent life elsewhere in the galaxy?

37. What is N_c if the values of all factors in the Drake equation are the same as in question 1, except f_l is 0.005?

38. What is the minimum length of time for sending a message and receiving a reply from an extraterrestrial on a planet 23 light-years away?

39. Assume that an RNA molecule has 144 nucleotides and that there are 4 different nucleotides. How many attempts will it take to produce one of these molecules, if it must be built from individual nucleotides one at a time?

40. Imagine that life that evolved on an exoplanet is based on 6 unique nucleotides instead of 4. An RNA molecule on that planet requires 138 nucleotides built from those 6 kinds of nucleotides. Building the RNA from individual nucleotides, how many attempts will it take to be sure to produce the necessary one?

41. Assume that the RNA molecule needed for life has 144 nucleotides, but it can be constructed out of a specific combination of 16-nucleotide subunits. Both can be built from 4 basic nucleotides. How many different combinations of the 16-nucleotide molecules are there?

42. A slowly rotating planet with an albedo of 0.5 orbits a star that has a luminosity 2 times that of the Sun at 5 AU. What are the inner and outer boundaries for the habitable zone? Is the planet in the habitable zone?

43. A slowly rotating planet with an albedo of 0.3 orbits a star that has a luminosity 5 times that of the Sun at 2.5 AU. What are the inner and outer boundaries for the habitable zone? Is the planet in the habitable zone?

44. A fast-rotating planet with an albedo of 0.4 orbits a star that has a luminosity 0.8 times that of the Sun at 0.5 AU. What are the inner and outer boundaries for the habitable zone? Is the planet in the habitable zone?

45. A fast-rotating planet with an albedo of 0.7 orbits a star that has a luminosity 3 times that of the Sun at 0.3 AU. What are the inner and outer boundaries for the habitable zone? Is the planet in the habitable zone?

 If your instructor assigns homework in **smartwork5**, access it at the Digital Landing Page for *At Play in the Cosmos*: **digital.wwnorton.com/cosmos**

THE SUN
AS A STAR

In this chapter you will learn about our Sun, its internal structure, and its outer environment. After reading through each section, you should be able to do the following:

10.1 Living with a Star
Describe the structure and composition of our star.

10.2 The Sun's Fusion Furnace
Explain the source of the Sun's energy and how we know about it.

10.3 Moving Energy
Explain the processes by which energy moves from the Sun's core outward to the surface.

10.4 The Active Sun: Photosphere to Corona and Beyond
Discuss the origin and dynamic nature of the Sun's magnetic field and its effect on solar behavior.

The sun as seen through the light of an emission line of once-ionized calcium. The blue light shows us a region of the Sun called the chromosphere, approximately 2,000 km above the Sun's "surface" (called the photosphere). This image was taken with an amateur telescope fitted with a special filter.

10.1
Living with a Star

THE SUN is the closest star to Earth; it is only 149 million kilometers (km) away. The next nearest star, Proxima Centauri, is nearly 30,000 times more distant. This proximity presents both benefits and problems. "We gather so much data about the Sun that putting it all together is a real challenge," says Lika Guhathakurta. "The Sun is such a remarkable object, and there are so many different questions to ask: Where does it get its power? What is its internal structure? What forces drive the incredible events happening on its surface?" As program manager for key NASA solar physics programs, Guhathakurta has brought together researchers from different fields—such as those who study the solar atmosphere or those who study internal solar rotation—and helped them integrate their findings.

In the 1990s, when space instruments began providing 24-hour views of the Sun, Guhathakurta and other scientists initiated a push to bring together all the different perspectives on the Sun. From the magnetic storms of its outer atmosphere to the nuclear fires of its core, the combined efforts taught scientists just how remarkable a star can be. "There is an old joke that stars look pretty simple because they are so far away," says Guhathakurta, "but the Sun shows us that up close, stars are astonishingly complex."

From a God to a Star

Many of our ancestors worshipped the Sun as a god—and for good reason. Even at a distance of millions of kilometers, the Sun is extremely powerful. On a clear summer day, looking anywhere near the Sun is enough to hurt your eyes, and staring directly into the Sun can cause blindness. The warmth that reaches Earth from the Sun can be either comfortable or oppressive, depending on the season. Up close, the blinding light and searing heat of the Sun reach scales unlike anything you've imagined or experienced. If humans tried to reach even the outermost layers of the Sun using existing technology, our machines would be vaporized and reduced to nothing more than stray atoms.

This description of the Sun as a source of vast energy is equivalent to its definition as a star. The Sun, like all other stars, is a massive sphere of

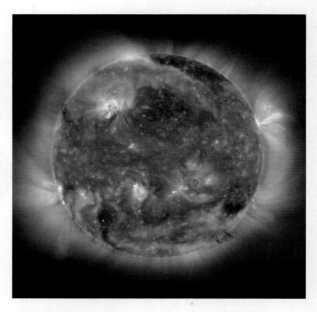

FIGURE 10.1 Our Sun
The Sun is a star—a massive sphere of superheated, electrically charged gas. This image, taken in three different wavelengths of ultraviolet light by the *SOHO* (Solar and Heliospheric Observatory) spacecraft, shows how active its surface can be.

superheated **plasma**, or electrically charged gas (**Figure 10.1**). And like all other stars, the Sun is also a forge in which light elements like hydrogen undergo **thermonuclear fusion**, which squeezes them together to create heavier elements like helium. Over time, these heavier elements have chemically enriched the galaxy, enabling life to form on Earth and perhaps other planets.

This chapter begins our exploration of stars and their life stories. Stars do not live forever but cycle through a process of birth, middle age, and decay, just as human beings do. The life story of a star involves a remarkable array of physical processes that shape the Universe and life. At the end of a star's life, some of the Universe's strangest inhabitants are created, including superdense neutron stars and enigmatic black holes. However, it makes sense to start at home with the most familiar star, the Sun, since it's the one we know the most about.

SECTION SUMMARY
From a God to a Star
- Because the Sun is our closest star, we can use it to study stars in general in great detail.

plasma Electrically charged gas.

thermonuclear fusion The transformation of lighter atomic particles into heavier ones, requiring high temperatures.

How We Know What We Know about the Sun

If it's impossible to send a probe into the Sun, how do scientists know anything about the Sun's deep interior? Much of the current knowledge comes from the remarkable study of **helioseismology**. As in seismological studies of Earth after an earthquake, scientists can use oscillations of the Sun's surface to infer properties of the solar interior. But to see those oscillations, astronomers must use spectroscopy, monitoring absorption lines in the Sun's spectrum for shifts in the wavelength. These changes in wavelength can then be tied to the motion of the Sun's surface. "We can see the Doppler shift of material on the solar surface," says Guhathakurta. "It's like watching the whole Sun ring like a bell. By tracking these solar oscillations, we have been able to work backward and build models of the different layers of the solar interior" (**Figure 10.2**).

Using helioseismology, astronomers from around the world in projects like the Global Oscillation Network Group (GONG) have tracked the Sun's vibrations in exquisite detail. With these data, they then constructed testable predictions about the Sun's structure and developed precise theories describing what makes stars tick. Let's begin with some basic facts and a quick tour of the Sun from its fusion-powered core to the vaporous violence of its outermost layer, the corona.

SECTION SUMMARY
How We Know What We Know about the Sun
- Helioseismology, combined with modeling, has provided astronomers with knowledge of the Sun's inner workings.

Anatomy of the Sun

"The Sun is far and away the most massive object in the Solar System," says Guhathakurta. It accounts for 99.86 percent of all the Solar System's material. The current mass of the Sun is about 2×10^{30} kilograms (kg). That means you would need 333,000 planets the size of Earth packed together to equal the mass of the Sun. And the Sun is not just massive, it's physically enormous. The radius of the Sun is 695,000 km, 100 times the radius of Earth. You could put 1.3 million Earths in the volume occupied by the Sun and still have room for quite a few more. Even Jupiter is dwarfed by the Sun, which could swallow at least 1,000 planets the size of the Solar System's largest gas giant.

"To put it into more human terms," says Guhathakurta, "a typical passenger jet would take nearly half a year to circle the 4.4-million-km equatorial circumference of the Sun."

Remarkably, all that mass is pretty stable, and the Sun's radius changes significantly only over very long timescales. The Sun shows this stability because it's in what astronomers call **hydrostatic equilibrium**, meaning that the inward crush of gravity is balanced by the outward push of pressure. The pressure has its origin in the energy released by nuclear fusion in the core. As long as the fusion continues, the Sun will continue its balancing act of pressure and gravity.

Like everything else in the Solar System, the Sun is spinning. It takes approximately 26 days for the Sun's equator to complete one revolution. But near the poles it's a different story: there, material takes almost 8 more days to complete one rotation. "We call this difference in the spin rates of different parts of the Sun **differential rotation**," says Guhathakurta. "It's not just a curiosity," she explains. "Differential rotation is essential for generating the kind of powerful magnetic field we find on the Sun, which controls so much of what happens on its surface." While not all magnetic fields require the presence of such differential rotation, for the Sun's dynamo it's a crucial ingredient (**Figure 10.3**).

At present, the elemental composition of the Sun (by mass) is 71.5 percent hydrogen and 27.1 percent

FIGURE 10.2
Helioseismology
Astronomers use helioseismology (the study of the Sun's oscillations and pulses) to investigate conditions in the solar internal structure. This image, taken from a theoretical model, is a snapshot of rising (blue) and falling (red) plasma in the Sun (the rising and falling regions switch as the oscillation progresses). The actual oscillations seen on the Sun are a combination of many such patterns.

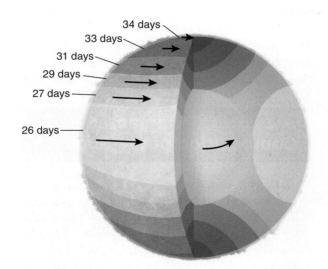

FIGURE 10.3 Differential Rotation
The core of the Sun rotates as a solid body with a period of 27 days, but the outer layers experience differential rotation. In other words, different latitudes and depths have different rotation periods, ranging from 26 days at the equator to 34 days at the poles.

helioseismology The study of oscillations in the Sun and their propagation. Helioseismological studies are used to develop precise models of solar structure.

hydrostatic equilibrium The condition in which gravity and gas pressure are balanced.

differential rotation The rotation of different regions of an object (different latitudes or depths) at different rates.

Lika Guhathakurta

Neither of Lika Guhathakurta's parents were scientists, but that did not stop them from instilling in their young daughter a scientist's most important quality: curiosity. "There was lots of very general discussion in the family, and I think both my parents were very open-minded individuals."

Born in Kolkata, India, Guhathakurta moved with her parents to Mumbai when she was 8. From an early age she was introduced to astronomy through books. "I remember these wonderful picture books about stars, comets, and planets," she says with a smile. Guhathakurta's early fascination deepened as she learned to read and began absorbing the story behind the pictures. "I remember poring through all those books in my native Bengali. I never lost that fascination."

Eventually, Guhathakurta applied to the University of Colorado to study astrophysics. There, in the Laboratory for Atmospheric and Solar Physics, she began her concentration in the study of the solar atmosphere. "There was so much going on then: eclipse observations and analyses of new satellite data. One of the images we took in X-rays of the solar corona made it onto the cover of *Time* magazine."

But Guhathakurta was always interested in seeing the big picture. As her career progressed, she became involved in one major solar observation program after another. The different interests of each project made her determined to keep the big picture of the Sun in mind. "I really like bringing people together from the different subdisciplines because it's getting the overview that makes science really fun! People should always try to remember that gaining a grasp of the whole is why we do what we do as researchers."

i interactive
Explorable Sun

helium; other elements, including oxygen, carbon, iron, neon, nitrogen, silicon and magnesium, each make up a fraction of a percent of the Sun's mass (**Table 10.1**). These numbers for the solar composition have been determined by spectroscopy. It is important to talk about the *present* composition of the Sun, because that

core The central region of a star where high temperatures and high densities allow thermonuclear fusion reactions to occur, releasing energy.

TABLE 10.1 ••• Elemental Composition of the Sun

Element	Percentage of Sun's Mass
Hydrogen	71.50
Helium	27.10
Oxygen	0.60
Carbon	0.25
Iron	0.14
Neon	0.13
Nitrogen	0.07
Silicon	0.07
Magnesium	0.07
All others	0.07

FIGURE 10.4 Structure of the Sun
Using helioseismology and other data, scientists have constructed detailed models of the Sun's principal layers. From the center outward, these are the core (where fusion reactions release energy), the radiative zone, the convective zone, the photosphere (the solar "surface"), the chromosphere, and the corona.

composition is changing. Fusion reactions in the Sun's core are, right now, converting hydrogen into helium, though the rate for significant changes in solar composition must be measured in billions of years. "If we could wait long enough," explains Guhathakurta, "we would see the helium fraction of the Sun's mass slowly increase and the hydrogen fraction decrease."

The Sun's composition is not uniform throughout, however. "The Sun is made up of different layers, one inside the other," says Guhathakurta. "Each of these layers has different properties, and each is controlled by different physical processes." Starting from the center, these layers are the core, the radiative zone, the convective zone, the photosphere, the chromosphere, and the corona (**Figure 10.4**).

CORE. "The **core** is the region where energy is produced," explains Guhathakurta. It is composed primarily of ionized hydrogen along with helium "ash" (left over from fusion). Extending from the center out to about 150,000 km (20–25 percent of the Sun's full radius), the core is the densest and hottest region of the Sun. Core densities are as high as 160,000 kg per cubic meter (kg/m^3), 160 times denser than water and 14 times denser than lead. Temperatures in the core can be as high as 15 million kelvins (K), and these temperatures are critical for the energy-generating fusion reactions to occur.

RADIATIVE ZONE. "After energy is generated in the core, it has to work its way to the surface," says

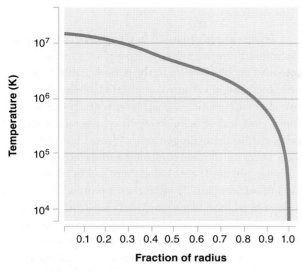

A Temperature of Sun

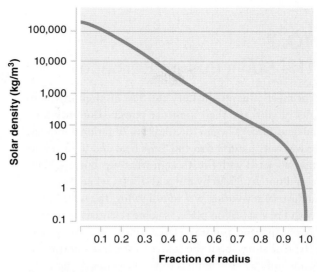

B Density of Sun

FIGURE 10.5
The Sun's Temperature and Density
The layers of the Sun exhibit decreasing (A) temperature and (B) density from the core to the photosphere.

Guhathakurta. "The **radiative zone** marks the first step in that journey." Extending from just above the core out to 500,000 km (70 percent of the Sun's radius), the radiative zone is a region, composed primarily of ionized hydrogen and helium, where the energy is transported outward by means of electromagnetic radiation (the scattering of photons off particles of mass). The density drops by more than a factor of 100—from 20,000 kg/m^3 to only 200 kg/m^3 across this region. The temperature, however, drops by less than a factor of 4, from 7 million K to 2 million K.

CONVECTIVE ZONE.   Although radiation is the means of energy transfer in the radiative zone, in the **convective zone** the motion of gas (convection) is what moves energy around. The motions in the convective zone originate when plasma at the bottom of the zone absorbs heat and then rises to the surface. Once on the surface, it cools by emitting radiation into space and then sinks again, and the process begins anew. The convective zone stretches from 500,000 km above the Sun's center all the way to its surface. The density drops from 200 kg/m^3 down to just a fraction of a kilogram per cubic meter just below the photosphere. The temperature falls dramatically, from nearly 2 million K near the border of the radiative zone to 5,800 K at the Sun's surface (the outer edge of the convective zone).

PHOTOSPHERE.   The **photosphere** is the closest thing the Sun has to a surface. The photosphere is where light from the solar plasma escapes into space. That means it's where the blackbody spectrum of the Sun is emitted. "This is the part of the Sun we actually see," says Guhathakurta. Temperatures at the photosphere are 5,800 K, and particle densities are 10^{-4} kg/m^3, only 0.37 times that of Earth's atmosphere at sea level. The temperature falls with height above

the photosphere, and it is in these colder regions that absorption lines form.

CHROMOSPHERE.   Extending 2,000 km above the photosphere, the **chromosphere** is considered the lower layer of the Sun's atmosphere. In the bottom parts of the chromosphere, the temperature drops to 4,100 K. "Once you get to the upper layers of the chromosphere," says Guhathakurta, "the temperature actually begins to rise again." These regions of chromosphere, being hotter than the underlying photosphere, produce an emission line spectrum. The density, however, continues to drop with distance from the photosphere, until the chromosphere merges with the ultratenuous corona above.

CORONA.   With temperatures rising as high as 1 million K, the **corona** is an extended region of extremely hot, low-density gas surrounding the Sun. The corona typically reaches as far out into space as 2 times the radius of the Sun, though gas densities there can be as low as 10^{-14} kg/m^3, a hundred trillion times less than the density of air at Earth's surface. The progression of temperatures and densities from the Sun's core to its photosphere is shown in **Figure 10.5**.

SECTION SUMMARY
Anatomy of the Sun

- The Sun comprises the majority of mass in the Solar System.
- The Sun rotates differentially, slower at the poles than at the equator.
- Hydrogen (71.5 percent) and helium (27.1 percent) are the primary elements making up the Sun.
- Six distinct layers constitute the Sun and its atmosphere. Density and temperature decrease in each successive layer except the corona, where temperature increases dramatically.

radiative zone The region of the Sun above the core, where energy is transported outward by means of electromagnetic radiation (photons scattering off of particles of mass).

convective zone The region of the Sun above the radiative zone, where energy is transported outward by means of the motion of gas (convection).

photosphere The thin, "surface" region of the Sun, from which radiation escapes into space.

chromosphere The lower layer of the Sun's atmosphere.

corona The extended region of extremely hot, low-density gas surrounding the Sun.

10.2
The Sun's Fusion Furnace

For most of human history, the Sun inspired mythologies of gods and their immense powers. But once the ancient Greeks began the tradition of rational inquiry, it was only natural that the Sun's heat and light would be associated with fire, though what was burning they could not say. Even by the mid-1800s, when the industrial revolution was in full swing, the solar link with fire was still in place, as some scientists hypothesized that the Sun was a large sphere of burning coal. The true story about the source of the Sun's power came only when scientists learned to unpack the secrets of atomic nuclei.

The Birth of Nuclear Physics

By the early 20th century, spectroscopy had enabled astronomers to deduce that the Sun was composed mostly of hydrogen gas. The question then arose whether the Sun's energy might come from gravity squeezing down on the large gaseous sphere. To the British scientist William Thomson (Lord Kelvin; **Figure 10.6**), such **gravitational contraction** was the answer to the mystery of the Sun's energy source. Kelvin believed that the inward pressure of gravity forced the solar gas to heat up. When some of that heat energy was radiated into space at the solar surface, gravity could squeeze a little tighter, shrinking the Sun a bit more and liberating a bit more energy in the process. The process would continue until the Sun was just a cold cinder.

Lord Kelvin calculated the timescale for this slow release of energy via contraction and showed that it could power the Sun for about 20 million years. Although this seemed like a long time, independent estimates based on geology were already putting Earth's age at more than a billion years. Since the Sun couldn't be younger than Earth, it soon became clear that some other kind of physics, unknown at the time, must also be operating to allow the Sun to shine for billions of years. The solution to this dilemma was not found until the science of nuclear physics was born in the late 1920s and early 1930s. "Nuclear physics studies the cores, or nuclei, of atoms," says Falk Herwig, an astrophysicist at the University of Victoria who specializes in the nuclear physics driving stellar evolution. "Early in the 1900s," he explains, "physicists discovered that all atoms were composed of negatively charged electrons orbiting a central positively charged nucleus."

Hydrogen, as discussed in Chapter 4, has a single electron orbiting a nucleus composed of a single proton. Take one step up in the periodic table, however, and things get more complicated. Helium has two electrons orbiting a nucleus with two protons. But the nucleus of helium weighs approximately *four* times that of the hydrogen atom. "Since electrons have a very small mass compared with protons," says Herwig, "physicists concluded there had to be another kind of nuclear particle other than a proton." Thus helium must have two *neutral* particles living inside its nucleus. In 1932, this uncharged particle, called the neutron, was discovered, and nuclear physics began in earnest. Physicists found that the nuclei of most elements contained a mix of protons and neutrons.

Although the number of protons is always fixed for a given element, the number of neutrons can vary, producing different isotopes of that element (see Going Further 6.1). A nucleus with 6 protons and 6 neutrons (carbon-12, or ^{12}C) is called carbon, but so is a nucleus with 6 protons and 7 neutrons (carbon-13, or ^{13}C).

As physicists probed the nucleus, they began mapping out a wide range of nuclear reactions between two or more nuclei. "Nuclear reactions are interactions that transform one nucleus into another," says Herwig. Within just a few years after the advent of nuclear physics, scientists like Hans Bethe and Enrico Fermi began to see how reactions called thermonuclear fusion might be the source of the Sun's energy. Then, over the course of the next two decades, the pieces of a full theory of stellar nuclear power were put into place. By the early 1960s—just five decades ago—one of the oldest of human questions had been answered. We finally knew how stars shine.

FIGURE 10.6
Lord Kelvin
William Thomson (the first baron of Kelvin; 1824–1907) assumed that the Sun was powered by the conversion of gravitational potential energy into thermal energy, which was then radiated into space. Given the Sun's present luminosity, he calculated that the Sun could be no more than 20 million years old.

gravitational contraction The shrinking of an object's radius and accompanying increase in its density that is due to the force of its own gravity.

binding energy The energy needed to break up an object being held together by a given force. The same amount of energy will be released in the formation of the object.

strong nuclear force One of the four fundamental forces of nature. It binds particles together in the nucleus of an atom.

nuclear fission The splitting apart of a nucleus into two or more nuclei of lower atomic weight.

SECTION SUMMARY
The Birth of Nuclear Physics
- Nuclear physics is the study of the nuclei of atoms and their reactions.
- Atoms of each element have a unique number of protons, but isotopes of that element can have different numbers of neutrons.
- Thermonuclear fusion in the Sun transforms one nucleus into another and produces energy in the process.

Thermonuclear Reactions: The Curve of Binding Energy

In transforming one type of nucleus into another, nuclear reactions also release energy. The concept of *binding energy* is crucial to understanding why and when nuclear reactions can give up more energy than they take to initiate and so act as a power source for stars. For this reason, the curve of binding energy is one of the most important diagrams in all of astrophysics (**Figure 10.7**).

"The curve of binding energy tells us how effective nature is at keeping protons and neutrons packed into the nucleus," says Herwig. **Binding energy** is the energy needed to break up an object being held together by a given force. The same amount of energy will be released in the formation of the object. Note that any attractive force can do the binding, such as gravity or electromagnetism (which is attraction between opposite charges). In the nucleus of an atom, protons and neutrons are held together by the **strong nuclear force**. (It needs to be strong to overcome the electromagnetic repulsion associated with the positively charged protons in the nucleus.) The strong force is one of the four basic forces in nature, along with gravity, electromagnetism, and the weak nuclear force. As you can infer from its name, the strong nuclear force is the strongest of the four forces.

The curve of binding energy rises as the number of protons and neutrons in an atom's nucleus increases, but this continual increase with the size of a nucleus goes only up to the element iron. The increase tells us that the energy associated with holding the nucleus together is increasing as the nuclei get larger. Therefore, iron nuclei, having the greatest binding energy, are the most stable atomic nuclei in nature. To the right of iron, the curve begins to fall, because the strong force (which acts only over short distances) has a harder time binding the protons and neutrons in a larger, heavier nucleus.

The curve of binding energy helps us understand the two most important kinds of nuclear reactions: fusion and fission. Fusion reactions, which occur when two nuclei are slammed together and produce a larger nucleus, generate energy by fusing nuclei that are lighter than iron. **Nuclear fission** involves large (heavy) nuclei, which fall beyond iron on the curve of binding energy. A heavy nucleus made of many protons and neutrons can be driven to fission—it can be split apart—if it is bombarded with neutrons. "With fission," explains Herwig, "you begin with heavy nuclei, such as isotopes of uranium. Since they are unstable, you just need to tickle them a little and they split up into two less heavy nuclei, giving up energy in the process."

All nuclear power stations on Earth utilize fission reactions (**Figure 10.8**). Most often, uranium-235

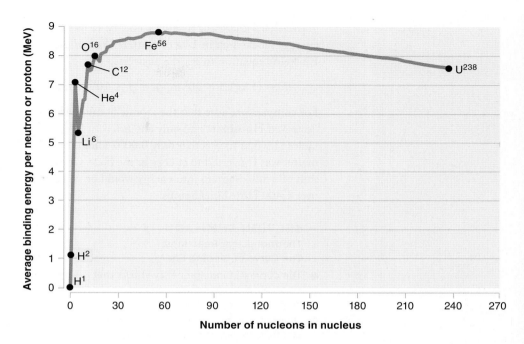

FIGURE 10.7 The Curve of Binding Energy
The curve of binding energy shows how tightly each element's nucleus is held together. The total binding energy in a nucleus rises as the number of protons and neutrons in the nucleus increases. Up to the element iron (Fe), energy can be released through fusion (bringing two smaller nuclei together to make a larger nucleus). Iron represents the peak of the curve of binding energy; for elements larger than iron, it takes more energy to create the larger nucleus than is released. Beyond the peak in the curve of binding energy, very large nuclei can release energy by being broken apart through nuclear fission. Binding energy in nuclei is usually measured in mega-electron-volts (1 MeV = 10^6 electron-volts). C, carbon; Fe, iron; H, hydrogen; He, helium; Li, lithium; O, oxygen; U, uranium.

(92 protons and 143 neutrons) is split into smaller nuclei. In the process of breaking the heavy nucleus apart, the fission reaction gives back more energy than was supplied to it via the bombarding neutrons.

But where does the extra energy in nuclear fission come from? "It's locked up as mass in the original heavy nucleus," says Herwig. "There is an absolutely tremendous amount of energy locked up even in a little ball of mass." Einstein's famous equation $E = mc^2$ tells us that mass (m) is really another form of energy (E) just like kinetic energy (energy of motion) or

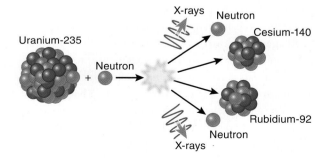

FIGURE 10.8
Nuclear Fission
Fission is the opposite of nuclear fusion, the source of energy in stars, and it is the process used in nuclear reactors. In fission, the large nucleus of a heavy element is split into lighter nuclei and neutrons, releasing energy in the process. The instability of uranium-235 (U-235) makes it an ideal fuel for fission.

gravitational energy or magnetic energy. How much energy is locked up in 1 kg of matter? We can use Einstein's equation to get the answer:

$$E = (1 \text{ kg})(3 \times 10^8 \text{ m/s})^2 = 9 \times 10^{16} \text{ joules}$$

For comparison, note that the atomic bomb that destroyed Hiroshima had only about 6.3×10^{13} joules (J). That means that only 0.0007 kilograms of matter was converted to energy in the blast. Conversion of a full kilogram into energy would be equal to over 1,400 Hiroshima bombs.

SECTION SUMMARY
Thermonuclear Reactions: The Curve of Binding Energy

- The curve of binding energy shows that fusing elements together gives up more energy than is put in only up to the element iron.
- Nuclear fission, the process used in nuclear power plants, splits apart heavy nuclei by bombarding them with neutrons. Energy is created from some of the mass of the nuclei.
- Einstein's equation $E = mc^2$ tells us that mass (m) is really another form of energy (E) and can be used to calculate the energy locked up in matter.

Thermonuclear Fusion in the Sun: The P-P Chain

Nuclear fusion reactions are the source of the Sun's power. Every second, the Sun is consuming 6.1×10^{11} kg of hydrogen gas through fusion in the core. But what exactly is fusion? Thermonuclear fusion is the process of combining light nuclei to form heavier ones. The mass of a nucleus coming out of a fusion reaction is always a little less than the total mass of the lighter elements going in. Thus, by binding the protons and neutrons together in the heavier nucleus, the fusion reaction converts some matter into energy. "As we add nuclear particles to a light nucleus through the process of fusion," says Herwig, "we're releasing energy, allowing them to bind together."

Throughout a star's evolution, the fusion of lighter elements into heavier ones is the fundamental energy generation mechanism that enables the star to support itself against its own gravity. Specifically, the Sun (and other stars like it) converts hydrogen gas into helium gas. Scientists use the term *nucleons* to refer to protons and neutrons in a nucleus. Thus, hydrogen has one nucleon (a proton), and helium has four nucleons (two protons and two neutrons). It might seem as though it should take four hydrogen nuclei to make a single

helium nucleus, but the fusion reaction from hydrogen to helium is not so direct. A chain of intermediate steps is needed to turn a group of H nuclei into a single He nucleus. Together these steps are called the **proton-proton chain** (or **P-P chain**).

"The P-P chain starts with collisions between two hydrogen nuclei," explains Herwig. When two hydrogen nuclei collide with sufficient speed to overcome their mutual electromagnetic repulsion, a nuclear reaction occurs that transforms one of the protons into a neutron. In the process, a new nucleus is born called a **deuteron**, containing one proton and one neutron, which is considered an isotope of hydrogen. "The deuterons then react with another hydrogen nucleus to create a helium-3 nucleus (two protons and one neutron)," says Herwig. "Finally, you get two helium-3 nuclei colliding, and that produces a helium-4 nucleus" (**Figure 10.9**).

The transformation of a proton into a neutron in the P-P chain is an essential aspect of nuclear physics and other branches of quantum physics. Particle identities are not fixed in the subatomic world. A proton can turn into a neutron as long as certain laws are obeyed. One of these laws is conservation of electric charge, which says that the proton's positive charge can't just disappear. The nuclear reaction that turns a proton into a neutron has to create another particle

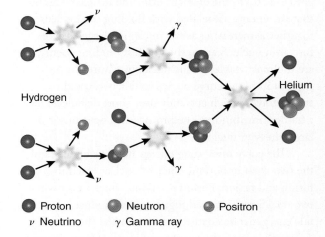

Proton   **Neutron**   **Positron**
ν **Neutrino** γ **Gamma ray**

FIGURE 10.9 The P-P Chain
Stars like the Sun use the proton-proton (P-P) chain to convert hydrogen to helium. The reaction begins with two protons combining to form a nucleus of deuterium (1 proton and 1 neutron), which is an isotope of hydrogen. Deuterium nuclei then collide with protons to produce tritium (2 protons and 1 neutron), an isotope of helium. Tritium collisions then yield a stable helium nucleus (2 protons and 2 neutrons). Note that the mass of the final products is less than the mass of the original protons, and the difference represents energy released in the fusion process.

---

proton-proton chain or **P-P chain** The sequence of reactions required to convert hydrogen nuclei into helium nuclei.

deuteron An isotope of hydrogen with one proton and one neutron.

that carries away the positive charge, called a **positron**. "Positrons are positively charged particles with the mass of an electron," says Herwig. "They're an example of what physicists call **antimatter**. If a positron and an electron meet, they annihilate each other, leaving only energy (in the form of photons) behind." Note that the proton does not split into a neutron and positron, but instead it is instantaneously and discontinuously transformed into these other particles. Such transformations are part of the behavior of the quantum world, upon which everything we experience is built.

Another particle, called a **neutrino**, and some energy in the form of gamma-ray photons are also by-products of the reaction. Neutrinos are not antimatter but are strange in their own way. "Neutrinos are like ghost particles," says Herwig. "They have very little mass, have no charge, and travel at close to light speed. Most important, they barely interact with the kinds of matter our bodies are made of (protons, neutrons, and electrons)." In fact, it would take a slab of lead (the densest metal) a light-year thick to slow down a bunch of neutrinos. "And that lead would only stop half of them," adds Herwig.

Fusion reactions, like the P-P chain that keeps the Sun shining, can occur only under very extreme conditions. "Normally, protons repel each other via the electromagnetic force because they have like charges," says Herwig. "That means the only way to get them to fuse (via the strong nuclear force) is by having them collide at very high speed to overcome the repulsion." Since temperature is a measure of gas particles' random motion, high velocity translates into high temperature. "You need really high temperatures to get fusion going, and the only place you get those is in a star's core." Temperatures at the center of the Sun reach 15 million K, high enough for the P-P chain to operate. "As you go farther out from the core," says Herwig, "the temperatures decrease. Fewer particles have high enough energy to overcome the electromagnetic repulsion, and fusion stops."

Other conditions in the core, such as density, also contribute to the energy generation process. High densities aren't needed for individual fusion reactions, but the *frequency* of fusion reactions depends on how often particles collide. High densities mean higher collision rates—the number of collisions per second. In the Sun's core, matter is packed so tightly that more than 10^{38} protons can be converted into helium every second. That seems like a big number, but how can we relate it to energy production? Recall that in fusion reactions, a fraction of the mass in the constituent particles is converted to energy. "The majority of the hydrogen mass is converted into helium, not energy," says Herwig. "In fact, only 0.7 percent of every kilogram of hydrogen is converted into energy in the fusion process." Thus, for the specific case of H burning in stars like the Sun, we can modify Einstein's formula to $E = 0.007mc^2$. Explaining how this number is derived requires detailed calculations, but in a nutshell, the number is tied to the concept of binding energy. While 0.7 percent may not sound like much, the Sun represents a very large reservoir of mass. Every second, more than 4.3 million tons of hydrogen is converted into helium, releasing 3.8×10^{26} J of energy—the same amount of energy as in 9.1×10^{10} megatons of TNT (see **Going Further 10.1**, page 260).

Energy can take many forms, ranging from magnetism to motion. What form of energy does fusion produce? As **Figure 10.10** illustrates, the P-P chain not only yields a nucleus of helium but also produces photons and neutrinos. In terms of a total energy budget, the photons (in the form of gamma rays) carry away most of the energy. But these fusion-generated light particles can't just escape immediately. The solar interior is far too dense for a photon to travel very far before it is absorbed again by a particle of matter. The neutrinos, however, suffer no such restriction.

While the neutrinos represent a small share of the fusion energy budget, their "ghostly" nature does provide physicists with direct evidence of fusion in the solar core. Because neutrinos interact so rarely with other matter particles, they immediately escape the solar core. Traveling through the entire Sun at

positron The antimatter version of the electron, having the mass of the electron but a positive charge.

antimatter A form of matter that annihilates on contact with normal matter. Particles and antiparticles of the same type have the same mass but opposite charge.

neutrino An electrically neutral, weakly interacting elementary subatomic particle often created in nuclear reactions.

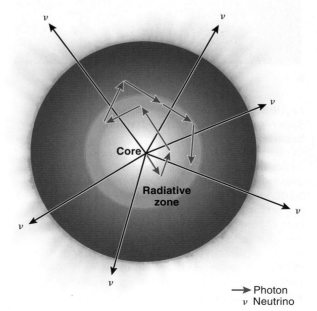

FIGURE 10.10
Solar Neutrinos
Neutrinos (ν) readily escape from the Sun's interior, since they interact very weakly with matter. Photons, however, are scattered many times as they encounter particles of matter before eventually escaping. It takes roughly 100,000 years for a photon to make it out of the Sun via its many interactions with matter.

GOING FURTHER 10.1   THE LANGUAGE OF THE COSMOS
Fusion, Matter, and $E = mc^2$

Let's see how the equation $E = mc^2$ is applied to fusion in the Sun. The efficiency of hydrogen fusion in the Sun is such that

$$E = 0.007\, m_H c^2$$

Note here we have replaced m with m_H to show that hydrogen is being converted to energy.

First let's compare how much energy is produced when 1 kg of matter is *fully* converted to energy, versus how much energy is produced when 1 kg of hydrogen is converted into helium in fusion.

$$E = (1\ \text{kg})(3 \times 10^8\ \text{m/s})^2 = 9 \times 10^{16}\ \text{kg m}^2\ \text{s}^{-2} = 9 \times 10^{16}\ \text{J}$$

$$(9 \times 10^{16}\ \text{J})/(4.18 \times 10^{12}\ \text{J/kiloton TNT}) = 21{,}531\ \text{kilotons TNT}$$

$$E_H = 0.007(1\ \text{kg})(3 \times 10^8\ \text{m/s})^2 = 6.3 \times 10^{14}\ \text{J} = 151\ \text{kilotons TNT}$$

You can see how comparatively little energy is released in a fusion reaction (remember that most of the hydrogen mass is converted not into energy but into helium mass).

Now let's answer a more complicated question. Given that we know how much energy the Sun produces every second (by measuring its light output), we can figure out how much mass is being converted into energy every second in the Sun. The energy output (luminosity) of the Sun is 3.84×10^{26} J/s. That means every second, the Sun produces 3.84×10^{26} J. Let's put this into our formula and solve for mass:

$$m_H = E_H/(0.007c^2) = (3.84 \times 10^{26}\ \text{J})/[0.007(3 \times 10^8\ \text{m/s})^2]$$

$$= 6.1 \times 10^{11}\ \text{kg}$$

All the energy we see coming from the Sun originates in fusion reactions. Since we know the efficiency of the fusion reactions in liberating energy (0.007), we can convert what we see (the Sun's energy output) into what we can't see (the Sun's mass consumption).

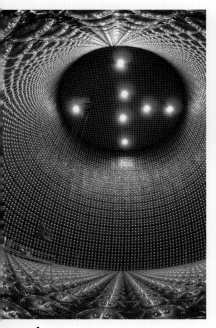

B

A

FIGURE 10.11
Super-Kamiokande Neutrino Detection Experiment
(A) Located 1,000 meters below Mount Kamioka in Japan, the Super-Kamiokande (or SK) experiment is a neutrino observatory consisting of a tank holding 50,000 tons of ultrapure water. The inner surface of the tank is covered by detectors sensitive to the flash of light that occurs when a neutrino interacts with an electron or nucleus in a water atom. Because neutrinos interact very weakly with matter, scientists must gather enormous quantities of them in hopes of capturing one interaction. (B) Using the SK detector, scientists actually used the elusive neutrinos to image the Sun.

just below the speed of light, the neutrinos quickly emerge into interplanetary space. Some of these solar neutrinos cross Earth's path, where physicists can use huge detectors to capture a few of the elusive particles. These solar neutrinos have been used to directly test theories of solar fusion and the P-P chain. It's worth noting that these experiments take enormous effort. Most detectors require thousands of tons of target mass (in the form of water or other substances) to catch even a few neutrinos (**Figure 10.11**).

For decades, every effort to match solar fusion theory with solar neutrino experiments failed. There were always 65 percent fewer neutrinos than expected. This **solar neutrino problem** remained unsolved, until some physicists imagined that the problem was not with the Sun or with their understanding of nuclear fusion physics but with the neutrinos themselves. In 2002 it was finally discovered that the ghostly neutrinos could transform from one type into another as they traveled through space.

There are actually three types of neutrinos, each associated with how it interacts with different kinds of subatomic particles. In addition to the so-called electron neutrinos, there are muon neutrinos and

FIGURE 10.12
Conduction of Heat
The conduction of thermal energy in a heated spoon is a familiar example of energy transport in everyday life. Holding a spoon over a flame causes rapid vibrations in the metal atoms, which are propagated down the spoon's length to the handle, which then becomes too hot to hold. There are many ways in nature to transport energy. In the Sun, the principal mechanisms are radiation and convection.

tau neutrinos (the muon and tau particles are each cousins of the electron type). The original solar neutrino detectors were designed to look for the electron neutrinos—the kind emitted in solar nuclear reactions. But as physicists learned in experiments carried out in 2002, all neutrinos can switch identities. By the time the electron neutrinos reached Earth, they had already transformed (the technical term is *oscillated*) into either tau or muon neutrino versions. In this way, advances in basic particle physics, including the discovery of neutrino oscillations, solved a vexing and long-standing problem in basic nuclear astrophysics.

SECTION SUMMARY
Thermonuclear Fusion in the Sun: The P-P Chain

- Using nuclear fusion, the Sun and all stars combine light nuclei into heavier ones and release mass that is converted to energy.

- The proton-proton (P-P) chain is a multistep fusion process at work in the Sun in which hydrogen nuclei are converted into helium nuclei; in the process, 0.7 percent of the mass in the original hydrogen is converted into energy.

- Nuclear fusion reactions require high temperatures and velocities to overcome the repellent electromagnetic forces of the charged protons.

10.3
Moving Energy

The majority of the energy released by fusion reactions at the Sun's core makes it all the way to the surface and from there escapes into space (minus the fraction that immediately escapes as neutrinos). How do we know? "Well, we can see it shining," says Falk Herwig. "Energy is always flowing *off* the Sun into space. That means the Sun is always losing energy from its surface." To understand how the energy that is generated in the solar core via fusion makes its way to the surface, you need only consider that most mundane of objects: the metal spoon.

If you hold one end of a long metal spoon over a burner on your stove, the tip of the spoon in the flame is, obviously, going to heat up as it absorbs energy. In a fairly short time, your end of the spoon will heat up too, becoming so hot that you will have to let go of it.

How did the heat get from one side of the spoon to the other in such a short time? "The heat moves," says Herwig. "It gets conducted from the hot end to the cold end. The process is called energy transport, and it's why you can't hold on to the heated spoon for very long" (**Figure 10.12**).

Energy transport is any process by which energy is moved from one place to another. In the spoon example, the heat moves from one tip of the spoon to the other as the flame sets off vibrations in the lattice of metal molecules. The vibrations at the end that is close to the flame then set nearby metal molecules in motion. In this way, the heat is moved (transported) from the heated tip all the way down the length of the spoon to the part you're holding.

The form of energy transport occurring in the metal spoon is called **conduction**, in which increased vibrational motion of atoms (heat) in the spoon's metal is propagated along its length. But the Sun and other stars use different processes to transport energy from the core to the surface. "A star like the Sun," says Herwig, "is always transporting heat from the inside out. If that transport becomes difficult in some region of the star, then the star will find an alternative mechanism of getting that energy moving again." As discussed in the first section of this chapter, detailed studies of the Sun's interior (via methods like helioseismology) have revealed that two mechanisms of energy transport operate within a star: radiation and convection.

The Radiative Zone

The fusion reactions in the core release the bulk of their energy as light, especially gamma-ray photons. When these photons emerge from a fusion reaction, they begin traveling at the speed of light in random directions. "But those gamma rays don't make it very far," says Herwig. The regions above the core constitute a soup of electrons and (mostly) ionized hydrogen and helium. Each encounter between a gamma ray and one of these charged particles **scatters** the light particle, sending it in a different random direction. Scattering occurs when light interacts with matter and has its direction of propagation changed. "The average distance a photon can travel before it interacts with a charged particle is just a few centimeters," says

solar neutrino problem The apparent discrepancy between the expected number of neutrinos being emitted from the Sun and the number measured experimentally, which has been resolved with the discovery that neutrino species can change form.

conduction The transfer of thermal energy from a hotter region to a cooler one via the collisions of atoms or molecules.

scatter To change a photon's direction of propagation.

Falk Herwig

It wasn't the lure of the night sky and its mysteries that led Falk Herwig into astronomy; it was the simple desire to find a field of study that was as far as possible from anything practical. "When I was younger, I think I must have had a big chip on my shoulder or something," he says. "I just couldn't imagine using my time to do anything applied to real life. I wanted to find the most abstract endeavor I could and dedicate myself to it. Now, of course, I am not so resolute about the importance of doing something useless."

When he started his PhD studies in Germany, Herwig explored a number of disciplines. "Areas like geophysics were interesting, but they still seemed too applied," he says. "Really abstract fields like string theory sounded great, but I had a feeling I would never be able to find a job if I made that my specialty. Astrophysics, however, seemed to have just the right mix of distance from day-to-day reality and a reality that included job prospects," he says, laughing.

Whatever his motivation, Herwig's skill as a scientist led to a successful career modeling the evolution of stars like the Sun. His work has placed him at the frontiers of understanding what happens deep within stars as they reach the end of life. From the vantage point of this success, Herwig now realizes that it's the methods he loves most. The specific field those methods are applied to comes second. "Now I think I could have gone into geophysics and it would have all been fine. It's the process of science and the tools like the computer simulations I use that are exciting. That's what I really love doing. The exact application doesn't matter that much."

Herwig. Only after trillions and trillions of these random scatterings does the photon (and its energy) slowly wander outward from the core.

The movement of energy through this repeated scattering of photons is called **radiative energy transport** and gives the radiative zone of the Sun its name. In radiative energy transport, a given quantity of energy is transported entirely by photons (electromagnetic radiation) as opposed to some other means. "On average," says Herwig, "it will take the energy bundled into a single photon at least tens of thousands of years to make its way out of the radiative zone."

The direction of the photon is not all that changes in these scattering events. Recall that gamma rays are at the high-energy end of the electromagnetic spectrum (high energy = high frequency = short wavelength). As the photon moves outward, it encounters solar plasma at ever-lower temperatures. The cooler ions and electrons in these layers change the photon's energy as they scatter it, shifting the light to longer wavelengths. "One gamma ray gets converted into two lower-energy X-rays," says Herwig. "Each of those X-ray photons will then be converted into multiple UV photons." Later, those UV photons are

radiative energy transport The transfer of thermal energy via electromagnetic radiation.

---

also "downshifted" in wavelength when they scatter off lower-temperature gas farther from the core. In this way, energy originally locked up in a single gamma ray that was released in a fusion reaction at the core is divided into almost 2,000 photons of lower energy that diffuse outward through the outer layers of the Sun. In other words, the photons that emerge from the photosphere are not the same as the ones produced by fusion reactions in the core.

SECTION SUMMARY
The Radiative Zone
- In the radiative zone, high-energy gamma rays are scattered numerous times and eventually are converted into lower-energy photons as the energy is transported outward.

The Convective Zone

"The solar plasma can transport energy in different ways, depending on the local conditions," says Herwig. "Moving energy by radiation is one option, but it's very inefficient because it takes so long. If the conditions are right, then the solar plasma can switch to a second option: moving the heat by moving the gas. That is what starts happening at the top of the radiative zone." In this next layer of the Sun, large-scale flows of matter accomplish the energy transport. Hot gas rises, taking its thermal energy up to the top of the convective zone, which is also the Sun's effective surface. This mode of energy transport via mass motions is called convection (see Chapter 6), and the layer of the Sun where it occurs is called the convective zone.

To understand how convection works in the Sun, recall that the temperature drops continuously from the Sun's core to its surface. As photons squirm outward through their multiple scatterings, they eventually encounter plasma that has cooled to the point where much of the hydrogen gas has become neutral rather than ionized. In this state, instead of simply scattering the light particles, the hydrogen and helium atoms now absorb them and trap their energy. Since the energy flow from the core can't stay backed up forever, an alternative method of transport comes into play: motion.

"A pot of water on a stove continually absorbs heat from the burner," says Herwig. "At some point, however, the water becomes so hot that the only way to move the energy absorbed at the bottom is to set the water in motion through boiling. That's sort of what happens in the Sun." The energy absorbed at the top of the radiative zone allows blobs of plasma in the convective zone to expand slightly, making them buoyant (lighter than surrounding gas). The buoyant blobs begin to rise through surrounding material just as a helium balloon rises into the sky. When the hot, rising plasma blobs reach the surface, blackbody radiation from the hot gas escapes into space. The gas cools and sinks back down toward the convective zone's base, starting the process over again. Thus, energy generated by fusion reactions in the core finally is released as starlight at the Sun's surface (the photosphere).

SECTION SUMMARY
The Convective Zone
- In the convective zone, gas is no longer fully ionized, and the photons become trapped by neutral atoms.
- Energy transport is accomplished by convection, the cyclical motion of gas as it heats, rises, cools, sinks, and is heated again.
- Blackbody radiation from the hot gas is released at the Sun's surface, the photosphere.

Solar Granulation

"It's important to remember that the convection zone reaches all the way to the photosphere," says Lika Guhathakurta, "and all that churning leaves its mark there." High-resolution photographs of the Sun's surface show it to be almost uniformly tiled with a network of bright patches surrounded by thin, dark outlines. This pattern is called **solar granulation** (**Figure 10.13**). "Each bright spot is a plume of hot material that's made it to the surface from deep inside the convection zone," says Guhathakurta. "Once that hot gas hits the photosphere, it spreads out, cools, and then sinks down again." The dark outlines of a solar granule show where the cool gas has begun its downward journey.

The solar surface is extremely dynamic; it's always changing. Using Doppler shift measurements, astronomers can see rising and descending plumes of plasma moving at approximately 0.4 km/s. A typical solar granule extends across 1,000 km (about the size of Texas). But despite this enormous size, a granule lasts only 10–20 minutes before it fades and the surface pattern changes.

The reason for granules' characteristic bright centers and dark edges lies in the Stefan-Boltzmann law for blackbody radiation (see Chapter 4). Recall that the power (energy emitted per second) from a section of blackbody surface can be expressed as

$$L \propto T^4$$

This formula tells us that even a small change in temperature results in a large change in brightness (that is, light-energy output). Imagine, for example, that the temperature in the center of a granule is just twice as high as in the cooler gas at its edge. According

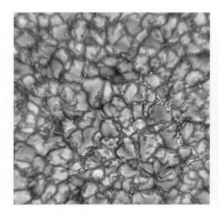
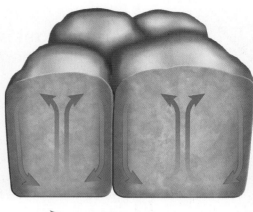

Hot gas rising
Cooler gas sinking

A   **B**

to the Stefan-Boltzmann law, the granule's center will be $(2)^4 = 16$ times brighter than its edge.

"That difference might not seem like much," says Guhathakurta, "but try staring into a 10-watt lightbulb versus a 160-watt bulb and see what happens. The 160-watt bulb would be painful to your eyes. Small differences in temperature really mean a lot for how bright surface features appear on the Sun." Guhathakurta notes that in reality, the temperature at the granule edges is only a few hundred degrees lower than at the granule center, but this difference is enough to create the light-and-dark pattern.

SECTION SUMMARY
Solar Granulation
- Solar granules on the photosphere result from convection cells rising from below, each showing brighter, hotter rising gas in the center and darker edges where the cooler gas is sinking back down.
- Differences in brightness on the Sun's surface are explained by the Stefan-Boltzmann law.

10.4
The Active Sun: Photosphere to Corona and Beyond

While we have to rely on helioseismology to "see" what happens inside the Sun, what happens *outside* is on display every day. "The Sun's surface, called the photosphere, is not really a surface in the sense of either a solid or a liquid," Lika Guhathakurta explains. "It's the place where light emitted by solar matter can escape into space." Gas in the photosphere has a temperature of approximately 5,800 K. Gas layers above the photosphere are transparent to the (blackbody) radiation the

FIGURE 10.13
Solar Granulation
(A) Granules, which live typically 20 minutes or less, are the primary solar surface feature. (B) Each granule is a convection cell in which hot gas rises toward the granule's lighter center and cooler gas sinks toward the outer edges.

solar granulation The pattern of cellular features on the solar surface, showing bright interiors and dark edges.

GOING FURTHER 10.2   THE LANGUAGE OF THE COSMOS
Sunspots and Temperature

The surface of the Sun shows a number of features that appear quite dark, such as sunspots and the edges of solar granulation cells. But are these features really dark, or is it just a matter of relative brightness? The solar surface radiates as a blackbody, so we can use all the formulas we learned for blackbodies to answer this question.

According to the Stefan-Boltzmann law, the luminosity of a blackbody's surface will depend very sensitively on its temperature. If we had two same-sized square plates of metal and heated them up to two different temperatures, we could use the Stefan-Boltzmann law to understand how much energy one plate would radiate relative to the other. To see this, let's write out the full Stefan-Boltzmann equation:

$$L = A\sigma T^4$$

where L is the luminosity (energy radiated per time), A is the area of the plate surface, σ is the Stefan-Boltzmann constant, and T is the blackbody temperature in kelvins. If we take the equations for both plates and divide them by each other, the area A and the Stefan-Boltzmann constant σ drop out. In the end, we get

$$L_1/L_2 = (T_1/T_2)^4$$

So if $T_1 = 400$ and $T_2 = 200$, then the ratio of luminosity between the two will be

$$L_1/L_2 = (400/200)^4 = (2)^4 = 16$$

This is quite a difference in energy radiated per second for such a small difference in temperature.

For sunspots, the temperature differences are smaller. A typical temperature in the sunspot will be about 4,000 K, while the rest of the solar surface has a temperature of 5,800 K. Using these numbers, we can calculate the difference in brightness between the sunspot and the gas surrounding it:

$$L_{spot}/L_{surface} = (4{,}000\ K/5{,}800\ K)^4 = 0.2$$

Thus, the sunspot produces only about 20 percent as much light as the surrounding hotter gas produces. This difference is enough for optical images (or your eye) to see the sunspot as dark relative to its surrounds. In reality, however, the spot is still producing a lot of light.

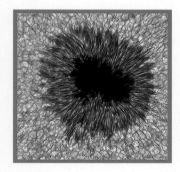

FIGURE 10.14
Sunspots
Each sunspot has a darker, cooler umbra (center) and a brighter, hotter penumbra. The umbra is approximately 1,000 K cooler than the surrounding solar surface, making the sunspot appear dark by comparison. The small-scale details reflect the structure of the magnetic fields within the sunspot.

photosphere creates. "That's why we 'see' the Sun as a 5,800-K ball of plasma," says Guhathakurta. "If radiation could escape freely from deeper inside the Sun, then we would see a hotter photosphere."

What we do observe in the photosphere and the regions above it tells us that the Sun's outermost domains host activity of extreme power and violence. From sunspots that cover regions the size of planets to flares that release the equivalent of 100 billion megatons of TNT, the final layers of the Sun give stark testimony to the vast reservoirs of energy flowing through a star. While these forms of solar activity are fascinating in and of themselves, they have come to present significant dangers to us as we take our steps in becoming a space-faring race.

Sunspots and Their Cycles

While solar granulation constitutes a kind of ever-shifting carpet of structure on the solar surface, far larger and more dramatic features exist there as well. **Sunspots**, in particular, are giant structures that play a critical role in our understanding of the Sun. Sunspots are huge, dark regions that come in a range of sizes. A typical spot can extend across 10,000 km of solar surface (almost the diameter of Earth). "Structurally, an individual sunspot is composed of a central dark region called an *umbra* and a surrounding area that is brighter and is called the *penumbra*," says Guhathakurta (**Figure 10.14**).

Like solar granules, sunspot umbras appear dark because gas there is cooler than surrounding material and therefore emits less blackbody radiation (see **Going Further 10.2**). "Sunspots are only dark in contrast to the bright face of the Sun," says Guhathakurta. "If you could cut an average sunspot out of the Sun and place it in the night sky, it would be about as bright as a full Moon."

Unlike solar granules, sunspots are not convective plumes of rising gas. Instead, powerful magnetic fields are the agents creating sunspots. "Giant columns of the Sun's magnetic field well up to the photosphere from deep in the convective zone to form spots," says Guhathakurta. Just as magnetic fields are generated in the interiors of many of the Solar System's planets, the

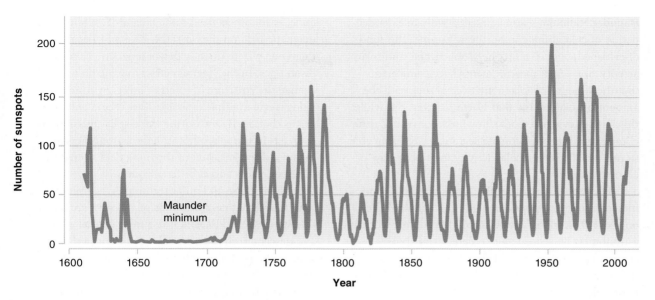

FIGURE 10.15
Sunspot Cycle
Historical records of sunspot data show that the sunspot cycle has an average period of about 11 years. Note that the number of sunspots at the peak of the cycle varies by a factor of 4 or more. Note also the period in the later 1600s called the "Maunder Minimum," when the number of sunspots was greatly reduced.

Sun also creates its own magnetic field. However, surface fields on the Sun are many times more powerful than the strongest regions of the strongest planetary fields. "All that energy locked up in the Sun's magnetic field," says Guhathakurta, "drives the breathtaking solar activity occurring on and above its surface, including sunspots."

Powerful columns of magnetic fields rise up in a sunspot from below the solar surface. The fields exert force on the solar plasma and thus create their own form of pressure, which pushes plasma out from the sunspot's center (recall that plasma is an ionized gas, so plasmas will feel forces exerted by magnetic fields). The lower gas pressure in the sunspot translates into lower gas temperatures and a darker appearance (again, as explained by the Stefan-Boltzmann law). "A large sunspot might have a temperature of about 4,000 K," explains Guhathakurta, "much less than the 5,800 degrees Kelvin of the bright photosphere surrounding it."

Like granules, sunspots are transient. "Sunspots form over periods lasting from days to weeks. They can even last for months," says Guhathakurta. But unlike the granules, sunspots appear and disappear following a regular pattern. "The average number of spots on the face of the Sun is not always the same; it goes up and down in a cycle." Historical records of sunspot counts show that this sunspot cycle has an average period of about 11 years. Large numbers of spots appear on the Sun at the beginning of each cycle, and over the next 11 years their numbers slowly decrease (**Figure 10.15**). The location of the spots also varies with time. Groups of sunspots first appear at middle latitudes on the solar globe (about 30° from the equator). Then, as the cycle progresses, the spots make their appearance closer and closer to the solar equator.

This progression from many spots to a few and from high latitudes to the equator is cleanly captured in so-called Maunder butterfly diagrams, named after Edward Maunder, who first "visualized" sunspot data in this way (**Figure 10.16**). These diagrams plot the location and number of spots as a function of time. The butterfly pattern that emerges constitutes the raw data of solar science. Any theoretical explanation of sunspots must explain how this pattern is created.

In addition to this 11-year cycle, the magnetic behavior tied to sunspots shows even longer-term changes. The number of sunspots occurring at the maximum of the cycle can vary considerably, from more than a hundred to just a few. Looking back at observations spanning 350 years, the record shows a remarkable period in the late 1600s when the number of sunspots appears to have remained extremely low from one cycle to the next. This period—called the

interactive
Sunspot Cycle

sunspot A large, strongly magnetized region of the Sun's surface that appears darker because it has a lower temperature than its surroundings have.

DAILY SUNSPOT AREA AVERAGED OVER INDIVIDUAL SOLAR ROTATIONS

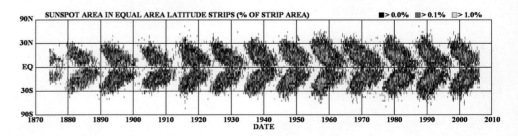

FIGURE 10.16 Maunder Butterfly Diagram
Detailed observations of the sizes of sunspots and the latitudes on the Sun where they appear show a predictable pattern. When plotted as latitude versus time, the pattern resembles a butterfly, with sunspots first forming at midlatitudes. Later in the cycle, sunspots appear toward the equator. The pattern was first recognized by Edward Maunder, a British astronomer. In this graph the x-axis represents time, and the y-axis represents the latitude of the sunspots.

Maunder Minimum The period in the late 1600s when the number of sunspots remained extremely low from one sunspot cycle to another.

Maunder Minimum—also coincided with an extended period of cold climate in northern Europe. This kind of evidence suggests a link between climate and solar activity. Spacecraft measurements have confirmed that the strength of solar activity correlates with the amount of solar energy that Earth receives. But while this kind of evidence does show links between Earth's climate and sunspots, climate scientists have been able to rule out the role of solar activity in present-day climate change.

It's also worth noting that spots may be a common astronomical phenomenon. "Our Sun isn't the only star with spots," says Guhathakurta. "Just recently, astronomers have been able to detect 'star spots' appearing on other stars." This discovery means that the stellar physics of spots, including the role of magnetic fields, is likely a universal process.

SECTION SUMMARY
Sunspots and Their Cycles

- Sunspots on the photosphere are regions of powerful solar magnetic field that create irregularly shaped areas with central, dark umbras and brighter penumbras.
- Sunspots follow a repeating 11-year cycle, during which they decrease in number and move from higher to lower latitudes steadily over the cycle.
- Sunspot cycles also show variances over time in terms of the number of spots that appear in each cycle.

Solar Magnetic Cycle

Sunspots and their 11-year cycle are only one part of an even longer cycle in which the Sun's magnetic field rebuilds itself over and over again. To study the solar cycle, astronomers need some means of measuring the magnetic field on the Sun. The *Zeeman effect* gives them one way to directly detect magnetic fields in distant gas. In the Zeeman effect, magnetic fields change the structure of electron orbits in atoms. The fields create multiple emission lines where only one existed before. By measuring these split emission lines from gas in the solar photosphere, astronomers can obtain a direct measure of the Sun's field strength in different places and at different times.

Using these and other techniques, astronomers have found that nearby groups of sunspots are linked by magnetic-field lines that rise from one set of spots and descend into the other. Thus, one sunspot constitutes the northern pole of the field, and another constitutes the southern pole. Sunspot groups in both the northern and southern hemispheres of the Sun show these kinds of magnetic links.

"But there are differences between the behavior of sunspot groups in the two hemispheres," explains Guhathakurta. "And the differences shift from one cycle to the next." In any particular cycle, all sunspot groups in the northern hemisphere show the same field direction. For example, trailing spots (in terms of the Sun's direction of rotation) will be the source of an emerging field (the north polarity), and leading spots have the south-field polarity, where the field dives back down. Throughout a single sunspot cycle, this pattern holds true in the northern hemisphere. "In the southern hemisphere you get the opposite story," says Guhathakurta. "There the polarity between trailing and leading spots is reversed. The trailing spot acts as the field's 'south pole' and the leading spot is the 'north pole'" (**Figure 10.17**).

"Then," Guhathakurta continues, "things get really weird." As a cycle ends, the number of sunspots gradually goes almost to zero, and as the new cycle rises, new groups of spots appear. "But now things have flipped," says Guhathakurta. The new groups of leading and trailing sunspots have the opposite polarity compared with those of the previous cycle. In the new cycle, the orientation of magnetic north and south for the spots in the northern hemisphere is the same as what was seen in the southern hemisphere in the previous cycle. The same flip occurs for southern-hemisphere spots in the new cycle. Southern spots now have fields that look like the old northern-hemisphere spots in terms of magnetic orientation. "But it's not just the sunspots that change their magnetic orientation," says Guhathakurta. As the Sun moves from one sunspot cycle to the next, the polarity of the entire Sun reverses. The solar north magnetic pole becomes the south pole and

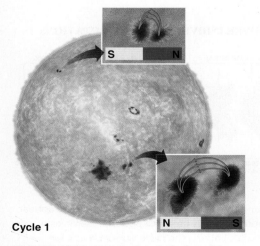
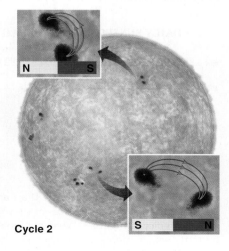

Cycle 1   Cycle 2

FIGURE 10.17 Alternating Polarity of Sunspots
As the polarity of the entire Sun reverses with each sunspot cycle, the polarity of sunspot pairs by hemisphere reverses as well.

the south magnetic pole becomes the north pole. "The Sun's entire field has flipped!" exclaims Guhathakurta.

This large-scale reorientation of the Sun's magnetic field is called the **solar magnetic cycle**. Note that it takes two full sunspot cycles (22 years) for the Sun to go through a single magnetic cycle, returning the "north" magnetic pole to its starting orientation.

SECTION SUMMARY
Solar Magnetic Cycle
- The Sun's magnetic field exhibits changes linked to the solar cycle, including reversals in the polarity of sunspot groups.
- As sunspot activity goes to near zero at the end of each 11-year cycle, the entire Sun's magnetic polarity reverses.

Solar Dynamo

What is the explanation for the Sun's strange magnetic behavior? We've already seen how planets generate their own magnetic fields. "Magnetic fields are produced by electric currents," says Guhathakurta. "These currents are generated within the Sun by the flow of the Sun's hot, ionized gases. There are strong flows of plasma on the Sun's surface and within its interior. Nearly all of them contribute in some way to the production of the solar magnetic field."

Anytime a charged fluid creates a magnetic field, astronomers say that a dynamo is at work. The solar dynamo works in a similar way to the dynamos in planets, but the timescales for dramatic changes, such as field reversals, are much shorter. "The Sun's field reverses every 22 years," says Guhathakurta, "but the same process can take 100,000 years or more for Earth."

The origin of the solar dynamo rests in the Sun's differential rotation. Recall that different regions of

the Sun rotate at different speeds. Material near the solar poles completes one full rotation in 35 days, but material at the equator makes a complete circuit in just 25.6 days. Not only are there differences in rotation rate from pole to equator, but material at different distances from the Sun's center also rotates with different speeds. In particular, plasma at the outer edge of the radiative zone rotates faster than material just above it in the convective zone.

"Magnetic fields within the Sun get stretched out and wound around the Sun by this differential rotation," says Guhathakurta. "It can be useful to think of magnetic-field lines as rubber bands. They're continuous lines of force which, like rubber bands, can be strengthened by stretching and twisting." The magnetic fields increase in strength as they are stretched because energy in the motion of the gas gets converted into energy of the magnetic field.

For a better picture of what's happening, imagine a magnetic-field line that runs from one pole to the other along the solar surface. Because of differential rotation, the plasma at the equator rotates faster than the plasma near the poles, so the initially straight field line will soon be stretched out along the Sun's waist (**Figure 10.18**). In other words, the line that used to run pole to pole now has a section that runs parallel to the equator. As the Sun continues its differential rotation, this winding of the field line becomes more pronounced. "The Sun's differential rotation," says Guhathakurta, "can take a north-south-oriented magnetic-field line and wrap it around the whole circumference of the Sun in just about 8 months."

Much of this winding occurs at the base of the convective zone. As the field is wrapped tighter and tighter, bundles of field lines gather together into structures called **flux tubes** and become buoyant. The

solar magnetic cycle A large-scale reorientation of the Sun's magnetic field occurring every 22 years.

flux tube A bundle of magnetic-field lines.

FIGURE 10.18
Wrapping of the Sun's Magnetic Field
Because of the Sun's differential rotation (gas at the equator rotates more rapidly than gas at the poles), the magnetic field is wrapped up during the course of a cycle. The field lines become buoyant and rise to the surface as loops called flux tubes, which then define the Sun's large-scale field. Eventually, the field structure's complexity collapses and another dynamo cycle begins.

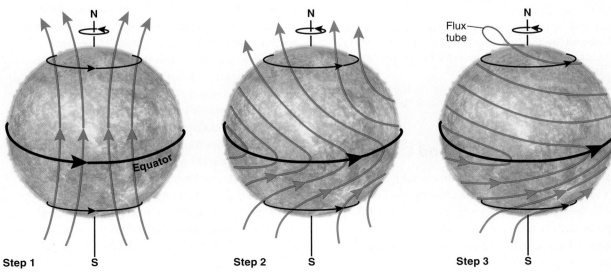

Step 1   S   Step 2   S   Step 3   S

FIGURE 10.19
Solar Prominence
The *SOHO* spacecraft captured this solar prominence in ultraviolet light in 2008. An arc from the Sun's magnetic field erupted from the surface, briefly suspending ionized gas along with it.

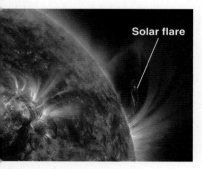

FIGURE 10.20
Solar Flare
Powerful magnetic forces above an active region on the Sun act on a region of plasma above the Sun, driving it into space in this solar flare in 2014. The plasma in this flare comprised a volume larger than several Earths.

prominence A giant arc of magnetism anchored in the photosphere.

flare An eruption that launches material from the solar surface.

reconnection The breaking and simultaneous re-joining of magnetic-field lines.

coronal mass ejection (CME) A large-scale ejection of mass and magnetic energy from the Sun's corona.

buoyancy occurs because the magnetic pressure in these tubes pushes out plasma, which makes them less dense than their surroundings. Like helium balloons in the air, the flux tubes feel an upward buoyant force and rise through the convective zone. On their way to the surface, the tubes twist around each other like shoelaces. "The twisting," says Guhathakurta, "comes from the effect of the Sun's rotation on the tubes as they rise up from deep within the convection zone." When the magnetic-field lines break through the surface, they form sunspots. "The twisting of those flux tubes becomes evident as arcs of the field that rise out of one sunspot (with one polarity) and plunge back down again into another sunspot (with a different polarity)." These arcs are also the source of magnetic activity such as prominences and flares in the solar atmosphere.

This solar dynamo model predicts that the winding of field lines by differential rotation would cause the reversal of the sunspot magnetic-field orientation from the northern hemisphere to the southern hemisphere. Eventually, the ongoing winding via differential rotation makes the solar field so complex that it can no longer maintain itself. The large-scale magnetic field of the Sun becomes unstable because of the complex overlaps created by all the field windings. Each disruption of the large-scale field marks the end of the 11-year sunspot cycle and either the midpoint or the end of the 22-year solar magnetic cycle. Since the solar dynamo model so neatly predicts what is observed, astronomers feel that, despite many unanswered questions, the model gives them a good basis to explain solar magnetism.

SECTION SUMMARY
Solar Dynamo
- The Sun generates magnetic fields through a dynamo process, which is driven by the differential rotation of different regions of the Sun in both latitude and depth.
- As the magnetic field lines are wound ever tighter around the Sun, areas within the convective zone become buoyant, rise up, and create sunspots on the surface.
- At the end of the 11-year solar cycle, the magnetic field has become so complex that it can no longer maintain itself.

Prominences, Flares, and CMEs

Sunspots come and go in a matter of weeks, and granules appear and fade in 10–20 minutes. "From just these two types of phenomena," says Guhathakurta, "you can see that the photosphere is *active*." But there are forms of solar activity with far more power and violence than either granulation or sunspots have. Prominences,

flares, and coronal mass ejections (CMEs) are three related classes of activity. While they differ in behavior, they are all driven by powerful solar magnetic fields.

Prominences (**Figure 10.19**) are giant arcs of magnetism whose legs are anchored in the photosphere. Two classes of prominences are observed on the Sun: *Eruptive prominences* explode from the photosphere only to collapse back down in just a few hours. *Quiescent prominences*, in contrast, maintain their arched magnetic fields for many days. Smaller prominences can be tens of kilometers long and extend into the chromosphere. The largest prominences can stretch across much of the Sun's disk, as well as rise far into the corona.

"Most of the solar atmosphere is so hot that the gas is primarily in the plasma state," says Guhathakurta. "That means electrons are no longer bound to atomic nuclei. The gas is now made up of separated charged particles, mostly protons and electrons." Recall from Chapter 6 that magnetic forces trap charged particles by redirecting the particle motion to spiral around the magnetic-field lines. "When arcs of field erupt from the surface, the plasma is carried along with them. Charged particles making up the plasma reach the top of the magnetic arcs and then drain back down, emitting radiation as they get heated by collisions." The collisions typically drive temperatures in a prominence to values of 70,000 K. Emission lines produced by this hot gas provide one means for astronomers to observe and analyze prominences.

Flares (**Figure 10.20**) differ from prominences in that they launch material away from the solar surface into space. "Solar flares are essentially huge explosions on the Sun," says Guhathakurta. In a prominence, the magnetic fields remain tied to the surface, but in a flare, a process called **reconnection** (**Figure 10.21**) occurs in which the old field lines are simultaneously broken and re-joined with other field lines. Reconnection frees matter and magnetic energy from their anchor points in the photosphere. "Flares occur when the intense magnetic fields on the Sun become too tangled," explains Guhathakurta. "Like a rubber band that snaps when it is stretched and twisted too far, the tangled magnetic fields break and reconnect, releasing huge energies in the process."

In a reconnection event, magnetic-field lines running in opposite directions (north and south) are squeezed close enough together that they cancel each other out. The energy that was locked up in the magnetic field is rapidly converted into light, heat, and motion. Flares are remarkably violent events. "Just for perspective," says Guhathakurta, "the energy emitted by even a small solar flare is more than a million times

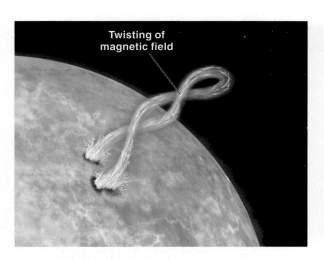

Twisting of magnetic field

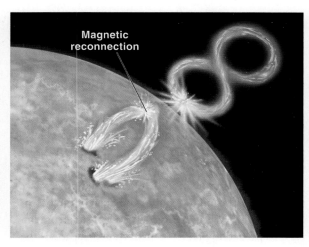

Magnetic reconnection

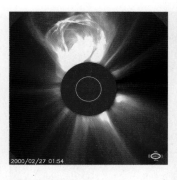

2000/02/27 01:54

FIGURE 10.22
Coronal Mass Ejections
Powerful coronal mass ejections (CMEs) can spew a billion tons of matter into space at speeds of millions of miles an hour. This CME occurred in February 2000 and was captured by the *SOHO* satellite. The disk of the Sun is blocked so that the light from the corona is visible. The size of the Sun is marked by the white circle.

FIGURE 10.21 Magnetic Reconnection
Magnetic reconnection—the breaking and reconnecting of oppositely directed magnetic-field lines in a plasma—is the mechanism that drives solar flares. As magnetic fields anchored in the Sun move with the solar plasma, field lines can become tangled, leading to reconnection events that release matter and energy to escape into space.

greater than the energy from a volcanic explosion on Earth." Much of the energy in a flare is converted to radiation across a wide range of wavelengths. "Although solar flares can be visible in white light," says Guhathakurta, "they are often studied via their bright X-ray and ultraviolet emissions."

While much of the energy in a flare travels upward (some of which will escape into space), the flare also affects the photosphere. Astronomers have observed powerful "sunquakes" propagating away from flares in the same way an earthquake travels away from a slipped fault line.

Coronal mass ejections, or **CMEs** (**Figure 10.22**), are particularly potent phenomena associated with solar flares. "Coronal mass ejections are explosions in the Sun's corona that spew out huge amounts of matter and energy," says Guhathakurta. "And they can be particularly dangerous when they hit Earth." This danger is ever present, even though only a small fraction of CMEs hit Earth.

CMEs appear as a kind of superflare, but their exact connection with the lower-energy phenomena is still unclear. "Coronal mass ejections often accompany solar flares, but not always," explains Guhathakurta. Astronomers are sure of the connection between CMEs and solar magnetic fields, however. "CMEs and solar flares explode outward from the vicinity of magnetically active regions on the Sun. They are most common during times of peak solar activity, the so-called **solar max** years of the sunspot cycle." A typical CME will drive 10 billion kg of plasma into space, along with energy equivalent to that of a flotilla of

220 aircraft carriers moving at 500 km/s. "That," says Guhathakurta, "is a whole lot of energy."

Explosions on the Sun that prove devastating to Earth may sound like the stuff of cheesy disaster movies, but they can pose a real threat. **Space weather**, a new term in the lexicon of science, refers to the CMEs blown off the Sun that periodically sweep across interplanetary space (**Figure 10.23**). A particularly potent example occurred on March 13, 1989, when a severe space storm caused a system-wide power failure in Quebec. The time from the onset of the storm to total blackout was 90 seconds, and 6 million people were left without power.

solar max The period of peak activity in the sunspot cycle.

space weather Coronal mass ejections sweeping through interplanetary space, which produce auroras when they hit Earth and may pose threats to astronauts and technological systems.

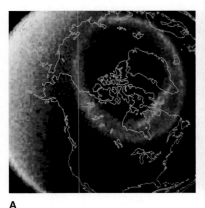

A

B

FIGURE 10.23 Space Weather
Coronal mass ejections traveling through interplanetary space constitute storms in what is now called space weather. (A) Viewed from high orbit, charged particles from a CME strike Earth and work their way into the atmosphere along magnetic-field lines to create a large oval aurora. (B) The auroral glow from charged particles flowing in the atmosphere, as seen from the International Space Station.

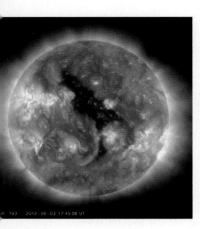

FIGURE 10.24
Coronal Holes
Coronal holes appear dark in X-ray light because of their lower densities relative to the surrounding bright regions. In this X-ray image, the bright regions indicate hotter areas of the solar corona, mainly above the sunspot regions. A large, dark region represents a coronal hole imaged in June 2012.

coronal hole A region of low density where the solar wind is generated.

heliosphere A cavity, created by the solar wind, in the interstellar gas surrounding the Solar System.

heliopause The outer boundary of the heliosphere.

Before we became a high-tech culture, the collision of a CME with Earth was no cause for alarm. The result would, at best, produce beautiful auroral displays. But now that we have come to depend so strongly on the space environment for everything from communications to weather prediction, what happens on the Sun can affect us all. In a worst-case scenario, the ionizing radiation could knock out orbiting satellites and space stations, possibly proving fatal to anyone working on them.

To deal with the problem, NASA and other space agencies have begun to monitor the Sun continuously. As soon as a CME is observed, powerful supercomputers are engaged to predict its path through space. If the storm of matter and magnetism appears to be headed toward Earth, precautions can be taken, such as bringing astronauts in from spacewalks or putting satellites into "safe mode" so that their electronics will not be harmed.

Space weather represents a potent example of how science and technology raise new issues for politicians and policy makers. The growth of our high-tech society—and, in particular, our reliance on satellites—means that as a society we now need to consider the dangers that space weather poses. Policy makers who are not scientists themselves must decide whether and how to dedicate scarce resources to prepare for such risks. In this kind of situation, scientists and politicians must find ways to work together to find optimal solutions to new problems.

The hardest part, however, is to make accurate predictions. "Those supercomputer calculations of a CME's path can take longer to run than the time it takes the storm to reach Earth," says Guhathakurta. "That means predicting space weather is going to remain an inexact science for some time to come." Still, it's a measure of how far we have come as a species that now even the weather in space has become our daily concern.

SECTION SUMMARY
Prominences, Flares, and CMEs
- Prominences are giant arcs erupting from the solar surface that carry ionized plasma.
- Flares are violent events that launch material out into space when magnetic-field lines are broken via reconnection.
- The most powerful solar eruptions are called coronal mass ejections (CMEs).
- *Space weather* is the term used to describe geomagnetic storms that can create auroras but can also damage communications satellites and other equipment on Earth and in space.

Solar Wind

Flares and CMEs are distinct events in which matter from the Sun is launched into space. Though they are dramatic, they are not the only way the Sun loses mass. Every moment of every day, a continuous stream of particles called the solar wind is blown off the Sun. But don't let the word *particles* fool you. All those solar protons and electrons add up. "The Sun is flinging more than a million tons of matter into space every second," says Guhathakurta. "And with typical speeds in the solar wind ranging from 200 to 500 km/s (that's about a million miles per hour), it all adds up so you get a lot of matter and a lot of energy plowing into interplanetary space every day."

Like so much else in the Sun, the solar wind is related to magnetic fields and their ability to trap charged particles. "Near the Sun's equator, magnetic-field lines are more likely to loop back on themselves and reenter the Sun's surface," says Guhathakurta. "These 'closed' field lines trap hot coronal gases that emit X-rays. The particles are trapped within the magnetic field and can't escape into space." In X-ray observations, the closed field lines show up as bright regions of intense emission clustered near the Sun's equator.

The situation is reversed near the Sun's poles. There, the large-scale magnetic field does not arc back to the Sun but extends out into interplanetary space. "The solar wind is primarily launched through regions called **coronal holes** near the solar poles," says Guhathakurta (**Figure 10.24**). As in a deflating balloon, thermal pressure is high enough in these regions to launch hot gas out along the open magnetic-field lines and produce the solar wind. The term *holes* comes from X-ray observations, in which the coronal holes appear dark. That means the X-ray coronal holes have low emission because of their low gas densities. Note that coronal holes and sunspots are not, in general, connected, since the scale of the magnetic-field variations and their locations are much different.

The matter particles composing the solar wind blow through the entire Solar System. By our terrestrial standards, however, the wind is not very dense. "The solar wind is pretty thin," says Guhathakurta. "It only has about six particles per cubic centimeter near Earth." Even though that number might seem small, the solar wind packs enough punch to carve out a vast cavity in the interstellar gas that surrounds the Solar System. Over the course of the Sun's lifetime, the solar wind has pushed all interstellar gas out to a radius of 15 billion km. Astronomers call this solar wind–dominated bubble the **heliosphere** (**Figure 10.25**), and its boundary is called the **heliopause**. Only recently (August 2012) did the *Voyager I* spacecraft, launched in the late 1970s, reach the heliopause, transmitting back valuable data about this last stand of the Sun at the edges of interstellar space and providing proof of its location, 18 billion km from the Sun.

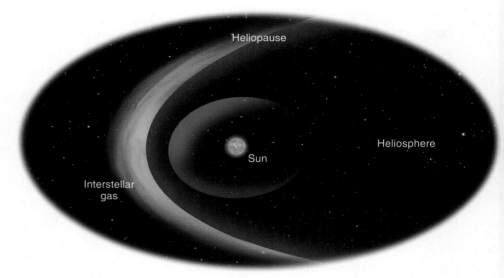

FIGURE 10.25 The Heliosphere
The heliosphere is a cavity in the interstellar gas surrounding the Solar System that has been created by the pressure of the outward-flowing solar wind. The heliopause marks the boundary between solar wind and the swept-up interstellar gas. The inner bubble around the Sun is a region where the solar wind begins decelerating as it pushes against the interstellar gas farther out.

┌─ **SECTION SUMMARY**
Solar Wind

- The Sun continually blows particles from its surface at high velocity; most particles near the equator are trapped in magnetic-field lines, but open magnetic lines near the solar poles allow particles to escape through coronal holes.

- The region in space created by the solar wind is called the heliosphere, and its boundary is the heliopause.

●→ chapter summary

10.1 Living with a Star

The Sun, our closest star, contains more than 99 percent of the Solar System's mass. It rotates differentially—faster at the equator than at the poles. The Sun and its atmosphere comprise six layers: core, radiative zone, convective zone, photosphere, chromosphere, and corona. Density continually decreases from the core to the outer layers, and the temperature decreases as well, except in the hot corona. Most recently, helioseismology has been used to study the inner workings of the Sun.

10.2 The Sun's Fusion Furnace

It took advances in nuclear physics to understand the source of the Sun's energy. Stars generate energy by means of nuclear fusion in their cores. Fusion requires high densities and temperatures of tens of millions of degrees. In the proton-proton chain characteristic of Sun-sized stars, hydrogen is converted to helium, releasing 0.7 percent of the hydrogen's mass as energy. Neutrinos are also released, and studying them has provided insight into the Sun's nuclear engine.

10.3 Moving Energy

The Sun uses two processes to transfer energy from the core to its photosphere. In the Sun's radiative zone, gamma-ray photons created in the core are scattered many times, with each scattering leading eventually to the creation of multiple photons of lower individual energy. In the convective zone, gas heated from below (the top of the radiative zone) rises to the solar photosphere (the visible surface of the Sun), where it cools by emitting photons and then sinks downward again. Granulation on the photosphere is caused by smaller convection cells with cooler, darker edges.

10.4 The Active Sun: Photosphere to Corona and Beyond

Sunspots are caused by the Sun's magnetic-field lines and follow a predictable pattern over their 11-year cycle, and the Sun's polarity also reverses on a 22-year cycle. The solar dynamo is caused by the Sun's differential rotation. Magnetic activity on the Sun's surface also causes eruptions of varying size and energy, characterized as flares, prominences, and coronal mass ejections. Surges of matter and magnetic energy from the Sun, called space weather, can cause power failures and danger for satellites and astronauts in space. The constant solar wind, a stream of particles blown off the Sun's surface, escapes through coronal holes near the Sun's poles. The solar wind clears out a large, bubble-shaped region of interstellar gas around the Solar System known as the heliosphere.

•→ questions and problems

Narrow It Down: Multiple-Choice Questions

1. Which of the following statements about differential rotation is/are correct? Choose all that apply.
   a. The average speed varies with the sunspot cycle.
   b. The spin rate is slower at the poles.
   c. The Sun rotates at different speeds at different times of the year.
   d. Different layers of the Sun rotate at different speeds.
   e. The Sun's outer layers rotate at different speeds at different latitudes.

2. Choose the correct order of the Sun's layers from the center outward.
   a. corona, chromosphere, photosphere, convective zone, radiative zone, core
   b. core, magnetosphere, heliosphere, atmosphere
   c. atmosphere, heliosphere, magnetosphere, core, solar wind
   d. corona, chromosphere, convective zone, photosphere, radiative zone, core
   e. core, radiative zone, convective zone, photosphere, chromosphere, corona

3. Which layer(s) of the Sun does *not* decrease in temperature as distance from the core increases? Choose all that apply.
   a. radiative zone
   b. convective zone
   c. photosphere
   d. chromosphere
   e. corona

4. How will the composition of the Sun change over the next billion years?
   a. It will not change appreciably.
   b. There will be more hydrogen, more helium, and less of the heavier elements.
   c. The proportions of carbon and iron will increase
   d. There will be less hydrogen and more helium.
   e. There will be less hydrogen and more of all the heavier elements.

5. Which of these statements about temperature, pressure, and density in the various layers of the Sun is correct?
   a. The higher the temperature, the higher the pressure.
   b. Temperature decreases from the innermost layer to the outermost.
   c. The deeper below the Sun's surface the layer is, the higher the pressure.
   d. The higher the temperature, the higher the density.
   e. The values of these parameters are unknown for some layers of the Sun.

6. If a star like the Sun did not have the ability to create energy by nuclear fusion,
   a. thermal energy from gravitational contraction would have sustained it until now, but its life expectancy would be much shorter.
   b. it would have exhausted its energy long ago.
   c. it would last longer because it would not be radiating away its stored energy.
   d. its antimatter and matter would have recombined, making it disappear.

   e. it could have sustained itself by nuclear fission as long as it was not a first-generation star.

7. What particle is the antimatter counterpart of an electron?
   a. positron
   b. boson
   c. neutrino
   d. proton
   e. neutron

8. Nuclear fission is possible on Earth with current technology, but nuclear fusion is not, because
   a. temperatures required for fission can be achieved and sustained, but those for fusion cannot.
   b. there is not enough hydrogen on Earth to adequately fuel fusion.
   c. fission reactions do not release deadly amounts of energy, but fusion does.
   d. nuclear fusion requires the reunification of matter and antimatter, but no antimatter is available on Earth.
   e. heavier elements required for fusion are not abundant on Earth.

9. Which statement about energy from nuclear fusion is correct?
   a. Matter is annihilated and completely converted to energy in nuclear fusion.
   b. The proportion of hydrogen converted to energy in nuclear fusion is 7 percent.
   c. The proportion of hydrogen converted to energy in nuclear fusion is 0.7 percent.
   d. Nuclear fusion requires more energy put in than it gives back.
   e. Fusion converts radio photons into gamma-ray photons.

10. The "solar neutrino problem" is accurately described in which of the following statements?
    a. The Sun lacks enough neutrinos to sustain fusion.
    b. Too many neutrinos are produced in solar fusion, and they are damaging to Earth.
    c. No neutrinos were found coming from the Sun.
    d. Experimental results found more neutrinos than were expected from models of nuclear fusion.
    e. Experimental results found fewer neutrinos than models of nuclear fusion suggested would exist.

11. A gamma ray is produced in the core of the Sun. What happens after that?
    a. It emerges intact from the photosphere.
    b. It is scattered many times as a gamma ray before emerging from the photosphere.
    c. It is scattered and converted into lower-energy photons many times before emerging from the photosphere.
    d. It remains intact in the Sun's interior.
    e. It undergoes nuclear fusion.

12. Which of the following statements about sunspots is true?
    a. They occur in predictable cycles.
    b. They are permanent features on the Sun's surface.
    c. They are a primary cause of climate change.
    d. They exist in the convective zone.
    e. Each sunspot is unrelated to any nearby sunspots.

13. The Zeeman effect, in which some individual lines in a stellar spectrum are split into multiple lines, is caused by
    a. gravity.
    b. nuclear fission.
    c. nuclear fusion.
    d. gravitational contraction.
    e. magnetism.

14. Once the Sun's magnetic north pole is located at its geographic north pole, on average how many years will pass before it shifts back to being near the geographic south pole?
    a. 11 years
    b. 22 years
    c. 33 years
    d. 44 years
    e. The answer is unknown.

15. Which statement(s) about nuclear fusion and nuclear fission is/are true? Choose all that apply.
    a. Both involve molecular reactions.
    b. Both release energy that had been in the form of mass.
    c. Both convert 100 percent of their fuel to energy.
    d. Both were unknown phenomena until about 100 years ago.
    e. Both require large magnetic fields.

16. What property of neutrinos allows them to mostly pass right through matter?
    a. their very small size
    b. their zero mass
    c. their low probability of interaction with other particles
    d. their high temperature
    e. their immutability

17. A solar feature that lasts about 10 minutes is most likely to be a
    a. granule.
    b. sunspot cycle.
    c. sunspot.
    d. prominence.
    e. CME.

18. Which statement about the inputs to and outputs from nuclear fusion in the Sun is true?
    a. Outputs have smaller atomic weights than their corresponding inputs.
    b. Only elements up to iron can be inputs.
    c. The total mass of outputs is less than the corresponding inputs.
    d. Outputs are always radioactive.
    e. Any element can be an output of nuclear fusion.

19. True/False: Nuclear fusion requires extremely high temperatures so that collisions of protons and neutrons occur at velocities high enough to overcome the strong nuclear force.

20. True/False: Nuclear weapons with enough destructive power to annihilate a city require several tons of nuclear fuel.

To the Point: Qualitative and Discussion Questions

21. What is the P-P chain?

22. What are the characteristics of a plasma?

23. The Sun was once considered to be perfect and changeless. Explain how that belief has been disproved.

24. Compare and contrast energy transport in the radiative zone and in the convective zone.

25. How has the Sun's composition changed from when it was a newly formed star?

26. What layer of the Sun is visible to the naked eye?

27. What conditions of the hydrogen in the core of the Sun are necessary to create the enormous amount of energy produced there?

28. What phenomenon within Earth is similar to how energy is transported in the convective zone, and in what layer of Earth's structure does it occur?

29. Describe the cause of the sunspot cycle. In your explanation, discuss how it can be considered both an 11-year cycle and a 22-year cycle.

30. How does a solar flare compare with a CME?

31. Describe the relative quantity and movement of sunspots over the solar cycle.

32. What is the underlying cause of the solar dynamo?

33. What is helioseismology, and what useful information does it yield?

34. Would you expect that nuclear fusion of *any* element to another would convert 0.7 percent of the mass to energy? Explain.

35. What other phenomena are like the Sun's energy in that they are life-giving and also potentially harmful and destructive?

Going Further: Quantitative Questions

36. What is the Sun's mass in units of Mars masses? in units of Saturn masses? in units of Earth's-Moon masses?

37. How many Mercury diameters are there in the equatorial diameter of the Sun?

38. How long, in years, would it take a passenger jet to arrive at the Sun's photosphere, assuming it travels at 800 km/h?

39. How much energy, in joules, is released when 50 kg of hydrogen is converted into helium by nuclear fusion?

40. If 210 kg of hydrogen could be entirely converted to energy, how many joules would be produced?

41. How much brighter is the area surrounding a star spot with a temperature of 5,000 K than the area surrounding a star spot at 3,500 K? (You can assume the two areas are equal in size and differ only in temperature.)

42. You are viewing two light sources of the same size at the same distance. One is 1,000 K and the other is 2,400 K. How many times brighter is the hotter light source?

43. The center of a solar granule is 25 times brighter than an equal-sized area at its edge. By what factor is the temperature of the center greater than that of the edge?

44. The temperature of a sunspot is 0.66 as high as the surrounding photosphere. What is the ratio of its brightness to that of an equal-sized area around it?

45. Four kilograms of hydrogen is converted to helium by nuclear fusion. How much of it, in kilograms, remains as matter (and is thus not converted to energy)?

 If your instructor assigns homework in **smartw⚙rk5**, access it at the Digital Landing Page for *At Play in the Cosmos*: **digital.wwnorton.com/cosmos**

MEASURING THE STARS

The Main Sequence and Its Meaning

● **In this chapter** you will learn about the discoveries and tools leading to our understanding of stars and their evolution. After reading through each section, you should be able to do the following:

11.1 The Life of the Stars
Explain how the discovery of stellar nuclear fusion allowed astronomers to conclude that stars have life cycles from birth to death.

11.2 Measuring Stars
List the measured properties needed to understand stellar evolution, and describe the methods astronomers use to determine each of them.

11.3 From Observations to Explanations
Explain the importance of the main sequence on the HR diagram, and explain what it tells us about stellar evolution.

11.4 The March of Time: Stellar Evolutionary Tracks
Describe how astronomers employ stellar models tested using globular clusters to further their understanding of stellar structure and evolution.

Globular Clusters like M13 (the Hercules Cluster), shown here, are gravitationally bound collections of thousands of stars all born at the same time but with different masses. Because a star's mass determines how fast it evolves, a snapshot of a globular cluster like this shows stars and many stages in their evolution.

11

11.1
The Life of the Stars

Jim Kaler remembers what astronomy was like before computers controlled telescopes and Google put all the Universe's information at our fingertips. Back in those days, if you wanted to find an object on the night sky, you couldn't just plug the coordinates into a digital machine; you had to find it yourself.

"That was how I watched a star explode one night," says Kaler. Now retired from the University of Illinois, Kaler spent his career as a professor of astronomy studying stars and their life cycles. He remains one of the world's experts on extracting stellar properties from stellar spectra.

"I had been following this one object for years," recalls Kaler. "It was a pair of stars orbiting so close to each other that they exchanged mass. We call it a *symbiotic binary system*" (**Figure 11.1**). Symbiotic systems change their brightness in regular ways over time, as mass from one of the stars flows onto the other. By tracking the response of the star that was gaining mass, Kaler hoped to learn something quite general about the physics governing all stars.

"One night," Kaler continues, "I'm at the telescope and something just doesn't look right. The system was much too bright. I was sure something was wrong, so I closed the dome and climbed down into the pit where the telescope's instruments were mounted to make sure everything was working." Kaler then reopened the observatory dome and let the stars' light flow once more through the telescope. To his astonishment and chagrin, he found the new measurements were the same as before. The instruments were fine. It was the binary star system that had gone crazy.

"Usually stars change over very, very long timescales," says Kaler. "To understand how they evolve, we have to measure lots and lots of them at different points in their lives and then deduce their individual stellar life stories. But on that night, stellar evolution was happening at high speed. The atmosphere of one of my binary stars had exploded, and I just happened to be there to see it."

Stellar Stories

For most of human history, people assumed that the stars were eternal. The idea that stars had life stories—that they were born, lived, and eventually died—seemed too absurd even to consider. But the human perspective on the stars slowly changed, and with that change came the recognition that even the stars might age.

This idea caught hold in the mid-1800s with the science of thermodynamics, the study of energy and its transformations. Steam engines—driving everything from trains to factories—had taught scientists the value of thinking about energy and the evolution of a "system." It didn't matter what the system was; any collection of interacting parts fit the bill. Thermodynamics provided laws that described how transformations of energy from one form to another forced systems to evolve. In a steam engine, for example, the chemical energy in coal was transformed into heat energy in steam and then transformed yet again into the train's motion.

By the beginning of the 1900s, astronomers using spectroscopy had discovered that stars are nothing more than giant systems of atoms. As such, stars must also be subject to the laws of thermodynamics. Forty

FIGURE 11.1
Binary Star Systems
Examples of binary stars. (A) Bright Sirius A and its white dwarf companion, Sirius B. (B) Gamma Caeli, a complex multiple-star system in the Caelum constellation, consisting of one pair with an orange giant whose companion is much fainter, and a second pair with a yellow-white giant and a fainter companion star. (C) Albireo, a favorite with amateur astronomers for its contrasting blue and yellow pair.

A

B

C

years later the energy source powering the stars was discovered in nuclear fusion. Once it was recognized that stars "burn" hydrogen fuel to keep themselves alive, it was also understood that sooner or later the fuel would run out and the stars would have to die. Stars *must*, therefore, have life stories.

But learning those stories proved a long and difficult journey for astronomers. As is always the case in science, the road began with data. To understand how stars evolve, astronomers first had to find links between the properties they could measure and the unknown story of stellar evolution. Before taking that step, however, they had to acquire accurate measurements of the important stellar properties and figure out which measurements mattered.

Our own journey through the life and death of stars will begin with that most basic of questions: Where, in terms of measurements and data, do we start?

SECTION SUMMARY
Stellar Stories
- Stars were once thought to be changeless, but astronomers now understand that they evolve and have life cycles, including birth and death.
- Measurements of the properties of stars form the basis for understanding their life cycles.

11.2
Measuring Stars

"There are a handful of properties we need to build up the life story of stars," says Jim Kaler. "Some of them you get directly from measurements. Others get *derived* from measurements. That means you calculate them using other measurements as inputs." Specifically, in their attempts to understand stellar evolution, astronomers focus on six critical properties: luminosity, distance, temperature, size, mass, and composition (**Table 11.1**). One additional property—the velocity of a star through space—is also useful as astronomers attempt to build a picture of stellar evolution in the context of the Milky Way Galaxy as a whole.

How do astronomers get these measurements? You can't stick a thermometer in a star to get its temperature, and you can't run a tape measure around its equator to find its size. In every case, astronomers must rely on their ingenuity and their knowledge of the laws of physics to make these measurements. And, in every case, they must use only the stellar light collected in their telescopes.

Measuring Luminosity and the Magnitude Scale

Among a star's most important properties is its luminosity. "A star's luminosity is really a measure of its energy generation rate," says Kaler. "Luminosity is measured in energy per time. That means if we know a star's luminosity, then we know how fast it's burning its nuclear fuel." But astronomers can't measure luminosity just by looking at how bright a star appears on the sky. Remember from Chapter 4 that brightness B decreases with increased distance:

$$B = L/4\pi D^2$$

The luminosity L is the intrinsic energy output of the star. Brightness is measured as the energy per second per unit area falling on a detector. Measured at a distance D from the star, brightness is the fraction of the star's energy that reaches Earth—the object's brightness that we see in the night sky. What this formula tells us is that once we know the distance to a star, we can convert a simple measurement of brightness into a measurement of luminosity and hence stellar energy generation.

To make the actual measurements of brightness, astronomers use a scale that dates back to the ancient Greeks. When Hipparchus made his first maps of the stars, he categorized their brightness using what's now called the **magnitude scale**, in which faint objects are given higher numbers than brighter objects. In Hipparchus's original formulation, each star was assigned to one of six magnitude classes. He assigned a magnitude of 1 to the 20 brightest stars. The stars in the next-brightest group were assigned a magnitude of 2. He continued this categorization until he reached stars at the limit of detection by the naked eye, which he gave a magnitude of 6.

magnitude scale
A categorization of stellar brightness in which faint objects are given higher numbers than brighter objects.

TABLE 11.1 ⬤⬤⬤ Properties of Stars Key to Evolution	
Stellar Property ⬤	**How Determined**
Luminosity	Apparent brightness and distance
Distance	Parallax; standard candles, other methods
Temperature	Blackbody spectra; Wien's law
Size (radius)	Luminosity and temperature (Stefan-Boltzmann law)
Mass	Binaries, spectral lines
Composition	Spectral lines
Velocity	Doppler shift

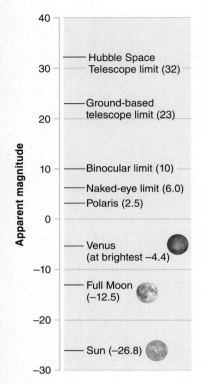

FIGURE 11.2
The Magnitude System
Astronomers use the magnitude system to measure brightness. Higher numbers imply fainter objects.

Remarkably, astronomers still use this system, but they allowed the scale to extend in both directions, since Hipparchus did not include objects, like the Sun, that look much brighter to us than distant stars, or objects that can be seen only through telescopes, which are much fainter. The Sun has a magnitude of about −27 (the scale had to go negative for bright objects), and the faintest objects detectable by the Hubble Space Telescope have a magnitude of about 32 (**Figure 11.2**).

The magnitude scale used today by astronomers is logarithmic. In other words, every step up in magnitude means a decrease in energy flowing into the detector (like your eye) such that the object appears smaller, and therefore dimmer, by a factor of 2.512. Astronomers chose this number so that two objects separated by five magnitudes—such as the brightest star in the night sky compared with the dimmest star visible to the naked eye—differ in brightness by a factor of 100, which equals 2.512.

Astronomers call the brightness we observe from Earth *apparent magnitude* because it's the brightness that the object appears to have, given its distance. To determine the *absolute magnitude*, which is a more direct measure of an object's luminosity, astronomers have to know the object's distance. The absolute magnitude tells us how bright the star would appear (in the magnitude scale) if it were seen from 10 pc away. **Table 11.2** gives the apparent magnitude, the absolute magnitude, and the distance for different objects in the night sky, including the Sun. Notice that the apparent magnitude of the Sun is a whopping −26.72 because it's so close. Placed at 10 pc away, the Sun would appear on the sky with a brightness of just 4.8 in the magnitude scale—not vey bright at all.

SECTION SUMMARY
Measuring Luminosity and the Magnitude Scale
● A star's luminosity is its energy generation rate, which can be calculated from its apparent brightness if the distance is known.
● Brightness is measured on a logarithmic magnitude scale in which brighter objects are assigned lower numbers than dimmer objects.

Measuring Distance

The inverse square law for brightness provides a simple way to determine a star's luminosity once its distance is known, but how do astronomers measure those vast distances? "The distance to a star is often the most important measurement we make," says Kaler. "Many of the properties we need for understanding stars are derived from combinations of other measurements. Distance appears in many of those formulas. If you get that wrong, then lots of other things will be wrong too."

"Distance is not really a property of stars in the same way as, say, temperature," continues Kaler. What he means is that some properties of stars are *relational*: they make sense only relative to other stars. Other properties, however, are *intrinsic*: they relate to the star itself. Distance is a relational property, while temperature is intrinsic. Luminosity is an intrinsic property of a star, but its apparent brightness from Earth is relational.

One easily understood unit of distance on interstellar scales is the light-year, which is the distance light travels in one year. Since light moves at $c = 2.99 \times 10^8$ meters per second (m/s) and there are 3.16×10^7 seconds in a year, some quick multiplication tells us that a light-year equals

TABLE 11.2 ••• Magnitude and Brightness			
Star	**Apparent Magnitude**	**Absolute Magnitude**	**Distance (pc)**
Sirius	−1.44	1.45	2.6
Procyon	0.4	2.68	3.4
Vega	0.03	0.58	7.7
Arcturus	−0.05	−0.31	11.1
Capella	0.08	−0.48	12.5
Achernar	0.45	−2.77	50
Canopus	−0.62	−5.53	100
Betelgeuse	0.45	−5.14	125
Rigel	0.18	−6.69	250
Sun	−26.72	4.8	$5 \times 10^{(-6)}$

9.46×10^{15} meters. However, astronomers tend to use another unit, the parsec, or pc (1 pc = 3.26 light-years = 3.09×10^{16} meters). The parsec is based on the most direct method used to find distances to the stars, called parallax.

USING PARALLAX TO DETERMINE DISTANCE.
Parallax is the apparent change in position of a foreground object relative to a more distant background object when the point of view changes. If that sounds confusing, try this simple experiment: Hold up your index finger about 6 inches from your nose. Now close your left eye and notice how your finger lines up relative to objects in the background. Then open your left eye and close your right. What happened? The relative position of the background has shifted to the left. The angular change in position of the background relative to your finger is the parallax that occurs as your point of view shifts from the right eye to the left eye (**Figure 11.3**). Thus, the parallax is directly proportional to the distance between your finger and your nose and the distance between your eyes. For example, if you move your finger to an arm's length from your nose, the amount of parallax shift will decrease.

Astronomers can use this principle to determine the distance to stars by making measurements 6 months apart—say, in June and January. Making these measurements at the maximum separation across the year is like shifting the point of view from one eye to the other as just described, but now the "baseline" over which the two measurements are made stretches across the diameter of Earth's orbit (2 astronomical units, or AU) rather than just a few inches (**Figure 11.4**).

If the object is very far away, such that the distance from it to Earth is much larger than the baseline distance of 2 AU, a simple formula for parallax relates the distance D to the object and its observed angular shift p relative to more distant background stars. If p is measured in arcseconds and D is measured in parsecs, we have

$$p = 1/D$$

"The parallaxes of stars are small because they are so far away," Kaler explains. "That is why it took so long to be able to measure them. You need a pretty good telescope to resolve their tiny shifts on the sky" (see **Going Further 11.1**, page 280). In 2013 the Gaia satellite was sent into orbit to measure the parallax of millions of stars with extreme precision, giving accurate distances (less than 10 percent error) out to thousands of parsecs. (Distances with a 20 percent uncertainty will be achieved out to 10,000 pc, or about 30,000 light-years.)

A Right eye closed   **B Left eye closed**

FIGURE 11.3 Parallax
Parallax is a change in the apparent angular position of a foreground object relative to a more distant background as the observing position changes. Here we see an everyday example, in which an observer views her index finger in the foreground, relative to the background of the more distant building while closing first the left eye (A) and then the right eye (B). Parallax is the angular change in position of the finger relative to features like the door and windows on the building.

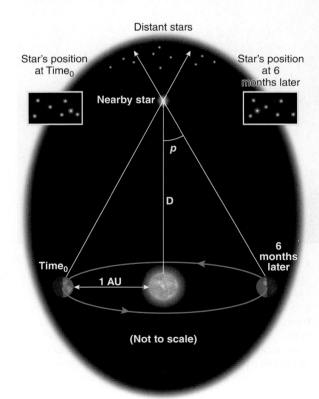

Distant stars

Star's position at $Time_0$

Star's position at 6 months later

Nearby star

p

D

$Time_0$

6 months later

1 AU

(Not to scale)

interactive
Parallax

FIGURE 11.4
Stellar Parallax
In making stellar parallax measurements, the observing position shifts from one side of Earth's orbit to the opposite position 6 months later. Basic trigonometry then allows an accurate calculation of the distance D to a star, based on its angular change in position relative to the more distant background stars (its parallax p).

GOING FURTHER 11.1 THE LANGUAGE OF THE COSMOS
Parallax

The formula for parallax comes from the basic geometry of triangles when one of the triangle's sides is much shorter than the other two. That is the situation we have in astronomy when we view a star at two points in Earth's orbit separated by 6 months, because the diameter of Earth's orbit is much smaller than the distance between Earth and any star other than the Sun.

The formula for parallax shows that there is an inverse relationship between the star's angular shift seen on the sky (p, measured in arcseconds, or ″) and the distance to the star (D, measured in parsecs):

$$p = 1/D$$

For a star that is 1 pc (3.26 light-years) away, we clearly have

$$1″ = 1/(1 \text{ pc})$$

Recall that the Moon is almost 1,500″ across, as viewed from Earth. Therefore, to see a star 1 pc away requires a telescope capable of resolving shifts that are 1,000 times smaller than the diameter of the Moon.

Let's put these calculations to practical use. Gliese 581 is a star with a planet known to be in its habitable zone. The parallax for Gliese 581 is 0.16″. How far away is this star (and its planets) from us?

To calculate its distance, we use the parallax formula:

$$D = 1/p = 1/0.16 = 6.2 \text{ pc}$$

Now convert parsecs to light-years to find

$$6.2 \text{ pc} \times 3.26 \text{ ly/pc} = 20.21 \text{ ly}$$

Thus, Gliese 581 lies about 20 light-years away. A spaceship traveling at 10 percent of the speed of light (something we're not even close to building) would take 200 years to reach its planets.

variable star A star with regular periodic changes in its temperature and luminosity. Periods tend to be on the order of years.

period-luminosity relationship The correlation between the luminosity and period of a variable star, allowing a star to act as a standard candle for determining distance.

FIGURE 11.5 Cepheid Variables
This Cepheid variable star in the Andromeda Galaxy, called V1, is shown here at four different levels of brightness, the result of instability occurring at certain times of a star's evolution. The time it takes for the variation in brightness to cycle from low to high and back is called the Cepheid's period, which is directly related to the energy output or luminosity of the star.

USING STANDARD CANDLES TO DETERMINE DISTANCE.
To obtain precise distances to objects even farther than Gaia can measure, astronomers have to be more clever. One particularly powerful method relies on finding standard candles, objects whose luminosity is intrinsically related to another simply measured property (see Chapter 4). Standard candles are worth their weight in gold because they make distance measurements simple and yield very accurate results. The best example of a standard candle comes from the study of what are called variable stars.

While most stars evolve and change their properties over millions or billions of years, some stars change over much shorter periods. These are called **variable stars**, and they come in many forms, but all of them change their surface properties (temperature and luminosity) over periods of years or less (**Figure 11.5**). In most cases, these changes are quite regular and repeat themselves with well-defined periods. "Variable stars are usually in some stage where their surface layers have become unstable," explains Kaler. "Their surfaces are ringing like a gong. The outer layers swell and shrink in cycles. The luminosity and surface temperature of the stars change in step with each repeat of the cycle."

Cepheid variables constitute one class of changing star and are of vital importance to astronomers because of their role in distance determination.

FIGURE 11.6
Henrietta Leavitt's Contribution to Astronomy
Henrietta Swan Leavitt (1868–1921) was an astronomer who worked as a "computer" at Harvard College Observatory, earning $10.50 per week. She performed meticulous studies of images of Cepheid variable stars on thousands of photographic plates.

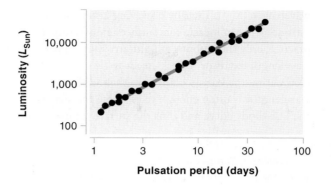

FIGURE 11.7 The Period-Luminosity Relationship
When Henrietta Leavitt plotted her observations of Cepheid variable stars in the Small Magellanic Cloud (SMC), she found an unmistakable relationship between the period of the star's variation and its luminosity at the brightest point in the cycle: the longer the period, the greater the luminosity.

Cepheids are bright stars that have already evolved off the main sequence and usually have 4–20 times the mass of the Sun. Over very well defined periods of days to months, an individual Cepheid will change brightness by a factor of about 10.

The special properties of Cepheid variables were discovered in 1908, when Henrietta Swan Leavitt (**Figure 11.6**) was studying stars in the Magellanic Clouds, two small satellite galaxies that orbit the Milky Way. Because the apparent brightness of an object decreases with distance and all the stars in the Magellanic Clouds are at roughly the same distance from us, Leavitt could directly compare their luminosities, leading her to a huge discovery.

Leavitt was officially not an astronomer, but a "computer" hired by the Harvard College Observatory to process the enormous amounts of data being collected at the time. Not content simply to repeat calculations time and time again, Leavitt began asking her own questions. After trolling through thousands of images, she found a relationship between the average luminosity of a Cepheid and how long it took the star to cycle through its pulsations. After exhaustively confirming her results, she published a paper announcing the **period-luminosity relationship** for Cepheid variables (**Figure 11.7**). Her studies gave astronomers a way of directly determining a Cepheid's intrinsic energy output (its luminosity) purely from an observation of its pulsation period.

Since Cepheids are fairly common throughout our galaxy and others, they soon became the quintessential standard candle. A simple measure of a Cepheid's period (P) yielded its luminosity (L). Comparing the star's luminosity with how bright it appeared on the sky

yielded an accurate measure of its distance. In addition, the fact that Cepheids are so bright means that they can be seen from very far away. If a Cepheid is found in another galaxy, astronomers can determine the galaxy's distance from us. Distance determinations with Cepheids became essential in the history of astronomy, providing the first true measure of the scale of galaxies and the Universe.

Other objects can act as standard candles as well. We will explore the topic of distance determination and its vital role in cosmology—the study of the Universe as a whole—in Chapters 16, 17, and 18.

SECTION SUMMARY
Measuring Distance

- Some stellar properties are absolute—intrinsic to the star itself; others are relational, meaning that they make sense only in relation to other stars.

- Distance, a relational characteristic, is measured for stars nearest Earth by parallax, the apparent motion of an object against the background of stars.

- Standard candles are objects whose luminosity can be determined by measuring other intrinsic properties. The luminosity can then be compared with the apparent brightness of the object to determine distance.

- The first standard candles to be discovered were Cepheid variables, whose luminosity can be determined through measurements of the variable star's period and the period-luminosity relation.

Measuring Temperature

After luminosity and distance, the next property astronomers need is a star's surface temperature, T_*. "A crude measure of temperature can be taken just from the color of the star," says Kaler (**Figure 11.8**). The trick, of course, is that stars are blackbodies. Recall from Chapter 4 that the spectrum of any blackbody is determined by its temperature, and Wien's law can be used to calculate the temperature from the peak

A Vega

B Arcturus

FIGURE 11.8
Wien's Law
Because stars radiate as blackbodies, there is a clear relationship between the wavelength where the emission peaks and the surface temperature (Wien's law). Thus we can "take a star's temperature" just by noticing its color. Here we see (A) Vega, which is hot and blue, contrasted with (B) Arcturus, which is cooler and red.

wavelength (λ_{max}). Thus, by observing a star's spectrum, astronomers can measure its surface temperature. In practice, only the amount of light arriving in a few wavelength bands or "colors" needs to be measured. Comparing the energy emitted in these different wavelengths is often enough to estimate the shape of the blackbody curve and thus the surface temperature.

SECTION SUMMARY
Measuring Temperature
- A star's peak spectral wavelength gives its temperature via Wien's law.

Measuring Size

The size of a star (its radius) is a key property needed for piecing together the story of stellar evolution because stars change their radii as they evolve. But how do astronomers determine the radius of a star that may be hundreds or thousands of parsecs away?

"In some cases," says Kaler, "you get a direct measure of size by observing a star's angular diameter and combining that with a measure of its distance. If you have both measurements, you can then use the small-angle formula to convert the star's angular size into a physical size" (see Going Further 2.1). Very few stars, however, are close enough and large enough to appear as anything more than a point of light in even the most powerful telescopes (**Figure 11.9**).

Even though astronomers rarely obtain direct measurements of size, they still have a powerful, indirect means of determining stellar radius. "Size can be directly derived from a star's surface temperature and luminosity," explains Kaler. To see Kaler's point, let's start with the Stefan-Boltzmann law for the luminosity of a blackbody: $L = A\sigma T^4$ (see Chapter 4), where A is the surface area of the blackbody. Since stars are spheres and the area of a sphere is calculated as $A = 4\pi R^2$, we can rewrite this equation as

$$L = 4\pi R^2 \sigma T^4$$

A little algebra shows how this equation yields R if T and L are known:

$$R = \sqrt{L/(4\pi\sigma T^4)}$$

For two stars with the same temperature T, the one with higher luminosity will be larger (have a larger radius) than the one with low luminosity. If the luminosity L of the two stars is the same, the star with the lower temperature will have the larger radius. Thus, the Stefan-Boltzmann law for blackbodies gives astronomers an indirect means for establishing the radius of a star.

This is a good time to stop for a moment and reflect on what you've just learned. It is nothing short of astonishing that Earth-based scientists can use just the laws of physics to measure the size of a star they will never visit that lies across trillions of kilometers of deep space.

SECTION SUMMARY
Measuring Size
- Size can rarely be obtained from observation, but it can be derived from temperature and luminosity using the Stefan-Boltzmann law.

Measuring Mass and Composition

At least two other significant properties of a star are needed to understand its evolution. "The mass and chemical composition of a star are both critical," says Kaler. "Taken together, those two properties pretty much set the entire life cycle of the star. Mass and composition will tell you the path a star takes from birth to death."

SPECTRAL LINES. The width of spectral lines can, in some cases, give a measure of a star's mass. The width of stellar absorption lines can be attributed to the thermal motions of atoms in the star's atmosphere. The higher the pressure that atoms experience, the

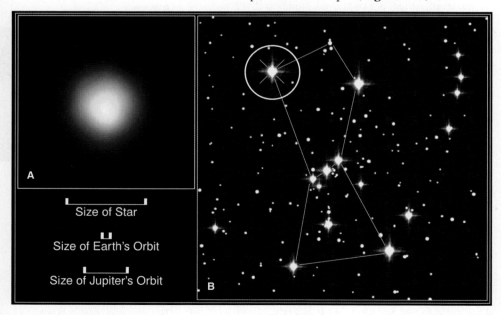

FIGURE 11.9 Betelgeuse
Most stars are so distant that they appear only as points of light, making direct measurement of their diameters impossible. A rare exception is (A) the red giant Betelgeuse, in (B) the constellation Orion. Although it is about 600 light-years distant, it is large enough to appear as an extended object, allowing its radius to be directly measured.

Size of Star

Size of Earth's Orbit

Size of Jupiter's Orbit

faster they bounce around. But the pressure on the star's surface is directly related to the gravitational pull of the star itself, which is proportional to the star's mass and radius. Since the radius can be determined if the luminosity and temperature are known, astronomers have learned to exploit the widths of spectral lines, and in some cases they are able to use them to estimate a star's mass.

Composition is also measured by spectral lines. Recall that every element produces a "fingerprint" in terms of its absorption or emission lines (see Chapter 4). Astronomers must first identify all the absorption lines in a stellar spectrum, connecting the lines to different elements in the atmosphere. The depth of the absorption lines (which indicates how much light has been absorbed) can be converted into a measure of how much of each element or molecule resides in the star's atmosphere. In this way, astronomers can build a complete census of the elements present in a star. "Most stars in our neighborhood of the galaxy have a chemical composition similar to the Sun's," says Kaler. "That means they are mostly hydrogen, a little bit of helium, and a tiny smattering of everything else."

Measurements of composition not only help astronomers build models of stellar evolution, but also allow them to explore the galactic *ecology* of stars. "Different parts of the galaxy show different chemical abundances," says Kaler. "Since stars move quite far from where they were born over time, observations of a star's chemical composition can tell where in the galaxy the star was born. That helps us understand the history of star formation in the galaxy as a whole."

BINARY STARS.   Another, more direct way to measure stellar mass comes from two stars orbiting each other, called **binary stars**. As Kaler's story at the beginning of this chapter suggests, not all stars are "loners." In fact, by taking a careful census of stars, astronomers have learned that at least half of all stars in the galaxy may be born as multiples, meaning two or more stars formed together that orbit their common center of mass. Binary star systems, in which two stars spend their lives locked in a gravitational dance, are by far the most common form of multiple system.

To determine the mass of a pair of binary stars, astronomers track the motion of the stars to obtain their orbital periods and orbital radii. They then use Newton's equations to calculate the total mass for the stars (see Chapter 3). From Newton's laws it can be shown that the ratio of the two stars' masses is the inverse of the ratio of their orbital radii. Therefore, if we observe two stars, A and B, orbiting about their center of mass, then $R_A/R_B = M_B/M_A$. This means that if the two stars' orbital radii are roughly equal, their masses are also roughly equal. And if one star's orbital radius is larger than the other star's, then its mass will be lower.

How astronomers track the stars' motion in the first place depends on the type of binary system. From an observational standpoint, binaries are classified as one of three types: astrometric, spectroscopic, or eclipsing (**Figure 11.10**). Note that these three categories relate to how we can observe the binary stars and don't say much about the stars themselves. In **astrometric** (or **visual**) **binaries**, one star can be seen moving back and forth around the other, and astronomers use images

binary star Two stars in orbit around their common center of mass. Binary stars are born together.

astrometric binary or **visual binary** Co-orbiting stars that have been detected by the proper motions of the components.

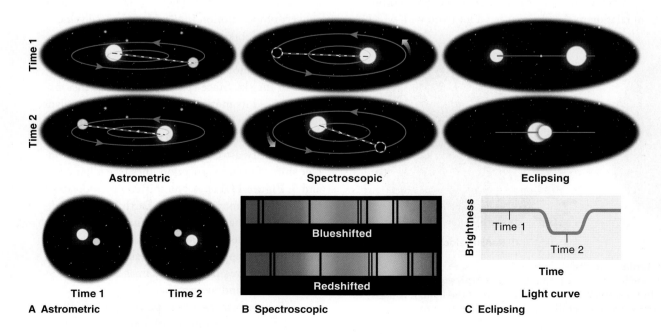

FIGURE 11.10
Three Types of Binary Systems
Astronomers describe binaries as one of three types. In astrometric binaries (A), both stars are visible and the proper motion of one star around another is directly observed. In spectroscopic binaries (B), Doppler shift of one star indicates its motion toward and then away from the observer as it orbits its unseen companion. In eclipsing binaries (C), the total light from the binary system decreases periodically as one star passes in front of the other to our line of sight.

Jim Kaler

When Jim Kaler was 8 years old, he asked his grandmother a simple question: "Grandma, why do stars all have points?" Kaler's grandmother told the young boy that stars didn't have points on them. "Just look up and see for yourself," she told him (like a good scientist). "I did look up," says Kaler. "But this time I really saw! I was looking into the sky at what turned out to be the giant star Aldebaran, and sure enough, stars didn't have points at all."

This first direct experience with astronomical observations set Kaler ablaze. "I remember Aldebaran was a slightly orange shade of red. The more I stared at it, the more I was fascinated by it. At that moment I fell in love with astronomy and, as crazy as it sounds, I never looked back."

Kaler eventually became one of the world's experts in the late stages of evolution for stars like the Sun. "I always viewed astronomy as a very aesthetic science," Kaler says. "It's as much an art form to me as it is a science. I remember just falling in love with the images I saw of what are called planetary nebulae—clouds of gas blown off dying solar-type stars."

But studying stars was not enough for Kaler. "Somewhere in middle life I decided I was having more fun writing about astronomy than actually doing it," says Kaler. "I did a sort of a career change which was unheard of back in those days. I went into public education." Kaler is now a well-recognized author of popular books on stars.

spectroscopic binary Co-orbiting stars that have been detected by the Doppler shift of absorption lines.

eclipsing binary Co-orbiting stars in which one star is observed passing in front of another.

proper motion An object's motion that is on the plane of the sky and can be detected by observations taken over time.

radial velocity An object's motion that is perpendicular to the plane of the sky along the observer's line of sight and can be detected via Doppler shift measurements.

taken over many years to determine the orbital periods and radii. The motions of stars in **spectroscopic binaries** can be detected via Doppler shifts, because absorption lines are blueshifted when either star is headed toward us in its orbit and redshifted when it moves away. In **eclipsing binaries**, one star passes in front of the other during an orbit, blocking out some of the light of the combined system. Astronomers use a combination of light curves and Doppler shifts to determine the motions of stars in this type of binary system.

Notice how the orientation of the binary orbit in relation to the observer has a lot to do with which

category might be assigned to an individual binary. Two stars can appear as an eclipsing binary, for example, only if their orbital plane is nearly lined up with the line of sight of our observations. In certain orientations, however, whether a system is classified as an astrometric or a spectroscopic binary depends on whether astronomers have a telescope powerful enough to resolve their orbital motions.

SECTION SUMMARY
Measuring Mass and Composition
- Astronomers use spectral lines to determine the mass and composition of stars.
- Stellar mass can also be calculated using binary stars.

Measuring Velocity

Astronomers are keenly interested in knowing a star's velocity through space (velocity, like distance, is a relational property). Although velocity isn't used directly in modeling the lifetime of an individual star, measuring thousands of stellar velocities can help astronomers understand the galactic context of stellar evolution. "Our galaxy is constantly in motion," says Kaler, "and knowing stellar velocities now can sometimes help us figure out the history of galaxy-wide star formation and evolution in the past."

To determine a star's velocity, astronomers need two measurements. If they can watch the star long enough, they can see it move across the sky. This movement is called **proper motion** and is measured in arcseconds per year. If the distance to the star is also known, the small-angle formula can be used to convert proper motion into an actual space velocity measured in kilometers per second. But proper motion can't be the whole story, because it tells us how a star is moving only in the two-dimensional plane of the sky. Since space has three dimensions, we also need its motion perpendicular to that plane, toward or away from us. Astronomers call this perpendicular motion the **radial velocity**. To measure radial velocity, astronomers must look for the Doppler shift of stellar absorption lines (see Chapters 4 and 5; **Figure 11.11**). A blueshifted line indicates that the star is moving toward Earth; a redshifted line indicates that the star is moving away. Obtaining radial velocity requires just a simple onetime measurement of Doppler shift, whereas proper motion measurements require multiple images taken over many years to find movement.

When information from proper motion (along with distance) is combined with Doppler shift data,

Moving toward us, light is blueshifted.

Moving away from us, light is redshifted.

Proper motion

Observer

A Proper motion

B Radial velocity

FIGURE 11.11 Proper Motion and Radial Velocity
(A) Proper motion is the change in position of objects in the plane of the sky over time. (B) Radial velocity is measured by the detection of the Doppler shift of the light; it is the velocity of the source perpendicular to the plane of the sky toward (blueshift) or away from (redshift) the observer.

astronomers can determine the full three-dimensional motion of the stars through space.

11.3
From Observations to Explanations

After astronomers measure all of these stellar properties, they can begin to deal with the theoretical side of stellar evolution. The data gathered from their observations provides the raw materials for identifying patterns and, ultimately, developing explanations for those patterns.

From Data into Insight: The HR Diagram

You can't explain the patterns in nature without finding them first. All good science begins with buckets of data, and once the buckets have been gathered, the work of sifting through all of them begins. Facing down those difficult hours of labor in search of patterns in the properties of stars was exactly the task given to the indefatigable Annie Jump Cannon back in 1884 (**Figure 11.12**).

While working at the Harvard College Observatory, Cannon was given the task of sorting through absorption spectra from hundreds of thousands of stars. Cannon used the presence and strength of different absorption lines to sort her stellar spectra. In the end, she managed to combine them all into a **spectral classification** of seven categories denoted by the letters O, B, A, F, G, K, and M.

In Cannon's scheme, O stars, for example, have strong Balmer-series hydrogen lines, as well as prominent helium, oxygen, nitrogen, and silicon lines. On the other end of the classification, M stars show no hydrogen lines but do have lines from many molecules (most notably, titanium oxide). Eventually it was recognized that Cannon's scheme made connections between stellar temperature and the underlying blackbody emission. The bluish O stars are the hottest stars in the classification, and the redder M stars are the coolest (see **Figure 11.13** and **Table 11.3**).

FIGURE 11.12
Annie Jump Cannon As an employee of Harvard College Observatory, Annie Jump Cannon (1863–1941) created a powerful system of classifying stellar spectra that is still in use today. The spectral classes are O, B, A, F, G, K, and M.

spectral classification
A fundamental grouping of stellar types, based on observed spectral lines, that specifies seven categories denoted by the letters O, B, A, F, G, K, and M.

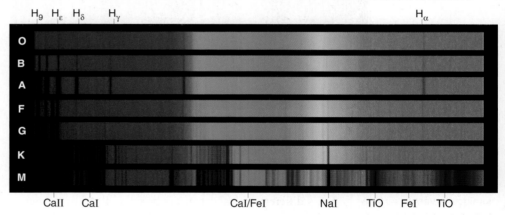

FIGURE 11.13 Stellar Spectra
Because each element or molecule has a unique set of absorption lines, the spectrum of a star reveals its chemical composition (as well as its temperature via Wien's law). A star's spectral class can be determined by the "fingerprints" of these lines. Here we see sample spectra (in the visible waveband) for stars from each spectral type. Note that the Sun is a class G star. Spectral lines of hydrogen are identified at the top of the figure and lines from other elements and ions are identified at the bottom. (Roman numbers indicate ionization level. Thus "CaII" is once ionized calcium.)

TABLE 11.3 ●●● Properties of Spectral Types					
Spectral Type	Typical Mass (M_{Sun})	Typical Radius (R_{Sun})	Typical Temperature (K)	Peak Wavelength (nm)	Main Spectral Lines
O	60	12	42,000	69 (UV)	Helium II
B	5.9	3.9	15,200	191 (UV)	Helium I, hydrogen
A	2.0	1.7	8,180	355	Hydrogen, calcium II
F	1.4	1.3	6,650	436	Calcium II, hydrogen, metals
G	0.92	0.92	5,560	522	Calcium II, iron, other metals
K	0.67	0.72	4,410	658	Metals, CH (methylidyne radical), CN (cyanogen radical)
M	0.21	0.27	3,170	915 (IR)	Titanium dioxide

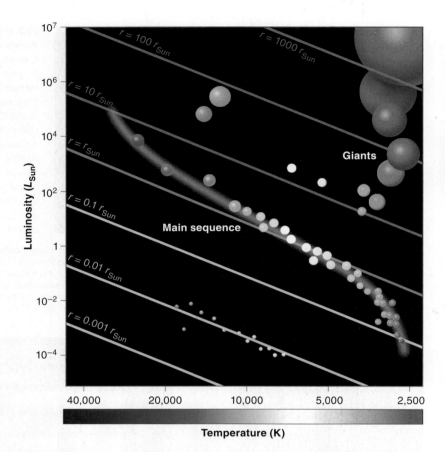

A Ejnar Hertzsprung   **B Henry Norris Russell**

FIGURE 11.15 Hertzsprung and Russell
In 1910, the Danish astronomer (A) Ejnar Hertzsprung (1873–1967) and the American astronomer (B) Henry Norris Russell (1877–1957) independently plotted the positions of stars on a graph of luminosity versus temperature (though they used magnitude and stellar classification). Their work led to a powerful tool for understanding patterns in stellar evolution: the HR diagram.

FIGURE 11.14 HR Diagram
Plotting many stars at many phases in their lifetimes on the HR diagram, which shows luminosity versus surface temperature, reveals that stars tend to "collect" in different regions. For example most stars will typically be found on the main sequence (the diagonal band running from upper left to lower right). Insight into why stars cluster in different regions made the HR diagram an essential tool in understanding the life cycles of stars. Note how radii stars can be found from their location on the HR diagram.

main sequence The diagonal band on the HR diagram representing core hydrogen-burning stars. High-mass stars appear on the upper left; low-mass stars, on the lower right.

HR diagram A plot of luminosity versus temperature (or spectral class) for a population of stars.

Cannon's spectral classification was a critical first step in understanding the story hidden in stellar properties. She provided a way of arranging stars into different logical categories based on those properties. The next step was to look for deeper patterns that would emerge from sorting stars into the categories. That task fell to Ejnar Hertzsprung and Henry Norris Russell, who both (separately) decided to relate Cannon's spectral classification to measurements of stellar luminosity. In 1910, each of these scientists made a graph in which spectral type (and therefore stellar temperature) was plotted on the *x*-axis and stellar luminosity was plotted on the *y*-axis (**Figure 11.14**). What these diagrams revealed changed astronomy and our understanding of stars forever.

Hertzsprung and Russell (**Figure 11.15**) were, essentially, looking for a correlation between luminosity and temperature in stars. A correlation means that as one property (such as *T*) changes, another property (such as *L*) changes too, in a regular and understandable way. The number of times you go to the ATM machine in a month and the amount of cash left in your bank account at the end of that month are correlated. In contrast, the number of times you go to the ATM machine in a month and the amount of rainfall in Brazil during that month are unlikely to be

correlated. Perfectly correlated data might look like a straight line (an increase in quantity X is perfectly correlated with an increase or decrease in quantity Y). Uncorrelated data would show points uniformly spread over the entire plot, as if someone had fired a shotgun at the page (scientists call uncorrelated data a *shotgun pattern*).

Before Hertzsprung and Russell made their graphs, it wasn't at all clear whether any correlation between stellar properties like temperature and luminosity would exist. "But they did see a correlation," says Jim Kaler. "Once they plotted spectral type versus luminosity, it was clear that the two quantities were strongly related for stars." Taking hundreds of thousands of stellar data points and plotting them on the *L*-versus-*T* curve, Hertzsprung and Russell saw the vast majority of stars falling onto a well-defined band that ran from high temperature and high luminosity to low temperature and low luminosity, which astronomers call the **main sequence**. Hertzsprung and Russell had no way of knowing why most stars "lived" on the main sequence in their graph, but they had clearly discovered a correlation—a pattern that later astronomers with a better understanding of stellar physics would have to explain.

Though the main sequence was the home for most stars on this graph, now called the **HR diagram**, significant numbers of stars still existed in different locations of the plot. Using the relationship of radius, temperature, and luminosity, astronomers were able to categorize these stars by their physical size. In particular stars above the main sequence luminous, indicating large radii. These stars were called *giants*.

With the HR diagram in hand, stellar astrophysicists now had a pattern that could help them create working models of stellar evolution. What had begun as a giant stack of spectra had been condensed down to a few simple facts (main sequence, giants, etc.) expressed as a single graph that demanded explanation.

SECTION SUMMARY
From Data into Insight: The HR Diagram

- Astronomers, notably Annie Jump Cannon, studied the spectra from many stars and categorized them into seven spectral classes, which also represented temperature classes.
- Hertzsprung and Russell independently correlated the spectral classes (representing temperature) with luminosity.
- The resulting graph, known as the HR diagram, showed the majority of stars falling in a narrow diagonal band called the main sequence.

Making Model Stars

The HR diagram revealed the presence of distinct patterns, such as the main sequence, in the properties of stars. To understand the meaning of these patterns, however, astronomers began developing theoretical models of stars. The goal of this theory was to take the data they gathered from their observations and construct mathematical models of stars at any point in evolution. If these theoretical models capture the reality of stellar evolution, they should provide insight into why a pattern like the main sequence appears in the HR diagram.

One basic fact about stars that is crucial for understanding their evolution is that they are forever at war with their own gravity, and it is a war they cannot win (**Figure 11.16**). "The whole history of a star is a battle between the inward pull of gravity and the outward push of gas pressure," explains Kaler. There is a lot of mass in a star, and all that mass produces a lot of gravitational force. "The gravity keeps trying to squeeze the star down to nothing," continues Kaler. "But it's the energy released by thermonuclear fusion deep within the star that provides the pressure to push back against gravity. Ultimately, however, gravity is going to win."

From the moment a star begins burning nuclear fuel, its days are numbered. The quantity of nuclear fuel in a star is limited (and set at birth), even though the star may switch from fusing one kind of element to another as it ages. "Eventually," says Kaler, "all the elements that can be fused are gone. The nuclear fuel runs out, gravity wins, and the star dies."

Tracking the history of a star as it fights this lifelong battle between gravity and fusion is the job of a theoretical astrophysicist. The tools required are a set of physics equations that describe the structure of a star and a computer to solve those equations. "We know the basic laws of physics," says Kaler, explaining how a mathematical model of a star is constructed. "What you need to do is take those equations and apply them to the star. One of the equations describes the balance between the inward pull of gravity and outward push of gas pressure." This crucial balance is called hydrostatic equilibrium. As the star evolves, its structure will change to keep that equilibrium. "We also have equations that describe the rates of thermonuclear energy generation and the way radiation generated through fusion gets transferred outward toward the surface of the star. Altogether, there are five different equations you have to solve at the same time. It's complicated, but in the end you've built a mathematical model of a star that should behave like a real one."

Specifically, a stellar model is a description of how the properties of a star—such as density, temperature, chemical abundance, and degree of ionization—change with distance from its center. The model also calculates what the star's surface temperature and luminosity would be if astronomers could observe it on the sky.

But it's important to remember that models are not reality. The mathematical equations solved on a computer are never perfect. Some effects are neglected because, for example, the computer may not be powerful enough to take them into account. "The first models of stars were calculated by hand," says Kaler. "With the early computers, things got better. But still it could take a week or more to finish a single model."

The models that took Kaler a week to compute back in his graduate student days might take only seconds on a smartphone today. Still, the state of the art is always moving on. "Nowadays," says Kaler, "the models are extremely sophisticated. But we're still stopped by processes we really don't understand well." Even with the most powerful modern supercomputers, scientists cannot produce a fully three-dimensional model of a living star from core to surface that accounts for all detailed physical processes. Still, these models have revealed a great deal about the different types of stars and the deeper patterns that underlie their positions on the main sequence.

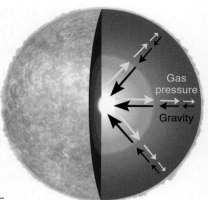

FIGURE 11.16
Opposing Forces in Stars
Stars are stable when they are in hydrostatic equilibrium, meaning that the inward force of gravity (due to the star's mass) is balanced by the outward force of gas pressure. The pressure is maintained by energy released by fusion reactions in the star's core.

SECTION SUMMARY
Making Model Stars

- By knowing the equations for the relevant physical processes, astronomers can build mathematical models of a star that describes its structure from core to surface.

From Insight to Theory and Back: The Meaning of the HR Diagram and Main Sequence

"If you took a random bunch of stars," says Harriet Dinerstein, a spectroscopist from the University of Texas, "you might guess that they would have a an arbitrary combination of properties like brightness, radius, and temperature. After all, why not?" Why shouldn't there be just as many very large, very cool stars as there are very small, very hot stars, as well as everything in between? "Well," she explains, "it turns out that just is not how nature set things up."

The HR diagram shows astronomers that some properties of stars, like luminosity and temperature, correlate very well. "The next question," says Dinerstein, "is what those correlations are trying to tell us. What, in other words, does the main sequence tell us about stars and their histories?"

The most important thing to remember about stars is that they evolve very slowly, taking millions or billions of years to change. In general, we don't have the opportunity to watch any single star evolve. We have only snapshots of many stars that are at different points in their life histories. The assumption that we're catching stars at different points in their trek from birth to death enables astronomers to turn the HR diagram into a map of stellar evolution.

"An analogy with people can be helpful," says Dinerstein. "Imagine you went to the mall and recorded people's heights and weights. Then you made a plot of your data with height on one axis and weight on the other. Most people's height and weight go roughly together with some variation, but generally increasing or decreasing together," says Dinerstein. "That means your plot would show them to be correlated. Tall people would tend to be heavier than shorter people."

But we can also ask what would happen if you repeated the same experiment at a preschool. In that case, you would see only short, low-weight data points. And if you went to a seniors' care seniors-care home, you would tend to see midheight, low-weight data. The point here is that age matters (**Figure 11.17**).

The HR diagram is powerful for a similar reason. By taking snapshots of many stars and plotting them on the HR diagram, we get a clear picture of the patterns associated with evolution at all points in stellar evolution. Where they spend most of their lives, in terms of temperature and luminosity, is where we expect to see most of them show up on the diagram. As it turns out, most stars appear on the main sequence. But the main sequence is most definitely not a sequence in evolution. Instead, it's a snapshot showing where most stars spend most of their lives. What the main sequence shows are stars of many different masses appearing at their own positions on the sequence—high masses toward the top and low masses toward the bottom. Note that the Sun is considered to fall in the low- or intermediate-mass stellar range.

"Plot the Sun on the HR diagram and it ends up right on the main sequence," says Kaler. From their stellar evolution models, astronomers came to understand that the Sun is a middle-aged star that uses hydrogen fusion to support itself against gravity. As astronomers became more adept at studying the structure of stars through their theoretical models, it became clear that all the stars on the main sequence are also hydrogen burners. Because hydrogen is the fusion fuel every star will rely on for most of its life, most of the stars we see are on the main sequence. (Similarly, in the height-versus-weight analogy, most people you would interview at the mall would be between the ages of 18 and 65.)

But why the does the main sequence appear as an extended band from hot and bright stars down to cool and dim ones? Why don't all stars burning hydrogen have the same luminosity and temperature and so appear in the same place on the HR diagram? The answer to that question came only as scientists learned more about the physics of nuclear fusion as applied to their model stars. "The mass of the star controls its luminosity," says Kaler, "and that is all about gravitational compression." The harder gravity squeezes down on the plasma at the star's core, the higher the temperature rises and the faster the star burns its nuclear fuel. "More fusion means more radiation," says Kaler, "and that means the 'sequence' in the main sequence is really about mass." Very luminous stars at the upper-left edge of the main sequence are also very massive hydrogen-burning stars. Stars with very low luminosity, at the

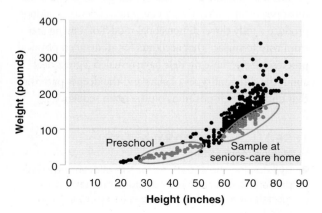

FIGURE 11.17

Height-Weight Correlation Like the HR diagram, a plot of height and weight for a group of humans shows distinct correlations, meaning that the points representing individuals lie along a curve rather than being randomly distributed across the plot in a shotgun pattern. Distinct areas on the plot would also come to be populated if the sample of people represented extremes in age (babies or old people).

interactive
The Age/Mass/ Luminosity Relationship

lower-right edge, are hydrogen-burning stars with very low mass (**Figure 11.18**).

The link between the main sequence and age goes beyond simply knowing that a star is in its hydrogen-burning phase. As astrophysicists learned more about stellar nuclear fusion, they found a simple relationship between the life expectancy of a star and its mass. "The high temperatures at the core control the thermonuclear fusion rate," says Kaler, "and that controls the lifetime of stars. So as you go up the main sequence, the stars get brighter, and their lifetimes get shorter." Astronomers have found a simple formula, the **life expectancy–mass relationship**, that expresses this relationship for stars (**Table 11.4**). **Going Further 11.2** (p. 290) explores the consequences of the life expectancy–mass relationship in more detail, but it's worth noting here that a star of 100 solar masses (M_{Sun}) lives for only 100,000 years (100,000 times shorter than the Sun), while a star of 0.08 M_{Sun} lives for 5.5 trillion years (more than 500 times longer than the Sun). That is quite a spread.

SECTION SUMMARY
From Insight to Theory and Back: The Meaning of the HR Diagram and Main Sequence

- All stars on the main sequence are hydrogen burners.
- Very massive, very luminous stars burn much hotter and thus use up their nuclear fuel much faster than less massive stars do, giving them shorter lifetimes.

11.4
The March of Time: Stellar Evolutionary Tracks

The life expectancy–mass relationship is just one of the fruits of using stellar evolution models to understand a pattern, like the main sequence, seen in the HR diagram. But other patterns in the HR diagram require explanations as well. For example, how do astronomers account for stars off the main sequence? How can the HR diagram be used to understand all of stellar evolution and not just stars' long middle age?

Modeling Stellar Evolution

Theoretical astrophysicists are a lot like kids with Lego robot kits. They love to take the models of things they study, turn them on, and then see what happens. We've already seen how these models gave astronomers insights that helped them understand the main

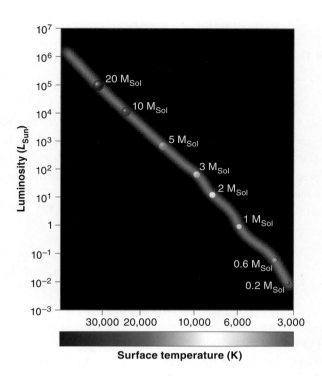

FIGURE 11.18
Mass and the HR Diagram
Stars spend most of their lives on the main sequence in the HR diagram as they fuse hydrogen into helium in their cores. The "sequence" part of the main sequence refers to stellar mass. High-mass hydrogen-fusing stars are at the top of the main sequence; they are bright and hot because their weight makes them burn hydrogen rapidly. Low-mass hydrogen-fusing stars are at the bottom of the main sequence.

sequence, but they can go much further than that. The comparison of the models and real data also helps astronomers understand, for example, how stars originally on the main sequence end up in the giant region.

The best way to answer these kinds of questions is to take a model of a single star with specified starting conditions (mass and chemical composition) and allow that star to evolve. Astronomers then watch the changes in that star's structure and properties, such as luminosity and temperature, as the star consumes its nuclear fuel over time. They could also follow how the star's surface properties L_* and T_* change over time. In this way, astronomers can create a **stellar evolutionary track** on the HR diagram for the star. The track would be the sequence of locations (L_*, T_*) that the star occupies on the HR diagram over its entire lifetime.

"Stellar evolutionary tracks come from the success of stellar models," says Jim Kaler. "To create a track, you build a model of a star on a computer starting with its initial composition: about 90 percent hydrogen and 10 percent helium in the core [**Figure 11.19**]. Then you allow it to burn a little bit of its hydrogen in steps, maybe going down to 89 percent hydrogen, 11 percent helium. Now you make a new model with these new core conditions. This new model gives you a different temperature and luminosity at the surface. As you keep calculating successive models with successively less core hydrogen, you can watch the star move

TABLE 11.4 • • • **Star Masses and Life Expectancies**	
Mass (M_{Sun})	● Life Expectancy (yr)
0.08	5.5 trillion
0.25	300 billion
0.5	60 billion
0.75	20 billion
1	10 billion
2	2 billion
4	300 million
8	60 million
20	6 million
100	100,000

life expectancy–mass relationship The mathematical connection between the life expectancy and mass of a star: the greater the star's mass, the shorter its life expectancy.

stellar evolutionary track A path on the HR diagram representing the evolution of a star through time.

GOING FURTHER 11.2 THE LANGUAGE OF THE COSMOS
The Life Expectancy–Mass Relationship

A swimmer at the bottom of a deep pool feels the pressure of the all the water above her. In the same way, the hydrogen plasma at the core of a star feels the gravitational force—the weight—of all the matter in the layers above it. The pressure of the overlying material heats up the core plasma to fusion temperatures, so that hydrogen burns into helium. The more massive the star, the higher the core temperature and the faster the store of hydrogen is fused into helium. This is the story behind the life expectancy–mass relationship:

$$t_{ms} = 10^{10} \text{ yr} \left(\frac{M_*}{M_{Sun}} \right)^{-2.5}$$

Let's play with this formula a bit to see if we can understand what it tells astronomers. First let's assume the star has a mass equal to the Sun's: $M_* = 1\ M_{Sun}$. If we plug this value into the formula, we find that the amount of time on the main sequence, or t_{ms}, is 10^{10} years. This tells us that a star like the Sun will spend 10^{10}, or 10 billion, years on the main sequence before it uses up its hydrogen and

turns onto the red giant branch. Now let's use the formula to find the main-sequence lifetime of a star that is 10 times more massive than the Sun and a star that is 10 times less massive than the Sun.

For the star with $M = 10\ M_{Sun}$, we get

$$t_{ms} = 10^{10} \text{ yr} \left(\frac{10\ M_{Sun}}{M_{Sun}} \right)^{-2.5} = 10^{10} \text{yr} \ (10)^{-2.5} = 3.2 \times 10^{7} \text{ yr}$$

This somewhat massive star lasts only 30 million years, which is less than the time elapsed since the age of the dinosaurs. For the star with $M = 0.1\ M_{Sun}$, we get

$$t_{ms} = 10^{10} \text{ yr} \left(\frac{0.1\ M_{Sun}}{M_{Sun}} \right)^{-2.5} = 10^{10} \text{yr} \ (0.1)^{-2.5} = 3.2 \times 10^{12} \text{ yr}$$

This star will last 3 trillion years, longer than the current age of the Universe. Thus we conclude that low-mass stars live for a very long time.

---

along the HR diagram. Then you simply connect the dots between the models."

Astronomers have built enough evolutionary tracks to lay out the basic story of stars. They have built evolutionary tracks for everything from stars of extremely low mass ($M \approx 0.08\ M_{Sun}$) to stars of extremely high mass ($M \approx 100\ M_{Sun}$). "The evolutionary tracks are our abstract way of 'keeping track' of how the changes inside the star look from the outside," says Harriet Dinerstein. "The star may be 'moving' its

position on the HR diagram, but it's not actually going anywhere. It's the properties of the star that are moving, not the star itself."

SECTION SUMMARY
Modeling Stellar Evolution
- Astronomers can create a stellar evolutionary track by modeling the conditions inside the star as it continually converts hydrogen to helium.

Exit on the Right:
Leaving the Main Sequence

In the next two chapters we will explore the evolution of stars, starting with their birth from interstellar clouds of gas all the way to their spectacular deaths. In all cases, the evolutionary tracks computed via the detailed stellar models enable us to translate what we observe in different stellar populations into an understanding of the physical processes guiding the stars' changes.

For now, however, we need to focus on one important change: the stars' departure from the main sequence. We already know that stars on the main sequence are hydrogen burners: they fuse hydrogen nuclei into helium nuclei and release energy to support

FIGURE 11.19
Stellar Model for a 3-M_{Sun} Star
Astronomers use stellar models to predict the evolution of stars over time. As hydrogen fusion in the core proceeds, the models describe how the relative fractions of hydrogen and helium change. Here we see a star that is 3 times more massive than the sun at three points in evolution. By 3 billion years, the core is almost entirely composed of helium.

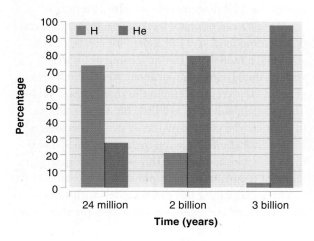

themselves against their own weight in the process. But recall that the process cannot go on forever. There is only so much hydrogen in the core, and once it has been mostly replaced with helium, the star runs into trouble. The temperatures in the core are not high enough to start helium fusion (slamming helium nuclei together to make either carbon or oxygen). Therefore, by the time the core has become mostly helium, the star is not producing enough energy to maintain its current configuration. Something has to give.

We will explore post-main-sequence evolution in detail in Chapter 13, but for our purposes here we need to notice only one thing: that there *is* a post-main-sequence evolution. In other words, when a star stops getting its energy from core hydrogen burning, the crush of gravity forces a rearrangement of the star's distribution of density, temperature, and so forth. As a consequence, the surface properties of temperature and luminosity change, and the star leaves the main sequence (**Figure 11.20**). It's worth noting that stars barely change their mass during their time on the main sequence. While a star like the Sun loses mass in a wind, the mass loss rate in the wind on the main sequence is so small that the stellar mass hardly changes at all. In the later stages of life, however, as the star moves off the main sequence, these winds become much more powerful.

What is most remarkable about the evolutionary tracks provided by computer models is that they make very specific predictions about how long it takes for a star to burn up its hydrogen in the core and turn off the main sequence. That timescale—the main-sequence turnoff time—depends only on the star's mass, which determines the pressures, densities, and temperatures in the core. These factors, in turn, determine the star's rate of nuclear burning.

The one-to-one relationship between the mass a star begins with and the time it takes for it to leave the main sequence turns out to be of no small consequence. It is, in fact, the key to validating much of the entire stellar evolution model-building enterprise. In other words, it's how astronomers know they've gotten their models right. To elucidate that point, we need to travel to the edge of the galaxy and meet some of astronomy's most beautiful players: the globular clusters.

SECTION SUMMARY
Exit on the Right: Leaving the Main Sequence
- When stars end their hydrogen core burning, they leave the main sequence as they continue to evolve.
- The main-sequence turnoff time for a star depends on its mass.

Harriet Dinerstein

"I grew up in the middle of Manhattan," laughs Harriet Dinerstein. "I barely even saw the night sky. I did look through binoculars at the Moon when I was a senior in high school." How did the young Dinerstein, living in the middle of a large city, ever catch the astronomy bug? "The Hayden Planetarium!" she says. "Every time there was a new show at the Hayden, my mom would take us."

The interest that began at the planetarium deepened in the library. "I remember reading book after book by Isaac Asimov." Dinerstein says. "One chapter he wrote about the Solar System's weirdest objects—like how Uranus rotates on its side. Then, when I was 14 came the *Apollo* Moon landing."

In college, Dinerstein was one of only a handful of female students determined to find a career in science. "I went to Yale when it had barely turned coed," says Dinerstein. "I plunged right into physics, chemistry, linear algebra, and astronomy my freshman year. I guess I never looked back."

Dinerstein eventually ended up at the University of Texas at Austin, where she studies the chemical composition of gases thrown off stars at the ends of their lives, to see how fusion transformed them. "Stars are always cooking up new mixtures of atoms through fusion. Then they spew those newly formed elements back into space," says Dinerstein. "What I see coming off the stars now will eventually end up in the next generation of stars or even some future planet."

Throughout her career, the possibility of discovery has always kept Dinerstein tied to astronomy. "I love the excitement of identifying things that weren't known before," she says. "Two emission lines were discovered long ago and remained unidentified for 25 years, until I recognized that they came from the elements selenium and krypton. It was really a thrill to understand something that had never been understood before."

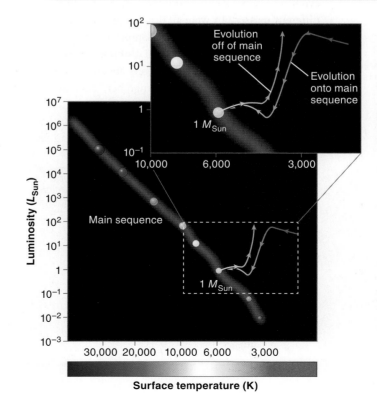

FIGURE 11.20
Evolutionary Track of a Sun-like Star The entire lifetime of a star can be traced through its changes in luminosity and temperature. This HR diagram shows the path that a star like the Sun traces from its early life as protostar, onto the main sequence where core hydrogen burning begins, and off the main sequence (almost 10 billion years later) once hydrogen burning in the core has ended.

A

FIGURE 11.21
Globular Cluster M80
(A) The globular cluster M80, shown in this Hubble Space Telescope image, is a gravitationally bound collection of more than 100,000 stars. (B) The HR diagram of a globular cluster allows astronomers to test their models of stellar evolution. The known hydrogen-burning lifetimes of stars at the cluster turnoff point indicate the age of the cluster. Stars with greater masses have already evolved off the main sequence.

globular cluster
A gravitationally bound cluster of 100,000 or more stars all born at the same time.

FIGURE 11.22
Metallicity of Globular Clusters
As successive generations of stars evolve, the interstellar medium from which new stars are formed grows increasingly populated with heavier elements (metals), enriched by the death of the earlier stars. Thus, one method for estimating how old star clusters are is to determine their metallicity and use it as a basis for extrapolating their age.

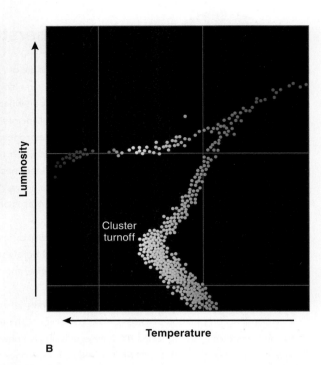

B

Testing the Tracks: Globular-Cluster HR Diagrams

"If we are doing science," says Kaler, "we had better be able to test our models." Since astronomers cannot watch individual stars evolve, they need to test evolutionary tracks on a group of stars that have different masses but were all born at the same time. That is exactly the stellar evolution playground astronomers find in the beautiful **globular clusters** (**Figure 11.21A**). "Globular clusters are large collections of stars that were all born together," explains

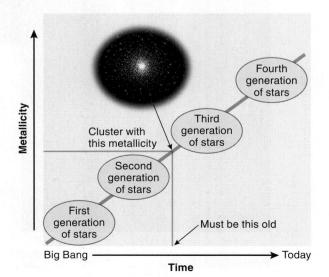

Kaler. "A typical globular cluster will contain tens of thousands of stars. It is really a magnificent sight."

Not only were all the stars in each globular cluster born at about the same cosmic moment (billions of years ago when the galaxy was just forming), but they also contain a wide range of stellar masses. Having so many stars with so many different masses all born together makes globular clusters much better for calibrating the evolutionary tracks than are other groupings of stars, such as those that are called open clusters associated with newborn stars within the disk of the galaxy. By making an HR diagram with a cluster's stars, astronomers can see the full range of stellar evolution at work (**Figure 11.21B**). "Having all those stars with different initial masses when the cluster is born is really important," says Kaler, "because it means when you look at a cluster now, you are seeing stars at very different stages of their evolution. The highest-mass stars have used up all their fuel and will be completely gone from the main sequence. The lowest-mass stars will still be on the main sequence. Somewhere in between there will be intermediate-mass stars that have just used up their core hydrogen and are starting to move off the main sequence."

Astronomers have a variety of ways to date the birth of a globular cluster and estimate its current age. The most important is to use the total amount of elements heavier than hydrogen and helium in the cluster's stars. These elements are, for the most part, formed through fusion in stellar evolution. Astronomers refer to all elements heavier than helium as *metals* and refer to the metallicity of a star as its fraction of these stellar fusion-processed elements. Since no new stars are forming in a globular cluster now, the amount of processed heavier elements in the cluster comes from the time of its formation. To make these measurements, astronomers must determine the abundance of elements in the stars making up the clusters. As we saw in Section 11.2, they do so by carefully measuring the spectra of many stars and using absorption lines to determine the elemental abundances, which yield a measure of the cluster's overall metallicity.

Because the metallicity of the Universe has been increasing steadily since the Big Bang (by means of fusion in the core of successive generations of stars), the metal content of the cluster can be used to determine when it formed (**Figure 11.22**).

After dating a cluster, astronomers create an HR diagram using only the cluster's stars. Once they see how the cluster is arrayed on the HR diagram, they can see at what mass the main sequence ends. Younger globular clusters should show a main sequence that still includes higher-mass stars. An older globular cluster should have lost most of its massive stars and show only intermediate- and low-mass stars still on the main sequence. In this way, astronomers get an observational "globular-cluster turnoff mass," which yields an age for the entire cluster. Now astronomers are in a position to test their stellar evolution models. Here's how they do it.

Suppose a measurement of a cluster's average metallicity indicates that the cluster is 8 billion years old, having formed nearly 6 billion years after the Big Bang. The metallicity of this cluster's stars gives an independent measure of the cluster's age. Astronomers can now run stellar evolution models for many stars of different masses, each for 8 billion years. Combining all these models allows them to make a theoretical HR diagram for an 8-billion-year-old cluster of stars.

Comparing the 8-billion-year-old stellar evolution models with the real 8-billion-year-old cluster HR diagram (whose age was derived from the metallicity measurements) enables astronomers to compare the theoretical and observational globular-cluster turnoff masses. If they match, astronomers can be confident that their models do a good job of predicting reality.

Using the comparisons of real cluster HR diagrams with predictions from evolutionary tracks in this way has enabled astronomers to steadily improve their stellar models to the point where now the match between prediction and data is remarkably close. "In every case," says Kaler, "you can see where the main sequence truncates and the stars go off the main sequence. It's a beautiful thing that we have learned to use pretty effectively."

SECTION SUMMARY
Testing the Tracks: Globular-Cluster HR Diagrams
- Star clusters known as globular clusters provide the means for astronomers to test their models of stellar evolution, since the stars were born at the same time with a variety of masses.

From Life Stories to Birth and Death

Now that we've seen how astronomers use measurements of stellar properties to understand the evolution of stars, we can turn to the more difficult question of stellar evolution's extremes. As strange as it may sound, at this point in human history we understand main-sequence stars pretty well. For all its power, a hydrogen-burning main-sequence star is a fairly stable object whose most important secrets have now been revealed through 100 years of scientific inquiry.

The birth and death of stars, however, are another story. While stars change slowly throughout the bulk of their lives, during their birth and final stages, evolution can be very fast and very violent. The speed and chaos associated with stellar birth and death have made them subjects whose investigation requires more exotic physics and more powerful tools. In the next chapter we begin at the beginning: the formation of stars.

●→ chapter summary

11.1 The Life of the Stars
Stars evolve over long timescales. Key observational properties are vital to understanding stars: distance, luminosity, temperature, radius, mass, composition. Astronomers measure all of these properties using the information that reaches Earth in the form of light. Information about star velocities helps astronomers understand stellar evolution in a galactic context.

11.2 Measuring Stars
Astronomers measure distances to nearby stars through parallax. For more distant stars, they use variable stars such as Cepheids as standard candles because of their period-luminosity relationship. Astronomers determine the surface temperatures of stars by applying Wien's law to observations of peak wavelength. Specific spectral lines indicate the composition of a star, while the width of those lines can translate into mass measurements for some stars. The motions of binary stars, which account for 50 percent of the stars in our galaxy, also enable astronomers to derive the stars' masses.

11.3 From Observations to Explanations
The identification of stellar spectral classes was an important step in understanding stars and led to the discovery that the luminosity and temperature of stars were highly correlated. In the HR diagram, most stars fall on a diagonal band called the main sequence during the majority of their lives, when they're fusing hydrogen into helium in their cores. The HR diagram led to an understanding of the life cycles of stars, since it shows stars of different masses in different evolutionary stages. To understand the structure and evolution of stars, astronomers use mathematical models that describe the opposing forces of gravity and gas pressure from nuclear fusion. High-mass stars burn their fusion fuel faster, as described

by the life expectancy–mass relationship. Low-mass stars smaller than the Sun live many billions of years. Intermediate-mass stars live 500 million to 15 billion years. The most massive stars live only millions of years.

11.4 The March of Time: Stellar Evolutionary Tracks

Astrophysicists have developed sophisticated computer models that help them understand how stars of various masses evolve through their lifetimes. The end of hydrogen burning in a star leads to changes in temperature, pressure, and density that take the star off the main sequence of the HR diagram. Because globular clusters contain many stars of different mass that are all born at the same time, they enable astronomers to test their models of stellar evolution. By creating an HR diagram of a globular cluster, astronomers can assess its age from the point where stars turn off the main sequence.

●→ questions and problems

Narrow It Down: Multiple-Choice Questions

1. What is the primary source of information needed to make stellar measurements?
    a. cosmic rays
    b. magnetic fields
    c. electric charge
    d. light
    e. atomic structure

2. True/False: A standard candle is useful for determining distance but not orbital velocity.

3. What is the definition of a standard candle?
    a. an object whose luminosity remains the same throughout its life
    b. an object whose age is known
    c. an object whose luminosity is equal to the Sun's
    d. an object whose luminosity is known
    e. a star within a cluster

4. Wien's law tells us that
    a. wavelength is related to frequency.
    b. blackbodies are essentially black.
    c. blackbodies radiate light at all wavelengths and absorb light at all wavelengths.
    d. temperature is regulated by spectral class and luminosity.
    e. in blackbodies, temperature and peak radiation wavelength (color) are related.

5. Two neighboring stars are seen with significant differences in composition. What is *not* likely to be true of these stars?
    a. They were born together.
    b. They are of different spectral types.
    c. One or both migrated from other locations.
    d. They are of different ages.
    e. They are from different generations of stars.

6. Which definition of proper motion is correct?
    a. motion due to the rotation of the galaxy
    b. observed motion of an object against very distant background objects
    c. motion that does not follow Newton's laws
    d. motion around the Sun
    e. motion due to the stretching of space-time

7. Astronomers must take many factors into account when modeling individual stars. Which of the following is *not* a factor?
    a. limitations of computers
    b. star mass
    c. star composition
    d. distance between stars
    e. stellar energy generation

8. Choose the correct list of spectral classes in ascending order of temperature.
    a. O B A F G K M
    b. A B F G K M O
    c. M K G F A B O
    d. O A G M B F K
    e. M G A O K F B

9. A star on the upper right of the HR diagram is
    a. a giant.
    b. cool and small.
    c. hot and small.
    d. a dwarf.
    e. hot and large.

10. What does the main sequence show?
    a. the entire life cycle of a star
    b. a population of $1\text{-}M_{Sun}$ stars
    c. the distribution of core hydrogen-fusing stars by mass
    d. the middle third of a star's lifetime
    e. the series of stages in a star's life

11. A star's position on the main sequence does *not* tell us
    a. its chemical composition.
    b. its mass.
    c. its luminosity.
    d. its temperature.
    e. its exact age.

12. The more massive the star,
    a. the longer it remains on the HR diagram.
    b. the lower it is on the HR diagram.
    c. the longer it stays on the main sequence.
    d. the hotter it is as a main-sequence star.
    e. the higher its apparent magnitude is.

13. A star's changing position on the HR diagram indicates
    a. its motion within the galaxy.
    b. its evolutionary track.
    c. how its mass changes as it ages.
    d. its changes as a live star.
    e. its degree of parallax.

14. Which of the following is true for low-mass stars? Choose all that apply.
    a. They may be spectral class O.
    b. They are not found on the main sequence.
    c. They are hotter than intermediate-mass stars.
    d. Their mass must be at least 8 percent of the mass of the Sun.
    e. Their life expectancies are in the billions or trillions of years.

15. What does the HR diagram of a cluster *not* tell astronomers?
    a. the age of the cluster
    b. the highest mass of cluster stars still on the main sequence
    c. the rotation of the cluster
    d. which stars are similar to the Sun in size and mass
    e. the highest luminosity among the remaining stars on the main sequence

16. Binary stars detectable by direct observation of the proper motion are called
    a. astrometric binaries.
    b. spectroscopic binaries.
    c. eclipsing binaries.
    d. visual binaries.
    e. a symbiotic binary system.

17. Which of the following is true of stars in binary systems?
    a. They may evolve without interfering with each other.
    b. They will eventually merge into a single star.
    c. They are rare.
    d. They may not be gravitationally related.
    e. They almost always have identical masses.

18. Choose the statement about variable stars that is *not* true.
    a. They have unstable outer layers.
    b. They change temperature.
    c. They change luminosity.
    d. They may be useful as standard candles.
    e. Their luminosity may be continually increasing.

19. How would parallax measurements of stars made from Jupiter differ from those made from Earth?
    a. The stars would appear to be at greater distances.
    b. The stars would appear to be closer.
    c. Parallax for a given star would be greater.
    d. Parallax for a given star would be less.
    e. All stars would appear to have greater radii.

20. True/False: A star with a mass 10 times that of the Sun would have a life span one-tenth as long.

To the Point: Qualitative and Discussion Questions

21. Define the following terms: *parallax, energy transport, luminosity, hydrostatic equilibrium, correlation*.

22. What are the six properties vital to studying the life stories of stars?

23. What is a standard candle, and how is it used to determine distance?

24. What information in a star's spectrum indicates the star's surface temperature?

25. What information in a star's spectrum indicates its chemical composition?

26. Describe the two approaches by which astronomers determine the mass of a star.

27. Doppler shift can provide information on which characteristics of a star's motion through space?

28. A star is observed with a repeating pattern of redshifted spectral lines followed by blueshifted lines. What does this pattern tell you about the star?

29. Explain what the main sequence on the HR diagram indicates. What is the difference between a star that falls higher and to the left on the main sequence and a star that falls lower and to the right?

30. What is going on in the core of a star that is on the main sequence?

31. How are globular clusters useful to astronomers studying stellar evolution?

32. What is meant by *cluster turnoff mass*?

33. Name and describe three categories of binary stars.

34. Which variable stars have helped astronomers understand the scope of the Universe, and how have they done so?

35. Describe how an astronomer develops a stellar model.

Going Further: Quantitative Questions

36. You observe that a star's wavelength of maximum intensity is a very reddish 770 nm. What temperature is the surface of that star, in kelvins, and how does it compare to the Sun?

37. The surface of a star has a temperature of 6,500 K, a bit hotter than the Sun. What is its wavelength of maximum intensity, in nanometers?

38. Find the expected main-sequence lifetime of a 4-M_{Sun} star, in years.

39. What is the ratio of the main-sequence lifetime of a 2.5-M_{Sun} star to that of the Sun?

40. A star has one-third the mass of the Sun. How many times longer will it live on the main sequence than the Sun will?

41. A star has a mass of 55 M_{Sun}. What is the ratio of its expected lifetime to that of the Sun?

42. What is the distance, in parsecs, to a star whose parallax is 2″? Compare this distance with the closest known star. Why would discovering a star with this parallax be a surprise?

43. How far from Earth, in parsecs, is a star whose parallax is 0.43″?

44. A star is known to be 25 pc from Earth. What is its parallax? Would such a change in position be easily visible to the human eye?

45. The star Vega is 25.05 light-years from Earth. What is its parallax?

 If your instructor assigns homework in **smartwork5**, access it at the Digital Landing Page for *At Play in the Cosmos*: **digital.wwnorton.com/cosmos**

NURSERY OF THE STARS

The Interstellar Medium and Star Formation

In this chapter you will learn about the material between the stars and the processes that turn this gas and dust into newborn suns. After reading through each section, you should be able to do the following:

12.1 Seeing in the Dark
Explain the nature and importance of the interstellar medium.

12.2 Anatomy of the ISM
Identify and compare the different phases of the material between the stars (called the interstellar medium, or ISM)

12.3 Molecular Clouds: The Birthplace of Stars
List the features of molecular clouds that make them the site of star formation and that enable astronomers to observe star formation in progress.

12.4 From Cloud to Protostar
Explain how a cloud becomes a protostar, and name the characteristics of protostars.

12.5 From Protostars to Fusion and Brown Dwarfs
Outline the process by which a protostar transitions to a hydrogen-fusing star, and describe how this process varies for stars of different masses.

12.6 Stellar Interaction
Describe the various ways that stars interact with one another and with their environments.

The Trifid Nebula is a large cloud of star-forming gas and dust located 1,600 parsecs from Earth. More than 150 new stars are forming in the cloud. This image, taken by the Spitzer Space Telescope, shows the Trifid in infrared light.

12

12.1
Seeing in the Dark

"IT TOOK A LONG, LONG TIME," says Judy Pipher, a professor of astronomy at the University of Rochester, telling the story of the Spitzer Space Telescope. "It started in 1983. That was the year a group of us proposed that NASA build a big, space-based infrared telescope. All of us were interested in studying how stars form in interstellar gas clouds." Recall from Chapter 4 that light of any given wavelength will be absorbed by objects with a size close to that wavelength, and the dust in most star-forming clouds is just the right size to absorb visible light. "The best way to see into those clouds and see what was going on with star formation was to use long-wavelength light in the infrared," says Pipher.

The infrared space telescope project was approved, but over the next two decades Pipher and her colleagues had to ride waves of interest and antagonism from both within and outside NASA. "A couple of times the project looked as if it were in trouble," she recalls. "And at one point the funding got scaled back really dramatically. All the astronomers working on the project had to sit down and decide what parts of the mission were most important and what parts we could let go. We knew we needed to get the details of what was happening in those star-forming clouds, so we focused on that as our primary mission."

interstellar medium (ISM) The gas and dust spread between the stars in different phases that depend on temperature and density conditions.

Despite funding and technical hurdles, the Spitzer Space Telescope (or SST) made it all the way from design to construction and final testing. On August 25, 2003, launch day finally arrived. Pipher and the rest of her team were there to watch the Delta 5 rocket, carrying the Spitzer Space Telescope, rise on a pillar of flame into the sky (**Figure 12.1**). "I was pretty nervous," she says. "Really, it had taken so much work I was just happy to finally make it to launch day." In the end there was nothing to be nervous about. The launch went off without a hitch, and within a few weeks Pipher and her colleagues confirmed that the telescope was working perfectly.

At the time, the SST was the only instrument capable of seeing so deeply, and with such accuracy, into interstellar clouds where the mysteries of star formation are hidden. Just a few months later, new data were pouring in, giving the waiting astronomers unprecedented views of giant clouds collapsing, new stars forming, and young planets hiding in surrounding disks of gas and dust. "It was definitely worth the wait," Pipher says with a wide smile.

In Space, Everyone Can Hear You Scream: The Interstellar Medium

"In space, no one can hear you scream," or so said the trailer for the 1978 space horror movie *Alien*. And indeed, most folks think of space as a vacuum, devoid of matter and energy. The truth, however, is a lot more interesting.

Interstellar space is far from empty. Floating between the stars is the **interstellar medium (ISM)**, a constantly moving mix of gas and dust between stars within the galaxy. The density of material in the ISM is certainly tenuous by Earth standards: a cubic centimeter of air at Earth's surface contains 10^{19} atoms, but a cubic centimeter of gas in the denser regions of the interstellar medium contains on average 10^6 atoms. That means that the ISM is 10 million million times less dense than the air you're breathing now. But 10^6 is not zero. Even at these low densities, if you added up all the matter between the stars, it would amount to more than billions of times as much material as the Sun contains. If we think on larger scales, space is full of stuff, and it is from this stuff that new stars are born (**Figure 12.2**).

The space between the stars is also not quiet or still. The ISM is constantly in a state of motion called turbulence. Turbulence is a very particular kind of motion in which a gas swirls, tumbles, and crashes against itself on many size scales. As in a fast-moving

A   **B**

FIGURE 12.1 Spitzer Space Telescope
The Spitzer Space Telescope (SST), launched in 2003, captures infrared light, which, unlike visible light, is not absorbed by dust grains. Thus, the SST is able to observe the inner regions of dusty components of the interstellar medium. (A) In this image, the SST departs Earth aboard a Delta 5 rocket. (B) The telescope trails behind Earth as it orbits the Sun, keeping its distance so that Earth's infrared emission doesn't overwhelm the emissions of objects being observed. The SST continues to drift away from us at about 0.1 astronomical unit (AU) per year.

river that swirls around obstacles, in the turbulent ISM there are huge "whorls" that are hundreds of parsecs across, and within those large-scale motions are smaller whorls that may extend over less than a single parsec.

It is unclear what drives the ISM into such turbulence, but many lines of evidence point to the combined action of supernovas. Because the ISM is a gas (though a very tenuous one), disturbances in the ISM create sound waves like the ones that occur in Earth's atmosphere. If the disturbance is violent enough (as in a supernova explosion), the extreme cousins of sound waves called shock waves form. These shocks can cross parsecs of space and sweep up everything in their path (**Figure 12.3**). Thus, supernova blast waves are a specific example of astrophysical shocks.

Over the course of millions of years, the explosions of many massive stars set local regions of gas into motion. When these supernova blast waves expand to large enough sizes, they collide with one another. Collisions tear each blast wave into shreds, sending gas tumbling in all directions. Computer simulations of this process indicate that the explosion of about one supernova every 10,000 cubic parsecs (pc^3) every million years is enough to keep the ISM in its turbulent state. Winds from young and middle-aged stars, and even the rotation of the galaxy itself, may also contribute to turbulence in the ISM.

SECTION SUMMARY
In Space, Everyone Can Hear You Scream: The Interstellar Medium

- The ISM is composed of a low-density mix of gas and dust, but it contains enough mass to enable ongoing star formation in the galaxy.
- The ISM is not static but is in a state of turbulent motion caused primarily by blast waves from supernovas.

12.2
Anatomy of the ISM

To understand the formation of stars, astronomers must first understand the different kinds of gases between stars. There's a sort of ecological system in the galaxy in which some kinds of interstellar material can collapse to form new stars, and then, as stars age, they return gas back to the ISM. Thus the ISM is a subject of great interest in its own right, and an important part of our story about the lives of stars and, eventually, the lives of galaxies themselves.

FIGURE 12.2 Interstellar Medium
This map of the whole sky indicates the location of ionized hydrogen gas in which all the hydrogen atoms have been stripped of their electrons (redder regions indicate more ionized gas). The map shows us that interstellar gas is broadly distributed throughout space, but that the brightest regions are highly localized and form distinct clouds.

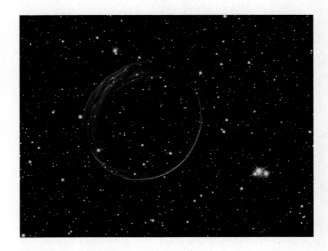

FIGURE 12.3 Supernova Blast Wave
A supernova remnant, bounded by an outward-expanding shock or "blast" wave, results from the explosion of a massive star. As the shock expands, it encounters and compresses nearby clouds of the ISM and eventually can collide with other supernova shocks to stir the interstellar medium into a turbulent state. Here we see SNR 0509, the visible remnant of a powerful stellar explosion in the Large Magellanic Cloud (LMC), a small galaxy near the Milky Way. Ripples in the shell's surface may be caused by subtle variations in the density of the ambient interstellar gas.

"The ISM is made up of different kinds of clouds," says Judy Pipher, "and they all have different conditions." Each kind of cloud tends to have a different density and different temperature. In addition, the gas takes on different microscopic states in terms

TABLE 12.1 ••• Components of the Interstellar Medium				
Phase	**Temperature (K)**	**Density (atoms/cm³)**	**State of Hydrogen**	**How Observed**
Molecular clouds	10	10^2 to 10^6	Molecular (plus many other molecules)	Molecular emission and absorption lines
H I clouds	10^2	10^2 to 10^3	Neutral atomic	21-cm-line absorption
H II regions	10^4	10 to 10^6	Ionized	Hα emission (656 nm, in red part of visible spectrum)
Warm interstellar medium (WISM)	8×10^3	0.2–0.5	Ionized	Hα emission
Coronal gas	10^6	10^{-4} to 10^{-2}	Highly ionized	X-ray emission

FIGURE 12.4 Star-Forming Nebulas
The Orion Nebula (also called M42), the region of massive star formation that is closest to Earth, displays glowing gas clouds and hot young stars (right). Note that the red color is due to the emission of H alpha photons. A smaller nebula, M43, appears near the center of the image, and the dusty, bluish nebula NGC 1977 appears on the left. Astronomers have identified what appear to be numerous infant planetary systems located in M42.

molecular cloud The low-temperature, high-density phase of the interstellar medium where molecules can form. Molecular clouds are the site of star formation.

H I clouds The phase of the interstellar medium that is composed of relatively low-temperature neutral hydrogen gas.

of ionization or chemistry (see **Table 12.1**). "In some clouds," says Pipher, "the gas will be primarily composed of molecules, and in others it will be atoms that have become highly ionized."

Phases of the ISM

Astronomers have identified different forms of the interstellar medium, which they refer to as *phases*. These phases include molecular clouds, H I clouds, H II regions, warm interstellar medium, and coronal gas. Astronomers think of these different phases the way physicists think about liquid water, ice, and water vapor. Each is a phase of water whose existence depends on conditions like temperature and pressure. Similarly, it has proved useful for astronomers to think about the different kinds of clouds as phases of the interstellar medium under different conditions of temperature, pressure, and chemical makeup.

MOLECULAR CLOUDS. The coldest and densest regions of the galaxy are vast clouds of gas in which stars form. These are called **molecular clouds** because their low temperatures (about 10 kelvins [K]) and relatively high densities (about 10^2–10^6 atoms/cubic centimeter [cm³]) allow complex molecules to form. Astronomers map out these clouds using emission lines from the molecules (**Figure 12.4**).

H I CLOUDS. Much of the galaxy's volume is occupied by extended clouds of atomic hydrogen gas called **H I clouds** (the *H* refers to hydrogen, and the Roman numeral I indicates the neutral state). H I clouds have temperatures of a few hundred kelvins and densities

FIGURE 12.5 H I Clouds

Cool H I clouds are typically found in the spiral arms of galaxies, as illustrated here in these images of NGC 6946 (A) and NGC 3184 (B), where the 21-cm emission line (seen here as red) traces the presence of the H I clouds. These clouds lie close to the midplane of the galaxies.

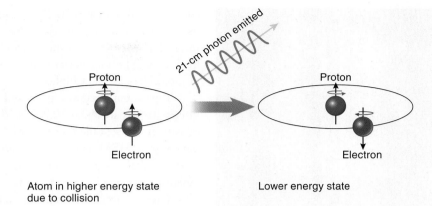

FIGURE 12.6 21-Centimeter Line

Neutral hydrogen (H I) clouds are too cold to emit light from transitions of electron energy states. They can be seen instead via emission of the 21-cm line, which occurs when a collision between two hydrogen atoms (not shown) leaves one of the atoms with the spins of both its electron and proton aligned. This higher-energy state eventually decays back to a configuration in which the spins are oppositely aligned. A 21-cm photon whose energy is exactly the difference between the aligned and unaligned states is emitted in the process.

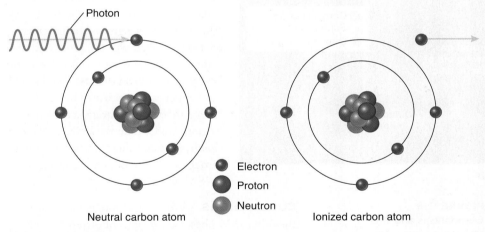

FIGURE 12.7 Photoionization

In photoionization, a photon of high enough energy collides with a neutral atom and ejects one of its electrons from its orbit. The resulting configuration is a singly ionized atom with one fewer electron than it has protons. Photoionization can also strip an already ionized atom of more of its electrons.

that range from on average tens of particles per cubic centimeter to as high as thousands of particles per cubic centimeter. In general, these cold clouds tend to be found in the spiral arms of our galaxy and are confined to a relatively thin region (300 pc in height) around the midplane of the galaxy's disk (**Figure 12.5**).

H I clouds are so cold that they do not emit light through transitions of electrons in hydrogen atoms between orbits. There simply is not enough energy for collisions between particles to kick an electron up from the ground state. How, then, do astronomers know these clouds exist?

All the particles that make up an atom have an intrinsic spin, like a top. Each charged, spinning particle (such as a proton or electron) is like a tiny magnet. Magnetic forces between the electron and proton tend to keep their spins oppositely oriented. When an occasional collision between particles knocks both spins into alignment, a quantum jump eventually flips them back into their lower-energy state. When this quantum "spin-flip" occurs, a photon is emitted with a wavelength of 21 centimeters (cm), which is in the radio band (**Figure 12.6**; see also Chapter 9). These long-wavelength radio photons can travel enormous distances without being absorbed. In this way, 21-cm radio observations enable astronomers not only to detect neutral hydrogen across the galaxy but also to map the entire galactic distribution of H I clouds.

H II REGIONS AND WARM INTERSTELLAR GAS.

"While H I clouds are neutral," says Pipher, "other regions of the ISM are defined by their ionization." **Photoionization** occurs when a photon with enough energy kicks an electron off an atom (**Figure 12.7**). Photoionizing radiation can come from many sources: young massive stars, evolved low- and intermediate-mass stars, or even the disks of gas surrounding supermassive black holes at the centers of galaxies.

Photons with energy E of or greater than 2.18×10^{-18} joule (J)—which corresponds to a wavelength of 91.2 nanometers (nm), putting it in the ultraviolet range—are capable of ionizing hydrogen. Once an electron is kicked off a hydrogen atom, it will

photoionization The stripping of an electron from an atom or ion via the absorption of a photon.

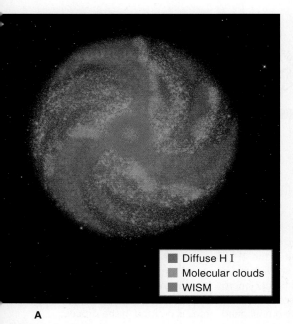

Diffuse H I
Molecular clouds
WISM

A

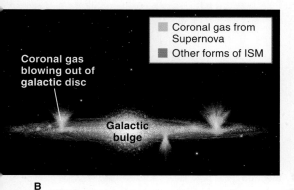

Coronal gas from Supernova
Other forms of ISM

Coronal gas blowing out of galactic disc

Galactic bulge

B

FIGURE 12.8
Coronal Gas and the Structure of the ISM
The distribution of different phases of the ISM is shown in (A) the top-down view of our galaxy. Threaded between molecular, H I, and warm phases, however, is hot coronal gas created by many supernova blast waves over galactic history. (B) High pressures in the hot material will also lead to blowouts above and below the plane of the galactic disk.

collide with other gas particles, sharing its energy and heating the gas. "That means ionized regions of the interstellar medium will be a lot hotter than neutral or molecular gas," says Pipher.

Typical temperatures in ionized interstellar gas are about 10,000 K. The density of this so-called warm component of the ISM depends on where it occurs. When a star produces the radiation-stripped hydrogen atoms, astronomers call the environment an **H II region**. The Roman numeral II indicates that the hydrogen atoms have each lost one electron. H II regions have typical densities of 100–10,000 atoms/cm^3.

Regions far away from any particular ionizing source, like a hot star, still have a diffuse background of ionizing radiation. This background flux of ionizing photons is the sum of all the sources in the galaxy. Although the background radiation is relatively weak, it will still produce ionizations to create what is called the **warm interstellar medium (WISM)**. The WISM has temperatures similar to those in H II regions, but lower average densities. Regions colder than the WISM, like H I clouds, maintain their low temperatures through emission lines not available to WISM material.

CORONAL GAS.   Worming through and between the other ISM phases are regions of extremely hot, low-density gas. Temperatures in this phase can reach as high as 10^7 K. Densities in this phase can be as low as 10^{-4} atom/cm^3.

"If these kinds of conditions seem familiar," says Pipher, "it's because we see them in the Sun's corona. That's why astronomers use the term 'coronal gas' to refer to this phase of the ISM." Despite the similar conditions, the source of **coronal gas** is not living stars but dead ones. The hot, low-density interstellar gas originates with supernova blast waves created at the end of a massive star's life. "When a star goes supernova, its explosion can sweep across hundreds of light-years of interstellar space," says Pipher. "All the heat released in the explosion drives swept-up ISM gas to such high temperatures that it gets strongly ionized." As Pipher explains, an element like iron can lose up to 13 of its 26 electrons.

At these temperatures and ionization states, the newly heated gas can't cool effectively by emitting radiation, because there are no electronic transitions that can be easily excited for the electrons that remain around the atoms. Thus, once ISM gas is driven into the coronal state by a passing supernova blast wave, it remains hot for billions of years. As time passes, the hot gas continues to expand, filling in regions between other phases of the ISM. Eventually, a kind of Swiss-cheese network of hot, low-density regions forms throughout the galaxy (**Figure 12.8**).

These ISM phases do not exhaust the forms of clouds floating between the stars. Astronomers have found other mixes of hot or cold, dense or tenuous gas in the ISM. However, the phases discussed in this section provide the main distinctions necessary for understanding the ISM and star formation.

SECTION SUMMARY
Phases of the ISM
- Molecular clouds are cold and dense and contain complex molecules.
- H I clouds are cold atomic clouds with a wide range of densities and are observed via the 21-cm line.
- Warm H II regions contain atoms ionized by nearby massive stars; the WISM includes regions similar to H II regions but ionized by weaker background radiation.
- Powerful blast waves from supernovas create regions of hot, low-density coronal gas that is highly ionized.

Many Clouds, One Pressure

When astronomers first began exploring the ISM, they noticed that while the various types of clouds seemed very different from one another, the pressures inside those clouds were in the same range. To understand the meaning of this result, remember that the ISM is a gas and therefore obeys the *ideal gas law*, which states that the pressure of a gas depends on its density and temperature.

Anyone who has ever played with balloons has an intuitive understanding of pressure. To blow up a balloon, you have to force air into it until its elastic skin expands. But if you inflate a balloon and then put it in the freezer for a couple of days, the balloon will deflate a little, even if no air escapes from it. What happened? "By cooling the trapped air, you lowered its pressure," says Pipher. "The air molecules are at a lower temperature, and that means they are no longer hitting the inside wall of the balloon with a high velocity." Temperature, velocity, and pressure are related. Recall

GOING FURTHER 12.1 THE LANGUAGE OF THE COSMOS
Pressure, Equilibrium, and the ISM

Temperature (T), density (d), and pressure (P) have a simple relationship in a gas, given by the equation

$$P = (k_b/m)dT$$

where k_b is a constant called Boltzmann's constant (we don't need to worry about its value here) and m mass of the particles in the gas. For our purposes, what is important is that pressure is proportional to the density and temperature, which we can write as

$$P \propto dT$$

The symbol $\propto$ means "is proportional to."

Let's use this equation to learn something about the relationship of the different phases of the ISM, which have different temperatures and densities. First we give this relationship for two clouds of different types or phases (labeled 1 and 2):

$$P_1 \propto d_1 T_1$$

$$P_2 \propto d_2 T_2$$

If we divide the two equations, we get a ratio:

$$\frac{P_1}{P_2} = \left(\frac{d_1}{d_2}\right)\left(\frac{T_1}{T_2}\right)$$

Now imagine that we observe a cloud of phase 1 embedded inside a larger cloud of phase 2. If we expect that the clouds are in pressure equilibrium, then

$$\frac{P_1}{P_2} = 1$$

which gives us a relationship between the densities and temperatures in the two clouds:

$$\left(\frac{d_2}{d_1}\right) = \left(\frac{T_1}{T_2}\right)$$

This relation tells us that if the temperature in cloud 1 is much higher than that in cloud 2, the density in cloud 1 must be much lower than that in cloud 2 to maintain pressure balance.

Imagine for a moment that we knew the temperature in cloud 1 was 10,000 K, the temperature in cloud 2 was 100 K, and the density in cloud 1 was 1,000 particles per cubic meter (m³). If we know that the clouds are in pressure equilibrium, we can calculate the density in cloud 2 by rearranging the equation as follows:

$$d_2 = d_1 \left(\frac{T_1}{T_2}\right) = 10^3 \text{m}^{-3}\left(\frac{10^4 \text{ K}}{10^2 \text{ K}}\right) = 10^5 \text{m}^{-3}$$

The density in the second cloud has to be 100 times higher than that in the first cloud to maintain equilibrium. This exercise illustrates how the concept of pressure equilibrium, when it applies, gives astronomers a powerful tool to understand the properties of clouds in the ISM.

When clouds are not in equilibrium, however, astronomers must also describe the motions that occur when pressure differences between two phases of the ISM set the gas in motion.

---

from our discussion of planetary atmospheres in Going Further 6.2 that higher temperatures in a gas mean particles moving at higher velocities. Since the pressure is just the effect of collisions with gas particles (balloons stay inflated because of atoms inside bouncing off the balloon's skin), high temperatures also mean high pressures. "The pressure outside the balloon that's been inside the fridge hasn't changed," says Pipher, "but the pressure inside the balloon has dropped with the temperature. That means the balloon has to shrink."

What applies to balloons should also apply to the ISM, if we think of the different kinds of ISM as clouds pushing against one another. When scientists calculated the pressure for all the different phases of the ISM, they found that many of the phases had the same pressure. When neighboring gas clouds have the same pressure, astronomers say they are in equilibrium. A cloud that is in equilibrium with its surroundings will neither expand nor contract. From these observations, astronomers concluded that gas in the ISM naturally expands or contracts until it reaches states that are all in pressure equilibrium. The net result of these findings was a multiphase model of the ISM. There is hot, low-density gas and cold, high-density gas. The different phases can coexist indefinitely because they all have the same pressure (**Going Further 12.1**).

interactive
Pressure Equilibrium

H II region The phase of the interstellar medium that contains ionized hydrogen gas.

warm interstellar medium (WISM) The phase of the interstellar medium that is warmed by the galaxy's diffuse ionizing radiation (the sum of all its ionizing stellar sources).

coronal gas The phase of the interstellar medium that consists of extremely hot, low-density gas resulting from explosions of high-mass stars (supernovas).

reflection nebula A dense, dusty gas cloud made visible by reflected starlight.

As better observations were obtained, astronomers realized that the clouds of the ISM were not in perfect pressure equilibrium. Many regions of the ISM were in motion. "In particular," says Pipher, "most phases of the ISM were turbulent, and that changed how astronomers had to imagine the ISM and its history." The turbulent motions of a gas can be thought of as another form of pressure. The roiling motions of a turbulent gas cloud push against other material, just as the random motions of individual atoms in a hot gas do.

SECTION SUMMARY
Many Clouds, One Pressure
- Observations of the ISM show pressure equilibrium across many regions.
- The multiphase model of the ISM assumes that regions expanded or contracted until the ISM reached this steady state.

Interstellar Dust: It's the Little Things That Matter

Individual atoms or molecules of gas are not the only inhabitants of interstellar space. Along with the gas are the smallest examples of solids known to astronomers: small grains of matter called dust. "Actually," says Pipher, "calling it *dust* is kind of a misnomer. Most of the grains we see in space are much more like soot from automobile exhaust or the smoke in a

smoke-filled room rather than the dust you find under a bed." What matters, explains Pipher, is the size of the grains.

Typical interstellar dust grains range from 0.1 to 10 microns (μm) in length (1 μm = 10^{-6} meter), though within molecular clouds the grains can be even larger. If you compare these sizes with the wavelength of optical light (0.390–0.700 μm), you will immediately see one reason that dust is so important for astronomers. "The size of dust grains is just right to absorb or scatter optical light," says Pipher. "The grain sizes and the light wavelengths are pretty much the same [**Figure 12.9**]. That means dust limits how far we can see across the galaxy. It also keeps us from seeing inside molecular clouds, where stars form."

On average, there is 1 kilogram (kg) of dust for every few hundred kilograms of gas in the ISM. Although dust accounts for only a small fraction of the mass in the ISM, it creates two effects that astronomers must be careful to account for in their studies. First, dust grains absorb certain wavelengths of incoming visible light, making objects appear dimmer than they actually are. Second, dust tends to scatter blue light. This scattering can create beautiful blue **reflection nebulas**, where starlight reflects off of relatively dense dust clouds, but it also makes objects beyond the dust appear redder than they actually are. Astronomers in the early 1900s didn't yet understand dust and its power to dim distant stars. By miscalculating the true brightness of these stars, they greatly underestimated the true size of our galaxy.

Dust is crucial for astronomers for another reason: the structure of a typical dust particle enables it to act as a chemical factory. "As the smallest bits of solid matter, dust grains act as a platform for complex chemistry," says Pipher. "The grains are also the first step in building larger solid structures like rocks, asteroids, and eventually planets."

Interstellar dust grains appear to be irregularly shaped crystals of silicates (silicon and oxygen) or carbon, with many other elements mixed in at lower abundances. The dust grains are often surrounded by a "mantle" or covering of water ice or other frozen volatiles. "When atoms collide with dust grains, they get stuck to the grain surface," explains Pipher. "Once an atom is on the surface, it's more likely to find other elements that are being bounced around the grain." When these surface atoms combine, they form molecules (**Figure 12.10**). "Building complex molecules could never happen in free space, because out there, atoms rarely collide. The dust grains act like flypaper for the atoms." By catching atoms on their surface, the grains can build

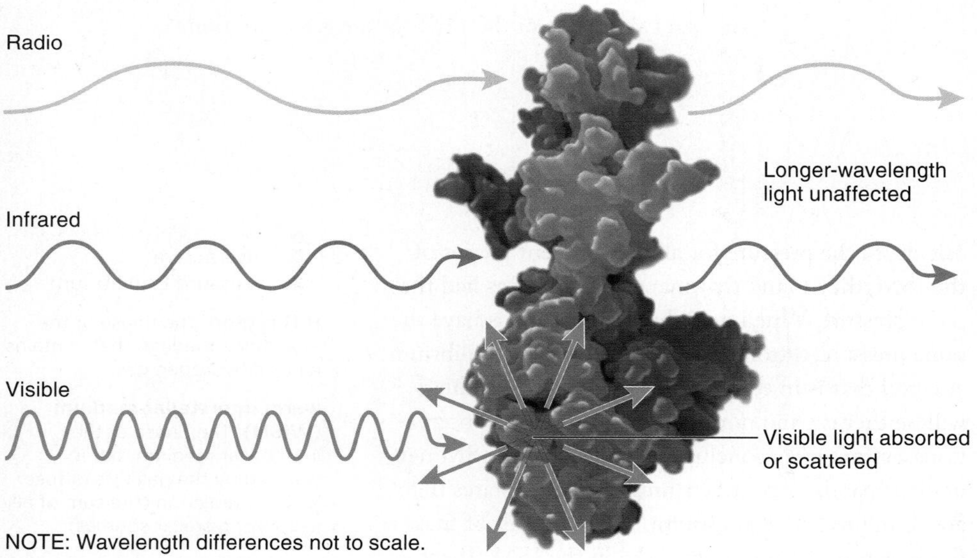

interactive
Interstellar Reddening and Extinction

Radio

Infrared

Visible

Longer-wavelength light unaffected

Visible light absorbed or scattered

NOTE: Wavelength differences not to scale.

FIGURE 12.9 Interstellar Dust
A dust particle absorbs light whose wavelength is less than or equal to the size of the particle. Since typical interstellar dust grains range from 0.1 to 10 μm, they absorb or scatter visible light, which has wavelengths of 0.390–0.700 μm. Longer-wavelength light, such as radio and infrared, can pass dusty regions with little absorption or scattering.

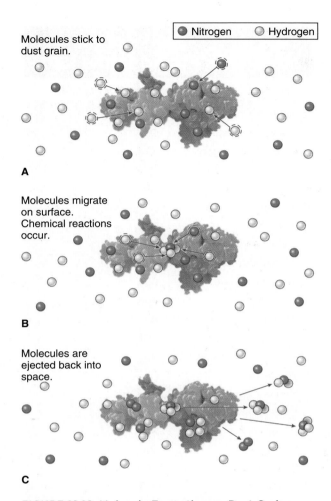

Molecules stick to dust grain.

Nitrogen · Hydrogen

A

Molecules migrate on surface. Chemical reactions occur.

B

Molecules are ejected back into space.

C

FIGURE 12.10 Molecule Formation on Dust Grains
(A) Atoms from interstellar gas first collect on the icy mantles that compose the surfaces of dust grains. (B) Atoms encounter one another as they move randomly along the grain surface, allowing chemical reactions to occur. Here, atoms of hydrogen and nitrogen combine to make ammonia (NH_3). (C) Eventually the newly formed molecules will be ejected back into the interstellar gas.

up high-enough atomic abundances for chemistry to take off and begin creating ever-more-complex molecules in relatively short times, astronomically speaking.

Astronomers can make such definitive statements about the properties of interstellar dust grains because the grains emit light in the form of blackbody radiation (they are solids), providing information about their temperature. More important, however, are the vibrations of the dust grains' atomic lattices. Light emitted by these vibrations produces clearly detectable spectral signatures, enabling astronomers to identify the detailed properties of dust grains, including size and composition.

Judy Pipher

"My mom was a biologist," says Judy Pipher, "so we never grew up with any biases about girls and science. I liked physics quite a bit, and that was my major at the University of Toronto." But in Pipher's junior year, an observational astronomy course changed her life's direction and changed the field of astronomy. "Even as a kid I can remember lying on the beach at my family's cottage and using a book to find different stars. That class in my junior year just let me see how physics was the basis of astronomy."

For graduate school, Pipher went to Cornell University, where she became involved in the young science of infrared astronomy. "We used to fly very primitive infrared detectors on rockets out at the White Sands Proving Ground in New Mexico," she explains. "The data would get collected on these long strips of graph paper. It was a nightmare trying to interpret what the detectors saw during their short flights. Every night the professors would tell us to have a report ready in the morning, and then they'd head off to bed."

Pipher was good at building the detectors and good at interpreting their observations. Eventually, she found her way to a professorship at the University of Rochester, where she became a leader in advancing generation after generation of infrared instruments. When the time came to consider the construction of a large-scale infrared space telescope, Pipher was ready to lead one of the teams. Twenty years later, that effort paid off with the successful launch of the Spitzer Space Telescope.

At 72, Pipher is now considered by some "the grandmother" of infrared astronomy. "I am not sure if I like being the grandmother of anything," she laughs. "I just went with what I loved to do, what interested me. I guess it worked out okay." In 2007 she was inducted into the National Women's Hall of Fame for her pioneering efforts in infrared astronomy.

Even better than spectral analysis, however, is the direct capture of interstellar dust grains. In 2004, after a 1999 launch, the *Stardust* space probe was sent into the coma of comet Wild 2. Approaching within 300 kilometers (km) of the comet, the spacecraft used a tennis racket–shaped collector to capture dust grains blown off the comet. Since comets are made of pristine material left over from the formation of the Solar System, the dust grains captured in the *Stardust* mission represent ancient material that once floated freely through interstellar space. After the *Stardust* mission returned to Earth, scientists began to harvest the grains from the collector and study their properties.

SECTION SUMMARY
Interstellar Dust: It's the Little Things That Matter

- On average, there is 1 kg of dust for every few hundred kilograms of gas in the ISM.

- Because of their size, dust grains effectively extinguish visible light, making background objects appear dimmer and redder.

- The surfaces of dust grains capture atoms and provide a platform for the formation of molecules.

12.3
Molecular Clouds: The Birthplace of Stars

For the story of star formation to make any sense, we need to delve more deeply into the composition and structure of molecular clouds. Only inside these dense, dusty clouds are the conditions just right for new stars to be assembled.

The Molecules in Molecular Clouds

The molecules that give molecular clouds their name form on dust grains within the clouds, but if these molecules were to form in other parts of the ISM they would quickly be dissociated (broken apart) by passing ultraviolet photons. Within a molecular cloud, however, the gas and dust are dense enough to shield the molecules from dissociating radiation. The high densities of dust and gas in molecular clouds thus enable them to act as stellar nurseries.

Given that hydrogen is the most abundant element in the Universe, it shouldn't come as a surprise that molecular hydrogen (H_2) is the dominant molecule inside molecular clouds. However, it's difficult for H_2 to radiate at the low temperatures inside molecular clouds ($T = 10$ K). "Hydrogen just doesn't emit a lot of light at conditions within a cloud," says Judy Pipher, "unless something violent happens to raise the temperature of the gas."

Because astronomers can't use hydrogen molecules to study these clouds, they need to find another molecular tracer of cloud structure. The term *tracer* is used because these molecules may not constitute much of a cloud's mass, but they are prevalent enough to trace out the cloud's structure. Luckily, these clouds contain other kinds of molecules, such as carbon monoxide (CO), that radiate more efficiently at low temperatures. Using CO as a tracer, astronomers can map molecular clouds in detail and determine important properties like cloud mass and internal motions.

"The chemistry in these clouds is pretty rich," says Pipher. "You get a lot more than just H_2 and CO molecules." Astronomers have identified hundreds of molecular species in star-forming clouds, including vast amounts of water (H_2O) and ethanol (C_2H_6OH). "One way to look at a molecular cloud is that it is like a giant martini," quips Pipher.

Table 12.2 lists some of the molecules currently known to exist in star-forming clouds. Notably, even a

giant molecular cloud (GMC) A large-scale region of gas in molecular form spanning tens to hundreds of parsecs and containing up to 10 million solar masses.

TABLE 12.2 ••• Some of the Molecules Found in the ISM	
Molecule	**Chemical Formula**
Molecular hydrogen	H_2
Molecular oxygen	O_2
Molecular nitrogen	N_2
Diatomic carbon	C_2
Ammonia	NH_3
Acetone	$(CH_3)_2CO$
Benzene	C_6H_6
Carbon monoxide	CO
Carbon dioxide	CO_2
Ethylene glycol	$C_2H_6O_2$
Ethanol	C_2H_6O
Formaldehyde	H_2CO
Hydrogen cyanide	HCN
Hydrogen peroxide	H_2O_2
Iron oxide	Fe_2O_3
Methanol	CH_3OH
Methane	CH_4
Nitrous oxide	N_2O
Ozone	O_3
Propanol	$CH_3CH_2CH_2OH$
Water	H_2O

number of amino acids have been discovered. This fact has provoked some astrobiologists to wonder whether life's building blocks might be produced in space first and then delivered to planets like Earth much later.

SECTION SUMMARY
The Molecules in Molecular Clouds
- The high densities in molecular clouds enable molecules formed on dust grains to remain shielded from radiation that would break them apart.
- Hydrogen is the dominant type of molecule, but many others occur—notably carbon monoxide, whose abundant emission allows the structure of the clouds to be studied.

Molecular Clouds: Shapes and Sizes

"There isn't just one kind of molecular cloud," says Pipher. "They come in a bunch of different sizes." On the largest scales, astronomers find **giant molecular**

clouds (GMCs) that can span 100 pc and contain enough mass to form a million stars. On the smallest scales are the so-called **dark globules** (sometimes called **Bok globules**), which extend only a parsec or less and contain enough mass for just a few stars (**Figure 12.11**). Densities within an individual cloud can vary as well. In their central regions, molecular clouds can be quite dense by ISM standards, with up to 10^{12} particles per cubic meter. At the edges of a cloud, however, the densities fall by a factor of a million as the cloud merges with other regions of the ISM.

This variety of sizes and masses may have a lot to do with the internal structure of molecular clouds. The largest star-forming clouds, GMCs, are not uniform but instead show a hierarchy of structure. "Hierarchy means there are big clouds embedded with smaller, denser clouds," says Pipher, "and within those smaller clouds are even smaller, denser structures."

A GMC 50 pc wide, for example, will show smaller, denser regions called clumps, which may extend across a few parsecs and contain enough mass for hundreds of stars. Within these clumps, astronomers find many denser conglomerations called cores, which may be 0.1 pc or less in diameter and contain enough mass for just a single star. Given these nested sets of ever-denser structures, it is likely that objects like dark globules were once chunks of a larger cloud but then separated from the parent body.

Although astrophysicists might assume a cloud to be spherical for theoretical simplicity, the GMCs that are home to the most active sites of star formation are not even close to spherically shaped (**Figure 12.12**). "Scientists like to picture things as spheres because it makes the math easier," says Pipher, "but what we see is

A NGC 1999

B NGC 281

FIGURE 12.11 Dark Globules
(A) The Snake Nebula contains these examples of dark (or Bok) globules. The globules, spanning about 5 light-years across, are regions of cold dense gas, molecules, and dust that block all starlight from behind. (B) NGC 281 includes several dark globules, such as the ones seen here.

really much messier. Many clouds appear as a network of long, dense filaments with less dense gas in between."

The shapes of clouds may give some indication of their history. For many years, astronomers believed that molecular clouds assembled slowly. The idea was that atomic gas in H I clouds would condense into molecular form once a chance bump or wiggle allowed the atomic cloud to become dense enough to shield the interior from ultraviolet light. According to this hypothesis, once the molecular clouds formed, they would persist for many millions of years. The inward-directed gravitational weight would also be balanced by either the clouds' thermal pressure or magnetic fields. This idea implied that molecular clouds were relatively quiet structures without any turbulence.

dark globule or **Bok globule**
A small molecular cloud containing a few solar masses' worth of material whose density and dust content render it opaque to light in visible wavelengths.

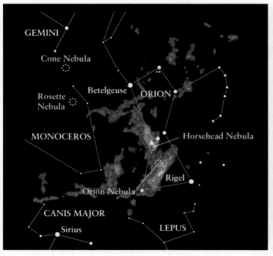

A

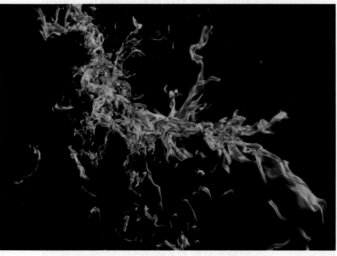

B

FIGURE 12.12
Giant Molecular Clouds
GMCs are not smooth spheres, but irregular clouds displaying filaments and clumps, as seen in this radio map of (A) the Orion molecular cloud and (B) computer simulation of the formation of structure within a molecular cloud.

As mentioned earlier, however, Doppler shift measurements have shown that many clouds are turbulent; that is, they are filled with fast-moving, randomly directed gas. Some astronomers hypothesize that the observed turbulence is an important clue to the formation and history of molecular clouds. They propose that clouds form from collisions between huge gas flows in the galaxy. These flows could come either from the galaxy's rotation or from some other mechanism, such as supernova blast waves. In either case, when these enormous streams collide, dense pockets of gas form, allowing molecular chemistry to get started. Clouds formed through this kind of violence "remember" their histories in the sense that they still show turbulent motions.

These two models of cloud formation—quiescent coalescence versus turbulent collisions—represent the cutting edge of thinking about star-forming clouds and their histories. They also represent the kind of split that often occurs at the frontiers of science. Two competing models (and their advocates) will battle it out for years until new data, often obtained through new observing technologies, allow the issue to be settled. "Controversy is the heart of the scientific process," says Pipher. "People love to fight over their favorite idea, but in the end it's data that wins the day."

Astronomers also understand that all molecular clouds are threaded by magnetic fields. The strength of these fields is still a subject of debate, but in general, the fields do not appear strong enough to keep most clouds from collapsing. The fields are strong enough, however, that they affect the way a cloud behaves as it collapses to small scales during star formation.

free-fall time The length of time required for a cloud to collapse under the force of its own gravity.

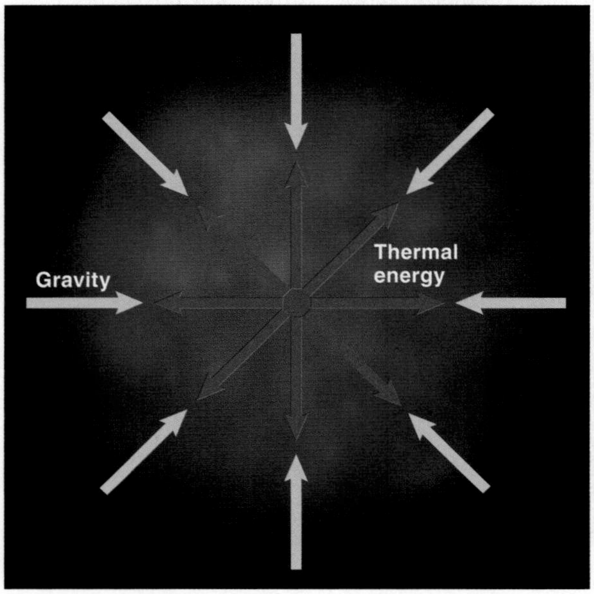

FIGURE 12.13
Thermal Energy Resisting Gravity
A molecular cloud may resist collapse for millions of years by balancing thermal pressure (which is due to collisions between gas particles) gained during its formation against the inward pull of gravity.

─ **SECTION SUMMARY**
Molecular Clouds: Shapes and Sizes
● Molecular clouds are composed of dense structures that take filamentary as opposed to spherical shapes.
● The turbulence in molecular clouds may be the result of collisions between large-scale gas flows.

12.4
From Cloud to Protostar

Now that we've explored the properties of molecular clouds, we're ready to take the next step and ask how stars form within these clouds. How does the mass in the clouds become so concentrated that densities and temperatures are high enough for fusion reactions to begin? The key to the story is gravity and its ability to drive the cloud, or at least a part of it, into collapse.

Collapse

Like the Sun and other stars, molecular clouds are at war with their own gravity. "The clouds, however, tend to lose that war a lot faster than stars," says Judy Pipher. Molecular clouds contain a large amount of mass in a relatively small region of space, and they are often close to the point where gravity can overwhelm gas pressure and cause the cloud to collapse on itself. Stars, in contrast, generate enough energy and pressure through nuclear fusion in their cores to counteract the force of gravity for most of their lives. "Molecular clouds can't generate any new energy," says Pipher. "What they can do is support themselves with thermal energy gained during their formation. If the cloud does not radiate its energy quickly, it can maintain a balance between gravity and pressure for millions of years" (**Figure 12.13**).

Molecular clouds can also get some limited support from internal magnetic fields (recall the discussion of magnetic pressure in sunspots from Chapter 10) or from the tumbling chaos of their own turbulence. In both cases, the fields or the turbulence provides some outward push to help fight against the inward pull of gravity. The problem with all of these support mechanisms is that they cannot last forever. Thermal energy is eventually lost as the cloud emits radiation and cools, magnetic fields also dissipate with time, and even turbulence eventually slows down. Without renewed energy deposition into the cloud, gravity will eventually win.

Even a cloud that reaches a balance between gravity and internal forces might not be stable. "You can balance a pencil on its point if it's in equilibrium," says Pipher, "but if you nudge the pencil, it will fall. That means the equilibrium wasn't stable. In the same way, a cloud can be in equilibrium between gravity and pressure, but a passing supernova shock can compress it enough to push it into collapse."

Gravitational collapse occurs when internal forces can no longer support some portion of a molecular cloud against its own gravity. Specifying that it's a *portion* of the cloud that begins collapse is important. Astronomers never see an entire GMC collapsing all at once. Instead, smaller regions of the cloud—the cores—are usually pushed to the point where gravity overwhelms pressure. "Once gravity starts winning," says Pipher, "the core begins falling in on itself."

To develop theories of cloud collapse, astronomers imagine following a "parcel," or blob, of gas on its fall toward the core's center. As many parcels drop toward the center, the density of gas increases: ever more material squeezing into an ever-smaller space. During the initial phases of collapse, the falling gas maintains its low temperature. Any heat generated as material falls inward and gets denser can still escape into space as radiation. Since pressure still plays no role in slowing the collapsing gas, during this phase astronomers speak of the gas as being in free-fall.

How long does this phase of collapse last? The **free-fall time** is a measure of how long it takes for a cloud of a given density to complete the free-fall phase. The remarkable thing about free-fall time is that it depends only on a cloud's density: denser gas collapses faster than less dense gas. The radius of the cloud does not affect the free-fall time. (**Going Further 12.2** explores in detail how different densities yield different free-fall times.) For typical conditions in a cloud core, however, the free-fall time turns out to be from 1 to 2 million years. "That is why a million years is the typical timescale astronomers think about for assembling a star," says Pipher.

SECTION SUMMARY
Collapse
- Like stars, clouds must have energy to resist the collapsing influence of gravity; the energy may be from thermal energy, magnetic fields, or turbulence.
- If gravity is able to overwhelm a portion of the cloud, that portion will free-fall inward.
- The timing of free-fall is based on cloud density and is typically from 1 to 2 million years for conditions in the cores of molecular clouds.

GOING FURTHER 12.2   THE LANGUAGE OF THE COSMOS
Star Formation and the Free-Fall Time

If the inward-directed force of gravity due to a molecular cloud's mass becomes greater than its outward-directed pressure, the cloud is destined to collapse on itself and form a star. How long does that collapse take?

The free-fall time t_{ff} is the time it takes for a cloud to fall in on itself when gas pressure can be completely ignored:

$$t_{ff} = \sqrt{\frac{3}{2\pi Gd}}$$

where G is the gravitational constant and d is density. Note that the free-fall time depends only on gas density. That means an extended cloud with a large radius R_L and a small cloud with a smaller radius R_S will take the same amount of time to collapse, as long as they begin with the same density.

The average density of particles in a molecular cloud is $n = 10^9$ m^{-3}. To get the mass density, we simply multiply n by the mass of a hydrogen atom, which is 1.67×10^{-27} kg. This gives us

$$d = (10^9 \text{ m}^{-3})(1.67 \times 10^{-27} \text{ kg}) = 1.67 \times 10^{-18} \text{ kg/m}^3$$

Let's use this number to calculate the average free-fall time in a section of cloud where gravity has won out over gas pressure:

$$t_{ff} = \sqrt{\frac{3}{2\pi Gd}} = \sqrt{\frac{3}{(2\pi)(6.67 \times 10^{-11} \text{ m}^3 \text{ kg}^{-1} \text{ s}^{-2})(1.67 \times 10^{-18} \text{ kg m}^{-3})}}$$

$$= 6.5 \times 10^{13} \text{ s} = 2.1 \times 10^6 \text{ yr}$$

This equation tells us that it takes a little more than 2 million years for the material in the cloud to collapse into a protostar.

How much shorter will the free-fall time be for a region of gas that is denser than average? Imagine that a passing shock wave compresses cloud material to 4 times the average density, which we will call d_{avg}. In other words,

$$d_* = 4d_{avg}$$

We can divide the free-fall times for both cases and see how they compare.

$$\frac{(t_{ff})_*}{(t_{ff})_{avg}} = \frac{\sqrt{\dfrac{3}{2\pi Gd_*}}}{\sqrt{\dfrac{3}{2\pi Gd_{avg}}}} = \sqrt{\frac{d_{avg}}{d_*}} = \sqrt{\frac{d_{avg}}{4d_{avg}}} = \frac{1}{2}$$

Our calculation tells us that it will take half as long for stars to form in the dense regions. The free-fall time for the denser region of the cloud is half that for the average-density region, so the dense regions collapse to form stars first.

protostar A dense central object, plus an accretion disk and envelope, that forms after cloud collapse but before nuclear fusion begins.

birth line The region on the HR diagram where protostars begin their descent onto the main sequence.

interactive
HR Diagram/
Evolutionary Tracks

Contraction to a Protostar

"Collapse can't last forever," explains Laura Arnold, a former graduate student at the University of Rochester who studied star formation. "As the density at the center of the collapsing core increases, any photons the gas emits begin to get trapped. The infalling gas starts heating up, and pressure begins to play a role again." Trapping radiation in the collapsing cloud's dense center means trapping heat, and that means pressure can begin to halt the collapse.

Where does this heat come from? In their studies of the Universe, physicists have found that energy is always conserved when a physical system evolves. That means the total energy at the beginning has to equal the total at the end. Energy can, however, be transformed from one form into another. In the case of star formation, we begin with an extended cloud with a lot of potential gravitational energy (like a book on a high shelf). As collapse proceeds, that gravitational energy is converted into the kinetic energy of motion—the energy associated with gas in free-fall toward the center of the core (like a book falling from the shelf). As the density at the center of the collapse increases, particles begin colliding more frequently. The kinetic energy of motion is then transformed into heat energy. In the collapse's early phases, this heat is effectively radiated away as photons. But in the later phases, density at the collapse center becomes very high and traps the emitted photons, turning them back into heat and thus raising the temperature and pressure of the gas.

The free-falling collapse of the central regions ends when the gas pressure at the center becomes high enough to decelerate infalling material. The central object is now a warm, dense, extended object that astronomers call a **protostar**. Further evolution of the protostar is governed by how fast photons can be radiated away from its surface. Only by losing energy in the form of light can the protostar continue to shrink more. This process of contraction dominates the subsequent evolution of the protostar.

In general, a protostar is an object with a radius of 50 million km, about the distance at which Mercury orbits the Sun. Densities within the protostar will be, on average, 10^{24} particles/m^3. Temperatures at the surface of the protostar will be 3,000 K, and at the center they will have climbed to 1 million K. Although this is much higher than the temperature in the original cloud (10 K), it is still too low to initiate nuclear fusion. Thus, a protostar is an object that shines because of escaping heat energy alone (the photons that escape are mainly in the infrared and therefore cannot be observed with optical-wavelength telescopes).

Protostars appear on the HR diagram to the right of and above the main sequence. This region is called the **birth line**. From this region, stars of different mass move toward the main sequence and core hydrogen fusion. The path each star of different mass takes down to the main sequence is called a Hayashi track after the Japanese astrophysicist who first calculated them (**Figure 12.14**). High-mass protostars appear at higher luminosities on the birth line than their low-mass cousins, just like the main-sequence stars they eventually become.

FIGURE 12.14
The Birth Line
Here, we see the evolution of stars of different masses on to the main sequence. Protostars enter the HR diagram on the upper right. They are cooler and more luminous than the stars they will eventually turn into. From this region, called the birth line, stars of different masses begin their evolutionary track toward the main sequence and core hydrogen fusion. Colors along the evolutionary tracks (called *Hyashi tracks*) represent the color of the (proto)star based on its temperature at the time.

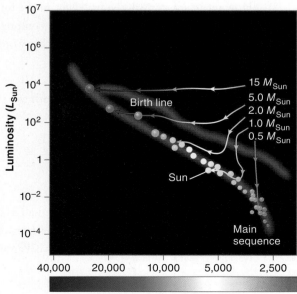

─ **SECTION SUMMARY**
Contraction to a Protostar

- The collapse of a cloud halts when its density and temperature are high enough to trap heat and generate pressure in its interior.

- Because the temperature for initiating nuclear fusion has not yet been reached, heat energy from contraction escaping at the surface of a protostar is the source of the star's energy radiated into space.

Why a Protostar Is More than a Star

In the process of collapse, more structures than just a star are formed. "The centrally condensed object is not the whole story," says Laura Arnold. "A protostar's structure includes the thing that is going to become a star and an accretion disk spinning around it."

Since molecular clouds are turbulent, a cloud's gas is in a constant state of rotation. "Each chunk of the cloud has some spin," says Arnold. Conservation of angular momentum tells us that the velocity of rotation increases as the mass approaches the axis of rotation. This is why an Olympic skater spins faster when she pulls her arms inward (see Chapters 3 and 5). Similarly, as a spinning cloud core begins its gravitational collapse, its rate of spin increases. Gas along the core's spin axis, however, is not rotating at all. It can just fall straight onto the forming protostar. The rest of the material takes a spiral path inward. These parts of the cloud fall inward toward the core's center while they rotate ever faster around the spin axis (**Figure 12.15**).

For gas located at the core's "equator"—the plane perpendicular to the spin axis—the increased rotational velocity about the spin axis prevents it from falling onto the surface of the forming star. "That stuff goes into orbit around the protostar," says Arnold. In this way, an accretion disk of gas and dust is formed. Astronomers consider this disk to be part of the extended object they call the protostar.

Material in the disk, however, doesn't orbit forever. Friction slows the disk material, allowing it to spiral slowly inward. Eventually, much of the gas in the disk makes it all the way to the newly forming star, where it falls—or *accretes*—onto the central object. These disks are also the locations of planet formation (see Chapter 5). "That's why accretion disks are so

Laura Arnold

Laura Arnold's story, like that of many other scientists, started with a love for science engendered by supportive parents. "My father's an engineer who helped design Gorilla Glass for electronics like iPhones. My mother works at the University of Rochester. When I was kid, she enrolled me in a bunch of science, math, and computer science programs they had for young girls." Soon Arnold was finding her own ways to engage with the subject. "My parents tell stories of me doing math on the sidewalk in chalk as a little girl, just adding and subtracting numbers while the other kids were drawing ponies."

Arnold's interest in physics began in high school with an AP class that let her carry forward her own research project. Astronomy, however, came from the night sky. "I grew up right on the beach, where the sky can be really thrilling. Later, I got to see Jupiter's moons through a telescope at the planetarium. I thought that was pretty cool, but when I came back later in the night, they had moved. They were swinging around in their orbit, and I could actually see it happening. That really blew me away."

Once she started college, Arnold was sure she was headed toward a PhD and a research career as an astronomer. But during a summer internship in Germany, she had the chance to focus on teaching rather than research. "I realized that research just wasn't really right for me. I'm really excited and kind of passionate, and I like to talk to people. A life focused only on research wasn't going to be a good fit for me."

On her return from Germany, Arnold threw herself into learning how to teach astronomy. "I helped teach some of the intro courses," she says, "and I loved it." After graduation she looked for science education programs and began to work in earnest toward becoming an astronomy educator. "This is what I love, and I am really excited that I can spend my life doing it."

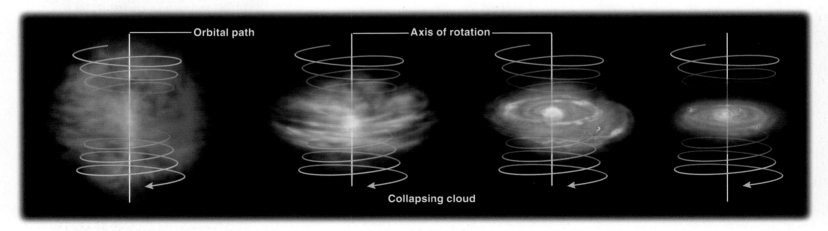

FIGURE 12.15 Formation of a Protostar
The spinning core of a forming star collapses along its axis of rotation, creating a central protostar surrounded by a flat accretion disk. As with a spinning skater who pulls her arms in and rotates faster, during the collapse of the cloud, conservation of angular momentum leads to higher rotation rates. Material along the poles, which is not spinning, can plunge all the way to the center. Material close to the equator eventually rotates fast enough to orbit the central mass that has accumulated. Thus, what began as an extended, slowly rotating spherical cloud ends up as a flattened, rapidly rotating disk at least 1,000 times smaller.

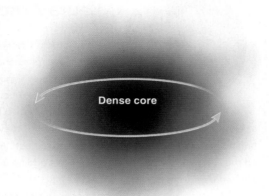

A Dark cloud

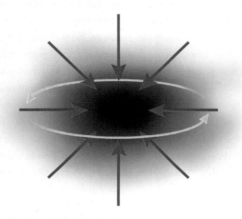

B Gravitational collapse

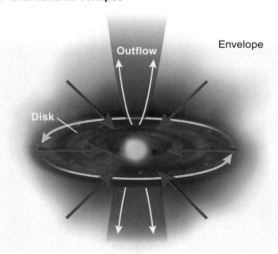

C Protostar

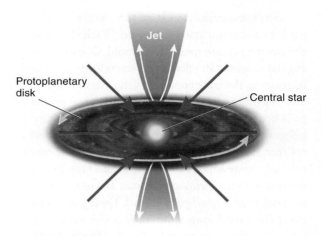

D T Tauri star

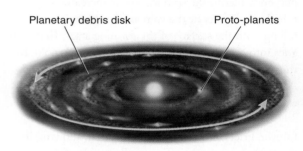

E Pre-main-sequence star

F Young planetary system

FIGURE 12.16 Stages of Formation of a Star and Planetary System
A protostar consists of the centrally condensed object (which will eventually begin nuclear burning) and its surroundings, which include an accretion disk and the surrounding infalling envelope of gas and dust. Beginning with (A) the initial cloud, the stages of evolution are (B) collapse and (C) formation of the protostar, which includes a central mass, accretion disk, and envelope. Outflows and jets will form during this stage. As the envelope thins and the central object continues to contract, the protostar enters (D) the T Tauri phase, and (E) planets begin to form within the disk. (F) Evolution leads to a main-sequence star and planetary system.

important," says Arnold. "They help build the star *and* they're the place where planets grow" (**Figure 12.16**). This dual role compels astronomers to expend enormous effort studying the properties of accretion disks, using both observations and theoretical tools such as computer simulation.

Finally, protostars—the dense central objects plus their accretion disks—are always surrounded by a third component, called the **envelope**. The envelope is made of material from the original gaseous core that is still free-falling inward. As the system evolves, the envelope will continually fall toward the center until eventually only the star and disk remain.

In the later stages of their evolution, protostars whose mass is less than 3 times the mass of our Sun enter the **T Tauri star** phase. The name *T Tauri* comes from the prototypical example of a young star discovered in the constellation Taurus. T Tauri stars are variable stars that emit X-rays created by both powerful flares on the star's surface and magnetic interactions between the star and its surrounding disk.

> ### SECTION SUMMARY
> #### Why a Protostar Is More than a Star
> - A protostar's spin is amplified by conservation of angular momentum during collapse, resulting in an accretion disk.
> - Disk material eventually falls onto the protostar, but some of its material may form planets in the process.

Jets and Outflows: Star Formation Fireworks

In the 1980s, just as astronomers were making the first positive identifications of newly forming stars, they were also beginning to understand the true role of an actor that had been discovered a few decades earlier. "In almost every star-forming region they looked at," explains Arnold, "astronomers also found these bright knots moving at really high speeds."

The knots were first discovered in the 1950s by the Mexican astronomer Guillermo Haro. The American astronomer George Herbig also observed the bright knots. At first these bright knots, called **Herbig-Haro (HH) objects**, were thought to be newly forming stars in themselves. Then, proper motion and Doppler studies showed that all HH objects were racing through space at more than 100 km per second (km/s). This was much too fast for a star, and the true nature of HH objects still remained a mystery. By the 1980s, however, better telescopes revealed gas between the knots. Thus the knots were just the brightest parts of beams of gas that were racing away from the stars.

With this new information, astronomers understood that the HH objects were part of the phenomena known as protostellar jets—beams of high-velocity plasma launched from the protostars into the surrounding molecular cloud (**Figure 12.17**). It is estimated that at least 50 percent of all young stars form protostellar jets. "In some cases the jets are enormous," says Arnold. "We can see them stretching across 5 pc or more." At the head of the jet, where fast-moving jet gas is overtaking slower-moving material, astronomers see bright bow shocks. These are curved shock waves, similar to the waves that form in front of the bow of a boat as it plows through water. The jets begin during the earliest protostellar phases, but they will continue as long as the disk is present and even after fusion has begun in the star.

envelope The region surrounding a protostar that is composed of free-falling gas.

T Tauri star A variable star of less than 3 solar masses that is in a later stage of star formation and exhibits bright X-ray flares.

Herbig-Haro (HH) object A protostellar jet that is distinguished by knots in the jet beam.

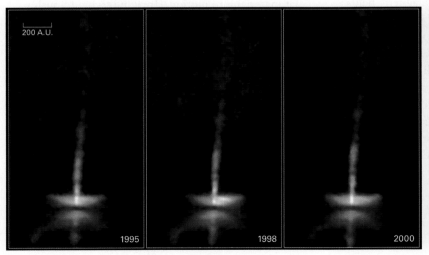

A

B

FIGURE 12.17
Herbig-Haro Objects
Herbig-Haro, or HH, objects are outflows of gas created during star formation. (A) HH 30 is shown here over three time periods as knots of gas are driven away from the star by pulses at the source of the jet. By tracking the motion of these knots over time, astronomers have measured the jets' speed. (B) Bipolar jets HH 901 emerge from a star embedded in the tip of a pillar of gas in the Carina Nebula.

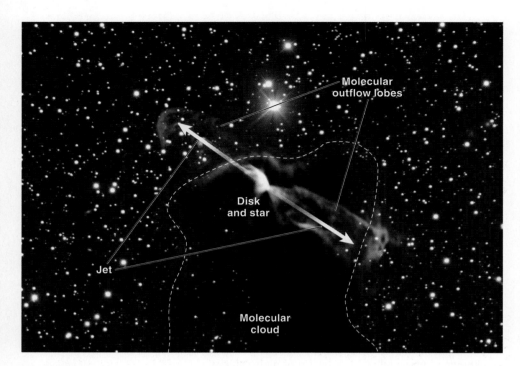

FIGURE 12.18
Bipolar Outflows and Jets
Two-sided, or "bipolar," outflows emerge from a star embedded in a molecular cloud. Here we see an HH jet moving to the right through the cloud, creating a cavity (one lobe of the bipolar outflow). The jet streaming to the left breaks out of the dense cloud, so that its swept-up cavity is not as apparent. This composite image of the object, which is called HH 46/47, shows light emitted by both atoms (the jets) and molecules (the cavities). Since not all bipolar outflows show internal jets, it's unclear whether the two phenomena are always connected.

molecular outflow The bipolar swept-up shells of molecular material, which may be driven by protostellar jets.

If we take 100 km/s to be the typical speed of an HH object, then a 5-pc-long beam implies a central star driving its jet for more than 100,000 years. Since the entire star formation process takes approximately a million years, this simple calculation demonstrates that jet formation occurs for at least 10 percent of the time it takes for a star to be assembled. This large proportion of the star formation time suggests that jets are a critical aspect of the star-making process. But what exactly are these jets, and how do they form?

"Most astronomers agree that magnetic fields have something to do with creating jets," says Arnold. In the most popular models, strong magnetic fields that emerge from the accretion disk act like rigid wires. As the disk rotates, the field lines rotate as well. Because charged particles feel magnetic forces, plasma becomes attached to the fields and is flung outward as it "rides" the field lines. Since jets are seen in environments other than young stars (for example, jets emanate from supermassive black holes near the centers of galaxies), this process may be quite common.

Along with the narrow jets and their bright knots (the HH objects), young stars also produce wider but slower-moving streams of matter called molecular outflows. As the name suggests, the gas in molecular outflows is primarily in the form of molecules, while the Herbig-Haro jets tend to be composed mostly of ions and atoms. The molecular outflows most likely form in the same way as the jets: via magnetic fields and rotation in the disk. The outflows, however, may form farther out in the disk, away from the central star. Cavities opened up by the outflows allow starlight to escape from the dusty inner regions of the protostar even after the jets have weakened or stopped flowing. These regions are one example of the reflection nebulas described earlier, since the light that we see is reflected off dust grains into our line of sight.

"There is a lot of mass in these molecular outflows," says Arnold. "Sometimes we see 0.1 to more than 1 solar mass in the flows." Protostellar jets, in contrast, contain only a fraction of a Sun's worth of material. Both jets and molecular outflows appear, for the most part, as bipolar outflows: material is observed moving away from the star in opposite directions along poles defined by the accretion disk's axis of rotation (**Figure 12.18**). "The fact that bipolar outflows are aligned with the disk's axis of rotation tells astronomers that the disks and the outflows are connected," says Arnold. "The magnetic fields in the disk are the likely link between the two."

Although astronomers have found these two dramatic forms of outflows emerging from young stars (often from the same stars), they do not yet understand the connection between them. An early proposal was that jets actually drive the molecular gas, just as a snowplow pushes snow in front of it. But astronomers soon found the jets' energy and momentum to be too weak to drive all the observed molecular material into motion. (Imagine trying to push a mountain of snow using a snowplow connected to a go-cart.) For now, the relationship between jets and outflows remains on the frontiers of what we do, and do not, understand.

SECTION SUMMARY
Jets and Outflows: Star Formation Fireworks
- Fast-moving protostellar jets, with their bright knots called Herbig-Haro objects, are observed emanating from young stars for a significant portion of their lives and are thought to be caused by magnetic-field lines from the disk and star.
- Slower-moving molecular outflows are also observed, but the relationship between the jets and outflows is not well understood.

From Protostars to Fusion and Brown Dwarfs

A protostar is a long way from generating its own energy via nuclear fusion. The density and temperatures in its core are simply too low for hydrogen nuclei (protons) to overcome electromagnetic repulsion and fuse into helium. Recall that the proton-proton (P-P) chain, which occurs in stars like our Sun, requires central temperatures of greater than 10 million K. How do protostars reach that magic temperature and graduate into full-fledged stars? Time and contraction are the answer, and along the way a lot more evolution has to occur. That evolution is closely tied to a star's mass.

P-P Chain and CNO Cycle

The protostar continues to shrink as heat energy escapes from its surface. With each decrease in radius, both the central density and the central temperature increase. The protostar's surface temperature, however, does not increase by very much. During this final contraction phase, the evolutionary track of the protostar moves straight downward on the HR diagram: the protostar decreases in luminosity but maintains an almost constant temperature.

As the core temperature increases, a series of fusion reactions is initiated (such as the fusing of two protons into deuterium), but these reactions are relatively ineffective at generating energy. Eventually, in protostars larger than about 0.08 solar mass (M_{Sun}), a full network of reactions begins that turns hydrogen nuclei into helium nuclei. This transformation occurs by way of the P-P chain in stars like the Sun.

In stars just 1.3 times more massive than the Sun and greater, where core temperatures exceed 16 million K, another route for fusion of hydrogen to helium dominates: the **CNO (carbon-nitrogen-oxygen) cycle**. In the CNO cycle, nuclei of the elements carbon, nitrogen, and oxygen play an intermediate role. The chain of reactions begins with a hydrogen nucleus fusing with a ^{12}C nucleus to create an unstable (radioactive) nucleus of ^{13}N. The ^{13}N then decays into ^{13}C, which combines with another hydrogen nucleus to form ^{14}N. The chain of reactions continues until, finally, a ^{4}He nucleus is created (**Figure 12.19**). The CNO cycle is very temperature sensitive. Small

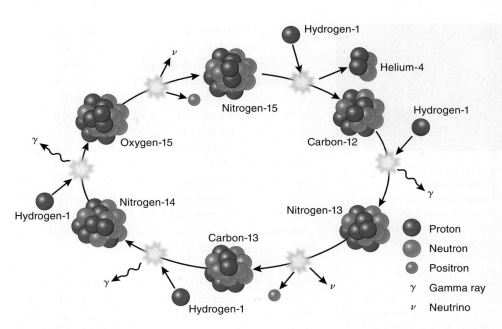

FIGURE 12.19 CNO Cycle
In the CNO cycle, massive stars convert hydrogen to helium through the decay of unstable isotopes and the fusion of carbon, nitrogen, and oxygen nuclei with hydrogen nuclei. It is the dominant source of energy in stars larger than about 1.3 M_{Sun}. It is most effective at temperatures of 16 million K and above. Each "flash" in the image represents a transformation of some kind—a decay or a reaction.

increases in temperature at the stellar core will lead to much larger increases in energy production in stars dependent on the CNO cycle than in stars fueled by the P-P cycle. (Chapter 13 explains the consequences of this sensitivity for the lifespans of stars.)

Of course, for the CNO cycle to occur, the star must already contain some atoms of carbon, nitrogen, and oxygen. Those elements must have come from earlier generations of stars. When the stars died, these elements were blown back into space, eventually ending up in the cloud that formed the protostars of today.

SECTION SUMMARY
P-P Chain and CNO Cycle
● Protostars contract over hundreds of thousands of years, increasing their internal temperatures.
● The mass of the protostar determines whether its future fusion process will be the P-P chain or the CNO cycle.

Brown Dwarfs

For protostars of very low mass ($M < 0.08\ M_{Sun}$), core temperatures never get high enough to burn hydrogen. With too little mass to drive further contraction, these

CNO cycle A chain of nuclear fusion reactions, involving carbon, nitrogen, and oxygen, that produces helium. It is the mechanism for powering stars of more than 1.3 solar masses.

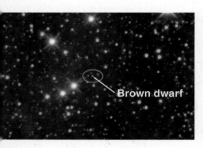

FIGURE 12.20
Brown Dwarfs
Brown dwarfs are too cool to radiate appreciably in visible light, but by using infrared telescopes, astronomers can see swarms of them in star clusters, such as the Trapezium star cluster in the Orion Nebula. This image, captured via the Wide-field Infrared Survey Explorer (WISE) survey, shows a single ultracool brown dwarf (the circled green star) among a field of stars.

brown dwarf A failed star that has too little mass to drive core temperatures high enough to initiate hydrogen-burning nuclear fusion.

initial mass function (IMF) The mathematical relation describing the relative abundance of stars of different masses.

objects, called **brown dwarfs**, can only initiate the fusion of deuterium. They spend billions of years slowly bleeding their stores of heat energy into space (**Figure 12.20**). "Brown dwarfs are essentially failed stars," says Laura Arnold. "They lie somewhere between a very massive planet like Jupiter and a full-fledged, fusion-powered, low-mass star."

The range of masses for brown dwarfs bookends the two categories that Arnold describes. Any object with a mass larger than $0.08\ M_{Sun}$ will drive hydrogen fusion in its core and must be considered a star. The lower end of the mass spectrum for brown dwarfs is less certain, though some researchers put it at $0.01\ M_{Sun}$, which is about 10 Jupiter masses.

"Using fusion to define the difference between a star and a brown dwarf is very clear," says Arnold, "but the line between a low-mass brown dwarf and a high-mass planet is less clear." Generally, astronomers believe that brown dwarfs form from gravitational collapse in molecular clouds. Planets, even the most massive ones, are believed to form within the accretion disks surrounding young stars.

<hr>

SECTION SUMMARY
Brown Dwarfs
● Brown dwarfs are "failed" stars that do not have enough mass to initiate fusion as they contract.

High-Mass Stars versus Low-Mass Stars

What brown dwarfs lack in mass they make up in number. Astronomers expect that 100 billion brown dwarfs may be floating freely in our galaxy (a smaller number may be members of binary or other multiple-star systems with fusion-powered stars). In total, the number of brown dwarfs is probably equal to the number of fusion-powered stars in the Milky Way.

Thus, the galaxy has far more brown dwarfs than 1-M_{Sun} stars, and in turn, there are far fewer high-mass stars than 1-M_{Sun} stars. This overall negative correlation between stellar population and stellar mass is neatly captured in what astronomers call the **initial mass function (IMF)**. The IMF tells astronomers how many 1-M_{Sun} stars they expect to form relative to the number of 0.5-M_{Sun} stars, how many 2-M_{Sun} stars they should expect relative to 1-M_{Sun} stars, and so on.

The shape of the IMF was discovered many years ago by astronomer Edwin Ernest Salpeter, who found a steep decrease in the number of stars with larger

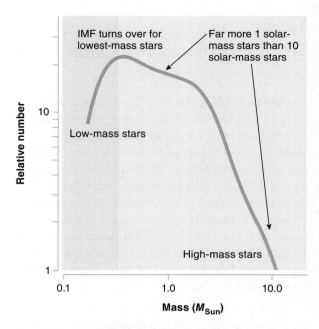

FIGURE 12.21 Initial Mass Function
Not all stellar masses are equally represented in our galaxy: lower-mass stars far outnumber high-mass stars. The distribution of stellar masses is set at birth as stars form from molecular clouds. The initial mass function (IMF), discovered by astronomer Edwin Ernest Salpeter, describes the steep decrease in the numbers of stars with larger and larger masses. Only at the lowest end of the stellar-mass range do we see the curve turn over, with the lowest-mass stars appearing in fewer numbers than those that have more mass.

and larger mass (**Figure 12.21**). From Salpeter's work, astronomers expect that for every thousand 1-M_{Sun} stars that form, only a single 10-M_{Sun} star should be created. Massive stars, which astronomers define as having a mass greater than $8\ M_{Sun}$, are thus relatively rare in the galaxy.

The reasons that stars form with such a particular distribution of masses (the IMF) is still a mystery to astronomers. They do understand that the gravitational collapse process favors fragmentation into smaller, rather than larger, protostars, but the detailed physics that sets the IMF into its exact mathematical relationship remains an open question.

<hr>

SECTION SUMMARY
High-Mass Stars versus Low-Mass Stars
● The initial mass function (IMF) tells astronomers how often stars of a particular mass should occur in a population of stars.
● There are many more low-mass stars than high-mass stars.

12.6
Stellar Interaction

So far, we have been discussing star formation as if each star formed in isolation. In most cases, however, stars form in relatively close proximity to other stars, and they can affect one another's evolution.

A Family Affair: Young Star Clusters

Stars are almost never born alone. Instead, they form in families called young star clusters. The dense structures called clumps within molecular clouds, with masses that are typically in the range of 1,000 M_{Sun}, are the forerunners of star clusters. The similarity between young star clusters and the globular clusters discussed in Chapter 11 is that in both cases the stars all formed together. The difference, however, is that eventually the stars in a young cluster will disperse; in other words, they will not remain gravitationally bound together as the stars in a globular cluster do.

The best-known young star cluster is the Pleiades, a collection of over a thousand stars visible from Earth's Northern Hemisphere (**Figure 12.22A**). The Pleiades is an example of an **open star cluster**, meaning that the stars are not as tightly packed together as they are in some clusters. Seven bright, massive stars called the "Seven Sisters" dominate our view of the Pleiades. (These stars are so noticeable that

they became the logo for auto maker Subaru, which is the Japanese word for "Pleiades.") **Figure 12.22B** shows an HR diagram for all the stars in the Pleiades although only six are visible with the naked eye. Notice that most of the stars appear on the main sequence. Only the most massive stars have begun to evolve away from hydrogen burning. "That means the Pleiades must be young," says Laura Arnold. "All of its stars were born less than tens of millions of years ago."

Other clusters, such as NGC 1333 in the Perseus molecular cloud, contain only a few hundred stars, none of which are high-mass O stars, though a few B stars may be part of NGC 1333's population. Some clusters, however, are far more densely populated. Star formation clusters appear to occur in two modes: high-mass (massive) and low-mass. "The number of stars that form within a cluster," says Arnold, "will depend on the size and conditions in the clump. Massive clumps make lots of stars (and some O stars too), while low-mass clumps tend to make clusters with fewer stars and mostly low-mass stars." Thus, massive stars appear only in massive clusters where many thousands of stars (most of them low- and intermediate-mass) are born. The Orion Nebula, an H II region at the edge of the Orion molecular cloud that can be seen even with a small-aperture telescope or binoculars, is one example of a so-called rich, young star cluster because it contains so many new stars. An even more spectacular example is the Carina Nebula, visible from Earth's Southern Hemisphere

open star cluster A loosely distributed group of stars that were born from the same cloud at the same time and remain bound for some period before dispersing.

A

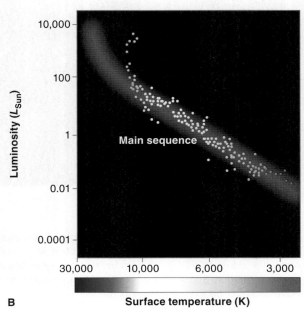

B

FIGURE 12.22
HR Diagram for the Pleiades
(A) The Pleiades is an open star cluster, readily spotted with the naked eye in winter skies from Earth's Northern Hemisphere. (B) Its HR diagram reveals its young age (only a few tens of million years), since only the most massive stars have evolved off the main sequence.

A

FIGURE 12.23
Molecular Clouds in the Carina Nebula
(A) Dark knots in the Carina Nebula are regions of dense molecular gas, seen here in optical light. (B) In infrared light, we see the complex structure of a large region of molecular cloud. Here, pink indicates dense dusty regions, and hot gas ionized by massive stars is colored green. Note the structures in the dense gas that look like pillars.

B

(**Figure 12.23**). The Carina Nebula contains many rich clusters, including one of the most massive stars ever studied (Eta Carinae, at approximately 125 M_{Sun}) and at least 14,000 stars in total.

Compared with the range of ages across the rest of the galaxy, the population of stars in a cluster can be thought of as all having the same birthday. Closer examination shows that stars within a cluster do have a range of ages, but it's narrow—on the order of a few million years. "Like triplets in a human family, some stars will be born before others," says Arnold, "even though on galactic timescales they all appear to be born at the same time."

The range of ages within a cluster may be a consequence of the turbulence occurring in the collapse process. Another contributing factor may be a process called feedback, in which the star formation in one region of a cluster may affect the star formation in another region. Some astronomers have argued that the jets, winds, and radiation from newborn stars can either suppress or enhance local star formation. Feedback takes its most dramatic form in a process called **triggered star formation**, in which a supernova blast wave, created by the death of a nearby massive star, sweeps through a molecular cloud, compressing gas and pushing it into gravitational collapse (**Figure 12.24**). If triggered star formation occurs in a molecular cloud, astronomers would expect to see successive generations of star clusters spread across a region of the cloud.

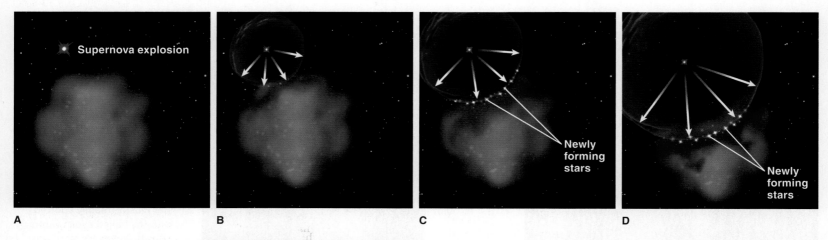

A   **B**   **C**   **D**

triggered star formation Star formation that is initiated by the compression of a molecular cloud by a supernova blast wave.

FIGURE 12.24 Triggered Star Formation
Triggered star formation is a dramatic example of how star evolution in one region of a cluster may affect star formation in another region. (A and B) The supernova blast wave from the death of a nearby massive star sweeps through a molecular cloud. (C and D) By compressing and gravitationally collapsing the cloud, the blast wave sets off a new cycle of star formation. Successive generations of this phenomenon can occur if massive stars are formed by the first blast wave and later become supernovas themselves.

┌─ **SECTION SUMMARY**

A Family Affair: Young Star Clusters

● Stars in clusters tend to be born at about the same time as their siblings.

● Triggered star formation results from supernova blast waves, an example of feedback in star formation.

High-Mass Star Formation: Mess with Your Surroundings

High-mass stars—in particular, O stars with masses of 25 M_{Sun} and above—dominate the regions of a molecular cloud where they form, and sometimes far beyond. Recall that O stars have high temperatures, of more than 30,000 K. "Look at a blackbody spectrum for a star like this," says Arnold, "and you'll see it generates lots and lots of ultraviolet photons." Recall that ultraviolet photons with a wavelength less than 91.2 nm can photoionize neutral H atoms. Since O stars produce a torrent of these photons, they are cosmic engines of ionization. "And it's not just neutral atoms that feel the heat," says Arnold. "Molecules get dissociated by the ultraviolet light of O stars too."

A single O star can ionize material out to 100 pc or so—an enormous volume of space. When the gas surrounding an O star becomes ionized, it is also heated. "Molecular material that was 10 degrees above absolute zero before the O star 'turned on,'" explains Arnold, "gets flash-ionized and heated. Its temperature jumps to 10,000 degrees. High temperatures mean high pressures, so the ionized gas expands." Like a rapidly inflating balloon, the hot gas sweeps up surrounding cloud material that has yet to be ionized. Eventually, the expanding bubble, now an H II region, reaches the edge of the molecular cloud and rapidly blows outward into the lower-density ISM. In this way, the ionized H II regions created by O stars are agents for dispersing the clouds.

As the ionizing radiation eats into cold molecular material, it also sculpts some of astronomy's most astonishing and beautiful structures. If knots of gas are dense enough, the ultraviolet radiation will ionize only the outer layers, while the interiors remain neutral (or molecular). Material behind the knots (relative to the hot star) is also shielded from ultraviolet photons. In this way, the molecular cloud material at the edges of the H II regions often shows long fingers pointing toward the hot stars that produce the ultraviolet radiation. These **elephant trunks** may have low-mass stars forming inside them that can be seen only via infrared

M16 ▪ Eagle Nebula

Hubble Space Telescope ▪ WFC3/UVIS/IR

Visible

Infrared

NASA and ESA

STScI-PRC15-01c

radiation (**Figure 12.25**). Eventually, however, the O stars' ultraviolet photons erode the trunks away. "If dense knots in the elephant trunk haven't formed stars by that time," says Arnold, "they get eroded too, and their evolution will be cut short."

The ionizing radiation affects more than just the molecular gas. A single newly formed O star will be surrounded by thousands of lower-mass stars (recall the IMF). The histories of these newly forming low-mass stars, which are still surrounded by their accretion disks, will also be altered by the O star's ultraviolet photons. When an accretion disk is irradiated by ionizing photons, its surface heats up. The pressures on the disk surface can be high enough to launch gas into space through a process called **photoevaporation**. So much mass can be lost from a disk via these photoevaporation disk winds that any ongoing planet formation within the disk can be cut short. If the flux of ultraviolet photons from the massive star is high enough, the entire disk can be boiled away in just 10,000 years.

An example of all the processes described above can be seen in the Orion Nebula. The Orion Nebula is a stellar hothouse. More than a thousand newly formed stars are crowded together in a cubic parsec (for comparison, note that the Sun's nearest neighbor is more than a parsec away from us). While most of these stars are of low or intermediate mass (less than

FIGURE 12.25
Elephant Trunks in the Eagle Nebula Ionizing radiation from a massive star (off to the top of this image) erodes dense molecular gas at the edge of an H II region. This phenomenon, sometimes called "elephant trunks," or in this case "pillars," occurs in many H II regions. The images here show pillars in (A) optical wavelengths and (B) infrared wavelengths. Star formation can occur within the densest region of the elephant trunks.

elephant trunk A linear feature at the edge of an H II region composed of molecular gas and pointing toward the ionizing star or stars.

photoevaporation The process by which gas is launched from the surface of an accretion disk or molecular cloud through heating by ionizing radiation from a young, massive star.

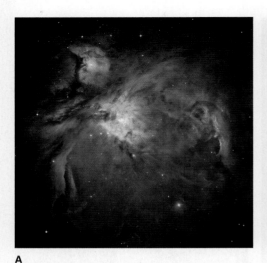

Infrared • NICMOS

A

B

FIGURE 12.26 Orion Nebula and the Trapezium
(A) The Orion Nebula, with more than 2,000 stars forming in a region just a few parsecs across, provides rich examples of the processes by which stars form from turbulent, collapsing clouds of gas and dust. (B) A cluster of four massive stars at the core of the nebula called the Trapezium provides the ultraviolet light that creates the extended H II region.

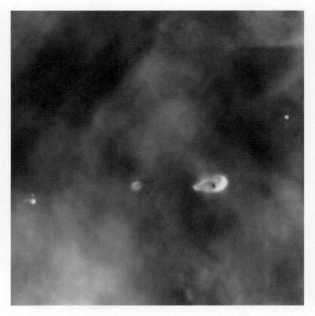

FIGURE 12.27 Proplyds in the Orion Nebula
Proplyds are accretion disks around young stars that are being photoionized by the ultraviolet light of a nearby massive star. The ultraviolet photons ionize the surface of the disk, heating it and creating a wind that rises off the disk. This "disk wind" subsequently encounters winds being driven from the massive star. A teardrop-shaped shock wave forms from the interaction of the disk wind and the stellar wind.

8 M_{Sun}), the evolution of the nebula is dominated by four massive stars (one O star and three B stars) called the Trapezium (**Figure 12.26**). The ultraviolet flux from these stars (mainly the single O star) illuminates, ionizes, and sculpts the Orion Nebula.

Accretion disks surrounding individual lower-mass stars can be seen in the Orion Nebula, and in some cases they appear in shadow as they block ionizing light from the Trapezium. In other cases, the photoevaporative winds rising from the disks' surfaces run into stellar winds driven by the O star, forming teardrop-shaped shock waves that are bright enough to be seen across the nebula. These objects, called **proplyds**, have been found in a number of star-forming regions with massive stars, such as the Orion and Carina Nebulas. Proplyds provide ample evidence of the effect just a few massive stars can have on a star-forming region (**Figure 12.27**).

In today's Universe, there appears to be an upper limit to the mass of a newly formed star. Astronomers don't see stars above about 150 M_{Sun}. The reason for this limit is the force exerted by the very light these stars create. High-mass stars are brighter than low-mass stars, and stars above about 125 M_{Sun} (no one knows the exact limit) may produce so much light pressure (force per area) that they cannot hold together.

proplyd A teardrop-shaped knot of gas formed by the interaction of stellar winds and the photoevaporation flow from a protostar's accretion disk.

SECTION SUMMARY
High-Mass Star Formation: Mess with Your Surroundings

- High-mass (and therefore very hot) stars produce enough ultraviolet photons to ionize large volumes of gas around them, creating hot H II regions that can expand into the ISM.
- The accretion disks of low-mass stars can be boiled away by the ionizing photons from high-mass stars in the same cluster.

●→ chapter summary

12.1 Seeing in the Dark

Infrared light is needed to study star-forming clouds and the star formation process, since visible light is absorbed by dust in the clouds. The Spitzer Space Telescope, launched in 2003, was one example of a tool used in that effort.

12.2 Anatomy of the ISM

The space between the stars (within galaxies) contains a turbulent mix of gas and dust called the interstellar medium (ISM). The ISM has different phases, including molecular clouds (where stars form), H I clouds of atomic hydrogen gas, H II regions, the warm interstellar medium, and coronal gas. All phases of the ISM tend to have pressures in the same range. In the ISM, dust acts as a platform on which molecules can form.

12.3 Molecular Clouds: The Birthplace of Stars

Molecular clouds are the birthplaces of stars, and they contain many molecules, such as water and alcohol. They vary widely in size and shape, and they tend to be turbulent, with clumpy, dense regions referred to as cores.

12.4 From Cloud to Protostar

Gravitational collapse occurs when internal forces can no longer support a cloud against its own gravity. When a cloud does collapse, its free-fall time depends inversely on its density. The gravitational potential energy inside the collapsing cloud is transformed into infalling motions and then internal gas pressure as its core heats up. Once the protostar is hot enough, its gas pressure stops the collapse, and further contraction requires the star to radiate away energy. The original collapse of the spinning cloud creates an accretion disk of material, some of which falls onto the protostar and some of which may form planets. An envelope of still-infalling gas surrounds the protostar. Bipolar outflows associated with star formation include protostellar jets with embedded knots of gas (called Herbig-Haro objects) and slower molecular outflows that surround the jets. Both kinds of outflow are believed to form with the help of the protostar's magnetic field.

12.5 From Protostars to Fusion and Brown Dwarfs

Continued contraction of the protostar causes it to heat up enough to initiate nuclear fusion. Hydrogen is fused into helium via either the P-P chain or, if the mass of the star is high enough, the CNO cycle. If the mass is not sufficient for hydrogen fusion, a brown dwarf will result. Stars form in a broad range of masses, from 0.08 to 150 M_{Sun}, but the most massive stars are very rare. Such massive stars radiate ultraviolet photons that ionize atoms and molecules around them, producing H II regions that expand and blow out into the less dense ISM. The ultraviolet radiation can also cause accretion disks around nearby smaller stars to evaporate. The initial mass function gives the relative proportions of stars of different mass that are expected to be created in the star formation process.

12.6 Stellar Interaction

Stars are born together over a few million years in families called young star clusters. Star formation clusters are found in two sizes—high-mass and low-mass—and the most massive stars are born in the high-mass star clusters. Young and fully formed stars can influence star formation around them via feedback processes. Supernova blast waves can cause triggered star formation.

●→ questions and problems

Narrow It Down: Multiple-Choice Questions

1. Which of the following characteristics is most likely to be the same among stars in a given cluster?
   a. age
   b. spectral type
   c. temperature
   d. mass
   e. presence of planets

2. Which phase of the ISM is most opaque to visible light?
   a. H I region
   b. H II region
   c. molecular cloud
   d. WISM
   e. coronal gas

3. Which of the following is/are *not* a source of turbulence in the ISM?
   a. rotation of the galaxy
   b. supernova explosions
   c. ultraviolet radiation from young stars
   d. jets from young stars
   e. infrared radiation from young stars

4. Given three forms of hydrogen, rank them correctly according to increasing temperature of the environments in which they are usually found.
   a. neutral, ionized, molecular
   b. molecular, neutral, ionized
   c. ionized, neutral, molecular
   d. molecular, ionized, neutral
   e. neutral, molecular, ionized

5. Which phase of the ISM is found near hot, young stars?
   a. H I region
   b. H II region
   c. molecular cloud
   d. WISM
   e. coronal gas

6. Which of the following is *not* a quality that all phases of the ISM share?
   a. absence of free electrons as part of gas
   b. more gas than dust
   c. more hydrogen than other elements
   d. turbulence
   e. much lower density than that of Earth's atmosphere

7. What kinds of photons are absorbed by typical dust grains?
   a. infrared
   b. visible
   c. ultraviolet
   d. radio
   e. microwave

8. Why does interstellar dust play a vital role in the chemistry of the galaxy?
   a. It allows gamma rays to penetrate into dark clouds.
   b. When it breaks apart, it provides new elements for the ISM.
   c. There is so much more of it than there is gas.
   d. Chemistry can occur on its surface.
   e. Its color changes the appearance of stars.

9. How do molecular clouds provide protection from ultraviolet photons that can dissociate molecules?
   a. Their thermal energy deflects the ultraviolet photons.
   b. Their temperature is too low for the photons to be effective.
   c. Water within the cloud absorbs the photons.
   d. Their turbulence keeps molecules in rapid motion.
   e. The dense gas and dust in their interiors shield molecules from interstellar ultraviolet radiation.

10. Which of the following is *not* a characteristic of molecular clouds?
    a. uniform structure throughout
    b. presence of several kinds of molecules
    c. high density, relative to other phases of the ISM
    d. presence of dust
    e. mass many times greater than the Sun's

11. Many astronomical objects have dense cores. Which of the following do *not*?
    a. stars
    b. protostars
    c. molecular clouds
    d. globular clusters
    e. coronal gas clouds

12. Which of the following places the objects in increasing order of radius?
    a. GMC, Solar System, Sun, protostar, brown dwarf
    b. GMC, protostar, Solar System, Sun, brown dwarf
    c. brown dwarf, Sun, protostar, Solar System, GMC
    d. Sun, brown dwarf, protostar, Solar System, GMC
    e. brown dwarf, Sun, Solar System, GMC, protostar

13. Which of the following is *not* a typical outcome for material in an accretion disk around a protostar?
    a. It accretes onto the protostar.
    b. It becomes part of a planet.
    c. It gets eroded by photoevaporation.
    d. It remains in the disk indefinitely.
    e. It returns to the ISM.

14. Two protostars have evolved to the point of nuclear fusion. One has a temperature of 12 million K; the other, a temperature of 17 million K. Which of the following statements is/are true about the stars? Choose all that apply.
    a. The 17-million-K star burns nuclear fuel faster.
    b. The 17-million-K star may be using the CNO cycle of nuclear fusion.
    c. The 17-million-K star will die sooner.
    d. The 17-million-K star can be using only the P-P process of nuclear fusion.
    e. The stars may have identical masses.

15. A star has a mass of 0.7 M_{Sun}. Which statement about it is true?
    a. It will live less than 10 billion years.
    b. Its temperature is greater than 16 million K.
    c. It may be using the CNO cycle for nuclear fusion.
    d. It is a brown dwarf.
    e. It is using the P-P process of nuclear fusion.

16. From a 1,000-M_{Sun} molecular cloud, one or more stars form. Which of the following is likely to be true?
    a. Many stars form that are almost all 1-M_{Sun} stars.
    b. One 1,000-M_{Sun} star forms.
    c. Since molecular clouds exist near massive young stars, none of the stars will have accretion disks.
    d. A number of stars will form with a range of sizes, and most will be less-massive stars.
    e. Astronomers will be able to observe all the stars that form in visible wavelengths throughout the star formation process.

17. The last supernovas were observed in the Milky Way Galaxy more than 400 years ago. Which of the following statements related to these events is/are true? Choose all that apply.
    a. The rarity of these events is explained, at least in part, by the IMF.
    b. Each one may trigger the formation of new stars.
    c. Turbulence will increase in the ISM surrounding these supernovas.
    d. Coronal gas will have been created in the process.
    e. Many more would have been observed in our galaxy with today's observing techniques.

18. Which of the following statements about interstellar dust is *not* true?
    a. It provides a platform for the formation of molecules.
    b. It emits blackbody radiation.
    c. It makes distant objects appear closer than they are.
    d. It shields star-forming regions from ultraviolet light.
    e. It is much rarer than gas in the ISM.

19. Carbon monoxide is useful as a tracer for molecular clouds because
    a. it is found in high concentrations.
    b. it can be excited to emit photons at lower temperatures than many other molecules do.
    c. it is an organic molecule.
    d. the light it emits is mostly ultraviolet light.
    e. it is found at higher temperatures than other molecules within molecular clouds.

20. An object has a mass that is less than 8 percent of the Sun's mass, and its elemental composition includes carbon, nitrogen, and oxygen. Which of the following is/are true? Choose all that apply.
    a. It could be a planet.
    b. It could be a brown dwarf.
    c. It must be on or moving toward the main sequence.
    d. It must not have reached 16 million K.
    e. It must be fusing protons into deuterium.

To the Point: Qualitative and Discussion Questions

21. Name and describe the phases of the ISM. How do these phases differ, and what do they have in common?

22. Where does the WISM's warmth come from?

23. In the Sun's structure, the corona is the outer layer of the atmosphere. Where is coronal gas found in the ISM?

24. How does the ideal gas law relate to the ISM?

25. What two measurement techniques provide astronomers with information about the temperature and composition of interstellar dust?

26. Astronomers believe most molecular clouds are made up of primarily molecular hydrogen, even though they locate most molecular clouds by using observations of CO emission lines. Why don't they just look for the molecular hydrogen?

27. Why do molecular clouds have limited lifetimes?

28. What is the birth line on the HR diagram, and where is it?

29. Describe the components of a protostar.

30. How does a disk form around a protostar?

31. Describe the differences in the formation of a brown dwarf versus a planet; then describe the differences in the formation of a brown dwarf versus a main-sequence star.

32. How do live, massive stars affect nearby stars and clouds? What about dying massive stars?

33. Define *proplyds* and *elephant trunks* as they relate to star formation and the ISM.

34. Describe the cause, location, and direction of jets in young stars.

35. The original mission of the Spitzer Space Telescope lasted from launch in 2003 to depletion of its helium coolant in 2009, but then the mission was redesigned to use the capabilities that remained. Do some outside reading and see whether you can identify other space missions that have done the same.

Going Further: Quantitative Questions

36. Molecular cloud A has a pressure of P. Cloud B has a temperature half that of cloud A and a density 3 times as great. What is the ratio of the pressure in cloud B to that in cloud A?

37. Two regions of the ISM have identical pressure, but the temperature of region 1 is 2.5 times that of region 2. How does the density of region 1 compare with that of region 2?

38. Radiation from a nearby star has heated a cloud in the ISM from 100 K to 550 K, and its density has dropped to 0.25 times its earlier density. What is the ratio of its current pressure to its pressure before these changes?

39. Region 2 of an interstellar cloud has 1.7 times the density of region 1. Assuming the pressure of the two regions is equal, what is the ratio of the temperature in region 2 to that of region 1?

40. The temperature in a region of coronal gas is 10^6 K, while a nearby H II region has a temperature of 10,000 K. Assuming the pressure of the two regions is equal, what is the ratio of the density of the coronal gas to that of the H II region?

41. The free-fall time of a gravitationally collapsing cloud is T. What would be its free-fall time in terms of T if its density were 3.2 times as high?

42. A cloud C_1 of density D has a calculated free-fall time of T_1. For a similar cloud C_2 whose free-fall time is twice as long, what is the ratio of its density to that of C_1?

43. Two separate cores of equal density in a molecular cloud have radii of 1 light-year and 1.7 light-years, respectively. How does the free-fall time of the larger cloud compare with that of the smaller one?

44. Free-fall time for region 1 of an interstellar cloud is 1.3×10^6 years; for region 2, free-fall time is 0.9×10^6 years. What is the ratio of the density of region 2 to that of region 1?

45. From Salpeter's projections, how many 10-M_{Sun} stars, on average, will form among 17,000 M_{Sun} newborn stars?

 If your instructor assigns homework in **smartwork5**, access it at the Digital Landing Page for *At Play in the Cosmos*: **digital.wwnorton.com/cosmos**

TO THE GRAVEYARD OF STARS

The End Points of Stellar Evolution

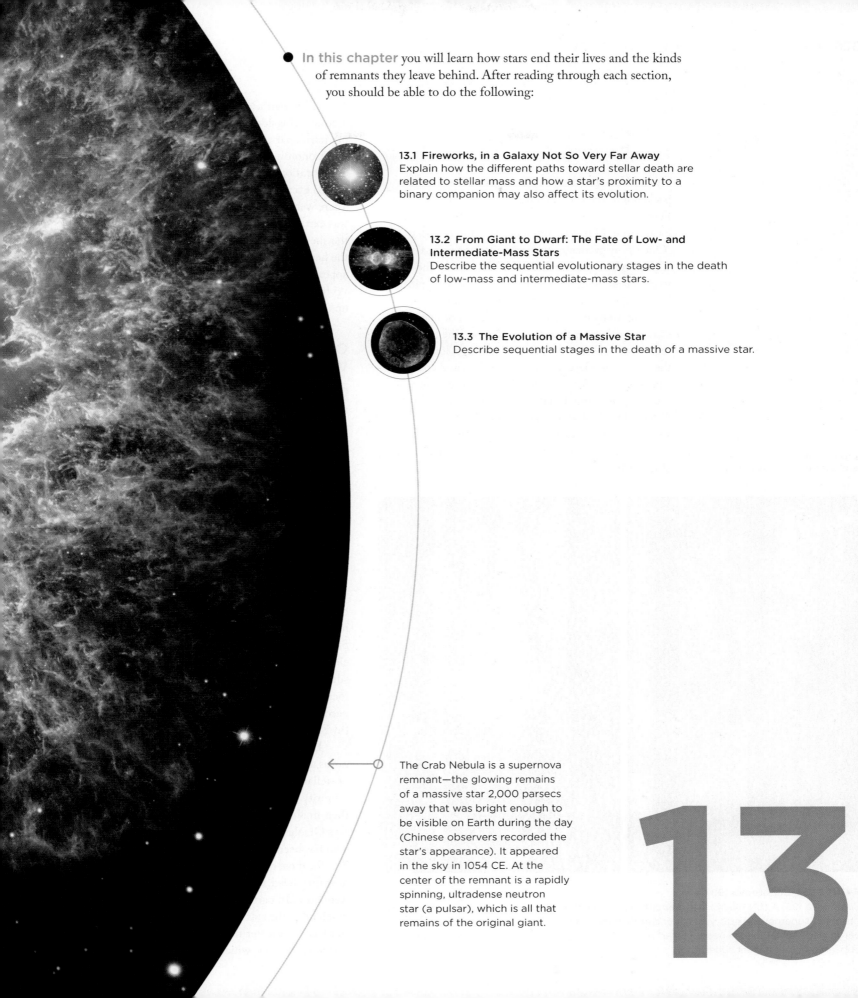

In this chapter you will learn how stars end their lives and the kinds of remnants they leave behind. After reading through each section, you should be able to do the following:

13.1 Fireworks, in a Galaxy Not So Very Far Away
Explain how the different paths toward stellar death are related to stellar mass and how a star's proximity to a binary companion may also affect its evolution.

13.2 From Giant to Dwarf: The Fate of Low- and Intermediate-Mass Stars
Describe the sequential evolutionary stages in the death of low-mass and intermediate-mass stars.

13.3 The Evolution of a Massive Star
Describe sequential stages in the death of a massive star.

The Crab Nebula is a supernova remnant—the glowing remains of a massive star 2,000 parsecs away that was bright enough to be visible on Earth during the day (Chinese observers recorded the star's appearance). It appeared in the sky in 1054 CE. At the center of the remnant is a rapidly spinning, ultradense neutron star (a pulsar), which is all that remains of the original giant.

13

13.1
Fireworks, in a Galaxy Not So Very Far Away

STAN WOOSLEY had all his family's ski equipment packed into the car when the call came. It was February 23, 1987, and the Universe, or at least our local corner of it, had different plans for Woosley. "One of my graduate students called me and said, 'Check your mail,'" explains Woosley, one of the world's leading experts on the titanic stellar explosions called supernovas. "At first I thought it was a joke."

Supernovas, which mark the death of massive stars, are as rare as they are spectacular. The last one seen in our galaxy was four centuries ago, long before the advent of telescopes, spectrographs, and computer analysis. But when Woosley checked his e-mail that morning, he learned that a supernova had just been observed in the Large Magellanic Cloud, or LMC, a satellite galaxy orbiting the Milky Way. The supernova was close enough to be seen with the naked eye (**Figure 13.1**).

supernova The highly energetic explosion of a massive star.

After waiting 400 years, astronomers were finally getting a ringside seat to watch a massive star tearing itself apart. "It turned out to be a golden weekend in astronomy," Woosley explains. "An international collaboration of teams from all around the world turned every instrument they had at the supernova. There were so many discoveries happening so fast, it was extraordinary. And of course there was also just the sheer excitement and joy at this new presence in the heavens." But not everyone was overjoyed at the appearance of supernova 1987A (as it was called). "The ski trip had to be canceled," laughs Woosley. "I made it up to my family later that year."

Out, Out, Brief Candle: Why Stars Die

Gravity pulls stars together, driving their central cores to extraordinary temperatures and densities at which nuclear fusion reactions become possible. In turn, only through the energy released by this nuclear fusion do stars manage to support themselves against the crushing force of their own gravity. But sooner or later, the raw fuel for fusion has to run out. When the fuel in the stellar core is gone, the star finds itself on the losing end of its gravity war and heads inevitably toward its death. This chapter takes up the story of stellar evolution at that point and brings the story to its conclusion.

The narrative of stellar death is also a story of renewal. In the long march to their end, stars return much of their mass back to the interstellar medium. Some of this recycled gas is enriched with heavy elements—such as carbon, nitrogen, oxygen, and silicon—in the cauldron of stellar fusion. These elements find their way back into interstellar clouds and become the raw stuff from which stars, planets, and (in Earth's case at least) life itself form.

But not all stars die the same way. While supernovas are spectacular, they represent the minority of stellar deaths because massive stars represent a minority of stars. Lower-mass stars like the Sun go to their final reward differently. "Stars are like people," says Orsola De Marco, an Italian-born astronomer who has been working on the death of stars like the Sun for most of her career. "How they die depends on how they lived, which for a star depends on how they were born. In particular, a star's death depends on how much mass the star was born with." The more mass a star has, the greater the gravitational weight crushing down on the core, which determines the temperature

FIGURE 13.1 Supernova 1987A
Supernova 1987A (SN 1987A), the dramatic explosion of a dying massive star, was the closest observed supernova in 400 years. It occurred in the outskirts of the Tarantula Nebula in the Large Magellanic Cloud, a nearby dwarf galaxy. The image on the left shows the star as it appeared before it went supernova.

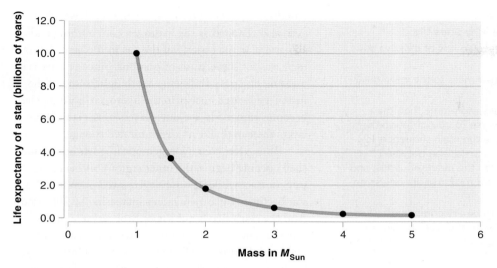

Mass (M_{Sun})	Life Expectancy (yrs)
0.08	5.5 trillion
0.125	1.8 trillion
0.25	320 billion
0.5	57 billion
1	10 billion
1.5	3.6 billion
2	1.8 billion
3	642 million
4	313 million
5	179 million
6	113 million
7	77 million
8	55 million
16	10 million
32	2 million

FIGURE 13.2

Stellar Mass and Life Expectancy

The length of a star's life depends on its mass: the larger the star, the faster it burns through its fuel. The most massive stars live only millions of years. A 1-solar-mass (1-M_{Sun}) star will live for about 10 billion years. This graph and table show the wide disparity in life expectancies as a function of mass.

in the core. The core temperature, in turn, determines the rate at which stars burn hydrogen fuel. Double the temperature in the core and the nuclear burning rate goes through the roof, increasing by a factor of 16 at least. The amount of time a star spends burning hydrogen on the main sequence therefore depends on the mass of the star (**Figure 13.2**).

Stars fall into different categories depending on their mass (note that there may be some variation in how astronomers arrange these categories). First there are what might be called very-low-mass stars, with masses ranging between 0.08 and 0.85 times the mass of the Sun. Then there are low-mass stars like the Sun, ranging between 0.85 and 2 solar masses. After that there are the intermediate-mass stars between 2 and 8 solar masses. High-mass stars include everything above 8 solar masses. Very-low-mass stars are so puny they basically burn hydrogen forever, relative to the age of the cosmos. "The Universe hasn't been around long enough for them to have consumed their core hydrogen and evolved off the main sequence," says DeMarco. Low- and intermediate-mass stars eat up their core hydrogen at a rate that lets them live on the main sequence for anywhere between half a billion years and 15 billion years. High-mass stars burn very quickly through their hydrogen, spending only a few million years or less on the main sequence. As you will learn in this chapter, a star's mass determines not only the length of its life on the main sequence, but also the way it leaves.

Mass is not the only factor that influences a star's evolution. Recall from Chapter 11 that at least 50 percent of the stars in our galaxy exist in binary systems with another star. Having a companion can

change a star's evolutionary track as well. Understanding the effect that being binary has on stellar evolution depends heavily on details of the two stars' orbit. "If the orbital separation between the two stars is always very large, then they don't really affect each other much," says De Marco. Astronomers call these systems *wide binaries*, and the orbital separations for these kinds of stars can be 1,000 astronomical units (AU) or more. "At the other end of the spectrum are stars that are so close that gravity distorts their shape and they are essentially touching. We call these *contact binaries*." In between are binaries with orbits large enough that they evolve separately for billions of years but then begin to interact when one of the stars evolves off the main sequence and becomes a giant. "The more massive of the pair will become a giant first," says De Marco. "Once it does, it can become so large that it interferes with the second star."

A number of things can happen once binary stars begin to interact. Perhaps the most important consequence of interacting binary evolution is that mass from the giant star can be pulled onto the smaller star. When mass is exchanged from one star to the other in this way, the process is called mass transfer. Mass-transfer binaries occur in many forms, depending on the kinds of stars in the binary. The important point to know now is that the accretion disks that form around these objects often emit enough radiation to enable detailed analysis of their properties. Thus, astronomers know quite a bit about the physics of mass transfer in binary systems and the ways it can affect the ultimate fate of the two stars.

We begin our story of stellar death close to home, with low- and intermediate-mass stars like our Sun.

SECTION SUMMARY
Out, Out, Brief Candle: Why Stars Die

- Supernovas, the explosive deaths of massive stars, are rare events.
- Stars return much of their mass to the interstellar medium as they die.
- Very-low-mass stars take more time than the current age of the Universe to burn through their core hydrogen and leave the main sequence, low- and intermediate-mass stars spend between 5×10^8 and 1.5×10^{10} years on the main sequence, and massive stars spend a few million years or less on the main sequence.
- Many stars are born with binary companions, and the first star that evolves to the giant phase can affect the subsequent evolution of the companion.

interactive
Stefan-Boltzmann Law

13.2
From Giant to Dwarf: The Fate of Low- and Intermediate-Mass Stars

Stars on the main sequence are temporarily winning their battle against gravity by releasing energy through hydrogen fusion. But what happens when the hydrogen in the core has all been converted into helium "ash"? No matter what a star's mass, the end of core hydrogen fusion marks a crisis point. The star's time on the main sequence has come to an end.

Life after Hydrogen Burning

For a low- or intermediate-mass star like the Sun, the end of hydrogen core burning is the beginning of a story of transformations both at the surface and in the star's interior as the battle between gravity and nuclear fusion is itself transformed. Once hydrogen in the core is gone, the helium core begins to contract under the force of gravity. As nuclear particles are squeezed ever tighter by the shrinking core, the temperature and pressure rise.

Although fusion stops in the core at the end of the main sequence, there is still a **hydrogen-burning shell**—a layer of hydrogen just outside the core where hydrogen burning continues. The energy produced by this shell is not, however, enough to stop the core from contracting. As the inner regions of the star shrink, the hydrogen-burning shell also heats up, causing its rate of fusion to rise dramatically. Prodigious energy generated at the shell flows outward, away from the core region and into the outer layers of the star.

Conditions in the outer layers of the star have, however, also changed as the hydrogen core burning ends. In particular, the opacity of the gas in the outer layer increases for some kinds of photons. Since opacity is a measure of a gas's ability to absorb light, some of the fusion-generated energy in the hydrogen-burning shell becomes trapped in the outer layers of the star. These layers then swell like an oversaturated sponge.

Thus, after core hydrogen burning has ended and shell burning begins, the inner regions of the aging star contract while its outer layers swell outward. The star becomes a giant, its radius increasing by 200 times. And the star's size is not the only observable property that changes. Even though the stellar core and hydrogen-burning shell are increasing in temperature, the temperature of the distended stellar surface layers decreases by a few thousand degrees. Using Wien's law (Going Further 4.1), we find that a stellar surface at this temperature yields a blackbody spectrum that peaks in the red part of the visible spectrum (**Going Further 13.1**). Thus the star is not just a giant; it is a **red giant**, a cool, evolved star with a radius much larger than that of the Sun (**Figure 13.3**).

Recall from Chapters 11 and 12 that the HR diagram serves as astronomers' principal tool for decoding the properties and life stories of stars. The evolutionary track of a star on the HR diagram is the snail's

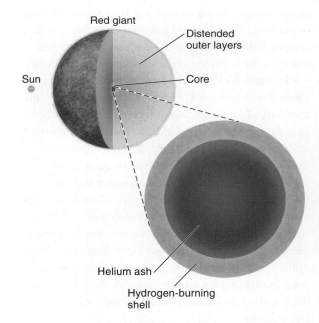

FIGURE 13.3 Hydrogen-Burning Shell
Once the hydrogen in a star's core has been converted to helium, nuclear fusion there ceases. Although there is a hydrogen-burning layer outside the core, its energy is not sufficient to stop the core from contracting.

hydrogen-burning shell The nuclear burning layer outside the core of a star where hydrogen continues to be fused into helium after such reactions have ended in the core.

red giant A cool star with a large radius that has evolved off the main sequence because hydrogen fuel in its core is depleted.

helium flash Rapid internal ignition of helium fusion in the core of an evolved star.

trail that an individual star makes on the diagram as it ages and its observable outer layers respond to internal changes. After the hydrogen core burning ends for low- and intermediate-mass stars, its evolutionary track heads up and to the right on the HR diagram, a region called the red giant branch (**Figure 13.4**). The movement upward on the HR diagram means that the star's luminosity (energy emitted per time) is increasing. The increase in luminosity is a direct result of the increasing energy generated at the hydrogen-burning shell. The move rightward on the HR diagram happens because of the decreasing temperature of the stellar surface—which, in turn, is a result of the star's distended size.

For a star like the Sun, the ascent up the giant branch on the HR diagram is not the end of the story. Eventually, the core temperature reaches the critical value of 100 million K—high enough for the helium nuclei to fuse into carbon. This is a new source of energy and the star, for a while at least, gets a new lease on life.

"When helium ignites, it can do so either in a quiet way or as a kind of internal, unseen explosion which astronomers call a core **helium flash**," explains Orsola De Marco. The difference depends on the mass of the core: heavier cores lead to the helium flash. "Either way, once core helium burning begins, stars enter a stable phase of their lives."

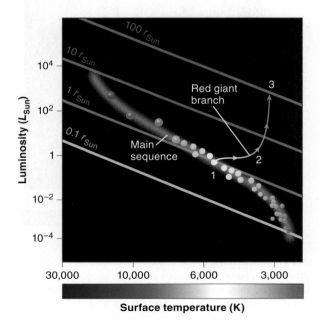

FIGURE 13.4 Red Giant Branch
Stars follow evolutionary paths that can be illustrated on the HR diagram. Here, a star that has run out of hydrogen in its core evolves off the main sequence and moves up the red giant branch.

GOING FURTHER 13.1   THE LANGUAGE OF THE COSMOS
Why Red Giants Are Red and Giant

Chapter 4 introduced the Stefan-Boltzmann law for stars. This is the formula for the luminosity of a star emitting as a blackbody (as all stars do). The star's luminosity L_* (that is, the energy radiated per second) depends on the star's surface area A_* (which, for a sphere, is $4\pi R_*^2$) and temperature T_*. Thus we have

$$L_* = A_* \sigma T_*^4 = 4\pi R_*^2 \sigma T_*^4$$

where σ is a constant called the Stefan-Boltzmann constant.

Using this formula, we can find out how large a star becomes after leaving the main sequence. Before we begin, it will be easier if we think about the problem relative to the Sun, comparing the star's luminosity, temperature, and radius to the Sun's current values. Let's begin by dividing the equation above by the same equation referring to the Sun. That means we replace the "*" with "Sun" for the Sun. When we cancel out the elements that are constant from the top and bottom, we get

$$\frac{L_*}{L_{Sun}} = \left(\frac{R_*}{R_{Sun}}\right)^2 \left(\frac{T_*}{T_{Sun}}\right)^4$$

If we already know the luminosity of the star and its temperature but want to know the radius, we can solve this formula for R_*/R_{Sun}:

$$\left(\frac{R_*}{R_{Sun}}\right) = \left(\frac{L_*}{L_{Sun}}\right)^{1/2} \Big/ \left(\frac{T_*}{T_{Sun}}\right)^2$$

Now let's look at the evolutionary tracks on the HR diagram in Figure 13.4 and find the luminosity and temperature of a low-mass star just before helium core burning begins. We find that the luminosity is already in "solar units," since that is the way the graph is written: $L_* = 10^3 L_{Sun}$. We also find that its temperature T_* is about 3,500 kelvins (K). Now let's plug these values into the equation, recalling the values for the Sun from Chapter 10 (we'll approximate the Sun's temperature as 5,800 K):

$$\left(\frac{R_*}{R_{Sun}}\right) = \left(\frac{10^3 L_{Sun}}{1 L_{Sun}}\right)^{1/2} \Big/ \left(\frac{3,500 \text{ K}}{5,800 \text{ K}}\right)^2 = 86.8$$

In other words,

$$R_* = 86.8 R_{Sun}$$

Just before helium burning begins, the star will be about 87 times larger than the Sun is today. That is why it's called a giant.

But why is the giant colored red? To answer this question, we need Wien's law, which tells us the peak wavelength at which a blackbody radiates energy. For our star with surface temperature $T_* = 3,500$ K, Wien's law tells us that the radiated light will have a peak wavelength λ_{max}, in nanometers (nm), given by

$$\lambda_{max} = \frac{0.0029 \text{ m K}}{3,500 \text{ K}} = 8.29 \times 10^{-7} \text{ m} = 829 \text{ nm}$$

This is the infrared part of the spectrum, so the giant star appears red in our optical telescopes (and to the naked eye).

Orsola De Marco

Orsola De Marco's initial scientific inclinations leaned more toward chemistry than astronomy. "I liked to mix things and make them explode," she says. "That didn't make my parents very happy." In an effort to turn her attentions elsewhere, De Marco's parents bought their ambitious elementary school student a telescope. "I used it a lot to spy at my neighbor's windows."

Near the end of high school, De Marco's interests did turn away from chemistry, toward astronomy. She traveled to England to pursue her studies at University College London, where she encountered the topic that would become her passion. "That was the time I began my love affair with planetary nebulas. They are so beautiful, and there is so much we don't understand about them."

De Marco became one of the leading astronomers to advocate for the new binary-accretion-disk-magnetic-field paradigm in planetary nebula studies. Pushing the envelope of what is known has provided some of her most difficult and exciting moments as a scientist. "At a meeting in Hawaii on planetary nebulas, I gave a review talk outlining the arguments for the new binary model. After the talk, I remember being pinned by some older scientist who kept telling me I was discrediting myself. I remember thinking 'Okay, if this is what it's like to push against the establishment, then I don't want to do this.' But later in the meeting, there was this amazing moment where someone stood up and made a key point about the need for binary stars and magnetic fields. All of a sudden you could feel people in the room get it. You could feel the tide turning. People were starting to be convinced, and I knew it. It was really quite incredible."

star becomes a red giant again, but this time with a vengeance. The star climbs onto the **asymptotic giant branch (AGB)**, the region of the HR diagram for highly evolved low- and intermediate-mass stars when their surfaces are cool and they have very large radii and very high luminosities. The AGB phase is the next-to-last stage of stars like the Sun. Unlike its massive cousins, a low- or intermediate-mass star does not have enough mass to raise the core's temperature as high as is needed to ignite carbon. The central engine has been shut down permanently (**Figure 13.5**; **Table 13.1**).

An AGB star is a behemoth. Its outer layers can swell far beyond the radius reached during the first red giant phase. When our Sun becomes an AGB star some 5 billion years from now, its expansion will spell the final end of Earth. Though detailed calculations have yet to provide an exact value for the size of the Sun during its last years, it is likely that our star will become so large that our planet will, in the worst case, become engulfed by its own bloated star. At best, the surface of the Sun will have expanded so close to Earth's orbit that the surface of our world will be baked

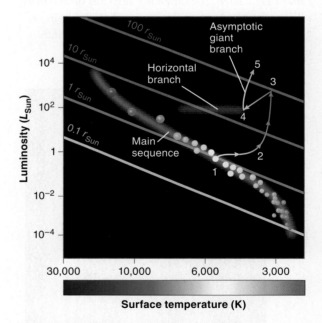

FIGURE 13.5 Horizontal Branch and AGB
When stars begin a period of stable helium burning in their cores, they move onto the horizontal branch of the HR diagram. But eventually, the core helium supply runs out, just as the hydrogen did. At this point, nuclear burning occurs only in shells—a helium-burning shell surrounding an inert carbon core and a hydrogen-burning shell above it. The star's luminosity and radius increase again as it moves onto the asymptotic giant branch (AGB) of the HR diagram.

Stars in this stage of evolution reach a region called the horizontal branch on the HR diagram. Horizontal branch stars are the core helium-burning versions of main-sequence stars. "But eventually, the helium in the core burns out too, just like hydrogen did," says De Marco. The helium core burning happens much faster than the hydrogen main-sequence burning, however. "A star like the Sun can spend just 50 million years on the horizontal branch," says De Marco. "Like the main sequence, more-massive stars spend less time there than lower-mass stars."

When all the helium in the core has been burned into carbon ash, the star faces the same crisis it faced at the end of the main-sequence phase. Its response mimics the first red giant transition. Once again, the inner regions of the star contract. These inner regions can now be defined as the carbon ash core and two nuclear-burning shells outside it: a helium-burning shell just outside the inert core, and a hydrogen-burning shell above it. As the inner regions contract, the temperature climbs again, driving a fury of nuclear burning in the shells. A new torrent of energy is sent coursing through the outer layers, which once again expand. The

asymptotic giant branch (AGB) The region of the HR diagram to which a low- or intermediate-mass star evolves when it has burned through the hydrogen and helium in its core. The AGB's external characteristics are a cool surface, large radius, and high luminosity.

TABLE 13.1 • • • Stages of Stellar Death for Low- or Intermediate-Mass Stars			
Evolutionary Phase	Duration (yr)	Size (R_{Sun})	Surface Temperature (K)
Post-main-sequence to red giant	100 million	3	4,500
Red giant	100,000	100	3,500
He burning	10 million	10	5,000
AGB star	100,000	500	3,000

to searing concrete as the oceans boil away and the atmosphere becomes too hot to breathe (though this is likely to happen during the Sun's first red giant phase).

SECTION SUMMARY
Life after Hydrogen Burning
- After core hydrogen burning stops, a low- or intermediate-mass star's core contracts and its envelope expands, creating a red giant.
- Temperatures rise in the star's core, and helium eventually ignites and burns to carbon.
- When helium core burning ends, the star's core contracts, the outer layers swell, and the star becomes an asymptotic giant branch (AGB) star.

Planetary Nebulas

The outer layers of red giants are so extended that they are not tightly bound by gravity to the star. With the flood of energy (in the form of photons) flowing outward through these layers, it does not take much for some fraction of the bloated stellar atmospheres to be blown into space. In this way, red giants, particularly AGB stars, produce dense stellar winds. The winds of AGB stars are powered by radiation impinging on tiny grains of dust that form in the cool atmospheres of the stars. As the grains are pushed

outward by the stars, they collide with gas particles, and in this way the entire outer atmosphere is accelerated away from the star to form a wind. "A star like the Sun will lose about half of itself, half its mass, as an AGB star," explains De Marco. "For more-massive stars, the winds are even more dramatic. An intermediate-mass star that's, say, 5 times the Sun on the main sequence, will lose almost 4 solar masses through its wind. That means four-fifths of the star gets blown into space."

The powerful winds of red giants contain heavy elements—carbon, oxygen, and nitrogen—by-products of billions of years of nuclear fusion. Some fraction of the processed elements is dredged up from the core of the star to the surface by deep convective flows during the AGB phase. These elements become part of the stellar winds. The winds then return their heavy elements to the galaxy in a kind of cosmic recycling. Eventually, the atoms find their way into the next generation of stars (and planets), perhaps setting the stage for the evolution of life. Closer to the dying star, however, the powerful winds set a process in motion that produces some of the Universe's most beautiful structures, called **planetary nebulas**—glowing, sculpted gas clouds surrounding dying, low- or intermediate-mass stars (**Figure 13.6**).

planetary nebula An evolutionary stage for low- and intermediate-mass stars that comes between the AGB and white dwarf phases, in which stellar wind material is driven off the hot star and ionized.

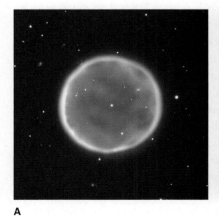 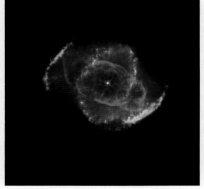 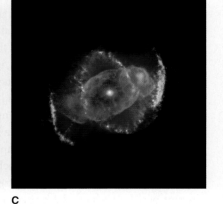

A   B   C

FIGURE 13.6
Planetary Nebulas
(A) When fast winds from the hot central star of a planetary nebula expand into material ejected earlier during the AGB phase, they sweep the gas up into a dense shell as seen here in the planetary nebula Abell 39. (B, C) The dense inner bubble of a planetary nebula, unlike the spectacular outer layers, is not visible in optical light but glows in the X-ray part of the spectrum. Here, NGC 6543 is shown in optical light (B) and optical plus X-ray light (C).

FIGURE 13.7
Charles Messier's Catalog
(A) Charles Messier (1730–1817) was a comet hunter who cataloged all the fuzzy things he saw on the sky so that he wouldn't mistake them for comets when he observed them later. (B) His catalog, now known to contain galaxies, star clusters, double stars, and nebulas, is prized for its usefulness by amateur and professional astronomers today.

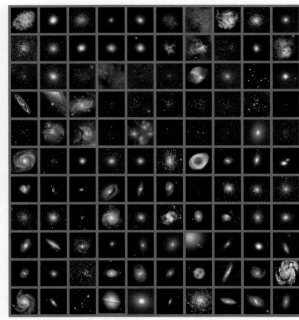

B

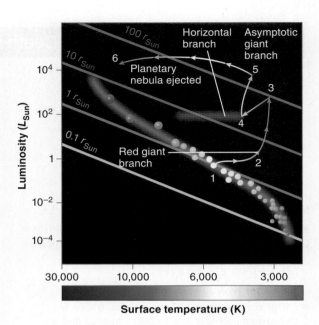

FIGURE 13.8 Planetary Nebulas on the HR Diagram
Planetary nebulas are the second-to-last phase of evolution for low- and intermediate-mass stars. After reaching the top of the AGB on the HR diagram, the giant star loses its outer envelope, exposing its hotter inner layers and moving leftward to higher temperatures at approximately constant luminosity. The ultraviolet photons from the hot star ionize the gas ejected and allow it to glow as a planetary nebula.

In 1764 the French astronomer Charles Messier was attempting to catalog all nonstellar features that can be seen in the night sky through a telescope (**Figure 13.7**). As his work progressed, he found a class of objects that looked like extended, spherical, glowing clouds. Because of the tiny size of his telescopes (their apertures ranged from 3 to 8 inches), he was unable to resolve much detail. Later, astronomer William Herschel named the objects "planetary nebulas" because the shapes and colors of these celestial clouds reminded him of the planets he saw in our own Solar System. It was an inaccurate choice of words, but it stuck, and astronomers still use it today. "Planetary nebulas have nothing to do with planets. That much is certain," explains De Marco. "But while we do know they are connected with the death of low- and intermediate-mass stars, much of our understanding of these beautiful objects is in a state of flux right now."

One thing astronomers know is that after a star has spent 100,000 years on the AGB, its stellar winds have stripped it almost down to its nuclear-burning shells. The surface above the shells and core is a violent scene of high-temperature gas releasing a torrent of energetic photons, mostly in the ultraviolet part of the spectrum. The increase in surface temperature takes the star off the asymptotic giant branch on the HR diagram and moves it leftward (**Figure 13.8**). The photons punch much of whatever atmosphere is left

into space, creating a tenuous, high-velocity wind. This "fast" wind, with speeds up to 1,000 kilometers per second (km/s), quickly overtakes the slower, dusty wind from the AGB star, which is moving at only 10 km/s. When the expanding fast wind reaches the inner edge of the AGB wind, it slams into it with the force of a trillion 1-megaton hydrogen bombs. That's when the fireworks begin.

As you learned in Chapter 12, a shock wave forms any time gas is pushed faster than it can react by getting out of the way. As a shock wave moves through a medium, it quickly and violently smashes together atoms in the gas like cars in a highway pileup. In the case of an AGB star, the collision of the fast and slow stellar winds produces two powerful shock waves. First a shock wave moves outward, accelerating and compressing the AGB wind, squeezing it into a dense layer of atoms and ions. At the same time, another shock wave rebounds off the AGB material, moving back through the fast wind toward the star. This rebound shock jerks the fast wind to a near stop, and the violent deceleration heats the material to more than

interactive
Expansion, Speed, and Observable Size of Planetary Nebulas and Supernovas

10 million K, creating a hot bubble of gas. Ultimately, the result is a kind of shock wave layer cake. The inner shock wave is closest to the star, surrounded by the hot bubble, which in turn is surrounded by the dense AGB gas layer and its outer boundary, the outer shock wave (**Figure 13.9**).

As the gas is heated and compressed by the shock waves, the ultraviolet light from the hot central star ionizes the surrounding material and causes it to emit light in the same way that occurs in H II regions around young massive stars. The high densities in the shell make it emit most intensely, creating the bright rims seen in planetary nebulas. The gas in the hot bubble is not dense enough to produce much optical light, despite its enormous temperatures. It does, however, produce enough X-rays to be seen with orbiting X-ray telescopes like the Chandra X-ray Observatory. The shell and the outer shock wave make the beautiful glowing forms we see when we view a planetary nebula from Earth.

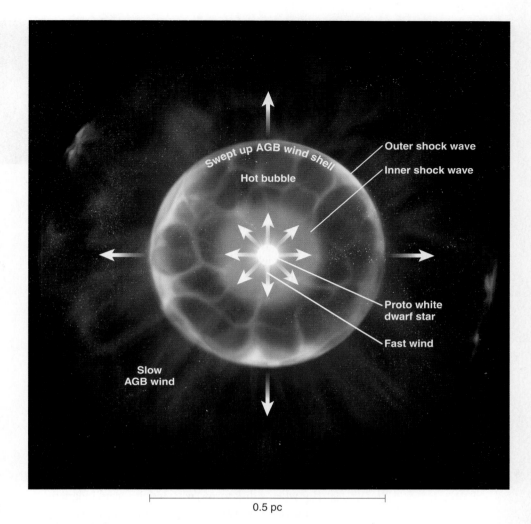

FIGURE 13.9 Shock Waves
The structure of a planetary nebula is driven by a sequence of spherical winds. An inner shock wave forms closest to the star. This shock wave heats the fast-wind gas, creating a hot bubble. The expansion of this bubble sweeps up a dense shell of previously ejected AGB gas and is bounded by an outer shock wave.

---SECTION SUMMARY
Planetary Nebulas

● The outer layers of gas and dust are blown off the AGB star as a dense, slow wind.

● Once the star is stripped almost to its core, the surface heats up and a fast wind is created; when the fast wind collides with the slow wind, shock waves form.

● Ultraviolet light from the central star ionizes the surrounding gas that has been sculpted by the winds, making them glow so that we observe a planetary nebula.

Planetary Nebulas: New Data, New Stories

Spherical winds produce spherical shock waves, so we might expect all planetary nebulas to be round. But when astronomers in the 1990s turned the Hubble Space Telescope (HST) to look at planetary nebulas, they got a shock of their own. "It's an absolute zoo," says De Marco, referring to the bizarre assortment of shapes that planetary nebulas take. Instead of concentric spheres, astronomers found some planetary nebulas that looked like narrow-waisted peanuts, some that looked like starfish, and others that looked like writhing snakes a light-year long (**Figure 13.10**). After years of work, astronomers slowly realized that the old idea of sequential interacting winds—a slow wind from the AGB star being overtaken by a fast wind from the hot core—could not be the whole story. To

explain the HST images, they would need to add more players to the drama of planetary nebulas. As often happens in science, new data forced astronomers to go beyond their old ideas.

"We reached a crisis," says De Marco. "The paradigm we had all been using just wasn't working anymore." Astronomers now have some advanced theories about what creates the amazing variety of planetary nebula shapes. "We are pretty sure that the outflowing gas in nonspherical planetary nebulas is shaped by jets from a magnetized accretion disk that forms around a star," says De Marco. "It's just like the jets you see with young stars. The spinning accretion disk whips magnetic-field lines around, and gas tied to

A B C D

FIGURE 13.10
Planetary Nebula Shapes
The incredible variety of planetary nebulas led astronomers to broaden their original models for the nebulas' formation. Note the bipolar (two-lobed) structure and the symmetry sometimes seen between the two lobes. Seen here are (A) NGC 6302, (B) HE3 1475, (C) Mz 3 (Ant Nebula), and (D) HD 44179 (Red Rectangle) in visible light, as imaged by Hubble Space Telescope.

Roche lobe The region around a star within which orbiting material is gravitationally bound to that star.

binary-accretion-disk-magnetic-field model A theory of planetary nebula formation that involves a binary system and accretion disks. Jets from the disk sculpt the shape of the planetary nebula.

Heisenberg uncertainty principle The quantum mechanical principle stating that as the position of a particle is specified to higher accuracy, its velocity becomes increasingly uncertain, and vice versa.

those field lines gets flung out into space. On the large scale of the nebula, the twisting magnetic-field lines can explain why we also get those amazing shapes."

But where do the accretion disks come from? That turns out to be the most interesting question for the new paradigm of planetary nebulas. In young stars, accretion disks form from the collapsing cloud that creates the stars in the first place. But a dying Sun-like star whose cloud dispersed 10 billion years ago can create an accretion disk only if it has a companion—another star, a brown dwarf, or even a giant planet. De Marco explains, "If you have a companion to your giant and the companion is close enough, then you can make an accretion disk."

Depending on the initial size of the binary orbit and the size of the two stars, an accretion disk can form around either the secondary or the primary star. But when a disk forms, wherever it forms, magnetic fields can drive material off its surface to create a jet. "In all cases," says De Marco," you form a disk around one of the stars, and that powers the outflow."

In one specific scenario of disk formation, the giant star expands to the point where it overflows what astronomers call its **Roche lobe**. When two objects orbit each other, each one defines regions of gravitational influence. Place an object close to the giant star in the system, and it orbits the giant star. Place an object close to the smaller star, and it orbits the smaller star. The Roche lobes define the limits of the two domains. There is always a point between the stars that marks the crossover from the gravitational dominance of one to the other. As a star swells to become a giant, its outer layers may expand past its own Roche lobe. When that happens, material cascades across to the second star. Because the two stars are already orbiting each other, the conservation of orbital angular momentum means that the transferring gas can't just fall onto the second star. Instead it goes into orbit, and as new material piles up, it forms an extended accretion disk. This process is called

Roche lobe overflow (**Figure 13.11**). Only as angular momentum is lost through friction within the disk can the gas slowly spiral inward and then drop onto the companion star.

But there is more than one way for a disk to form. If, for example, the primary star expands during its phase as an AGB star to a radius that is close to the radius of the orbit of its companion, tidal forces can drive the companion's orbit into decay. The companion is eventually swallowed by the giant, forming what is called a *common envelope system*. Once the smaller star is inside the common envelope, its orbit will continue to decay through aerodynamic drag (the same thing a skydiver feels). The process continues until the companion merges with the core of the giant, liberating enough energy to blow the envelope into space. In other cases, the giant star's gravity can completely tear the companion apart. In either of these situations, a disk made of the companion star's material will form around the now-exposed core of the primary.

This new paradigm for planetary nebulas is still controversial. Astronomers who favor the new **binary-accretion-disk-magnetic-field model** have a lot of work ahead of them to convince their more skeptical colleagues. But progress in science is not always serene; it can involve a lot of arguments and hot tempers. Out of these conflicts come new advances and deeper understanding.

SECTION SUMMARY
Planetary Nebulas: New Data, New Stories
● To account for the strange shapes of planetary nebulas, astronomers are currently debating a new binary-accretion-disk-magnetic-field theory, according to which binary companions create magnetized accretion disks that lead to powerful jets.

White Dwarfs

You might expect that once the nuclear burning in the core was completely exhausted, gravity would have its final say and crush the star completely. But what does "crushed completely" mean? Can all the mass in a star be crushed out of existence? Doesn't the star's mass have to exist somewhere? Are there any forces that can stop a dead star from being completely crushed by gravity? These were the kinds of questions physicists and astronomers wrestled with through the middle of the 20th century. Their answers, which relied on the most advanced theories of atomic and subatomic matter, led them into the bizarre world of stellar cinders—the forms a stellar corpse can take.

Just as the final years of a star depend on its original mass, so does its final fate. For low- to intermediate-mass stars like the Sun, during the planetary nebula stage the hydrogen nuclear-burning shell is just below the surface, and what little mass the star has above the core and the nuclear burning shells is almost all blown into space in the final fast wind. What's left is the inert carbon core from the long-past phase of core helium burning.

With no nuclear burning to create thermal pressure, why doesn't this inert core get crushed? The answer comes from quantum mechanics, the remarkable understanding of the micro world (molecules, atoms, and subatomic particles), which you first encountered in Chapter 4. One of the most important conclusions of quantum mechanics is something called the **Heisenberg uncertainty principle**, which tells us that particle properties in the micro world are essentially uncertain in very specific ways. For example, the Heisenberg uncertainty principle tells us that if we could specify the position of a particle to great accuracy, then its velocity would be highly uncertain. In turn, if we could specify the velocity of a particle to great accuracy, then its position would be highly uncertain. These uncertainties have nothing to do with our measuring instruments but are built into the very structure of matter itself. If you think this is weird, you're in good company. The world at the atomic scale of atoms behaves in very different ways than does the world at the human scale.

How does this principle relate to dying stars? As gravity crushes carbon nuclei and their electrons in a star's core, each particle has less space to occupy. Here, quantum physics rules, and the Heisenberg uncertainty principle determines the reaction of the nuclei and electrons to the big squeeze. In response to their confinement, the electrons (which have very little mass) take on an ever-larger range of velocities. The rapid oscillation of confined electrons corresponds to a pressure that supports the naked core against further

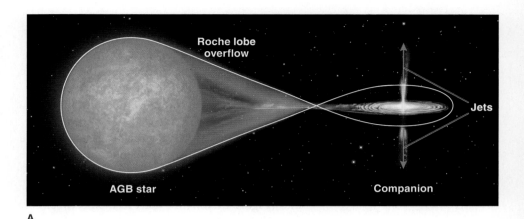

FIGURE 13.11 The Binary-Accretion-Disk-Magnetic-Field Model
A swollen red giant provides material that is then drawn in by a white dwarf as an accretion disk. Here a disk forms either (A) by Roche lobe overflow or (B) by the capture of the AGB wind. (C) In both cases, magnetic-field lines threading the disk then push material back into space as a jet that drives through the previously ejected AGB wind.

collapse. Physicists call this **degeneracy pressure**. Unlike the thermal pressure created by nuclear burning, the degeneracy pressure of the electrons does not require any material to be consumed, so the core of the star has won a permanent victory over gravity.

The almost-naked core, now supported by electron degeneracy pressure, is called a **white dwarf**—the compact remains of a dead intermediate-mass star. The *dwarf* part of the name is easy to understand, since a typical white dwarf, containing about half the mass of the Sun, will have a radius equal to about 1 Earth radius (**Figure 13.12**). (The exact relation between core mass and the mass the star began with depends on the history of its winds, which strip material away over the star's lifetime.) It is called a *white* dwarf because of its surface temperature. There is a very thin atmosphere on top of the carbon core, and it's kept very hot because of the intense gravitational field associated with cramming so much matter into so small a sphere. Temperatures on the surface of young or newly formed white dwarfs are typically 100,000 K or more. A quick calculation using Wien's law shows that a blackbody at this temperature would radiate most of its energy in the ultraviolet part of the electromagnetic spectrum. In the visible band, all wavelengths of the blackbody spectrum are almost equally represented, so the star appears white to us.

As time goes on, the white dwarf's atmosphere radiates away its energy and cools down. There is a well-defined white dwarf cooling track on the HR diagram, which astronomers have painstakingly confirmed through observational studies (**Figure 13.13**).

degeneracy pressure Pressure exerted because of quantum mechanical motions of particles resulting from the associated confinement to a small volume.

white dwarf The compact remnant of a low- or intermediate-mass star supported by electron degeneracy pressure.

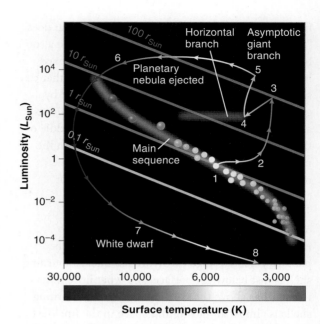

FIGURE 13.13 White Dwarfs on the HR Diagram
After the winds of the AGB and planetary nebula phases, the star has been stripped down to its core. Nuclear burning no longer continues, and the star, now a white dwarf, dims and cools over time as it moves downward and to the right on the HR diagram.

After billions of years, white dwarfs cool enough to barely emit visible radiation at all. The final fate of a star like the Sun is to become a black dwarf, spending eternity as a dark cinder.

Electron degeneracy has it limits, however. If you could pack more mass into a white dwarf, you would find that it actually shrinks as its mass increases. This makes sense from the point of view of the Heisenberg principle. Electron degeneracy pressure depends on squeezing particles into less space to increase their range of velocities. If more matter and hence more gravity are added, the star must shrink to squeeze its electrons harder and increase degeneracy pressure. Thus, more-massive white dwarfs have smaller radii than their lighter-weight cousins (**Figure 13.14**).

SECTION SUMMARY
White Dwarfs
- Once the planetary nebula phase is over, the inert carbon core supports itself as a white dwarf via electron degeneracy pressure.
- White dwarfs begin with very hot atmospheres but cool down over time to become black dwarfs.
- The more massive a white dwarf is, the smaller its radius is.

FIGURE 13.12 White Dwarfs
White dwarfs radiate approximately equally across all wavelengths of the visible spectrum and thus appear white. Although they contain masses equivalent to that of our Sun, their radii are at least 100 times smaller, comparable to that of Earth.

The Chandrasekhar Limit

Matter cannot be packed onto a white dwarf forever, and understanding what the limitations were required the innovations of both relativity and quantum physics. In July 1930, the physicist Subrahmanyan Chandrasekhar (**Figure 13.15**) boarded a boat from his native India to travel to England, having been awarded a scholarship to attend graduate school at Cambridge University. After overcoming his seasickness, the young physicist decided to use his week-long journey to work on the unresolved problem of the fate of stars. He started his calculations by trying to combine the new science of quantum physics with Einstein's theory of relativity. As the ship made its way to England, Chandrasekhar developed the basic outline of how degeneracy pressure could support a dead star and what limits that pressure would set in terms of the final mass of the stellar cinder. According to his calculations, no core of more than about 1.4 solar masses (M_{Sun}) could be supported against its own gravity by electron degeneracy pressure. When he arrived in England and began working out the details of his new theory, he often discussed his results with Sir Arthur Eddington, who was the most famous astrophysicist of his time. Eddington did not believe the new results, but he encouraged Chandrasekhar and invited him to present his results at an upcoming and highly prestigious congress of astronomers.

With some trepidation, Chandrasekhar delivered his talk on electron degeneracy pressure and the 1.4-M_{Sun} limit. Then, in one of the most famous betrayals in the history of astronomy, Eddington immediately followed with a talk of his own in which he savaged Chandrasekhar's work as completely misguided. To have a scientist of Eddington's stature take so strident a position against one's work would normally be the end of a young scientist's career. But though the experience was initially devastating to Chandrasekhar, it turned out that Eddington was wrong. History and astronomical data eventually proved the truth of degeneracy pressure and its limits. Chandrasekhar went on to win the 1983 Nobel Prize in Physics for his research on stellar evolution. Astronomers now use the term **Chandrasekhar limit** to describe the 1.4-M_{Sun} limit on dead cores supported by electron degeneracy pressure.

For cores with mass higher than the Chandrasekhar limit, electron degeneracy pressure is simply not strong enough to counteract the inward crush of gravity. Thus, more massive cores from more massive

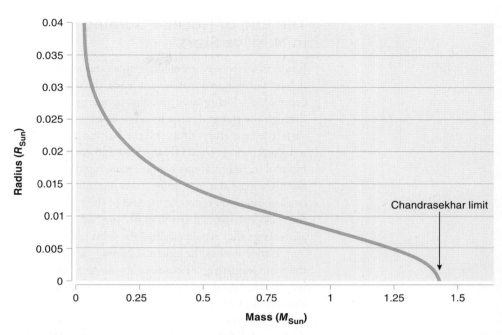

FIGURE 13.14 Mass and Radius in White Dwarfs
More-massive white dwarfs must shrink to increase their degeneracy pressure; thus they have smaller radii than their less-massive counterparts have.

stars must rely on a different form of support after fusion ends. That is the story we turn to now.

> **SECTION SUMMARY**
> **The Chandrasekhar Limit**
> • White dwarfs cannot be more massive than 1.4 M_{Sun}, the Chandrasekhar limit.

13.3
The Evolution of a Massive Star

It takes more than 100 million years for a low- to intermediate-mass star to evolve from the main sequence to a white dwarf, but more-massive stars evolve far faster and with more dramatic consequences. Once a star greater than 8 M_{Sun} completes its core hydrogen burning, it proceeds rapidly to its demise. The timescale for the evolution of a star depends on temperatures in its core. Core temperature determines both what can burn and how fast that fuel is spent. Even on the main sequence, massive stars produce high core temperatures, which shorten their main-sequence lifetimes to a small fraction of those achieved by their smaller cousins. Once the hydrogen in its core is exhausted, a high-mass star moves quickly through a series of transformations that hurtle it toward its eventual dramatic death.

FIGURE 13.15
Subrahmanyan Chandrasekhar
As a young man, Subrahmanyan Chandrasekhar (1910–1995) described the physics of degeneracy pressure in stars and discovered the maximum allowable mass supportable by that pressure. This limit is now called the *Chandrasekhar limit.*

Chandrasekhar limit The maximum mass (1.4 M_{Sun}) of a white dwarf; the mass that can be supported by electron degeneracy pressure.

The End of Nuclear Burning in Massive Stars

Just as in a low- to intermediate-mass star, when helium burning in a massive star ends, the core starts to contract, producing large amounts of energy that flow through the star and cause its outer layers to swell. The surface temperatures drop, and the star moves into the red giant region of the HR diagram. However, the internal transformations happen more rapidly and take the high-mass star into domains its intermediate-mass cousins can never enter. At the end of their lifetimes, the most-massive stars swell to become *supergiants* that are so large they would swallow the orbit of Mars.

Let's consider the case of a 25-M_{Sun} star. After the main sequence, the core of this massive star rapidly switches to helium burning. After only hundreds of thousands of years, the helium will all be consumed and turned into carbon. In massive stars, carbon is created through the so-called *triple-alpha process* (**Figure 13.16**), in which two helium nuclei (which, for historical reasons, are also called alpha particles) combine first to form a nucleus of beryllium (8 protons and 4 neutrons). The beryllium nucleus then fuses with another helium nucleus to create a nucleus of carbon with 12 protons and 4 neutrons.

Intermediate-mass stars never achieve temperatures high enough to ignite carbon (600 million K), but the core of a high-mass star will be squeezed by a tremendous force of gravity. Eventually, the temperature reaches 600 million K, at which point carbon begins fusing into neon. The carbon burning lasts only about a thousand years, however. After carbon is exhausted, neon begins fusing into oxygen; this stage lasts about a year. After the neon is spent, oxygen begins burning to silicon, which also lasts about a year. Things are happening fast now because each new form of fusion requires a higher temperature to ignite. As we've seen, higher temperatures mean that the star burns more rapidly through its store of nuclear fuel.

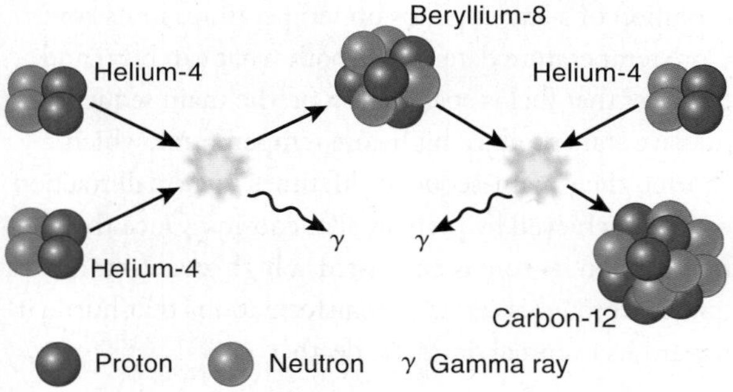

Eventually, silicon ignites and is fused into iron while the other elements are still fusing in their shells. But after a timescale on the order of Earth days (the exact timescales depend on the star's mass), all the silicon in the core is exhausted. This is the beginning of the star's crisis point (**Figure 13.17**).

Up to the creation of an iron core, the star was able to extract energy from the fusion reactions, which provided thermal pressure to resist gravitational collapse. But iron fusion requires more energy to be put in than it gives off (see the discussion of the curve of binding energy in Chapter 10, and especially Figure 10.7). Once the core is full of iron, there is no means of supporting the star against its own weight. The single 24-hour period of silicon burning becomes the day of reckoning for the star. "That's when the star has to die," says Stan Woosley, who has studied the fate of massive stars for more than 40 years. "It burns through carbon, neon, oxygen, and silicon until it finally ends up having sort of a spherical layer cake of elements, with iron down in the core and the untouched hydrogen on the surface. When a star gets to a point where it has made an iron core, you can't get any more energy out of fusing it to heavier elements, because iron is the tightest-bound state of matter. The neutrons and protons inside the iron nucleus are packed as tightly to one another as possible. If you try to release more

FIGURE 13.16
Triple-Alpha Process
The triple-alpha process is a set of nuclear fusion reactions by which three helium-4 nuclei (also called alpha particles) are transformed into carbon. Beryllium (^{8}Be) is formed in an intermediate step. This reaction occurs in massive stars as they become red giants. The triple-alpha process is similar to the P-P chain, which occurs during nuclear fusion in Sun-like stars (see Figure 10.8).

Helium-4

Beryllium-8

Helium-4

Helium-4

Helium-4

γ γ

Carbon-12

● Proton   ● Neutron γ Gamma ray

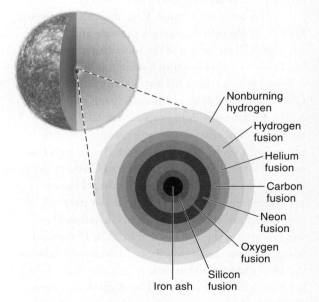

Nonburning hydrogen

Hydrogen fusion

Helium fusion

Carbon fusion

Neon fusion

Oxygen fusion

Silicon fusion

Iron ash

FIGURE 13.17 Structure of Massive Stars before Death
Because the fusion of iron into heavier elements takes more energy than fusion gives back, the creation of the iron core marks the beginning of the end of a massive star's life. Note the succession of nuclear-burning shells above the core.

energy by putting two iron nuclei together, it doesn't work. So the creation of iron is the end point of stellar evolution—the end point of core nuclear fusion" (**Table 13.2**).

---SECTION SUMMARY
The End of Nuclear Burning in Massive Stars
- Because of the higher temperatures achieved in their cores, massive stars can fuse elements heavier than helium after the main sequence, from carbon to neon to oxygen to silicon to iron.
- Once silicon burns into iron, nuclear burning in the core ends because energy cannot be extracted through the fusion of iron nuclei.

Supernova!

Once the iron core is complete and the nuclear engine dies, gravity doesn't take long to do its work. The core of a massive star will contain at least 1.5 M_{Sun} of iron at this point. Within a few milliseconds, it begins its catastrophic collapse.

The density in the core rockets to 10^{17} kilograms per cubic meter (kg/m^3) at the center, and the violence of the infall drives the temperature up to 10 billion K. Under these extreme conditions, the core is so dense and hot that neutrinos can be created from interactions within the iron nuclei and other particles in the core. In particular, protons and electrons are squeezed so tightly that they undergo a quantum transition to become neutrons. In the process, they emit a neutrino (**Figure 13.18**). Thus, few protons and electrons remain, and the core is made almost entirely of neutrons.

Recall from Chapter 10 that neutrinos are ghostly high-energy particles that rarely interact with matter. If you absolutely needed to capture just one neutrino, you would need to put up a wall of lead 22 light-years long in its path. But during the collapse of the core, neutrinos are produced prodigiously (including through thermal processes associated with the ultrahot gas). The core is so dense that the neutrinos cannot just escape, but have to worm or "diffuse" their way out, so in a sense, the core becomes a neutrino star. "The star is emitting a huge amount of energy in neutrinos," says Woosley. "In fact, the energy coming out in neutrinos exceeds the energy radiated by the rest of the entire visible Universe. All that energy has to come from somewhere, and the core gets it by shrinking, becoming more and more tightly bound. It's the gravitational collapse that powers the huge neutrino losses."

TABLE 13.2 ••• Stages of Nuclear Fusion for Massive Stars			
Burning Phase	Required Temperature (K)	Required Mean Density (kg/m³)	Duration
Hydrogen burning	4×10^7	5,000	7 million years
Helium burning	2×10^8	7×10^5	700,000 years
Carbon burning	6×10^8	2×10^8	600 years
Neon burning	1.2×10^9	4×10^9	1 year
Oxygen burning	1.5×10^9	10^{10}	6 months
Silicon burning	2.7×10^9	3×10^{10}	1 day

The core is now collapsing at about one-third the speed of light. It has less than a second to live before it collapses completely in its rush of gravity-powered neutrino radiation. "The core would collapse forever if it didn't reach a point where the strong nuclear force actually began to be repulsive," explains Woosley. "We think of a nuclear force as being attractive—and on the scale of ordinary nuclei, it is—but if you squeeze enough, a nucleus will actually push back, and if you try to compress matter beyond the density of nuclear matter, it also pushes back."

Once the repulsive nuclear force kicks in, the core suddenly stops collapsing and becomes rigid like a hard rubber ball. Parts of the star above the core, however, have no way of knowing about its sudden transition. As stellar layers above the core come crashing down, they are instantly halted, creating a powerful shock wave that begins to travel outward. For some years, astronomers were confident that this rebound-generated shock wave would travel all the way to the stellar surface and blow the star apart, resulting in the energetic explosion of a massive star called a supernova. Unfortunately, that simple idea met with an ugly reality. "The shock gets stalled in the collapsing layers above the core," says Woosley. "By the 1990s we all discovered that that this idea simply didn't work." Like their colleagues studying planetary nebulas, astronomers studying the death of massive stars were forced to look to other ideas.

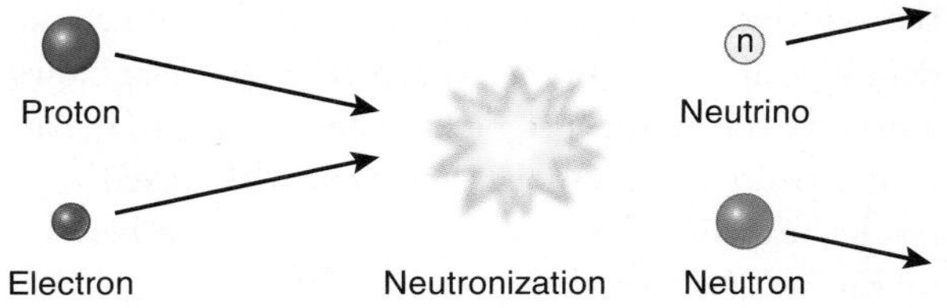

FIGURE 13.18
Neutronization
In the core collapse of a star, the quantum mechanical transformation of electrons and protons into neutrons is accompanied by the release of a neutrino.

Stan Woosley

Stan Woosley started college without a particularly strong passion for astronomy. "I was taking physics, and it was really hard. I would have changed majors, but I was just stubborn enough to refuse to quit," he explains. Then something happened that changed both his life and the course of history. It was 1961, and the Texas-born Woosley was a freshman studying at Rice University in Houston. By chance or fate, the young Woosley ended up sitting in the audience of President John F. Kennedy's famous speech at Rice that launched the US Moon-landing program. Kennedy's vision of astronomical science without boundaries lit a fire under the young Woosley. "The idea that space science was a big up-and-coming field that I could be part of was a critical influence for me."

Woosley stayed at Rice through graduate school but found his interest flagging again. A chance meeting with Don Clayton, however, an expert on the creation of elements and supernovas, once again ignited Woosley's interest and passion. "I always loved blowing things up," Woosley explains. Whatever doubts he had vanished as he plunged into the work at the frontiers of physics and astronomy. "It's amazing to me looking back now that I almost lost interest, because I love what I do now so much," says Woosley. "Sometimes I can't sleep I am so excited about working on a problem. It is such a privilege to be able to directly explore the nature of the world and learn about the how the Universe is constructed."

Woosley went on to study at Caltech before taking a professorship at UC Santa Cruz in 1975. Across his 40-year career, he has been one of the world's leaders in the study of supernovas, with more than 300 published papers to his credit. After all those accomplishments, he has earned the right to relax now and then. "I like margaritas and playing guitars by the pool these days," he confesses.

While astronomers may not know exactly how a star rips itself apart, they do know that eventually, some process within the star forms a shock wave powerful enough to blow everything above the core into space. These supernovas, which astronomers call **Type II supernovas**, are so bright that they can be seen in distant galaxies (**Figure 13.20**). Lasting for roughly a few months, Type II supernovas shine with the energy of a billion Suns—as much energy as an average-sized galaxy has. Along with this intense radiation, the explosions create **stellar ejecta**—material above the core that is blown outward at speeds of about 10,000 km/s. The stellar ejecta slams into the surrounding interstellar medium, sweeping it up and creating a **supernova remnant**—a vast, bubble-shaped shock wave that can grow to sizes of 100 light-years or more.

As the stellar ejecta and swept-up interstellar material move outward, they cool and decelerate, causing the supernova remnant to fragment. With their multicolored glowing networks of filaments and knots, these supernova remnants constitute some of the most beautiful and striking objects on the sky (**Figure 13.21**). Astronomers have become adept at using the remnants' observed structures to piece together the evolution of the explosion and its interaction with

To explain supernova explosions, astrophysicists turned to an earlier theory from late 1960s that the wild rush of neutrinos emanating from the core could be the power source. "If just a little bit of the energy the core radiates in neutrinos could be deposited in the material above it," says Woosley, "that would be enough to blow away the rest of the star." For the past 15 years, scientists have been building complex mathematical models of neutrino-driven explosions of massive stars using the world's most powerful supercomputers (**Figure 13.19**). The problem is immensely difficult: to get the physics right, the whole star must be realistically modeled in three dimensions with many different processes—the flow of neutrinos, the collapse of the stellar gas, the nuclear physics of the core—all happening at once. Although scientists think the neutrino idea is promising, they still do not have a definitive answer as to the exact mechanism that blows a massive star apart. "People have been trying to make neutrino-driven supernova explosions with mixed results," explains Woosley. "Sometimes the star blows up this way, but more often it doesn't."

Type II supernova The rapid collapse and violent explosion of a massive star.

stellar ejecta Material expelled in a stellar explosion such as a supernova.

supernova remnant An expanding shock wave that contains ejected material and swept-up interstellar material resulting from the explosion of a star.

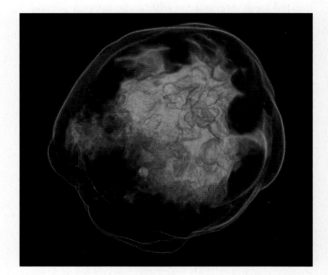

FIGURE 13.19 Computer Model of a Supernova
This image is taken from a three-dimensional simulation of neutrino-driven core-collapse supernova explosions. The image shows only the inner regions of the star, as some of the neutrinos produced in the contracting core are absorbed in the dense gas above it, providing extra pressure to drive the supernova blast wave. The process is not purely spherical, as plumes of neutrino-driven gas rise upward. Note that this image shows the supernova just 850 milliseconds into its evolution.

the surrounding gas. One aspect of supernovas and supernova remnants that astronomers have mapped out in great detail is the way these explosions seed the Universe with heavy elements.

"We are all made out of heavy elements," says Woosley. "Earth is made out of heavy elements too. That makes supernovas really important because most of the Universe's heavy elements are born in a supernova." We have already seen that a massive star just before its death is a kind of many-layered sphere of processed elements. The outer atmosphere is hydrogen, but below that are shells of nuclear-burning hydrogen, helium, carbon, neon, oxygen, and silicon, all surrounding an iron core. In a supernova explosion, all of these layers are blown into space. The violent conditions in the explosion itself are so extreme that all the naturally occurring elements heavier than iron are formed. Supernova explosions create all the truly heavy elements, such as nickel, lead, copper, and so forth, which enable life as we know it. As the astronomer Carl Sagan once said, "We are all starstuff looking at starstuff."

By combining the action of supernovas with the main-sequence burning of low-mass, intermediate-mass, and massive stars, astronomers can fully account for the range of elemental abundances in the Universe. Thus, the question of why carbon is relatively common in the cosmos compared with an element like scandium can be answered by tracking where and when each is fused in the course of stellar evolution. The origin of the elements, as well as their relative abundances, is an ancient question, so it is nothing short of extraordinary that you live among the first few generations of people to fully know its answer.

SECTION SUMMARY
Supernova!
- Once an iron core forms in a massive star, core fusion halts and the star quickly collapses. As it reaches nuclear density, it produces huge numbers of neutrinos.
- Repulsive forces in the core halt further collapse while overlying layers of the star continue to fall inward.
- Type II supernovas are explosions that mark the death of massive stars, but the exact mechanisms that cause core collapse are not fully understood yet.
- The exploding stellar ejecta sweeps up surrounding interstellar material, forming a supernova remnant that can extend across hundreds of light-years.
- Heavy elements fused in the core and in the supernova explosion are dispersed across the galaxy, seeding the next generation of stars, planets, and perhaps life.

A July 7, 2005   **B** June 7, 2011

FIGURE 13.20 Supernovas in the Whirlpool Galaxy
Supernovas are rare, but the nearby spiral galaxy M51, the Whirlpool Galaxy, has yielded three observable supernovas since 1995. The (A) 2005 and (B) 2011 occurrences of Type II supernovas are shown here. The short life spans of such massive stars mean that they die close to where they were born, in the galaxy's blue spiral arms.

FIGURE 13.21 Supernova Remnants
The bubble-shaped shock wave of SN 1006 is visible here in optical, radio, and X-ray wavelengths. The Type Ia supernova exploded more than a thousand years ago, driving stellar material into space and sweeping up the interstellar medium.

FIGURE 13.22
Nova
(A–B) Nova Cygni (labeled "V407 Cyg") brightens noticeably in these before and after images, captured in visible light by Japanese amateur astronomers in 2010. It was 10 times brighter in the second image, taken just a few days after the first. Novas occur when a white dwarf accretes mass from a binary companion onto its surface, taking the star above the physical limit for white dwarfs and causing brief nuclear fusion to burn the excess material. (C) The graph shows the changes in a nova's brightness over time. A nova can increase in brightness by a factor of 10,000 in just a few days.

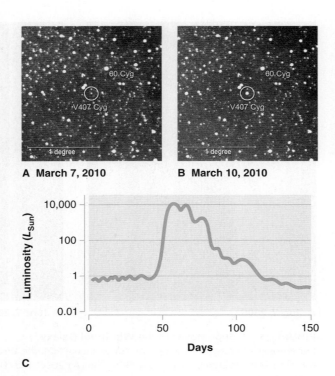

A March 7, 2010   **B** March 10, 2010

C

From Nova to Supernova: Exploding Stars in Binaries

From careful observations, astronomers studying supernovas have found that these stellar explosions come in different flavors. In addition to Type II supernovas, astronomers have discovered a variety called **Type Ia supernovas**, which are explosions related to binary stars. Type Ia supernovas have a very different story to tell than their single-star cousins.

Of course, for there to be supernovas, there must first have been regular **novas**. For hundreds of years, astronomers cataloged the recurring flaring of certain stars, which they called novas (*nova* is Latin for "new"). A typical nova will increase its brightness by a factor of 10,000 to 100,000 for a few days to a few weeks (**Figure 13.22**). The time between the nova outbursts can vary dramatically from a few decades (which can be observed) to 100,000 years (which can only be hypothesized). By the mid-1960s, astronomers had worked out a basic model for novas that explained these periodic flarings.

If the orbit of two stars in a binary system brings them close enough together that one of the objects extends outside its Roche lobe, their evolutionary pathways can be dramatically altered. "A nova happens if a white dwarf is in a binary star system and accumulates a little bit of matter from the companion star," says Woosley. A thermonuclear explosion of the accumulated mass on the white dwarf's surface produces the nova (**Figure 13.23**). Novas occur when the temperatures and densities on the surface of the white dwarf reach a **thermonuclear runaway**. In a thermonuclear runaway, the burning happens so fast that the matter can't react by expanding to slow the fusion. "In the case of a regular nova," says Woosley, "you're building up only about 10^{-5} of a solar mass of hydrogen and helium on the surface of the white dwarf before the nuclear runaway happens. So if the accretion rate from the companion is quite low like this, meaning that it's a solar mass every billion years instead of a solar mass every 10 million years, then you get into the case where the surface layer explodes before the white dwarf grows very much in mass."

Although the explosion is powerful enough to increase the brightness of the white dwarf 100,000-fold or more, it is not powerful enough to cause permanent damage to the binary star system. "When this layer, this 10^{-5} of a solar mass, explodes," says Woosley, "it has enough energy to push off everything

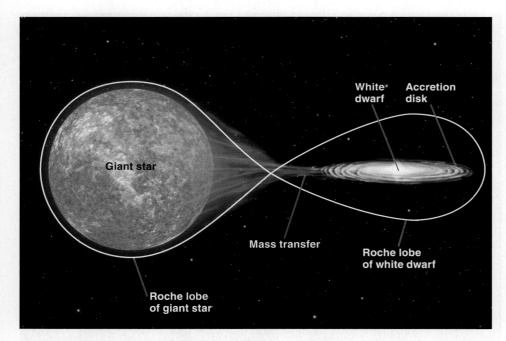

FIGURE 13.23 Mass Transfer and Novas
Mass transfer through the Roche lobes is the mechanism that triggers a nova. This image shows the Roche lobes of both the primary star and the secondary star (a white dwarf) in a binary system. When the primary star becomes a giant and overflows its Roche lobe, the material crosses over, forms a disk around the white dwarf, and eventually spirals onto the white dwarf's surface.

that had fallen onto the white dwarf and makes the star very bright for weeks, but it doesn't destroy the white dwarf." Once the nova explosion is over, the companion begins dumping matter onto the white dwarf again, setting the stage for another nova event sometime in the future.

If the accretion rate onto the white dwarf is more dramatic, however, the white dwarf can be entirely consumed. This is the origin of Type Ia supernova explosions. "If you put a white dwarf into a binary system and let the star next to it donate enough matter," says Woosley, "the mass of the white dwarf can grow to the Chandrasekhar limit. At that point, its central density rises and allows it to finally ignite carbon fusion. The star couldn't do this by itself, but the matter being dumped on it by the companion makes it happen."

In a Type Ia explosion, a "flame front" of carbon fusion begins in the center and races through the rest of the white dwarf. Once the flame gets going, it takes only a second for the carbon to burn all the way through the elements to iron. Unlike the steady burning in a massive star, here the fusion is part of an explosive detonation. The released energy tears the white dwarf and its companion apart (**Figure 13.24**). "A Type II supernova is powered by gravitational collapse. I call it a gravity bomb," says Woosley. "But a Type Ia—that is really all about fusion. It's basically a tremendous thermonuclear bomb."

Finally, it's worth noting that some researchers have argued that its not mass transfer at all but the merger of two white dwarfs that may be the source of Type Ia supernova. In the last decade or so, Type Ia supernovas have become extremely important to astronomers because they have a characteristic way in which their luminosity changes with time. When astronomers plot any object's energy output versus time, the resulting plot is a called a **light curve**. The shapes of these light curves are often quite different for different classes of objects that change their light output. The light curve of a Cepheid variable (which you learned about in Chapter 11) can be used to infer the distance to the Cepheid.

The light we see from a Type Ia supernova actually comes from the radioactive decay of elements, such as nickel-56, that were abundantly built up during the explosion. After looking at many Type Ia light curves, astronomers found similarities in their shape—for instance, how long the light curve remained at its peak (**Figure 13.25**). They recognized that the shape of the curve was tied to the total energy release in the

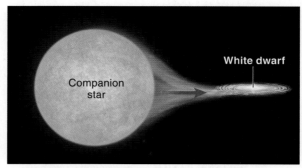

A

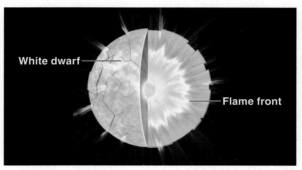

B

C

FIGURE 13.24 Type Ia Supernovas
(A) Through mass transfer, matter is dumped onto the white dwarf. (B) The additional mass ignites a wave of nuclear burning (called the *flame front*) within the white dwarf, (C) which then explodes as a supernova. Shown here are the stages of evolution of the binary system that lead to a Type Ia supernova.

Type Ia supernova The thermonuclear explosion of a white dwarf that has accreted mass from a binary companion, causing it to exceed the Chandrasekhar limit and experience thermonuclear runaway.

nova Nonterminal explosive nuclear fusion occurring on the surface of a white dwarf after it has accreted matter from a companion star.

thermonuclear runaway Nuclear burning that ignites and spreads catastrophically, consuming all available fuel.

light curve A graph of the light output versus time of an astronomical object.

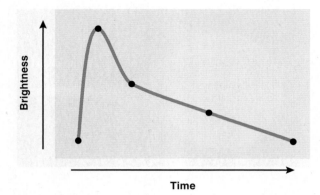

FIGURE 13.25 Light Curve of a Type Ia Supernova
Type Ia supernovas have characteristic light curves (the patterns of changes in their brightness over time) that help confirm their identity. The standard and incredible brightness of Type Ia supernovas, visible from across the galaxy, make them an ideal means of determining distances.

supernova, which is equal to the total luminosity. This discovery meant that astronomers had discovered a way to use Type Ia supernovas as standard candles, similar to Cepheid variables. The shape of each light curve could be decoded to extract the luminosity of the supernova. Comparing that luminosity with the apparent brightness of the supernova on the sky then enabled astronomers to find the supernova's distance using the inverse square law for brightness (see Chapters 4 and 11).

The fact that Type Ia supernovas can be used as standard candles was an unexpected boon for astronomers, who are always hungry for new distance determination methods. Supernovas are bright enough that they can be seen across significant fractions of the visible Universe, so Type Ia supernovas are extremely valuable standard candles.

SECTION SUMMARY
From Nova to Supernova: Exploding Stars in Binaries

- Ordinary novas occur when a white dwarf in a binary system pulls matter from its companion, causing periodic thermonuclear runaways on the white dwarf's surface.

- If too much matter falls onto the white dwarf, the increased mass enables its carbon to ignite, and the star is blown apart in a Type Ia supernova.

- Astronomers can derive the luminosity of Type Ia supernovas from their light curves. Type Ia supernovas thus serve as standard candles for determining distance.

neutron star A stellar remnant supported by neutron degeneracy pressure that results from the gravitational collapse of a massive star.

Neutron Stars

Type Ia supernovas completely destroy the white dwarf stars that gave them birth. But what remains after a Type II supernova explosion? When the torrent of light fades and the blast wave blows the bulk of the stellar material far into space, what is left at the center of the once-massive star? The answer, as with almost all questions related to stellar evolution, hinges on mass. The mass of the already ultradense core determines its final fate.

"The end of a big star's life is the beginning of something else," says Woosley. When the iron core collapses just before the supernova explosion, quantum physics once again comes into play, allowing one kind of subatomic particle to transform into another. Recall that as the core collapses, the charged electrons and protons get squeezed so tightly that they are transformed into neutrons, and this rapid creation of neutrons helps drive the intense production of neutrinos that may power the supernova. Turning so many charged electrons and protons into neutral neutrons enables the core to get rid of repulsive electromagnetic forces and draws it together ever tighter. The core continues to shrink, and in the end, only neutron degeneracy pressure can save it from complete collapse.

Neutron degeneracy pressure is similar to the quantum process of electron degeneracy pressure, which occurs in white dwarfs. In both cases, the squeezing of particles into a smaller space forces them to move at higher velocities. The high-velocity particles rattling around inside the dead star translate into a pressure force that supports the stellar corpse against its own gravity. In white dwarfs, the electrons provide the degeneracy pressure. In a neutron star, the neutrons play that role.

"At the end of the core's collapse, it has nearly the same density as matter in the nucleus of an atom," says Woosley. "Basically, the core *is* a giant nucleus, and that's the beginning of the life of a neutron star." What Woosley describes is one of two forms of a stellar "corpse" left over from the death of massive stars. **Neutron stars**—the compact remains of massive stars—have masses between 1.4 and 2.1 M_{Sun}, but they pack all that matter into a sphere just 10 km in radius. So much material is stuffed into such a small space that the gravitational field of a neutron star is almost 200 billion times stronger than Earth's. A mere teaspoon of neutron star material would weigh more than 1 billion tons (**Figure 13.26**), several times as much as all the people on Earth.

These extremes of gravity and density in neutron stars make them perfect laboratories for scientific investigation of the frontiers of physics. For example, the nuclear physics of matter at the center of a neutron star is extreme. Physicists believe that entirely new states of matter may exist in a neutron star core, where strange particles are created in such large numbers that they might form an ultradense kind of nuclear liquid. These calculations have proved extremely useful (though in some cases inconclusive) for ascertaining the nature of neutron stars.

When neutron stars appear as part of a binary system with a normal star, they can also help reveal how accretion works under extreme conditions. Binary systems known as **X-ray bursters** show intense outbursts of energy in the X-ray band of the electromagnetic spectrum. The cause of the bursts appears to be linked to a pileup of material on the neutron star's surface. As in a nova, when enough material builds up on the neutron star's surface, nuclear fusion is triggered, releasing a burst of X-ray photons.

These kinds of observations have helped astronomers unpack the remarkable physics of neutron stars. Nature, however, has proved to be even more generous in this domain, providing scientists with a most unusual and unexpected laboratory for studying neutron stars: the enigmatic cosmic lighthouses called pulsars.

FIGURE 13.26 Neutron Stars
Matter is so compressed in a neutron star that the mass equivalent to the entire human population on Earth would fit into a space the size of a sugar cube. A white dwarf is about the size of Earth, but although a neutron star's mass is greater than that of a white dwarf, the neutron star is many times smaller, closer to the size of a city than to that of Earth.

SECTION SUMMARY
Neutron Stars
- Neutron stars, the remnants of massive stars that have undergone Type II supernova explosions, have masses between 1.4 and 2.1 M_{Sun} and radii just 10 km across.
- Neutron stars are laboratories for extreme physics.

Pulsars

"A neutron star starts out very hot," explains Woosley, "but it doesn't stay that way for very long. Since hot gas radiates as a blackbody, the star produces lots of radiation in the visible part of the spectrum, just like a hot white dwarf. But as the star cools, it quickly becomes invisible in optical light." If that were all there were to the story, you might wonder how neutron stars have been studied so much by astronomers. In many cases, the conditions that create neutron stars also provide them with a means to be recognized far across the galaxy.

"Neutron stars are born rotating very rapidly," says Woosley. "That's because even if the massive star was rotating slowly on the main sequence, when it collapses and shrinks to such high density, conservation of angular momentum forces it to spin faster and faster." Recall that the principle of conservation of angular momentum appeared in the story of star and planetary system formation (see Chapters 5 and 12). This principle works the same way for a massive collapsing star as it does for giant star-forming clouds.

The reduction in size of the rotating core is so extreme that newborn neutron star rotation rates are mind-numbingly fast. A typical young neutron star—a 20-km sphere packing a Sun's worth of mass—will complete more than 600 rotations every second, giving it a rotation period of about 2 milliseconds (ms). That incredible speed of rotation is the drive belt for powering the phenomena known as **pulsars**—rotating, magnetized neutron stars whose radio waves periodically sweep across Earth, producing emission patterns that are observed as bursts of energy. "On top of their outrageously fast rotation rates," Woosley explains, "neutron stars are also born with incredibly powerful magnetic fields." Most stars have some degree of magnetism, but just as the collapse amplifies the spin of the massive star's core, it also drives even weak stellar magnetic fields to extreme values. Thus, newborn neutron stars also host hyperstrong dipole magnetic fields whose shapes are like Earth's field, only a trillion times more intense.

X-ray burster A binary system that shows intense X-ray outbursts caused by the pileup of material on the surface of a neutron star.

pulsar A rotating, magnetized neutron star whose radio waves periodically sweep across Earth, producing emission patterns that are observed as bursts of energy.

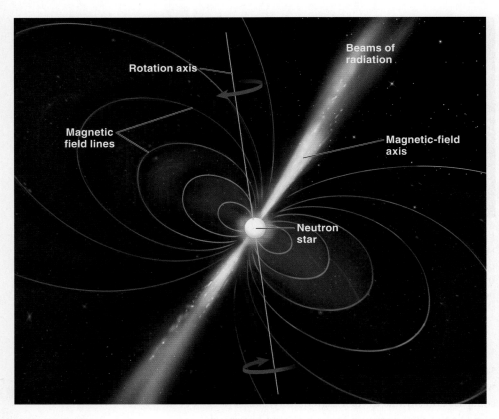

FIGURE 13.27
Pulsars
Pulsars result from the rapid rotation of neutron stars with misaligned magnetic-field and rotation axes. As the neutron star rotates, beams of radio waves, like the beams of light from a lighthouse, sweep around with the magnetic field. A pulsar will be observed from Earth only if the beam of radio waves passes across the planet.

interactive
Pulsar Observation:
Beam vs. Viewing Angle

The axis of the neutron star's magnetic field and the axis of its rotation need not be aligned. Significant misalignment between the neutron star's spin axis and its magnetic axis leads to dramatic, far-reaching effects, as the star's rotation sweeps the magnetic poles around. Electrons in the neutron star's atmosphere react to the magnetic field's movement, becoming rapidly accelerated and radiating energy in the process. The dipole magnetic-field lines funnel the electrons and the radiation along the field's axis. The resulting beams of radiation, which are intense enough to be detected across many light-years, emanate from the north and south magnetic poles (**Figure 13.27**). The radio beams rotate with the pulsar much like the light from a lighthouse rotates, swinging around so that they appear to pulse. Pulsars have also been observed through space-based X-ray and gamma-ray telescopes. When electrons in the region surrounding a pulsar interact directly with photons or the magnetic field, they produce emissions in these high-energy, short-wavelength bands of the spectrum.

In 1967, a Cambridge graduate student named Jocelyn Bell was using a specially built telescope to look for radio sources. While scanning the sky, Bell found what she first thought was "a bit of scruff" on the paper readout. Going back and observing again, she found, to her amazement, beautifully timed repeating pulses of radio emission coming at 1.3-second intervals (**Figure 13.28**). She brought the stunning results to her graduate adviser, Anthony Hewish. For a brief moment the pair considered the idea that the pulses might be evidence of extraterrestrial intelligence, naming the object LGM-1 for "Little Green Men-1." But they soon decided the phenomenon must have a natural origin and named it a pulsar. A year later, the American astrophysicist Thomas Gold and others provided the theoretical explanation for pulsars in terms of rotating, magnetized neutron stars. The story ends with some controversy, since Hewish alone was later awarded the Nobel Prize in Physics for the discovery. Many astronomers have argued that Jocelyn Bell deserved to share the prestigious prize.

In the years since, more than 1,700 pulsars have been discovered, having a range of periods from milliseconds to seconds. From decades of intense study, astronomers now understand that pulsars slow down as they age. The rotational energy of a pulsar is tapped as the radiation slowly drains its spin over millions of years. Thus, one of the most remarkable properties of pulsars is their ability to act as precise "cosmic time-pieces." Once a pulsar's spin rate and its rate of spin

FIGURE 13.28
Jocelyn Bell and the Discovery of Pulsars
(A) Jocelyn Bell was the astronomer who found the perfectly timed radio pulses of neutron stars. (B) The nonrandom cadence of the pulses initially caused Bell and her colleagues to ponder the possibility of extraterrestrials as the source, before they identified the true nature of the phenomenon.

decay are determined, astronomers can use the steady pulses as extremely accurate stopwatches to measure properties of the pulsar and its environment.

The Final Collapse

"When a massive star dies and the iron core collapses, there are only two things it can do," explains Woosley.

"It can make a neutron star, or it can make a black hole. Which kind of stellar corpse results depends on the mass of the thing that is finally left behind. If the core is more than about 2.5 times the mass of the Sun, then the strong nuclear force, the crowding of neutrons, and neutron degeneracy pressure just can't hold it up."

Scientists are still debating which massive stars fit into this category. Although they feel certain that a 10-M_{Sun} star on the main sequence will leave a neutron star (or pulsar) behind, the fate of a 30-M_{Sun} star is not so clear. The answer depends on how much mass is blown off the star by stellar winds as it evolves throughout its life, as well as how much mass is blown away in the supernova explosion. One thing is clear: If there is too much mass in the core, then nothing can stop gravity from winning a complete victory. The core collapses all the way until it becomes that most extreme and remarkable denizen of the stellar graveyard—a black hole.

●→ chapter summary

13.1   Fireworks, in a Galaxy Not So Very Far Away

Stars live on the main sequence for only as long as they are able to counteract their own gravity by producing energy through nuclear fusion of hydrogen. Once hydrogen for core fusion is exhausted, they begin a series of changes that eventually returns much of their mass to the interstellar medium. A star's mass determines how, and how quickly, these processes will proceed and what the end state will be. Another important factor is the presence of a binary companion and its proximity to the dying star.

13.2   From Giant to Dwarf: The Fate of Low- and Intermediate-Mass Stars

Once the core hydrogen is depleted in a low- or intermediate-mass star, the core contracts and heats up. A shell outside the core heats up and begins the shell hydrogen-burning phase, causing the outer layers to swell, so that the star becomes more luminous and its surface cools down. This phase corresponds to the red giant branch of the HR diagram. Contraction of the core creates a temperature high enough for helium to begin burning to carbon in the core (the horizontal branch on the HR diagram). Once helium is depleted, the carbon ash core contracts again, and a helium-burning shell above the core is created. The star, now very large and luminous, moves to the asymptotic giant branch of the HR diagram. Dense, relatively slow stellar winds will blow off the outer layers of the dying star, causing it to lose much of its mass. Eventually, a fast wind blown from the now high-temperature dwarf star overtakes the slower wind, creating violent shock waves. These shells are visible to us as planetary nebulas. The diverse shapes of planetary nebulas have led astronomers to postulate that outflows from an accretion disk around one of the stars in a binary system may be required for their formation. Stars with core masses under the Chandrasekhar limit of 1.4 M_{Sun} collapse into white dwarfs, smaller high-temperature stars able to withstand gravity through electron degeneracy pressure. White dwarfs cool over billions of years until they fade as black dwarfs.

13.3   The Evolution of a Massive Star

In a massive star, higher temperatures accelerate nuclear burning and enable it to proceed to heavier elements. The star moves to the red giant stage and begins fusing helium to carbon in the core through the triple-alpha process. It then proceeds through a series of contraction and increased-temperature stages in the core, fusing neon, oxygen, silicon, and finally iron. Each transition is accompanied by the creation of a new shell fusing the previous element. Since energy cannot be created by fusing iron, the star gravitationally collapses into an ultradense state consisting almost entirely of neutrons and resulting in the creation of a large quantity of neutrinos. A strong shock wave explosively blows away the outer layers of the star in a Type II supernova, bright enough to be seen from a distant galaxy for weeks or months. Material blown from above the core sweeps through the interstellar medium to become a supernova remnant. First appearing as a bubble, it will finally fragment. Elements heavier than iron are created through fusion in supernovas, and these elements are ejected into the interstellar medium via the explosion. In contrast, Type Ia supernovas involve a white dwarf with a binary companion that accumulates additional material until it exceeds the Chandrasekhar limit and violently explodes. Type Ia supernovas can be used as standard candles to determine distance. The neutron stars formed in Type II supernovas have strong magnetic fields and rotate many times per second initially, slowing as they age. The orientation of their magnetic fields with respect to their spin axes can create radio emission, seen as pulsars. Massive stars whose cores exceed the limit for neutron stars (whose exact limit is still unknown but believed to be about 2.1 M_{Sun}) collapse into black holes.

•→ questions and problems

Narrow It Down: Multiple-Choice Questions

1. A star is 10 billion years old. What final form may it take when it dies?
   a. white dwarf or black hole, but not neutron star
   b. neutron star or black hole, but not white dwarf
   c. only a white dwarf
   d. only black hole
   e. There's not enough information to answer the question.

2. How does doubling a star's core temperature affect its rate of nuclear fusion?
   a. It lowers it.
   b. It does not change it.
   c. It doubles it.
   d. It raises it by at least an order of magnitude.
   e. There's not enough information to answer the question.

3. Just after a star has exhausted the hydrogen in its core, the core properties differ from the those of the shell directly above it. Which of the following statements regarding the differences between the two layers is *not* true?
   a. The elemental composition differs.
   b. The density drops between the two layers.
   c. There is more gravitational pressure on the core.
   d. The hydrogen fraction is lower in the core.
   e. Nuclear fusion occurs in both the core and the shell.

4. As the hydrogen in the core of a star runs out, which of the following does *not* start to occur?
   a. Hydrogen burns in the shell above the core.
   b. Helium immediately burns in the core.
   c. The star's total energy output increases.
   d. The core contracts.
   e. The shell heats up.

5. A star is on the horizontal branch of the HR diagram. Which of the following describes nuclear fusion within the star?
   a. only hydrogen to helium in the core
   b. helium to carbon in the core; none in shells above the core
   c. no fusion in the core; hydrogen to helium in the first shell
   d. helium to carbon in the core; hydrogen to helium in the first shell
   e. no fusion in the core; helium to carbon in the first shell; hydrogen to helium in the second shell

6. Which of the following accurately describes changes to a star when it first moves off the main sequence of the HR diagram?
   a. increased radius and increased surface temperature
   b. increased radius and decreased surface temperature
   c. shorter wavelength of peak radiation
   d. decreased luminosity
   e. absence of nuclear fusion

7. A nearby star is observed to currently be in a stage of burning helium into carbon in a shell above the core. What mass can the star be?
   a. only low-mass   d. either intermediate-mass or massive
   b. only intermediate-mass   e. any size
   c. only high-mass

8. True/False: If a gas cloud were discovered that was composed only of elements up to and including the atomic weight of carbon, you could conclude with certainty that it did not originate from a supernova explosion.

9. Which of the following is/are true about an accretion disk around a star? Choose all that apply.
   a. Its star must be massive.
   b. Its star must be low mass.
   c. It is clear evidence of a companion star.
   d. It is almost certain to be composed of material left over from the birth of the star.
   e. It is almost certain to contain only the same elements as the star.

10. Which of the following is *not* a possible source of an accretion disk?
    a. the birth of a new star
    b. one star pulling material from a binary companion
    c. a larger star pulling a smaller binary companion apart
    d. a slowly spinning single main-sequence star
    e. All are possible.

11. What does the Heisenberg uncertainty principle imply about the behavior of an electron?
    a. The more confined it is, the greater will be the range of its velocity.
    b. When its velocity is greater, it requires more space.
    c. The greater the uncertainty in its position, the higher will be the range of its velocity.
    d. The electron can be squeezed out when matter becomes degenerate.
    e. The charge of the electron is reduced when it is confined.

12. Which of the following has the highest density?
    a. intermediate-mass main-sequence star
    b. white dwarf
    c. neutron star
    d. planetary nebula
    e. high-mass main-sequence star

13. Which of the following statements about white dwarfs is true?
    a. They can be the remains of massive stars.
    b. They don't allow any light to escape.
    c. They first appear at the center of a planetary nebula.
    d. They are smaller than neutron stars.
    e. The larger their mass, the larger their radius.

14. The Chandrasekhar limit refers to
    a. the minimum mass of a very-low-mass main-sequence star.
    b. the minimum mass of a white dwarf or neutron star.
    c. the maximum mass of a white dwarf.
    d. the maximum mass of a neutron star.
    e. the maximum mass of a black hole.

15. A star is currently creating the element silicon. Which statement about it is true?

   a. It must be a low-mass star.
   b. It must be an intermediate-mass star on the main sequence.
   c. It must be an intermediate-mass star on the giant branch.
   d. It must be a high-mass star on the main sequence.
   e. It must be a high-mass star on the giant branch.

16. Which of the following statements about the sequential nuclear fusion that occurs in a high-mass star is true?

   a. The duration of fusion increases as increasingly heavy elements are produced.
   b. Lighter end products are created in each successive stage.
   c. The last element that serves as input to nuclear fusion is iron.
   d. Higher temperatures are required for nuclear fusion of each successively heavier element.
   e. Fusion can occur in only one shell at a time.

17. What characteristic do the processes that produce planetary nebulas and Type II supernovas share?

   a. electron degeneracy pressure
   b. density of the remaining star
   c. shape of the outflows
   d. mass of the dying star
   e. the outer layers of the star being driven into space

18. Which of the following is/are *not* characteristics that Type Ia and Type II supernovas have in common? Choose all that apply.

   a. mass of the dying star
   b. fusion of elements up through iron prior to the massive explosion
   c. creation of elements heavier than iron
   d. enormous shock waves
   e. usefulness as standard candles

19. Which statement accurately describes a neutron star?

   a. It is an object created from the outer layers of a high-mass star.
   b. It is a star fusing iron in its core.
   c. It is an object that results from the death of a massive star.
   d. It is an object created by electron degeneracy pressure.
   e. It is the end product of a very-low-mass star.

20. A pulsar is also

   a. a neutron star.
   b. a white dwarf.
   c. a strong source of visible light.
   d. a dead low-mass star.
   e. the site of a nova.

To the Point: Qualitative and Discussion Questions

21. Explain the relationship between mass and core temperature in stars and how it determines a star's longevity.

22. Describe how different elements are created and recycled by stellar life cycles.

23. What event signals the beginning of the death process for all stars?

24. Why does a midsized star move up and to the right on the HR diagram relative to the main sequence as it approaches death?

25. For low- and intermediate-mass stars like the Sun, what keeps the inner core from collapsing after the star's hydrogen fuel is depleted and fusion stops?

26. Describe the correlation between the masses and radii of white dwarfs.

27. Describe the most current explanation of how planetary nebulas form.

28. Subrahmanyan Chandrasekhar determined that the limit for a dead star to be supported by electron degeneracy pressure is $1.4\ M_{Sun}$. What happens to stars more massive than that?

29. Describe the end-of-life stages in a massive star.

30. Explain why the stages of fusing progressively heavier elements move progressively more rapidly.

31. Why are massive stars unable to continue burning elements heavier than iron?

32. What causes the core of an intermediate-mass star to cease collapsing? How about that of a massive star?

33. Describe the differences in the causes and observations of Type Ia and Type II supernovas. Which type serves as a standard candle?

34. Would you expect to observe more planetary nebula remnants or supernova remnants? Why?

35. How does conservation of angular momentum explain the initial spin rate of neutron stars and the subsequent changes as they age?

Going Further: Quantitative Questions

36. A star's radius and temperature are both twice that of the Sun. How does its luminosity compare with the Sun's? Which had a greater effect on the final luminosity: the doubling of temperature or radius?

37. A star's temperature is 3 times as high as the Sun's, and its luminosity is 48 times that of the Sun. What is the ratio of the star's radius to the Sun's radius?

38. The temperature of a red giant is 3,300 K, and its radius is 60 times that of the Sun. What is its luminosity, in L_{Sun}? Does this result make sense, given the cooler surface temperature of the red giant?

39. A red giant has a temperature of 3,700 K and luminosity of $1.8 \times 10^3\ L_{Sun}$. What is its radius, in solar radii?

40. What is the wavelength of peak radiation, in meters, for a white dwarf with a temperature of 35,000 K? What kind of light is this?

41. A neutron star has a temperature of 50,000 K. What is the wavelength of its peak radiation, in meters? What kind of light is this?

42. What is the temperature, in kelvins, of a star with a peak wavelength of 6.7×10^{-7} meter?

43. A white dwarf has a peak radiation of 3.3×10^{-8} meter. What is its temperature, in kelvins?

44. A planetary nebula with a radius of 0.1 pc was created during the death of its star 3490 years ago. At what rate has it been expanding, in kilometers per second?

45. A supernova remnant has expanded at the rate of 6,000 km/s and now measures 4.1 pc in diameter. How many years ago did the supernova occur?

 If your instructor assigns homework in **smartwork5**, access it at the Digital Landing Page for *At Play in the Cosmos*: **digital.wwnorton.com/cosmos**

DOWN THE RABBIT HOLE

Relativity and Black Holes

In this chapter you will learn about the concepts of special and general relativity, as well as their implications for our understanding of space, time, and the remarkable phenomena called black holes. After reading through each section, you should be able to do the following:

14.1 The Black-Hole Shuffle
Describe the relationship between gravity and the limits of the physical processes supporting stellar remnants.

14.2 Special Theory of Relativity
Summarize the origin, postulates, and implications of Einstein's special theory of relativity.

14.3 General Theory of Relativity
Summarize the origin, postulates, and implications of Einstein's general theory of relativity.

14.4 Anatomy of a Black Hole
Describe the structure and characteristics of black holes.

14.5 Real Black Holes in Astronomy
Explain the evidence for the existence of black holes and the conclusions that astrophysicists have drawn from observing them.

An artist's conception of a black hole and an accretion disk of gas surrounding it. As material spirals inward through the disk, jets of matter and energy are driven outward at close to the speed of light along the axis of rotation.

14

14.1
The Black-Hole Shuffle

NOTHING WAS WORKING. In fact, nothing had worked for almost a decade.

"People were getting desperate," says Manuela Campanelli, one of the world's experts in using computers to explore the physics of black holes. "The idea should have been relatively straightforward: take the equations Einstein gave us that describe how matter bends space-time and use those to simulate something like two black holes orbiting each other." Einstein's theory of relativity—the theory behind black holes—is notorious for its difficulty, both in terms of understanding its mind-bending consequences and working with its complex mathematics. Still, in principle it should have been possible to convert it into a form that computers could use for simulations. After all, many other difficult astrophysical phenomena, including the turbulent interiors of stars, have been simulated with supercomputers. "There were some initial successes in the 1970s," says Campanelli, "and that gave everyone hope that you could do black-hole simulations. But then things started grinding to a halt."

For years, astrophysicists struggled to simulate even a single, isolated black hole moving through space. Using the codes the scientists developed, the computers would take a few steps but then spew errors and fail. "People even began to think that it could not be done," says Campanelli. "Maybe Einstein's equations governing black holes really were just too hard."

Then, in August 2005 at a conference in the mountains of Aspen, Colorado, one lone researcher took the stage and shocked the world. "His name was Frans Pretorius, and he was the one who first made it all work," says Campanelli. "He showed these beautiful simulations of binary black holes spiraling inward toward each other. It was amazing." Within months, two other groups, including Campanelli's, made their own leaps and were soon simulating the entire process

of black-hole mergers (**Figure 14.1**). "It was like a dam had broken," says Campanelli. "Now we can get a detailed understanding of these strangest of nature's creatures. It's a black-hole revolution happening right before our eyes."

Gravity's Final Victory

All stars are at war with their own gravity, and sooner or later they have to lose that war. As you have learned in the previous chapters, stars support themselves against their own gravitational weight by generating energy and pressure through nuclear fusion. Eventually, however, the fuel for nuclear fusion runs out. With nothing to balance gravity, dying stars must contract in on themselves. For stars with low enough mass, the collapse can eventually be halted by quantum mechanical properties associated with subatomic particles, such as electrons and neutrons. White dwarfs the size of Earth and neutron stars the size of Manhattan Island give testimony to how tightly more than 10^{30} kilograms (kg) of matter can be squeezed, as quantum effects step in to support these burned-out stars against their own crushing gravity.

But white dwarfs form only from stars with core masses of 1.4 solar masses (M_{Sun}) or less (this is the Chandrasekhar limit, which corresponds to stars with total mass of 8 M_{Sun} or less on the main sequence). For neutron stars, the exact upper limit on progenitor main-sequence core mass is not yet known. The best estimates, however, tell astronomers that only stars with cores of 2–3 M_{Sun} end their lives as ultradense neutron stars. These may correspond to stars with masses of about 8–20 M_{Sun} on the main sequence, though there is some uncertainty about these limits.

What fate awaits stars more massive than this limit? What can keep a 30-, 50-, or 70-M_{Sun} behemoth from infinite collapse? The remarkable answer is *nothing*. No known force in the Universe can balance the titanic gravity of the most massive stars when

FIGURE 14.1
Black Holes Merging
These images come from a computer simulation of the merger of two black holes. In the process, the structure of space and time around the black holes is distorted as predicted by Einstein's general theory of relativity. Moving from left to right, we see gravity waves (red) emitted more intensely as the two black holes spiral inwards towards each other. Eventually the holes merge into a single object.

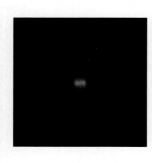  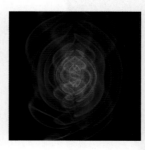 

they exhaust their fusion fuel. These stars collapse all the way down, crushing their matter into apparently infinite densities and zero radii. It is stars of such enormous mass that become the most enigmatic of astrophysical objects: **black holes**.

"Black holes are objects with gravity so powerful that even light cannot escape their pull," says Sean Carroll, a theoretical physicist at the California Institute of Technology. Black holes are so strange that they push our understanding of physics to new limits that might appear to be more the domain of science fiction than science fact. Using Albert Einstein's theory of relativity, scientists have been able to understand and model black holes, but as Carroll points out, even Einstein was unsure whether such outrageous objects could actually occur in nature. Only through the most intense study have astronomers built up a convincing case for the existence of black holes. The effort poured into the study of these objects not only has shown that black holes exist but also has provided new insights into their bizarre nature.

Before delving directly into a discussion of black holes, we must first explore the theory that led to their discovery: Einstein's radical reimagining of the physical world, known as the theory of relativity.

SECTION SUMMARY
Gravity's Final Victory
- The most massive stars collapse to black holes when they run out of fuel, because there is no force to stop their gravitational collapse.

14.2
Special Theory of Relativity

To understand a black hole, we must first understand how Einstein reimagined gravity. But understanding how Einstein reimagined gravity means we must first understand how he reimagined the very nature of space, time, and motion. This first step was the subject of his famous 1905 paper introducing what he called the special theory of relativity.

Electromagnetic Waves and the Speed of Light

Einstein was only 26 years old when he submitted his manuscript on special relativity to the most famous physics journal of his era. At the time he wasn't a physics professor, a researcher at a major laboratory, or even a graduate student. He was simply a clerk working at the Swiss patent office (**Figure 14.2**). This was the best job he could find after getting his degree in physics. "The patent office job," says Sean Carroll, "paid his bills and gave him a lot of time to think about physics."

Einstein was given the task of evaluating patents for new devices that could synchronize clocks across giant factories or immense cities. Time and space were the meat and potatoes of his daily work life. Ironically, he came to the job already engaged in a puzzle deeply linked to ideas of time and space that had been bothering him since he was a child. By the time he reached the patent office, he was determined to think his way to a solution.

Just eighteen years before Einstein was born, the British theoretical physicist James Clerk Maxwell gave physicists a set of equations linking electric phenomena (such as charge and current) with magnetic phenomena (bar magnets, magnetic fields from moving charges, and so forth). An explanation of light as an electromagnetic wave also emerged from these equations. According to this explanation, the visible light our eyes respond to is nothing more than waves of oscillating electric and magnetic fields traveling through space at the tremendous speed of 300,000 kilometers per second (km/s), or 670 million miles per hour (see Chapter 4).

But what exactly were these waves made of? "Scientists in Einstein's era knew about water and sound waves," says Carroll. "They also knew these waves needed a medium through which to propagate. So they assumed that the existence of electromagnetic waves also implied a medium for their propagation."

Since light crossed vast distances between stars, physicists imagined an all-pervading "luminiferous aether" filling space and supporting the passage of light waves. Although there was no direct evidence of the aether, physicists were certain it existed. "Once light was recognized as an electromagnetic wave," says Carroll, "the race was on to try to reconcile its behavior with the known principles of mechanics as Newton had laid them down." Researchers started designing experiments to study the propagation of light through the aether, expecting to detect the speed of light changing as Earth moved against the background aether.

"Before Einstein came along, adding up different velocities was pretty straightforward," says Carroll. "Imagine you are standing on the shore watching a boat move up a river. The water is flowing past you

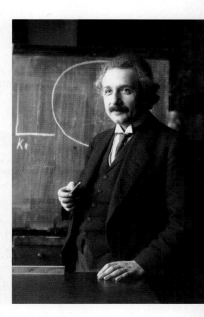

FIGURE 14.2
Albert Einstein
After graduating with a degree in physics, Albert Einstein (1879–1955), shown here in 1921, took a job as a Swiss patent clerk. Much of his work involved patents dealing with the transmission of electric signals and the electrical-mechanical synchronization of time. This period served as a background for the development of the special theory of relativity and its dramatic reimagining of the role of space and time in physics.

black hole A region of space-time whose strong gravitational distortion prevents anything, including light, from escaping.

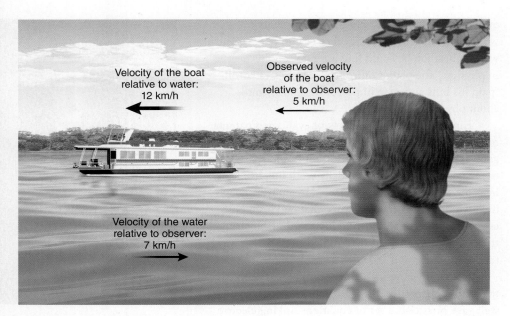

Velocity of the boat
relative to water:
12 km/h

Observed velocity
of the boat
relative to observer:
5 km/h

Velocity of the water
relative to observer:
7 km/h

FIGURE 14.3 Relative Motion
Watching a boat move on a river from the shore, you would measure its velocity to be the velocity of the water relative to you plus the velocity of the boat relative to the water.

frame of reference
An individual perspective from which observations are made.

special theory of relativity
Albert Einstein's first theory of relativity, stating that the laws of physics are the same for all observers that are in uniform motion relative to one another, and that the speed of light in a vacuum is the same for all observers, regardless of their relative motion.

at some fixed velocity, so the velocity of the boat that you measure is the velocity of the water *compared to you* plus the velocity of the boat *compared to the water*." Viewed from the shore, the boat *appears* to move slowly because its engines are fighting against a strong current. Physicists expected that light (like the boat) and the aether (like the river) would behave in the same way (**Figure 14.3**). "If you're trying to measure the velocity of an object like a light beam moving through the aether, then you'd expect the velocity you measure to be that of a light beam moving with respect to the luminiferous aether plus the velocity of the aether moving with respect to you."

But what physicists observed was not what they expected. No matter which way they set up their experiments, the observed speed of light always stayed the same. No effect of an "aether current"—the relative flow of the aether past a moving observer—could ever be found. In response to the experiments, physicists began inventing new ideas to save the aether. Einstein, however, declined to play by the rules. He did not believe in the aether, so he did not care about solving the aether problem. Instead, he changed the entire discussion.

"It was Einstein who took the leap," says Carroll. "It was Einstein who said, 'You know what? Forget the aether and just assume that the speed of light is always constant.' That changed everything."

Frames of Reference and Special Relativity

Einstein recognized that the problem lay in describing phenomena from different **frames of reference**—a key concept in physics. A frame of reference is the perspective that constitutes an observer's account of the world (an observer is anyone making measurements). "It's the way we measure length and time intervals according to a point of view," says Carroll. "A frame of reference is just the individual perspective of a person extended throughout the whole Universe." If you're standing in a field watching clouds pass overhead, you have one frame of reference. If you're sitting in an airplane a mile high and moving at 500 miles per hour as you watch the same clouds, you have a different frame of reference (**Figure 14.4**).

Since the age of Galileo and Newton, physicists had understood that the motion of a frame of reference could affect the outcome of experiments performed inside that frame. They learned how to reconcile the different ways that observers in different frames of reference would describe what they saw, and in doing so they developed a coherent account of Newtonian physics.

Though Einstein called his theory "relativity," what he was really after were invariants—the parts of physics that did *not* change from one frame of reference to another. To find nature's true invariants, Einstein first abandoned the aether and then let go of Newton's version of space and time as separate entities that were unchanging from one frame of reference to another. The **special theory of relativity** he arrived at rests on two postulates:

1. All motion is relative: no special frame of reference exists from which an absolute standard of motion can be judged.

2. The speed of light must be the same for all observers, no matter what their state of motion.

Taken together, the consequences of these simple postulates are stunning and very much against our intuition. "If someone shines a flashlight in our direction, and we measure the beam of light as it passes us," explains Carroll, "then we will get 300,000 km/s. Now imagine that same person is on a spaceship zipping toward us at 200,000 km/s. It would be the most natural thing in the world to believe that now we would see the light passing us by at 300,000 + 200,000 = 500,000 km/s. We would expect to add the velocity of the spaceship to the velocity of the light with respect to that spaceship. But Einstein and his postulates of special relativity say, 'Nope.' Despite the fact that the person on the spaceship sees the light leave them at 300,000 km/s and the spaceship is moving with respect to us at 200,000 km/s, we see the light pass by us at 300,000 km/s."

To see how strange this behavior of light is relative to our common experience, imagine two workers on a fast-moving mail train. Both workers are in a

Sean Carroll

When Hollywood film directors need to get their physics right for a superhero movie, Sean Carroll is one of the working scientists they call. Carroll has acted as adviser to films like *Thor* and *The Avengers*. "I see my role as helping them at least understand the limits of the science they want to include in the films," he says with a smile. "I don't get paid much, but it's really cool to hang out on the set."

Sean Carroll picked up his passion for science in childhood. "There was a book I found in my town's library," Carroll recalls. "It was called *High-Energy Physics*, and it explained everything they knew at the time about the fundamental structure of matter." That emphasis on fundamentals became the heart of Carroll's scientific career. After earning a BS at Villanova, Carroll was accepted to Harvard for his PhD and began his work looking into the nature of relativity. "The structure of relativity as a set of physical ideas has always interested me," he says. "In particular, I still get really excited thinking about the basic nature of time."

Along with a graduate textbook on general relativity, Carroll is the author of a number of trade books on physics and cosmology. In addition, Carroll was one of the first scientists to begin blogging, and he has appeared numerous times on television in venues as diverse as the History Channel and the *Colbert Report*.

For all his interactions with the public sphere, Carroll still loves his work as a scientist. "I have some of my best moments doing science on airplanes," he says. "Since there is nothing to do but work, I sometimes have these experiences where I am working on something and it all just falls together. Those moments of insight make everything else pale in comparison."

FIGURE 14.4 Frame of Reference
A frame of reference is the individual perspective of a person making observations—for example, using a clock to measure time, using rulers to measure distance, or watching a ball fall to the ground. The frame of reference for a person standing on Earth watching a passing airplane differs from that of a passenger in that plane.

moving boxcar with all doors and windows closed. The worker at the back end of the car heaves a bulky bag of mail forward to the other. The velocity of the bag as it flies through the air is, from the perspective of the other worker, just the velocity the first one gave it when he let it go. The velocity of the train does not affect the workers' experience. But now imagine that you're standing on a platform as the train roars by and a worker tosses the heavy mailbag to you from an open boxcar. Would you want to catch the mailbag? Not likely. The velocity of the bag would now be the velocity at which the worker released it plus the velocity of the train. From your frame of reference, the velocity of the bag and the velocity of the train have to be added together (**Figure 14.5**).

This addition of velocities is what physicists would have expected for light as well. Einstein, however, had a deeper vision. The second postulate of relativity is analogous to demanding that the mailbag thrown from the moving train travels toward you at the same velocity at which it leaves the rail worker's hands—as if the velocity of the train does not exist. This is how light behaves.

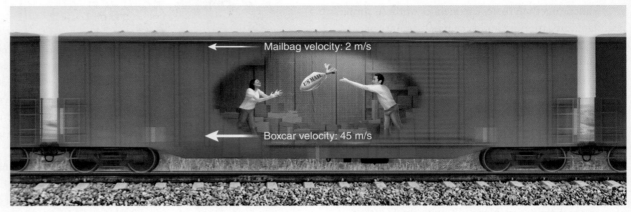

A

B

FIGURE 14.5 Addition of Velocities
(A) Within the train, the heavy mailbag moves at a low velocity and is easy to catch. (B) In the reference frame of the observer beside the tracks, however, the velocity of the train must be added to the velocity at which the mailbag was thrown by the worker, making for a much more risky catch.

SECTION SUMMARY
Frames of Reference and Special Relativity

- Taking frames of reference into account is essential to an accurate understanding of physics.
- Einstein created a new paradigm called special relativity whose postulates were that (1) there is no special reference frame from which to measure motion, and (2) the speed of light is the same for all observers regardless of their motion.

Relativity and Space-Time

Einstein further hypothesized that the speed of light is an upper bound on all cosmic motion. This fact alone changes the meaning of time and space. Einstein realized that if the Universe had a speed limit and light ran at this maximum, then something else had to give way in order for light to achieve its constancy from one frame to another.

Every measurement of velocity is really a mix of two other measurements—one measurement for a change in position and one for a duration, or change in time. In the language of math, this means

$$V = \frac{x_2 - x_1}{t_2 - t_1} = \frac{\Delta x}{\Delta t}$$

"If you need a constant light speed," says Carroll, "then something else needs to change, either your definition of space and the way you measure it, or your definition of time and the way you measure it, or both." If the speed of light is a constant (c), independent of its frame of reference, then measurements of length (Δx) and time (Δt) can't be independent.

Length (space) and duration (time) have to become flexible, changing from one frame of reference to another. Thus, the Δx that *you* measure and the Δx that *I* measure for the same event will be different, if we are in motion relative to each other. And our two measurements of the time between those events (Δt) will also be different.

Of course, for the practical purposes of our daily life, these differences are so small as to be imperceptible. Only when objects are in motion at a significant fraction of the speed of light do the predictions of relativity come into play in any sort of noticeable way. (An object must be moving at about 10 percent or more of c, or $V = 0.1c$, before physicists consider it to be moving at "relativistic" velocities.) That's why Newton's physics works fine in everyday life, where speeds are much less than c. What Einstein saw was that Newton wasn't wrong; it was just that Newton's physics relied on a low-velocity limit for describing reality.

Thus, the heart of Einstein's new vision for physics was the recognition that space by itself and time by itself were illusions. In Newton's physics, three-dimensional space was a kind of empty stage on which the drama of physics played out. Time just flowed uniformly through this empty stage; that is, the rate of time's flow was always the same for all observers. Einstein's special theory of relativity did away with Newton's separate and unchanging time and space. In their place, Einstein gave physics a new actor—a single unified **space-time**, which was four-dimensional (three dimensions of space and one dimension of time).

Einstein's conceptual revolution led, as well, to the merger of another long-standing distinction: energy and matter. The famous equation $E = mc^2$ is just a statement that mass can no longer be seen as something unchanging and unchangeable. Einstein's relativity showed that mass is a form of energy. The terrible power of atomic weapons and the gift of light from the Sun are both testaments of how, through $E = mc^2$, Einstein's relativity captured new fundamental truths about the world.

THE TWIN PARADOX. As a good example of relativity's remarkable consequences, consider the *twin paradox*. "The twin paradox," says Carroll, "brings home how different the world of relativity is from the world of Newton, which is the world of our learned 'common sense.'"

We start with a pair of identical twin girls. At age 20, the more adventuresome twin climbs into a

A Before launch   **B After return**

Twin 1: Age = 20 years   Twin 2: Age = 20 years   Space traveler is younger than twin   Twin 1: Age = 23 years   Twin 2: Age = 80 years

FIGURE 14.6 The Twin Paradox
The twin paradox illustrates the effect of "time dilation" in special relativity. For the twin who rockets away on a spaceship traveling at a substantial portion of the speed of light, time flows more slowly relative to the twin who stays home on Earth. Note, however, that each sister feels her own time passing at a normal rate. It's only the relative flow of time between the two frames of reference that changes.

spaceship and rockets away. She flies to a star 30 light-years away, always traveling at 99.9 percent of light speed. When she reaches the star, she turns around and returns to Earth at the same speed. The twin back on Earth has been waiting 60 years for her sister to return and is now 80 years old. When the space-traveling twin lands, she jumps out of the spaceship and her Earth-based sister is stunned to see that she's still a young woman, only 23 (**Figure 14.6**).

How can such a thing be possible? According to Einstein, the stay-at-home sister recorded 60 birthdays, but the space-traveling twin experienced only 3. That means the space-traveling twin's time did not flow at the same rate as her sister's did. The beat of time flowed faster for the stay-at-home sister as measured by everything from clocks in the town square to the pulse in her chest; thus, duration (Δt) was different for the two twins. The fact that time flows more slowly in some reference frames than in others is an effect called relativistic **time dilation**. "Neither twin perceives her own time to be moving faster or slower," says Carroll. "It's only the measurements of duration

interactive
Time Dilation

space-time A unified, four-dimensional geometry of the Universe that incorporates both space and time.

time dilation The effect of relativity in which time runs slower for objects near a strong gravitational source or for objects that are moving at relativistic speeds in the reference frame of distant observers.

GOING FURTHER 14.1   THE LANGUAGE OF THE COSMOS
The Simple Math of Time Dilation

Remarkably, you can prove time dilation for yourself.

Imagine you're riding in a spaceship at a significant fraction of the speed of light, and suddenly you have to sneeze. Your head snaps forward and you blow mucus particles across the space capsule. Now imagine you have a special kind of clock made of lasers and mirrors in your ship. The clock shoots a laser from the floor to the ceiling (at height H), where a mirror bounces the laser pulse back down to the floor. The time it takes for the laser to make a round-trip and the duration of your sneeze happen to be the same from your perspective (your frame of reference). Let's calculate how long your sneeze took. As **Figure 14.7** illustrates, the total round-trip time for the laser pulse must be

$$\Delta t_y = 2H/c$$

That means your sneeze (we'll call it Δt_y) also had a duration of $2H/c$. (Given the speed of light c and the likely size of your space capsule H, this would be a remarkably fast sneeze, but let's not worry about that for the moment.)

But how would your sneeze look to someone on a planet watching as you whizzed by? Would the time that person measures for the sneeze (let's call it Δt_p) be the same as the time you experience? Let's see if we can figure out what Δt_p is in terms of Δt_y.

Einstein said that the speed of light is the same for all observers, no matter whether or how they are moving. To see how this works in terms of Δt_p and Δt_y, let's draw a picture of what the observer on the planet sees as your spaceship shoots past (**Figure 14.8**). Notice that from that person's perspective, the light beam doesn't go straight up

What you see:
$\Delta t_y = 2H/c$

"achoo"

FIGURE 14.7
Sneeze as Seen by the Sneezer

and down. The sideways motion of the whole spaceship now has to be taken into account. We can use geometry (Pythagoras's theorem) to calculate how far the light beam travels in its round-trip as seen from the planet-based observer's frame of reference:

$$(c\Delta t_p/2)^2 = (V\Delta t_p/2)^2 + H^2$$

each makes about the other's reference frame that show something interesting happening." (**Going Further 14.1** shows how time dilation works.)

You may be wondering why there is a difference between the two twins if all motion is relative. Couldn't we say that the twin on Earth was moving and the twin on the spaceship was standing still? An important difference between the two sisters breaks the symmetry between their situations. Because the

twin in the spaceship had to accelerate and decelerate during her voyage to the star and back again, she is the one who experiences the slower relative clock ticks. The acceleration (no matter how brief) is what breaks the symmetry between the sisters.

The important point to digest in thinking about the twin paradox and relativity is that both twins have made proper and accurate measurements of length and duration. Einstein's fundamental insight was that no

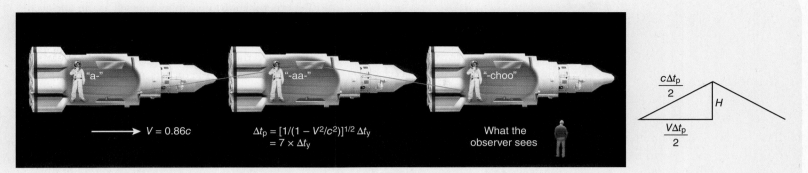

FIGURE 14.8 Sneeze as Seen by a Planet-Based Observer

This is Pythagoras's theorem for the triangle made by the laser beam as seen from the planet. But how do we get your sneeze duration (Δt_y) into the picture (and the equation)? We take the first equation above, solve it for H, and then substitute that value into the second equation:

$$H = c\Delta t_y/2$$

$$(c\Delta t_p/2)^2 = (V\Delta t_p/2)^2 + (c\Delta t_y/2)^2$$

Now we turn this into an equation that gives us Δt_p in terms of Δt_y. In other words, we have to solve for Δt_p. When the smoke clears, we find

$$\Delta t_p = [1/(1 - V^2/c^2)]^{1/2}\Delta t_y$$

What does this equation tell us? Notice that if the velocity of your spaceship is a lot less than the speed of light ($V << c$), then the factor in front of Δt_y, which is $[1/(1 - V^2/c^2)]^{1/2}$, comes out to about 1. In that case, the time you experience for the light pulse to bounce up and down (which is also the time you experience for your sneeze) is pretty much what the person on the planet experiences.

But if your spaceship is moving close to light speed, something amazing happens. Let's say you're moving at 86 percent of light speed. In that case the calculation works out to

$$\Delta t_p = 1.95\Delta t_y$$

In other words, your sneeze takes almost twice as long as seen by the planet-based observer. If you are traveling at 99 percent of light speed, then

$$\Delta t_p = 7.1\Delta t_y$$

Now your sneeze takes seven times as long. That would be really slow motion. Your head would not snap forward but would lazily loll from back to front, and those mucus particles would drift away from your nose in an agonizingly slow cascade.

There is no magic here. It's all a consequence of the constant speed of light. But those consequences are profound. There is no universal time flowing through the Universe. Your experience of time is literally your own and differs, even if by just a little, from anyone moving relative to you.

"right" answer exists for these kinds of questions about space and time because there is no absolute frame of reference in which to judge the answer. Understanding the correct physics requires abandoning the notion that space and time are totally separate entities. The problem, of course, is that the time we recognize is the one our brains evolved for us to experience. Since we never move close to the speed of light, we never have the chance to directly experience these domains of the real world.

SECTION SUMMARY
Relativity and Space-Time
- Since the speed of light is fixed, space and time must be flexible.
- Traveling at speeds close to the speed of light creates effects called time dilation and length contraction.

14.3
General Theory of Relativity

Einstein's first version of relativity was called "special" because it did not consider the role of acceleration produced by all possible forces. One force in particular, *gravity*, was so general and so prevalent that Einstein spent 7 years trying to understand how to encompass it in his new vision of physics. The result is called the **general theory of relativity**, and in its wake, the concept of gravity and the human understanding of space and time would be forever altered.

Space-Time and the New View of Gravity

Einstein often used thought experiments to work out his next steps beyond special relativity. To conduct a thought experiment means to imagine a set of circumstances and then work through its logical consequences using the laws of physics. To understand the effect of gravity in relativity, Einstein needed a thought experiment that would help him grasp how accelerations, gravity, and different frames of reference are related.

You can roughly replicate Einstein's thought experiment by imagining yourself as the sole occupant of a windowless enclosure, such as a rocket ship floating in space. When the rocket motor is turned on and the ship begins to accelerate, the "floor" of the capsule (where the rockets are located) comes rushing up to meet you, as the entire enclosure accelerates in response to the force of the rocket motor. The floor hits you and your equipment, sweeping you up and transmitting the rocket's incessant push. You feel as if you're being pulled toward the floor. At this moment in the thought experiment, Einstein came to a realization: A person inside an accelerating rocket has the same experience as a person in a rocket at rest sitting on a planet's surface. From a physics perspective, the two situations are equivalent.

Now think about what would happen if the capsule's motor was turned off. In that case, you would have no way to know whether the capsule was in motion. It might be stationary with respect to the stars, or it might be moving at constant velocity. In either case, you and any equipment on board would just float freely inside the enclosure. No experiment you performed would reveal any difference between standing still and moving at constant velocity, because both are inertial frames of reference.

general theory of relativity
Albert Einstein's second theory of relativity, which extends the special theory of relativity by incorporating gravity and states that matter and energy tell space-time how to curve while space-time tells matter and energy how to move.

FIGURE 14.9
Gravity and Acceleration
Einstein realized that moving in an accelerating frame of reference feels equivalent to the presence of gravity. (A) The astronaut and tools in the spacecraft moving at a constant velocity seem to be weightless. But when the astronaut is (B) in an accelerating spacecraft or (C) experiencing the gravity on the surface of a world, he and his tools react similarly.

A   B   C

As a final step in the thought experiment, imagine that the capsule is in a state of gravitational free-fall; for example, it is in a stable orbit around a planet. Einstein realized that in this situation, too, a person would float freely in the capsule, and the conditions would be indistinguishable from a state of constant motion. Therefore, gravitational free-fall must also be equivalent to an inertial state, even though the capsule and its inhabitant are accelerating (**Figure 14.9**). But how is it possible for an object to be in an inertial state while still experiencing an acceleration?

These questions inspired Einstein to make a bold conceptual leap. He did away with gravity as a separate force and substituted the *geometry of space-time*. Einstein saw that if the geometry of four-dimensional space-time became malleable in response to matter, it could play the same role as gravity.

For example, drop a ball on Earth and it accelerates toward the ground. Using his new vision of a flexible fabric of space-time, Einstein translated this acceleration into the ball's unforced movement through a curved space-time. According to Einstein, Earth's mass does not create a gravitational force that pulls on the ball. Instead, it distorts geometry, the very shape of space-time, around the ball. Remove all imposed forces (like the support of your hand) and the ball is free to do what space-time wants it to do. It is free to fall along the curve of space-time the way water flows down a children's slide. Thus, a ball falling from your hand and a spaceship orbiting Earth are both in free-fall, moving according to the shape of space-time around them (**Figure 14.10**).

This general theory of relativity cleanly extended Einstein's earlier special theory. Now, individual frames of reference moved on a flexible fabric of reality—four-dimensional space-time—that could be bent, stretched, and folded. The new postulates of general relativity can be captured in the following way:

1. Matter-energy tells space-time how to bend.

2. Space-time tells matter-energy how to move.

In special relativity, measurements of space or time depend on an individual frame of reference and its motion. But in general relativity, an observer's location relative to a large body of matter could also affect space and time measurements through the massive body's distortion of the space-time fabric. Like special relativity, general relativity has implications that challenge our everyday understanding of space and time. For example, clocks close to a planet's surface run slower than those far away (a phenomenon known

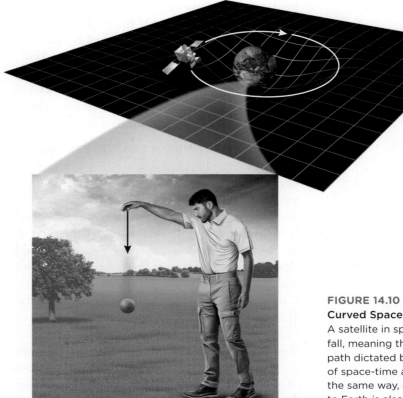

FIGURE 14.10
Curved Space-Time
A satellite in space is in free-fall, meaning that it follows a path dictated by the curvature of space-time around Earth. In the same way, a ball dropped to Earth is also in free-fall and moves according to the shape of space-time. Space-time can be thought of as stretched or curved toward the center of Earth by Earth's mass. (Note that the upper part of this figure is an attempt to represent the curvature of three-dimensional space in a four-dimensional space-time by showing curvature of the gridded two-dimensional sheet in a three-dimensional space.)

as gravitational time dilation), and lengths measured close to a planet's surface yield smaller values than do measurements taken far out in space. Time and space are relative, but this malleable geometry of space-time and the distributions of matter-energy provide the framework to understand them all as a single, unified whole. In this way, the entire human understanding of reality changed through Einstein's efforts.

SECTION SUMMARY
Space-Time and the New View of Gravity
- The general theory of relativity included gravity and the motions it produces.
- The general theory of relativity states that matter-energy tells space-time how to bend, and space-time tells matter-energy how to move.

How Do We Know?
Tests of Relativity

Einstein's ideas may seem so unintuitive that your initial response is "No way!" How can 1 year for someone flying past on a rocket ship moving close to light speed be 20 years for me? How can empty space bend and warp? You would be right to feel some skepticism, because good scientists seek experimental verification of theories before accepting them. As it turns out, both the special and general theories of relativity have been verified many times. Let's look at just a few ways in which they've been shown to be accurate descriptions of the world we inhabit.

PARTICLE LIFETIMES: A TEST OF SPECIAL RELATIVITY.

Physicists studying the structure of matter have used certain short-lived particles to provide evidence of relativistic time dilation. These particles, which move at speeds close to the speed of light, quickly decay into other types of matter (recall the discussion of radioactive decay in Chapter 6) and thus can serve as very exact clocks for physicists. One such particle is the **muon**. "The muon," says Sean Carroll, "is an elementary particle very much like the electron, but it's heavier and it decays in about one-millionth of one second. If you have a bag of muons, then one-millionth of a second later half of them are gone." Muons are routinely created in the upper atmosphere when fast-moving particles from space, called

muon An elementary, extremely short-lived particle that is heavier than an electron.

cosmic ray A high-speed subatomic particle (atomic nucleus) from space.

gravitational lensing A change in the path of light passing a massive object produced by the curvature of space-time.

cosmic rays, slam into atmospheric atoms. The newly created particles are sent flying downward at speeds close to the speed of light. On the ground, physicists can use detectors to catch these muons that started their journey high in the atmosphere.

How do these muons become a test of special relativity? "If we plug in how much time it takes a muon to travel from the upper atmosphere toward detectors here on Earth," says Carroll, "it is a lot longer than a millionth of a second." Let's do this calculation ourselves. The muons created 10 km high in the atmosphere ($D = 10$ km) are moving with a speed V of $0.99c$. That means it takes them 3×10^{-5} seconds to reach the ground ($t = D/V$). Since muons live only about one-millionth (10^{-6}) of a second and the trip takes 3×10^{-5} seconds, the muons should decay before ever reaching the ground. But scientists do find muons in their ground-based detectors, suggesting that these particles are not decaying at their normal rate.

The reason for this surprising finding is relativistic time dilation. Time is flowing more slowly for the high-speed muons relative to time measured on the ground. Thus, in the Earth-based frame, the muons can travel from the top of the atmosphere to the ground before they decay. The observed behavior of the muons exactly matches the predictions of special relativity (**Figure 14.11**).

BENDING STARLIGHT: A TEST OF GENERAL RELATIVITY.

Other experiments have confirmed the predictions of general relativity. "Like all good scientists," says Carroll, "Einstein thought about a number of different ways of testing general relativity." One of these involved comparing the positions of stars in the night sky (which can be measured very accurately) with their apparent positions during a solar eclipse. According to general relativity, when starlight passes near the position of the Sun, its path should be deflected by the massive object's curvature of space—an effect called **gravitational lensing**. However, scientists can observe this deflection only during a solar eclipse, when the Sun is obscured by the Moon.

"By bending space-time, the Sun acts as a distorting lens," says Carroll. "It bends the path of the light rays. A picture of the stars' positions when they were near the Sun during a total solar eclipse would look different from their relative positions at night," when those stars are far away from the Sun's position on the sky. General relativity gives a very definite prediction for the change in the stars' positions, and astronomers were looking for a way to measure that change (**Figure 14.12**).

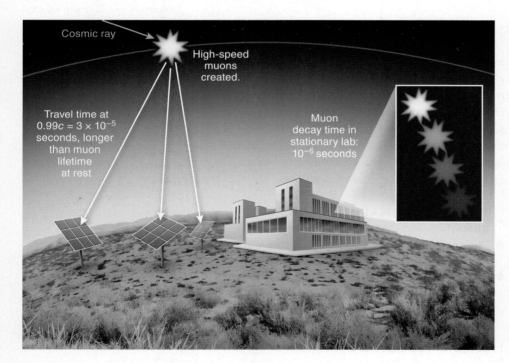

FIGURE 14.11 Time Dilation and Particle Lifetimes
Muons live only about one-millionth of a second, yet those created 10 km up in the atmosphere are able to reach Earth's surface—a trip that requires several multiples of the muon's life expectancy (its decay time). According to special relativity, the muon's time relative to observers on Earth is dilated (slowed down) because of its high speed. Thus, even though the travel time is long enough that the muon should have disappeared, in the muon's frame of reference less time has passed.

A solar eclipse is not an everyday event, but in 1919, just 5 years after Einstein finished his general theory, scientists were able to observe an eclipse and demonstrate that even light is subject to the bending of space-time by a massive object. Two separate teams of astronomers were sent to Brazil and Principe, an island off the coast of Africa, to observe the eclipse. Although the sky was cloudy in Brazil, the eclipse was visible from Principe. The deflection of starlight was observed just as Einstein predicted. General relativity had been experimentally confirmed. More recently, there has been some dispute about the degree to which the data taken that day matched the theory, but at the time the eclipse measurements were seen as a triumph for relativity. "That was really the event that catapulted Einstein into worldwide celebrity," says Carroll. Later experiments would provide many confirmations of the bending of starlight by curved space-time.

GPS: A GENERAL RELATIVISTIC TECHNOLOGY.

Every time you use a global positioning system (GPS) to find your position on Earth, you're conducting an experiment in general relativity. The GPS uses a network of 24 orbiting satellites. Each is equipped with an atomic clock that keeps hyperaccurate time, and each satellite continually broadcasts the exact time down to a billionth of a second. Ground-based receivers (like your cell phone) pick up the signals from four of these satellites and use them to compare signal travel times. Your cell phone compares its own clock with the time measured by the satellites. It then uses the difference between the times to calculate your distance from each satellite. Since light travels at 300,000 km/s, a time difference of one-thousandth of a second implies that the ground-based receiver must be 300 km from the satellite. This comparison of the receiver's time and the time signal sent by the four satellites enables accurate determinations of the receiver's position (**Figure 14.13**).

This explanation may seem completely straightforward, but a hyperaccurate measurement of time is not possible without incorporating our understanding of general relativity. Remember that general relativity tells us that space-time is distorted by the presence of matter, such that time passes more slowly for observers on the ground than it does for observers in space. General relativity predicts that GPS satellite clocks will run ahead of ground-based clocks by about 45 millionths of a second per day. While this difference might seem tiny, if it weren't accounted for in GPS technology, after only a week or so your GPS would locate you hundreds of miles from your real position

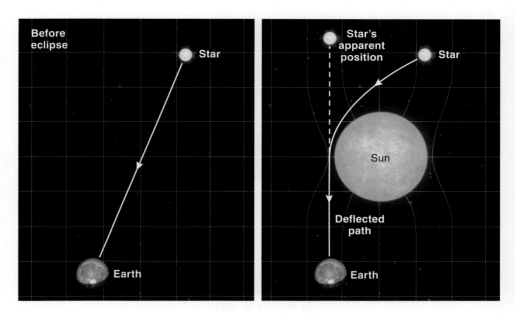

FIGURE 14.12 Bending of Light by the Curvature of Space-Time
The total eclipse of 1919 provided an accepted validation of general relativity at the time. The relative positions of distant stars were observed to change because of the Sun's gravitational distortion of space. Here we see the path of light from a star being bent as it passes the Sun, which alters the star's position as perceived by an observer.

FIGURE 14.13 Time Dilation and GPSs
Global positioning systems work by comparing time signals from orbiting satellites and the ground receiver. Both special relativistic (motion) and general relativistic (location relative to a massive body) effects must be taken into account. According to general relativity, the clocks on the satellites run faster than the ones on Earth's surface (gravitational time dilation).

FIGURE 14.14

Gravity Well of a Black Hole
(A) The gravity well of a normal star differs significantly from (B) that of a black hole. The space-time surrounding a black hole is stretched far more near the singularity at its center. If the Sun were replaced by an equivalent-mass black hole, Earth would continue to orbit as it currently does (although without heat or light).

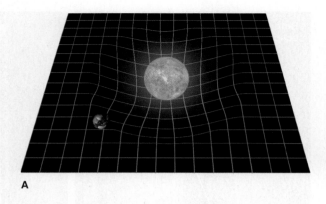

A

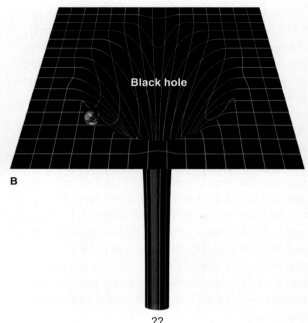

Black hole

B

??

and a couple of miles above the ground. Thus, the GPS has to take this relativistic time dilation into account or the system would never work.

PULSARS AND GRAVITY WAVES. In one particularly famous test of general relativity, the binary pulsar PSR 1913+16—two pulsars orbiting each other—was used to test Einstein's theory of gravity. The theory predicted that the two orbiting pulsars should produce gravity waves—ripples in the fabric of space-time—as they orbited each other. No one had ever directly detected gravity waves, so it was still unclear whether Einstein got this part of his theory right. Using the pulsar pulses to measure the orbit precisely, astronomers were able to demonstrate conclusively that the two pulsars were slowly spinning inward toward each other. Their orbit was shrinking as energy was lost to the elusive gravity waves. The exact amount of orbital decay matched perfectly the predictions of gravity wave generation. Einstein had been proved correct once again.

SECTION SUMMARY
How Do We Know? Tests of Relativity

- Because of relativistic time dilation, muons created at the top of the atmosphere can be detected on Earth's surface, apparently having traveled farther than should be possible during their short lifetimes.

- As light travels, its path will be bent by the curvature of space, as observed in the 1919 solar eclipse.

- Global positioning systems work with high accuracy only when relativistic effects are factored in.

- Astronomers used the binary pulsar PSR 1913+16 to infer the existence of gravity waves, ripples in the fabric of space-time.

14.4
Anatomy of a Black Hole

When the British geologist John Mitchell was contemplating Newton's formula for escape velocities back in 1783, he noticed something peculiar. If the mass M of the central object was large enough—or, more important, if its radius R was small enough—then the escape velocity would be greater than the speed of light: $V_e > c$. This meant that it was theoretically possible to have a "black" star from which light could not escape (**Figure 14.14**).

Like a cannonball blown straight upward with less than escape velocity, a light beam shot from such a black star would fall backward and never reach outside observers. "All light emitted from such a body would be made to return toward it by its own proper gravity," Mitchell wrote to a colleague. Only a few astronomers paid attention to the possibility of these black stars over the next century or so, and the idea was thought to be nothing more than a theoretical oddity.

But when the new physics of general relativity came to replace Newton's vision of gravity, the concept of such black stars was taken up again and pushed to its limit. These black stars were eventually reconceptualized as black holes, and the Universe as we understand it was forever changed.

How to Make a Black Hole: Massive Stellar Cinders

Chapters 12 and 13 tracked the evolution of high- and intermediate-mass stars through their main-sequence phases and beyond. When the hydrogen-burning phase ends for intermediate-mass stars ($M_* < 8\ M_{Sun}$), the cores contract enough to initiate helium fusion. Once the helium is gone, core fusion ends. From that point on, only the quantum mechanical degeneracy pressure of electrons keeps the burned-out stellar cinder alive as a white dwarf.

Higher-mass stars ($M_* > 8\ M_{Sun}$) burn through a sequence of elements up to iron after their core hydrogen is exhausted. But since iron requires more energy to fuse than it gives back, it represents the end of the line for the star. The star collapses on itself, creating a solid neutron core in some cases. The ultradense neutron star is another kind of stellar cinder that, like white dwarfs, is supported against gravity by the quantum physics of degeneracy pressure, though this time it's the neutrons that are degenerate.

"There's a limit to what quantum mechanics can do in terms of supporting a dead star against its own weight," says Manuela Campanelli. "If the core mass is too big, then even degeneracy pressure can't fight gravity's pull." Thus there can be no salvation for the highest-mass stars. With nothing to restrain the inward crush of gravity, the star collapses further and further toward an ever-smaller radius. Eventually the star becomes so small that its escape velocity is greater than the speed of light, just as John Mitchell first imagined. But the infall of the star doesn't stop there. "There's no known physical force that can resist gravity for these massive objects," says Campanelli, "so on the face of it, the star collapses all the way down toward zero radius."

Since a zero radius gives a zero volume, the stellar density (mass divided by volume) shoots toward infinity as the radius shrinks to nothing. Physicists call this kind of infinite limit a **singularity**, and its creation is the hallmark of the beast we call a black hole.

Astronomers are not certain how massive a star has to be to end up as a black hole. Many factors, including the rate of rotation, determine whether a newly formed neutron star will be stable or will survive for just a brief time before collapsing into a black hole. Truly massive stars may even form a black hole at their center before a neutron core can form. Astronomers currently estimate that stars greater than 20–30 M_{Sun} become black holes, but future research may revise that limit.

SECTION SUMMARY
How to Make a Black Hole: Massive Stellar Cinders
- Very massive stars (probably greater than 20–30 M_{Sun}) collapse to black holes when their fusion fuel is gone.

Schwarzschild Radius and Event Horizon

You don't need Einstein's general relativity to see that a massive enough star will eventually collapse completely under its own weight. You do, however, need general relativity to understand what is left after the total collapse is over.

One of the important features of a black hole is its **Schwarzschild radius** (R_s), the distance from the center to where the escape velocity equals the speed of light, just as Mitchell first described. What Mitchell could not see, however, is that this distance—the Schwarzschild radius—defines a point of no return for all objects falling into a black hole (**Figure 14.15**).

"Even Newton could have figured out that a massive and compact enough star would have an escape velocity equal to light speed," says Campanelli. "But general relativity tells us much more. It also shows us that once you cross the Schwarzschild radius, no amount of force can get you back out. Newtonian gravity would not have given you that conclusion."

interactive
Schwarzschild Radius

singularity The central point of a black hole, where density and space-time curvature is infinite.

Schwarzschild radius The distance from a black-hole singularity where the escape velocity equals the speed of light.

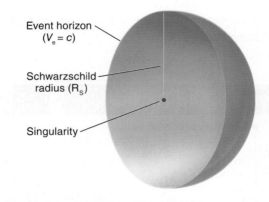

FIGURE 14.15 Structure of a Black Hole
A black hole has a simple structure: the event horizon at the Schwarzschild radius, where the escape velocity is the speed of light, and a dimensionless singularity at its core, where all mass appears to be concentrated. It is expected that an as yet undiscovered theory of quantum gravity will articulate the physics of the singularity.

Manuela Campanelli

"I was always interested in science," says Manuela Campanelli, who grew up in Switzerland. "When I was a little girl, I read about everything scientific I could get my hands on. I was interested in biology, but I didn't want to do experiments with animals, so I knew this was not the field for me. Then I tried chemistry. I liked chemistry a lot. But then one day I did an experiment in my room and everything caught fire. I stopped doing chemistry after that," she says. "This is when I realized that I needed to work on the theory side of science."

Campanelli says that even from her youngest days, she was interested in answering "big" questions. "I remember watching all astronomy shows and documentaries. They were so fascinating. The questions about the nature of time and of space really fired my imagination. Those questions never left me, and I still get to think about them in my work with black holes." Campanelli went on to become one of the leaders in the field of gravitational astrophysics. Her group, along with another led by Joan Centrella of Drexel University, helped open the door to computer simulations of binary black holes.

"My parents were not scientists," she says. "They work in fields aligned with economy and business. My brother is an economist too. My whole family is made up of very pragmatic people. I guess I am the exception to the rule."

event horizon The boundary of a black hole, located one Schwarzschild radius out from the singularity.

quantum gravity A theory (yet to be fully worked out) that unifies general relativity and quantum mechanics, describing how space-time becomes granular ("quantized") at the smallest scales.

Planck length The size scale at which space-time should begin showing its quantum nature, about 10^{-44} meter.

Once an object drops below the Schwarzschild radius, the structure of space-time demands that it continue its fall inward toward the center. That is why the sphere defined by R_s is also called the **event horizon**. Just as you can't see distant objects that lie over the horizon on Earth, objects closer to the center of a black hole than the Schwarzschild radius lie over a "horizon" for observers in the rest of the Universe. Inside a black hole—meaning inside the event horizon—is a region of space-time that is entirely cut off from the rest of the cosmos.

A simple formula gives us the Schwarzschild radius (see **Going Further 14.2**):

$$R_s = 2GM/c^2$$

TABLE 14.1 ••• The Schwarzschild Radius

Object	Mass	Schwarzschild Radius (R_s)	R_s Comparison
Human being as black hole	70 kg	10^{-28} km	Radius of an atom
Sun-like star as black hole	1 M_{Sun}	2.9 km	Width of Manhattan Island, NY
Massive star as black hole	10 M_{Sun}	29 km	Length of Long Beach Island, NJ
Supermassive black hole progenitor	1 billion M_{Sun}	20 AU	Neptune's orbital radius

Using this equation, we can see that the event horizon for a 1-M_{Sun} black hole would be only about 3 km in radius! That is basically the distance from the US Capitol to the Lincoln Memorial. Imagine trying to hide 2,000 billion billion billion kg of matter (1 M_{Sun} = 2 × 10^{30} kg) on the national mall!

Moving to larger masses, the Schwarzschild radius of a billion-solar-mass black hole, like those that exist at the center of some galaxies, would be only 20 AU, about equivalent to Neptune's distance from the Sun. Going in the other direction, if all the mass in your body were squeezed down to the point where a black hole formed, it would have an event horizon smaller than an atom (**Table 14.1**). "While it might sound crazy," says Campanelli, "when the Universe was born in the Big Bang, the conditions were so violent that these kinds of micro–black holes may have formed quite easily."

SECTION SUMMARY
Schwarzschild Radius and Event Horizon
- A black hole has a Schwarzschild radius (R_s) where the escape velocity is c; this defines the event horizon, beyond which nothing can be seen and nothing can escape.
- The Schwarzschild radius equals about 3 km for each solar mass in the black hole.

The Singularity, Quantum Gravity, and the No-Hair Theorem

General relativity provides precise descriptions of the physics of the event horizon, but the singularity is an untamed and unknown monster. The problem for astrophysicists trying to build models of black holes is that the density, temperature, and curvature of space-time become infinite at the singularity. "The singularity is essentially a geometric point," says Campanelli. "That's where all the mass and energy of the black hole live." But geometric points have no volume, so how can the entire mass of a star occupy zero volume? Clearly there must be more to the story.

"Of course, the real density can't actually go all the way to infinity, just as the radius of the star can't shrink all the way to zero," Campanelli continues. "There must be some form of new physics which occurs before the radius reaches zero and the density goes to infinity. Unfortunately, we don't yet know what that new physics is, so we work within the limit of what Einstein's theory of relativity tells us is possible."

The fact that quantities like density and temperature approach infinity at the singularity really tells us that the physics we do understand—in this case Einstein's general theory of relativity—must break down at the center of a black hole. In that way, the singularity is a kind of road block for physicists.

Physicists know from their studies of matter that, on small enough scales, all of reality must be described by quantum physics, the central lesson of which is that on the smallest sizes nature always becomes granular. As Campanelli puts it, "Nature breaks up into chunks on small enough scales. But the fabric of space-time as Einstein described it is smooth and continuous like a bedsheet. There is nothing quantum about it, and that's the problem."

To describe what happens inside the singularity, physicists need a theory of **quantum gravity** that explains how space-time itself breaks up into discrete bits. Physicists even know when quantum gravity should kick in: Physicists call the size scale at which space-time should begin showing its true quantum nature the **Planck length**, and they estimate that it is somewhere around 10^{-44} meter. "That's about a hundred million billion billion billion billionth of a meter," says Campanelli. "It's beyond small, and much, much smaller than the nuclei of atoms."

What, then, happens when a singularity is forming at the center of black hole and collapsing material reaches the Planck length? At that point the laws of quantum gravity should begin to operate, modifying the collapse. Unfortunately, after decades of trying, physicists still have not developed a complete theory of quantum gravity. Progress appears to have been made through various approaches to the problem, such as *string theory* and *loop quantum gravity*. Some of this work indicates that the collapse may be halted at the smallest scales, as the tiny quanta of space-time resist further compression. This is a useful idea, but without a complete description of a quantized space-time, the singularity remains a fundamental mystery—a tear in the fabric of reality.

Since singularities are places where the known laws of physics break down, some physicists see them as a kind of dangerous abomination. Some, like the great cosmologist Stephen Hawking, propose that singularities can never occur in nature without an event horizon around them. In a sense, the event

GOING FURTHER 14.2   THE LANGUAGE OF THE COSMOS
Black-Hole Edges: The Event Horizon

One of the strangest attributes of black holes comes from one of the simplest ideas in physics. The event horizon represents the point of no return for an object falling toward a black hole. Mathematically, the event horizon is defined as the point where the black hole's escape velocity equals the speed of light. For a black hole of mass M, we can use this simple idea to determine the event horizon's radius—the distance from the black hole's center to the event horizon, also called the Schwarzschild radius (R_s).

To see how the Schwarzschild radius is derived, we simply write the equation for escape velocity:

$$V_e = (2GM/R)^{(1/2)}$$

If we let $V_e = c$, square both sides of the equation, and solve for R_s (because $R = R_s$ when $V_e = c$), we get

$$c^2 = 2GM/R_s$$

which can be solved for R_s as follows:

$$R_s = 2GM/c^2$$

Since the speed of light (c) and Newton's constant (G) don't change, we can write this equation in a way that is much easier to use because it shows us how R_s changes (or *scales*, as astronomers say) with mass:

$$R_s = 2.9 \text{ km} \times (M/M_{Sun})$$

Thus a 1-M_{Sun} black hole has a 2.9-km event horizon, a 10-M_{Sun} black hole has a 29-km event horizon, a 100-M_{Sun} black hole has a 290-km event horizon, and so on. How large, then, is the event horizon of the black hole at the center of the Milky Way Galaxy? If we know that the black hole's mass is 3.7 million M_{Sun}, we can quickly calculate the answer:

$$R_s = (2.9 \text{ km})(3.7 \times 10^6) = 1.1 \times 10^7 \text{ km}$$

If we now convert to astronomical units (AU), we find

$$R_s = (1.1 \times 10^7 \text{ km})/(1.5 \times 10^8 \text{ km/AU}) = 0.07 \text{ AU}$$

Thus, the black hole containing as much mass as 3.7 million Suns has a size that is much less than the distance from Mercury to the Sun.

horizon forms a protective membrane for the Universe, since nothing can ever cross the event horizon *going out*. Hawking even proposed a principle of **cosmic censorship** in which **naked singularities** (singularities without an event horizon around them) would never occur. "Hawking never explained the physics controlling cosmic censorship," Campanelli explains. "He simply stated that nature will find a way to keep naked singularities from forming. Only time is going to tell us if he was correct."

For all their weirdness, black holes are actually rather simple. Once matter and energy fall through the event horizon, all properties like color, size, shape, and so forth are lost. That strange fact gives rise to the so-called **no-hair theorem** for black holes, which states that every black hole can be described by just three quantities: mass, spin, and electric charge. Thus, they have no extra properties—no "hair." "These three quantities are the only ones which don't get 'erased' when matter crosses the event horizon," says Campanelli. "That's why we like to say black holes are simple objects."

One last point is worth noting about black-hole structure. From science fiction movies, people often get the idea that a black hole has infinite gravitational power and will inexorably suck in everything across the cosmos. The reality is much more benign. "If the Sun turned into a black hole tomorrow," explains Campanelli, "Earth would still happily go on in its orbit (though in darkness, of course). It's only relatively close to the event horizon that the strange general relativistic effects appear." Of course, as Campanelli reminds us, our Sun does not have enough mass to become a black hole and will end its life as a white dwarf.

What exactly are these "strange relativistic effects"? There is only one way for us to find out. We must embark on a very dangerous and, happily, very imaginary plunge downward into a black hole.

cosmic censorship The conjecture that the Universe will not allow the existence of a singularity without an event horizon.

naked singularity A singularity that lacks an event horizon.

no-hair theorem The description of a black hole as having only three properties: mass, spin, and electric charge.

SECTION SUMMARY
The Singularity, Quantum Gravity, and the No-Hair Theorem

● To fully understand the nature of a black hole's singularity requires still-unknown laws of quantum gravity.

● A black hole has the same gravitational pull as a star or other body of the same mass (although a much smaller radius) and is characterized by just three properties: mass, spin, and electric charge.

The Longest Fall: Plunging into a Black Hole

The nature of a black hole will change if it is rotating or carries an electric charge. For our journey, we're going to keep things simple by making the plunge into a black hole that has no spin and no charge.

Imagine you're in a spaceship orbiting a black hole at a safe distance and preparing to enter it. You are thousands of Schwarzschild radii away from the hole, so space-time is gently curved and you feel no unusual effects of the black hole, except for its gravitational attraction (as described by Newton's version of gravity).

A friend in her own ship is orbiting with you as you make preparations for your trip into the black hole. The two of you routinely exchange signals by radio, ensuring that your clocks are perfectly synchronized. Since the two ships are close together and far from the event horizon, your clocks appear to run at the same rate. The only thing that looks strange is the view of space in the direction of the black hole (via a telescope, since you are so far away from it).

"Because the black hole distorts space-time so strongly," Campanelli explains, "you can actually see objects directly behind the black hole. The space-time around the hole acts like a lens bending the path of starlight. Objects like stars and galaxies that are directly behind the hole would appear to a distant observer as a bright band that runs around the edge of the event horizon" (**Figure 14.16**). This is another example of gravitational lensing.

The time arrives for you to start your journey. With some misgivings, you fire the ship's retro-rockets in the opposite direction of your orbital motion to slow down. The gravity of the black hole begins to pull you inward, and your path becomes a long spiral that will end at the event horizon. Note that by firing the rockets *to speed up*, you can still stop your inward spiral and get back into a stable orbit.

But you are brave and continue your spiral inward, watching as the black maw of the event horizon and its gravitationally lensed rim of stars grow in your field of view. From a safe distance, your friend tells you that the closer you get to the hole, the slower your clock appears to run relative to hers. The radio signals that you send her are also getting redshifted. To receive your messages, she must keep tuning her receiver to longer and longer wavelengths. Both of these effects become stronger as you drop deeper into the black hole's well of stretched space-time.

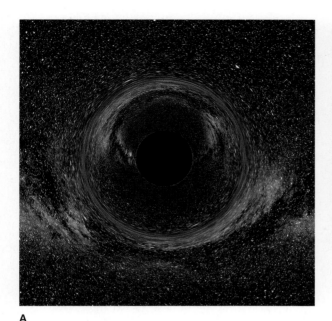

A

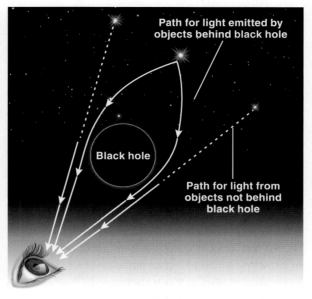

B

FIGURE 14.16
Gravitational Lensing by Black Holes Because of the strong curvature of space-time, the light from stars behind a black hole (relative to our line of sight) will be gravitationally lensed into a ring around the event horizon. These images show (A) what you would see if you were close to a black hole and (B) the path that light takes to reach you.

"The slowing of your clock relative to your friend's is called gravitational time dilation," says Campanelli. "Time in your frame of reference will run slower relative to that of someone far from the black hole. The second effect is called a gravitational redshift. Radiation emitted close to a black hole gets shifted to longer wavelengths as it climbs out of the hole's gravitational well." Both effects occur around any mass, not just black holes. Gravitational time dilation and redshift occur around Earth too, but they are so weak that only extremely precise instruments (like a GPS) can detect them.

As you get closer, the blackness of the event horizon fills your field of view. Suddenly your ship lurches and begins to plunge directly toward the blackness. The alarms are blaring. You ramp up your rockets to speed up, but it's no use. You can't get back into orbit. You're falling, and falling fast. "What you just discovered," says Campanelli, "is that no stable orbits are possible around a black hole past a certain point. The shape of space-time is such that a balance between orbital motion and gravity is not possible anymore." The location of this last stable orbit is at $R = 3R_s$ for a nonrotating black hole; for a rotating hole, the position will be farther out.

Time dilation is getting very strong now. Even though you feel time passing at its normal rate, your friend sees you slowing down to a crawl. You are feeling strange in a different way, though. Your feet seem to be pulling away from your head as if you were being stretched on a medieval rack. At the same time, your insides are getting squeezed, and you feel like you're being crushed in a giant fist. It's becoming too painful to bear.

What you're now experiencing is an extreme form of tidal forces. "From Newton, we know that tidal forces result because the effect of gravity decreases with distance from a central object," explains Campanelli. "In general relativity, we think of tidal forces as differences in the stretching of space-time from one point to another. Near a black hole, space-time is so distorted that your feet are being stretched toward the black hole much more strongly than your head." Scientists sometimes call the effect you're feeling *spaghettification*. Objects falling into a black hole are pulled into strands of spaghetti until even intermolecular forces can't hold up and are pulled apart. This is your eventual fate (**Figure 14.17**).

Even after you and your spaghettified spaceship are

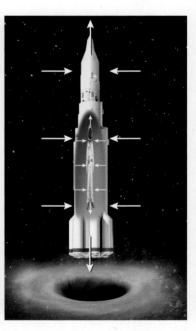

FIGURE 14.17 Spaghettification
Tides—which in general relativity are caused by the fact that space-time is more strongly curved on the side of an object closer to a large mass than on the side farther away—become greatly amplified near a black hole. An unfortunate person approaching the black hole would therefore be pulled apart (or "spaghettified") as his feet were pulled away from his head, as in this figure.

pulled apart, the remains fall toward the event horizon. From your friend's perspective, however, your time is running so slowly that it stops entirely when you reach the Schwarzschild radius. All signals become redshifted to infinity, so your friend can't actually see an image of you anymore, but if she could, you would appear to be forever frozen at the lip of the event horizon.

Your remains, which are now just a cloud of subatomic particles, pass right through the event horizon. It's important to note here that the horizon is not actually a physical boundary like the surface of a lake. It's just a location where the stretching of space-time becomes so extreme that light and everything else are permanently bound to the black hole.

What happens inside the horizon? "According to Einstein's equations, space and time reverse roles in a very strange way," explains Campanelli. "Just as we can only travel into the future in normal space-time, now there is only one direction in space you can travel and that is *into* the singularity. You can't help but move toward it."

Thus, your journey ends as you are drawn toward that giant question mark that is the singularity. What happens when you reach it is more than anyone can say right now.

SECTION SUMMARY
The Longest Fall: Plunging into a Black Hole

- The space-time distortion near a black hole causes light to be redshifted and the flow of time (relative to distant observers) to be dilated (slowed down).
- An object falling toward a black hole is pulled apart (spaghettified) by tidal effects.

14.5
Real Black Holes in Astronomy

The formation of black holes through stellar collapse was inferred from Einstein's general relativity as far back as 1939. But at that point most scientists didn't imagine that black holes were real. "Even Einstein believed that event horizons were too strange to exist," says Manuela Campanelli. "He thought the star would somehow stop collapsing before one formed."

As the decades passed, however, researchers explored general relativity's description of event horizons and other aspects of black-hole physics in ever-greater detail. "All that work built up confidence that nature might actually create something like black holes," says Campanelli. "But until you go out and find one through observations, you can't know for sure."

From the 1960s through to the 1990s advances in observational techniques helped build the case that real black holes exist. Astronomers generally group black holes into two categories: stellar-mass black holes, with masses between 1 and (approximately) 100 M_{Sun}, and **supermassive black holes**, with masses between 10^6 and 10^9 M_{Sun}. In principle, there is no reason why an intermediate-mass black hole of, say, 10,000 M_{Sun} can't exist. At this point, however, astronomers do not think that a large population of such objects has naturally formed.

Black Holes and X-Ray Binaries

The search for stellar-mass black holes got a strong push forward when NASA launched a satellite called Uhuru in 1970, one of its first X-ray space telescopes. Scouring the sky for sources of bright X-rays, Uhuru scientists detected strong signals emerging from a region associated with a blue giant star (of spectral class B) located in the constellation Cygnus. The object, christened Cygnus X-1, was immediately tagged as a candidate for a black-hole binary system when analysis of the B star's motion showed it in orbit with a massive but unseen object. The binary orbital period was just 5.6 days. Using Newton's laws, astronomers quickly determined the mass of the invisible companion to be at least 10 M_{Sun}. "That was big," says Campanelli, "and way too massive for the object to be a 'normal' compact object like a white dwarf or a neutron star."

The X-rays that Uhuru detected were also considered strong evidence for a black hole. Theoretical predictions told astronomers that material pulled off the companion B star by the black hole would naturally form an accretion disk as it spiraled inward toward the event horizon. "Gas in any accretion disk gets heated as it drops deeper into the gravitational well of the central object," says Campanelli, "and the deeper the well, the higher the temperature of the gas as it spirals around." Since black holes represent the deepest gravity well possible, astronomers expected temperatures in the inner regions of a black-hole disk to reach many millions of degrees (**Figure 14.18**). At these temperatures, dense gas radiating as a blackbody would hit peak emission in the X-ray region of the

supermassive black hole
A black hole of millions or billions of solar masses that is found in the center of a galaxy.

spectrum. Thus, any accretion disk around a black hole was expected to be glowing in X-rays.

Given the mass of the unseen companion and the X-ray emission from the orbiting gas, you might think the identification of Cygnus X-1 as a black hole would have been an open-and-shut case. Scientists are skeptical by nature, however. They expect every explanation for a given set of data to be explored, not just the one that gets everyone excited. Throughout the 1970s and 1980s, a number of alternative explanations were suggested for Cygnus X-1 that didn't involve a black hole. In the meantime, other X-ray binaries were discovered. Some of these had companions that turned out to be neutron stars. For others, as with Cygnus X-1, a black hole was still considered a viable possibility. Only later did new data from advanced X-ray telescopes finally give astronomers the proof needed to convince even the most hardened skeptics that stellar-mass black holes had, in fact, been discovered.

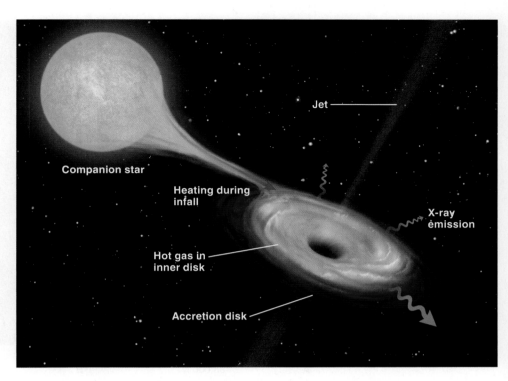

SECTION SUMMARY
Black Holes and X-Ray Binaries
- Because of their extremely high temperatures, accretion disks in binaries with a black hole can be observed using X-ray telescopes.
- X-ray observations have provided strong evidence for the existence of black holes in binary systems.

Supermassive Black Holes and Galaxies

A galaxy is a vast collection of stars, gas, and dust with a total average mass of about $10^{12}\ M_{Sun}$. Beginning in the 1960s, it became clear that the centers of many galaxies were home to wildly energetic phenomena (**Figure 14.19**). Torrents of light were pouring from the central regions of these galaxies, and many of them were driving energetic jets of gas into space. Over time, astronomers came to realize that only a black hole of titanic proportions could account for these observations. They hypothesized that supermassive black holes, with masses more than a billion times that of the Sun, must exist at the core of these galaxies.

These supermassive monsters turned out to be good targets for black-hole research. "Using the Hubble Space Telescope and other high-resolution instruments," Campanelli explains, "astronomers could zoom in on material spiraling into these larger black holes." The light from this material was emitted by gas orbiting far enough from the event horizon that

it could easily escape to reach Earth-based telescopes. By analyzing the light's Doppler shift, astronomers determined the speed of the orbiting gas and therefore could derive the mass of the central object (see Going Further 3.3).

"The mass was so large and was all packed into such a small region of space that a black hole was really the only explanation," says Campanelli. The discovery of these supermassive black holes ushered in a new era of black-hole studies. It is perhaps ironic that some of the black holes we've come to know best are the ones Einstein and the early pioneers of general relativity never imagined. In fact, astronomers now think that *all* galaxies have black holes at their centers.

SECTION SUMMARY
Supermassive Black Holes and Galaxies
- Some of the first black holes discovered were supermassive black holes in the centers of galaxies, observed via the light emitted from orbiting and infalling material.

Jets from Black Holes

While most people think of black holes as giant cosmic vacuum cleaners, the truth is more complicated. Many environments hosting black holes also show material flowing outward from the central object. In particular, many black holes show evidence for

FIGURE 14.18
X-Ray Binaries
Astronomers expected temperatures in the inner regions of a black-hole accretion disk to reach many millions of degrees, as gas in the disk would be drawn deeper into the black hole's gravity well. At these temperatures, any gas that radiated as a blackbody would reach peak emission in the X-ray region of the spectrum.

FIGURE 14.19

Supermassive Black Holes
(A) Supermassive black holes exist at the center of many galaxies. Here, we see the galaxy NGC 4261 with its brilliant radio jets (driven by the black hole at the center) extending well beyond the limits of the galaxy. (B) This HST image shows the dusty disk 400 light-years in diameter that surrounds the black hole and the radio jets aligned perpendicular to the disk's major axis. HST spectral observations of gas in the nucleus suggest a supermassive black hole of $5 \times 10^8 \, M_{Sun}$.

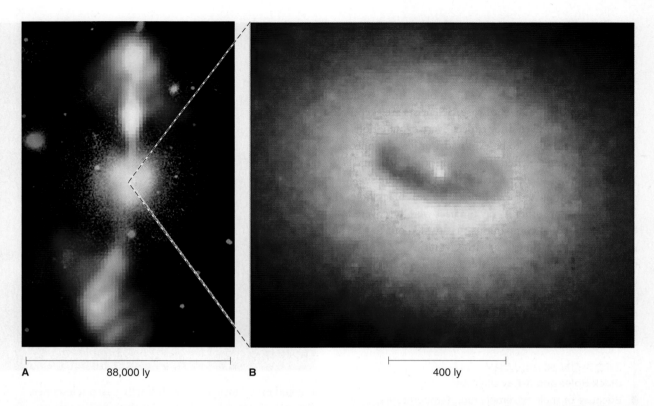

| A | 88,000 ly | B | 400 ly |

powerful jets blasting across space at velocities just a hair shy of light speed.

Microquasars are one example of a system that includes jets associated with black holes. Microquasars are binaries in which one of the companions is a black hole or a neutron star. (Quasars, phenomena associated with supermassive black holes at the center of galaxies, are discussed in detail in Chapter 16.) Jets of material moving at velocities of 99 percent of light speed are seen hurtling away from microquasars and their larger galactic cousins.

Galactic jets are another example of a beam of relativistic matter created by a black hole. The engines of galactic jets, which can extend many thousands of parsecs into space, are supermassive black holes in galactic centers. These black holes will also be surrounded by accretion disks, which form as material falls in toward the central region of a galaxy.

In both cases, these relativistic jets are composed of material that has been lifted off the inner regions of an accretion disk surrounding a black hole. Magnetic fields tied to the rapidly spinning disk fling material out into space in a process resembling the way jets are created from young stars (see Chapter 12). "Magnetic-field lines extending from the disk act like wires," explains Campanelli. "Plasma tied to those field lines gets whipped around with the disk's rotation at relativistic speeds. When the gas eventually gets flung out into space, it's already moving close to the speed

of light" (**Figure 14.20**). The relativistic speeds of the jets tell astronomers that the launching must occur in the inner regions of the disk, where the orbital speeds are also relativistic (roughly 10 percent of the speed of light or greater). Gas rotating at 200 km from a $10\text{-}M_{Sun}$ black hole moves at about 81,000 km/s, or 27 percent of light speed.

If the black hole itself is spinning, then something new and remarkable can occur. First, a spinning black hole will actually drag space-time around with it. This process is called **frame dragging**, and it is caused by any massive rotating object. Frame dragging due to Earth's rotation has been detected by a spacecraft called *Gravity Probe B*. For spinning black holes, the effect is much stronger and can change conditions in the inner regions of the accretion disk by eliminating the zone of stable orbits close to the event horizon. Once material in the disk spirals into these regions, frame dragging forces it to plunge directly into the event horizon.

If the spinning black hole carries an electric charge, magnetic fields will also extend out of the event horizon. When the black hole spins, these magnetic-field lines spin as well, twisting up like tangled wires sending jets of magnetic energy flowing out into space. Note the phrase "jets of magnetic energy" in the previous sentence. Unlike the beams of plasma that come from young stellar objects, for example, jets that form directly from rotating black holes have very little matter in them and are composed almost entirely of magnetic-field

frame dragging The process by which a massive, spinning object pulls the surrounding space-time around with it.

energy. Such magnetically dominated jets can, however, push on matter in the environment as they expand away from the black hole, creating shock waves that astronomers have observed in a variety of wavelengths.

> **SECTION SUMMARY**
> **Jets from Black Holes**
> - Some black holes are observed to have jets moving at relativistic speeds.
> - Rotating black holes drag space-time around as they spin, in a process called frame dragging.

The End of Stellar Death

Just as a very massive star exhausts its fuel and collapses to form a black hole, with our exploration of black-hole physics we have now come to the end of our exploration of stars and their lifetimes. Stars live by the billions in vast, swarming structures called galaxies, and the Universe, it turns out, is teeming with galaxies. Now that we've learned about planets and stars, it is time to take the last step on our journey outward and learn to see the cosmos first through its multitude of galaxies and then through its own fundamental history. It is time to go to the edges and beginnings of the Universe itself.

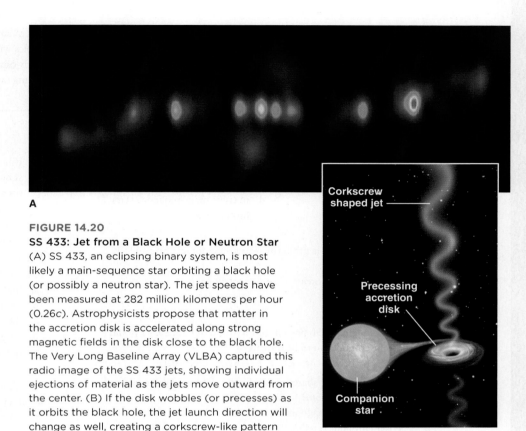

A

FIGURE 14.20
SS 433: Jet from a Black Hole or Neutron Star
(A) SS 433, an eclipsing binary system, is most likely a main-sequence star orbiting a black hole (or possibly a neutron star). The jet speeds have been measured at 282 million kilometers per hour (0.26c). Astrophysicists propose that matter in the accretion disk is accelerated along strong magnetic fields in the disk close to the black hole. The Very Long Baseline Array (VLBA) captured this radio image of the SS 433 jets, showing individual ejections of material as the jets move outward from the center. (B) If the disk wobbles (or precesses) as it orbits the black hole, the jet launch direction will change as well, creating a corkscrew-like pattern like the one seen in the radio image.

Corkscrew shaped jet
Precessing accretion disk
Companion star

B

●→ chapter summary

14.1 The Black-Hole Shuffle

Black holes are objects whose properties defy our everyday experience, and scientists have expended great effort to understand their physics. Dying massive stars collapse into black holes when thermonuclear fusion no longer offsets gravity's inward pull. Not even light can escape from a black hole.

14.2 Special Theory of Relativity

Einstein proposed (and experiments later proved) his special theory of relativity, which considers how measurements of space and time differ in different frames of reference. Two important consequences of the special theory of relativity are: (1) Since the speed of light must be the same for all observers, then measurements of space and time must be different for different frames of reference. (2) If a person is moving close to the speed of light relative to another observer, the observer will see the moving person's time dilated (flowing more slowly), although the person in motion will perceive his or her own time and space to be unchanged.

14.3 General Theory of Relativity

Albert Einstein reached some of his most radical conclusions through thought experiments, realizing, for example, that the physics of a body at rest in a gravitational field is equivalent to that of a body moving at constant acceleration. His general theory of relativity implies that (1) matter-energy tells space-time how to bend, and (2) space-time tells

matter-energy how to move. According to the general theory of relativity, gravity is the effect of curved space-time. One effect of general relativity is gravitational time dilation: clocks run slower when they are closer to large masses. Observations of the extended lifetimes of muons confirmed special relativity, and general relativity was initially confirmed by the bending of starlight as observed in the 1919 eclipse. Today's GPSs would not be accurate if they did not account for the effects of general relativity.

14.4 Anatomy of a Black Hole

Black holes are objects so compact and so massive that even light cannot escape. They can result from the deaths of very massive stars or may be located at the centers of galaxies. All black holes have a central point called the singularity, where our current understanding of physics breaks down. A theory of quantum gravity is assumed to be the key to understanding what happens at the singularity, but no such theory has been fully developed. Black holes are bounded by an event horizon at the distance of the Schwarzschild radius from the center, where the escape velocity equals the speed of light. The Schwarzschild radius can be calculated as simply 3 km multiplied by the number of solar masses in the black hole. Black holes have only three parameters: mass, electric charge, and spin. An object falling into a black hole becomes "spaghettified" and torn apart by the strong tidal forces. Light from the object becomes greatly redshifted. The flow of time near the horizon, relative to a distant observer, becomes strongly dilated.

14.5 Real Black Holes in Astronomy

At least two classes of black holes are observed to exist. Supermassive black holes at the centers of galaxies are evidenced by the rapid motions of nearby stars and gas. Black holes resulting from stellar death were first observed as part of a binary pair with a visible companion star. Accretion disks around black holes will have high temperatures and will radiate in X-rays. Jets are frequently observed flowing at relativistic speeds from stellar-remnant black holes, flung out by the black hole's magnetic-field lines.

●→ questions and problems

Narrow It Down: Multiple-Choice Questions

1. The Chandrasekhar limit applies to which kinds of objects?
   a. neutron stars
   b. white dwarfs
   c. black holes
   d. massive main-sequence stars
   e. brown dwarfs

2. Which force or process is primarily responsible for creating a black hole?
   a. the strong nuclear force
   b. inertia
   c. gravity
   d. centrifugal force
   e. gas pressure

3. An observer moving at 90 percent of the speed of light $(0.9c)$ would observe light around him to be moving at
   a. $0.9c$.   b. $0.1c$.   c. c.   d. $0.81c$.   e. $-0.1c$.

4. Which of the following shares your reference frame while you're reading this book in your room?
   a. an astronaut on the Moon
   b. a geosynchronous satellite directly over your location
   c. an airplane 30,000 feet up in the atmosphere traveling 400 miles per hour
   d. an Indy driver speeding on the course
   e. a person in an adjacent room of your residence

5. Two spaceships are moving in opposite directions, spaceship A traveling at $0.9c$ and spaceship B traveling at $0.95c$. Which of the following statements is true?
   a. B measures A moving at the speed of light.
   b. A measures B moving at the speed of light.
   c. Both measure light passing them at c.
   d. Only A is actually moving.
   e. Only B is actually moving.

6. In the twin paradox, the twin who travels in the spaceship experiences
   a. time passing slowly.
   b. time passing quickly.
   c. a normal flow of time in her frame of reference.
   d. no sense of time passing.
   e. time passing at the same rate as her twin experiences it, as measured by the same distant observer.

7. Regarding special relativity and general relativity, which of the following is true?
   a. General relativity alone deals with gravity.
   b. Special relativity alone has c as a constant.
   c. Special relativity incorporates the effects of gravity.
   d. In special relativity alone, space and time are flexible.
   e. "General relativity" refers to objects in uniform motion.

8. Which of the following is true about gravitationally induced motion, according to general relativity?
   a. It is a result of an object's spin.
   b. It occurs as objects fall freely following the curvature of space-time.
   c. It does not follow Newton's laws.
   d. It is the same as uniform motion.
   e. It exists only near extremely massive objects.

9. Which of the following does/do *not* represent a successful test of relativity? Choose all that apply.
   a. the detection of muons created by cosmic rays
   b. the inference of gravity waves via binary pulsars
   c. the discovery of the wave nature of light
   d. weightlessness experienced by astronauts in space
   e. the accuracy of GPS devices

10. If GPS satellites orbited higher, which of the following would be true regarding the general relativistic adjustments currently required for accuracy of positioning on Earth's surface?
    a. They would be greater because of the difference in the satellites' speed.
    b. They would be smaller because of the difference in the satellites' speed.
    c. They would be the same, since the surface positions remain the same.
    d. They would be greater because of the greater difference in the curvature of space-time between the ground and satellite.
    e. They would be the same because general relativity can be ignored in the near-Earth environment.

11. Which of the following is the densest?
    a. core of a $1\text{-}M_{Sun}$ star
    b. core of a $100\text{-}M_{Sun}$ star
    c. white dwarf
    d. neutron star
    e. singularity of a black hole

12. Which of the following is/are *not* measurable properties of black holes? Choose all that apply.
    a. mass
    b. elemental composition
    c. electric charge
    d. spin
    e. color

13. A patent clerk in a spaceship observes that time on the clock of an astronaut on a spaceship passing at $0.25c$ runs slower than does time on his own clock. This phenomenon is called
    a. relativistic time dilation.
    b. time travel.
    c. gravitational time dilation.
    d. curvature of space-time.
    e. time warp.

14. As an object approaches the event horizon of a black hole, the light from it is observed to become
    a. longer in wavelength.
    b. shorter in wavelength.
    c. slower and slower.
    d. faster and faster.
    e. unchanged.

15. If you could be 475 km from the center of a 200-M_{Sun} black hole, which of the following would be true?
    a. You would see an accretion disk present.
    b. You would be orbiting at a constant speed.
    c. You would no longer be able to communicate with the rest of the Universe.
    d. You would be in the process of being accelerated into the black hole.
    e. You would be able to speed up to move away from the black hole.

16. Two spacecraft (*X101* and *X102*) with different velocities relative to each other pass in interstellar space. Which of the following *cannot* be true?
    a. *X101* measures its own velocity as zero and measures a nonzero velocity for *X102*.
    b. *X102* measures its own velocity as zero and measures a nonzero velocity for *X101*.
    c. Each measures both its own velocity and the velocity of the other as nonzero values.
    d. Both measure light passing them at exactly *c*.
    e. A distant observer measures zero velocities for both.

17. Which of the following statements about Einstein's theory of gravity, compared with Newton's law of gravity, is true?
    a. It concluded that the effect of mass is less than expected, while that of distance is more than expected.
    b. It concluded that the effect of distance is less than expected, while that of mass is more than expected.
    c. It proved Newton's law entirely wrong.
    d. It made the use of Newton's law unnecessary.
    e. It showed that Newton's laws were a "limiting case" of general relativity that applied only when velocity and gravity were relatively low.

18. Which of the following statements is/are consistent with Einstein's special theory of relativity? Choose all that apply.
    a. There is no privileged frame of reference.
    b. The speed of light is constant, except when affected by relativistic time dilation.
    c. The speed of light is an upper bound to motion in the Universe.
    d. Time is not the same in all frames of reference.
    e. Space and time exist independent of each other.

19. True/False: Spaghettification upon approaching the event horizon of a black hole occurs because gravity varies with distance from its source.

20. True/False: The 1919 solar eclipse provided support for relativistic length contraction.

To the Point: Qualitative and Discussion Questions

21. Explain the difference between core mass and main-sequence mass in determining the end product of star death.

22. What were some of Einstein's thought experiments?

23. Describe the evidence for the existence of black holes.

24. What was the "luminiferous aether" believed to be, and why did physicists propose that it existed?

25. What is a frame of reference? What and who currently share your frame of reference?

26. Describe a situation in which the "addition of velocities" gives a correct result.

27. Why did Einstein conclude that space and time must be unified and measurements of each depend on an observer's frame of reference?

28. What is the twin paradox?

29. For what kinds of situations is special relativity relevant? For what kinds is general relativity relevant?

30. What are some of the effects that occur near black holes?

31. How does the detection of muons on the surface of Earth, after they are created high in Earth's atmosphere, provide evidence of special relativity?

32. What phenomenon defines the event horizon of a black hole?

33. What are the differences between supermassive black holes and stellar-mass black holes in terms of how they form?

34. What evidence suggests that massive black holes exist in galaxies?

35. Think about any movies or TV shows in which black holes have been portrayed. According to what you now know about black holes, were they reasonably accurate?

Going Further: Quantitative Questions

36. What is the Schwarzschild radius, in kilometers, of a 13-M_{Sun} black hole? of a 2,500-M_{Sun} black hole?

37. What is the mass, in M_{Sun}, of a black hole whose Schwarzschild radius is 990 km?

38. A black hole has a Schwarzschild radius of 7×10^6 km. What is its mass, in kilograms?

39. What is the Schwarzschild radius, in meters, of a black hole whose mass is 4×10^{22} kg?

40. A planet's orbital radius is 43 AU. If the star were a black hole instead, at what radius would the planet orbit?

41. Sophia (on Earth) sees her twin sister, Stella, passing on her spacecraft at 0.4*c* and observes that Stella's clock runs slower than her own. Sophia's favorite movie runs 2 hours. How long by Sophia's clock, in hours, would the show run for Stella?

42. While riding on a jet from a distant space station at 0.9*c*, the occupants of a spaceship take 35 minutes to prepare their dinner. How long, in minutes, does the dinner preparation take according to the clock of a NASA scientist on Earth who is observing the activity and the clock of the spaceship?

43. An American football game runs 60 minutes (not counting time-outs and commercials). You are on Earth observing a game being played aboard a starship traveling at 0.79*c*. How many minutes does the game last on your clock?

44. An event takes 2 hours on the clock of a spacecraft at relativistic speed, and 8.4 hours on the clock of an observer on a nearby planet. At what speed is the craft traveling? (State your answer in terms of *c*.)

45. What is the escape velocity from a location that is 92 km from the singularity of a black hole with mass 25 M_{Sun}? (State your answer in terms of *c*.)

 If your instructor assigns homework in **smartwork5**, access it at the Digital Landing Page for *At Play in the Cosmos*: **digital.wwnorton.com/cosmos**

OUR CITY OF STARS

The Milky Way

In this chapter you will learn about our home galaxy, the Milky Way. After reading through each section, you should be able to do the following:

15.1 A Hard Rain: Gaining a New Vision of Galactic Studies
Recount the major steps that led to an understanding of the Milky Way's size and structure.

15.2 Anatomy of the Milky Way
List the observable components of our galaxy and describe their important characteristics, including the types of stars and other objects found in each component.

15.3 Spiral Arms: Does the Milky Way Have Them?
Summarize the evidence for the presence of spiral arms in our galaxy, as well as theories about their formation and nature.

15.4 The Galactic Center
Describe the structure and characteristics of the central region of the galaxy, and explain why astronomers think that a massive black hole resides there.

15.5 Dark Matter and the Milky Way
Explain the evidence for dark matter in galaxies and the way physicists think it interacts with the four known fundamental forces.

15.6 Constructing a Galaxy: Evolution of the Milky Way
Relate the characteristics of the various components of the Milky Way to current theories about the galaxy's formation and evolution.

The beautiful Andromeda galaxy, with its spiral arms and central bulge, is thought to be much like our own Milky Way.

15

FIGURE 15.1
Galactic Fountain
These composite radio-light and visible-light images of our galaxy show clouds of gas above the plane of the Milky Way. The cloud circled in (A) is falling (or "raining" down) onto the Milky Way disk, seeding it with material for star birth. (B) In this diagram of the galactic fountain concept, supernovas drive material rich in heavy elements out of the plane of galaxy, which then rains back down in other locations.

15.1
A Hard Rain: Gaining a New Vision of Galactic Studies

YOUNG STUDENTS don't usually get unexpected calls from world-famous scientists, but that is exactly what happened to Bob Benjamin one morning after he submitted his first scientific paper for review. "I was a graduate student and had written a short paper on clouds of gas falling down onto the disk of the Milky Way Galaxy," says Benjamin, now a professor at the University of Wisconsin. "I was really excited about the idea because we can see these clouds of hydrogen gas high above the Milky Way's disk, and they're moving

FIGURE 15.2
Lyman Spitzer
Lyman Spitzer (1914–1997) was an American theoretical astrophysicist known for his research in many areas of astronomy and for conceiving the idea of telescopes operating in outer space. NASA's infrared Spitzer Space Telescope is named in his honor.

downward at pretty high velocities. I thought I had a neat, and pretty simple, idea for explaining them."

In his paper, Benjamin hypothesized that supernova explosions could blow material out of the disk of the Milky Way, but the gas would have too low a speed to escape the gravitational pull of our galaxy. The phenomenon was called a *galactic fountain*. Like a rocket with too little fuel, the gas would rise and then begin to fall back (**Figure 15.1**). "I had written down some simple equations that compared the falling gas clouds to raindrops," says Benjamin. "The speed of the cloud is determined by the balance between gravity and aerodynamic drag, just like rain." If correct, Benjamin's idea would help explain one piece of the puzzle of galactic recycling: how gas from one part of the galaxy ends up in another.

And then Lyman Spitzer (**Figure 15.2**), arguably one of the most important astronomers in the world, called him. "I've been given your paper to referee, and it's very interesting," said Spitzer. "However, I have some concerns about equation 1." Benjamin was stunned. "Oh dear!" he said to himself. "I'm in for it now." Equation 1, Benjamin knew, was just a version of $F = ma$.

It turned out that the only thing Benjamin was really in for was a long and fruitful discussion with the great scientist. Eventually, Benjamin's work was published with Spitzer's blessing. "This paper presents a new and potentially important theory of interstellar dynamics," the older scientist wrote. Benjamin's work as a researcher studying our galaxy, the Milky Way, had officially begun.

Life in a Universe of Galaxies

Every day the Sun rises and sets, and every night the stars appear and create their patterns of illumination against the backdrop of empty space. But where are all these stars located? Is there any structure to their

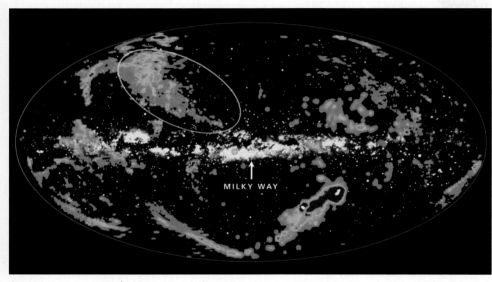

A

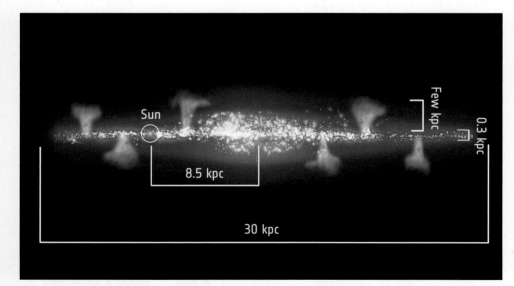

B

FIGURE 15.3
The Milky Way on the Sky
The Milky Way appears as an arc of diffuse light reaching from one horizon to the other. For most of human history, it was a familiar part of the nighttime sky, but the widespread use of electric lights has made it a rare sight for anyone living in highly populated areas.

distribution in space, or is the cosmos just an infinite array of stars? Does the night sky provide any clues about where, exactly, we reside in the architecture of the Universe?

For most of human history, a hint to this giant question could be seen overhead every night in the **Milky Way**, an arc of light reaching from one horizon to the other (**Figure 15.3**). These days, most of us see the Milky Way only on occasional trips out of the cities and suburbs where we live. "So few people actually live in a place where they can see more than a few stars," says Benjamin. "But if you get far away from city lights, you'll see many, many, many more stars than you ever realized were up there. And in particular, you'll see a sort of a smooth band of light stretching across the sky. That band is the Milky Way, and it's where most of the stars in the galaxy are located."

For hundreds of millennia, however, humans looked at the Milky Way in wonder, creating explanations first in myth and later in science. Through the efforts of generations of astronomers, the Milky Way was gradually recognized for what it is: a galaxy—a teeming "city" of a few hundred billion stars, all gravitationally bound together.

The last step outward in building our story of the Universe is essentially a story of galaxies, the building blocks of the Universe. And what better way to understand the structure, formation, and history of galaxies than to start with our own vast system of stars, the Milky Way?

SECTION SUMMARY
Life in a Universe of Galaxies
- All the stars that we see at night exist within a single galaxy called the Milky Way.
- Galaxies are gravitationally bound systems of stars that are the building blocks of cosmic structure.

The Long Road to the Milky Way

Ancient astronomers marveled at the nightly spectacle of the Milky Way, but without telescopes it was impossible for them to determine its true nature. "It was Galileo with his telescope who first realized this Milky Way was not just some fuzziness in the sky," says Benjamin. "He saw it was a collection of thousands of stars whose light was blurred together. I guess in that way you could call Galileo the first galactic astronomer."

After Galileo surmised the true starry nature of the Milky Way, the race was on to find its true shape. Understanding the real structure of the Milky Way meant understanding the true distribution of visible matter (which at that time meant stars) in the Universe. By the time of Isaac Newton, many astronomers believed that an infinite number of stars were distributed smoothly across infinite space. But as telescopes became more powerful, some astronomers resolved to answer the question of the Milky Way's size and structure by using the best tool at their disposal: direct observation.

By the late 1700s, astronomer William Herschel and his sister Caroline had set out to directly measure the distribution of stars in the Milky Way (**Figure 15.4**). Their strategy was simple: count them. "Herschel was already an expert at building big telescopes, and he had used these instruments to make amazing discoveries," says Benjamin. "He was the first person in history to discover a planet, Uranus, and that made him an instant celebrity."

In time, however, William Herschel turned his attention to the Milky Way. Beginning with the assumption that the Sun was located at the center of the Milky Way, Herschel and his sister spent hours each night at their telescope counting stars in different directions of the sky. "Herschel would begin by pointing the telescope at a certain location in the sky," explains Benjamin. "He'd

A

B

FIGURE 15.4
Star Counts
(A) Caroline (1750–1848) and (B) William Herschel (1738–1822) made labor-intensive counts of the number of stars in all directions of the sky in an effort to discover the structure of the Milky Way Galaxy.

Milky Way The spiral galaxy in which our Sun is located.

FIGURE 15.5
Herschel's
Grindstone Diagram
Demonstrating a remarkable ability to interpret data, the Herschels correctly deduced from the distribution of stars in the star counts that the Milky Way had a disklike shape. William Herschel likened it to a grindstone, a stone wheel used to sharpen metal blades.

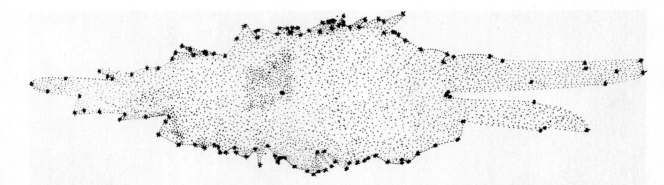

count up the number of stars he could see through the eyepiece and call the numbers out to Caroline at the base of the big instrument. Then he'd reset the telescope, point it in another direction, count up those stars, and call them out again." It was exhausting work. Together the Herschels spent many months meticulously observing and building up their map of star counts.

The effort bore fruit: the Herschels' work enabled them to accurately infer that the Milky Way was shaped like a grindstone, or disk (**Figure 15.5**). Their assumption that the Sun was located at the grindstone's center was, however, flawed for reasons they could not have suspected at the time.

By the early 20th century, astronomers not only had a better determination of the Milky Way's shape, but also had begun establishing limits on its size. "By then, people had photographic plates to make their star counts," says Benjamin. "That was a huge advance over staring into an eyepiece." In 1906, the astronomer Jacobus Kapteyn used the star count method, together with maps of stellar motions, to infer that the Milky Way was a flattened disk 10,000 parsecs (pc) across and 2,000 pc high (**Figure 15.6**). These distances deduced by Kapteyn would have been unimaginably large to astronomers just a few centuries earlier. The study of the Milky Way was pushing researchers ever farther out into the cosmos. The need for a new distance unit, the kiloparsec (kpc), equal to 1,000 pc, is testament to the fact that they had left the realms of stars behind and were considering far larger cosmic structures.

Ironically, further progress determining the true (and enormous) size of the Milky Way was blocked by the tiniest pieces of solid matter—interstellar dust. Recall from Chapter 12 that these tiny flecks of material affect observations in two ways. First, when starlight passes through dust clouds, some of its light will be scattered or absorbed. "That means the stars appear dimmer," says Benjamin. Since astronomers were not taking the intervening dust into account in their calculations, they thought the stars were farther from Earth than they really were.

A

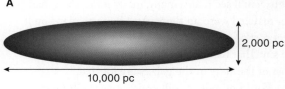

2,000 pc

10,000 pc

B

FIGURE 15.6 Kapteyn and the Size of the Milky Way
(A) Jacobus Kapteyn (1851–1922) counted stars, as the Herschels had before him, but he used photographic plates instead of observations through an eyepiece. (B) Kapteyn inferred that the Milky Way was a flattened disk 10,000 kpc across and 2,000 kpc high, greatly expanding previous estimates of the galaxy's size.

Dust clouds can also completely obscure light from some regions of the Milky Way, so that no stars can be seen. Thus, any star counts carried out in those directions will be highly inaccurate. "Dust really limits how far you can see," says Benjamin. "If you're looking toward the inner galaxy through the Milky Way's disk, dust blocks out almost everything more than a few kiloparsecs away. If you don't know about dust, then you are going to get a lot of aspects of the galaxy dead wrong."

interactive
Explorable Milky Way

It was some time before astronomers understood how to take interstellar dust into account in their calculations of the Milky Way's size. A crucial step came several years after Kapteyn's model, in 1914, when Harlow Shapley, an astronomer at Harvard, decided to focus not on counts of individual stars, but on the jewel-like collections of stars called globular clusters. Shapley was able to use the largest telescope of the day to resolve variable stars that could be used as standard candles (see Chapter 11). In this way, he mapped out the distances to and positions of the globular clusters, which he found took the shape of a sphere, not a disk. Just as important, he determined that the center of the sphere lay more than 16 kpc away from the Sun, in the constellation Sagittarius as seen from our point of view (the true distance is about 8 kpc). This, he concluded, was the true center of the galaxy (**Figure 15.7**).

"What made Shapley famous was he managed to go outside the plane of the Milky Way's disk," says Benjamin. "When Herschel and Kapteyn did their star counts, they weren't seeing very far because dust in the disk blocked their view. That meant they were getting a very biased picture of the galaxy." Shapley trained his telescope away from the dust and toward clusters that lay above and below the disk. Doing so enabled him to see farther, find the Milky Way's center, and discover that the disk was not the only component of the galaxy's structure. His work set the stage for the modern era of Milky Way studies. In the decades that followed Shapley's studies, the true structure of the Milky Way would be mapped out in exquisite detail, revealing the complex anatomy of our home galaxy.

SECTION SUMMARY
The Long Road to the Milky Way
- Many early astronomers studying the Milky Way recognized that it had a disklike component.
- Star counts and studies of stellar motions were used to determine the shape of the galaxy, but the presence of interstellar dust, which obscures and reddens distant objects, limited the effectiveness of these methods.
- The study of globular clusters above and below the galactic disk made possible better models of the size and shape of the Milky Way.

15.2
Anatomy of the Milky Way

Even though we can see the Milky Way only from the inside, years of study reveal that it has three basic directly observable structural components: the galactic disk, the stellar halo, and the bulge. **Figure 15.8** shows

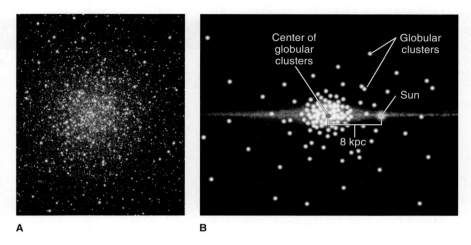

A   **B**

FIGURE 15.7 Finding the Galactic Center
Harlow Shapley (1885–1972) studied globular clusters such as (A) M3 in our galaxy and determined that they were distributed spherically around the galaxy's center. (B) Using the globular clusters, he also estimated the position of the Sun from that center.

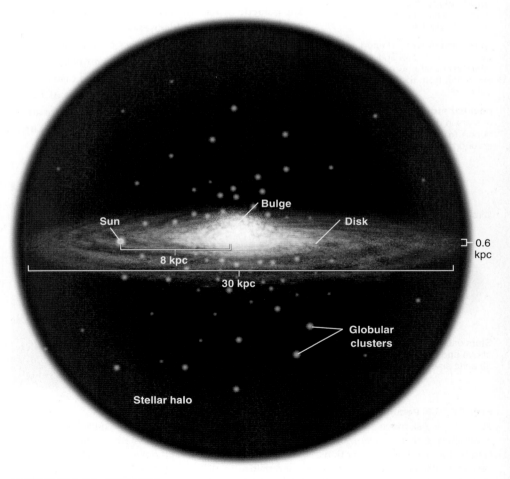

FIGURE 15.8 The Milky Way
A schematic view of the Milky Way Galaxy shows its three primary luminous components (components made of "normal" or nondark matter): the galactic disk, bulge, and stellar halo.

TABLE 15.1 • ••	Visible Components of the Milky Way Galaxy		
	● Bulge	● Galactic Disk	● Stellar Halo
Location in the Galaxy	Central region	Equatorial plane	Surrounding the disk
Size	Radius: ~5 kpc	Radius: ~15 kpc Height: ~0.3 kpc	Radius: ~15 kpc
Shape	Flattened sphere	Flat disk; may have spiral arms	Sphere
Features	Very dense stellar population	Star-forming regions in spiral arms	Old, widely dispersed globular clusters and stars

galactic disk A flattened distribution of stars and interstellar material within a galaxy.

spiral galaxy A pinwheel-shaped galaxy that may or may not have spiral arms.

spiral arm A spiral-shaped distribution of stars and interstellar material, often extending from the bulge outward.

peculiar motion The motions of stars in the Milky Way that depart from pure revolution around the galactic center.

the Milky Way's structure, and **Table 15.1** gives the basic properties of each component.

The Galactic Disk

The **galactic disk** is the most recognizable of the three fundamental and directly observable galactic structures, since the Milky Way appears to be a classic example of a **spiral galaxy**, much like our neighbor the Andromeda Galaxy. The disk is a flattened distribution of stars and interstellar material with a radius of 15 kpc and a height of approximately 0.3 kpc. There are somewhere between 200 billion and 400 billion stars in the disk (implying a mass of approximately 10^{11}

solar masses, M_{Sun}). The amount of interstellar material is lower but still significant (in the tens of billions of solar masses). The Sun is located at a distance of approximately 8 kpc from the center of the galaxy and 0.03 kpc above the midplane of the disk.

Although material is distributed throughout the disk, there is strong evidence for the existence of **spiral arms**, where the density of stars and gas increases. The exact number of distinct spiral arms remains in dispute, but most astronomers agree that there are up to four major spiral arms in our galaxy, and they tend to be the location of molecular clouds where stars are forming. Later in this chapter, we will spend some time understanding the spiral arms' nature. With the exception of the arms, however, the density of stars and gas clouds decreases fairly smoothly with both radius and height in the disk. The fraction of heavy elements in both the stars and the stellar clouds also decreases with distance from the disk's center.

Material in the disk revolves around the galactic center on orbits that are generally circular. "There's an overall circular revolution. At the position of the Sun, the galactic revolution period is roughly 233 million years," says Bob Benjamin. "But stars move in other ways as they go around the Milky Way too. We call these **peculiar motions**." Gravitational encounters between stars and clouds cause stars to "scatter," similar to the way planets can scatter within solar systems (see Chapter 5). These scattering encounters cause disk material to bob up and down in an oscillating vertical motion that takes stars and gas above and below the disk midplane. "In other words," says Benjamin, "if you look at a star, it may be going slightly upward or slightly downward relative to the disk, and it may be somewhat above or below the disk" (**Figure 15.9**).

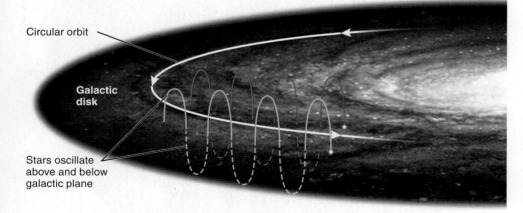

FIGURE 15.9 Peculiar Motions
Stars in the disk move on circular orbits in the plane of the disk around the galactic center. In addition to that circular motion, gravitational encounters between stars and clouds lead to so-called peculiar motions, in which the stars bob up and down in an oscillating vertical motion that takes them above and below the disk's midplane. Stars also oscillate radially around their circular orbits, meaning that they move closer to and farther from the galactic center.

In general, the older a star is, the more opportunity it has had for encounters and scatterings. That means the oldest disk stars will be found at the greatest distances above and below the Milky Way's midplane.

The age distribution of stars in the disk is quite mixed. There are many older stars with ages of about 8 billion years. Stars tend to become redder as they evolve off the main sequence, so the old stars contribute red light to the overall spectrum of the galaxy. There are also newly formed, extremely massive young stars, which are found close to where they were born, in molecular clouds. Since these clouds tend to be found in the spiral arms, the arms have higher concentrations of the massive blue stars.

"As time goes on, the amount of metals, or the **metallicity**, of interstellar gas increases as each generation of stars dumps processed heavy elements back into the galaxy," Benjamin reminds us. "That means you can divide stars into metal-poor and metal-rich classes." While the overall fraction of heavy elements decreases with distance from the disk center, most stars in the disk contain high enough concentrations of metals to indicate that they were formed from material that had already been processed through earlier generations of stars. Thus, astronomers put disk stars in what they call the **Population I** category. This classification distinguishes them from other populations of stars of different ages found in other locations in the galaxy, such as the stellar halo (see **Table 15.2**).

TABLE 15.2 ••• Stellar Populations		
Characteristic	Population I	Population II
Location	Disk; spiral arms	Stellar halo (including globular clusters); bulge; small proportion of disk stars
Age	Younger	Older
Temperature	Hotter and bluer	Cooler and redder
Metallicity (abundance of elements heavier than helium)	High	Lower than Population I

SECTION SUMMARY
The Galactic Disk

- Most of the observable matter of the Milky Way Galaxy is found in the galactic disk, which appears as a thin circular distribution of several hundred billion stars along with tens of billions of solar masses of dust and gas.

- The disk is thought to have two to four spiral arms.

- The disk contains a mix of young, blue, metal-rich stars and older, redder stars that primarily orbit around the galactic center but have additional peculiar motions.

- The relatively high metal content of disk stars puts them in the Population I category.

The Stellar Halo

"There is some fraction of stars like the Sun that are on this nice merry-go-round about the galactic center," says Benjamin. "Then there are these stars on more extreme orbits, tipped in all different directions. They make up the **stellar halo** of the galaxy." Most, but not all, of these halo stars reside in globular clusters (there are approximately 150 globular clusters in the halo). Globular clusters are gravitationally bound collections of up to 100,000 stars that were all born at the same time. These clusters in the stellar halo are what enabled Shapley to determine the true center of the galaxy. The radius of the halo is about 15 kpc. That is about the same size as the galactic disk, though the halo is spherical. A few outlying globular clusters, such as Pal 14, lie at much larger distances, extending out as far as 72 kpc from the galactic center.

X-ray observations indicate that the stellar halo may be full of gas at a temperature of more than a million degrees. If this is true, the mass in the halo would be comparable to that of the galactic disk. Like the corona of the Sun, the density of the Milky Way's hot gas appears to be quite low, so it escaped detection for some time. The study of this hot halo is ongoing, as astronomers work to understand its properties.

As Benjamin emphasizes, the motion of the halo's globular clusters is very different from the stellar motions in the disk. Disk stars (and gas clouds) revolve around the galactic center on roughly circular orbits, but globular clusters tend to dive inward and then swing around the galactic center on highly elliptical paths. These paths, called plunging orbits, provide evidence that whatever formed the globular clusters was not rotating very quickly; if it had been, these stars would have a far more significant component of circular motion. Just as important, to the degree that the globular-cluster orbits do show rotational motion around the galactic center, though on highly elliptical orbits, many of them have a sense of rotation that

metallicity The abundance of elements heavier than helium in an object.

Population I Stars found primarily in the disk and bulge, which have higher metallicity and therefore must have been born relatively recently in cosmic history.

stellar halo The spherical distribution of Population II stars that are found mainly in globular clusters.

FIGURE 15.10
Globular-Cluster Orbits
Many globular clusters travel far above and below the plane of the galaxy in highly elliptical orbits. Some move on retrograde orbits relative to the rotation of the disk.

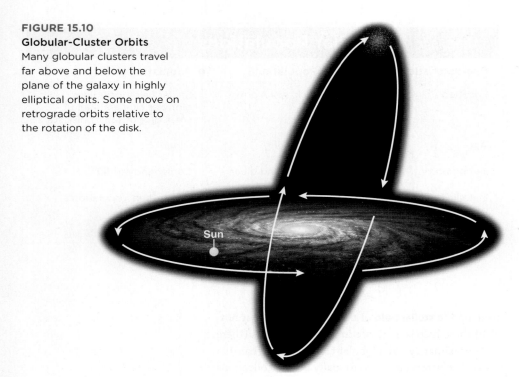

category. All Population II stars must have formed at an earlier epoch of cosmic history, before many generations of star formation and nuclear processing. The material from which they formed was relatively pristine, compared with the metal-rich material found in Population I stars.

Astronomers recently created a third category of stars, *Population III*, to describe the first generation of stars born after the creation of the Universe. These stars would have had no or extremely low metallicity, and all of them are expected to have met their stellar death in one form or another by the current cosmological epoch.

┌─ **SECTION SUMMARY**
The Stellar Halo

● The old, metal-poor stars in the stellar halo (Population II) form a spherical distribution around the galactic disk and move on highly elliptical orbits.

● Star formation is not observed in the stellar halo.

The Bulge and the Bar

The central region of the Milky Way is dominated by the aptly named **bulge**, a roughly spherical distribution of stars and dust about 5 kpc in radius. Stars are packed tightly together in the bulge, with typical stellar densities reaching as high as 1,600 stars per cubic parsec (pc³), more than 10,000 times higher than the densities of stars found near the Sun, where the density is much less than 1 star/pc³.

The bulge is composed mainly of Population II stars. Thus, stars in the bulge are, on average, older and redder than the Population I stars composing the disk. This observation might tempt you to think of the bulge as a dense version of the stellar halo, but that would be a mistake. The bulge is a separate component of the galaxy with its own unique features and history. The gas and dust content of the bulge is quite high, making it very different from the stellar halo. The presence of the gas allows new stars to form, and there are scattered examples of hot, blue stars and supernovas in the bulge.

The presence of so much gas and dust also makes the bulge much harder to study. There is so much interstellar material in this innermost component of the galaxy that seeing inside is almost impossible without using very-long-wavelength telescopes. In a few places, such as the region called Baade's Window, a path through the gas clouds has opened up, affording

is opposite that of disk material. Thus, astronomers consider these globular clusters to be on retrograde galactic orbits, which serves as another important clue for constructing models of the galaxy's history (**Figure 15.10**).

The composition of the stellar-halo stars and interstellar medium is also very different from what is found in the disk. One difference of note is the lack of star-forming gas in the halo. "The disk shows a rich interstellar medium of gas and dust, including gas clouds that are forming new stars," says Benjamin. "But the amount of gas that could form stars in globular clusters is comparably pretty small." Given the absence of star-forming regions, it should come as no surprise that the halo does not contain newborn stars. "Every now and then when I am observing, I will look for star formation in the halo," says Benjamin. "But really, there's nothing." All of these findings suggest that the halo is an older system of stars than the disk.

The conclusion that the stellar halo is older than the galactic disk is also strengthened by studies of the chemical compositions of halo stars. While disk stars are relatively abundant in heavy elements, halo stars show distinctly lower processed-element compositions. Because of their low metal abundances, astronomers group halo stars into what they call the **Population** II

Population II Stars found primarily in the stellar halo, which have lower metallicity and therefore must have been born relatively early in cosmic history.

bulge The roughly spherical distribution of stars and interstellar material in the central region of a galaxy.

bar A thick, rectangular region of stars extending across the center of some spiral galaxies.

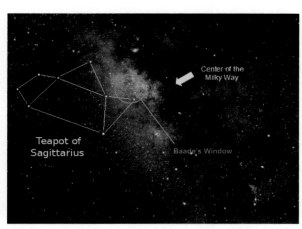

A

B

FIGURE 15.11 Baade's Window
The density of gas and dust near the galactic center makes it very difficult to study the bulge. (A) An opening in the clouds known as Baade's Window in the constellation Sagittarius affords astronomers an optical view into (B) the bulge's central regions.

Bob Benjamin

Bob Benjamin's journey into astrophysics began with a single high school math teacher. "She was always telling me, 'You've got to take more math,'" he recalls. Once Benjamin had arrived at Carleton College, he not only took his teacher's advice, but found he had real talent in the subject. A single astronomy class drew Benjamin's interest but was still not enough to set the fire roaring. "When I reached the end of my undergraduate career, I really didn't know what I wanted to do with my life. I had a pretty intense debate with myself the summer before senior year about graduate school in physics versus astronomy. On impulse, I decided astronomy."

After starting the astronomy program at the University of Minnesota, Benjamin soon knew he'd made the right choice. "There was this moment that really stayed with me. This other graduate student named Kim and I were sitting in our office talking. First we started comparing our favorite bands and where they came from. Then we started talking about O stars, which are the biggest and brightest stars in the sky. Both of us were interested in them in our research. At some point I said, 'Kim, do you realize that we just talked for 20 minutes about O stars, and it was as interesting as the stuff about the bands?' She smiled and said, 'Yeah. Wasn't that weird?'"

As a professor of astrophysics, Benjamin has become an expert on the structure and evolution of galaxies. "I remember in the beginning of graduate school wondering, 'Well, how do you think of a research project?' Then somewhere in the middle of graduate school I was thinking of my own exciting research projects, and I realized I couldn't stop thinking about new research projects I wanted to do." And Benjamin still can't stop thinking up new research projects. "I know that shouldn't be a problem," he laughs. "But I already have way too many things on my to-do list."

astronomers a valuable optical view into the bulge's central regions (**Figure 15.11**).

"People started thinking the inner region of the galaxy was a bulge when, basically, they could only see through these little windows," Benjamin explains. "But now, with infrared astronomy, we can see much more of the Milky Way, and we realize that this thick, squarish region of stars down in the center of the Milky Way may be a **bar** and that the formation of the bar and bulge might go together." Many other spiral galaxies also show bars in their centers with spiral arms emanating from the tips of the bars (**Figure 15.12**). We will explore the formation of bars and

FIGURE 15.12 The Milky Way's Bar
Infrared astronomy has enabled astronomers to postulate that a thick, rectangular region of stars in the center of our galaxy may be a bar, similar to the ones in other spiral galaxies, such as NGC 1300, shown here in visible light.

their importance to galactic structure in more detail in Chapter 16, when we compare galaxies of different types and histories.

Determining the true shapes of the Milky Way's bulge and bar is not the only reason astronomers are keen to see deep into the galactic center. A bright radio source was discovered in a region close to the center of the bulge in 1974, and it appears to be associated with a supermassive black hole (see Chapter 14). The presence of an enormous, relatively nearby black hole was enough to quicken the hearts of many astronomers, and the central region of the bulge has remained a topic of intense study ever since.

Local Group A gravitationally bound collection of galaxies within which the Milky Way resides.

SECTION SUMMARY
The Bulge and the Bar
● The bulge-shaped center of the galaxy is very hard to study because of the dense gas and dust it contains.

● Although the stars in the bulge are mostly Population II stars, some star formation occurs there.

● A bar-like structure may stretch across the bulge, similar to bars seen in other galaxies.

Our Neighborhood of Galaxies

If we could step outside the Milky Way and look at it from afar, we would see that it resides in a small collection of galaxies, called the **Local Group**, which contains more than 50 galaxies in all (**Figure 15.13**). Although these other galaxies are not part of the Milky Way's "anatomy" per se, many of them may be influencing our galaxy's structure and evolution. In fact, the current understanding of the Universe suggests that every large galaxy is surrounded by a multitude of smaller "satellite" stellar systems, which are involved in ongoing interactions, collisions, and even cannibalism.

The Milky Way is orbited by at least 14 confirmed satellites. The largest and most massive are the Large Magellanic Cloud (LMC) and the Small Magellanic Cloud (SMC), two irregularly shaped galaxies with diameters of 4 kpc and 2 kpc, respectively. (They are so named for the Portuguese explorer Ferdinand Magellan, who noted them on his historic circumnavigation of the globe in 1519–22. They are visible only from Earth's Southern Hemisphere.) Many of the

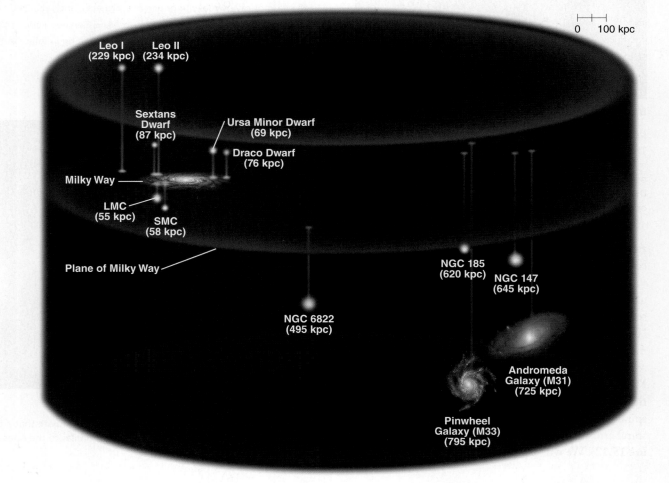

FIGURE 15.13
The Local Group
Galaxies rarely appear in isolation but occur in groups (or in larger clusters). The Milky Way is part of the Local Group, which encompasses two other spiral galaxies and a host of smaller satellite galaxies. The Milky Way has a number of satellite galaxies, including the Large and Small Magellanic Clouds.

smaller galaxies are on orbits that will lead to a collision with the Milky Way in the future.

These galaxy motions also imply that collisions with the Milky Way have happened in the past. Evidence of such collisions can be seen in so-called tidal streams of stars and hydrogen gas formed as an orbiting satellite galaxy is pulled apart by the Milky Way's much larger gravity. For example, astronomers have found two tidal streams associated with the Sagittarius Dwarf Galaxy, torn off by the Milky Way's huge gravitational pull (**Figure 15.14**). Sagittarius may once have been one of the brightest of the Milky Way's satellite galaxies, but the Milky Way's immense gravity has ripped it apart, dispersing half of Sagittarius's stars and virtually all of its gas over the last billion years.

While the larger gravitational pull of the Milky Way tends to dominate all satellite interactions, the satellites themselves sometimes collide. The so-called Magellanic Stream represents evidence of a near-collision between the LMC and SMC. The stream is composed of high-velocity neutral hydrogen (H I) clouds stretching 180° across the sky. Detailed computer simulations indicate that the stream formed almost 2.5 billion years ago when the LMC and SMC passed close enough to each other for tidal forces to become significant. The near-collision not only pulled out the Magellanic Stream, but also triggered a burst of star formation in both galaxies, as well as deforming them into their current shapes.

FIGURE 15.14 Galactic Interaction
Evidence of an ancient galactic interaction, the Magellanic tidal stream formed billions of years ago, when the Large and Small Magellanic clouds passed close to one another. This composite radio image shows the Magellanic stream in pink.

SECTION SUMMARY
Our Neighborhood of Galaxies

- Large galaxies like the Milky Way typically have smaller, gravitationally bound satellite galaxies, and these galaxies can collide, merge, and tear one another apart.
- These interactions change galaxy properties and are therefore an important part of galactic evolution.
- The Milky Way has two significant satellites: the Large and Small Magellanic Clouds. Gravitational interactions between the Milky Way and these satellites have produced elongated tails of gas and dust.

15.3
Spiral Arms: Does the Milky Way Have Them?

One of the mysteries astronomers in the 20th century faced was whether our galaxy hosts spiral arms.

Unfortunately, it's not easy to resolve details of the galaxy's structure from the inside. "We're sitting in this disk, partway out from the center," says Debra Elmegreen, a galactic astronomer at Vassar College. "When we look up at that band of light in the sky we call the Milky Way, how are we supposed to see spiral structure in that?"

In other spiral galaxies considered to be similar to our Milky Way, however, spiral arms are often observed as part of the disk. These spiral patterns are some of the most beautiful structures the night sky has to offer, and their origin was poorly understood by astronomers for more than a century. The presence of spiral arms within the Milky Way itself took some time to confirm. Finding spiral arms from the inside meant obtaining detailed maps of the Milky Way's stars, star clusters, and gas clouds.

Evidence for Spiral Arms

The first evidence for spiral arms in the Milky Way came from studies of young star clusters that included bright, massive stars. When astronomers mapped the locations of the most massive stars (those with spectral classifications O and B), they found that they did not occur alone, but in groupings called **O and B associations**. More important, their distribution is not uniform in space. Instead, the O and B stars tend to

O and B association A group of massive stars of O and B spectral types found tracing out a spiral arm.

cluster in long strips (**Figure 15.15A**). In time, these strips were inferred to be segments of spiral arms. Our own Sun is inferred to be part of the so-called Orion-Cygnus Arm. At larger distances from the Milky Way's center, the O and B associations trace out the Perseus Arm, while at distances closer to the Milky Way's center, astronomers find evidence of the Carina-Sagittarius Arm (**Figure 15.15B**). These names are based on the directions on the sky and the constellations in which the O and B stars appear.

The fact that spiral arms can be traced out by the massive stars they contain already provides some information about their nature. Because they do not live very long compared with low-mass stars, massive stars have not had time to migrate far from where they formed. Thus, the presence of massive stars in the arms suggests that these stars were born *within* the arms. Whatever process creates the arms must also trigger star formation.

Direct evidence for the existence of spiral arms in the Milky Way came with the advent of radio astronomy in the 1950s. Because of their very long wavelengths, radio waves emitted by interstellar gas can propagate unobstructed from one end of the galaxy to the other. Radio astronomers' attempts to map the galaxy were aided enormously by the fact that cold, neutral hydrogen atoms naturally emit at a radio wavelength of 21 centimeters (cm). Since hydrogen is the most abundant material in the cosmos, the 21-cm line became the go-to tool of astronomers seeking to map galactic structure. Maps of the galaxy in the 21-cm line gave astronomers their first direct evidence for the presence of spiral arms in the Milky Way.

As we have seen, the presence of O and B associations suggests that star formation and spiral arms are linked. That link is strengthened when radio astronomers use emission lines of the carbon monoxide (CO) molecule to map the galaxy. Recall from Chapter 12 that molecules form only in the densest, coldest regions of the interstellar medium, called molecular clouds, which are also the site of star formation. In their CO maps of the galaxy, astronomers find that these molecular clouds are not distributed randomly across the Milky Way. Instead, molecular clouds are found in bands or strips separated by wide regions in which no clouds exist. Just like the O and B stars, the molecular clouds appear to trace out spiral arms in the Milky Way. Again, star formation and spiral arms

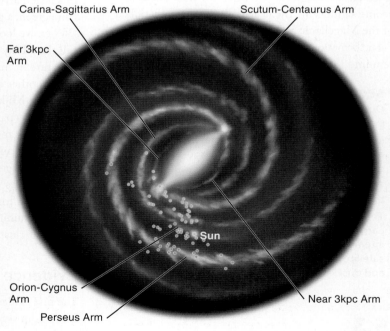

FIGURE 15.15 Spiral Arms in the Milky Way
Young star clusters with bright, massive stars light up the spiral arms of the Milky Way Galaxy. These massive O and B stars typically occur in groupings called O and B associations, which tend to cluster in long strips, now seen as evidence of spiral arms. (A) The cluster shown here, NGC 6231, is found in a region called the Scorpius O and B association. (B) This schematic diagram shows the names of the Milky Way's spiral arms and the position of the bar. The dots show the locations of O and B associations.

appear linked. Any working model of the physics of spiral arms must explain this connection.

┌─ **SECTION SUMMARY**
│ **Evidence for Spiral Arms**
● The positions of young, hot stars (O and B stars) provide evidence for spiral arms in the Milky Way.
● Spiral arms can be seen using the 21-cm line of neutral hydrogen.
● Mapping molecular clouds via carbon monoxide also shows evidence for spiral arms.

Spiral Arms: From Windings to Waves

Tie a string to a stick. Now begin rolling the stick between your fingers in one direction. What happens to the string? With each rotation, the string gets wound once around the stick. Is this a good model for what is happening with spiral arms? Are the arms composed of material that is winding up with the galaxy's rotation?

This idea for the origin of spiral arms was one of the first to occur to astronomers. But measurements of the galaxy's rotation rate soon killed the hypothesis. Observations of stellar orbits about the galactic center show that the period of the galactic rotation varies with distance from the center. At the distance of the Sun, the galaxy rotates about once every 233 million years. Measurements of the age of the oldest stars in the galactic disk show, however, that the galaxy itself is nearly 13 billion years old. If we divide the total age of the galaxy by the time it takes to make one orbit, we find that the number of times the Milky Way has spun completely around is approximately 50. But if we return to our string-and-stick example, we realize that if the spiral arms were really material that was winding up with the Milky Way, then we would expect to see just about as many windings of the spiral arms as there have been galactic rotations. In other words, we would expect to see very tightly wound spiral arms, like the grooves on a vinyl record. "Over the lifetime of our galaxy there should have been many, many, many windings. And so we ought to see 40 or more, but that is not what we observe in our galaxy or others," says Elmegreen. Observations suggest that the Milky Way hosts between two and four well-defined spiral arms.

Astronomers now have strong evidence that spiral arms result from a process described by **spiral density wave** theory. Instead of material winding up with each rotation of the galaxy, spiral density wave theory sees the arms as patterns moving through the disk at a

Debra Elmegreen

If you're ever on a train and you see a fellow passenger building a telescope out of cardboard, you might have just run into Debra Elmegreen. "A little while ago my husband and I were on a train coming down from Alaska," she explains. "Venus was about to pass in front of the Sun [a rare event called a transit]. We didn't want to miss it, so we got out a piece of cardboard and stuck a pinhole in it. I had binoculars, which we turned backward and held up so we could project the Sun's image. A couple of people on the train got really excited. 'This is crazy!' they were saying. 'What is it?'"

That kind of do-it-yourself enthusiasm has been a hallmark of Elmegreen's astronomical career going all the way back to when she was a kid. "I grew up in Indiana, where the skies were very dark. We lived along the river, and there were no other houses around. My parents got me a telescope when I was five, one of those 2-inch refractors. I was hooked from that moment on." In school, Elmegreen joined the science club and built her own telescope. Then came local science fairs and the Westinghouse Science competition.

These days, the responsibilities of teaching, grant proposal writing, serving on national committees (including recently being president of the American Astronomical Society, AAS), and the nuts-and-bolts details of doing science can wear her down. "You know, the day-to-day computer drudgery can get to you sometimes," she says. But a view of the sky is all she needs to fire up her imagination again. "Every now and then I have to step outside. Once I look up and remember why I'm doing this, it's all okay." That sense of excitement can also come from new data. "You get this new plot and you realize no one else has ever seen what you are finding before. I'm living my dreams. It's fun. I still love it. I can't imagine doing anything else."

speed different from that of the disk's physical material (stars and the interstellar gas). Like sound waves on Earth, spiral waves are **compression waves** that move through their supporting medium. For sound waves, the medium is the gas molecules of the atmosphere. For spiral waves, it's the gas molecules of the interstellar gas and the distribution of stars in the disk. In a compression wave, material piles up and is squeezed inside the "crest" of the wave and then spreads out in the wave's trough.

You've undoubtedly been part of this kind of compression wave many times while driving in traffic. Cars are initially traveling at high speed and then are forced to slow down for a time as they move through a region of congestion. The region of congestion ends, and then the cars are able accelerate back up to their cruising speed. The key point is that from a helicopter flying above, the congestion looks the same even as the speed of individual cars passing through it continually changes. The congestion is a stable pattern of compression waves on the highway. The density of cars (number of cars per kilometer) increases at the peak of

spiral density wave A wavelike pattern of compression that moves through the physical material of a galaxy's disk. Spiral density waves are seen in spiral arms.

compression wave A wave of increased density that passes through background material. The density returns to normal after the wave passes.

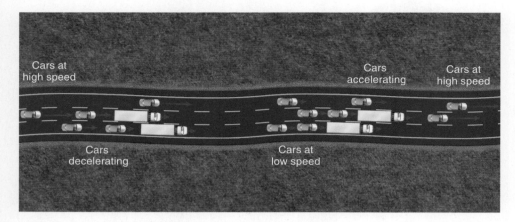

FIGURE 15.16 Spiral Density Waves
The spiral arms of a galaxy are often described as density waves, a type of compression wave similar to sound waves. Gas and dust passing through the spiral arms are compressed, thus increasing the density. In everyday life, we experience a similar wavelike phenomenon in stop-and-go traffic, where cars bunch together as they slow down and then spread out as they resume cruising speed.

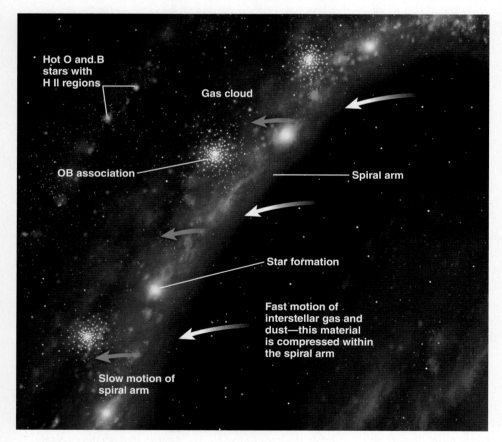

FIGURE 15.17 Star Formation in the Spiral Arms
Spiral arms represent regions of high density and higher gravity, and they rotate more slowly than the stars and low-density gas clouds that sweep through them. As a low-density cloud overtakes and moves into the spiral arm, the stronger gravity leads to a compression of the cloud, setting the stage for star formation. Thus, we see molecular clouds in spiral arms and star formation occurring within the clouds. O and B associations will be seen in or downstream of the spiral arms.

the wave and then decreases on either side of the peak (**Figure 15.16**).

Seeing spiral arms as compression waves explains many of the features observed in the Milky Way. In particular, the association between star formation and spiral arms can be easily explained. The spiral arms rotate more slowly than the stars and gas clouds. Imagine a low-density gas cloud orbiting the center of the Milky Way: as it overtakes the spiral arm and is swept into the arm's gravitational well, the gas in the cloud is compressed. Molecules then form in the dense, cold interior regions that are self-shielded from interstellar ionizing radiation. Thus, spiral density wave theory would predict that molecular clouds should occur more frequently in spiral arms than elsewhere in the galaxy, and this is exactly what astronomers have observed (**Figure 15.17**).

Because massive stars live for only a few million years, we should also expect to see them close to where they formed. The compression produced in spiral arms should act as a trigger for the formation of new stars, and therefore we expect to see O and B stars preferentially in the arms. Once again, this is exactly what is observed.

Spiral density wave theory provides a robust answer for the question, "What are spiral arms?" It also provides a reasonable explanation of how spiral arms form, because the spiral wave pattern is the result of what are called **instabilities** in the galactic disk. Any small perturbation that disturbs the smooth pattern of rotation will naturally organize itself into spiral waves running around the disk. These perturbations might come from a globular cluster passing through the disk or even from a collision with a satellite galaxy. In either case, just a small nudge is enough to get the spiral pattern going.

SECTION SUMMARY
Spiral Arms: From Windings to Waves
- Spiral arms cannot be caused by the winding action of galactic rotation.
- Spiral arms appear to be compression waves moving through the disk that were created as a result of instability, and this theory is consistent with star formation occurring in the arms.

15.4
The Galactic Center

Located in the constellation Sagittarius, the galactic center is a place like no other in the Milky Way. It's an environment dense with stars and giant arcs of magnetic fields stretching across parsecs. Most remarkably, the galactic center is home to a black hole with millions of times the mass of the Sun.

The Galactic Hub

We have already seen that the bulge contains a higher density of stars than any of the galaxy's other components. The galactic center contains even higher densities than the rest of the bulge. Within the central parsec of the galaxy, more than 10 million stars swarm together. Compare this value of 10^7 stars/pc^3 with our local value of 0.14 star/pc^3. The galactic center has a stellar density that is 100 million times denser than the local neighborhood of the Sun (**Figure 15.18**).

The galactic center also plays host to an enormous quantity of gas, most of which resides in a dense ring of molecular material. This **molecular ring**, or **molecular disk**, orbits the galactic center at a distance of about 2 pc. While some stars are forming in the ring now, astronomers predict that the future holds a much more spectacular event. Within 200 million years, gravitational forces in the ring may trigger what astronomers call a **starburst**, a rapid increase in star formation in which an appreciable fraction of the molecular gas will quickly be converted into stars. Such starbursts are seen in many other galaxies and may be the way the majority of stars in the Universe have formed. The star formation rate in our galaxy now is puny compared to what happens in the brief "pulse" of a starburst.

The nature of stars in the galactic center is surprising. From their observations of the rest of the bulge, astronomers expected to find high densities of older, redder stars. Instead, the density of red stars drops in the central regions, forming a kind of "hole" in the distribution of highly evolved stellar objects. Astronomers do not yet understand why there are so few older stars, and they refer to this problem as the *conundrum of old age*. Equally surprising is the number of young stars, including a large population of massive O and B stars. Since star formation is not prevalent in the rest of the bulge, the presence of these short-lived young stars came as a surprise to astronomers, who refer to their presence as the *paradox of youth*. One explanation for the presence of these young stars may be the collapse of a massive gas cloud as it orbited the central black hole.

Along with gas, dust, and stars, the galactic center hosts highly energetic phenomena, some of which are

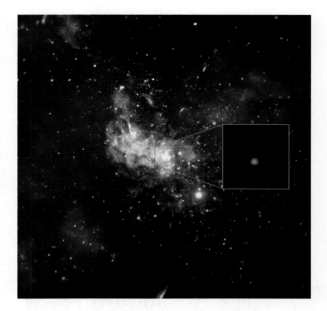

FIGURE 15.18
Black Hole at the Center of the Milky Way
Seen in X-ray wavelengths, the highly energetic processes and objects in the center of our galaxy become visible, including the tracers of star-forming regions, supernova remnants, and the supermassive black hole Sgr A* (inset). Here, low-energy X-rays appear in red, medium-energy in green, and high-energy in blue.

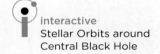

interactive
Stellar Orbits around Central Black Hole

connected with the black hole. Radio observations of the center show vast arcs of magnetic fields strung across the region. The arcs are 50 pc in length and contain fields that are more than 10 times stronger, on average, than the field in the rest of the galaxy. Their structure is similar to that of the prominences seen on the solar surface, but unlike the solar fields, the galactic center has no surface on which the fields can anchor. That means these great arcs are generated by large-scale flows of charged gas swirling around and through the center.

SECTION SUMMARY
The Galactic Hub
- The galactic center has the highest density of stars in the galaxy, as well as a dense molecular ring that is expected to be the source of a future starburst event.
- Near the center, there are fewer old stars and more young stars than expected.
- Large-scale magnetic arcs are also observed at the galactic center.

Monster at the Center: The Galactic Black Hole

In 1974, two graduate students, Bruce Balick and Robert Brown, were using the National Radio Astronomy Observatory's big antenna in West Virginia for their thesis project. They were looking at a complex region of radio emission in the constellation Sagittarius that includes a supernova remnant. As the radio antenna accumulated signals from the distant

instability A complex pattern of gas flow resulting from a disruption in an originally smooth flow.

molecular ring or **molecular disk** A dense ring of molecular material orbiting the galactic center.

starburst A rapid increase in star formation in which an appreciable fraction of the molecular gas is quickly converted into stars.

region, the two scientists saw a bright object pouring out radio energy off to the side of the supernova remnant.

Balick and Brown dubbed their new object Sagittarius A* (abbreviated Sgr A*). As recognition of its strange properties grew, the object soon became a subject of intense study, and astronomers slowly built the case that this bright radio source was none other than the Milky Way's own supermassive black hole.

Twenty-eight years later, a swarm of stars that orbit Sgr A* provided the final proof that it was a black hole. "There are lots of stars near that black hole that are going around very rapidly," explains Debra Elmegreen. "They allow astronomers to deduce the mass of the black hole, which of course we don't see."

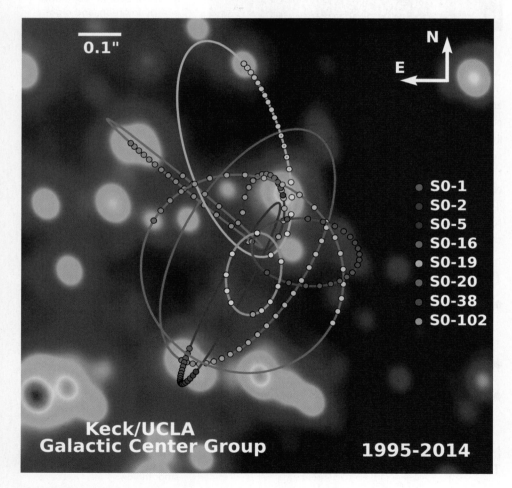

FIGURE 15.19 Star Tracking
Tracking the stars orbiting Sgr A* confirmed the existence of the Milky Way's supermassive black hole. Here, we see high-resolution observations taken over time (labeled as points). Once a number of star orbits had been defined, such as the one for S0-2, astronomers used Newtonian physics to calculate the mass of the central object. Combining the mass with the size of the region implied a density possible only in a black hole. Here we see a representative image and an orbit compiled from a series of such images for the star labeled S0-2 (note the observation dates).

Using hyperprecise radio observations, astronomers accurately mapped the microarcsecond paths that these stars made on the sky (**Figure 15.19**). After watching for 10 years to track the stars' complete orbits, astronomers could use Newton's laws to measure the mass of the central object binding the stars to their orbits: $4.3 \times 10^6\ M_{\text{Sun}}$. Since many of the stars were orbiting the central object at distances of just 45 astronomical units (AU), only one interpretation made sense. Only in a black hole can millions of Suns' worth of material be crammed into a volume smaller than our Solar System (see **Going Further 15.1**).

The existence of a supermassive black hole in the galactic center of the Milky Way explained a wide range of high-energy phenomena in the region, such as X-ray flares in Sgr A* that come from clouds of gas as they are torn apart by the black hole. Astronomers have also seen evidence for higher-energy gamma-ray emission, possibly caused by positrons created outside the black hole's event horizon and funneled into relativistic jets. Recall from Chapter 10 that positrons are the antimatter version of electrons; when a positron meets an electron, they annihilate in a flash of gamma rays. Thus, the detection of such high-energy radiation from Sgr A* implies exotic processes that can occur only near a black hole.

The confirmation of a supermassive black hole at the galaxy's center closed the books on one question, but it raised another that astronomers have yet to answer. Compared with monster black holes in other galaxies, which spew vast amounts of energy into space, our black hole seems relatively inactive. "The region around the black hole at the center of our galaxy does not emit much radiation," says Elmegreen. "Usually, we expect these kinds of systems (a black hole and matter surrounding it in an accretion disk) to be extremely luminous. The ones in other galaxies can be seen from enormous distances because they are so bright. But our Milky Way black hole puts out about one-billionth of their energy. No one is really sure why."

"While we do have a supermassive black hole," continues Elmegreen, "it's just not as enormous as those in some galaxies. The really bright ones appear to have masses that are billions of times the mass of the Sun. Because it's got relatively low mass, our black hole in the Milky Way may never have gone through a really luminous phase, as we see in other galaxies."

However, the black hole at the center of the Milky Way may be able to flare a bit brighter every once in a while. In a sense, our black hole may have been put on a temporary diet with little material around to "eat" just now. But astronomers have evidence that infalling material may have made the black hole brighter in eons past. In fact, from the current position and motion of nearby molecular clouds, astronomers can even predict that the black hole will flare up again in a few centuries as new material drops onto it.

SECTION SUMMARY
Monster at the Center: The Galactic Black Hole
- Observations of stellar orbits confirmed the enormous mass (4.3 million M_{Sun}) and small radius of the radio object Sgr A*, showing that it had to be a black hole.
- Sgr A* emits less radiation than do other supermassive black holes observed in other galaxies, possibly because of its smaller mass and lack of available infalling material.

15.5
Dark Matter and the Milky Way

When you look out into the night sky, you see a lot of empty space punctuated by some bright dots. Through a telescope you can see those dots for what they are: stars. You will also see other objects among the stars, such as interstellar clouds and other, distant galaxies. Using techniques such as spectroscopy, astronomers have been able to probe the light from these distant objects, determining that they are made up of the same kinds of physical constituents as you and me—protons, neutrons, electrons, and other subatomic particles. But what if there's other stuff out there that we cannot see? What if there are other kinds of matter that simply do not interact with light at all—neither emitting nor absorbing it? How can we tell whether that kind of stuff is out there?

The Footprints of Dark Matter

These were exactly the questions astronomers studying our own galaxy found themselves facing as they began mapping out the orbits of stars in the Milky Way's disk. The idea was to measure the mass of the Milky Way by first measuring the orbital speed of stars at ever-larger distances from the galactic center. Recall from Going Further 3.3 that Newton's law of gravity can be used to determine the mass of a central object by measuring the period and radius of satellites orbiting that object. The formula looks like this:

$$M = 4\pi^2 R^3 / G P^2$$

GOING FURTHER 15.1   THE LANGUAGE OF THE COSMOS
How to Bag a Black Hole and Survive

How did astronomers use the orbits of stars in the Milky Way's center to determine that a supermassive black hole was there? As we've seen a number of times before, Newton's laws can be used to determine the mass of an object if the orbital properties of its satellite can be observed. We can do a simple calculation to see how close we get to the results astronomers find when using more detailed and sophisticated calculations.

We'll focus on the star labeled S0-2 in Figure 15.19. Its orbit is quite elliptical ($e = 0.88$), but the average orbital radius projected on the sky, as measured in arcseconds (let's call it a), is found to be approximately 0.1″.

First we need to recall the small-angle formula:

$$a = 206{,}265 \times (R/D)$$

In this case, R is the radius of the orbit, and D is the distance to the star at the galactic center, which has been determined to be 8.42 kpc. Since both a and D have been measured, we can rearrange the small-angle formula to find R:

$$R = a \times D/206{,}265$$

If we put in the values we already have and express D in meters (1 kpc = 3.08×10^{19} meters), we find

$$R = 0.1 \times [8.42 \text{ kpc} \times (3.08 \times 10^{19} \text{ m/kpc})]/206{,}265$$

$$= 1.26 \times 10^{14} \text{ m}$$

To find the mass of the black hole, we also need to know the period of the orbit, which in this case is just 15.78 years. Now we can use Newton's version of Kepler's third law to determine the black hole's mass:

$$M_{bh} = \frac{4\pi^2}{G} \frac{R^3}{P^2} = 2.4 \times 10^6 \, M_{Sun}$$

This is just about one-half of the value found from a more detailed analysis using all the details of the star's orbit around the black hole. Thus, with just a few lines of math using Newton's powerful law, we get pretty close to the right answer.

In the 1970s and 1980s, astronomers hoped to use the stars in the outer regions of the disk to find the mass of the Milky Way (M_{MW}). The expectation was that most of the galaxy's mass would be close to the center, and measuring ever-more-distant stars would yield better and better estimates of the galaxy's total mass.

But how could astronomers be sure that their assumption about the distribution of mass in the galaxy was correct? Recall that in the Solar System,

GOING FURTHER 15.2   THE LANGUAGE OF THE COSMOS
Dark Matter and Galaxy Rotation

Astronomers measuring how the rotational velocities of stars in the Milky Way's disk change with distance didn't get the result they were expecting. The idea was simple: For stars orbiting the galactic center, Newton's laws show that only matter inside the star's orbit can affect its motion. That means stars orbiting at large distances from the galactic center should feel the gravitational force from the bulk of the galaxy's mass. In terms of the rotation curve of the stars, this situation should look a lot like what we see in the Solar System, the curve for objects orbiting a central mass. But how, exactly, do orbital velocities change with distance for objects in the Solar System?

Chapter 3 introduced a compact formula describing the circular orbital velocity of satellites orbiting a central mass:

$$V_c = \sqrt{\frac{GM}{R}}$$

This equation predicts the shape of the rotation curve for planets in our Solar System. Since neither the mass of the central object (the Sun) nor Newton's constant G changes, the rotation curve for the Solar System looks like this:

$$V_c = \frac{C}{\sqrt{R}}$$

Note that C (not to be confused with c, the speed of light) is now a constant: $C = \sqrt{GM}$.

This equation says that the farther out you go, the slower the orbital (or rotational) velocity of the planets should be (**Figure 15.20**, blue line). That kind of pattern is what astronomers expect for any system of orbits, once most of the mass has been enclosed within those orbits.

But this is not what astronomers see for the rotation curve of the Milky Way, which is represented in Figure 15.20 by the orange line.

The fact that the observed rotational velocity for the galaxy never decreases, even at large distances from the center, tells us something very important.

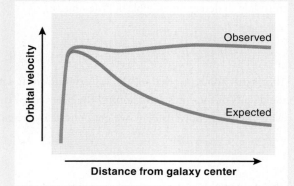

FIGURE 15.20

If we assume that Newton's laws hold at these distances (and we have every reason to expect they do), then there must be more matter in the Milky Way than we can see, and it must extend well beyond the visible limits of our galaxy. Scientists therefore think that the increasing mass with distance comes from a spherical distribution of dark matter that extends far beyond the visible galaxy's boundaries.

most of the mass resides in the Sun. Thus, as we look at larger radii in the Solar System, we see objects (planets, asteroids, and so forth) orbiting with slower velocities. Because gravity gets weaker as you move farther away from a massive object, the change in orbital velocity versus radius (called the **rotation curve**) of the Solar System shows velocities decreasing with increasing radius. Astronomers expected to see the same effect in the disk of the galaxy. Stars at the outer edges of the disk should be orbiting with a velocity much slower than those closer to the center. This is not, however, what observations showed.

Instead of seeing orbital velocities drop with radius, astronomers found that the orbital velocities of stars were constant across most of the disk, even out to the largest distances they could observe (see Figure 15.20 in **Going Further 15.2**). The galaxy was not behaving at all as expected. The rotation curve made it look as though the galaxy's matter was far more spread out than astronomers had thought, with enough material at larger radii to tug on the distant stars and keep their velocities high. But no matter how closely they looked with their telescopes, this extra matter was nowhere to be found.

At first, astronomers were certain they had made a mistake. But as more and more studies were completed, all showing the same result, it became clear that something truly unexpected was going on. Faced with an anomaly—something that does not make sense in terms of the best theories available—scientists always have two choices: they can abandon the theory, or they can imagine that an unseen process is occurring that was not part of the original assumptions of the calculation supporting that theory. "It really comes down to Occam's razor," says Debra Elmegreen. "You try to go for the simplest explanation." But when we're staring into an anomaly, which choice is the simplest?

Abandoning Newtonian gravity—the first option—meant doing away with 400 years of successful science, including Einstein's relativity, which had been built on Newton's work. For the most part, scientists were unwilling to take that option to explain the rotation curve anomaly. "The astronomer Vera Rubin [**Figure 15.21**] was one of the first people to work on rotation curves in galaxies," explains Elmegreen. "She was a Vassar alumna and used to return to campus to meet the current astronomy students. A student once asked Vera about the idea of modifying gravity, and she didn't scoff. But she did say we don't need to do that to explain what we see." What Rubin meant was that there was a second option: dark matter.

Rubin and others imagined that there was far more mass in the galaxy than we can directly observe. While this was a remarkable claim, it did less violence to our understanding of the Universe than abandoning Newton's laws did. The unseen mass would produce no light, but if it were distributed differently than the luminous matter, it would produce gravity. In this way, the extra mass would provide the needed gravitational force on the orbiting stars, raising their orbital velocity to give the observed flat rotation curve. Because Newton's laws allowed the distribution of this material around the galaxy to be inferred, astronomers soon embraced the idea, calling the new mass **dark matter**. The label *dark* meant simply that they knew nothing of its properties other than its ability to generate a gravitational force. "The only thing missing is we don't know what this dark matter is," says Elmegreen, "but as an explanation for the rotation curves, it makes sense."

By assuming the existence of dark matter, astronomers could use the velocities they measured for disk stars and work backward. Newton's laws gave them both the amount of dark matter and its distribution in space. The astonishing results suggested that for every kilogram of "normal," luminous stuff in the Milky Way, there are 10 kilograms (kg) of invisible dark material (**Figure 15.22**). As Rubin put it, "The ratio of dark-to-light matter is about a factor of ten. That's probably a good number for the ratio of our ignorance-to-knowledge. We're out of kindergarten, but only in about third grade."

If there is so much dark matter in the galaxy, where is it all? "Most of the dark mass can't be concentrated at the center of the galaxy," says Elmegreen, explaining that if it were, the rotation curves would still decline with distance. "It's got to be in a big extended halo around the disk of our galaxy in order to account for the rotation curves." This halo is theorized to be a spherical region whose density decreases with distance from the center. The dark matter halo does not have a sharp edge, but astronomers estimate its radius (based on its gravitational influence) to be approximately 100 kpc. "That means the dark matter halo extends several times past the stellar halo and the disk," says Elmegreen (**Figure 15.23**).

FIGURE 15.21
Vera Rubin
Vera Rubin (b. 1928) was one of the pioneers in the field of rotation curves in galaxies and their implications for the existence of dark matter.

SECTION SUMMARY
The Footprints of Dark Matter
- Astronomers attempting to calculate the mass of the Milky Way from stellar orbits found an unexpected result: stars at greater distances from the center do not orbit slower than those nearer the center.
- Astronomers speculated that a nonluminous form of mass called dark matter, accounting for roughly 90 percent of the galaxy's mass, could be the cause of the unexpectedly high rotational velocities.

What Is Dark Matter? The Four Forces

If the Milky Way is constructed primarily of dark matter, a nagging little question remains: What *is* dark matter?

rotation curve The change in rotation speed as a function of distance for objects orbiting around a common center.

dark matter Nonluminous mass that is evident only by its gravitational influence.

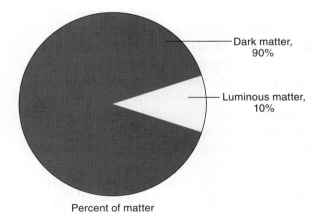

FIGURE 15.22 Dark versus Luminous Matter in the Milky Way
Luminous matter—the kind we can see through its interaction with light—constitutes only 10 percent of all matter in our galaxy, while 90 percent is dark matter. The dark matter is observed only through its gravitational effect on the luminous material.

FIGURE 15.23 Dark Matter Halo
The dark matter halo is believed to be a spherical region of decreasing density with an estimated radius of 100 kpc, extending several times past the galactic disk and the stellar halo.

Characteristic	● Weak Nuclear Force	● Strong Nuclear Force	● Gravity	● Electromagnetic Force
Range of influence	10^{-18} m	10^{-15} m	Infinite (declining with square of distance)	Infinite (diminished by shielding)
Effects	Decay of nuclei and fundamental particles	Holds nuclei together	Creates stars and planets; holds us on Earth's surface	Holds atoms together; includes light

TABLE 15.3 ••• **Fundamental Forces**

FIGURE 15.24
Large Underground Xenon Detector
The Large Underground Xenon (LUX) detector, installed in a former gold mine in South Dakota and activated in 2013, is looking for evidence of dark matter. Loaded with 350 kg of liquid xenon, LUX uses ultrasensitive photon detectors to watch for possible collisions between the xenon and WIMPs. The underground location protects the detector from both surface noise and cosmic rays.

massive compact halo object (MACHO) A proposed form for dark matter that includes objects such as black holes and extremely faint neutron stars.

weakly interacting massive particle (WIMP) A proposed form for dark matter consisting of particles that interact only via gravity and the weak nuclear force.

Astronomers once thought that dark matter might be something made of ordinary matter that was simply extremely dim—like faded white dwarfs, neutron stars, and, of course, black holes. They called these objects **MACHOs**, for **massive compact halo objects**. Astronomers carried out exhaustive searches for MACHOs by looking for so-called microlensing, the gravitational focusing of light that would occur as a MACHO passed in front of a more distant star. These microlensing surveys did find some candidates, but far too few to account for the total amount of dark matter. That meant dark matter must be made of particles that did not respond to electromagnetic force (light).

If dark matter doesn't interact with light, how can we determine whether it exists? To answer that question, we first have to ask an even broader one: How does *anything* in the Universe interact with anything else in the Universe? A more explicit way of posing this question is to ask what kinds of forces there are in the Universe. After all, Newton taught us that only through forces does matter accelerate or decelerate. Thus, without forces there can be no interaction among bits of matter.

From their studies of subatomic particles, physicists have constructed an extremely detailed understanding of matter and forces. What they've discovered is that only four fundamental forces exist in the cosmos (see **Table 15.3**). "You have gravity, the electromagnetic force, the strong nuclear force, and the weak nuclear force," says Elmegreen. "You need all of the forces to account for structures we see in the Universe on different size scales." Gravity holds stars and planets together and keeps your feet on the floor. Electromagnetism holds atoms together and makes chemical reactions possible; you are able to read this book because of light, which is an electromagnetic phenomenon. The strong nuclear force binds protons and neutrons together inside atomic nuclei,

and the weak nuclear force drives certain kinds of radioactive decay.

We already know that dark matter must be made of particles that do not interact with the electromagnetic force. These particles also don't respond to the strong nuclear force; if they did, astronomers would have seen tracers of their strong interactions. Thus, two of the four forces were ruled out. Dark matter was assumed to feel gravity, so all that was left was the weak nuclear force.

Neutrinos are a classic example of a particle that responds to other particles mainly through the weak force. Given that a light-year of solid lead would stop only half the neutrinos fired into it, the weak force must be weak indeed. Therefore, dark matter might very well be *weakly interacting*. Thus, astronomers and physicists began focusing on **weakly interacting massive particles (WIMPs)** as their best candidate for the mysterious unseen stuff composing most of the Milky Way. (The acronym *WIMPs* was a direct response to the acronym *MACHOs*, demonstrating that scientists do have a sense of humor.)

Today, physicists are hard at work looking for direct evidence of WIMPs. If dark matter is composed of these kinds of particles, then with large enough detectors and long enough time frames, astronomers should be able to capture at least a few WIMPs in the act of interacting with normal (luminous) matter. Dark matter detectors need to be located far below Earth's surface to shield them from cosmic rays and other forms of "noise" that could confuse the extremely sensitive detectors. A number of these detectors exist, such as the Large Underground Xenon detector, 1.5 kilometers (km) below the surface in a former gold mine (**Figure 15.24**). In addition to the underground dark matter searches, astrophysicists have launched satellites such as the Fermi Gamma-ray Space Telescope, which is designed to see photons resulting from the very rare interaction of a dark matter particle slamming into a particle of normal stuff.

To date, however, dark matter has not been convincingly detected. Whatever makes up the bulk of the Milky Way Galaxy, for now it remains invisible and unknown. We can only infer its properties from the gravitational effects produced on that small fraction of the Milky Way whose matter we can see. Luckily, the luminous matter presents us with a rich storehouse of questions all by itself.

SECTION SUMMARY
What Is Dark Matter? The Four Forces

- Dark matter, if it exists, must interact via gravity but must not interact via the electromagnetic force or the strong nuclear force.

- Astronomers think dark matter may be made of particles that feel the weak nuclear force (WIMPs), but no dark matter particles have yet been directly detected.

15.6
Constructing a Galaxy: Evolution of the Milky Way

To piece together the history of the galaxy, like good detectives we have to stick with the facts we can gather now and see what they tell us about the past. But which facts matter? Which features of the galaxy shed light on its evolution in time and space? A good start might be to focus on the different populations of stars in the Milky Way and use their locations to piece together the galaxy's story.

Early Stages of Cloud Collapse

To begin, let's quickly review stellar populations. There are metal-rich Population I stars and metal-poor Population II stars. The abundance of processed elements (referred to as metals) in any population acts as a measure of age. Stars with high proportions of metals must have formed relatively recently, so Population I stars are young. Stars with few metals must have formed a long time ago, so Population II stars are old. The locations of these populations tell us that the disk, which contains mainly Population I stars, is young, while the halo, containing mainly Population II stars, is old. This is our first key piece of the history: the stellar halo itself must have formed earlier than the disk.

Stellar populations are not the only evidence for concluding that the halo is older than the disk. Gas and dust offer another clue. The disk is rich with gas and dust, but significant, dense distributions of interstellar material are rare in the stellar halo. Whatever gas and dust existed once in the halo now appear to have been converted into stars or dispersed long ago. It also makes sense that we don't see any massive stars in the halo: if star formation stopped in the halo billions of years ago, all the massive stars that formed would have blown themselves up as supernovas long ago, leaving only a population of low-mass, low-metal Population II stars.

So far, so good. But how can we turn the relative ages of the halo and disk into a realistic model for the formation of the Milky Way? As usual, everything begins with primordial clouds and gravity. "We can go back to the very early Universe and see how all structures get started," says Debra Elmegreen, who explains that after the Big Bang, gravity began to pull small concentrations of luminous mass together to create ever-denser structures. Filaments of gas began to form, and in those filaments the seeds of galaxies originated. "The buildup of galaxies over time occurs primarily through this inflow of gas," says Elmegreen.

Using the oldest stars in the stellar halo, astronomers can place the birth of the Milky Way at about 12.6 billion years ago. The Milky Way most likely began as a relatively dense region (denser than the background) within a filament. In this sense, the Milky Way began as a large, extended gas cloud, similar to the denser regions *within* galaxies that become centers of star formation (see Chapter 12).

Unlike the clouds where star formation occurs, however, the bulk of the proto–Milky Way would have been made of dark matter. From the initial large-scale distribution of dark matter, a region of higher-than-average density occurred via random fluctuations. This dense mix of mostly dark matter and a smaller fraction of luminous mass acted as a seed, allowing local gravity to begin pulling material inward and eventually forming into a well-defined dark matter halo. This structure began collapsing in on itself through its own gravitational force.

There is, however, an important distinction between the dark and luminous components of this proto–Milky Way. Because dark matter is so weakly interacting, it can never experience the friction that slows down particles of normal matter and leads them to clump together. Thus, the galaxy's dark matter collapsed to the point where it formed an extended halo. Any further collapse, however, was impossible.

FIGURE 15.25

How the Milky Way Formed
The Milky Way began at the intersection of large (cosmic-scale) filaments of collapsing gas. There, an extended cloud, composed of more dark matter than luminous matter, gathered via the force of gravity. Random fluctuations resulted in a region of higher-than-average density, which acted as a seed for gravity to pull material inward and form it into a proto-dark-matter halo. The luminous material collapsed down to where stars begin to form, with globular clusters and the stellar halo forming first. The rest of the gas continued to fall toward the cloud's center, forming a disk through conservation of angular momentum. Star formation began in the disk, accelerated by compression waves of the spiral arms. In the central regions of the newly formed Milky Way, the bulge, bar, and black hole formed.

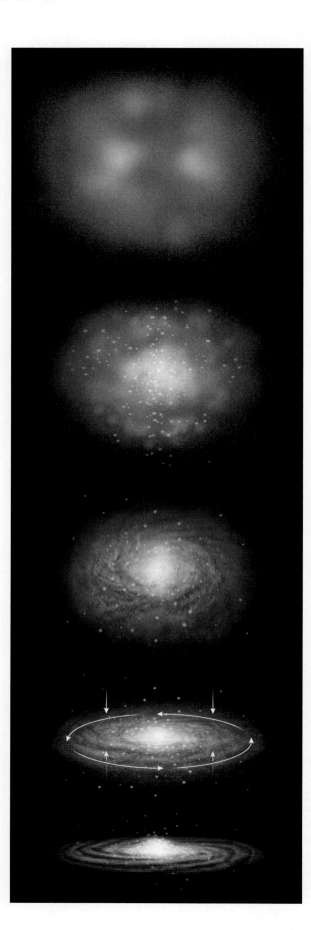

The luminous matter suffered no such restriction. Since luminous material experiences all four fundamental forces, the infalling, pregalactic gas could collapse all the way to the point where stars began to form. This process didn't occur all at once or all in the same place. Instead, early on, as the extended cloud of gas began falling inward, smaller fragments or cloud-lets must have formed. These collapsed on themselves to form the globular clusters. These early clusters were the birthplace of the Population II stars that we see today in the stellar halo.

SECTION SUMMARY
Early Stages of Cloud Collapse
- Clouds of gas and dark matter condensed by gravity in the early Universe to form the Milky Way.
- The dark matter remained in a spherical distribution, but some star clusters also formed and created the stellar halo.

Formation of the Disk, Bulge, and Black Hole

As the rest of the gas in the infalling protogalaxy continued to fall toward the cloud's center, it formed a disk in much the same way that an accretion disk forms around a young star. Conservation of angular momentum increased the rotation speeds around the new galaxy's center, while collisions within the material at the cloud's midplane flattened the mass distribution into a disk. Note that the rotation speeds of the stars and globular clusters that had already formed also would have increased because of conservation of angular momentum, but their distributions would not have been flattened by collisions. Because the distance between stars is much larger than the typical radius of a star, stars rarely physically collide the way gas clouds do. (Like clouds, stars do feel each other's gravitational force from large distances.) In contrast, the distance separating interstellar clouds can be of the same size scale as the clouds. This is why astronomers expect (and observe) cloud collisions.

The material in the newly formed disk eventually began its own sequences of star formation, leading to the formation of the younger Population I stars. Spiral waves that were generated in the disk material sped up the process by providing regions of extra compression, where clouds were more likely to collapse into stars.

In the central regions of the newly formed Milky Way, the bulge built up as material that fell straighter inward piled up at the center. This gas in the bulge still had some rotational motion, and later a bar may have been generated from the same general kinds of instabilities that form the spiral arms. As the mass of the central regions grew larger, material streamed inward. Mass there built up until it collapsed under its weight to form the central supermassive black hole. In this way, over time, the dark matter halo, stellar halo, disk, and bulge bar all formed, creating the Milky Way Galaxy we know today (**Figure 15.25**). In addition, the supermassive black hole at the center of the galaxy may have formed before the disk or built up through inward migration of massive star-forming clumps that contained black holes. The whole process of formation was over in just 5 billion years.

The preceding narrative fits the classic version of the story of galaxy formation. A more modern update would include the role of smaller galaxies that formed and then collided and merged with the Milky Way. "Our galaxy might have eaten many, many tiny neighbors that were a tenth or a hundredth of the mass of the Milky Way," says Elmegreen. "Those probably became part of the stellar halo."

There is one more very important detail that we must add to the Milky Way's story. A few billion years after the galaxy's formation, a small gas cloud lying about 8 kpc from the galactic center collapsed to form a very average yellow G-type star. After the chaos of the star's initial planet formation period, eight worlds settled down into stable orbits. On the third planet from that star, a blue-green world rich in liquid water, life, and oxygen emerged and then … well, you know how that story goes.

SECTION SUMMARY
Formation of the Disk, Bulge, and Black Hole
- Luminous matter collapses into a disk as a result of collisions at the rotating cloud's midplane, creating the galactic disk.
- In the central region, the bulge, bar, and black hole formed.

●→ chapter summary

15.1 A Hard Rain: Gaining a New Vision of Galactic Studies

Galileo was the first to realize that the band of light on the sky known as the Milky Way was made up of thousands of stars. In the late 1700s, William and Caroline Herschel used star counts to determine that the Milky Way is disk-shaped. Harlow Shapley's observations of globular clusters helped pinpoint the galactic center.

15.2 Anatomy of the Milky Way

The Milky Way's structure includes a disk that contains several hundred billion stars moving on roughly circular orbits around the galactic center. The stars are mostly younger Population I stars, which contain more heavy elements. The disk also has between two and four spiral arms. The Milky Way's stellar halo has a much smaller population of primarily old, red Population II stars, most of which are contained in globular clusters. The globular clusters' orbits follow highly elliptical paths with random orientations relative to the disk. The bulge, which is believed to include a bar structure, is a roughly spherical central region of the galaxy with extremely high densities of primarily old Population II stars.

15.3 Spiral Arms: Does the Milky Way Have Them?

Astronomers used various methods, such as mapping the locations of massive O and B stars, to determine whether spiral arms are present in the galaxy. Radio emission from cold, neutral hydrogen via the 21-cm line and emission lines from carbon monoxide have provided additional evidence. Spiral arms are created by spiral density waves that propagate through the disk to create bands of compression. Stars then form in the compressed regions.

15.4 The Galactic Center

The galactic center has been particularly difficult to study because of the extreme density of stars, gas, and obscuring dust. Astronomers have determined that a molecular ring orbiting near the very center may be the site of future starburst activity. The center also boasts a supermassive black hole whose existence has been demonstrated by the motions of orbiting stars. Also found in the galactic center are giant arcs of magnetic fields many times stronger than the average galaxy's field.

15.5 Dark Matter and the Milky Way

From its gravitational influence on the velocities of orbiting stars, astronomers deduce that an unseen halo of dark matter surrounds the galaxy and represents the majority of the galaxy's mass. Although dark matter responds to gravity, it does not interact with the electromagnetic force or the strong nuclear force. Astronomers are conducting experiments to see whether dark matter particles respond to the weak nuclear force, but to date they have found no conclusive results.

15.6 Constructing a Galaxy: Evolution of the Milky Way

The structure and motions of the galaxy provide insight into its formation. The stellar halo formed from an initial large sphere of gas and dust. Infalling luminous matter formed the disk and bulge, but the dark matter did not collapse, remaining instead as an extended halo. First generations of stars in the disk were metal-poor (Population II), but later generations were increasingly metal-rich (Population I). Smaller galaxies may have been consumed by the Milky Way through collisions.

●→ questions and problems

Narrow It Down: Multiple-Choice Questions

1. Using star counts, William and Caroline Herschel concluded that
    a. the Milky Way is made of stars.
    b. stars differ in mass.
    c. stars are not evenly distributed in our galaxy.
    d. the Milky Way galaxy is disk-shaped.
    e. there are fewer massive stars than smaller stars.

2. What role did interstellar dust play in the quest to determine the shape and size of our galaxy?
    a. It magnified the light from stars, making them appear closer.
    b. It obscured some regions and made others appear dimmer and redder.
    c. It outlined the Milky Way's structure.
    d. It played no role; dust is not important on a galactic scale.
    e. It made stars appear bluer.

3. The shape and orientation of globular-cluster orbits in our galaxy are most similar to those of
    a. long-period comets in the Solar System.
    b. planets in the Solar System.
    c. the Moon around Earth.
    d. asteroids near Jupiter.
    e. dwarf planets in the Solar System.

4. With which of the four fundamental forces must dark matter interact? Check all that apply.
    a. gravity
    b. electromagnetism
    c. weak nuclear force
    d. strong nuclear force
    e. none

5. The galactic disk is *not* characterized by
    a. spiral arms.
    b. a flat shape.
    c. stars, gas, and dust.
    d. mostly old, red stars.
    e. some massive, young stars.

6. In which component of the Milky Way do gas and stars bob slightly above and below the galactic plane?
    a. disk
    b. bulge
    c. stellar halo
    d. dark matter halo
    e. bar

7. Which statement about Population I and Population II stars is true?
    a. Population I stars are lower in metallicity.
    b. Population I stars are found in the halo.
    c. Population I stars are older.
    d. Population II stars may be supermassive.
    e. Population II stars are older.

8. A 0.5-M_{Sun} red giant star with high metallicity
    a. can be younger than the Sun.
    b. is composed of primordial material.
    c. is probably a member of Population I.
    d. may be found in a globular cluster.
    e. could not be located on a spiral arm.

9. Which of the following is *not* evidence that globular clusters formed very early in the history of the Milky Way?
    a. They all formed at the same time.
    b. They have low metallicity.
    c. Some have young, blue stars.
    d. Some have retrograde galactic orbits.
    e. They are nearly as old as the Universe.

10. What can be seen through Baade's Window?
    a. the nearest large spiral galaxy
    b. a spiral arm
    c. the central black hole
    d. the central region of the galaxy
    e. a globular cluster

11. Which statement about dark matter is true?
    a. It is denser toward the outer regions of the Milky Way.
    b. It represents 10 percent of total galactic matter.
    c. It does not respond to gravity.
    d. It ends at the average orbital radius of the globular clusters.
    e. Its nature is not yet understood.

12. Which of the following is *not* evidence of spiral arms in the Milky Way?
    a. the path of the Sun on the sky
    b. patterns of O and B stars
    c. CO maps of the galaxy
    d. the distribution of hydrogen seen via the 21-cm line
    e. the appearance of spiral arms in similar galaxies

13. Our Solar System orbits the center of the galaxy with approximately what period?
    a. 1 year
    b. 24 hours
    c. 200 million years
    d. 4.5 billion years
    e. 13 billion years

14. If there were no dark matter, the orbital period of the Sun around the galactic center would be
    a. shorter.
    b. longer.
    c. the same.
    d. impossible to predict.
    e. the same as that of every other star in the galaxy.

15. The term *starburst* refers to
    a. the death of a massive star.
    b. a collision between stars that annihilates both.
    c. a rapid increase in star formation.
    d. the merger of many stars into a black hole.
    e. the implosion of a Sun-like star.

16. Evidence for a central black hole in our galaxy includes all of the following *except*
    a. strong radio emission.
    b. stars orbiting rapidly around the potential black hole's position.
    c. the scarcity of brown dwarfs.
    d. a large apparent concentration of mass in a small volume of space.
    e. X-ray flares.

17. The age of the Milky Way Galaxy is about 13 billion years, but our Solar System is less than 5 billion years old. If the Sun had been one of the very first stars, how would the Sun and its evolution differ? Choose all that apply.
    a. The Sun would have lower metallicity.
    b. The Sun would already have reached the end of its fusion lifetime.
    c. The Sun would likely be in the bulge.
    d. There would be a lack of terrestrial planets because of the Sun's low metallicity.
    e. The Sun would be a more massive star.

18. Compare two Milky Way disk stars—one 3 billion years old and one 5 billion years old. The younger one is more likely to
    a. be farther from the midplane of the Milky Way.
    b. be from Population II than from Population I.
    c. have lower metallicity.
    d. reach smaller heights as it oscillates above and below the disk.
    e. be redder.

19. Which statement(s) about galaxy groups is/are true? Choose all that apply.
    a. The largest galaxies tend to have satellite galaxies.
    b. Interactions between massive galaxies and their satellites tend to affect the satellites more.
    c. Galaxy mergers are rare in galactic evolution.
    d. Tidal streams can result from galaxy interactions.
    e. The Local Group is one of the largest galaxy groups.

20. True/False: Stars near the galactic center of the Milky Way are generally members of Population I.

To the Point: Qualitative and Discussion Questions

21. What method did each of the following astronomers use to study the galaxy? Galileo, the Herschels, Kapteyn, Shapley.

22. How did Kapteyn's estimate of the galaxy's size compare with the estimate made by today's astronomers? Can you explain why his estimate differed from modern ones?

23. In what ways did dust inhibit early astronomers from obtaining an accurate picture of our galaxy?

24. How is dark matter believed to be similar to luminous matter, and how does it differ? How do we know this?

25. Where in the Milky Way is our Solar System located?

26. What are the so-called peculiar motions of disk stars?

27. Describe the differences in age, color, metallicity, and location of Population I and Population II stars. Can you explain why we would expect the age, color, and metallicity of stars to be interrelated?

28. How does the arrangement of globular clusters in our galaxy provide a clue to the timing of their formation?

29. How did the study of the galactic rotation curve help lead to the concept of dark matter?

30. How does the study of dark matter's interaction with the four fundamental forces provide evidence for its nature?

31. What would our galaxy look like if spiral arms were solely the result of winding due to rotation?

32. How do O and B associations trace out the locations of spiral arms? What other methods have astronomers used to verify the presence of spiral arms?

33. Describe how star orbits near the center of the galaxy provide additional evidence for the existence of a massive black hole.

34. "You can't see the forest for the trees." How is this saying relevant to the study of the Milky Way? What are some other situations you've experienced when it has been difficult to discern the big picture because you were immersed in it?

35. The nomenclature for Populations I and II is somewhat counterintuitive, since Population II is older. What other examples of astronomical nomenclature have you learned that are similarly counterintuitive?

Going Further: Quantitative Questions

36. The average orbital radius of a star around a galactic black hole has an angular size of 0.25 arcsecond when observed from a distance of 6.2 kpc. What is the orbital radius in kilometers? in astronomical units?

37. The orbital radius of a star orbiting Sgr A* is 3.45×10^{11} kilometers. Observed from a distance of 7.46 kpc, what is its angular size in arcseconds?

38. Calculate the mass of a black hole, in solar masses, based on the orbit of a star with period 7.8 years and orbital radius 2.8×10^{11} km.

39. A star orbiting Sgr A* has a semimajor axis of 1.4×10^{15} meters and a semiminor axis of 7.6×10^{14} meters. Using a mass for the supermassive black hole of $4.3 \times 10^6 \, M_{Sun}$, find the orbital period of the star, in years.

40. Calculate the orbital velocity, in kilometers per second, of a star orbiting 15,000 pc from the center of a galaxy whose mass is 130 billion M_{Sun}.

41. Star A orbits at 5,000 pc from the Milky Way's center; Star B orbits at 7,500 pc from the center. In the absence of dark matter, what would be the expected ratio of A's velocity to that of B? Assume that all of the luminous matter can be considered to reside in the galactic center.

42. How many times faster would a star's orbital velocity be at the Sun's position, compared with that of a star orbiting at 3.7 times the distance from the galactic center, assuming that dark matter is not a factor and that all the mass of the luminous matter can be considered to reside in the galactic center?

43. A star orbits in its galaxy at the same orbital radius as the Sun, but the mass of the galaxy is 2.6 times that of the Milky Way Galaxy. Assuming the ratio of the galactic mass lying within its orbit is the same as that of the Sun, how does the orbital velocity of this star compare to that of the Sun?

44. A globular cluster has an orbital radius of 25,000 pc. Using a galactic mass of $1.0 \times 10^{12} \, M_{Sun}$ for both luminous and dark matter combined (and assuming that all the mass lies within the cluster's orbit), what is the orbital velocity of the globular cluster, in kilometers per second?

45. What is the Schwarzschild radius, in kilometers, of a black hole whose mass is $6.7 \times 10^7 \, M_{Sun}$? (See Going Further 14.2.)

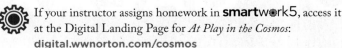 If your instructor assigns homework in **smartwork5**, access it at the Digital Landing Page for *At Play in the Cosmos*: **digital.wwnorton.com/cosmos**

A UNIVERSE
OF GALAXIES

In this chapter you will learn about the structure and evolution of galaxies as well as how astronomers study them across great distances from Earth. After reading through each section, you should be able to do the following:

16.1 The Great Debate and the Scale of the Universe
Explain how astronomers discovered that our galaxy was one of many in the Universe.

16.2 Galactic Zoology
Compare and contrast the different types of galaxies and locate each on the Hubble tuning-fork diagram.

16.3 The Cosmic Distance Ladder
List in order the rungs of the cosmic distance ladder, and explain what Hubble's law, in particular, tells us about the dynamics of the Universe.

16.4 Monsters in the Deep: Active Galactic Nuclei
Describe the model of active galactic nuclei that astronomers have proposed, and explain how it accounts for various phenomena.

16.5 Galaxies and Dark Matter
Summarize the evidence for dark matter in galaxies.

This dramatic image of the galaxy M51 shows atomic hydrogen in red marking many "HII regions," where massive stars are forming in spiral arms. M51 is interacting with its companion, the galaxy NGC 5195 seen at the top of the image.

16

16.1
The Great Debate and the Scale of the Universe

THE APRIL NIGHT was cold and blustery in Washington, DC, as men in spats, fedoras, and bowler hats filed into the National Academy of Science's Johnson Auditorium. The year was 1920, and the United States had recently emerged triumphant from the First World War. With the horrors of that great conflict behind them, the astronomers in the auditorium were concerned with a different kind of conflict: a contentious debate over the nature of the cosmos. The fate of galaxies, or at least our view of them, was hanging in the balance.

The event was called the Great Debate for good reason. For more than 100 years, astronomers had been building catalogs of the pinwheel-shaped forms called spiral nebulas. "People had observed these so-called spiral nebulas in the sky for decades," explains David Law, a galactic astronomer at the University of Toronto. "But in all that time their nature hadn't been resolved. That is what the Great Debate was really all about" (**Figure 16.1**).

As work progressed on measuring the size and shape of our own galaxy, some astronomers began to suggest that these spiral clouds were, in fact, vast and complete systems of stars similar to our Milky Way, but entirely separate and located at what were then inconceivable distances from us. The spiral nebulas, they claimed, were island universes strung throughout the vast emptiness of space. Other scientists scoffed at the "island universe theory" of spiral nebulas. The Milky Way, they claimed, was all there was to the Universe. The spiral nebulas were just that—spiral clouds of gas embedded within the Milky Way.

The debate pitted two eminent scientists against each other. Arguing against the island universe theory was Harlow Shapley, a leader in the study of the Milky Way. Shapley claimed the Universe was simply a single "enormous all-comprehending galactic system" encompassed in the Milky Way. Heber Curtis, an astronomer from Lick Observatory in California, defended the island universe theory, presenting multiple lines of evidence that spiral nebulas were distant, separate galaxies.

Each researcher fought hard for his vision of the Universe, but the debate that April night ended without a knockout. "No one really won the Great Debate," explains Law. "Each side had points in its favor which ultimately turned out to be true. But each side had biases and incorrect assumptions." Interstellar dust, for example, was affecting observations in ways that no one at the time could accurately predict. What makes the Great Debate so remarkable, however, is that as late as 1920, when airplanes and radios were becoming commonplace, science had yet to determine the nature of our own galaxy or to prove the existence of others. The true dimensions of galactic space, and of the Universe itself, would have to wait a few more years to be resolved.

The Island Universe Theory Confirmed

After serving in World War I and then studying at Oxford University on a Rhodes Scholarship, Edwin Hubble arrived at Mount Wilson Observatory in California in 1919, just a year before the Great Debate (**Figure 16.2**). Hubble favored the island universe theory and mounted his own attack on the issue.

Mount Wilson was the home of the 100-inch Hooker Telescope, the largest astronomical instrument in the world at the time. With it, Hubble could make out individual stars in the larger spiral nebulas. On October 5, 1923, Hubble spent the night searching the great spiral nebula called Andromeda. The next day he compared the night's work with previous observations, and to his surprise and delight, he found the Great Debate's resolution staring at him on fresh photographic plates (**Figure 16.3**).

Hubble found an all-important Cepheid variable star in Andromeda. Recall from Chapter 11 that Cepheids are standard candles that brighten and dim in a regular pattern. "There is the famous period-luminosity relationship for Cepheids," says Law. "Just

FIGURE 16.1
The Great Debate
(A) Harlow Shapley and (B) Heber Curtis were combatants in the 1920 Great Debate. Their evidence focused on the nature and distance of the spiral nebulas, but the deeper issue was the scale of the Universe.

A Harlow Shapley

B Heber Curtis

by observing how long the star takes to cycle from dim to bright and dim again, you can read off its intrinsic luminosity. Compare that with how bright it appears on the sky, and a distance measurement just falls right out." Through his discovery, Hubble had found astronomical gold that night. After taking more images over subsequent nights, Hubble determined the Cepheid's period and thus its intrinsic energy output (luminosity). Then, with a few simple lines of math, Hubble used the newly discovered star to sweep away 100 years of debate. "Isolating that Cepheid in Andromeda," says Law, "is what allowed him to end the Great Debate."

Using the Cepheid variable's period, Hubble calculated that the Andromeda Galaxy was almost a million light-years from Earth (we now know it to be 2.5 million light-years, or 770,000 parsecs, distant). This was far larger than any other astronomer's estimate for the Milky Way's outer boundary and implied that the spiral nebula in Andromeda could not be located within the Milky Way. It was, without doubt, a spiral *galaxy*, not a spiral nebula. More important, Hubble's result showed that the Universe was far larger than anyone had imagined.

Shapley had by this time moved on to Harvard, but he had not given up on his hypothesis that spiral nebulas were part of the Milky Way. On receiving news of Hubble's discovery, the defeated Shapley grimly told a student, "Here is the letter that has destroyed my Universe."

SECTION SUMMARY
The Island Universe Theory Confirmed

● In the Great Debate, scientists expressed conflicting views about the distance to spiral nebulas and its implications for the size of the Universe.

● Edwin Hubble subsequently used his observations of Cepheid variables to resolve the question and determine that ours is one of many galaxies.

A   B

FIGURE 16.2 Edwin Hubble and the Hooker Telescope
(A) Edwin Hubble (1889–1953) used (B) the 100-inch Hooker Telescope at Mount Wilson Observatory to resolve the Great Debate.

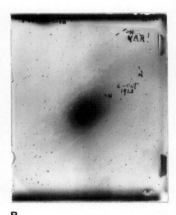

A   B

FIGURE 16.3 Confirmation of the Island Theory
(A) Hubble observed the Andromeda Galaxy (then known only as one of the spiral nebulas), establishing its true distance by identifying Cepheid variables in the galaxy via (B) photographic plates. Once the period of the star is determined, the period-luminosity relationship allows a derivation of the luminosity, which can then be compared with the apparent brightness to estimate distance. Given the great distance between the Milky Way and its neighbor, Hubble identified Andromeda for what it really was: another galaxy similar to our own.

16.2
Galactic Zoology

With Hubble's discovery, galaxy studies began in earnest. Astronomers started building up extensive catalogs of galaxy observations—the kind of pure data gathering that marks the first step in any new science. "There were all these different kinds of galaxies, and no one really knew what was going on," says David Law.

"So astronomers started by dividing the galaxies into bins. They created classifications based on the galaxies' observed properties and hoped those categories would expose underlying patterns in the physics of galaxy evolution and structure."

In galaxy classification, Hubble once again led the pack, helping to create the basic classification scheme still used today. Hubble found that galaxies could be sorted into four main types: spirals, barred spirals, ellipticals, and irregulars.

Pinwheels on the Sky: Spiral Galaxies

"The main thing that sets spirals apart," laughs Law, "is how pretty they look in Hubble Space Telescope images." But spirals are not important for their looks alone. Their beauty belies a wealth of physics, and they make up at least 70 percent of all galaxies in the cosmos. The Milky Way is an example of a spiral galaxy, and studies of it indicate that the basic (normal matter) components of a spiral galaxy are a disk, a bulge, and a stellar halo with globular clusters.

Fortuitously, one of the closest large galaxies to us, Andromeda (or M31), is also a spiral galaxy, so it provides us with an easily observed model for this class of galaxy. Both the Milky Way and Andromeda are called **grand design spirals** because they contain well-defined spiral arms. Other examples include the Whirlpool Galaxy (M51) and M81 (**Figure 16.4**). Law reminds us that whenever a galaxy has spiral arms in any form, how tightly bound the arms appear is a fundamental characteristic telling astronomers about the physics of spiral density waves that occur in the galactic disk (see Chapter 15).

But not all spiral galaxies have such well-defined arms. **Flocculent spirals** such as NGC 4414 show only patches of spiral arms in the disk (**Figure 16.5**). "In a flocculent spiral, sometimes it's difficult to see the spiral pattern at all," explains Law. "The spiral pattern is often fairly loosely wound, and it's broken up into lots of individual pieces. It's a very chunky spiral." In

grand design spiral A spiral galaxy with well-defined spiral arms that may be traced all the way from the bulge/bar to the outer regions of the disk.

flocculent spiral A spiral galaxy with poorly defined spiral arms.

FIGURE 16.5 Flocculent Spiral
As seen in this example of a flocculent spiral galaxy, NGC 4414, such galaxies have patchy, poorly organized spiral arms.

some cases, no spiral arms can be seen at all, and the disk component of the galaxy appears entirely smooth, as in the Sombrero Galaxy (M104).

Spiral galaxies come in a wide range of sizes, with disk diameters as small as 5 kiloparsecs (kpc) at the low end and as large as 100 kpc at the high end (the Milky Way has a diameter of 30 kpc). They also come in a wide range of masses, from 10^9 solar masses (M_{Sun}) to 10^{12} M_{Sun}. In the present epoch of cosmic history, "dwarf" spiral galaxies (very small, low-mass versions) are quite rare, though dwarfs are commonly found in other types of galaxies. The lack of true dwarf spiral galaxies turns out to be an important clue for understanding the history of all galaxy classes.

SECTION SUMMARY
Pinwheels on the Sky: Spiral Galaxies
- Spiral galaxies have a disk, bulge, and halo, and they may or may not have spiral arms that result from spiral density waves.
- Dwarf spiral galaxies are rare.

Pinwheels with a Twist: Barred Spirals

"When you look at the central regions of a spiral galaxy," says Law, "you basically see one of two things. Either the spiral arms extend pretty much directly into the center of the galaxy where they end in the

FIGURE 16.4 Grand Design Spiral
Prominent spiral arms characterize the classic grand design spiral galaxies, such as M81, which is found in the constellation Ursa Major and is shown here in optical light.

FIGURE 16.6 Barred Spirals
One-third of all spirals show a central bar, a structure that plays a major role in star formation. The barred spiral galaxy NGC 6217, shown here, lies about 2 Mpc away in the north circumpolar constellation Ursa Major.

bulge, or the arms don't curve smoothly into the very center but join onto a linear feature that cuts across the central bulge." The second kind are the so-called **barred spirals**, and their central bars are bright because the rate of star formation within them is quite high, just as in the spiral arms themselves (**Figure 16.6**). Fully 33 percent of all galaxies classified as spirals include a bar.

The fact that the spiral arms in a barred spiral galaxy begin at the bar, as Law describes, is an important clue to the bars' origin. This strong morphological link (*morphology* means "the study of structure") tells astronomers that bars are formed from the same kinds of instabilities that form spiral arms. That means they are also density waves—moving patterns of high density that flow through the background stars and

gas. "The bars are a feature reflecting some instability within the galaxy itself," says Law. "A bar will evolve over time in terms of its appearance. They are often a fairly transient feature. In computer simulations, they may stick around for many billions of years and then disappear and then reappear."

The computer simulations of spiral galaxies that Law mentions show that bars form when the orbits of stars near the bulge are perturbed enough to make their paths around the galactic center more elliptical (**Figure 16.7**). The bar pattern thus represents an increase in the local density of stars and gas as the entire galaxy rotates. The extra density in a bar can, in turn, pull interstellar gas from larger distances inward. As the bar "funnels" gas inward, conditions become ripe for enhanced star formation.

SECTION SUMMARY
Pinwheels with a Twist: Barred Spirals
- One-third of spirals exhibit bars across their centers.
- Like spiral arms, bars are thought to be caused by instabilities in the disk, and the presence of a bar drives enhanced star formation.

Footballs in the Sky: Elliptical Galaxies

Elliptical galaxies account for about 15 percent of the cosmic galactic population. As the name implies, they are elliptical in shape. "Elliptical galaxies have much less in the way of visible structure in contrast to the spiral galaxies," says Law. "They don't have the dominant disk with all of its star formation." Ellipticals appear on the sky as fairly smooth distributions of stars. They usually have one axis that is longer than the other, and the density of their stars increases from edge

barred spiral galaxy A spiral galaxy with a central rectangular bar.

elliptical galaxy A galaxy that appears elliptical on the sky. Elliptical galaxies are generally triaxial, meaning that the width is different in each dimension.

FIGURE 16.7
Bar Formation
The sequence of changes leading to the formation of a bar across the center of a galaxy, as seen via a computer simulation. Bars appear to form in galaxies because of both instabilities and resonances in stellar orbits. The Milky Way is believed to host such a bar.

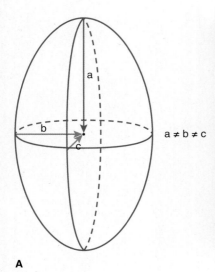

A

FIGURE 16.8
Elliptical Galaxies
Depending on their
orientation relative to our line
of sight, elliptical galaxies
may appear to be spheroid
or elliptical, but many are
(A) triaxial, meaning that
there is no equality between
any two axes. (B) NGC 1316 is
an example of a triaxial galaxy.

cD galaxy The largest, most
massive example of an elliptical
galaxy.

B

to center. Detailed studies of ellipticals show, however, that they are *triaxial*, meaning that the size of each of the three perpendicular dimensions can be different (**Figure 16.8**). A football, for example, is biaxial, but a lima bean—which is flattened by different amounts in each dimension—is triaxial.

Ellipticals show very little interstellar medium of any form. Radio studies using the 21-cm line of neutral hydrogen, for example, find almost no H I clouds in ellipticals. While spiral galaxies play host to active star formation, in elliptical galaxies stellar nurseries are practically nonexistent. "In general, ellipticals seem devoid of gas," explains Law. "That means they don't have much in the way of newly formed stars either." Most of the stars in elliptical galaxies, therefore, are old and red, whereas spiral galaxies have both red, older stars and blue, young, massive stars.

Most ellipticals do exhibit a halo of low-density, high-temperature gas, however. Space-based X-ray observations using instruments like the Chandra space telescope show that these extended halos are filled with gas as hot as 10^7 kelvins (K; **Figure 16.9**). The presence of hot halos indicates that ellipticals must have experienced violent events in their past that could drive the gas to such high temperatures. And according to many scientists, it also indicates that collisions with other galaxies must be an important part of elliptical galaxies' evolution.

Stellar motions within elliptical galaxies are also quite different from those in their spiral cousins. "Instead of stars orbiting on nice, circular orbits in a disk," says Law, "the stars in an elliptical galaxy are fairly chaotic. They are all plunging toward the galaxy center with randomly oriented directions." We've already seen that the globular clusters in the Milky Way move on similar plunging orbits. In fact, globular clusters and elliptical galaxies show a number of important similarities, including the old, red stellar population and the lack of an interstellar medium. Elliptical galaxies also tend to contain more globular clusters than spirals do.

The range of sizes and masses is much wider for elliptical galaxies than for spiral galaxies. The smallest elliptical galaxies, called *dwarf ellipticals*, are not much larger than the largest globular clusters. These galaxies have an average size of 1 kpc and a mass of 10^6 M_{Sun}. They represent the majority of elliptical galaxies. At the high end of size and mass are the giant **cD galaxies** that live at the center of dense galaxy clusters (explored in Chapter 17). These cD galaxies contain upwards of 10^{13} M_{Sun} and can extend across hundreds of kiloparsecs.

FIGURE 16.9
Galactic Halos
(A) The Chandra X-ray image
of NGC 4555 reveals that
this large, isolated, elliptical
galaxy is embedded in a cloud
of 10-million-K gas (blue).
(B) The hot gas cloud has a
diameter of about 120,000 pc,
roughly twice that of the
visible galaxy shown in this
image.

A X-ray image of NGC 4555

B Optical image of NGC 4555

TABLE 16.1 • • • Galaxy Types

Type	Shape	Structures	Star Formation	Special Classes	"Tuning Fork" Designations	Notes
Spiral	Disk	• Disk • Bulge • Stellar halo with globular clusters • Most have spiral arms	Prevalent in the spiral arms	• Flocculent • Grand design • Lenticular	Sa, Sb, Sc	Dwarfs are rare
Barred spiral	Disk	• Disk • Bulge • Bar • Stellar halo with globular clusters • Most have spiral arms	Prevalent in the spiral arms and bar	• Flocculent • Grand design • Lenticular	SBa, SBb, SBc	Dwarfs are rare
Elliptical	Spheroid to triaxial ellipsoid	Stellar halo with globular clusters	Very limited	• Dwarf • cD • Lenticular	E0 through E7	• cDs are found at the center of some galaxy clusters • Widest range of radii
Irregular	Amorphous	May have some disklike features	Varies; active in some	• Irregular I • Irregular II	Irr I, Irr II	Often found as satellites to larger galaxies

According to Law, there is a lesson to be learned in the difference between the mass and size ranges of spirals and ellipticals. "The spiral galaxies are more limited in size," says Law. "They tend to be on the middle and lower ends of the mass spectrum. You can certainly have reasonably high-mass spiral galaxies, but these are not as massive as the most massive elliptical galaxies. The range of masses and sizes shows us that the different kinds of galaxies must form in different ways and have different star formation histories" (**Table 16.1**).

SECTION SUMMARY

Footballs in the Sky: Elliptical Galaxies

- Elliptical galaxies have triaxial shapes, a smooth distribution of stars, and increasing stellar density toward the center, as well as a broader range of sizes than other galaxy types have.

- Stars in elliptical galaxies have highly elliptical orbits that can take them close to the galactic center.

Galactic Blotches: Irregular Galaxies

Spirals, barred spirals, and ellipticals represent relatively clean morphological "bins" into which astronomers can categorize galaxies. But there are galaxies with amorphous shapes, which astronomers call **irregular galaxies**, that don't fit into any of these categories. "They are, essentially, train wrecks of galaxies," says Law. At the simplest level, irregulars look like messy blobs with no easily categorized geometry. Taken a step further, though, astronomers see enough pattern to distinguish two types: the truly amorphous Irr I galaxies and the Irr II galaxies, which seem to retain some features of a disk.

The range of sizes and masses for the irregulars places them between the dwarf ellipticals and the normal spirals. The smallest irregulars have masses of $10^7 \, M_{\text{Sun}}$; the largest, $10^{10} \, M_{\text{Sun}}$. As with ellipticals, the dwarf irregulars represent the bulk of the class: most irregular galaxies have low mass and small diameter.

irregular galaxy An amorphous galaxy lacking common structures.

A   B   C

FIGURE 16.10
Irregular Galaxies
Irregular galaxies have amorphous shapes and lack features that are common across examples. They are typically small and are often found as satellites of spiral or elliptical galaxies. Seen here are (A) the Large Magellanic Cloud, (B) the Small Magellanic Cloud, and (C) NGC 1427.

"The irregulars are a population that's reasonably uncommon in the modern Universe," says Law. Just as important, many irregulars are companions to other, larger galaxies. Because they are typically low in mass, these irregulars can be considered satellites of the larger galaxies. The best known of all the irregular galaxies are, in fact, satellites of our own Milky Way. The Large Magellanic Cloud (LMC) and Small Magellanic Cloud (SMC) are a pair of dwarf irregulars (type Irr I) that can be seen from Earth's Southern Hemisphere (**Figure 16.10**). The LMC contains 10^{10} M_{Sun} of normal (nondark) matter and is about 4.3 kpc across, while the SMC contains only 7×10^9 M_{Sun} and is about 2 kpc in diameter. Stars continue to form in both systems. The fact that large galaxies can have smaller galaxies such as these trapped within their orbital influence is central to our understanding of galactic evolution.

SECTION SUMMARY
Galactic Blotches: Irregular Galaxies
- Irregular galaxies are amorphous, although some have features of a disk.
- Many dwarf irregulars are satellites of larger galaxies.

The Hubble Tuning-Fork Diagram

After establishing that galaxies can be sorted into four basic categories (spirals, barred spirals, ellipticals, and irregulars), the next step for astronomers was to try to find a relationship between the categories that exposes an underlying order. Recall how the stellar spectral classification OBAFGKM was given physical meaning once stars were plotted on the HR diagram. The diagram revealed how stellar temperature and luminosity combine (via the deeper principles of stellar structure and nuclear burning) to order stars by mass on the main sequence. "Galaxies were no different," says Law. "Once the categories (the galaxy types) were defined," says Law, "someone needed to work out the relationship between them." That someone was, once again, Edwin Hubble.

After providing the basic galaxy characterizations, Hubble refined the categories and organized them into a schematic diagram that he believed brought order to the galaxy classification. The fruit of his effort did, in fact, enable astronomers to make significant progress in understanding galaxies, but not for the reasons Hubble originally imagined.

For Hubble, the shapes of elliptical galaxies could be smoothly laid out on a spectrum running from circular to extremely elliptical (cigar-shaped). He designated the roundest ellipticals E0 galaxies. As the ellipticity of the galaxy increases, the integer following the letter E increases, with the very cigar-shaped galaxies labeled E7. Note that shape alone does not determine the appearance of elliptical galaxies; the observer's visual perspective matters too. A football seen end on appears circular. Look at the same football from the side, and it appears elliptical. Hubble was aware that his measure of ellipticity would be affected by both the actual shape of the galaxy and its orientation relative to the viewer. He hoped, however, that after averaging over many galaxies with many orientations, his schematic would provide insights into the underlying physics.

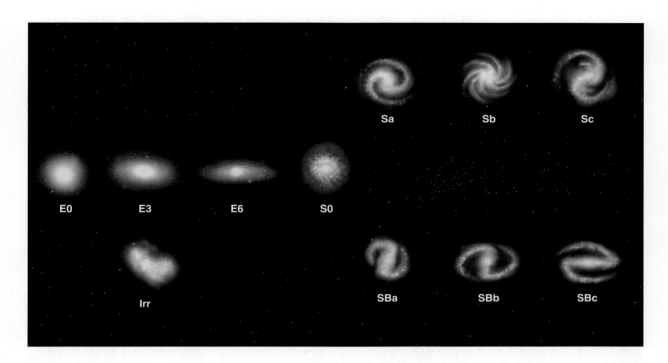

FIGURE 16.11
Hubble Tuning-Fork Diagram
Edwin Hubble classified the galaxies he observed and developed a classification scheme in the shape of a tuning fork (a common instrument of his day). Elliptical galaxies are placed on the left end of the sequence, which then moves through spirals or barred spirals. Although Hubble's assumption that his sequence represented the evolution of galaxies from young to old has been disproved, the classification scheme remains useful.

Hubble needed to account for two properties in spirals. The first property was the size of the central bulge, and the second was how tightly the spiral arms were wrapped around the galaxy. Luckily, the size of the bulge appeared to be correlated with how tightly the arms were wound, enabling Hubble to use a single classification scheme for both. Hubble used the term *Sa galaxies* for those with the largest bulges and the tightest, most circularly wrapped spiral arms. In contrast, *Sc galaxies* have the smallest bulges and spiral arms that are fairly open, and *Sb galaxies* have intermediate bulges and intermediate spiral arm windings.

Hubble used the label *SB* to denote a spiral with the presence of a bar. As with the regular spirals, there are three subcategories for the size of the bulge and angle of the spiral arms: *SBa*, *SBb*, and *SBc*. Note that orientation also matters for barred spiral galaxies. If a spiral galaxy is seen edge on, astronomers have no way of determining whether a bar exists.

From his extensive observations, Hubble found a category of galaxy that appeared intermediate between the last ellipticals (E7) and the first spirals (Sa). These objects appear to be almost all bulges with just a hint of evidence for the presence of a disk, and Hubble called them **lenticular galaxies**. "Lenticulars can get a little complicated," says Law. "They look elliptical, but you can often see they've got a disk. Still, those disks aren't necessarily forming a great deal of new stars." Hubble classified these objects as *S0*.

With this inventory of galaxies, Hubble was able to create what became known as the Hubble sequence, which could be cleanly represented on the so-called **Hubble tuning-fork diagram** (**Figure 16.11**). "Hubble placed the ellipticals on a line running from E0 to E7," explains Law. "The line ends with the S0 lenticulars and then splits off into two branches." The spiral galaxies run along one branch from Sa to Sc. The barred spirals run along the other branch from SBa to SBc. "So the entire sequence looks like a tuning fork," Law continues. "The irregulars don't get included because they are just a mess."

For Hubble, the sequence exhibited on the tuning fork appeared to be an *evolution in time*. He believed his classification scheme laid galaxies out in their different stages from birth onward. With his sequence, Hubble was suggesting that an individual galaxy would begin with the amorphous shape of an elliptical and then evolve into the grand pattern of spiral arms, and perhaps a central bar as well.

Two aspects of the Hubble sequence stand out. The first is that it appears *Hubble was entirely wrong about the evolution of galaxies*: there is no evidence that ellipticals ever turn into spirals; in fact, the truth seems to be just the opposite. "We now know elliptical-type galaxies tend to be bigger and older than the spiral galaxies," says Law, "which tend to be both lower-mass and more actively forming stars." So rather than ellipticals turning into spirals, astronomers now understand that

lenticular galaxy A galaxy with a bulge and limited evidence of a disk; it has no spiral arms.

Hubble tuning-fork diagram A scheme for representing galaxy types developed by Edwin Hubble that shows a progression from highly elliptical to spiral or barred spiral.

FIGURE 16.12
Galactic Collisions
Galactic evolution can be driven by collisions between galaxies. This composite image shows a variety of interactions between spirals. While each image shows a separate interacting pair of galaxies, the composite shows steps in evolution that are similar to what is seen in simulations of a single pair of colliding galaxies as gravitational interactions result in loss of spiral structure and evolution toward elliptical shapes.

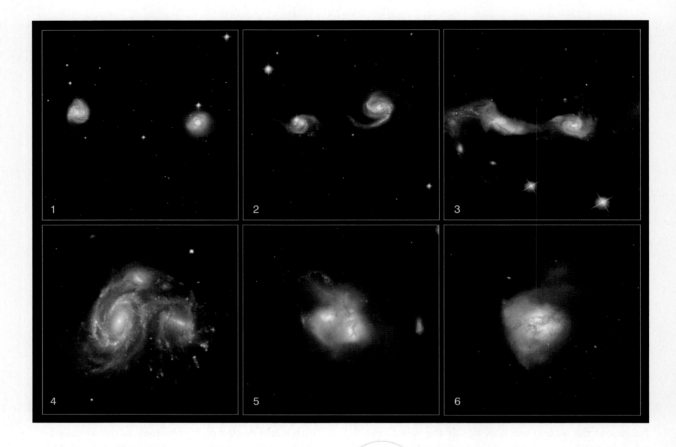

megaparsec (Mpc) A unit of distance equal to 1 million parsecs.

calibration The use of a known parameter value obtained by a proven method to validate a measurement of that parameter by an unproven method.

cosmic distance ladder The sequence of methods for determining distance to astronomical objects; arranged by increasing distance over which each method is effective.

spirals may turn into ellipticals through galaxy collisions (**Figure 16.12**). It's important to note, however, that the processes that shape galaxies remain at the frontier of research. Processes occurring during the birth of galaxies in the early Universe were also an important factor in determining the shapes we see today.

The second remarkable aspect of Hubble's scheme is how useful it turned out to be, despite being wrong. Categorizing galaxies by their main morphological types, as Hubble suggested, proved to be a good way for astronomers to sort through links between observed properties and important physical aspects like gas content and star formation. In this way, the Hubble sequence, as a classification scheme, has outlived the organizing idea (galaxy age) underlying its creation.

SECTION SUMMARY
The Hubble Tuning-Fork Diagram
- In Edwin Hubble's tuning-fork diagram, ellipticals are categorized by increasing ellipticity, and spiral galaxies and barred spirals are categorized according to bulge size and how tightly wound the spiral arms are.
- Hubble believed galaxies evolved from elliptical to spiral, but astronomers later discovered that the reverse is more likely to be true because of collisions.

16.3
The Cosmic Distance Ladder

Studying galaxies beyond our own Milky Way means dealing with an enormous increase in size scale. Distances are measured in parsecs for stars and in kiloparsecs for the Milky Way. The space between galaxies, however, takes us up to size scales that are thousands to millions of times larger. The distance between the Milky Way and Andromeda, the nearest spiral galaxy, is about 800 kpc, or 800,000 pc. The separation between galaxies must therefore be measured in units of **megaparsecs (Mpc)**, equal to 1 million pc.

As you begin to think about astronomical distances, keep in mind that light has a finite speed of 300,000 kilometers per second (km/s). Although this is fast enough to seem instantaneous in our everyday life, the light from objects at distances of megaparsecs takes millions of years to reach Earth. Thus, as we look at a galaxy that is 65 million light-years (19.9 Mpc) away, we're seeing it as it looked 65 million years ago. That was when the age of the dinosaurs ended.

Given these staggering astronomical distances, the first question must be, How do we know? How do we measure distances on these immense scales? "The answer to that question turns out to be remarkably hard," says David Law. "We're stuck here on Earth, and we can't take out a ruler and measure how far away an object is. We're really limited in terms of how we can determine astronomical distances. No one method for determining distances works over all ranges. So we have to come up with lots of different methods and hopefully get enough overlap between the ranges in which they're valid that we can calibrate them all against one another."

Calibration means using a known distance determination method to tune and prove a new method. When astronomers discover a new method, they first test it and calibrate it against proven methods to ensure its usefulness and accuracy. They can then use the new method to step outward and survey even larger volumes of cosmic space. Each new method for determining astronomical distances essentially represents a rung on a ladder extending out to ever-larger distances, so the entire process is referred to as the **cosmic distance ladder** (**Figure 16.13**). "What we are doing," says Law, "is pulling ourselves up by our own bootstraps. We combine lots of different methods one after the other to get to ever-larger distances."

Lowest Rungs of the Ladder

"For the very nearest stars, we can use parallax," says Law. "Parallax is just the yearly back-and-forth shift in the apparent positions of nearby stars on the sky." Parallax is the most direct method for determining astronomical distances because it relies only on basic trigonometry and the observed displacement of a star on the sky as Earth moves from one side of the Sun to the other during its orbit. "The problem is, for traditional ground-based telescopes, parallax works only out to about 100 pc or so," says Law. Space-based observatories give astronomers the ability to use parallax for

FIGURE 16.13

Cosmic Distance Ladder
The cosmic distance ladder is made up of a series of methods for determining distance to faraway astronomical objects, with each successive rung (represented by the curved horizontal bars) extending to greater distances. As astronomers discover new distance determination methods that can serve as new rungs on the ladder, they must first validate those methods' accuracy and calibrate the results. The overlap between each rung and the one preceding it provides the means to perform the validation and calibration.

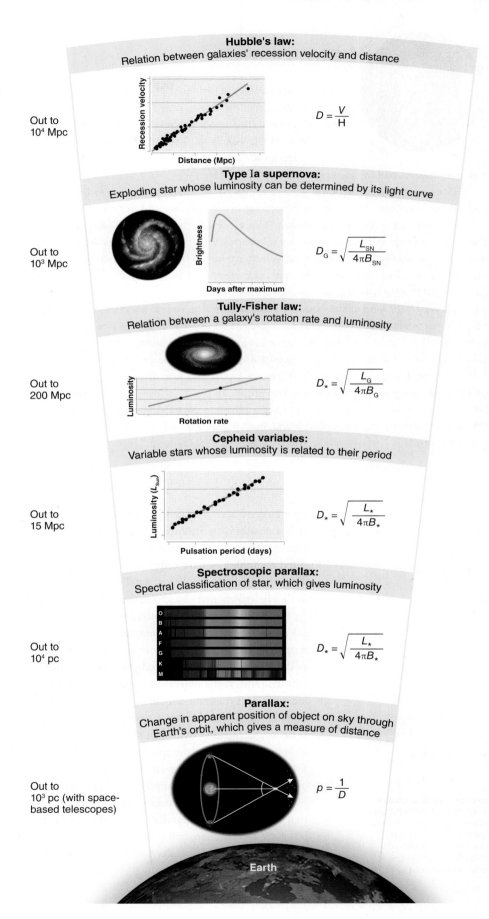

Hubble's law:
Relation between galaxies' recession velocity and distance

Out to 10^4 Mpc

$$D = \frac{V}{H}$$

Type Ia supernova:
Exploding star whose luminosity can be determined by its light curve

Out to 10^3 Mpc

$$D_G = \sqrt{\frac{L_{SN}}{4\pi B_{SN}}}$$

Tully-Fisher law:
Relation between a galaxy's rotation rate and luminosity

Out to 200 Mpc

$$D_* = \sqrt{\frac{L_G}{4\pi B_G}}$$

Cepheid variables:
Variable stars whose luminosity is related to their period

Out to 15 Mpc

$$D_* = \sqrt{\frac{L_*}{4\pi B_*}}$$

Spectroscopic parallax:
Spectral classification of star, which gives luminosity

Out to 10^4 pc

$$D_* = \sqrt{\frac{L_*}{4\pi B_*}}$$

Parallax:
Change in apparent position of object on sky through Earth's orbit, which gives a measure of distance

Out to 10^3 pc (with space-based telescopes)

$$p = \frac{1}{D}$$

Earth

David Law

One of David Law's best memories of astronomical research began by accident. "My colleague and I were just looking through some of the Hubble Space Telescope data we had recently taken," Law explains. These were images of very distant galaxies 10 billion light-years or so away. "Most of them looked weird, blobby, and irregular."

Since Law and his collaborator were seeing so far back in time, these galaxies were caught in the act of their formation, so it made sense that they looked like a mess. "But then we just saw one that looked totally different," Law continues. "It looked like a regular spiral galaxy. It had a bulge and two or three nice spiral arms. We figured we must have gotten the distance wrong." But checking and rechecking confirmed that the galaxy was clearly 10 billion light-years away—more than two-thirds of the way back to the dawn of time. They had found the oldest spiral galaxy anyone had ever seen.

Law wasn't the kind of star-happy kid who was sure he was going to be an astronaut or astronomer when he grew up. Instead, his fascination with the field grew slowly. "I took a physics course in high school, and I really liked the part about planetary orbits and gravity," Law explains. Once he started his undergraduate studies, he signed up for more physics courses. But the event that really pushed him over to the dark side (astronomical observing) was a summer program at the Haystack radio observatory in Westford, Massachusetts. "Even though I found out I didn't want to have anything to do with radio antenna design, I did get a lot of really useful experience and, most important, I learned that I really, really liked astronomy."

"But you can't use any new distance determination method until you calibrate it," Law reminds us. Thus, astronomers took stars whose distance had already been determined through geometric parallax, and they compared those distances with their findings using the new method—in this case, spectroscopic parallax—to confirm that it could, in fact, produce accurate distances. With this confirmation, they could push outward in space, using the new method to measure objects at greater distances.

But the usefulness of spectroscopic parallax does not extend forever. It is accurate out to only about 10,000 pc. To measure even larger distances, yet another new method needed to be found.

SECTION SUMMARY
Lowest Rungs of the Ladder

- Different methods of distance determination, collectively referred to as the cosmic distance ladder, are useful out to successively greater distances, and each is used to calibrate the next method.

- Parallax, the apparent shift in a distant object's position due to the Sun's annual orbital motion, can be used to determine the distance to the nearest objects.

- Spectroscopic parallax uses the correlation between stellar spectral class and luminosity to determine the distance to more distant stars.

Cepheid Variables

Cepheid variables come into the picture as the next step outward on the ladder. As mentioned earlier in this chapter, Cepheid variables are the classic example of a standard candle, an object whose luminosity can be inferred through some basic measurement of the object's properties. Measure the pulsation period of a Cepheid, and you can find its luminosity. Compare this luminosity to the brightness you see on the sky, and you can use the inverse square law for brightness (see Chapter 4) to determine the object's distance.

If an individual Cepheid can be resolved in a distant object like a galaxy so that its pulsation period can be tracked, the period-luminosity relationship can be used to calculate the distance to that object. In just this way, Hubble used Cepheid variables to measure the correct distance to the Andromeda Galaxy and prove that the spiral nebulas were, in fact, spiral galaxies. Current telescope technologies allow Cepheids to be used for distance determination out to 15 Mpc.

even larger distances. The GAIA satellite can measure stellar positions to distances as far as the galactic center, almost 10 kpc away. Of course, such sophisticated technologies have only recently become available. To go beyond parallax before satellite technology was available, astronomers had to find a different method.

In the 1940s, such a method was found in what is known as **spectroscopic parallax**. With spectroscopic parallax, astronomers use spectral classification to determine a star's horizontal placement on the HR diagram. If the object is a hydrogen-burning star, locating it on the main sequence gives an estimate of its luminosity. If, however, the star is a giant that is no longer burning hydrogen in its core, the widths of absorption lines in the spectrum provide a good measure of the luminosity. In all cases, measurements of spectra yield values of luminosity that can be compared with apparent brightness to find distance. (Note that there is no actual parallax involved in this method, but since astronomers had one distance measure called parallax, they added *parallax* to the name of the new method as well.)

spectroscopic parallax
A method for determining distance to an astronomical object that is based on obtaining the width of absorption lines in the object's spectrum.

SECTION SUMMARY
Cepheid Variables
- Astronomical distances can be measured using standard candles such as Cepheid variables and the application of the inverse square law for brightness.

Tully-Fisher Law

"Astronomers expect that high-mass galaxies should have higher energy outputs than low-mass galaxies," says Law. "That's because most of the visible mass will come in the form of stars." Thus, knowing a galaxy's mass could, in principle, provide a way of estimating its luminosity. This is the principle behind the **Tully-Fisher law**. While astronomers can't directly measure a distant galaxy's mass, they can indirectly measure it using the galaxy's average rotation speed. As we have seen, orbital revolution speeds provide some measure of the galaxy's mass via Newton's law of gravitation.

"If you measure the rotation speed of a galaxy using a spectrograph," explains Law, "then you can use the Tully-Fisher law to calculate how intrinsically bright the galaxy itself must be, and that gives

you an estimate of its distance" (**Figure 16.14**). The Tully-Fisher law is not very accurate (the errors in any individual measurement may be as large as 50 percent), but it does provide a relatively quick way of determining distances for galaxies as far away as 200 Mpc.

SECTION SUMMARY
Tully-Fisher Law
- The Tully-Fisher law uses the rotation speed of a galaxy to infer the galaxy's mass and associated luminosity.

Type Ia Supernovas

The brighter a standard candle is, the farther away it can be seen and used to determine distance. Therefore, it's easy to imagine why astronomers became giddy with delight when, in the 1990s, it seemed that one type of supernova could be useful as a standard candle. "The discovery of supernova standard candles was fantastic," says Law. "It gave us a method of getting distances out to incredibly large scales."

Recall from Chapter 13 that Type Ia supernovas are created in binary systems when a companion dumps so much mass onto a white dwarf that

Tully-Fisher law A relationship of galactic rotation speed to galaxy mass that is used to infer the intrinsic brightness of a galaxy.

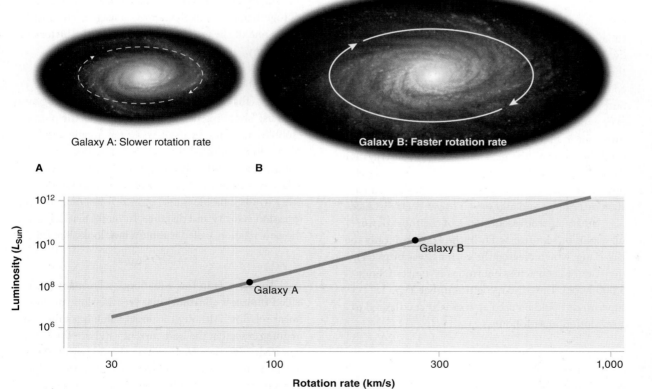

A

B

Galaxy A: Slower rotation rate

Galaxy B: Faster rotation rate

C

FIGURE 16.14
Tully-Fisher Law
The Tully-Fisher law defines the linear relation between galactic rotation speed (related to galactic mass) and the luminosity of the galaxy. (A) Stars in the disk of a less massive galaxy rotate more slowly. (B) Stars in the disk of a more massive galaxy rotate more quickly. This greater mass also means more stars, which means greater luminosity. (C) The luminosity compared with observed brightness provides the distance measurement.

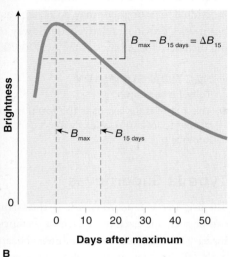

A   **B**

FIGURE 16.15 Type Ia Supernovas
(A) A supernova (SN 2011fe) in the Pinwheel Galaxy. This is a Type Ia supernova, the explosion of a white dwarf star that has exceeded its allowable mass by accretion from a companion star. (B) Light curves from these events follow a distinctive pattern, enabling astronomers to determine the supernovas' luminosity. In particular, astronomers have found that a measurement of difference between the peak brightness and the brightness 15 days later (ΔB_{15}) yields a value of a supernova's luminosity at its peak. In this way, Type Ia supernovas can be used as standard candles to determine accurate distances to objects (such as a host galaxy) out to 3,000 Mpc.

FIGURE 16.16
Milton Humason
Working with Milton Humason (1891–1972), pictured here, Edwin Hubble made the remarkable discovery that galaxies are receding from us at speeds proportional to their distances. The implication of this observation is that the Universe itself is expanding.

runaway nuclear fusion is ignited, blowing the white dwarf apart. In the mid-1990s, researchers discovered that the shape of a Type Ia supernova's light curve could be used to infer the total intrinsic light-energy output of the explosion (**Figure 16.15**). "The period over which the Type Ia supernovas brighten and fade turns out to be very closely linked to the intrinsic brightness during the peak of the explosion," says Law. "That means the light curve alone tells you precisely how much energy they put out. That makes them a standard candle and allows you to get their distances."

Since supernovas are some of the most powerful explosions known so far in the Universe (only the Big Bang, which created the cosmos, was brighter), they can be seen across enormous distances (about 3,000 Mpc). Thus, the recognition that Type Ia supernovas could be used as standard candles led to a revolution in astronomy.

SECTION SUMMARY
Type Ia Supernovas
● The light curves of Type Ia supernovas correlate with their luminosity.

Hubble's Law

In the mid-1920s, Hubble was exploring the motions of galaxies through space. To determine their speeds, he needed accurate Doppler shift data, which he obtained from measurements of galaxy spectral lines. "These come from the combined starlight of each galaxy," says Paul Green, an astronomer at the Harvard-Smithsonian Center for Astrophysics. "The spectra gave him velocities, but he also needed distance measurements so he could understand if there was a pattern in the way different galaxies were moving."

To get each galaxy's position in space, Hubble continued using Cepheid variables. Accumulating the measurements (velocity and distance) demanded many frigid nights sitting in the "cage" atop Mount Wilson's 100-inch telescope, keeping the 100-ton instrument trained on the right position as the faint light from the distant galaxy slowly exposed photographic plates at the telescope's focus. "I did this kind of thing as an undergraduate," says Green, "and it's really, really cold in the cage."

Working with Hubble was Milton Humason, a former mule-train driver and a mechanic for the observatory, who turned out to be a gifted astronomer (**Figure 16.16**). Together, Humason and Hubble painstakingly gathered the needed data. Putting it all together, the unlikely pair of colleagues stumbled across one of the greatest discoveries in the history of astronomy.

First, Hubble discovered that nearly every one of the 500 distant galaxies he had observed showed redshifts. That meant all the galaxies were moving away from us; "shunning us like the plague" was the phrase Hubble used. This, by itself, was a remarkable result, but when Hubble plotted the galaxies' velocities versus their distance, he found a clear pattern. The farther away a galaxy was from Earth, the faster it was receding. In other words, Hubble found a *linear relationship* between recession velocity and distance. In mathematical form, the new relation, called **Hubble's law**, looks like this:

$$V = HD \text{ (km/s)}$$

The term H in this equation is a constant (now called **Hubble's constant**) that sets the slope of the linear relation. Although its exact value is a subject of ongoing research, the best current estimate for H is 69 km/s/Mpc (**Figure 16.17**). "You should pay attention to the funny units on Hubble's constant (km/s/Mpc)," says Green. "That combination comes because Hubble's law can take measurements of galactic recession

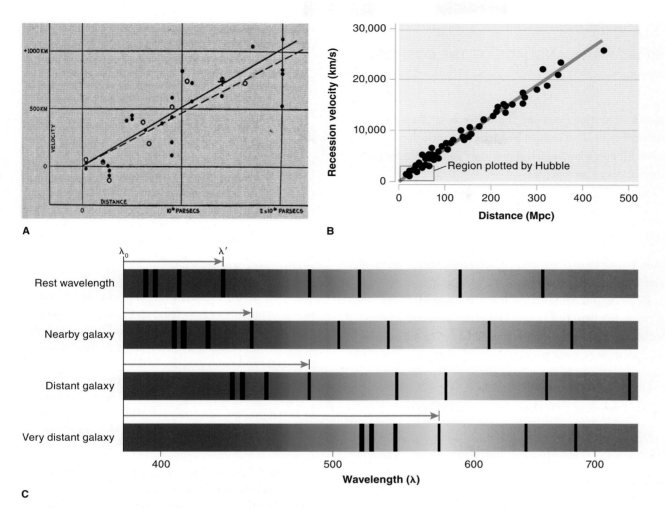

A

B

C

FIGURE 16.17
Hubble's Law
Hubble's law correlates galactic distance with recession speed. (A) Hubble plotted this relation for galaxies out to a distance of about 2 Mpc, as his original figure shows. (B) More recent studies of galaxies out to distances of hundreds of megaparsecs have produced an estimate of $H = 69$ km/s/Mpc. (C) Astronomers determine recession speed from the redshift of absorption lines in a galaxy's spectrum. The farther a known absorption line is shifted from its rest wavelength, the faster the galaxy is moving away and the greater its distance.

velocity (measured in kilometers per second) and turn them into distances (in megaparsecs) or the other way around" (see **Going Further 16.1**, p. 418).

Hubble's discovery that every galaxy is moving away from us had profound theoretical implications. It was the first direct evidence that the Universe had a beginning and was not infinite in time, as Aristotle and others had suggested. A full discussion of that topic will be the focus of Chapters 17 and 18; for now, it's enough to notice how Hubble's law functions as a rung on the cosmic distance ladder.

"Using Hubble's law, distance comes straight from velocity," says Green. "All an astronomer needs to do is make a measurement of a galaxy's redshift (which is relatively easy). Then the redshift gets converted into velocity and the velocity then becomes a distance via Hubble's law." In this way, Hubble's discovery opened up the vast distances between galaxies to astronomical exploration. The size of the Universe, once again, was exploding to ever-larger scales.

SECTION SUMMARY
Hubble's Law
- By combining measurements of velocity and distance for each of 500 galaxies, Hubble and Humason found a velocity-distance relation that came to be known as Hubble's law: $V = HD$.
- Hubble and Humason's discovery showed that all galaxies are receding from us.

interactive
Hubble's Law Generator

16.4
Monsters in the Deep: Active Galactic Nuclei

Hubble laid out the basic categories of galaxies (elliptical, spiral, irregular) using only optical telescopes. Once other wavelength windows were opened on the Universe, beginning with radio in the 1950s, the study of galaxies was swept in entirely new directions. Radio,

Hubble's law The linear relationship between a galaxy's recession velocity and its distance.

Hubble's constant The slope of Hubble's law relating recession velocity and distance, which is expressed in kilometers per second per megaparsec (km/s/Mpc).

GOING FURTHER 16.1   THE LANGUAGE OF THE COSMOS
Hubble's Law

Hubble's discovery that all galaxies are receding from one another represents one of the most significant scientific discoveries in all of human history. His conclusion is cleanly summed up in the so-called Hubble's law, which shows how galaxy velocity (as measured with redshift) and galaxy distance go together.

Expressed with the modern value of $H = 69$ km/s/Mpc, Hubble's law takes the form

$$V = 69 \text{ km/s/Mpc} \times D$$

Thus, if two galaxies have recession velocities of 6,900 and 13,800 km/s, we calculate their distances from us as 100 and 200 Mpc, respectively.

Note that if you double the distance, you double the velocity, because Hubble's law expresses a linear relation between the two. This is why Hubble's law is so useful to astronomers: a simple measure of redshift (velocity) can be translated into a measure of distance by a straightforward manipulation of the equation:

$$D = V/(69 \text{ km/s/Mpc})$$

Therefore, if a galaxy is moving away from us at 3,000 km/s, we can calculate its distance as follows:

$$D = (3,000 \text{ km/s})/(69 \text{ km/s/Mpc}) = 43 \text{ Mpc}$$

Again, doubling the velocity would double the distance, as we would expect for a linear relationship.

But how do we get from the actual measurement of redshift to velocity? Imagine that a galaxy shows a spectral line of a particular element that appears in the lab (at rest with respect to us) at a wavelength of $\lambda_1 = 500$ nanometers (nm). In the galaxy spectrum, however, the same spectral line appears at $\lambda_g = 505$ nm. To convert from spectrum to speed, astronomers use the formula

$$V = (\lambda_g - \lambda_1)/\lambda_1 \times c$$

where c is the speed of light (note that this equation is valid only when V is much less than c). Plugging in the values above, we find

$$V = [(505 \text{ nm} - 500 \text{ nm})/500 \text{ nm}] \times (3 \times 10^5 \text{ km/s}) = 3,000 \text{ km/s}$$

That's the power of Hubble's law. All you need is a galaxy's spectrum to determine its velocity, and from there you get distance. Keep in mind, however, that at distances the size of the Universe itself, even Hubble's law is no longer accurate. That story will have to wait until Chapter 18.

infrared, X-ray, and other wavelength bands yielded a treasure trove of new insights, but the most surprising discovery was the presence of extremely energetic and violent activity in the centers of many galaxies. Studies of these **active galactic nuclei (AGNs)** eventually led to the recognition that almost all galaxies have supermassive black holes residing in the deep gravitational wells at their centers. Although only 1 percent of galaxies are AGNs, these active galaxies have proved to be a window on black holes and on the evolution of all galaxies. "What's so cool about these AGNs," says Paul Green, "is that they show how looking at objects that have extreme, even bizarre, observational properties can lead to discovering entirely new kinds of physics in the Universe." Active galactic nuclei and their black-hole engines remain at one of the most vibrant frontiers of astrophysical research.

Early Findings

The study of active galactic nuclei began even before new instruments such as radio and X-ray telescopes were available. But the development of these technologies pushed the field forward, leading to the detection of quasars and radio galaxies in the 1960s. As is typical, the process of discovery and the process of recognizing the meaning of discoveries did not move in a straight line. Scientists sometimes find that a new discovery sheds light on phenomena that may have been seen many years before.

SEYFERT GALAXIES.   The earliest recognition of what would eventually be called AGNs came in the 1940s, when optical studies by astronomer Carl Seyfert unveiled a class of galaxies, now called **Seyfert galaxies**, whose spectra show a distinct pattern of broad emission lines. "These lines show up as tall peaks in the spectrum from elements like hydrogen, oxygen, neon, sulfur, and nitrogen," says Green. "The thing that's unusual about Seyfert galaxies is that they show lines from highly ionized versions of these elements. That means there must be very intense ultraviolet or X-ray radiation stripping the elements of their electrons."

The fact that the emission lines are so broad tells us that they come from gas moving in the galaxy core at high velocities. "Whatever is going on in there must include clouds with orbital speeds of up to 10,000 km/s," says Green. In addition to the high velocities indicated by the spectral lines, Seyfert galaxies show rapid changes in brightness not seen in any other object. "Taken together, the high ionizations, high velocities, and rapid variability all tell you something remarkable is happening in the nuclei of these galaxies." Seyferts are categorized as Type I or Type II. Type I Seyferts show the broadest emission lines, indicating fast-moving gas. Type II Seyferts show

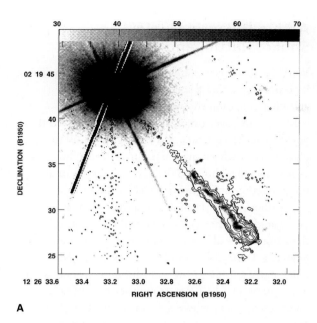

A

Laboratory spectrum (for comparison)

Hδ   Hγ   Hβ

Spectrum of 3C 273

B

FIGURE 16.18
Quasars
(A) In 1963, astronomer Maarten Schmidt determined the nature of 3C 273, which was the first quasar discovered. In this composite HST and radio image, we see light from the central quasar, as well as a relativistic jet being driven from its nucleus. (B) Its spectrum was perplexing because its emission lines were shifted so far into the red that they were almost unrecognizable, implying great distance, and its brightness exceeded expectations for an object so far away. In the spectrum shown here, the three bracketed lines at the top show the emission lines observed in the lab (no redshift). These correspond to the three lines in the redshifted spectrum of 3C 273 at the bottom.

narrower emission lines, indicating that the material is moving less rapidly.

QUASARS. In 1963, Maarten Schmidt, a professor of astrophysics at the California Institute of Technology, was working with a group of radio astronomers trying to identify optical counterparts of newly discovered radio sources. "Radio astronomy was still a relatively new science at this point," says Green. "Stars don't normally give off huge amounts of radio waves in the first place, and only one in 10,000 optical objects in the sky will give off radio emission that you could detect back then." Just as important, radio telescopes at the time had lower resolution than optical ones, so it could be quite difficult to find which object on a photographic plate corresponded to an object detected via radio telescope. Schmidt, however, had become quite good at his work, and he compiled an impressive list of galaxies that could be linked to specific radio observations.

Then along came 3C 273 (whose name stood for the 273rd object in the third Cambridge University radio survey). As well as Schmidt could determine, 3C 273 was associated with an unremarkable, bluish-looking star. When Schmidt looked at the object's spectrum, however, he was shocked. The spectral lines emitted by the object made no sense. No matter how hard he tried to fit the spectral lines to the usual library of elements, he always failed. Either he had discovered new forms of matter (an unlikely option), or something very strange was going on. Sinking deep into the problem for months, Schmidt finally came up with the answer: the spectral lines were just normal

hydrogen being redshifted to an enormous degree (**Figure 16.18**).

Racing through his analysis, Schmidt found that 3C 273 was receding from Earth at a whopping 16 percent of light speed. Using Hubble's law, Schmidt then calculated the object's distance to be 750 Mpc, making it the farthest object ever seen up to that point. With the distance in hand, Schmidt calculated 3C 273's luminosity. "That is when the real shock hit him," says Green.

At a luminosity of approximately 10^{40} watts (W), 3C 273 was pumping out more than a trillion times the Sun's energy output and was hundreds of times more luminous than the entire Milky Way Galaxy. While supernovas produce more light, they shine for only a few weeks at most. The energy output from 3C 273 stayed high no matter how long astronomers looked. Whatever was powering the distant object was something entirely new and entirely unknown to astrophysics at the time.

Other sources like 3C 273 were quickly discovered and given the name **quasar**, for *quasi-stellar radio source*. Their great distance and power made quasars an enigma for more than 20 years, but by the 1980s a scientific consensus had emerged that quasars were a type of energetic galactic nucleus. As extragalactic studies matured, however, a new and closely related phenomenon was recognized to be living in our own neck of the woods (on relative cosmic scales).

RADIO GALAXIES. Radio astronomical studies in the 1960s found, in addition to quasars, another new class of galaxies. These **radio galaxies**, as they were called, showed powerful jets from their central regions pushing into the intergalactic medium. "Radio galaxies are spectacular," says Green. "They show huge, bipolar lobes extending thousands of parsecs in both directions away from the galactic nucleus." The shape of the radio spectrum tells astronomers that the lobes

active galactic nucleus (AGN)
A highly luminous, energetic galactic nucleus powered by a supermassive black hole.

Seyfert galaxy A galaxy whose nucleus shows broad emission lines, indicating highly ionized material moving at great speed.

quasar Quasi-stellar radio source; an extremely luminous and distant active galactic nucleus.

radio galaxy A galaxy whose nucleus is the origin of relativistic bipolar jets.

Paul Green

"I started out wanting to be in a rock band," says Paul Green. "Astronomy came much later." Green can recall the exact moment of the shift. "There was this one night when I pretty much decided that astronomy was the coolest thing out there. During the winter term at Oberlin, they let you get out of town and do something different for a month," he explains. "I decided to go to the National Observatory in Arizona. I got hooked up on this project using what was then a big telescope: the 4-meter at Kitt Peak."

Green was testing different kinds of photographic plates for astronomical images. Carrying out his tests meant that the giant telescope had to be tipped over so that Green could climb into a cage mounted at the top. Then the telescope was pointed back at the sky with Green packed in the cage 20 feet above the ground. "I would be up there all night," he says. "First I'd put in a photographic plate. Then I'd take an image of the Andromeda Galaxy with the enormous telescope. Then I'd change out the plates again." It wasn't easy work, and it definitely wasn't comfortable. "The observatory is 7,000 feet up on a mountain," says Green. "You needed very warm clothes if you were going to sit there all night."

Despite the cold, the project was a revelation for Green. "It was dead quiet, except for the clanking of the machinery and the buzz of the electronics. And the sky . . . oh, man! The sky was perfect, just the black of space and the sharp light of the stars. I was in heaven."

Those nights at Kitt Peak put Green on a track that led him to an undergraduate degree in physics at Oberlin, a PhD in physics from the University of Washington, and a Hubble Fellowship, and finally landed him at the Harvard-Smithsonian Center for Astrophysics, where he works with the Chandra X-ray space telescope and studies quasars. "But I still have a band, okay?" he says. "Make sure you get that in."

FIGURE 16.19
Radio Galaxies
Radio galaxy Centaurus A is shown here in visible, X-ray (blue), and radio (orange) wavelengths. The lobes above and below the dusty disk are created by the powerful radio jets driven from the galaxy's nucleus.

must be filled with energetic electrons spiraling around strong magnetic fields (**Figure 16.19**).

With the advent of interferometric methods (many radio telescopes working in concert), the resolution of radio observations vastly improved. The jets and their lobes were mapped in exquisite detail, and astronomers were able to trace the source of the jets to central regions of their host galaxies just a few light-years across. Step by step, astronomers were zeroing in on the mysterious engines of AGNs.

SECTION SUMMARY
Early Findings
- Seyfert galaxies, first identified in the 1940s, are characterized by high ionizations, high velocities, and rapid variability.
- In 1963, using radio astronomy and spectral analysis, Maarten Schmidt discovered an enigmatic, extremely bright, extremely distant object, subsequently called a quasar.
- Radio galaxies, also discovered in the 1960s, have powerful jets that emanate from a very compact center.

Active Galactic Nuclei: Shake, Rattle, and Roll

With the discovery of an ever-larger census of galaxies, it became clear that a small fraction of the total galactic population showed remarkably strange behavior. And over time, astronomers recognized that Seyfert galaxies, quasars, and radio galaxies all displayed essentially similar behavior. By the 1980s, astronomers had lumped the driving engines of all three together under the label AGNs.

Other subclasses of AGN activity were also discovered, and often classes were distinguished by the amount of radio emission coming from the AGN. Seyfert galaxies, for example, are considered radio-quiet active galaxies. Low-ionization nuclear emission line regions, or LINERs, are another type of radio-quiet active galaxy whose spectra show gas with weak emission lines at the galactic cores. Radio galaxies, with their jet emissions, are an obvious example of radio-loud active galaxies, as are quasars. So-called blazars are also radio-loud; they show rapid variability in their brightness in radio, optical, and X-ray wavelengths but also show few emission lines.

In general, AGNs have some combination of six principal characteristics:

1. Compact size
2. High luminosity
3. Strong continuum emission from the core (radiation emitted at a wide range of wavelengths, from radio to X-ray)
4. Strong emission lines
5. Variability in both continuum and spectral-line emission
6. Strong emission at radio wavelengths

"Turning these six characteristics into a physical model for the AGN represented a major challenge for astronomers," says Green. Despite the differences among the types of AGNs (radio-loud, radio-quiet, strong emission lines, weak emission lines), astronomers were motivated to find a single model that could unite all the behaviors they observed. The model they eventually found was based on the assumption that the principal actors driving all the remarkable behavior in AGNs were the enigmatic supermassive black holes.

SECTION SUMMARY
Active Galactic Nuclei: Shake, Rattle, and Roll
- Seyfert galaxies, quasars, radio galaxies, LINERs, and blazars all have active galactic nuclei, now determined to be the result of supermassive black holes.

Linking AGNs and Black Holes

Chapter 14 introduced you to the black holes, with masses of more than 10^6–10^9 M_{Sun}, that reside at the centers of galaxies. Even before firm proof could be obtained for the existence of supermassive black holes, astronomers were convinced they were the engines of AGNs. The jets discovered in the radio galaxies were one of the early findings that led astronomers to suspect that black holes existed. "There are three properties of radio jets from radio galaxies (or radio-loud AGNs) that tell us they're coming from a very very compact, very very powerful source," explains Green. "First, the source at the base of the jet is generally very compact. Whatever is making the jet has a radius of a few astronomical units or less. Second, the power in the jet can be extreme, telling us that whatever is creating it must have lots of energy at its disposal. Finally, speeds in the jets, which can be measured by tracking individual knots or blobs as they move across the sky, tell us the jet plasma moves at speeds approaching the velocity of light. All these properties point to a black hole" (**Figure 16.20**).

The relativistic speeds of AGN jets are a particularly compelling piece of evidence (relativistic speeds mean that an object is moving faster than 10 percent of light speed, or $V = 0.1c$). They imply that whatever launches the jet must itself be relativistic. A jet formed just outside the event horizon of a black hole must have a velocity nearly the speed of light if it is to escape into space. In some cases, astronomers calculate jet speeds that appear to be even faster than the speed of light. These so-called **superluminal motions**, however, are simply optical illusions that occur when jets are oriented so that they're moving toward us. However, such superluminal motion will occur only for relativistic jets moving close to light speed.

Perhaps the most important early evidence for a connection between AGNs and supermassive black holes is the timescale for AGN brightness variations. Depending on the wavelength, X-ray observations show that AGNs can have increased light output by a factor of 2 or more on timescales of hours to days. How do these rapid changes in luminosity offer a clue to the size of AGN central engines? "The argument works like this," says Green. "Nothing can travel faster than the speed of light, right? So that means whatever physical process is responsible for increasing an AGN's brightness can't propagate across it any faster than the time it takes for light to make it from one side to the other."

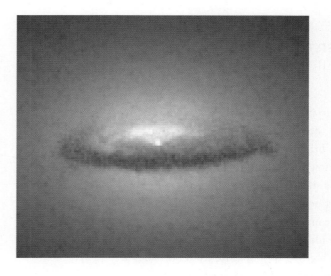

FIGURE 16.20
Supermassive Black Holes and Galaxies
This Hubble Space Telescope image shows the central region of NGC 7052, an elliptical galaxy 200 million light years away in the constellation Vulpecula. The disk is circling a 300-million-solar-mass black hole. Such supermassive black holes are the source of the behaviors associated with active galactic nuclei.

The reasoning Green outlines means that if an AGN increases its brightness in 3 days, then the size of the AGN must be 3 light-days, or just 80 billion km (see **Going Further 16.2**, p. 422). "That's a pretty small region of space," says Green, "especially when you think about all the energy released in a quasar, blazar, or the other forms of AGNs. With so much energy being liberated in such a small volume, it's hard to imagine that anything other than a black hole could be at work."

Understanding that AGNs and black holes were linked did not, however, explain the bewildering array of behavior from the different kinds of AGNs. Why were some radio-loud and others radio-quiet, for example? For astronomers, the next step was to come up with a model that used black holes and other actors to develop a single unified model for all AGNs.

SECTION SUMMARY
Linking AGNs and Black Holes
- Astronomers determined that black holes drive the strange, powerful features of AGNs because only the presence of a black hole could explain the compact size of the core, relativistic speeds of the jets, and rapid variability.

The Power of One: A Unified Model of AGNs

The deep gravitational well of a black hole is the most effective means known for converting matter into radiation. When an object falls onto a gravitating body, astronomers say it is accreted. In the process of accretion, the object's initial gravitational potential energy (when it is far away) is converted into kinetic energy of motion. Thermal energy (heat) is also generated if the infalling material experiences any kind

superluminal motion The illusion that an object's speed exceeds the speed of light, which is created by the object's motion toward the observer.

GOING FURTHER 16.2 THE LANGUAGE OF THE COSMOS
AGN Variability

The rapid timescales of AGN variability provided a critical clue for astronomers trying to pin down the nature of AGN engines. Using the time it takes for an AGN to brighten significantly and some simple geometric arguments, astronomers can calculate the central engine's size.

Imagine that the AGN engine is a spherical region, as in **Figure 16.21**. There are two ways to think about the argument. First, we could think of the time it takes for a physical process that causes brightening to propagate from one side of the engine to the other, as described in the main text. Another approach is to imagine that the whole engine suddenly brightens at the same time. Let's explore that case in more detail now.

If the entire spherical engine brightened all at once, then observers on Earth would detect the change from a point X on the sphere (closest to Earth) before they would detect it from point Y, which is farther from Earth on the sphere. The path difference between point X to Earth and point Y to Earth can be expressed using simple trigonometry. Take R to be the radius of the sphere, P_1 to be the path from Y to Earth, and P_2 to be the path from X (the near edge of the sphere) to Earth.

Because Earth is so far away that the two paths are pretty much parallel, there is a simple relation between their size:

$$P_1 = (P_2 + R)$$

so

$$R = P_1 - P_2$$

The difference in arrival time for the signal at X versus for the signal at Y is

$$\Delta T = (P_1 - P_2)/c = R/c$$

Thus,

$$R = \Delta T \times c$$

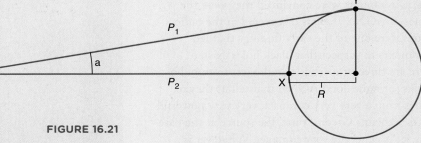

FIGURE 16.21

Since astronomers see AGN variations within a matter of hours (1 hour = 3,600 seconds), this last equation tells us that the engine must have a radius of

$$R = 3,600 \text{ s} \times (3 \times 10^5 \text{ km/s}) = 1.08 \times 10^9 \text{ km} = 7.2 \text{ AU}$$

That is a very small region. Now we can ask what the mass of the sphere would be if R was the Schwarzschild radius of a black hole. Using the formula from Chapter 14, we find

$$R = 2GM/c^2$$

Thus,

$$M = c^2 R/2G$$

Note that in using this formula, we will have to convert radius R from astronomical units into meters (1 AU = 1.5×10^{11} meters). Thus, we get

$$M = [(3 \times 10^8 \text{ m/s})^2 \times 7.2 \text{ AU} \times (1.5 \times 10^{11} \text{ m/AU})]/ [2 \times (6.67 \times 10^{-11} \text{ m}^3/\text{kg s}^2)]$$

$$= (7.3 \times 10^{38} \text{ kg}) \times [1 \, M_{Sun}/(2.0 \times 10^{30} \text{ kg})]$$

$$= 3.6 \times 10^8 \, M_{Sun}$$

This is the right size for a supermassive black hole.

of friction on its way down (as can happen when something is pulled apart by tidal forces). Heat released through accretion is turned into radiation, which makes accreting regions (such as accretion disks) emit light. A simple calculation shows that the more compact a massive object is, the more energy can be converted into light as material falls down its

gravitational well during accretion. Black holes, being the most compact objects possible, can convert the largest fraction of the energy locked up in an accreting mass m into light (via $E = mc^2$).

In itself, however, the black-hole scenario could not yet account for all the kinds of behavior seen in different types of AGNs. "There are such big differences between the different types of AGNs," says Green. "Even though everyone was sure a black hole was involved, people still needed to explain those differences with as simple and efficient a mechanism as possible." It took years of study, but eventually astronomers developed a coherent framework for understanding all AGNs. "Now we have the so-called **unified model of AGNs**. It starts with a few basic actors and then throws in the crucial element of viewing angle."

The first actor in the unified model is, of course, the supermassive black hole. The second actor is an accretion disk surrounding the black hole composed of galactic gas captured by the black hole's gravity. This relatively **thin disk** extends out to a radius of about 1 pc. At larger distances is a thick **torus**, or doughnut, of dusty molecular gas. The torus extends out to 100 pc or more. The fourth component is a halo of small clouds (sometimes called *clumps*) surrounding the black hole and accretion disk and extending out to a radius about the size of the torus. The fifth and last component is an **AGN outflow** from the combined black hole–accretion disk system. The outflowing gas includes the relativistic jets coming from the innermost regions. Outflowing material is also emerging from the central regions at slower speeds, a "mere" 1,000–10,000 km/s.

With these players in place, the differences between AGNs reduce to nothing more than the inclination of the entire system relative to our telescopes. "Much of the model is about the obscuration of the area closest to the black hole," says Green. "If the dense molecular and gas clouds block much of our view of the area closest to the black hole, then we get one kind of AGN," says Green. "If they don't, then we get another." If, for example, we're viewing the torus–disk–black hole system from the side, we can clearly see the jets, but our view (even using radio telescopes) of the innermost regions is obscured by the dense torus. These systems will be considered radio-quiet, so this viewing angle would explain the Seyfert galaxies. If we see the system more from the poleward direction, looking down the jet, then the system will be radio-loud, and we will see rapid variability because we're seeing down into the central regions. This viewing angle would produce blazars. Depending on the viewing angle, other objects will show a mix of properties, such as those seen in classic quasars (**Figure 16.22**).

"It's pretty cool how many of the phenomena associated with AGNs can be explained with the unified model," says Green. But the success of the unified model teaches us about more than just the

nature of AGNs and the central regions of galaxies. At the center of science's mission is the idea of unity in diversity—the belief that phenomena that appear to be distinct can often be attributed to a few basic principles operating together. Once those principles are understood, the bewildering buzz of experience becomes a symphony of order and unity.

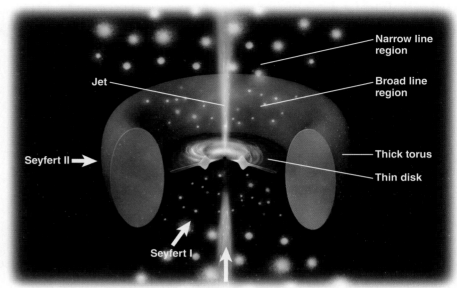

FIGURE 16.22
Unified Model of AGNs
Every AGN has the same basic components: a central black hole, a thin disk, a thick torus, small clumps of clouds, and an AGN outflow that includes relativistic jets. Over time, astronomers have come to recognize that that the different types of AGNs they observe are due to the orientation relative to us. If the orientation of an AGN leaves us looking down the barrel of the jet, we see a blazar. If we look directly from the side, the torus obscures the view of the innermost regions and we see a Seyfert galaxy.

SECTION SUMMARY
The Power of One: A Unified Model of AGNs
● The unified model of AGNs explains observed differences in the various AGN types as the result of our viewing position relative to the structures that make up the AGNs: black hole, accretion disk (thin disk and torus), and outflowing gas.

Chickens and Eggs, Black Holes and Galaxies

While only a tiny fraction of galaxies host AGNs, astronomers now believe that all galaxies host supermassive black holes. The reason not all supermassive galactic black holes create AGNs is simple: if there is not enough matter falling into the black hole, then luminosity will be relatively weak, as appears to be the case for the black hole at the center of our own Milky Way.

But there is more to the black hole–galaxy relationship than simply having a black hole in every galactic center. As astronomers became more adept at "weighing" black holes through the motion of surrounding material, they were able to build a fairly

unified model of AGNs A consolidated depiction of the morphology of the various active galactic nuclei.

thin disk An accretion disk surrounding a supermassive black hole; an element of the unified model for AGNs.

torus (pl. tori) A doughnut-shaped ring surrounding an accretion disk and. supermassive black hole; an element of the unified model for AGNs.

AGN outflow Material driven away from the central regions of an active galactic nucleus.

FIGURE 16.23

***M*-Sigma Relation**

The discovery of the *M*-sigma relation, correlating the velocity dispersion (velocity range) of bulge stars with the mass of the black hole, provided evidence that supermassive black holes may play an important role in galaxy formation and early evolution.

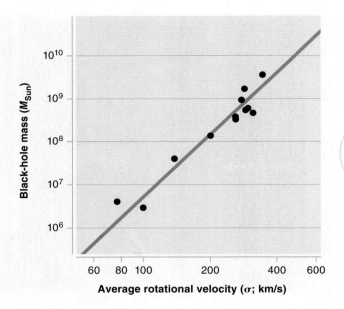

Average rotational velocity (σ; km/s)

***M*-sigma relation** The correlation between the average rotational velocity of stars in the outer region of a galaxy and the mass of the central black hole.

SECTION SUMMARY

Chickens and Eggs, Black Holes and Galaxies

● Spiral galaxies exhibit a strong correlation between the range of stellar velocities and the mass of the central black hole, pointing to an evolutionary cause that has yet to be explained.

16.5
Galaxies and Dark Matter

Chapter 15 described how studies of the Milky Way's rotation curve led astronomers to conclude that a significant component of the galaxy's mass is in the form of dark matter. Further evidence for the existence of dark matter came from explorations of other galaxies. "It's easier to study galaxies from the outside," says Paul Green. "The strongest early evidence for dark matter came from extragalactic studies—studies of galaxies other than our own."

Extragalactic Evidence for Dark Matter

In the late 1960s, Vera Rubin began using a new generation of high-resolution spectroscopes to map out the rotation curves of relatively nearby spiral galaxies. Her spectroscope was sensitive enough to get Doppler shifts for stars at different radii within the galactic disk. When Rubin and her collaborator Kent Ford plotted out the Doppler-measured rotation velocities versus the stars' distance from the galactic center, they found that every galaxy showed the same kind of rotation curve that the Milky Way has. "In each case," says Green, "the rotational velocity increased sharply near the center and then became flat all the way to the edge of the visible disk" (**Figure 16.24**). According to Newtonian physics and the distribution of the normal (luminous) matter, the rotational velocities of stars farther from the center would have been expected to drop (see Going Further 15.1), so Rubin knew she had a major discovery on her hands.

"Rubin forced people to accept that the gravitational influence of a galaxy *as a whole* on its stars is much larger than you could possibly account for by adding up the mass of all the stars and gas and dust," says Green. "For every galaxy she studied, there had to be a dark component of mass that was much larger than the luminous stuff." Astronomers now generally accept that every spiral galaxy, like the Milky Way, has an enormous spherical "halo" of dark material and

complete comparison of black-hole mass to the host galaxy's mass. In particular, spiral galaxies show a tight correlation between the mass of the bulge and the mass of the black hole. Bulges are almost always found to be approximately 700 times more massive than the central black hole.

Even more surprising is the correlation between the range of stellar velocities (called the velocity dispersion) in the bulge and the mass of the central black hole (M_{bh}). This is the **M-sigma relation** (the velocity dispersion is denoted by the Greek symbol sigma, or σ; **Figure 16.23**). The strong relationship between velocity dispersion and the black hole's mass points to an evolutionary process in which the early formation of the black hole created feedback that altered the formation of stars in the bulge. The *M*-sigma relation is important because it implies that the supermassive black holes at galactic centers are not an afterthought of the galaxy formation and evolution process but are, perhaps, important players in that drama.

One proposed evolutionary scenario that could help explain the *M*-sigma relation is the collapse of giant gas clouds into many black holes before most of the bulge mass has turned into stars. Thus, the black holes eventually merge into a single entity as the bulge is forming. As matter flows into this now supermassive black hole, the intense radiation produced drives a wind that pushes back on the gas and eventually stalls further accretion flow, limiting further growth of the black hole. "Still," Green concedes, "no one really understands the physics behind the relation. But it clearly points us to a much deeper connection between the black holes and their host galaxies."

that stars, gas, and dust inhabit only the inner regions. While dark matter halos vary in mass and size from one galaxy to another (just as the visible components of the galaxy do), other halo properties inferred from the rotation curves are remarkably similar. For example, the shape of the dark matter distribution (how it varies as a function of distance) seems to be common to all spiral galaxies.

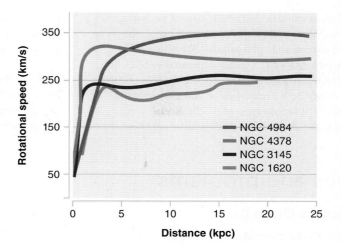

FIGURE 16.24

Galactic Rotation Curves and Dark Matter

Vera Rubin's work with galactic rotation curves demonstrated that rotational velocities in spiral galaxies do not drop off with distance from the center, as expected according to Newtonian physics. Thus, additional unseen mass (dark matter) must be present, keeping the stars rotating faster than the total mass of luminous material alone would lead us to expect.

SECTION SUMMARY
Extragalactic Evidence for Dark Matter

● The mapping of galactic rotation curves showed an unexpected result: stars farther from the galactic center don't rotate slower than stars near the center, thus indicating that a halo of unseen matter is present.

Dark Matter in Elliptical Galaxies

So far, we have discussed dark matter only in terms of spiral galaxies. Is there evidence for dark matter in elliptical galaxies too? The answer is a definitive yes. In fact, it's easier to determine the presence of dark matter in ellipticals than in spirals precisely because the orbits in the former are highly elliptical. Instead of a uniform circular motion, the orbits in an elliptical galaxy make up a kind of buzzing swarm of stars. So rather than mapping out an entire rotation curve from inside to outside, astronomers can look for the average orbital velocity V_{avg} in elliptical galaxies and compare

that with the escape velocity V_e deduced from all the luminous matter. In all cases, V_{avg} is greater than V_e, meaning that there is not enough luminous matter to hold the elliptical galaxy together. "You just measure what we call the **velocity dispersion** of the stars (basically V_{avg})," says Green, "and you can derive the mass from there." From these kinds of studies, astronomers infer that in ellipticals, as in their spiral cousins, dark matter composes more than 90 percent of the galaxy's mass.

SECTION SUMMARY
Dark Matter in Elliptical Galaxies

● The additional mass, called dark matter, is present in both spiral and elliptical galaxies.

velocity dispersion The characteristic range of velocities of bodies in an astronomical object, such as stars in an elliptical galaxy.

●→ chapter summary

16.1 The Great Debate and the Scale of the Universe

The true nature and scale of galaxies were not known until well into the 20th century. The Great Debate of 1920, on the distance to observed spiral nebulas, framed the question but left it unresolved. Soon after that, Edwin Hubble used Cepheid variables as the key to determining that our galaxy is one of many in a vast Universe of galaxies.

16.2 Galactic Zoology

Hubble classified galaxies into four main types. Spiral galaxies vary in size and exact configuration, but all exhibit a disk, bulge, and stellar halo, while many also show some form of spiral arms. Barred spirals show all of these characteristics, as well as a central bar, where star formation is accelerated. Elliptical galaxies vary widely in ellipticity and size. Irregular galaxies are amorphous, some showing disklike features. Hubble arrayed the various galaxy types into a tuning fork–shaped diagram and defined subclasses. He assumed that the arrangement represented an evolutionary track from elliptical to spiral morphology, but this assumption has since been shown to be inaccurate.

16.3 The Cosmic Distance Ladder

The cosmic distance ladder enables astronomers to determine ever-greater distances to objects. Most steps rely on comparing the observed brightness with the known luminosity of a class of objects to determine their distance, and each rung is used to calibrate the rung that follows. In order of increasing distance, the rungs are parallax, spectroscopic parallax, Cepheid variables, the Tully-Fisher law, Type Ia supernovas, and Hubble's law. Using Cepheid variables to measure brightness, combined with distance measurements, Hubble determined that the faster a galaxy is receding from us, the more distant it is. This discovery implied that the Universe was expanding, and it was later recognized to be evidence for the Big Bang theory.

16.4 Monsters in the Deep: Active Galactic Nuclei

The advent of radio astronomy yielded the discovery of galaxies with extremely energetic central engines, called active galactic nuclei (AGNs). A wide variety of these galaxies have been observed, but astronomers have determined that they all have supermassive black holes at their centers.

Evidence for the black hole is found in the compact size of the energy source and the short period of variability in brightness observed. The group includes Seyfert galaxies, quasars, and radio galaxies. According to the unified model of AGNs, each AGN has a central supermassive black hole, an accretion disk, a thick torus of dusty molecular gas, and a halo of small clouds. AGN outflows include both a relativistic jet and a slower wind. Differences in characteristics of various types of AGNs are due to the orientation of the galaxy to the observer. While all galaxies are believed to have supermassive black holes, not all have AGNs. Astronomers determine the mass of the black holes via the *M*-sigma relation, which correlates galactic rotational velocity to black-hole mass.

16.5 Galaxies and Dark Matter

More than 90 percent of the mass of galaxies is dark matter. Its presence is implied by the rotation curve of spiral galaxies, as first discovered by Vera Rubin, and by the velocity dispersion observed in elliptical galaxies.

●→ questions and problems

Narrow It Down: Multiple-Choice Questions

1. The factor that makes Cepheid variables particularly useful as standard candles is
   a. their distance.
   b. their spectral class.
   c. the inverse square relation.
   d. their position on the HR diagram.
   e. the period-luminosity relationship.

2. Which of the following structures is/are found in *all* spiral galaxies? Choose all that apply.
   a. disk   d. bar
   b. bulge   e. spiral arms
   c. jets

3. A galaxy is observed to be spheroid and to have increasing stellar density toward the center. What is its galaxy classification?
   a. spiral
   b. barred spiral
   c. lenticular
   d. elliptical
   e. irregular

4. A galaxy seen edge on is observed to have a disk and a bulge. What is its galaxy classification?
   a. spiral
   b. barred spiral
   c. either elliptical or barred spiral
   d. either spiral or elliptical
   e. either spiral or barred spiral

5. Which of the following galaxy types is most likely to be clearly identifiable, regardless of orientation?
   a. E0   d. E6
   b. Irr   e. SBc
   c. SBa

6. Which of the following is/are *not* among the subcategories of spiral galaxies? Choose all that apply.
   a. elliptical   d. barred
   b. halo   e. grand design
   c. flocculent

7. Which of the following galaxy types has no discernible structure?
   a. Irr II   d. lenticular
   b. E7   e. SBb
   c. Irr I

8. Which of the following cosmic distance ladder methods use(s) actual or apparent motion of objects to determine distance? Choose all that apply.
   a. Tully-Fisher law   d. parallax
   b. spectroscopic parallax e. Cepheid variables
   c. Type Ia supernovas

9. Which of the following cosmic distance ladder methods uses cyclical changes in star's brightness over time to determine distance?
   a. Tully-Fisher law   d. parallax
   b. spectroscopic parallax e. Cepheid variables
   c. Type Ia supernovas

10. Which method of determining distance would be most appropriate for a ground-based observer trying to find accurate distances out to approximately 100 pc?
    a. Tully-Fisher law   d. parallax
    b. spectroscopic parallax e. Cepheid variables
    c. Type Ia supernovas

11. Hubble's law says that
    a. all galaxies are expanding.
    b. the more distant the galaxy is, the faster it appears to be receding from us.
    c. larger galaxies rotate faster.
    d. the larger the galaxy is, the faster it is receding.
    e. all galaxies that will ever exist already do exist.

12. Galaxy A is receding from us at *x* km/s, while galaxy B's recession velocity is 3*x* km/s. Based on Hubble's law, which statement is true?
    a. B is 1/3 as far from us as A.
    b. A is 1/3 as far from us as B.
    c. B is 9 times as far from us as A.
    d. The relative positions of A and B depend on the types of galaxies they are.
    e. The relative positions of A and B depend on their size.

13. Which of the following is *not* true of 3C 273?
    a. It is very distant.
    b. It is very luminous.
    c. It is receding from the Milky Way.
    d. It is believed to have a black hole at its core.
    e. Its spectrum reveals no hydrogen.

14. Which of the following is/are included in the AGN designation? Choose all that apply.
    a. Seyfert galaxies
    b. radio galaxies
    c. quasars
    d. lenticular galaxies
    e. blazars

15. Which of the following describe(s) characteristics of AGNs that point to a black hole as the central engine? Choose all that apply.
    a. compact size of the core
    b. abundance of dark matter
    c. relativistic velocity of jets
    d. short timescale for variability in brightness
    e. rapid star formation

16. Superluminal motions of the jets of AGNs are
    a. jets moving faster than the speed of light.
    b. jets that appear to move faster than the speed of light because of their orientation.
    c. optical illusions due to interference from dust.
    d. light from the jets that has been accelerated to greater than light speed by the energy of the black hole.
    e. measurement error from our own galactic motion.

17. The unified model of AGNs
    a. says that the various AGNs have different physical characteristics.
    b. says that the orientation of the observer accounts for the differences among the types of AGNs.
    c. says that all galaxies have AGNs.
    d. explains all variations in observations of the various types of AGNs, except those of quasars.
    e. is based on observations of the Milky Way's AGN.

18. True/False: The defining characteristics of spiral galaxies include disk and bulge.

19. What led astronomers to infer the existence of dark matter?
    a. infrared images showing galactic halos
    b. the presence of far more satellite galaxies than expected
    c. observations of dark nebulas
    d. unexpected results in galactic rotation curves
    e. the discovery of black holes

20. Astronomers determined that elliptical galaxies have dark matter by
    a. calculating that the average stellar orbital velocity is greater than the expected escape velocity based on the luminous matter.
    b. calculating that the expected escape velocity based on the luminous matter is greater than the average stellar orbital velocity.
    c. measuring star rotation speeds near the axis of the galaxy.
    d. measuring the galaxies' rotation curves.
    e. measuring the observed brightness versus the luminosity.

To the Point: Qualitative Questions

21. What was the controversial topic debated in the Great Debate, and what position did each participant argue?

22. Speculate on the role that dust played in determining the distance to faraway objects such as the spiral nebulas.

23. What are the four main classes of galaxies? Describe the characteristics of each type.

24. Which of the four galaxy types commonly occur as dwarfs?

25. Which types of galaxies show active star formation?

26. Describe the stellar orbits in spiral versus elliptical galaxies.

27. What does the Hubble tuning-fork diagram describe?

28. What is the cosmic distance ladder, and what does it mean that each rung of the ladder has been used to calibrate the next-higher rung?

29. Describe each of the following methods for determining astronomical distances: parallax, spectroscopic parallax, Cepheid variables, Tully-Fisher law, Type Ia supernovas. How did each technique rely on the previous one for calibration?

30. Hubble's law describes a correlation between what two quantities?

31. What did the spectrum of 3C 273 indicate? Why was it a surprise?

32. Define the following terms: *AGN, Seyfert galaxy, radio galaxy, quasar*.

33. What evidence do astronomers have that AGNs are associated with supermassive black holes?

34. What is the evidence for dark matter in spiral galaxies? in elliptical galaxies?

35. How might people's outlook and understanding have changed upon learning that ours was not the only galaxy in the Universe?

Going Further: Quantitative Questions

36. A galaxy is observed moving away from ours at 2,415 km/s. What is its distance from us, in megaparsecs?

37. A galaxy is measured receding at 5,800 km/s. How far away is it, in megaparsecs?

38. A galaxy has been determined to be 162 Mpc from us. At what velocity, in kilometers per second, is it receding?

39. What is the recession velocity, in kilometers per second, of a galaxy 47 Mpc from us?

40. A spectral line whose wavelength in the lab is 500 nm appears at 540 nm in the spectrum of a galaxy. What is the distance to the galaxy, in megaparsecs?

41. The wavelength in the laboratory for a particular spectral line is 580 nm. It is observed in the spectrum of a galaxy at 660 nm. What is the recession velocity of the galaxy, in kilometers per second?

42. An AGN brightens significantly in a period of 5.5 hours. What is the maximum radius of the black hole, in kilometers? Estimate the mass of this black hole if it is this maximum size.

43. An astronomer has calculated the radius of a black hole to be 68 AU by observing its period of variability in brightness. What period, in hours, did she observe?

44. What is the mass, in kilograms, of a black hole whose Schwarzschild radius is 4.3×10^9 km?

45. Calculate the mass, in M_{Sun}, of a black hole with Schwarzschild radius 7.0×10^{11} km.

 If your instructor assigns homework in **smartwork5**, access it at the Digital Landing Page for *At Play in the Cosmos*: **digital.wwnorton.com/cosmos**

THE COSMIC WEB

The Large-Scale Structure of the Universe

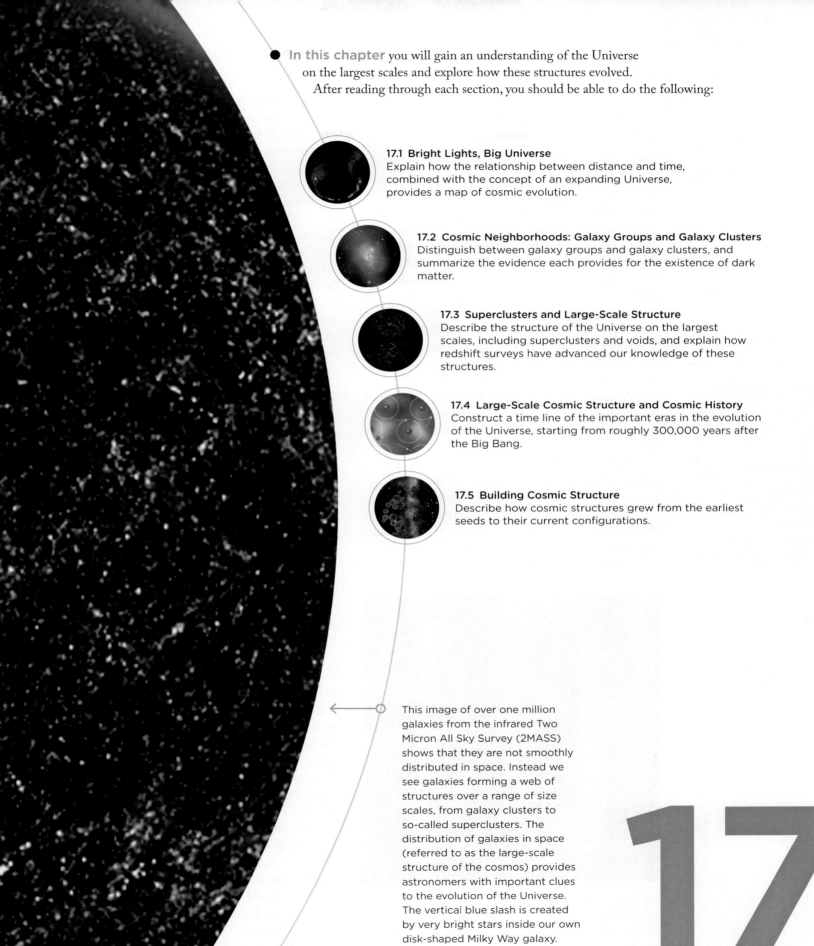

In this chapter you will gain an understanding of the Universe on the largest scales and explore how these structures evolved. After reading through each section, you should be able to do the following:

17.1 Bright Lights, Big Universe
Explain how the relationship between distance and time, combined with the concept of an expanding Universe, provides a map of cosmic evolution.

17.2 Cosmic Neighborhoods: Galaxy Groups and Galaxy Clusters
Distinguish between galaxy groups and galaxy clusters, and summarize the evidence each provides for the existence of dark matter.

17.3 Superclusters and Large-Scale Structure
Describe the structure of the Universe on the largest scales, including superclusters and voids, and explain how redshift surveys have advanced our knowledge of these structures.

17.4 Large-Scale Cosmic Structure and Cosmic History
Construct a time line of the important eras in the evolution of the Universe, starting from roughly 300,000 years after the Big Bang.

17.5 Building Cosmic Structure
Describe how cosmic structures grew from the earliest seeds to their current configurations.

This image of over one million galaxies from the infrared Two Micron All Sky Survey (2MASS) shows that they are not smoothly distributed in space. Instead we see galaxies forming a web of structures over a range of size scales, from galaxy clusters to so-called superclusters. The distribution of galaxies in space (referred to as the large-scale structure of the cosmos) provides astronomers with important clues to the evolution of the Universe. The vertical blue slash is created by very bright stars inside our own disk-shaped Milky Way galaxy.

17

17.1
Bright Lights, Big Universe

"WHEN ONE DOOR CLOSES, somewhere a window opens," or so the old saying goes. When it came to measuring the density of gas between distant galaxies, astronomer Eric Wilcots was pretty sure he had found just the portal he needed. "It's often really hard to directly measure gas density between galaxies," says Wilcots, a professor of astronomy at the University of Wisconsin. Intergalactic gas usually does not produce emission lines that can be used to make density measurements. This difficulty in directly measuring density posed a problem for Wilcots, since he was interested in seeing how the density between galaxies changes in different environments. "Some galaxies are packed fairly close together in big structures called galaxy clusters," says Wilcots, "and others come in less dense structures called galaxy groups. What we wanted was an *indirect* way of finding the background gas density between the galaxies in these different structures."

That's when Wilcots had his good idea. "Some galaxies produce jets," he explains, "and when those galaxies and their jets move through the background intergalactic medium, the jets get blown backward" (**Figure 17.1**). This effect is the same as what happens when you spray the flow from a garden hose into a strong wind. The galactic jets feel a *ram pressure* of the

large-scale structure of the Universe The hierarchy of cosmic structures on ever-larger scales from galaxy groups up to and beyond superclusters and voids.

cosmic expansion The expansion of space-time since the Big Bang.

Big Bang The event initiating cosmic expansion.

cosmology The field of astronomy that studies the Universe as a whole.

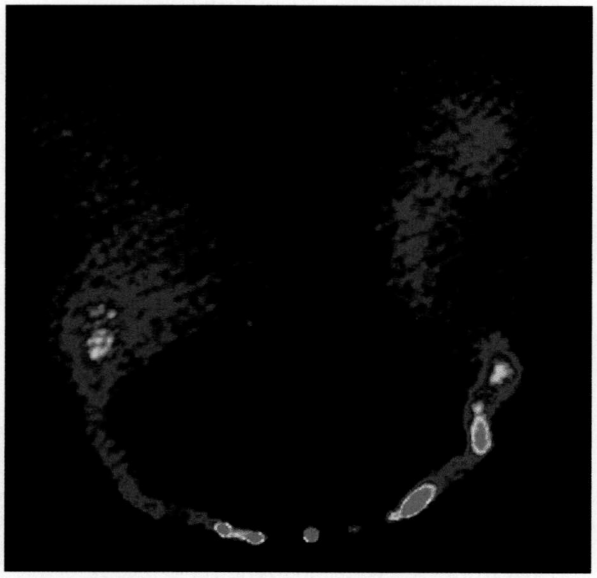

FIGURE 17.1 Galactic Jets
Galactic jets emanating from NGC 1265, the red dot at bottom center, interact with and are deflected by the intergalactic medium.

intergalactic medium as they fly through it, and that pressure bends the jets away from the direction their host galaxy is traveling.

By measuring the deflection of the galactic jets and the speed of the host galaxy, Wilcots and his team could infer the density of the intergalactic gas. "It worked out really well," says Wilcots with a broad smile. "We were able to see that the density between galaxies in the groups was lower than the density between galaxies in the clusters—exactly what we expected based on our understanding of how these large-scale structures form across cosmic history."

Wilcots's discovery was just one thread in a tapestry of studies focusing on the distribution of matter across the Universe at immense distances—the so-called **large-scale structure of the Universe**.

Large-Scale Structure and the History of Everything

When Edwin Hubble wrote his paper on the relationship between galaxy redshifts and distance (what we now call Hubble's law), he never tried to interpret the monumental discovery's meaning. His paper reported that galaxies were receding from us with a speed that depended on their distance—end of story. But he and almost every other astronomer knew what the galaxy redshifts implied: if all the galaxies are flying away from each other, then the entire Universe must be expanding.

Cosmic expansion was a radical idea that suddenly had solid observational support. At the time of Hubble's discovery, most astronomers assumed the Universe was static, stationary, and eternal—an idea that found its roots in the writings of Aristotle (**Figure 17.2**). Individual stars, planets, and people come and go, but the Universe in its entirety was immune to movement and change, or so it was thought. The discovery of Hubble's law changed those assumptions overnight. In time, the recognition that the Universe is expanding led to the idea that the Universe had a beginning. In other words, Hubble's law was the first major piece of evidence for what would come to be called the **Big Bang**.

Today, after more than 80 years of study, astronomers have compiled an overwhelming case that the Universe began in a single event that sent all space, time, matter, and energy flying apart. Astronomers call that event the Big Bang, and it is the standard model for **cosmology**, the field of astronomy that studies the Universe as a whole. Cosmology is such a rich branch of astrophysics that we will spend the next two full chapters exploring its ideas and consequences. This chapter focuses on the later stages of cosmic history and the ways that cosmic expansion affected the evolution of the large-scale structure of the Universe.

Using the world's largest telescopes and most powerful supercomputers, astronomers have mapped out the evolution of matter across space and time as it formed not only galaxies but also much larger structures called groups, clusters, and superclusters of galaxies. By studying these maps, astronomers have gained insights into the Universe's history going back billions of years. Hidden in these maps they've also found clues to the Universe's earliest moments. We will use those clues in our study of the Big Bang itself in Chapter 18, the final chapter of our cosmic exploration.

SECTION SUMMARY
Large-Scale Structure and the History of Everything

- Hubble's discovery that all galaxies appear to be receding from us was the first step in understanding that the Universe has been expanding ever since its beginning at the Big Bang.
- On the largest scales, the structures that make up the Universe—galaxies, galaxy groups, galaxy clusters, and superclusters—have also been evolving since their genesis in the Big Bang.

Cosmic Expansion: Looking Out, Looking Back

From the standpoint of the Universe's history, Hubble's law tells us that the Universe as a whole is expanding. And according to Einstein's general theory of relativity (see Chapter 14), that means the fabric of space-time itself is expanding. As Wilcots puts it, "It's not that all the galaxies in the Universe are rushing away from each other through space. Instead, it is space (or space-time itself) which is stretching. It's all the points of space that are pulling away from each other."

One way to visualize this cosmic expansion is to imagine a grid of galaxies glued to a vast rubber sheet (**Figure 17.3**). If the sheet is stretched uniformly in all directions, each galaxy moves away from all the others, even though no galaxy ever moves across the sheet. As Wilcots explains, "From the point of view of any randomly chosen galaxy, all the neighbors appear to be speeding away from it like the plague."

Just as important, the speed at which the neighbor galaxies appear to be receding depends on their distance from the observer galaxy. That means every

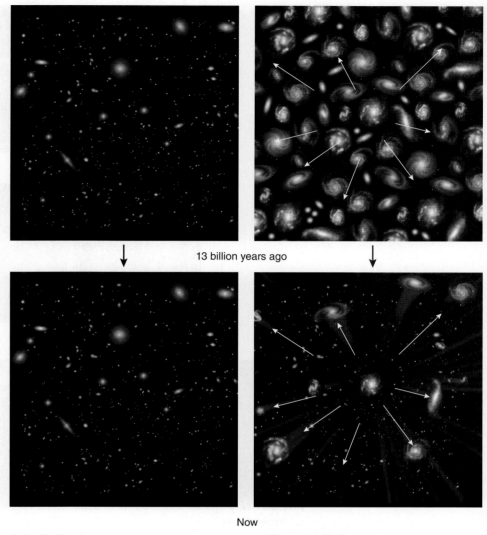

13 billion years ago

Now

A Static Universe

B Expanding Universe

observer in every galaxy sees the same Hubble's law for the motion of surrounding galaxies.

Now imagine making a movie of this cosmic expansion and then running the movie in reverse. The galaxies would all draw closer together as time wound backward. "Eventually," says Wilcots, "a point would be reached where each galaxy (or each point on our imaginary rubber sheet) would be squashed against its neighbors." Thus, we can imagine going so far back that the distance between points on the sheet will *approach zero*. Everything everywhere will be infinitely compressed together. In this way, the expansion of the Universe that we see today can, in principle, imply a cosmic beginning: the Big Bang. For now, we will focus on the amazing *long-term* consequences of the Big Bang: the origin and evolution of all the structure we see in the Universe today. To do so, we need to

FIGURE 17.2
An Expanding Universe
(A) Philosophers and astronomers alike had long assumed that the Universe was unchanging and had existed in its current state for eternity. (B) The discovery that galaxies are all moving apart disproved the old "static" view and opened up the possibility that the Universe was evolving, meaning that it was in a different state early in its history than it appears today.

FIGURE 17.3
Expansion of Space-Time
The expanding Universe can be visualized as a vast rubber sheet being stretched in all directions, such that the galaxies are swept along with it as it expands.

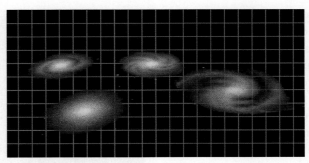

A Early universe

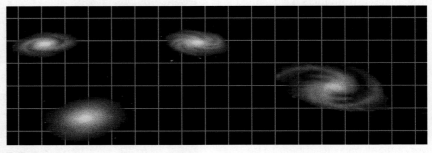

B Today

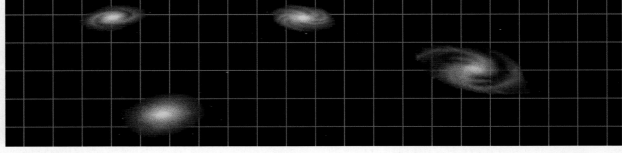

C Distant future

interactive
Look-Back Time

cosmological look-back time
A measure of time elapsed from the moment photons are detected on Earth back to the moment at which they were emitted by a distant object.

understand a few key cosmological ideas, which can be stated in terms of Hubble's law.

Recall the equation for Hubble's law:

$$V = H \times D$$

The first critical idea hiding in Hubble's law is the (approximate) age of the Universe. Hubble's constant H is just the slope of the relationship between distance and recession velocity (redshift). That means H tells us how fast space is expanding now. But we can flip that information on its head and use H to get an idea of how long ago the points were almost on top of each other.

Using Hubble's law and our best observational value for Hubble's constant, approximately 69 kilometers per second per megaparsec (km/s/Mpc), we can estimate that the Universe is about 14 billion years old (see **Going Further 17.1**). This number is roughly consistent with estimates based on other, more accurate,

measures. As we will see in Chapter 18, however, this simple estimate ignores any acceleration or deceleration that may have occurred during the evolution of the Universe. Taking those factors into account, most scientists now agree that the age of the Universe is closer to 13.7 billion years.

The second critical idea implied by Hubble's law (and the Big Bang) is that by looking out into space, we're looking back in time toward earlier epochs of cosmic history. Even if the Universe were not evolving, the fact that the speed of light is not infinite means that when we look outward, we always look back in time. So when we observe a star 1,000 light-years away, we are seeing that star as it was 1,000 years ago. The same relation is true for any galaxy.

The Big Bang and Hubble's law added something new to this picture. If the Universe has evolved, when we look into the past we're looking at something different from the present. In a static Universe, we would see a galaxy that is 7 billion light-years distant as it existed 7 billion years ago, but the Universe as a whole would still be the same then as it is now. "The cosmos itself would not have changed in any significant way," says Wilcots. If, however, the Universe began about 14 billion years ago and has been evolving ever since, seeing a galaxy 7 billion light-years away means looking a significant fraction of the way back to the beginning of everything! In an evolving Universe that began from a Big Bang, the cosmic conditions 7 billion years ago would have been different from what exists today.

Thus, because Hubble's law describes how a galaxy's recession velocity increases with its increasing distance from us, redshift (denoted in cosmology by the letter z) becomes a kind of clock. An object with redshift of $z = 0.1$ is being seen as it was when the Universe was 90 percent as old as it is today (about 1 billion years ago or 1 billion light-years away). An object with redshift of $z = 1$ is being seen as it was when the Universe was half as old as it is today (about 7 billion years ago). If you're looking at an object with a redshift of $z = 5$, you're seeing almost 90 percent of the way back to the dawn of time. In this way, we can associate every redshift with a **cosmological look-back time**—the time measured from where we are now in cosmic history back to the moment when the photons we see today were radiated (**Figure 17.4**).

For the Universe as a whole, the mathematical connection between redshift and look-back time is not simple, because light from distant objects has been moving through a space-time that has also been expanding.

─ **SECTION SUMMARY**
Cosmic Expansion: Looking Out, Looking Back
- Observers in all galaxies would see other galaxies receding from them as space-time expands.
- Astronomers trace the evolution of the Universe back to the Big Bang by imagining the expansion running backward.
- Hubble's law provides a method for estimating the age of the Universe.
- Measurement of an object's recession velocity (referred to as redshift) tells astronomers the look-back time of an object.

17.2
Cosmic Neighborhoods: Galaxy Groups and Galaxy Clusters

Climbing up the ladder of cosmic structure means that we're also climbing outward on the cosmic distance

GOING FURTHER 17.1 THE LANGUAGE OF THE COSMOS
Hubble's Constant and the Age of the Universe

The exact value of Hubble's constant is enormously important to astronomers because it controls the derivation of many quantities in the study of the Universe—most notably, the age of the Universe itself. To see how this works, we need to start by rearranging the equation a bit. If $V = H \times D$, then

$$V/D = H$$

and further,

$$D/V = 1/H = 1/(69 \text{ km/s/Mpc}) = 1 \text{ s} \times 1 \text{ Mpc}/69 \text{ km}$$

Now recall from Chapter 3 that velocity is a measure of distance over time ($V = D/T$), and therefore time is a measure of distance over velocity ($T = D/V$), so we can replace D/V in the equation above with T:

$$T = 1 \text{ s} \times 1 \text{ Mpc}/69 \text{ km}$$

Here, T represents the time it has taken any galaxy to reach its current distance from Earth, given its current velocity. In other words, T is equal to the age of the Universe, or T_U.

Next we need to cancel out the units of distance by converting megaparsecs to kilometers (1 Mpc = 3.08×10^{19} km) and then converting seconds to years (1 year = 3.16×10^7 seconds):

$$T_U = 1 \text{ s} \times (3.08 \times 10^{19} \text{ km})/69 \text{ km} \times 1/(3.16 \times 10^7 \text{ s/yr}) = 1.41 \times 10^{10} \text{ yr}$$

The current value of Hubble's constant tells us that cosmic expansion began about 14 billion years ago.

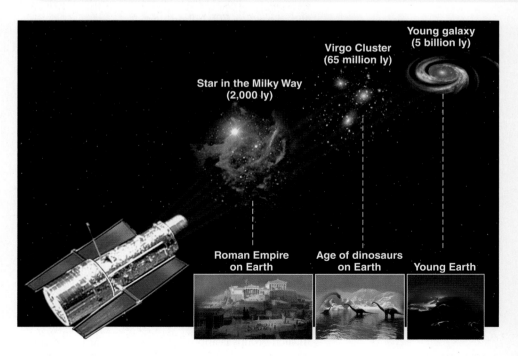

FIGURE 17.4 Look-Back Time
Looking back in space means looking back in time. When we see a distant object 1 million light-years away, we are viewing it as it was when the photons were emitted 1 million years ago. Here we see the light travel times to a few classes of objects and the events occurring on Earth when light from those sources was emitted.

A   B

FIGURE 17.5
The Local Group
Like other galaxy groups, the Local Group includes a mix of galaxy types, gravitationally bound as a whole, with dwarf galaxies clustered around the more massive spiral galaxies. The Milky Way is one of only three spiral galaxies in the Local Group. The other 50 or so are elliptical and irregular. Shown here are (A) Triangulum Galaxy (M33), the third, smaller spiral in the group; and (B) M110, a dwarf elliptical satellite to the Andromeda Galaxy.

interactive
Large-Scale Structures and Galaxy Clusters

galaxy group A gravitationally bound collection of galaxies, usually containing 50 or more members.

regular rich cluster A spherical, centrally condensed galaxy cluster with more than 1,000 galaxies.

scale. Beyond the neighborhood of the Milky Way and its small satellite galaxies, the next rungs we encounter are galaxy groups and galaxy clusters.

Galaxy Groups

Recall from Chapter 15 that the Milky Way is a member of the Local Group, a collection of gravitationally bound objects that also includes two additional large spiral galaxies and a number of dwarf galaxies (**Figure 17.5**). The smallest spiral galaxy in the Local Group is M33, the Triangulum Galaxy. The Milky Way comes next in size, and the Andromeda Galaxy, M31, is the largest galaxy in the Local Group. All told, the Local Group contains at least 50 galaxies and more than 700 billion stars.

"All of these galaxies are tied together by their mutual gravity," says Eric Wilcots. "They orbit around a common center of mass, and over time, many of them experience collisions with their neighbors." We've already seen that many of the Milky Way's satellites have interacted with each other or the Milky Way. Those interactions extend even to the largest members of the Local Group. "From their current relative motions toward each other," says Wilcots, "astronomers predict that the Milky Way and Andromeda will collide in 4 billion years." Thus, even at the largest scales, galaxy collisions are an important part of galaxy evolution.

The Local Group is not unusual. Many, if not most, galaxies can be found in gravitationally bound aggregates called **galaxy groups**. The typical group consists of fewer than 50 galaxies and has a diameter of about 2 megaparsecs (Mpc). The total mass in such groups is approximately 10^{13} solar masses, or M_{Sun} (which would equal 10 Milky Way galaxies). Like the motion of stars within galaxies, the motion of galaxies within groups indicates the presence of

unseen material whose gravity is binding the visible matter together. That means galaxy groups show evidence for dark matter. Typically, galaxy groups show ratios of dark to luminous matter of 5.5 to 1, providing more evidence that dark matter dominates the Universe's mass.

SECTION SUMMARY
Galaxy Groups
- Galaxies are typically found in gravitationally bound groups with up to 50 members, which interact with one another.
- The movement of galaxies within groups provides evidence for dark matter.

Galaxy Clusters

Groups are the smallest association of galaxies that have formed in the Universe. When astronomers make larger three-dimensional maps of galaxies and their distribution in space, however, they find larger and far richer structures. Galaxy clusters are the next step upward in cosmic structure. "Galaxy clusters contain anywhere from hundreds to thousands of galaxies in aggregates that stretch across 2–10 Mpc of space," says Wilcots.

The Virgo Cluster in the constellation Virgo, located 16 Mpc from the Milky Way, is a relatively nearby example of what astronomers call a **regular rich cluster**. Astronomers group clusters by the number of galaxies they contain and their shape. Regular clusters are spherical and centrally condensed, meaning that the density of galaxies (the number per volume) increases toward the cluster's center. Rich clusters are those containing more than 1,000 galaxies. With somewhere between 1,300 and 2,000 members, Virgo is a moderately rich cluster (**Figure 17.6A**). It's noteworthy that the Virgo Cluster has a diameter of about 5 Mpc, about the same size as the Local Group. "That means the cluster packs almost 50 times as many galaxies into the same space as is occupied by the Milky Way, Andromeda, and their 50 neighbors," says Wilcots.

The outer parts of the Virgo Cluster host a fairly even mix of the different galaxy types: spirals, ellipticals, and irregulars. "In the inner, denser parts of Virgo it's a different story," says Wilcots. Virgo's core shows far more elliptical than spiral galaxies. "Ninety percent of the galaxies in Virgo are dwarfs, meaning it's the little guys that really make up the cluster," Wilcots explains.

Another well-studied example of a galaxy cluster occurs in the constellation Coma Berenices. The Coma Cluster is a spectacular example of a regular rich cluster, situated about 99 Mpc from the Milky Way. The Coma Cluster contains at least 2,000 galaxies (the total may be as high as 10,000) in a volume of space more than 6 Mpc across (**Figure 17.6B**). At the heart of the cluster are two giant (cD) elliptical galaxies, each of which contains more than $2 \times 10^{13} M_{Sun}$ (more than 10 times the mass of the Milky Way). Beyond these two monsters, most of the bright galaxies in the central regions of the Coma Cluster are elliptical or S0-type galaxies.

Unlike the situation in the Virgo Cluster, only 15 percent of the brightest 1,000 galaxies in Coma are spirals. "The deficit of spirals in a rich cluster like Coma, compared with less dense collections of galaxies like the galaxy groups, points to an important feature of galaxy evolution," says Wilcots. "Density matters because that is what drives collisions. The higher the density of galaxies in a region, the higher the likelihood of galaxies running into each other in ways that disrupt spirals and turn them into ellipticals or irregulars."

SECTION SUMMARY
Galaxy Clusters
- After galaxy groups, galaxy clusters are the next level of structure, containing hundreds to thousands of members.
- The galaxies in denser clusters experience more collisions, resulting in a higher proportion of elliptical galaxies near cluster centers.

Eric Wilcots

Eric Wilcots was born in Philadelphia, which is not exactly dark-sky territory. When his parents gave him a telescope at age 8, he knew something had happened. "You could see the rings of Saturn!" he recalls. "I mean, how cool is that?"

The arrival of Wilcots's new telescope just happened to coincide with the *Voyager* mission passing by Jupiter. "The guys at JPL," says Wilcots, "they just looked like they were having fun. I mean there was Io and its volcanoes. You can't argue with that kind of thing." The images from the space probe, combined with his own backyard observing, set Wilcots on a course to a career in astronomy from which he never wavered.

"But the thing is, I didn't know anything about astronomy. I was taking all these classes, and they were great, but I didn't know what astronomers really did." That gap in knowledge led Wilcots to ask his astronomy professor at Princeton, "So how much do they pay you to do this?" It may have been an impolite question, but Wilcots asked it innocently. "I guess I wasn't thinking it might be rude to ask how much someone makes. I was just a kid and really wanted to know." Without missing a beat, however, Wilcots's professor answered, "Enough to go to the opera."

That was good enough for Wilcots. After earning a PhD at the University of Washington in Seattle, Wilcots found himself in the astronomy department at the University of Wisconsin, where he studies galaxies and their environments. "It's still fun," he says. "I wanted to do this when I was 8, and now here I am. Doing science, doing astronomy, thinking about galaxies . . . I still get excited thinking about getting new data and figuring out ways to answer a question no one else has asked."

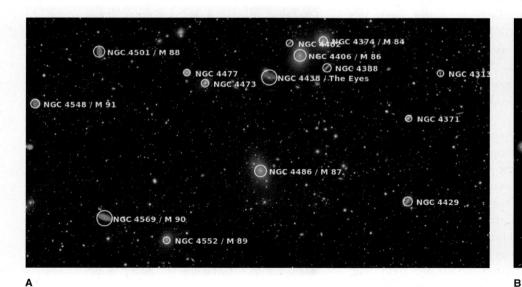

A

B

FIGURE 17.6 Galaxy Clusters
(A) Some of the Virgo Cluster's thousand-plus galaxies are shown in this image of the central region, including its dominant giant elliptical galaxy, M87. The cluster lies 16 Mpc away. (B) This false-color mosaic of the central region of the Coma Cluster combines infrared and visible-light images to reveal thousands of faint dwarf galaxies in green. Two large elliptical galaxies, NGC 4889 and NGC 4874, dominate the cluster's center.

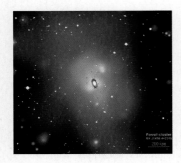

FIGURE 17.7
Cluster X-Ray Halos
This image of galaxy cluster RX J1416.5+231 (captured by the XMM-Newton satellite) shows its hot halos in X-ray light (blue). The temperatures of such halos reach tens of millions of degrees and constitute a substantial fraction of the total mass of the cluster.

X-ray halo A large, quasi-spherical region of hot gas surrounding a galaxy or galaxy cluster.

intracluster medium Thin gas in the space between galaxies in a galaxy cluster.

X-Ray Halos of Galaxy Clusters

Galaxy clusters are made of more than just galaxies. As astronomers became more proficient at lifting X-ray telescopes above the atmosphere, they used these space-based instruments to discover that every galaxy cluster has a halo of hot, X-ray-emitting gas.

These **X-ray halos** surrounding galaxy clusters are large, roughly spherical regions of gas at temperatures of 30–100 million kelvins (K). The halos are composed mainly of ionized hydrogen, meaning free protons and electrons. Detailed observations of the halos, using high-resolution space-based X-ray instruments such as the Chandra space telescope, reveal that each galaxy in a cluster has its own bright X-ray halo (just as the Milky Way does). "The space in between the galaxies is also filled with hot, X-ray-emitting gas, which forms an **intracluster medium**," explains Wilcots (**Figure 17.7**). The cluster halos contain at least 9 times as much mass as all the galaxies in the cluster.

The origin of the X-ray halos is still a subject of intense debate. Evidence has convinced most astronomers that the halos formed when the clusters themselves formed early in cosmic history. As galaxies were drawn together and collided, the individual halos created megamergers that led to a single X-ray halo for the entire cluster. Images from the Chandra telescope provide evidence of these megamergers where relatively cool 50-million-K clouds of gas can be seen falling into much larger and hotter clouds. This evidence for the buildup of X-ray halos via the merger of smaller subsystems provides further evidence for the critical role of interactions. "From galaxies upward in size," says Wilcots, "astronomers have come to see how important collisions are in shaping the structures we see today."

One of the most important characteristics of the cluster X-ray halos is their pressure. The halo is like a giant balloon. Without the gravity of the cluster, the hot, X-ray-emitting gas would expand away into space. Direct measurements of the density and temperature of the X-ray gas enable astronomers to derive its gas pressure. What they find is that the X-ray halos have far too much pressure to be held by the visible mass of the cluster. "All of the galaxies and the X-ray-emitting gas itself do not produce enough inward gravitational pull to constrain the outward push from the hot X-ray-emitting gas," says Wilcots. There has to be more mass to keep the X-ray gas in equilibrium—that

is, to keep it from expanding away. The extra mass must be in the form of dark matter. "That means cluster halos provide yet more indirect evidence for the existence of dark matter." Only by adding 10 times more mass in the form of dark matter can astronomers explain why the X-ray halos remain confined.

Thus, as in galaxy groups, both the X-ray gas and the galaxies themselves are gravitationally bound within galaxy clusters. That means the collective gravity of all the mass in the cluster is large enough to keep the galaxies and the material between them from escaping. Another way of thinking about this is that clusters are *long-lived structures*.

Further evidence that clusters are not just random, short-lived configurations comes from a general principle in astronomy, which states that the present moment we live in, the moment when we're making our observations, is not in any way special. Thus, if we see galaxies in a cluster now (which means hundreds of millions of years ago if the cluster is hundreds of millions of light-years away), it is improbable that they are just a random collection of objects passing close enough at the moment of their observation to give the appearance of being a cluster (**Figure 17.8**). Instead, the clusters must be orbiting a common center of mass created by the gravitational pull of the total mass

FIGURE 17.8 Cluster Abell 1689
The galaxies in a cluster are gravitationally bound, orbiting a common center of mass, as seen in the dense grouping of galaxies in the galaxy cluster Abell 1689.

GOING FURTHER 17.2   THE LANGUAGE OF THE COSMOS
Gravity and the Binding of Clusters

Imagine placing a bunch of marbles in a transparent box and then taking snapshots as you shake the box hard. On reviewing the images, you might find that as the marbles flew randomly around the box, at a certain moment the camera might have just happened to catch most of the marbles bunched up at the center. If that one image were all you had, how would you know whether the "cluster" of marbles was just a random occurrence or represented a more permanent structure formed by forces pulling the marbles together? This is the question astronomers must face when trying to understand galaxy clusters. To answer it, they apply a little bit of physics and a little common sense.

Using Doppler shift measurements, astronomers can find the velocity of any galaxy within a cluster. For the Virgo Cluster, typical galaxy motions clock in at V_g = 900 kilometers per second (km/s). Given this velocity, how long would it take for a galaxy to cross the diameter D of the cluster, which in this case is about 5 Mpc? The crossing time t_c is calculated as follows (being careful about the unit conversions):

$$
\begin{aligned}
t_c &= D/V_g = (5 \text{ Mpc})/(900 \text{ km/s}) \\
&= (5 \times 10^6 \text{ pc}) \times (3.1 \times 10^{13} \text{ km/pc})/(900 \text{ km/s}) \\
&= (1.55 \times 10^{20} \text{ km})/(900 \text{ km/s})/(3.15 \times 10^7 \text{ s/yr}) \\
&= 5.47 \times 10^9 \text{ yr}
\end{aligned}
$$

The idea is now to compare this crossing time t_c with the age of the Universe T_U, which is 1.37×10^{10} years. If t_c is much less than T_U, then there has been plenty of time for the galaxy to go from one end of the cluster to the other in the history of the Universe. Therefore, if

the galaxy were not bound to the cluster and you looked at a random time (which is essentially what happens when any astronomer makes an observation), you would not expect to see that galaxy bunched up with others in the cluster. Since you do see a cluster and you find out that $t_c << T_U$, it must be true that the cluster is being held together by a force. Thus, the cluster is a bound structure, not a random occurrence.

Note also that we can use the average galaxy velocity in a cluster, along with cluster size, to get a measure of the cluster mass. If we assume, for convenience, that the galaxy is on a circular orbit around the cluster center, we can begin with the circular-velocity formula:

$$
V_g = V_c = \sqrt{\frac{GM_{cluster}}{R}} = \sqrt{\frac{GM_{cluster}}{(D/2)}}
$$

and rearrange it to get

$$
M_{cluster} = (V_g)^2 \frac{D}{2G}
$$

Using the values above—V_g = 900 km/s and D = 5 Mpc—we have

$$
M_{cluster} = (900 \text{ km/s})^2 \times \{5 \text{ Mpc}/[2(6.67 \times 10^{-11} \text{ m/kg/s}^2)]\}
$$

Carrying out the unit conversion and performing the calculation, we find that $M_{cluster} = 4.7 \times 10^{14} \, M_{Sun}$, which is far more than what is observed from the luminous material in the galaxies and X-ray halos. Thus, even this simple calculation points to the existence of dark matter in clusters.

---

of the cluster, including its dark matter (see **Going Further 17.2**). Another way of understanding this is to recognize that we see similar structures at various different distances, which means different look-back times. This also convinces astronomers that there is nothing particularly special about our time. If we saw certain structures (such as clusters) only at certain times, then we could indeed say that those times must have been special.

SECTION SUMMARY
X-Ray Halos of Galaxy Clusters
- Galaxy clusters have large, massive halos of hot, ionized X-ray gas that appear to have formed from the individual halos of member galaxies as they merged.
- Halos and galaxies are gravitationally bound within a cluster, such that the total mass of a cluster indicates the presence of additional dark matter.

Galaxy Clusters and Gravitational Lenses

Unlike individual galaxies, galaxy clusters provide a means for astronomers to investigate the amount and distribution of dark matter through gravitational lenses—a remarkable effect predicted by Einstein's general theory of relativity (see Chapter 14). General relativity teaches astronomers that space-time can be curved by the presence of matter. Any object moving freely through space (not being acted on by a force such as a rocket motor) must move along paths defined by the shape (the curvature) of space-time. Photons traveling through empty space are perfect "sensors" of this curvature. Photon paths are deflected from straight lines to the degree that the geometry of space-time is distorted by mass.

"Just like a glass lens in a telescope can bend the path of light, collecting it into an image," explains Wilcots, "the distortion of space-time by matter can deflect the path of light, creating an image of background objects." **Figure 17.9A** shows how this process works. The dark matter of a galaxy cluster creates a large-scale deformation of space-time. As light from a more distant object (such as a quasar or young galaxy) passes near the cluster, it is deflected from its original path. Like a lens, the curved space gathers the light into an image. When the cluster's dark matter is arranged in a highly symmetric distribution, **Einstein rings** will be formed, in which the image of a background object appears in a thin, circular region around the cluster (**Figure 17.9B**). But since a cluster's dark matter tends to be asymmetric, the lens can rarely gather the light from the background object into a single image ring. Instead, multiple images of the background object usually form and can be seen as arcs.

"Since the properties of the image depend on the shape of space-time in and around the cluster, and since space-time is distorted by mass," explains Wilcots, "gravitational lenses give astronomers a powerful tool for exploring dark matter in clusters." Using the shape of the images, as well as their brightness, astronomers can not only determine the amount of dark matter but also probe its distribution. In this way, gravitational lenses have yielded some of the most compelling evidence for the existence of dark matter. As with other methods (galaxy rotation curves, X-ray halos, and so forth), studies of lenses show that the ratio of dark to luminous material is about 10:1. "Thus, for every kilogram of stuff like the kind you are made of," says Wilcots, "there are 10 kilograms of the mysterious dark stuff." (Note that the ratio of dark to

normal matter is not the same for different kinds of structures: Galaxy groups show ratios of about 5.5:1, while clusters show ratios of 10:1). The reason for these differences is still a point of active research, but astronomers hypothesize that the differences are due to evolutionary processes.

> **SECTION SUMMARY**
> **Galaxy Clusters and Gravitational Lenses**
> - Gravitational lensing occurs when a large concentration of mass bends space-time, thus magnifying and distorting the light from background objects.
> - Studies of gravitational lensing find further evidence for dark matter in galaxy clusters.
> - Einstein rings occur when dark matter is distributed symmetrically.

17.3
Superclusters and Large-Scale Structure

Before we can get to the largest and grandest perspective on cosmic structure, we must understand the difficulty involved in gaining that perspective. "Galaxy groups and galaxy clusters can both be found in relatively nearby regions of space," explains Eric Wilcots. "For example, for all its majesty, the Coma Cluster is 'only' about 100 Mpc away. While that might seem like an enormous distance, it constitutes only 1 percent of the distance to the edge of the visible Universe."

To get a sense of structure on the largest scales, astronomers need to build maps of the distribution of galaxies in space that reach out to billions of parsecs. Creating accurate maps means establishing the three-dimensional positions of hundreds of thousands or millions of galaxies, and that means determining distances for hundreds of thousands of objects. To make these maps, astronomers must once again turn to Hubble's law and measurements of redshift.

Redshift Surveys

Astronomers focus on redshift when studying cosmic structure simply because it is, in principle, easy to measure. Just take a spectrum of a galaxy, locate some familiar spectral lines, and determine how much they have been shifted to longer wavelengths. Using that measured redshift, Hubble's law then enables a quick

Einstein ring The ring-shaped appearance of a distant background object that is caused by a gravitational lens formed via symmetric distribution of dark matter in the foreground.

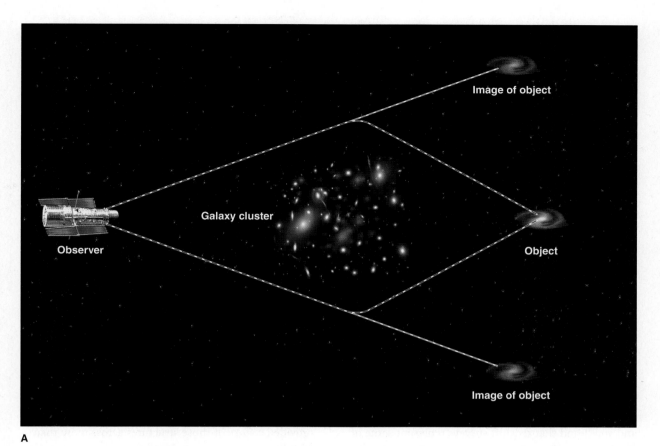

A

B

FIGURE 17.9
Gravitational Lensing
(A) The general theory of relativity predicts that light's path through space-time is bent by the presence of mass. In this way, a massive galaxy cluster (including its dark matter) can act like a lens, bending the light of a background quasar, causing it to appear as a ring or multiple arcs. (B) In this image, a foreground galaxy cluster, Abell 2218, has "lensed" the light of a distant (13.4 billion light-years away), extremely young galaxy in the background into many arcs, shown by red arrows.

FIGURE 17.10
Sloan Digital Sky Survey
The SDSS used this dedicated 2.5-meter telescope at Apache Point Observatory, New Mexico, to survey a large portion of the sky.

determination of distance. By measuring hundreds, thousands, or even millions of galaxy redshifts, astronomers can take the galaxies' positions on the two-dimensional bowl of the night sky and create a three-dimensional map of matter in the Universe over a tremendous scale. In this way, redshifts become the key to revealing the Universe's large-scale structure. That is why, as far back as the 1980s, astronomers began the difficult task of compiling **redshift surveys**.

As the name implies, a redshift survey is a compilation of spectra from many, many galaxies. "It's a time-consuming task because, as we know from the inverse square law for brightness, even very bright objects like galaxies will appear faint if they are far away," says Wilcots. "Catching enough light to create a spectrum that is high enough quality for accurate redshift determination requires long exposures." And long exposures for hundreds of thousands of galaxies require lots and lots of telescope time. Because the whole point of the project is to map out the three-dimensional locations of galaxies across space, astronomers also need to make their observations over as wide a patch of sky as possible. That means even more telescope time must be obtained. Put it all together, and you'll find that redshift surveys represent some of the most complicated and time-consuming undertakings for the astronomical community.

One of the most famous redshift surveys was the Sloan Digital Sky Survey (SDSS), which ran from 2000 to 2005 in its initial phase. The goal of the survey was to map out the three-dimensional structure over a large portion of the sky. To accomplish this, a special-purpose 2.5-meter telescope was built in the mountains of New Mexico (**Figure 17.10**). Then, to facilitate rapid data acquisition, special filters were constructed for each region of the sky to be observed by the SDSS. Holes were drilled into a metal plate at the exact locations where galaxy images would appear. Fiber-optic lines were then attached to the holes so that the light from the galaxies could be quickly and efficiently collected and analyzed. The SDSS imaged more than 350 million celestial objects during its operation. Additional spectra were then taken from 930,000 galaxies, 120,000 quasars, and 460,000 stars included in the full sample. Other surveys, such as the Australian 2DF project, collected data from almost 350,000 objects, none of which were covered in the SDSS survey.

These redshift surveys tell us a great deal about the future of astronomy, as well as our ever-expanding narrative of cosmic history. The first important point

about redshift surveys is that they are large-scale projects requiring the collaboration of many scientists over the course of many years. Although people often think of astronomers as lone wolves working in solitude at their telescopes, the redshift surveys show that modern astronomical research can involve small armies of researchers and technicians—a trend sometimes called *big science*. "A typical scientific paper by the SDSS team can have 30 authors," says Wilcots.

The redshift surveys also demonstrate the importance of large-scale computation in modern astronomy. Redshift surveys accumulate so much information that special software for data storage and analysis had to be developed. The largest redshift surveys represent the incursion of *big data* into astronomy. All in all, the redshift surveys, by demanding very large collaborative efforts and very large sets of data, are creating new ways of doing astronomical science.

What emerged from the redshift surveys were, however, not just new models for doing science but fundamentally new results. The three-dimensional maps of the Universe that SDSS and other surveys produced showed just how far away and how large cosmic structures could be. In the process, they also provided a story of those structures across more than 13 billion years of cosmic history.

SECTION SUMMARY
Redshift Surveys

● The SDSS and other redshift surveys provide three-dimensional maps of the Universe on the largest scales.

● Such surveys must cover large areas of the sky and capture the dim light of hundreds of thousands of galaxies or more, resulting in massive data collection that requires collaboration among many scientists.

● Redshift surveys provide clear evidence of the evolution of cosmic structure.

Galaxy Superclusters

The first attempts to produce large-scale redshift surveys were completed in the 1980s. Although they were crude by today's standards, their results stunned astronomers of the time.

"It has always been assumed that on large enough scales, the distribution of matter in the Universe would become nice and uniform, or what we astronomers think of as 'smooth,'" says Wilcots. "Astronomers had expected that hundreds of millions of parsecs would be large enough to see such uniformity." The first redshift survey, however, produced by Harvard astronomer Margaret Geller and her collaborators (**Figure 17.11A**), showed astronomers just how wrong they could be. With just 1,100 galaxies observed in a thin strip of sky, Geller's team was able to probe the distribution of mass out to more than 200 Mpc. Although the map was sparse, it showed two important features: First, the galaxies were not distributed uniformly, but were arranged into long filaments. Second, the filaments were separated by vast voids where no galaxies appeared (**Figure 17.11B**). Even on these enormous scales, matter still showed structure: galaxy superclusters had been discovered.

A galaxy supercluster is a chain of clusters and groups bound together by their mutual gravitational attraction. The Milky Way lies near the edge of the Virgo or Local Supercluster, an oblong structure containing at least 100 groups and clusters (**Figure 17.12**). The Local Supercluster is typical of most superclusters in mass and luminosity, at $10^{15}\ M_{Sun}$ and $10^{12}\ L_{Sun}$. The Virgo Cluster makes up the bulk of its mass. One way the Local Supercluster differs from other clusters, however, in how its galaxies are distributed from the center to the outer regions. The Local Supercluster is considered to be "poor" in the sense that it does not have a central condensation of galaxies. Note that the definitions of *poor* and *rich* are the same for superclusters and clusters, where "rich" structures have a high density of galaxies at their centers.

The Perseus-Pisces Supercluster is an example of a rich supercluster. More than 76.7 Mpc away, it appears as a particularly dense filamentary concentration of galaxies 45 Mpc in length. It is dominated by a centrally condensed region containing the Perseus Cluster, which is one of the most massive galaxy clusters known.

While Perseus-Pisces shows up as a notable example of a supercluster on redshift survey maps, it borders a similarly notable example of a **void**. The Taurus Void is a circular region 33 Mpc in diameter that is bounded by walls of galaxies on all sides (the Perseus-Pisces supercluster forms one of those walls). Across the vast expanse of the Taurus Void, astronomers can find only a handful of galaxies. It appears almost completely empty of structure. "The voids are the complements of the superclusters," says Wilcots. "Superclusters appear as elongated regions where the density of matter is higher than the cosmic average, and voids appear as regions where the density is below that average."

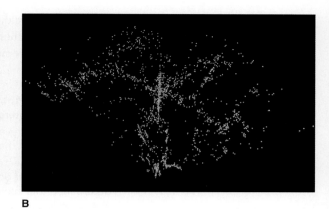

A **B**

FIGURE 17.11 Margaret Geller and "Stickman"
(A) Margaret Geller and her team conducted early redshift surveys and stunned the astronomy community by showing that, contrary to earlier assumptions, the distribution of matter in the Universe did not smooth out on the large scales they studied, but showed filaments and vast voids. (B) Their map was dubbed "Stickman" for the distinctive shape of large-scale structure it revealed.

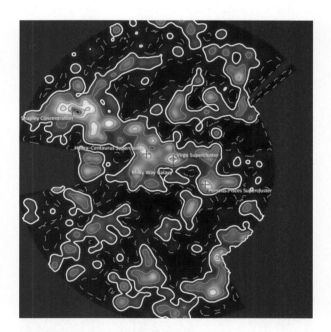

FIGURE 17.12 Galaxy Superclusters
This three-dimensional density map of galaxy superclusters, developed by a team of scientists from Canada and France, spans about 2 billion light-years (600 Mpc). Regions with higher densities of galaxies appear lighter on the map, and red areas represent the densest regions.

After completing their redshift surveys, astronomers were faced with the reality that structure still exists on hundred-million-parsec to billion-parsec scales. More important, such structure took the remarkable form of alternating empty voids and denser supercluster filaments. That discovery rocked

void A region between superclusters, spanning tens to hundreds of megaparsecs, where galaxy density is much lower than average.

the astronomical world, but the next crucial step in understanding why these structures exist at such large scales could not be completed without adding motion to the story.

─┐ **SECTION SUMMARY**
│ **Galaxy Superclusters**
● Redshift surveys revealed enormous superclusters separated by vast, nearly empty voids.
● Astronomers expected that cosmic structure would smooth out on very large scales, but they determined that such smoothing is not seen on the size scale of superclusters.

Rivers of Galaxies

Redshift surveys did more than tell astronomers where galaxies were located in space; they also showed astronomers how those galaxies were moving. Since redshifts represent velocities (which are then converted to distance via Hubble's law), astronomers could see that in some cases the galactic motions included components that differed from simple expansion. "From Newton's laws of gravity, we know that where there is motion, there must be mass causing that motion," says Wilcots.

Astronomers call the basic pattern of galactic expansion the **Hubble flow**. Remember that the motion it represents is not that of the galaxies through space, but the expansion of space itself. Galaxies

Hubble flow The pattern of large-scale galaxy motion associated with the expansion of space-time.

peculiar velocity The motion of a galaxy through space that is due to gravitational attraction, distinct from motion due to the expansion of space-time.

Great Attractor An unseen concentration of mass whose gravity affects the motion of galaxies in a large volume of space that includes the Milky Way.

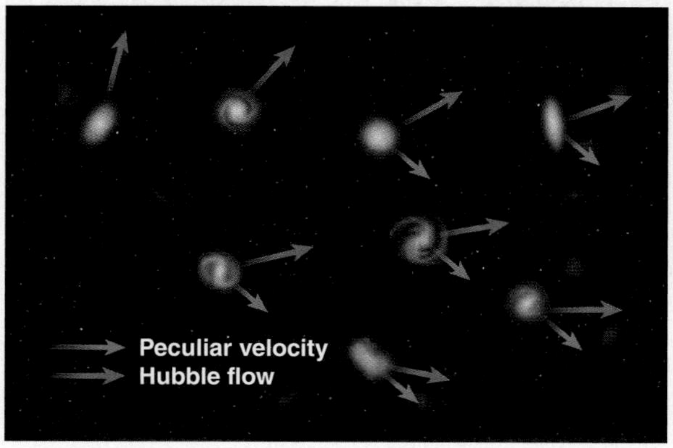

are simply being carried along with space-time as it expands in the aftermath of the Big Bang. Redshift surveys can, however, detect deviations from the pure expansion pattern of the Hubble flow, called **peculiar velocities** (**Figure 17.13**). "When astronomers see peculiar velocities, it tells them that galaxies are not just flowing *with* space because of the Big Bang but also flowing *through* space," explains Wilcots. "That means we are seeing the galaxies streaming toward each other because of gravitational attraction."

One of the most dramatic examples of a peculiar flow is caused by the so-called **Great Attractor** (**Figure 17.14**). Deviations from Hubble expansion of as much as 700 km/s are detected in galaxies in a region of sky centered on the constellation Centaurus. "What astronomers are, in essence, seeing in this region of the Universe is a vast river of galaxies all flowing in the same direction," says Wilcots. "Unfortunately, the plane of the Milky Way blocks our direct view

FIGURE 17.14 The Great Attractor
Obscured by our own galaxy, the Great Attractor cannot be directly observed. This image covers part of the Norma Cluster (Abell 3627), as well as a dense area of the Milky Way. The Norma Cluster is the closest massive galaxy cluster to the Milky Way. The enormous mass concentrated in the region of space that the Norma Cluster occupies produces so much gravitational attraction that it dominates the motions of our local region of the Universe, leading astronomers to call the region the Great Attractor. Observing the Great Attractor is difficult at optical wavelengths because the plane of the Milky Way outshines (with stars) and obscures (with dust) many of the objects behind it.

FIGURE 17.13 Hubble Flow versus Peculiar Velocity
Galaxy motions based on the Hubble flow (shown here in red) occur because of the overall expansion of space-time associated with the expanding Universe. Peculiar motions differ from the Hubble flow because they are responses to the gravitational pull of large nearby masses (on cosmic scales). In this diagram, the peculiar motions of most galaxies are being influenced by an unseen large mass in the direction indicated by the blue arrows (lower right).

of the massive object that causes the river to flow." Despite this obstructed view, astronomers named the unseen collection of mass the Great Attractor, since it appeared to be pulling everything across a region of space more than 77 Mpc in diameter toward it. Using X-ray observations, astronomers were eventually able to map out the Great Attractor's location and dimensions, finding its core associated with the Norma Cluster, an unusually massive galaxy cluster.

Although astronomers originally associated the river of galaxies with the Great Attractor, it soon became apparent that an even larger concentration of mass at even greater distances was also playing a role. "Beyond the Great Attractor lies what seems to be the most massive supercluster known, the Shapley Supercluster, at 500 Mpc," says Wilcots. The Shapley Supercluster packs the equivalent of 20 Virgo clusters into the same volume as the Virgo Cluster. "That is a lot of stuff in one place," Wilcots adds.

The motion of the Milky Way (and surrounding material) toward the Shapley Supercluster is not the only example of large-scale peculiar motions known. Other such "rivers of galaxies" flow through space, and each one represents the ongoing assembly of superclusters. In all cases, the mass needed to account for the motions of the material we can see (the galaxies) is far too large to be accounted for by other collections of observable luminous matter. Thus, peculiar motions represent yet another strand of evidence that dark matter dominates the dynamics of the cosmos. In fact, some astronomers have come to call peculiar motions "dark flows" since, for the most part, they are assumed to be made up of dark matter and are driven by dark matter.

SECTION SUMMARY
Rivers of Galaxies

- Redshift surveys reveal the movements of galaxies, resulting both from the Hubble flow and from peculiar motions caused by the presence of large-scale concentrations of material (most of which is dark matter).
- Massive objects such as the Great Attractor and the Shapley Supercluster create "rivers of galaxies" that flow through space.

The Cosmic Web

As we've seen so far, exploring the Universe on ever-larger scales has been a history of astronomers *discovering structure* on ever-larger scales. Astronomers found galaxy groups first (on the scale of megaparsecs), then galaxy clusters (tens of megaparsecs), and then

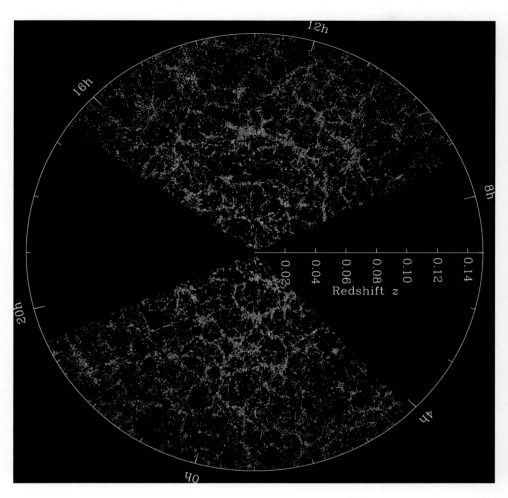

galaxy superclusters (hundreds of megaparsecs). "There is an obvious question," says Avi Loeb, a theoretical astrophysicist at Harvard. "Does it ever stop?" What Loeb means is this: if we pull back far enough, does the distribution of matter in the Universe ever begin to smooth out?

The answer is yes. As astronomers mapped the large-scale Universe in greater and greater detail, it became clear that no new distinct structures appear beyond a few hundreds of megaparsecs. "For astronomers, *smooth* has a very particular meaning," says Loeb. "What we are looking for is the length scale at which one patch of the Universe looks the same, on average, as any other patch." **Figure 17.15** shows one of the deepest and most complete maps of the three-dimensional Universe provided by the SDSS. What this stunning image reveals is that, on billion-parsec scales, the pattern of superclusters and voids begins to repeat itself. The individual superclusters now appear as a network of interconnected filaments with the densest regions (the dense clusters) showing up as the

FIGURE 17.15
Cosmic Web
The interconnected filaments of superclusters and the voids between them have the appearance of foam on a cosmic scale. This SDSS three-dimensional map shows Earth at the center, and each point outward represents a galaxy. The outer circle is at a distance of 2 billion light-years. The region between the wedges was not mapped by the SDSS, because dust in our own galaxy obscures the view of the distant Universe in these directions.

Avi Loeb

It wasn't astronomy or even physics that got Avi Loeb started on his career studying the earliest cosmic structure. It was philosophy.

"Originally, I was mostly interested in philosophy because philosophy addresses the most fundamental questions we have in life," says Loeb. But philosophy has its own problems. "It doesn't provide unique answers," he explains, "so it's difficult to feel like progress is being made." In addition, after high school Loeb was eager to continue his studies during the obligatory 2 years of military service in his native Israel. "They wouldn't let people pursuing the humanities forgo the service," he explains. "But by being admitted to the special Talpiot program, which allowed me to pursue a PhD in physics while working in defense-related research, I could engage in intellectual work. I preferred to do that rather than run around in the field with a heavy pack on."

As a student, Loeb had the chance to travel to Princeton to meet John Bahcall, one of the most respected astronomers of the 20th century. "Bahcall said he would offer me a postdoc position after I graduated," Loeb recalls, "but only under one condition. I had to convert from the physics I was doing at the time to astrophysics."

That was a fateful step for Loeb. As his research became more enmeshed with questions about the early Universe, he found he was returning to his first passion, philosophy. "After some months I found that I had really closed the circle," says Loeb. "In our work we address questions touching the realm of philosophy. For example, we consider how the first light was produced, which appears in the very first chapter of the Bible. For me, it's a fulfillment of my childhood dream. I can actually address the big questions I started with, but this time I'm using scientific tools and I feel like I'm making progress."

cosmic web Structure on the largest cosmic scales appearing as a network of interconnected filaments, surrounding voids, joined at nodes where dense clusters are found.

homogeneous Having the matter distribution property of being uniform throughout space. A homogeneous distribution has the same average properties at every point in space.

isotropic Having the matter distribution property of being uniform in terms of its appearance when viewed from different angles. An isotropic distribution looks the same no matter which direction a viewer chooses to observe.

nodes of the network where the filaments join. Taken together, these filaments become walls surrounding the voids, and the entire pattern takes on the appearance of a dense network of structure or, as some astronomers call it, the **cosmic web**.

The idea that the hierarchy of structure ends with the cosmic web has important consequences for the study of cosmology. It means that, at the largest scales, the Universe is **homogeneous**, meaning that the distribution of matter is the same, on average, in every part of space. For the cosmic web, that means the overall pattern of voids plus filaments remains constant across different regions of space in a statistical sense. Even though one individual void might be larger than another, the *average* properties of the web—things like density or void size—are the same from one place to another. "Of course, this means you must consider large enough volumes of space for such averages to make sense," says Loeb. "Many voids and

filaments must be included when average properties get calculated."

The cosmic web also shows us that, on the largest scales, the Universe looks the same in whichever direction you point your telescope: the same average pattern of voids and filaments always occurs. The Universe is therefore **isotropic**.

The fact that, on these largest scales, the appearance of the Universe becomes uniform, meaning it is both isotropic and homogeneous, makes the development of grand, all-inclusive cosmological theories much easier. "If every region of the Universe looked different at even the largest scales, it would be impossible, or at least very difficult, to embrace the whole Universe as a simple system that is statistically predictable," says Loeb.

It is a remarkable fact of astronomical history that when Einstein and others began developing the first cosmological models 100 years ago, they assumed isotropy and homogeneity because they had to. Any other choice made the mathematics too difficult to solve. It was these early models from which the theory of the Big Bang would eventually emerge. A century later, astronomers have discovered that the Universe has exactly the right large-scale properties to make models like those of Einstein and his followers possible. The cosmologists, it seems, got lucky.

SECTION SUMMARY
The Cosmic Web

- On the largest scales, the Universe's structure takes the form of a cosmic web, built from filaments of superclusters surrounding voids.
- The cosmic web is both homogeneous and isotropic.

17.4
Large-Scale Cosmic Structure and Cosmic History

As discussed earlier in this chapter, when astronomers look outward to distances vast enough to see large-scale cosmic structure, they are also looking backward into the depths of cosmic evolution. Thus, the vast superclusters of galaxies and the cosmic-scale flows associated with them give us one thread in the narrative of the Universe's history. Since the dawn of time, the Universe has been building ever-larger structures (galaxy superclusters) out of smaller ones (galaxy clusters, galaxy groups, and galaxies themselves).

If we want to piece together the full story of the Universe from the Big Bang to the current epoch (and beyond to the Universe's future), we need to understand which mechanisms can take the Universe from small to large structures across cosmic time. But to answer that question we must, once again, rely on observations that look as far back in time as possible. In addition to mapping the entire sky, we need to map backward along *individual lines of sight*, constructing a time line for the most basic building blocks of cosmic structure. In this way, we come to one of astronomy's holy grails—the quest to understand when the first stars and galaxies were assembled.

The Era of Quasars

One of the earliest clues to the history of galaxies came from the study of quasars. Recall that quasars are a highly luminous form of active galactic nuclei (AGN). The prodigious light pouring out of a quasar is directly connected with material pouring into a supermassive black hole at the center of a galaxy. Quasars are so far away that, in general, only the superabundant light from regions around the black hole is visible (**Figure 17.16A**). Light from the rest of the galaxy gets overwhelmed by the emission from the nucleus. But the enormous distances to quasars mean that we're seeing these galaxies in the earliest epochs of cosmic history.

Since the discovery of quasars in the 1960s, astronomers have pushed to find ever-more-distant examples. "The first quasars observed had $z = 0.3$," says Avi Loeb. "That puts them around 25 percent of the way back to the Big Bang." Currently, the farthest quasars observed have a redshift of $z = 5.7$ or more, placing them 95 percent of the way to the Big Bang. Merely seeing that far back into cosmic history is a remarkable achievement, but as they pushed back the frontiers of quasar observation, astronomers also made an unexpected discovery: a cosmic era of quasars had come and gone.

Figure 17.16B shows the density of quasars in the Universe as a function of time (meaning how many quasars existed per unit of volume of space as the Universe evolved). The figure tells a simple story of quasar history. The number of quasars begins low, rises to a peak, and then drops off sharply. This graph tells us that, by our epoch of cosmic history, no quasars were

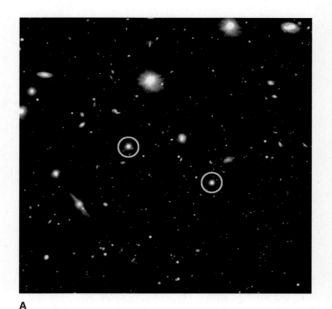

A

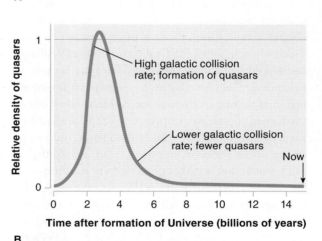

B

FIGURE 17.16
The Age of Quasars
(A) Shown here are a pair of quasars (denoted QP0110-0219) at redshift $z \approx 1$ (which implies a look-back time of 7.7 billion years ago). (B) This graph of the number of quasars versus the age of the Universe shows that there are no quasars in the current epoch of cosmic history. Quasars formed early in cosmic history because their birth required the accretion of huge amounts of material flowing into the center of a newly forming galaxy. Thus, the number of quasars grew in the early Universe, peaked, and then declined.

left, and therefore there are no quasars in the nearby Universe. Thus, the epoch of quasars ended long ago.

But why was there a quasar era in the first place? The answer to that question gives us two key pieces of information for understanding the history of galaxies in their proper cosmic context. "On the early side of the quasar era, it must have taken some time for the black holes (and the galaxies that host them) to form," says Loeb. "The end of the quasar era tells us that the massive infall of material into the supermassive black holes at these galaxies' centers could not last forever. There simply was not enough dense material in the center of the young galaxies to keep feeding the monsters at such a high rate." Thus, quasar history and galaxy history are closely tied together, and the fact that

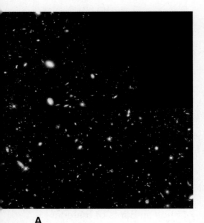

A

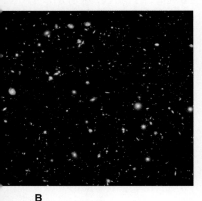

B

FIGURE 17.17
Hubble Deep Field Images
(A) The Hubble Deep Field (HDF) and (B) the Ultra Deep Field (HUDF) projects carried out long-time exposures on a small patch of dark sky to image very faint, very distant galaxies. In this way, the deep-field images provided a direct observation to the epoch when galaxies were first being assembled by gravity, such that we see galaxies in various stages of their early evolution.

Hubble Deep Field Data resulting from a week-long exposure in 1995 by the Hubble Space Telescope of a tiny, dark patch of sky, revealing thousands of galaxies at great distances. Since the original, other deep-exposure projects have been carried out.

Dark Ages The early era of cosmic evolution, when neutral hydrogen filled space and absorbed all ultraviolet and optical photons.

quasars came and went gives us an important glimpse into the evolution of galaxies over cosmic time.

SECTION SUMMARY
The Era of Quasars
- The era of the quasars, which are well-fed supermassive black holes at the center of galaxies, came early in cosmic history.
- The era of quasars ended when the abundance of available material to feed supermassive black holes was used up.

Hubble Deep Field

To study the evolution of galaxies across cosmic history, astronomers need to see the galaxies directly. But gaining such a direct view means taking the best instruments, like the Hubble Space Telescope, to extremes.

Gaining access to the HST isn't easy, and there is always fierce competition for even a few hours of observing time. Given the level of demand, it was a shock when, in 1996, HST's director, Robert Williams, announced that a whole week of telescope time was being dedicated to staring at one postage stamp–sized region of the sky. Even more incredible, the part of the sky he had chosen was empty! At least, there were no known objects located at those coordinates. But by just staring at one region of empty space for a week, the HST would slowly gather photons from the faintest objects in that patch of sky. This was the origin of the **Hubble Deep Field**. All the objects discovered by the telescope that week were indeed very faint, which meant they were very, very far away. The Hubble Deep Field had captured the first portrait of galaxies still in the agonies of birth (**Figure 17.17**).

The 1996 Deep Field was soon followed by other, similar projects, including the Hubble Ultra Deep Field, from images captured in 2003 and 2004, and the Hubble Extreme Deep Field, which was created in 2012 from overlaying 10 years' worth of images. Working in concert with the Spitzer (infrared) and Chandra (X-ray) space telescopes, the HST also produced the Great Observatories Origins Deep Survey (GOODS). Taken together, these projects enabled astronomers to see galaxies back to 13.2 billion years ago. That is a redshift of 11.9 and constitutes some of the earliest cosmic structures ever resolved with a telescope.

What astronomers find in their deep fields are galaxies that look like they're still being put together. "There are many irregular-looking objects with hints of spiral structure," says Loeb. "Also, any of the recognizable spiral galaxies appear remarkably blue,

and that means they are being overwhelmed by the formation of bright, massive stars." Close cousins of such starbursts have been seen in the local Universe near the Milky Way, but the deep fields make it clear that in the early stages of assembly, entire galaxies can be gripped by the starburst phenomenon. In addition, clearly recognizable in the images are interactions between the new galaxies. Everywhere, galaxies appear to be running into other galaxies, merging, or tearing their neighbors apart. It's stark evidence that space was so much more crowded when the Universe was young and that collisions were an essential process in making the galaxies we see today (**Figure 17.18**).

SECTION SUMMARY
Hubble Deep Field
- The Hubble Deep Field images are ultralong exposures of small sections of sky that reveal myriad galaxies far back into cosmic history.
- The deep-field images show early phases of galaxy evolution, including details such as collisions and starbursts.

The Dark Ages

Using data like the Hubble Extreme Deep Field, astronomers can see galaxies already in the process of forming, but going much farther back than about 13 billion years presents its own problems. For galaxies, the most important milepost of cosmic history comes about 300,000 years after the Big Bang. That's when all the normal matter evolved into the form of neutral hydrogen gas. Chapter 18 will explore this important transition in detail. For now, the critical point is that because the Universe had evolved to a state where it was filled with neutral hydrogen, it was also dark, and in this darkness the first stars and galaxies were born.

"Neutral hydrogen is very good at absorbing photons in the optical and UV bands," explains Loeb. "That means UV and optical light could not travel very far without being absorbed. If you were around back then, you couldn't have seen very far out into space before everything was lost in a gray fog." The situation that Loeb describes is very different from today's cosmos, in which a red or blue photon can travel across billions of parsecs of intergalactic space without being absorbed. Astronomers therefore call the era when space was full of neutral hydrogen the **Dark Ages** (**Figure 17.19**).

But how did the Universe go from the Dark Ages of early cosmic history to the transparent, well-lit conditions we see today? "It's the formation of stars and galaxies themselves that ended the Dark Ages," says

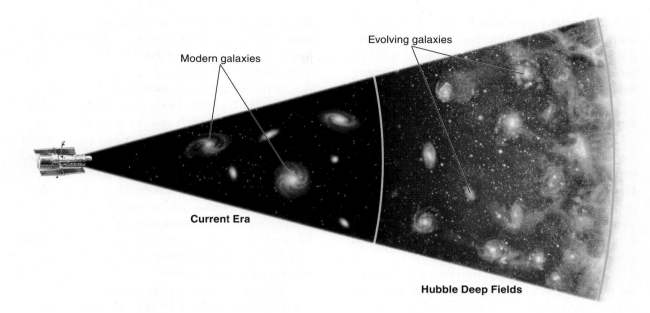

Modern galaxies

Current Era

Evolving galaxies

Hubble Deep Fields

FIGURE 17.18
Evolution of the Universe
To understand the evolution of the Universe, astronomers need to work backward along lines of sight, constructing a time line for the most basic building blocks of cosmic structure and the stages that created them. The HUDF represents the farthest (and therefore most ancient) galaxies directly observed to date.

Loeb. "That is why we are so interested in this epoch of cosmic history. It's where stars and galaxies begin."

—SECTION SUMMARY
The Dark Ages
● The emergence of stars and galaxies marks the end of the Dark Ages, a period following the Big Bang when neutral hydrogen had already formed and absorbed all visible-wavelength photons.

denser-than-average gas and makes them into the seed of all the structure we see today," says Avi Loeb. Since these perturbed regions have a little more density (and more mass) than their surroundings, they lie at the bottom of shallow "gravitational wells." Once neutral hydrogen has formed, gravity pulls surrounding material inward, toward the center of the perturbations. As more mass is pulled in, what started as a small ripple soon becomes a distinct cloud of gas.

density perturbation A small region where the density of hydrogen gas is slightly higher or lower than that of the surrounding regions.

17.5
Building Cosmic Structure

Using powerful computer simulations, astronomers have been zeroing in on the earliest stages of cosmic structure formation. Their conclusion can be simply summarized: All cosmic structures began forming in earnest at the outset of the Dark Ages. But even these beginnings hinged on holdovers from the truly "early" Universe: ripples in the sea of matter and energy called perturbations.

The First Stages: The Rich Get Richer

Density perturbations are small regions where the density of hydrogen gas is slightly higher or lower than the surroundings. (As we will see in Chapter 18, these perturbations formed in the barest instants after the Big Bang.) "Gravity takes these regions of slightly

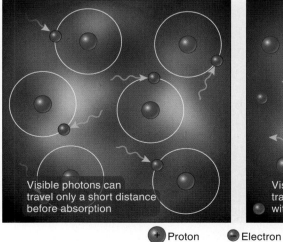

Visible photons can travel only a short distance before absorption

Visible photons can travel long distances with no absorption

⊕ Proton   ⊖ Electron   〰➤ Photon

A Dark Ages (neutral hydrogen)   **B Today (ionized hydrogen)**

FIGURE 17.19 The Dark Ages
(A) The dense clouds of hydrogen gas during the Dark Ages reabsorbed any ultraviolet and optical photons as soon as they were emitted, making the Universe opaque during that era. (B) In the Universe today, all of the intergalactic hydrogen is ionized, so it is transparent to optical and ultraviolet photons.

In contrast, regions that started with slightly lower-than-average densities find themselves becoming evacuated. "It's a case of the rich getting richer and the poor getting poorer," says Loeb. "The low-density regions eventually become voids, and the high-density regions are where stars, galaxies, and galaxy clusters form."

But which structures form first? According to computer simulations, the evolution of large-scale structure represents a process of **hierarchical formation**. The smallest structures are the first to emerge via gravity. Only later do the larger structures form. Thus, stars and low-mass black holes are the first macroscopic objects to make their appearance on the cosmic stage, and galaxies come afterward.

The first generation of stars that formed out of the Dark Ages was, however, unlike anything we see today. When astronomers simulate the formation of the first stars, what they find are monsters. Somewhere around a redshift of $z = 20$, clouds of a million or so solar masses have already formed from the sea of neutral hydrogen. Within these clouds, smaller pockets of gas collapse under their own gravity to become stars.

hierarchical formation An ordered sequence of formation, from smallest to largest structures.

But instead of the stars we see today, with an average mass less than the Sun, astronomers believe that the first generation of stars was dominated by behemoths with masses of 100–300 M_{Sun}.

In the current epoch, astronomers have never seen a star with a mass of more than 150 M_{Sun}, and all massive stars are relatively rare. So why were the first-generation stars so big? "There were no heavy elements around because there had been no stars to make them," says Loeb. "Heavy elements like carbon and oxygen can be much more effective at emitting radiation that allows gas to cool than lighter elements. That tends to allow star-forming clouds in today's Universe to fragment into smaller pieces and produce smaller-mass stars. Without heavy elements, the first generation of stars ended up really massive."

The monster first-generation stars produced powerful fluxes of ultraviolet light, which then created ionized bubbles in the otherwise neutral sea of hydrogen atoms. And when the tremendous stars ended their lives as supernovas, their powerful blasts created more ionizing radiation that further eroded neutral hydrogen. When these stars were close enough together, their surrounding ionized regions merged, creating more space free of neutral hydrogen. In this way, the first-generation stars helped end the cosmic Dark Ages. But they weren't alone; the young black holes were also doing their part.

As larger and larger clouds were gathering, gravity pulled them together to form the first protogalaxies. While stars formed in some regions, in others so much material was falling in that the first progenitors of supermassive black holes were created with 10,000 M_{Sun} or more in mass. Once these intermediate-mass black holes formed, they quickly became surrounded by accretion disks that continued to feed them new matter from the surrounding gas. Thus the first black holes became the first protoquasars.

Enough ionizing radiation was created by these young quasars that they, too, became surrounded by bubbles of ionized gas. As the number of quasars grew, the ionized bubbles overlapped, just as was happening with the first stars. "About a billion years after the Big Bang, all the neutral hydrogen was gone," says Loeb. "The entire Universe became, for the most part, transparent to optical photons." Only within the galaxies themselves could gas recombine to form clouds of neutral hydrogen. Thus the cosmic Dark Ages finally ended (**Figure 17.20**).

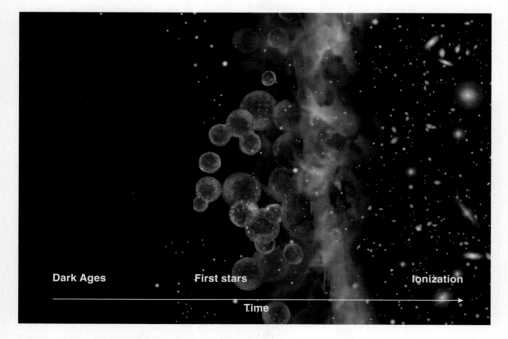

FIGURE 17.20 The End of the Dark Ages
Ionizing bubbles were created first from massive stars, then from those stars' supernova blasts, and finally from young quasars. As these bubbles overlapped, the vast stores of neutral hydrogen clouds were eroded until at last the Universe became transparent and the Dark Ages ended.

┌─ **SECTION SUMMARY**
│ **The First Stages: The Rich Get Richer**
● Density perturbations, which are small differences in mass in one place versus another, formed the seeds around which all larger structures could be created by gravity.
● The first stars were many times larger than stars that exist today, because of the absence of heavier elements to limit their growth.
● The first black holes were the progenitors of quasars.
● Ionizing radiation from the first stars and black holes ended the Dark Ages.

Dark Matter and the Secret to Large-Scale Structure

As we've climbed up the ladder of cosmic structure—from galaxies all the way to galaxy superclusters—we've seen that at every rung, dark matter dominates over luminous matter. Taking a giant step back and trying to simulate the evolution of all this structure from just after the formation of neutral hydrogen ($z = 2,000$) all the way up to the present epoch ($z = 0$), astronomers find that this dominance of dark matter extends to almost every feature of the Universe's evolution. More important, using these simulations as a guide, astronomers have been able to infer how the properties of dark matter shape the history of cosmic structure formation.

Modern observations enable astronomers to place fairly strict constraints on the composition of matter in the Universe. Taken as a whole (as opposed to what is seen in specific individual structures), 87 percent of all the matter in the Universe is composed of the dark component, while only 13 percent is the luminous kind that makes up stars and galaxies and human beings. (Note also that most of the Universe's luminous matter is in the form of interstellar gas.) This ratio represents the starting condition for simulations of cosmic structure formation. But since nothing is known from observation about the form that dark matter takes, astronomers had to make a choice: Should they treat dark matter as particles moving at high speeds (even close to the speed of light), as neutrinos are known to travel? Or should they treat dark matter as slow-moving particles like the majority of luminous matter particles (protons, neutrons, electrons, and so forth)?

"It turns out that the difference between such 'hot' and 'cold' dark matter scenarios really matters," says Loeb. When astronomers begin their simulations with **hot dark matter**, gravity has a hard time getting the fast-moving particles to collapse into structures like those seen on the sky. Only by assuming a **cold dark matter (CDM)** scenario do simulations show gravity grabbing hold of the initial perturbations and drawing sufficient material together to re-create the beautiful patterns of the cosmic web observed in redshift survey maps. "Whatever dark matter is made of, it must be moving slowly," explains Loeb. "That's an extremely important bit of information." Knowing that dark matter moves relatively slowly limits the kinds of candidates that particle physicists (the scientists who study the nature of matter) can propose to make up the invisible but dominant form of cosmic mass. Relativistic particles, for example, can't make up dark matter.

With the question of cold versus hot dark matter resolved, astronomers have turned to building ever-more-detailed and highly resolved simulation-based explorations of structure formation. What they see in these supercomputer studies is an initially smooth cosmic gas (other than tiny perturbations) emerging from the era when neutral hydrogen forms. Gas slowly collects in and around the perturbations, creating clouds that become ever denser and ever more massive. Long, dense filaments grow to link the densest regions. As we have seen, as the gas pours into these deepening gravitational wells, it fragments into clouds that form the first stars and the progenitors of the supermassive black holes (and hence quasars). As gas is drawn together, any rotational motion it had on large scales is amplified as it collapses, leading to the disks of spiral galaxies.

With time, more material is drawn together, lowering the density in the voids and increasing the mass of proto–galaxy groups and proto–galaxy clusters (**Figure 17.21**). Galaxies within these structures are gravitationally bound together, but they don't necessarily interact in a graceful way. "The orbits of galaxies through the clusters are often highly elliptical," says Loeb. "They dive through the cluster centers, rise to the cluster edges, and then dive back again like bees swirling around a hive." Collisions between the young galaxies are inevitable. Spiral galaxies often run into each other, stripping away the organized disks of stars and gas. Although the stars rarely hit each other in these galaxy-galaxy collisions, vast clouds of molecular gas are driven together, creating tremendous starbursts. These spikes in the stellar birth rate last only a few tens of millions of years. In that brief time, however, the rate

hot dark matter Nonluminous matter moving at relativistic speed.

cold dark matter (CDM) Nonluminous matter moving at slower-than-relativistic speed.

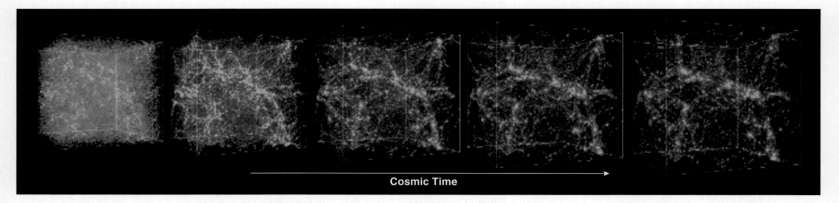

FIGURE 17.21 Evolution of Cosmic Structures
Computer simulations of cosmic history show how structures emerge from small-scale perturbations imposed at the earliest moments after the Big Bang. Gravity from the small, overly dense regions allows them to become amplified as more mass from surrounding regions is drawn in. Eventually the perturbations grow to become the hierarchy of large-scale structure we see today. In particular, these computer simulations show the emergence of the filamentary structure of matter associated with superclusters and voids.

of star formation can be 100–1,000 M_{Sun} per year (the current rate in our galaxy is about 1 M_{Sun} per year).

In the aftermath of the collision of two spiral galaxies, existing stars are tossed about but not lost. They eventually fall back, creating a new, larger galaxy that will be distinctly elliptical. In this way, the history of hierarchical large-scale structure formation explains the higher percentage of ellipticals in the centers of rich clusters. As gravity draws galaxies and their groups together into the clusters, the density of galaxies (the number per volume) increases thereby also increasing the likelihood of collisions. A collision between ellipticals always produces another elliptical (or perhaps an irregular). "Since collisions between spirals also produce ellipticals," explains Loeb, "you can see why the more galaxy collisions that occur, the larger the fraction of ellipticals we astronomers expect to observe."

> **SECTION SUMMARY**
> **Dark Matter and the Secret to Large-Scale Structure**
> - Dark matter played a fundamental role in cosmic evolution, and astronomers have learned through simulations that it must be "cold" (moving at nonrelativistic speeds).
> - In galactic evolution, spirals evolve into ellipticals through collisions.

Final Questions for the Beginning of Everything

Models for the evolution of cosmic structure that are based on cold dark matter have been remarkably successful at predicting everything from the distribution of galaxy types to the statistics of supercluster density (how much more dense superclusters are than their surroundings). They give us a coherent and consistent story of how the Universe went from a relatively structureless gas of neutral hydrogen 300,000 years after the Big Bang to the rich and diverse structures we see today, 13.7 billion years after the Big Bang.

But where did the initial perturbations that form such an important part of this story come from? And what happened before the formation of neutral hydrogen? Finally, and perhaps most important, what *was* this event called the Big Bang and how did the whole shebang we call the Universe get set into motion?

Attempting to answer these questions takes us to the end of our cosmic journey by bringing us to the beginning and, perhaps, the deepest questions science can ask. How did time begin, and why is there something rather than nothing? Chapter 18 explores these ultimate questions of cosmology: the birth of the Universe and its eventual fate.

chapter summary

17.1 Bright Lights, Big Universe

The discovery that the Universe is expanding was the first step in refuting the long-held belief in a static cosmos. It eventually led to the conclusion that the Universe has evolved, meaning that it is different today than it was in the past. The expansion of the Universe is apparent from observed receding motions of all galaxies, and it suggests that the Universe was once extremely compact. Using Hubble's law alone, astronomers can estimate the age of the Universe to be about 14 billion years. Using more detailed methods, they find the Universe to be 13.7 billion years old. An object's recession velocity is measured in terms of the redshift of its spectrum, which correlates with look-back time: how far back in time we're seeing the object.

17.2 Cosmic Neighborhoods: Galaxy Groups and Galaxy Clusters

The Milky Way is part of the so-called Local Group of galaxies. Galaxy groups are themselves members of clusters, some of which have more than 1,000 member galaxies. Clusters differ in their relative abundances of specific galaxy types. Galaxy clusters include X-ray halos of hot gas, which appear to result from the merging halos of individual galaxies. Evidence for dark matter comes from the fact that more mass is required to hold galaxy clusters (including their halos) together than can be observed as luminous matter. Gravitational lensing—in which large collections of matter cause space-time to bend and focus light from distant background objects—also demonstrates the presence of dark matter in galaxy clusters.

17.3 Superclusters and Large-Scale Structure

Redshift surveys provide a three-dimensional picture of galaxy distribution on the largest scales. These "big science" projects can require collaboration among large groups of scientists and enormous quantities of data. Through these surveys, vast superclusters of galaxies have been discovered, with similarly vast voids between them, showing that structure exists even on very large scales. The surveys have also identified that, along with the "Hubble flow" of cosmic expansion, galaxies show so-called peculiar motions driven in response to the gravitational pull of other large collections of mass, giving further evidence of dark matter. On the largest scales, the Universe's structure smooths out into a homogeneous and isotropic cosmic web, consisting of interconnected filaments of superclusters and the voids between them.

17.4 Large-Scale Cosmic Structure and Cosmic History

Studying the history of the Universe requires moving backward in time along individual lines of sight. Quasars came into existence at a relatively early time in cosmic history as large quantities of matter were funneled into supermassive black holes at the centers of galaxies. As the reservoirs of matter ran out, the quasars dimmed.

17.5 Building Cosmic Structure

Deep fields, in which astronomers collect light from very distant (and hence faint) cosmic objects, have advanced the study of the Universe's history by revealing galaxies in their early stages of construction. Visible in these deep fields are many galaxies in various stages of evolution, as well as evidence of frequent collisions. Stars and galaxies first started forming during the Dark Ages, when normal matter was mostly neutral hydrogen that absorbed all ultraviolet and optical photons. Massive neutral hydrogen clouds formed first, and then massive stars. Intermediate-mass black holes that formed the cores of galaxies came next. By 1 billion years after the Big Bang, almost all neutral hydrogen had been ionized, and the Universe became transparent to visible and ultraviolet light. The dominance of dark matter over luminous matter is seen throughout cosmic evolution and is central to the formation of galaxies, galaxy clusters, and superclusters. Simulations agree with observations in showing that the dark matter must be "cold" (moving at nonrelativistic speeds).

questions and problems

Narrow It Down: Multiple-Choice Questions

1. Select the correct sequence for cosmic structures, from smallest to largest.
   a. galaxy, galaxy group, galaxy cluster, supercluster, cosmic web
   b. cosmic web, galaxy, galaxy cluster, galaxy group, supercluster
   c. galaxy, galaxy cluster, galaxy group, cosmic web, supercluster
   d. galaxy, galaxy cluster, galaxy group, supercluster, cosmic web
   e. cosmic web, galaxy, galaxy group, galaxy cluster, supercluster

2. Which of the following provide(s) evidence of dark matter? Choose all that apply.
   a. the Solar System
   b. galaxies
   c. galaxy clusters
   d. superclusters
   e. black holes

3. Which statement about the observed expansion of space-time is true?
   a. Observers in all galaxies should see farther galaxies receding from them faster than nearer galaxies.
   b. Since we see all galaxies receding from the Milky Way, we must be near the center of the Universe.
   c. On average, all galaxies appear to be receding at the same speed.
   d. The farther away galaxies are, the more slowly they appear to recede.
   e. Space-time is expanding out toward the limits of the Universe.

4. The term *Local Group* refers to a group of
   a. planets.
   b. stars.
   c. galaxies.
   d. galaxy clusters.
   e. astronomers working on a redshift survey.

5. If H were 90 km/s/Mpc, how would the estimated age of the Universe change from current estimates?
    a. The value would be the same.
    b. The value would be larger.
    c. The value would be smaller.
    d. The value can't be determined from the information given.
    e. The value would depend on the density of the Universe.

6. From what you've learned about the stages of evolution of the Universe, and given the very best instruments available to detect light in visible wavelengths on Earth, what would be the earliest time frame or event that could be captured?
    a. the instant of the Big Bang
    b. within 300,000 years after the Big Bang
    c. the birth of the first galaxy
    d. the end of the Dark Ages
    e. none of the above

7. What does a redshift of $z = 0$ mean?
    a. The object is almost at the edge of the Universe.
    b. The object is nearly 14 billion years old.
    c. The object is not moving.
    d. The object shows no redshift.
    e. The object is obscured from our view by the plane of the Milky Way.

8. Which of the following is *not* characteristic of rich galaxy clusters?
    a. Their diameters are measured in megaparsecs.
    b. They can have more than 1,000 members.
    c. They contain many galaxy groups.
    d. They are gravitationally bound.
    e. They contain mostly spiral galaxies.

9. Which of the following is *not* indicated by the observation of an Einstein ring?
    a. the presence of dark matter
    b. a large mass between the observer and the source object for the ring
    c. a source object with a circular shape
    d. dark matter with a symmetric distribution
    e. curved space-time

10. Redshift surveys are difficult and time-consuming to obtain. Which of the following factors does *not* contribute to that difficulty?
    a. lack of evidence showing that redshifts imply expansion
    b. dim light from galaxies observed
    c. number of objects to be observed
    d. need for specialized equipment
    e. need for a large amounts of telescope time

11. Which of the following statements about the cosmic web is *not* true?
    a. It is filamentary.
    b. It is isotropic.
    c. It is homogeneous.
    d. It shows no voids.
    e. It resembles a sponge.

12. Which of the following is true of an object with a redshift of $z = 5$?
    a. Its spectrum shows longer wavelengths than is expected for specific elements.
    b. It is receding faster than an object with $z = 6$.
    c. It is only a few light-years away.
    d. It may be significantly closer than a second object with $z = 5$.
    e. It is farther away than an object with $z = 7$.

13. Which of the following were the earliest to form in the evolution of the Universe?
    a. $1\text{-}M_{Sun}$ stars
    b. galaxies
    c. supermassive stars
    d. protoquasars
    e. intermediate-mass black holes

14. Which of the following is/are expanding as the Universe expands? Choose all that apply.
    a. the Solar System
    b. galaxies
    c. galaxy clusters
    d. galaxy superclusters
    e. space-time

15. X-ray halos of galaxy clusters consist primarily of which of the following?
    a. neutral hydrogen gas
    b. hot, ionized gas
    c. brown dwarfs
    d. supernova remnants
    e. X-ray radiation

16. If the Local Group had begun forming with more total mass and at a higher density than it did, which of the following would *not* be true?
    a. More galaxy mergers would likely have occurred.
    b. The largest galaxy (Andromeda) might have been even larger.
    c. More satellite galaxies would likely be found around the spiral galaxies.
    d. The galaxies would have started moving relative to each other at higher velocities.
    e. The ratio of luminous to dark matter would likely have been greater.

17. If the rate of expansion of the Universe had been increasing since the Big Bang, how would the calculated age of the Universe differ from the value attained using the currently observed Hubble constant?
    a. It would be less than with constant expansion.
    b. It would be greater than with constant expansion.
    c. Without knowing the rate of change, it is not possible to determine.
    d. There would be no change, because the rate of expansion is taken into account in the current calculation.
    e. It would be a meaningless calculation, since astronomers now have definitively determined the true age of the Universe.

18. What specific piece(s) of evidence for dark matter do astronomers find? Choose all that apply.
    a. They calculate more mass in various large-scale structures than is visible as luminous matter.
    b. They observe large dark regions between galaxies.
    c. Galactic rotation curves do not match expectations based on visible mass and radius.
    d. X-ray halos in galaxy clusters have too much pressure to be gravitationally bound by the observed luminous matter in the clusters.
    e. Galaxies appear dimmer than expected.

19. Which statement(s) about redshift and look-back time is/are true? Choose all that apply.
    a. The greater the redshift, the greater the look-back time.
    b. The smaller the value of z, the shorter the look-back time.
    c. Objects with greater redshift appear younger than they actually are today.
    d. Look-back time is limited by the speed of light and the age of the Universe.
    e. The greater the redshift, the closer the look-back time approaches the Big Bang.

20. True/False: Gravitational lensing means that objects can appear shifted from their expected positions.

To the Point: Qualitative and Discussion Questions

21. Explain what *redshift* means as it applies to the study of galaxies.

22. What galaxy types are found in the Local Group, and how are they arranged?

23. The Virgo Cluster is a regular rich cluster. What do the terms *regular* and *rich* tell you about its structure?

24. How do astronomers explain the formation of X-ray halos in galaxy clusters?

25. What is required for the formation of an Einstein ring?

26. What are redshift surveys? Name some notable examples. How are their results used?

27. Explain the meanings of *big science* and *big data*.

28. How do peculiar velocities differ from the Hubble flow? How are they affected by dark matter?

29. Define the terms *homogeneous* and *isotropic*.

30. Explain why there are no nearby quasars.

31. How do deep-field images contribute to our understanding of the evolution of the Universe?

32. How do small perturbations form the seeds of larger structures?

33. Explain the term *cold dark matter*.

34. The first *Star Wars* movie series began with the phrase "Long, long ago in a galaxy far, far away. . .". What do you understand now about the scope of how long ago and how far away that could have been?

35. Considering that galactic evolution requires thinking about time periods far beyond human experience, if our species survives several billion years, where might we expect our own evolution to take us?

Going Further: Quantitative Questions

36. Estimate the age of the Universe, in years, using a value of 55.4 km/s/Mpc for Hubble's constant and assuming that the rate of expansion has remained constant.

37. Estimate the age of the Universe, in years, using a value of 81.5 km/s/Mpc for Hubble's constant.

38. An earlier estimate of the age of the Universe found it to be 7.5 billion years old. What value of Hubble's constant would be consistent with this age?

39. The average velocity for galaxies in a galaxy cluster is 850 km/s. The diameter of the cluster is 3.2 Mpc. How long, in years, would it take a galaxy to cross the cluster?

40. Calculate the mass, in M_{Sun}, of a cluster where the average galactic velocity is 970 km/s and the diameter of the cluster is 8.5 Mpc.

41. What is the average galactic velocity, in kilometers per second, within a cluster that has mass $2.5 \times 10^{14}\ M_{Sun}$ and diameter 2.7 Mpc?

42. A star orbits a black hole whose mass is $4.7 \times 10^{6}\ M_{Sun}$ at a radius of 1.9×10^{13} meters. What is its velocity, in meters per second? (See Going Further 3.1 and 15.1.)

43. What is the recession velocity, in kilometers per second, of a galaxy that is 790 Mpc from Earth (using the value of H given in Going Further 16.1)?

44. In the spectrum of a distant galaxy, a spectral line whose expected value is 510 nm is observed at 710 nm. If you ignore any relativistic effects, at what velocity, in kilometers per second, is the galaxy receding from the observer? (See Going Further 16.1.)

45. An astronomer calculated that the recession velocity of a galaxy was 65,000 km/s, by noting the redshift of a spectral line whose rest wavelength is 420 nm. At what wavelength, in nanometers, did he observe the spectral line? (See Going Further 16.1.)

 If your instructor assigns homework in **smartwork5**, access it at the Digital Landing Page for *At Play in the Cosmos*: **digital.wwnorton.com/cosmos**

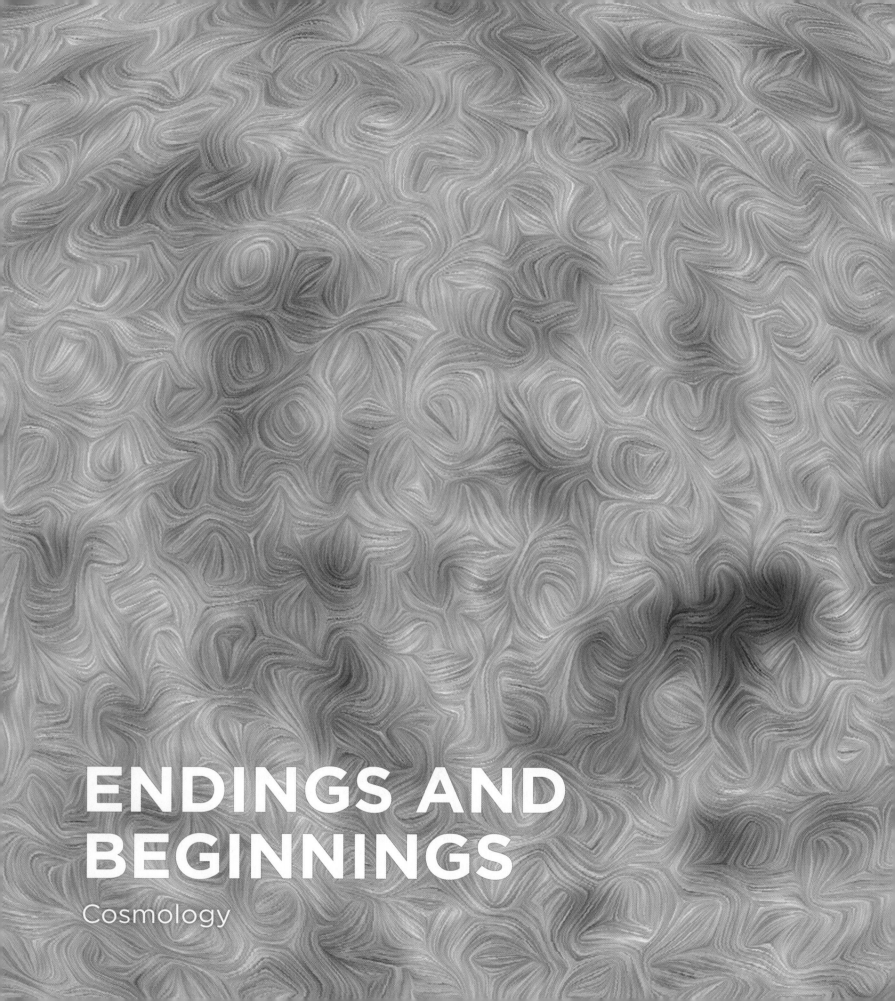

ENDINGS AND BEGINNINGS

Cosmology

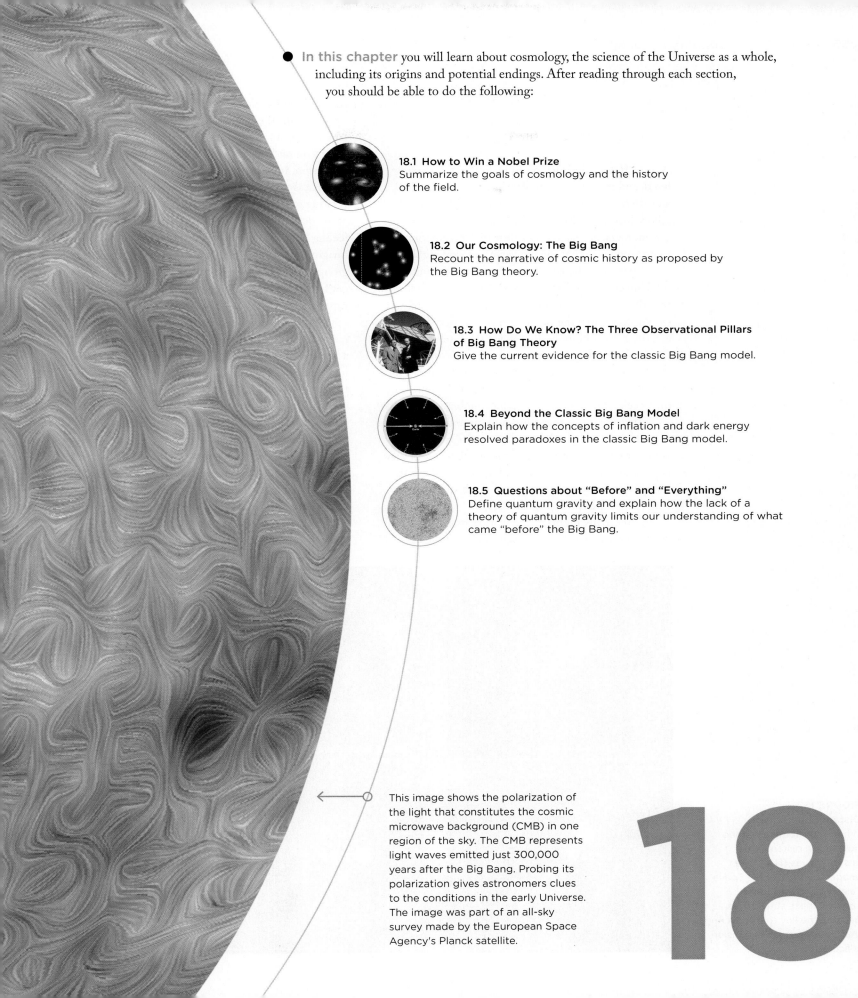

In this chapter you will learn about cosmology, the science of the Universe as a whole, including its origins and potential endings. After reading through each section, you should be able to do the following:

18.1 How to Win a Nobel Prize
Summarize the goals of cosmology and the history of the field.

18.2 Our Cosmology: The Big Bang
Recount the narrative of cosmic history as proposed by the Big Bang theory.

18.3 How Do We Know? The Three Observational Pillars of Big Bang Theory
Give the current evidence for the classic Big Bang model.

18.4 Beyond the Classic Big Bang Model
Explain how the concepts of inflation and dark energy resolved paradoxes in the classic Big Bang model.

18.5 Questions about "Before" and "Everything"
Define quantum gravity and explain how the lack of a theory of quantum gravity limits our understanding of what came "before" the Big Bang.

This image shows the polarization of the light that constitutes the cosmic microwave background (CMB) in one region of the sky. The CMB represents light waves emitted just 300,000 years after the Big Bang. Probing its polarization gives astronomers clues to the conditions in the early Universe. The image was part of an all-sky survey made by the European Space Agency's Planck satellite.

18

18.1

How to Win a Nobel Prize

ADAM RIESS knew he might be holding a winning hand, and it scared him to death. The year was 1998, and Riess was a young scientist with a postdoctoral fellowship at UC Berkeley. With a team of other astronomers, he was trying to use distant supernovas to find evidence that the Universe was slowing down.

Ever since the Big Bang (the event that initiated the Universe) was first proposed, astronomers assumed that the mutual gravitational pull of all cosmic matter would slow down the cosmic expansion first discovered by Edwin Hubble. That made the rate of "cosmic deceleration" a key factor in cosmology. Everyone wanted to know its value. But it was the big fish that always got away, as previous methods used to measure the deceleration had always yielded ambiguous results.

The supernova that Riess and his collaborators were using represented a new approach. "Supernovas are so bright," says Riess, "you can see them from very far away, which also means looking far back in time. Using supernovas, we see more of the history of the Universe's expansion" (**Figure 18.1**).

After many nights at the world's most powerful telescopes, Riess and his colleagues had all the data

they needed to calculate the exact value of cosmic deceleration. There was only one problem: it wasn't happening. "I found just the opposite," says Riess. "The data said the Universe wasn't slowing down in its expansion. It was speeding up. It was accelerating."

Riess was sitting on the scientific version of a very big grenade, and he knew it. "I remember being at my desk at UC Berkeley looking at my results and feeling kind of sweaty-anxious about it," he says. "I was imagining having to explain this to the whole research group I was part of. I knew the grilling I was going to get for bringing in this really weird result that no one expected."

Riess didn't have to justify the result only to his research group. Finding out that the Universe was accelerating meant he had also found evidence for a new and invisible form of energy—a "dark" energy—that was powering the cosmic acceleration. If Riess was right, he was in Nobel Prize territory. If he was wrong, he was about to make a fool of himself in the most public way possible. "It was like being in a big poker game and thinking you're going to play for relatively small stakes," he says. "But then you get dealt the most amazing hand ever, and you realize you can't just bet with pennies and nickels. You need to go all in."

Riess and his collaborators checked and rechecked their calculations many times as they

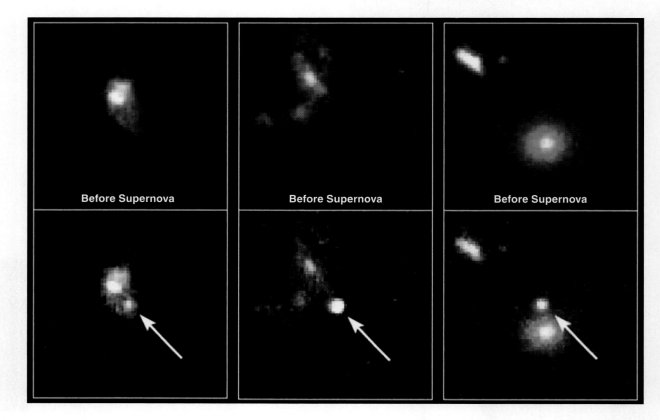

FIGURE 18.1
Distant Supernovas
When astronomers learned how to use distant Type Ia supernovas (like these imaged by the Hubble Space Telescope) as standard candles, these supernovas became the key to discovering that the Universe's expansion is accelerating.

prepared to announce their results. "Brian Schmidt, who led the supernova search, and I were planning what we would do if we had to leave the field in disgrace because we were proved wrong," Riess chuckles.

But they weren't wrong. Another group using a similar method had found the same result. The two teams ended up announcing their discoveries at the same time. Dark energy, it seemed, was here to stay. Fourteen years after that night worrying at his desk, Riess would take an all-expenses-paid trip to Sweden to pick up his Nobel Prize.

The Last Step in the Cosmic Journey

Just seventeen chapters ago, we began a journey through time and space. We started that journey by considering our own human evolution and the evolution of science from Galileo to Newton's laws and on through to the properties of light and the structure of matter. Using these laws of physics, we pushed outward, exploring planets, stars, and galaxies. Now, finally, we find ourselves at the end of the journey. Putting everything we've learned together, we can finally ask the biggest and most all-embracing questions possible. We are finally ready to consider *cosmology* and *the Universe as a whole*.

Beginnings have always held a special place in human curiosity, of course, but endings are equally important. In this chapter we will use the laws of physics to ask not only about the beginning of our Universe, but also about the ultimate cosmic ending. What is the long-term future of the Universe? Will it end? Will it continue forever?

More than almost any other science, cosmology takes us to the limits of what it means to ask scientific questions. Does the theory of the Big Bang really tell us why there is something rather than nothing? Does it really tell us what time is or how time began? Does it make infinite space easier to understand? As we tell this final story in the narrative of humanity's understanding of astronomy, we need to pay close attention not only to what we know, but also to what we don't know and the difficulties we face in asking the deepest kinds of questions.

> ## SECTION SUMMARY
> ### The Last Step in the Cosmic Journey
> - Astronomers now have answers to many—but not all—important questions in cosmology, the study of the Universe and its evolution.

What the Big Bang Isn't

We have already seen how Hubble's discovery of galaxy redshifts led some astronomers to the idea that the Universe must have a history, meaning it looked much different in the past than it does today. "If you imagine running the 'movie' of cosmic expansion backward," says Riess, "then it would seem obvious that as you go backward in time, the Universe must have been much denser than what we see today."

Taking that idea to its extreme means running the "movie" of cosmic expansion backward as far as possible. In doing so, we would find that at some point in the past, all the matter in the Universe (which is now expanding) must have been infinitely squeezed together, implying that the Universe began with an infinite density (**Figure 18.2**). However, an infinite matter density makes no sense from a physics standpoint. An infinite amount of matter can't be squeezed

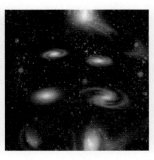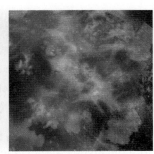

FIGURE 18.2 Running Expansion Backward
"Running the movie backward"—reversing the expansion of space-time seen today in the recession of galaxies—leads first to an ever-higher density of galaxies. Eventually, the stars and gas making up the galaxies must merge and mix. As we go further back in time and increasing density, even the matter making up individual stars must have all been crushed together. This "thought experiment" shows how the currently observed expansion of the Universe can imply very different cosmic conditions at earlier epochs of cosmic history from those we see today.

A George Gamow

B Ralph Alpher

FIGURE 18.3
Alpher and Gamow
With his adviser (A) George Gamow (1904–1968), (B) Ralph Alpher (1921–2007) published a paper in 1948 on the Big Bang theory titled "The Origin of Chemical Elements." In this paper and others, they introduced the idea that only specific light elements could have been produced in the Big Bang, and they predicted the discoveries that now support the theory.

steady-state model A theory, advanced by astronomer Fred Hoyle in the 1950s, stating that although the Universe is expanding, it always appears the same because new matter is slowly created between galaxies to maintain the average density.

into a cubic meter. So while cosmic expansion might seem to imply that the Universe had a beginning—a time when time itself started—most physicists and astronomers were not willing to go that far when Hubble made his discovery.

The modern theory of the Big Bang began in the late 1940s and 1950s with Russian-American physicist George Gamow and his graduate student Ralph Alpher (**Figure 18.3**). Their calculations suggested that the early Universe might have begun very dense and very hot and then cooled and rarefied with cosmic expansion. As we will see, this idea led to specific predictions that could be tested against observations (though it took another 12 years before that happened), becoming the basis for all our cosmological models today. What Gamow and Alpher did not explain, however, was where all this matter, energy, space, and time came from in the first place.

Also in the 1950s, astrophysicist Fred Hoyle proposed a **steady-state model** of cosmology, in which the Universe had always been expanding and always would be expanding. In Hoyle's theory, new particles of matter slowly appeared between the galaxies, such that the average cosmic density remained the same (**Figure 18.4**). When other scientists objected to Hoyle's continuous creation of matter, he asked

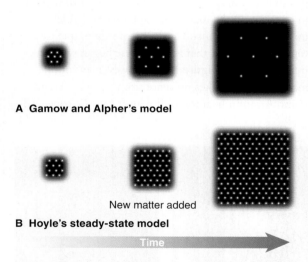

A Gamow and Alpher's model

New matter added

B Hoyle's steady-state model

Time

FIGURE 18.4 The Big Bang Model Compared with the Steady-State Model
(A) In Gamow and Alpher's Big Bang model, expansion leads to lower densities over time.
(B) In Hoyle's steady-state model, new matter is slowly created, such that new galaxies form and fill in the space left by the expansion of older galaxies. Thus the density of galaxies never changes.

whether having everything created in a "big bang" was any better. And so it was that Hoyle invented the name *Big Bang*, using the term to distinguish it from his own ideas.

The eventual fate of the steady-state model is a wonderful example of the scientific process at work. Many scientists at the time viewed Hoyle's idea favorably. But by the mid-1960s, new discoveries made clear that the Universe in the past looked very different from the Universe today. Those discoveries meant that Hoyle's steady-state model made predictions (the past should look the same as the present) that were not backed up by the evidence. In fact, what the new data pointed to was a Universe that must have been much hotter and denser in the past than it is today, just as Gamow and Alpher had predicted.

Still, although the theory of the Big Bang prevailed, it is important to remember that it is not a theory of the origin of the cosmos; it's a theory of what happened after the Universe began. The Big Bang never tells us how the cosmos, or existence itself, came into being. For now, that question remains unanswered by science.

SECTION SUMMARY
What the Big Bang Isn't
- Early scientific cosmological theories included Fred Hoyle's steady-state model, in which the Universe did not evolve but always looked the same from one epoch to the next.
- Today, the generally accepted model is the Big Bang, which describes what happened *after* cosmic expansion began.

18.2
Our Cosmology: The Big Bang

Now that we know that the Big Bang *isn't* a theory of the origin of the Universe, we are ready to explore what the Big Bang *is*: a theory of what happened from just after expansion began up to the present.

Part I. The Stage and the Actors

The first step in laying out our modern narrative of cosmic evolution is to set the stage for this scientific drama and introduce the actors, some of whom will be new while others are old friends. Let's start with the stage.

SPACE-TIME. The Big Bang is a general relativistic cosmological theory. It begins with the flexible space-time at the heart of Einstein's general theory of relativity (explored in Chapter 14). In a sense, cosmology was waiting for Einstein. There had been early attempts to build scientific models of the entire Universe. Newton had certainly thought about it, but previous efforts were hobbled without Einstein's great insight into the nature of space-time as a four-dimensional continuum (meaning a smooth fabric of reality). By allowing physics to describe space-time in terms of geometry and by assuming that the Universe was highly symmetric (that is, homogeneous and isotropic, as described in Chapter 17), Einstein was able to formulate equations describing the Universe *as a whole*.

With general relativity, describing the cosmic history meant describing the geometry (or curvature) of space-time and the evolution of that geometry with time. "Geometry is the stage," says Adam Riess, "and the play must take place on the stage. Knowing what kind of a stage you're on—meaning knowing the geometry of cosmic space-time—really changes how you interpret what you're seeing."

The geometry of the Universe is determined by the matter and energy within it. The only way to distinguish which geometry the Universe actually has is by making measurements that are either direct (looking for effects of the geometry) or indirect (measuring the Universe's store of matter and energy). Remarkably, these experiments have been successfully carried out and, as we will see, current evidence points to our living in a Universe with a flat geometry, but future research could indicate that the situation is more complex.

Of course, the Big Bang can't just describe the expanding stage, meaning the space part of space-time. It must also describe what's happening *on* that stage. Modern science has a theory to do just that, and it is a crowning achievement of the human intellect called the standard model of particle physics.

THE STANDARD MODEL OF PARTICLE PHYSICS. During the 1950s and 1960s, physicists expended enormous effort and resources to develop giant **particle accelerators**, machines designed to bring the smallest specks of matter to the highest speeds (which really means the highest energies) possible. By smashing the particles into each other and scanning the remains, physicists gleaned hints of the internal constitution of the particles. The American physicist Richard Feynman described the effort this way: "It's like smashing two watches together and using the debris to figure out how they worked." By the mid-1960s, accelerator-based studies had produced a remarkably successful description of all known subatomic particles, known as the **standard model of particle physics** (**Figure 18.5**).

Physicists found that there were just two classes of fundamental matter particles: **leptons** and **quarks**. The electron is the most familiar of the leptons, but two other lepton "generations"—the *muon* and *tau*—fill out the family. Quarks, which make up protons and neutrons, constitute the other family of matter. There is a significant difference between these two classes of particles in that only quarks can feel the strong nuclear force, whereas both quarks and leptons feel the weak and electromagnetic forces. And both quarks and leptons feel the force of gravity. Every particle comes with an *antimatter* twin, such that the "anti-" and "normal" particles take opposite electric charges (although particles that don't have an electric charge can be their own anti-particles). Thus, there are electrons and antielectrons (also called *positrons*), protons and antiprotons, and so forth. As discussed in Chapter 10, antimatter and matter are essentially mortal enemies. When matter and antimatter particles collide, they completely annihilate each other, transforming into photons.

interactive
Explorable List of
Particles and Forces

particle accelerator
An experimental apparatus used by physicists to study the structure of matter by colliding atoms and subatomic particles together at very high velocities (high energies).

standard model of particle physics A description of subatomic particles and forces.

lepton A class of fundamental particles of matter that includes electron, muon, and tau particles.

quark A class of fundamental particles of matter that make up other particles, such as protons and neutrons.

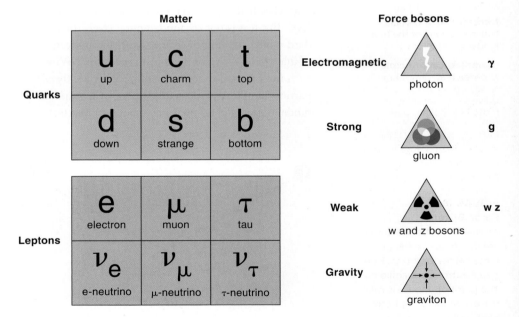

FIGURE 18.5 Standard Particle Model
The standard model of particle physics accounts for all known subatomic particles making up normal (nondark) matter. Leptons and quarks are the basic classes of matter particles, each of which occurs in six different forms. The standard model also accounts for three of the four fundamental forces between the particles, which are "felt" by the exchange of a so-called force boson particle.

interactive
Explorable Big Bang
Timeline

force boson A particle
that carries one of the four
fundamental forces.

cosmological principle
The idea that, on average, the
conditions in one part of the
Universe do not differ from
those in any other part.

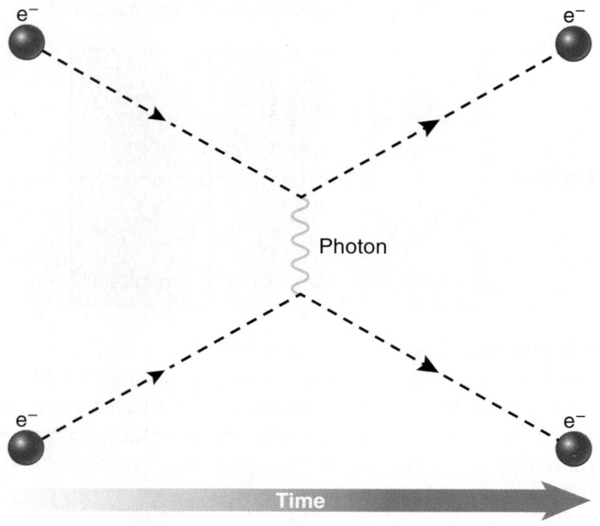

FIGURE 18.6
Force Bosons
All the forces between matter
are due to an exchange of
force carrier particles. Here
two electrons are colliding.
The point of impact is really
the exchange of a photon—a
light particle—which is the
carrier of the electromagnetic
force. The exchange of the
photon is what actually
changes the electrons'
direction as they move
through space.

But taking a census of all the Universe's particles and antiparticles was not the only job of the standard model. Particles "feel" each other by exerting forces. Recall that only four forces are at work in the Universe: gravity, electromagnetism, the strong nuclear force, and the weak nuclear force.

The standard model describes three of these forces and their effects on subatomic particles. The forces themselves are mediated by a separate class of particles called **force bosons**. The photon, for example, is the particle that carries the electromagnetic force. Exchanging photons is the way charged particles "feel" the electromagnetic force (**Figure 18.6**). The magnets hanging on your refrigerator are kept in place by unseen photons dashing back and forth between magnet and metal refrigerator door. Force bosons called *gluons* carry the strong nuclear force, while so-called W and Z bosons carry the weak nuclear force between particles. Scientists expect that the gravitational force is also carried by a particle, which they call a *graviton*. So far, however, there's been little progress toward a theory for the graviton because of the extremely difficult hurdles in trying to bring quantum physics and general relativity together.

This exquisitely articulated web of particles, antiparticles, and forces was the triumph of particle physics and is known as the standard model. With it, physicists had a grand map of matter at a fundamental level. But it was not *the* fundamental level. The standard model could not explain a long and important list of questions. Why are there just four forces? Why does each force have different strengths? Why are there two particle families: quarks and leptons? "The standard model doesn't tell us anything about dark matter

either," says Riess. "Since most mass in the Universe is in the dark form, that is a pretty big omission."

A particular concern for physicists was the host of numbers that had to be fed into the standard model from experimentation. More than 20 of these "constants" in the model did not arise from the theory itself but had to be specified by experiment. Without any way of knowing which "law" set those constants to the values we observe, physicists had to assume that the Universe could have ended up very different from what we observe. If, for example, the strength of the electromagnetic force were tweaked just a tiny bit, then biologically important molecules could never have formed. In general, any small change to the constants of nature in the standard model would lead to a Universe unlike what exists today. Physicists call this the "problem of fine-tuning," since it appears that the constants in the standard model must be tuned to the exact values we have or we could not exist.

Questions like fine-tuning are still unresolved and constitute the exciting frontiers of particle physics. Even with these open questions, however, enough has been learned from the standard model to enable a coherent description of cosmic evolution all the way back to fractions of a second after the Big Bang.

There is one last point to consider before we begin telling the story of cosmic history. In order to treat the entire Universe as an entity that can be studied using the mathematical laws of physics, astronomers invoke what they call the **cosmological principle**, which states that, on average, the conditions in one part of the Universe do not differ from those in any other part. This concept greatly simplifies the treatment of cosmological problems. In the end, however, astronomers must observationally test the truth of the cosmological principle. But recall from Chapter 17 that at the largest scales, the matter in the Universe does appear to become uniformly distributed.

e^- e^-

Photon

e^- e^-

Time

SECTION SUMMARY
Part I. The Stage and the Actors
- The Big Bang theory is built on the concept of flexible space-time (and the possibilities it entails for space-time geometry).
- The standard model of particle physics encompasses an understanding of matter and energy at a fundamental level.

Part II. The Story

Having set the stage and introduced the actors in our cosmic drama, we are now ready to begin the story.

Scientists break up the narrative of the Big Bang into a sequence of phases based on the dominant physical processes that operated during each phase. The most important point to remember is that as soon as we step forward from the "moment of creation," or $t = 0$, the Universe will be expanding, and the matter in that space dilutes. Thus, as we progress from one era of cosmic evolution to the next, we are also moving through a sequence of ever-decreasing average temperatures and densities.

Let's begin, then, just after the beginning.

THE SINGULARITY (t = 0).   Simply tracing the history of the expanding Universe backward suggests that the Universe would have begun with a state of infinite density, infinite temperature, and infinite space curvature. This is what physicists call a **cosmic singularity**, and it is similar to the kind of state that occurs in the center of a black hole. To fully understand what happens in a singularity, however, physicists need a theory of quantum gravity, which they don't have yet. Thus, the current version of the Big Bang is really only a theory of *after* the Big Bang. "That," says Riess, "is what we mean when we say our theory of the Big Bang really starts a short interval after all the players are already in place."

Physicists expect that at the incredible temperatures just after the Big Bang, all the forces were unified into a single force via what they call a "theory of everything." As the Universe expanded and cooled, these forces separated into the four we experience today. Gravity is thought to have separated first, and then the strong, weak, and electromagnetic forces separated one by one. So-called **grand unified theories** (or **GUTs**) focus just on the unification of these last three forces and aim to explain how all the particle interactions we see today emerged from the early history of the Big Bang. There are, however, no generally accepted grand unified theories yet, and the ones that have been proposed remain highly speculative.

INFLATION (t = 0 TO t = 10^{-33} SECOND).   In the first bare instants of cosmic history, the Universe was filled with quantum energy fields (or just **quantum fields**). These are distributions of energy that fill all space but appear as different kinds of fundamental particles. (Remember that in quantum mechanics, particles may behave as either particles or waves.)

When the Universe was just $t = 10^{-33}$ second old, many theorists think that something remarkable happened in post–Big Bang space as a quantum field found itself in an unstable, excited state (much like an electron in an upper orbit of an atom). The

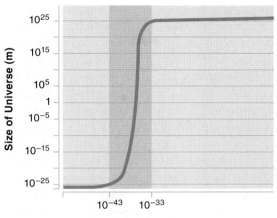

FIGURE 18.7
Inflation
During inflation, post–Big Bang space-time underwent a brief period of hyperrapid expansion such that the distance between any two points grew rapidly. Inflation began just a fraction of a second after the Big Bang and lasted even less time. Without the addition of inflation to the story of the Big Bang, the theory contains a number of paradoxes. After inflation ended, the Universe settled into the rate of expansion we see today, and cosmic evolution continued. Note that the observable Universe is that part from which light can reach us after 13.7 billion years of evolution.

energy locked up in that elevated state was enormous. Then, suddenly, the quantum field dropped back down to its ground state. As the field made its transition, the energy released pushed space into a kind of expansion on steroids. This period is called **inflation** (**Figure 18.7**).

"In Newton's gravity, there were only attractive forces," says Riess, explaining how inflation works. "But in Einstein's general relativity, gravity can actually be repulsive if there is the right kind of energy around. This is what happens during inflation when the quantum field drops back to its ground state."

During inflation, the scale of the post–Big Bang Universe blew up from the size of an atom to the size of a softball. By just a bit after 10^{-33} second following the Big Bang, the scale of the inflated region increased by a factor of 10^{60}. But inflation ended as quickly as it started, leaving behind a vastly enlarged space, which settled back into an expansion that was much slower (on the order of what we see today). Keep in mind that the Big Bang and inflation represent a stretching of space-time itself, not an expansion into a void. The phrase "the scale of the Universe" therefore refers to the part of it undergoing inflation that would become the observable Universe we see today.

Among the inflation era's most important legacies for the future of cosmic history were the small bumps and wiggles it left in the cosmic bath of matter-energy. These perturbations were the result of quantum fluctuations in the energy fields driving inflation, and they would persist through all the epochs that followed.

PARTICLE ERA (t = 10^{-33} SECOND TO t = 0.001 SECOND).   After inflation ended, the Universe was ultrahot (with a temperature of 10^{16} kelvins, or K) and

cosmic singularity The initial condition of the expanding Universe as a state of infinite density, temperature, and curvature from which the Big Bang was initiated. The presence of the infinities implies that a yet-to-be-discovered theory will provide a more complete account of initial condition.

grand unified theory (GUT) A yet-to-be-discovered theory that would explain how the strong and weak nuclear forces along with the electromagnetic force emerged from a unified force in the early instants following the Big Bang.

quantum field The distribution of energy that fills space, within which fundamental particles appear as excited states.

inflation A hyperrapid expansion of space-time that occurred extremely early after the Big Bang.

ultradense (with a density of 10^{16} kilograms per cubic meter, or kg/m^3). But cosmic density and temperature continued to fall with the expansion of space during this **particle era**. The matter and energy filling the Universe now went through a sequence of transitions with cooling and rarefaction. Many of these changes, particularly those at the earliest times and highest temperatures, are not fully understood by scientists today. What physicists are sure of, however, is that whatever emerged from the era of inflation eventually gave rise to the particles that make up the standard model (quarks, leptons, and force bosons).

To understand how these particles formed, physicists tend to think in terms of what are called **phase transitions**, which occur when matter moves from one form to another as temperature changes. The most familiar everyday example of a phase transition is the switch from liquid water to ice. As the temperature in a pool of water drops, freely moving H_2O molecules in the liquid become locked into ice crystals. Physicists think of the early Universe as a soup of particles (and the quantum fields associated with them). In this

phase of matter, particles could interact and switch from one form to another in specific ways. As the temperature dropped, some kinds of particles could no longer participate in the back-and-forth transformations, so their populations were locked in their current form and number for the rest of cosmic history. For example, when temperatures were high enough, quarks were able to move freely through the early Universe. But as the cosmic temperatures dropped, all quarks became bound together by the strong nuclear force into particles called baryons (such as protons and neutrons), consisting of three quarks, and mesons, consisting of a quark and an antiquark. This phase transition from free quarks at high temperatures to bound quarks at lower temperatures was essential to the emergence of the Universe we experience today (**Figure 18.8**).

One unanswered question in this process is how the Universe ended up with more matter than antimatter. Physicists expect that both forms of matter would have been produced in equal amounts during the early parts of cosmic history. But through a mechanism called **baryogenesis**, which is still poorly understood, the Universe ended up with far more protons and neutrons than antiprotons and antineutrons.

BIG BANG NUCLEOSYNTHESIS (t = 0.001 SECOND TO t = 3 MINUTES). When the temperature of the Universe fell to 10^9 K and the density fell to 10^4 kg/m^3, conditions were just right to turn the entire Universe into a nuclear reactor. In this epoch, known as **Big Bang nucleosynthesis (BBN)**, the light elements like helium and lithium could first be forged.

Big Bang nucleosynthesis must be distinguished from stellar nucleosynthesis, which you learned about in Chapter 10. Elements could not be "cooked" in the center of a star until the first stars formed, about a billion or so years after the Big Bang. Big Bang nucleosynthesis, however, began when the mix of protons, neutrons, electrons, and photons emerging from the particle era had the right range of temperatures and densities to allow fusion to begin. "It's important to distinguish between the relatively constant density and temperature in the core of a star and the rapidly dropping density and temperature in the Universe in the early moments after the Big Bang," says Riess. "That's what makes the BBN era so remarkable. It only lasts about 3 minutes."

Collisions between protons and neutrons led to deuterons, the same heavy form of hydrogen nuclei that is produced during fusion reactions in the Sun. As cosmic deuterium built up, continued collisions added

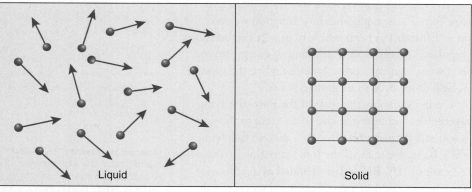

A Phase transitions of water

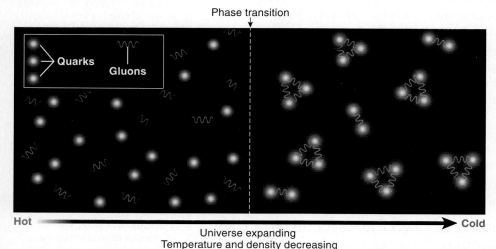

Phase transition

Quarks

Gluons

Hot

Cold

Universe expanding
Temperature and density decreasing

B Particle era

neutrons and protons until helium-4—with its two protons and two neutrons—became abundant.

Nuclear reactions depend on temperature and density, so as the cosmic temperature dropped, some reactions slowed to a crawl, like a cake that stops baking as the oven cools. In this way, the Universe's expansion carried the soup of matter through successive stages where once again some nuclei stopped transforming via nuclear reactions and were frozen out of the ongoing reactions. Within 3 minutes after $t = 0$, temperatures and densities dropped low enough that any further nuclear reactions were impossible. At that point fusion ended, and the elemental abundances of the lightest elements and isotopes were set. A nuclear census of the Universe by mass just a few minutes after creation would show it to be 75 percent hydrogen nuclei, 25 percent helium-4 nuclei, 0.0028 percent deuterium, 0.0015 percent helium-3, and about 10^{-8} percent lithium (**Figure 18.9**).

RECOMBINATION ($t \approx$ 380,000 YEARS).

As exotic as the early Universe might have been, it was still a collection of hot, dense matter. In this way, the entire early Universe was a blackbody. We have encountered blackbodies and blackbody radiation many times in our journey from planets to stars and on to galaxies. When matter in an object is dense enough that the distance between a photon's emission and its absorption is small (compared with the size of the object), then that object behaves as a blackbody. We've seen that blackbody radiation has a distinct spectral energy distribution with peak energy occurring at a wavelength determined by Wien's law ($\lambda_{max} = 0.0029$ m K/T).

This equilibrium of matter and radiation was the rule for the entire early history of the Universe. Matter particles jostled along with photons that were absorbed and spat out again in endless reactions. All the while, the young Universe remained in blackbody conditions. Then, about 380,000 years after the Big Bang, the blackbody photons were kicked out of the party.

The key event was the capture of electrons by protons to form the first hydrogen atoms. This period is called the **recombination era**. "This is a terrible misnomer," says Riess, "because it's the first time protons and electrons could form long-lived hydrogen atoms. We really should call it *combination*" (**Figure 18.10**).

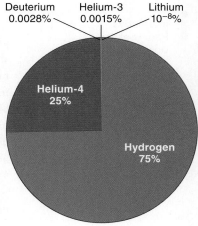

Deuterium 0.0028% Helium-3 0.0015% Lithium 10^{-8}%

Helium-4 25%

Hydrogen 75%

FIGURE 18.9
Big Bang Nucleosynthesis
In the epoch of Big Bang nucleosynthesis, falling temperatures and densities in the expanding Universe were briefly in the range to allow fusion of protons (hydrogen nuclei) into light nuclei such as deuterium, helium, and lithium. The proportions of each nucleus are collectively referred to as the "primordial abundance of elements," and its close match with theoretical predictions is one of the three primary pieces of evidence for the Big Bang.

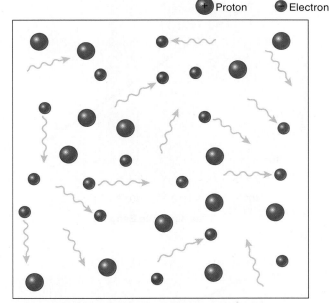

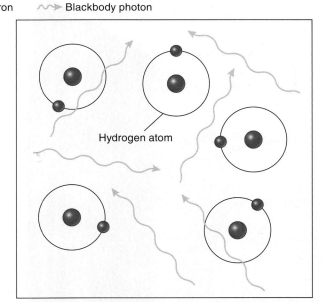

⊕ Proton   ⬤ Electron   〜➤ Blackbody photon

Hydrogen atom

A Before recombination: photons absorbed after traveling a short distance

B After recombination: photons can travel freely without being absorbed

FIGURE 18.10 Recombination
In the epoch of recombination, lower temperatures meant that protons and electrons were moving slowly enough for them to bind together, forming the first hydrogen atoms. (A) Before recombination, matter and photons were in equilibrium because of their rapid interactions, giving the radiation a blackbody spectrum. (B) After recombination, the newly formed atomic hydrogen was unable to absorb the blackbody photons, leaving the radiation "decoupled" from matter. The blackbody photons became fossils that can be observed today as the cosmic microwave background. Note that when decoupling occurred, the blackbody radiation had wavelengths in the ultraviolet range.

particle era A period in cosmic evolution when the particles described by the standard model of particle physics emerged.

phase transition A change in the state of matter based on changes in temperature (or other quantity).

baryogenesis A process that produced an asymmetry between matter and antimatter in the early Universe.

Big Bang nucleosynthesis (BBN) Nuclear fusion of light elements in the first 3 minutes after the Big Bang.

recombination era The era approximately 380,000 years after the Big Bang in which the first electrons were captured by protons to form hydrogen atoms, thus decoupling matter from the cosmic blackbody radiation.

interactive
Radius of the
Universe vs. Time

One of the most important changes that came with recombination was the fate of the blackbody photons. The newly formed hydrogen atoms could not absorb the still-present blackbody electromagnetic radiation. The physics of hydrogen's electron orbits made interactions between hydrogen atoms and the blackbody photons all but impossible. Thus, recombination also meant a *decoupling* of matter (the now ubiquitous hydrogen atoms) from the bath of blackbody photons.

The blackbody photons were left without dance partners. Just a few thousand years earlier, they could not travel more than a tiny distance through the Universe without being absorbed. After decoupling, they were left orphaned, free to wander space almost unimpeded. "It's like when you are stuck in a fog," says Riess. "If you turn on your headlights, you can't see anything. The light just scatters around. But once the fog clears, the headlight beams can travel a long distance through the air without interruption." As time marched forward, the only change this fossil light would experience was a stretching in wavelength directly tied to the expansion of space itself.

The decoupling of matter and radiation via recombination had other consequences, the most important of which are related to cosmic structure. As long as radiation and matter were strongly linked, the tiny perturbations left over from inflation could not grow. Before recombination, if gravity tried to draw material into a region of higher density, the region simply grew hotter. Increased heat meant increased pressure, which pushed back against gravity, smoothing out the perturbations via sound waves. After matter and the blackbody radiation went their separate ways, gravity could begin its work because contracting bodies could cool by emitting radiation that escaped from the infalling gas. Thus, dense regions could become denser and build up all the large-scale structure we explored in the previous chapter. In this way, recombination marked the end of the Universe's early history. **Figure 18.11** summarizes the phases following the Big Bang.

Time since Big Bang

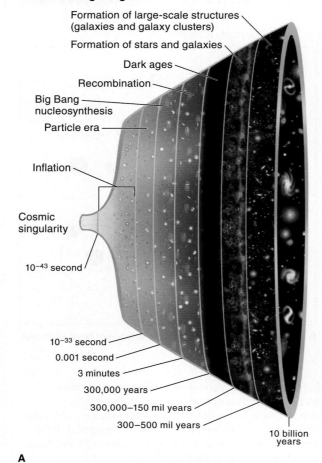

FIGURE 18.11
The Post–Big Bang Universe
As the Universe expands, the temperature and density of matter-energy within it drop, leading to transitions from one epoch of cosmic evolution to the next. (A) Here we see the main stages of the evolution of the Universe as it expands: inflation, particle era, Big Bang nucleosynthesis, recombination, and the formation of large-scale structure. Cosmic expansion causes both (B) density and (C) temperature to decline with time.

A

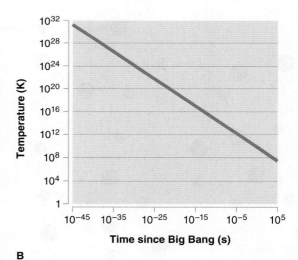

B

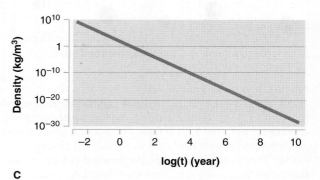

C

SECTION SUMMARY
Part II. The Story

- Tracing cosmic history backward takes us to a cosmic singularity, which implies infinite density, temperature, and space-time curvature. This state is similar to the singularity at the center of a black hole, and a complete theory for it does not yet exist.
- Inflation was a period of hyperexpansion of a small region of post–Big Bang space-time.
- During the particle era, the particles of the standard model formed as the temperature and density of the Universe dropped.
- During the era of Big Bang nucleosynthesis (BBN), the nuclei of the lightest elements formed by fusion, and their primordial abundances, became fixed.
- During the era of recombination, matter and its partner blackbody photons decoupled, allowing the photons to travel unimpeded through space.

18.3
How Do We Know? The Three Observational Pillars of Big Bang Theory

It's quite a story really: the Universe begins in an ultradense, ultrahot state and immediately begins expanding. What observations, however, do astronomers have to convince themselves that the Big Bang actually happened?

Three key pieces of evidence support the classic narrative of the Big Bang. Here, *classic* means everything except inflation, which is a more recent addition to Big Bang theory. Let's deal with the classic part of the story first and come later to the evidence for inflation.

Pillar 1: Cosmic Expansion

You should now be quite familiar with the first pillar. Hubble's discovery of galaxy recession (see Chapter 16) was the first evidence of a Universe that was not static. Note, however, that if expansion were the only evidence we had, it would still be possible to develop a model like Fred Hoyle's steady-state Universe, in which space always expanded but always looked the same. The Big Bang model, in contrast, tells us that the Universe looked very different in the past than it does today. "It's kind of like seeing a cake that is rising as it bakes," says Adam Riess. "It's natural to infer that it was all much smaller and more compact when it started out."

Adam Riess

Adam Riess is not that interested in astronomy, though he won the Nobel Prize for work he did at a telescope (a lot of telescopes actually). "I was just not one of those kids who was into it," he says. "I always liked science. At first I liked biology. Then I did a summer program in high school that convinced me it was really physics that I liked." Riess majored in physics in college, but it wasn't until graduate school that astronomical science appeared in his personal universe. "That was the point when I discovered astrophysics and cosmology. It was cosmology that really got me excited."

Riess was born with the kind of curiosity that lands a kid either in science or in the hospital. "I was always blowing stuff up," he recalls. "I would put a wire across those two prongs on an electrical outlet to see what would happen. But I did less dangerous stuff too, like taking radios apart." Along with this practical experimental side, Riess held deeper philosophical questions close to his heart. It was the possibility of pursuing those questions that sent him to the "dark side" of astrophysical studies.

"When I started learning about cosmology," Riess explains, "I thought, 'Really? You can really ask basic questions like, How old is the Universe? or What's its ultimate fate?' I was even more amazed that you could actually design quantitative experiments that answer those questions. When I stumbled on this aspect of cosmology, I was just thrilled!"

Since then, Riess has never lost interest in the big questions. "I would say the questions that I have—or the questions that seem important to me, that I want to answer—are all physics questions. But I have to use astronomical tools to answer them."

SECTION SUMMARY
Pillar 1: Cosmic Expansion

- The first evidence for the Big Bang came from the expansion of the Universe, as seen in the recession of galaxies described by Hubble's law.

Pillar 2: Big Bang Nucleosynthesis (BBN)

The brief period lasting a few minutes after the Big Bang, when the Universe was hot enough to drive fusion reactions, left a mark on the cosmos that we can see today. The relative abundance of nuclei such as hydrogen, deuterium, helium-3, helium-4, lithium-6, and lithium-7 can be directly measured and compared against detailed models of the Universe during its nucleosynthesis epoch. In this way, the light elements act as a kind of fossil record of the Big Bang, enabling us to infer conditions in the early Universe.

Astronomers can measure the **primordial abundances of light elements** in a number of ways. For example, to measure the primordial helium-4 abundance, they must find relatively primitive

primordial abundance of light elements The proportions of hydrogen, helium, deuterium, and other light elements created in the first 3 minutes after the Big Bang.

FIGURE 18.12
Arno Penzias and Robert Wilson

Arno Penzias and Robert Wilson, standing in front of the radio telescope they used to discover the cosmic microwave background—fossil photons left over from the early Universe.

cosmic microwave background (CMB) Blackbody radiation released during recombination, with a peak wavelength in the red part of the electromagnetic spectrum. The expansion of the Universe now puts the peak of CMB in the microwave region of the electromagnetic spectrum.

systems—astronomical environments that haven't been tainted by helium-4 formed within stars. Dwarf galaxies make good candidates for such "primitive" systems because they are especially poor in oxygen and nitrogen (elements that must be processed within stars). Thus, measurements of helium-4 in dwarf galaxy gas clouds give astronomers a measure that is close to the abundance it exhibited after BBN finished.

The measured values of light-element abundances correspond well with the predictions from BBN theory. This match between theory and observations makes it clear to astronomers that the Universe must have been hot enough and dense enough early in its history for nucleosynthesis to occur. In other words, the Universe *then* looked very different from the Universe *now*.

SECTION SUMMARY
Pillar 2: Big Bang Nucleosynthesis (BBN)
- The observed abundances of the lightest elements in the Universe are well matched by models predicting the proportions that Big Bang nucleosynthesis would have produced.

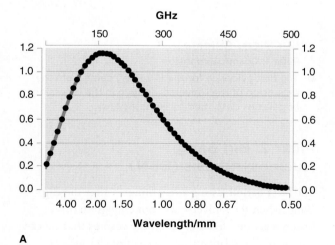

FIGURE 18.13
CMB Spectrum
(A) The spectrum of the CMB is that of a blackbody that has a peak radiation consistent with a temperature of 2.7 K. (B) As the Universe expanded, the wavelength of all the electromagnetic waves making up the CMB was stretched, and correspondingly, the CMB's temperature dropped.

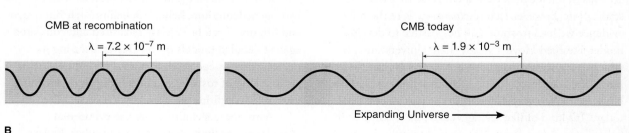

Pillar 3: The Cosmic Microwave Background (CMB)

In the spring of 1964, two Bell Laboratory scientists, Arno Penzias and Robert Wilson, managed to trip over one of the greatest cosmological discoveries ever made and win themselves a Nobel Prize in the process (**Figure 18.12**). At the time, the two astronomers were working on the new technology of satellite telecommunications via a massive horn-shaped microwave antenna situated in a field outside of Holmdel, New Jersey. (Recall that microwaves are electromagnetic waves with wavelengths between 1 meter and 1 millimeter [mm].) Bell Labs was initially interested in microwave transmissions to and from the orbiting satellites that were just beginning to populate the sky. In time, however, the lab (which was a powerhouse of both industrial and pure scientific research) allowed the two scientists to use the antenna for basic astronomy. The plan had been to look at distant galaxies in microwave radiation. Unfortunately, that plan kept failing.

The problem was an annoying, low-level "noise" that persisted regardless of which direction the antenna was pointed—a microwave hiss that refused to go away. For weeks, Wilson and Penzias struggled to root out the problem. They rebuilt the electronics. They cleaned layers of pigeon poop from the antenna surface. Nothing changed. Then, through painstaking work, they came to understand the problem: there was no problem.

The signal was real, and it was yet another "fossil" left over from the Big Bang. The light waves flooding the microwave antenna were radiation released more than 13 billion years ago during the epoch of recombination. When the full spectrum of Wilson and Penzias's mystery microwaves was measured, it turned out to be a *perfect blackbody*. That meant the electromagnetic waves they discovered were the blackbody light left over from the era of recombination 380,000 years after the Big Bang.

The peak wavelength λ_{max} of their blackbody radiation occurred in the microwave portion of the spectrum at 1.9 mm (1.9×10^{-3} meters). This was almost exactly what Gamow and Alpher's theory had predicted back in 1954. With some subsequent fine-tuning, the temperature of the **cosmic microwave background (CMB)** was found to be just 2.7 K. The reason it was so cold was that the wavelength of the blackbody radiation had been stretching with the expansion of space since recombination had decoupled it from the cosmic gas. This spatial stretching translated into an ever-decreasing temperature for the waves. Thus, although the blackbody's temperature had been about 3,000–4,000 K at the time the radiation was released more than 13 billion years earlier, now it was just barely above absolute zero (**Figure 18.13**; **Going Further 18.1**).

"The CMB tells us that the Universe had to have been much hotter than it is now," says Riess, "because you still see the glowing radiation. It's as though you step into a room a moment after somebody pulled a grenade pin, and you see shrapnel flying apart and you feel heat. So even though you didn't see somebody trigger the grenade—you didn't actually see the explosion—you can infer what's going on in the room."

SECTION SUMMARY
Pillar 3: The Cosmic Microwave Background (CMB)

- The discovery of the cosmic microwave background was a key piece of evidence for the Big Bang, as it constitutes fossil radiation from the time of recombination.

18.4
Beyond the Classic Big Bang Model

One of the most important consequences of the CMB studies was the development of support for the existence of inflation as part of the Big Bang. The idea of inflation was first fully developed by Alan Guth of MIT in the 1980s. It was soon picked up by others and became a major addition to the story of cosmic

GOING FURTHER 18.1   THE LANGUAGE OF THE COSMOS
The CMB and the Expansion of the Universe

How much has the Universe expanded since the cosmic microwave background (CMB) radiation decoupled from matter?

Recall that the temperature of the plasma filling the Universe had dropped to $T = 4{,}000$ K when electrons and protons were finally moving slowly enough for electromagnetic attraction to allow them to find each other and bind together into neutral hydrogen atoms. In other words, just before recombination the Universe was, essentially, a blackbody with a temperature of $T = 4{,}000$ K. Now let's calculate the wavelength where this blackbody radiation peaked. We can do this using our old friend Wien's law:

$$\lambda_{max} = 0.0029 \text{ m K}/T$$
$$= 0.0029 \text{ m K}/4{,}000 \text{ K} = 7.2 \times 10^{-7} \text{ m} = 720 \text{ nm}$$

This tells us that at the moment of decoupling, the whole Universe was glowing with a peak at 720 nm, in the red part of the visible spectrum.

But how does this number tell us anything about the Universe's expansion? Recall that after decoupling, the electromagnetic waves making up this blackbody radiation stopped interacting with matter. The only thing affecting the radiation was the expansion of space, which explains why the CMB seen today has a peak wavelength that is so much larger than 720 nm. Thus, the expansion of the Universe is encoded in the CMB radiation.

If we call the size scale of the Universe at decoupling R_d and the size scale of today's Universe R_o, understanding that the wavelength of the CMB is directly proportional to the size scale, then we have

$$R_o/R_d = (\lambda_{max})_o/(\lambda_{max})_d$$

As we have seen, the peak of microwave radiation in the current cosmic epoch is $(\lambda_{max})_o = 0.0019$ meter $= 1.9 \times 10^{-3}$ meter. Plugging these values in, we see that

$$R_o/R_d = (1.9 \times 10^{-3} \text{ m})/(7.2 \times 10^{-7} \text{ m}) = 2.638 \times 10^{3}$$

For an accurate result, the details of how recombination occurred do matter, but our answer is good enough to tell us that the Universe has expanded by a factor of about 1,000 between the era of recombination and the current epoch. That is a lot of new space that has been added to the cosmos.

evolution. But astrophysicists needed to find some solid evidence before they could accept the possibility that it had, in fact, occurred.

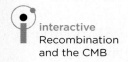
interactive
Recombination
and the CMB

Inflation, the CMB, and the Origin of Structure

When Wilson and Penzias stumbled on the CMB, they found that every region of the sky was filled with blackbody radiation at a temperature of 2.7 K. But astrophysicists knew that structures such as today's galaxy clusters must have originated in small-density perturbations present at the era of recombination, when the CMB radiation was released. According to the physics governing the plasma at the time of recombination, regions that were slightly denser would be slightly hotter, while regions that were lower in density would be lower in temperature. Thus, small perturbations in density implied that small perturbations (or fluctuations) in temperature must have existed as well. Recall from Chapter 17 that the large-scale structure of clusters and superclusters that we see in the Universe now must have come from some kind of early fluctuations.

These temperature fluctuations were expected to show up in maps of the CMB. But when physicists estimated the expected changes in temperature, the challenge ahead was clear. The fluctuations led to a change in temperature of only one part in 100,000. Taking these measurements required overcoming enormous technical difficulties, which forced the observations to be made from space. After years of effort, the Cosmic Background Explorer (COBE) satellite, launched in 1989, successfully mapped the CMB across the entire sky with enough sensitivity to find the cosmic temperature fluctuations (**Figure 18.14**). It was yet another triumph for the Big Bang theory. Other CMB satellites followed, such as the Wilkinson Microwave Anisotropy Probe (WMAP) in 2001 and Planck in 2009, each mapping the sky at ever-higher resolution.

One significant prediction of inflationary Big Bang cosmology was the so-called **spectrum of perturbations**. Quantum fluctuations imposed on matter during inflation became the tiny bumps and ripples that persisted into the era of recombination as the seeds of all current large-scale structure. Inflation made very definite predictions for the distribution, or spectrum, of those perturbations in terms of their size and strength. With higher-resolution maps of the CMB and its temperature fluctuations, astronomers

spectrum of perturbations
The range and strength of variations in density imposed on cosmic matter during inflation.

A

B

C

FIGURE 18.14 Mapping the CMB
A succession of satellites have mapped fluctuations in the CMB with ever-greater detail and accuracy beginning with (A) the COBE mission (1989), continuing with (B) WMAP (2001), and ending (to date) with (C) Planck (2009). The fluctuations correspond to small regions in the early Universe with different temperature and density conditions. These small fluctuations were the gravitational seed for the large-scale structure of the Universe we see today.

were able to show how closely inflation's prediction for the spectrum of perturbations matched what was actually observed. This was the first real test of inflation theory using data, and it passed that test with flying colors.

SECTION SUMMARY
Inflation, the CMB, and the Origin of Structure
- Large-scale structures must have formed from density fluctuations in the early Universe, and the fingerprints of those fluctuations were found in temperature variations in the CMB.

The Paradox of Universal Uniformity

Why were astronomers so keen on the idea of inflation? "Inflation ended up solving a number of problems or paradoxes that were appearing in the classic version of the Big Bang," says Clifford Johnson, a theoretical physicist at the University of Southern California.

One of these paradoxes was the recognition that the Universe, as shown by the CMB, was remarkably uniform. Statistically, the temperature fluctuations on one side of the sky looked just like those on the other side of the sky. It was like reading the paper one morning and finding out that every city on the planet had exactly the same temperature. Since these different regions corresponded to seeing different parts of the Universe at early times, the uniformity implied that all regions of the entire Universe at recombination must have already been able to share information, smoothing out any large-scale differences (**Figure 18.15**). "They should have been in contact with each other," says Johnson, "but that just wasn't possible with the

Clifford Johnson

If you're going to be a scientist, it helps to have a secret lab. "There was this place under the stairs in the house I lived," says Clifford Johnson. "The house was still being built, so you had to sort of climb up into this spot. It was the perfect hideout. I had my little bottles of pond water and this old, badly functioning microscope I had fixed."

While growing up on the island of Montserrat in the Caribbean, Johnson was vaguely aware of where he was going with his work. "I knew that there was this thing called 'scientist' and didn't know exactly what it was. I did know I wanted to be one. In fact, I remember a family friend once asking me, 'What kind of a scientist do you want to be?' and me being really angered by the question. I just wanted to know about everything, so I didn't want to choose." Eventually, Johnson realized that there were different kinds of scientists, and sooner or later he would have to make that choice. "I actually just got a dictionary and tried to find all the '-ists' and '-ologists' to see what they studied. In some sense, I chose to be a physicist because whoever wrote that entry said it was the study of everything."

After moving back to England, Johnson excelled in his high school studies. Despite his continuing interests in everything, however, he ended up specializing in theoretical physics. These days he is one of the world's experts in string theory. Since strings are considered a good candidate for a TOE, or theory of everything, perhaps Johnson has not strayed too far from his dreams in the lab under the stairs.

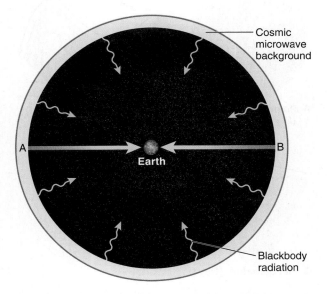

FIGURE 18.15 The Paradox of Universal Uniformity A paradox in cosmology was the recognition that the Universe, as shown by the CMB, was too uniform, with temperature fluctuations on one side of the sky looking statistically like those on the other side of the sky. If light from points A and B on the CMB is reaching Earth only now, then there has not been enough time since the epoch of recombination for light waves (the fastest signals possible) to smooth out the different parts of the cosmos. The brief period of inflation implies that the different parts of the Universe we now see widely separated were, in fact, once close enough together before inflation happened to be connected by light signals.

classic version of the Big Bang because the expansion of the Universe was too slow." According to classic Big Bang theory, there had not been enough time from $t = 0$ to the era of recombination for light waves (the fastest signals possible) to reach from one side of the cosmos to the other, bringing different regions into contact and hence smoothing out their differences.

By assuming there was an early period of inflation, cosmologists were able to resolve this and other paradoxes in classic Big Bang theory. The brief period of inflation implies that the different parts of the Universe we now see widely separated were, in fact, once close enough together to be connected by light signals before inflation happened.

SECTION SUMMARY
The Paradox of Universal Uniformity
- Inflation explains how the Big Bang produced a Universe with statistically uniform structure.

Density and Cosmic Destiny

Inflation solved one other important problem for cosmologists that, in a way, led to headaches of its own. In the classic Big Bang theory, Einstein's equations imply a critical link between the density of the

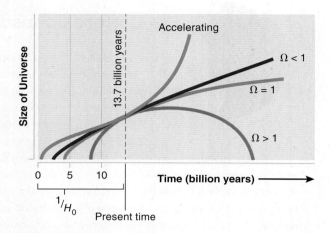

FIGURE 18.16 The Fate of the Universe
There is a critical link between the density of the Universe and its geometry. With high enough density of matter-energy, the gravitational attraction would slow, stop, or reverse the expansion of the Universe. Omega (Ω) is the ratio of the measured density to the critical value, where an Ω value of 1 will slow the expansion such that even though the expansion proceeds indefinitely, the expansion speed approaches zero more and more closely as time passes. An Ω value of greater than 1 will result in a Big Crunch, while an Ω value of less than 1 will mean that the Universe keeps expanding forever. However, the addition of dark energy allows for more complicated relationships between the Universes' geometry and it's fate.

Universe and its geometry. If the Universe began with a high enough density of matter-energy, then gravitational attraction should, eventually, overwhelm the expansion initiated at the Big Bang. The gravitational attraction of everything to everything else eventually stops cosmic expansion and turns it around into an overall cosmic contraction, leading to the opposite of the Big Bang: a Universe ending in a **Big Crunch**. "It's like firing a missile that doesn't reach escape velocity," says Johnson. "The rocket eventually falls back to Earth. With the Universe, if the initial Big Bang isn't strong enough, then the Universe collapses back on itself."

If the matter-energy density of the Universe is too low, however, gravity can never decelerate cosmic expansion enough to halt the stretching of space, and the Universe goes on expanding forever. The final possibility is that the Universe has just the right value of matter-energy density—what cosmologists call the **critical density**. In that case, the Universe will slow down in such a way that even though the expansion proceeds indefinitely, the expansion speed approaches zero more and more closely as time passes. Thus, everything depends on the density of the Universe

Big Crunch A theoretical contraction of the Universe that would occur if the combined matter-energy exceeded critical density.

critical density The matter-energy per unit volume that is just sufficient to slow the expansion of the Universe such that while the expansion proceeds indefinitely, the expansion speed approaches zero more and more closely as time passes.

omega (Ω) The ratio of the observed density of matter-energy in the Universe to critical density.

relative to the critical density. Astronomers express this relationship in terms of a parameter they call **omega (Ω)**: the ratio of the measured density to its critical value (**Figure 18.16**). Each of the possibilities just described is associated with its own form of geometry.

Einstein's equations also predict that the Universe can have different kinds of geometry, meaning different kinds of curvature, depending on density. If $\Omega > 1$, we have a Universe with positive curvature. "That gives the same kind of three-dimensional spherical geometry as we have for the two-dimensional surface of Earth," Adam Riess says. Like the surface of Earth or the surface of a beach ball, a Universe with positive curvature would be finite (only so much space) but have no boundaries. If $\Omega < 1$, we have a Universe with negative curvature, yielding a saddle-shaped geometry. An Ω value of 1 means a flat Universe with no curvature (**Figure 18.17**).

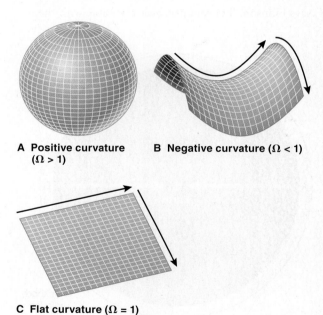

A Positive curvature ($\Omega > 1$)

B Negative curvature ($\Omega < 1$)

C Flat curvature ($\Omega = 1$)

FIGURE 18.17 Omega and the Curvature of Space-Time
Each value of Ω can be related to a specific geometry of the Universe. (Note, however, that the images here are just two-dimensional representations of the three dimensions of space.) (A) $\Omega > 1$ means positive curvature, like that on the surface of a sphere. (B) $\Omega < 1$, a Universe with negative curvature, yields a saddle-shaped geometry. (C) $\Omega = 1$ means a flat Universe with no curvature. Note that in all cases, space is not expanding "into" anything. The fabric of space-time represented by these figures (even when it stretches to infinity) represents all that exists.

One of the problems cosmologists faced before the inflation theory was developed was that observations implied Ω was close to 1 (typical values were about 0.2 or 0.3). But according to the equations governing the evolution of the Universe, if Ω started even a little bit below 1 (say, 0.9999), it would evolve into incredibly low values by our current epoch (on the order of 10^{-22} or less). Likewise, if Ω started greater than 1 at the Big Bang (say, 1.0001), it should now be greater than 1 by many orders of magnitude. In other words, unless Ω started *exactly* at 1 at the Big Bang, cosmic history should have pushed it very far from 1 as the Universe evolved. But observations were showing Ω close to but not exactly 1, meaning either that astronomers were missing something or that somehow the Universe had been fine-tuned early on to give this nearly-but-not-quite-1 value that we see today.

Inflation solved this problem. By pushing space-time into hyperexpansion, inflation flattened the geometry of space, driving it toward 1 rather than away from 1 (**Figure 18.18**). Because the rapid expansion stretches space, in turn reducing the curvature, inflation always takes the part of space we can see and drives its value of Ω toward the critical value. Thus, inflation theory allowed cosmologists to breathe a sigh of relief, since their values of Ω that were close to 1 ($\Omega \approx 0.3$) no longer seemed like a miracle.

But hiding in this result was another dilemma. As cosmologists got a better look at the CMB, they found their best models predicting a value of Ω not around 1, but exactly equal to 1. If the matter-energy detected (including the dark matter) added up to only 0.3, then what was making up the rest of Universe that was giving a total Ω actually equal to 1? The answer came as a surprise and represented one of the most important cosmological discoveries of the 20th century, after the discovery of the CMB.

Although the geometry of space-time might seem like an abstract property it does, in fact, have observable consequences. For in a flat space the sum of the angles in a triangle must equal 180°. But in a space with positive curvature ($\Omega > 1$) the angles would sum to more than 180°, while a space with negative curvature ($\Omega < 1$) would have angles that sum to less than 180°. Remarkably, the fluctuations in the CMB provide a direct way of distinguishing effects like these. Rather than providing giant triangles on the sky, it's the size of the CMB fluctuations that is sensitive to the kind of curvature our

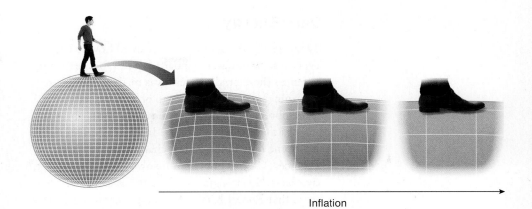

FIGURE 18.18 Inflation Explains Flatness
Each panel here shows a sphere inflated by a factor of 3 from the previous panel. By the third panel, the sphere's curvature is practically undetectable. Similarly, inflation provides the explanation for the flattening of the Universe's geometry, consistent with the observed value of $\Omega = 1$.

Universe has. Measurements of the size of the CMB brightness variations carried out with the satellites WMAP and Planck have confirmed that the Universe has a flat geometry.

Finally, it is important to understand that the space-time of our Universe (as represented by the different geometries discussed in this section) does not expand *into* anything. It can be very hard to wrap our minds around thinking about the space-time of the Universe as a whole, but the two-dimensional curved surfaces in Figures 18.17 and 18.18 represent the whole of physical reality, which is itself expanding. There is no other area for it to expand into.

SECTION SUMMARY
Density and Cosmic Destiny

- The fate of the Universe depends on the density of matter-energy and the resulting gravitational effects.

- The Universe has three possible geometries. $\Omega > 1$ indicates a space that is finite but unbounded, whose expansion will eventually stop and begin contracting. $\Omega < 1$ indicates a space with a saddle shape that will expand forever. $\Omega = 1$ indicates an infinite flat space (no curvature) whose expansion proceeds indefinitely, but the expansion speed approaches zero more and more closely as time passes.

- Current observations support a value close to 1 for the density parameter Ω.

- The hyperexpansion that occurred during inflation was crucial to setting the value of Ω at 1.

Dark Energy

By the 1990s, astronomers had learned how to use Type Ia supernovas as standard candles for measuring distance. These bright supernovas made it possible to determine accurate distances to objects far across the entire observable Universe. One of the most obvious projects that made use of Type Ia supernovas set out to determine how Hubble's law changed at very large distances. "That is how I got involved with the project," says Riess. "I was a grad student at Harvard, and I heard that Robert Kirshner had a project using Type Ia supernovas to test Hubble's law at huge distances. So I asked him if I could join his team." The rest, as they say, is history.

Recall that astronomers had always expected that the mutual gravitational attraction of the Universe's matter-energy would slow cosmic expansion to some degree. A value of $\Omega > 1$ meant lots of deceleration and a Big Crunch, whereas $\Omega < 1$ meant almost no deceleration and infinite expansion. By using Hubble's law and measuring distances (D) and velocities (V) of objects very far away, the deceleration could be determined by plotting V versus D on a graph and seeing how their relation departed from a straight line. "Objects far away (deeper in the past)," explains Riess, "would be moving faster than the linear Hubble's law predicted because they had not yet been decelerated."

But, as you already know, Riess and his colleagues' painstaking multiyear project to push the V-versus-D plot deeper into space and farther back in time found acceleration rather than deceleration. The expansion of the Universe wasn't slowing down; it was speeding up (**Figure 18.19**).

Cosmic acceleration implies that a force of some kind is pushing space apart, and since forces require energy, astronomers began speaking in terms of a **dark energy** that pervaded all space. The term *dark* simply meant that we could not measure it directly but could only measure its effects on luminous matter (just as with dark matter). This new form of dark energy, however, had to be different from the "normal" matter-energy we usually encounter. Whereas normal matter-energy always leads to gravitational attraction, dark energy has an "antigravity" effect in pushing space apart. "Whatever the dark energy is," says Riess, "it's having a repulsive gravitational effect."

Because of this difference between normal matter-energy and dark energy, astronomers now specify different forms of the density parameter Ω. There's Ω_m for normal matter, Ω_{dm} for dark matter, and

dark energy Energy driving the acceleration of the expansion of space-time.

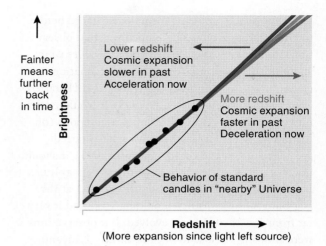

FIGURE 18.19 Acceleration of the Universe's Expansion
By using Type Ia supernovas as standard candles, astronomers can look back across large distances and large look-back times to see how cosmic expansion has changed over time. This figure shows the relationship between brightness and recession velocity for standard candles in the nearby and distant Universe. Since standard candles all have the same luminosity, objects observed to be fainter are farther away and hence are further back in time. In cosmologically nearby objects, the brightness-recession velocity plot is linear, consistent with Hubble's law. But at large distances sampled by Type Ia supernovas, either the curve will bend downward, signifying faster expansion in the past and hence deceleration now, or it will bend upward, signifying slower expansion in the past and hence acceleration now. The Type Ia data favor acceleration and hence the existence of dark energy.

Ω_{de} for dark energy. The total matter-energy density of the Universe then becomes the sum of the different kinds: $\Omega_T = \Omega_m + \Omega_{dm} + \Omega_{de}$. Astronomers soon gathered other evidence for the existence of dark energy, including direct measures of the geometry of the Universe via the size of CMB temperature fluctuations. All of this led to the remarkable discovery that $\Omega_{de} = 0.7$ and therefore $\Omega_T = 1$ (**Figure 18.20**). "It's pretty remarkable that as late as 1998, we still didn't know what most of the Universe was made out of," says Riess. "Now it seems pretty clear; dark energy is the dominant constituent of the cosmos."

Astronomers and physicists still do not know what this dark energy is made of, though many suggestions have been put forth. Einstein himself proposed a form of dark energy in 1917 in his first attempts to derive cosmological models from his equations. To keep his model Universe static, he added a term called a cosmological constant (referred to by the simple label Λ). Once the Universe was discovered to be expanding, Einstein abandoned the cosmological constant and called it his greatest blunder. But since this term represented exactly the kind of antigravitational energy needed to explain cosmic acceleration, today many astronomers have resurrected the cosmological constant. "The cosmological constant means that every cubic centimeter of space has the same amount of vacuum energy, and that amount doesn't change with time," explains Riess. "As space expands, you get more volume (more cubic centimeters) and hence more dark energy."

A second possibility is that dark energy does change with time. "This idea sees what is happening with the dark energy now very much like inflation in that there must be invisible fields in space that are decaying (though this time very slowly) and giving up their energy to cosmic acceleration."

Many other ideas about the nature of dark energy have been proposed and are being actively explored as you read these words. Without a doubt, learning the origin and nature of dark energy is one of the greatest challenges for modern physics and cosmology. Its discovery shows just how much there still is to learn about the Universe we inhabit.

SECTION SUMMARY
Dark Energy
- The discovery that the expansion of the Universe is accelerating led to the identification of dark energy, which represents 70 percent of the total matter-energy.

Cosmology as a Precise Science

"Even as recently as when I was an undergraduate, cosmology used to be the butt of a lot of jokes," says Clifford Johnson. "People made fun of how poorly known most of its basic measured quantities were. The accuracy was pretty bad."

That's not the situation anymore. From the CMB to Big Bang nucleosynthesis to the statistics of large-scale structures, astronomers now have many ways to test the modern narrative of cosmology. With the addition of inflation and dark energy, the story of the Universe's evolution from the Big Bang has become richer and more detailed than the early version of the classic Big Bang. Most important, our ability to test key features of that story has grown. Just 50 years ago, cosmological parameters like Hubble's constant H or the density parameter Ω were either poorly known or not known at all. Today, astronomers are able to pin these numbers down to within less than a few percentage points. This is what scientists sometimes mean when they talk about cosmology as a precise science. "We've also got a lot more statistics now," says Johnson. "That allows us to really bring down the uncertainties. We know what we know pretty accurately."

One measure of precision is the ability to use the observed data to distinguish between variations of a cosmological model. For example, all versions of Big Bang models make predictions for the number of temperature fluctuations of a given size that should be imprinted in the CMB. As we've seen, these fluctuations come from perturbations imposed by inflation, but they were also modified by sound waves running through the cosmic gas when recombination occurred.

The number of large-scale temperature fluctuations (in terms of how much space they occupy) versus small-scale fluctuations will be different for different models of the Big Bang. For example, more large-scale fluctuations occur in models with more dark energy relative to dark matter. Astronomers have computed many "model" universes in this way, and each model has a different mix of dark energy, dark matter, normal matter, and so forth. Each model yields different predictions for the statistics of temperature fluctuation in the CMB. "By comparing these predictions with the actual data from something like the Planck satellite," explains Johnson, "we've been able to eliminate a lot of possibilities for the constitution and history of the Universe."

SECTION SUMMARY
Cosmology as a Precise Science
- The parameters of cosmic evolution can now be measured with great precision.

18.5
Questions about "Before" and "Everything"

As we've seen, the Big Bang is not really a theory about "creation" but one about what happens "after time

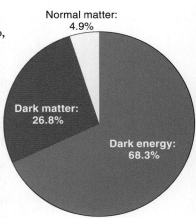

FIGURE 18.20
Dark Energy
Atoms, or luminous matter, of which we, planets, stars, and mobile phones are made, constitute only a small portion (about 4.9 percent) of the matter-energy of the Universe. Dark matter makes up 26.8 percent, and dark energy represents the majority, 68.3 percent.

began." Most scientists, however, were never quite satisfied with the idea of the Universe just popping into existence. The theories all led back to questions of what happened before the first instants of the Big Bang and the possible existence of more than one universe.

Before the Big Bang: The Hunt for Quantum Gravity

As mountains of data piled up showing the early Universe to be hotter and denser than it is today, researchers assumed that someone, at some point, would figure out what $t = 0$ really meant. "People always knew there were several options for the beginning of space and time," explains Clifford Johnson. "One idea is that the Universe always has been. It just seems to start at zero at the beginning of the Universe, but really there was no beginning. Another option is that there was no 'pre-Universe' in the sense of there being a large, smooth space in which one could move around with everything widely separated from everything else." What Johnson means is that before the Big Bang, the kind of space and time we're familiar with may not have existed yet. "Maybe both space and time are just something completely different fundamentally from what we think of," says Johnson. "Maybe they require a different language to describe them such that creation of the Universe means taking some kind of protospace and prototime and allowing them to evolve into the smooth space and time we know today."

The problem with describing the Big Bang's true beginning is that the further back in time scientists push, the hotter and denser the Universe gets, until space-time itself becomes so compressed that infinities appear in Einstein's general relativity. In this cosmic singularity, a space-time that's so strongly curved on such tiny scales, scientists believe that general relativity is no longer applicable. "The equations themselves become singular," says Johnson, "meaning infinities appear in particular ways that tell us that we just can't use those equations anymore." To make any further progress, scientists must switch over to a different theory—a *quantum description* of space-time.

Quantum mechanics describes nature at the smallest scales. At the Universe's very beginning, even space-time itself should become quantized, appearing as a kind of "foam" of location and duration. Unfortunately, no one has ever developed a full working theory for these kinds of "atoms" of space-time. Without such a theory of *quantum gravity*—a merger of Einstein's general theory of relativity with quantum mechanics— scientists cannot really describe the Universe's beginning or address whether something might have come before the Big Bang.

> **SECTION SUMMARY**
> **Before the Big Bang: The Hunt for Quantum Gravity**
> ● Discovering what happened at $t = 0$ requires a theory of quantum gravity, or quantized space-time, which scientists have not developed yet.

Journey's End

We have come a long way on our journey. We began learning about small bands of hunter-gatherer humans looking in wonder at the night sky and followed our species' progress as we initiated the agricultural, city-building, industrial, and digital revolutions. With each step, we humans learned more about the celestial drama occurring over our heads—including the fact that we lived on a spherical planet and the drama was playing out below our feet as well.

As human culture matured, our understanding of the sky grew richer. Our journey through the modern story of astronomy in this book has reflected that richness. From the birth and evolution of planets to the birth and evolution of stars, we have seen that the cosmos is in a constant state of change and evolution. In these last chapters we have seen how even at the largest galactic, extragalactic, and even cosmological scales, change and evolution are the rule as well. It is, indeed, a grand and sweeping story playing out across billions of years and billions of parsecs.

But what makes this narrative of cosmic history we have followed even more remarkable is that it's a story we have learned for ourselves. No book was dropped by aliens telling us what happens inside a star or how to interpret the fossil cosmic microwave photons left over from the Universe's early history. We figured it all out for ourselves using our own capacity to reason and to measure, as well as the remarkable tools that, taken together, are called science. It's a remarkable accomplishment, requiring endless hours spent squinting through telescopes or hunched over lab benches and calculations. For scientists of the past—and those working today—there is no better work than to hear the world's own voice speak to us through our investigations. That effort and adventure is ongoing because the story of the Universe and our place in it is not yet complete. In that way, we remain at play in cosmos.

●→ chapter summary

18.1 How to Win a Nobel Prize

Cosmology is the study of the Universe as a whole. Early attempts at cosmological theories included Fred Hoyle's steady-state model, which said that the Universe has always expanded and will continue to do so forever. Although this model and others were replaced by the Big Bang theory after new data and observations were gathered, not even the Big Bang explains how the Universe originated.

18.2 Our Cosmology: The Big Bang

The classic Big Bang model relies on the concept of a flexible space-time from general relativity, along with the standard model of particle physics, which defines all the elementary particles and the forces they experience. The classic Big Bang theory begins from a cosmic singularity, the infinitely dense and hot source of all matter and energy, that began expanding and cooling 13.7 billion years ago. The post–Big Bang space-time underwent an accelerated expansion called inflation, and this region became the observable Universe. The important epochs in the evolution of the Universe were the particle era, during which the currently dominant particles emerged; Big Bang nucleosynthesis, when the original proportions of the light elements hydrogen, helium, and lithium formed; and the recombination era, at about 380,000 years after the Big Bang, when matter and blackbody photons separated.

18.3 How Do We Know? The Three Observational Pillars of Big Bang Theory

Three primary pieces of evidence support the Big Bang theory: the observed expansion of the Universe, the primordial abundance of elements produced in the initial Big Bang nucleosynthesis epoch, and the blackbody radiation called the cosmic microwave background, or CMB (the fossil remnants of photons emitted during recombination).

18.4 Beyond the Classic Big Bang Model

Several aspects of the classic Big Bang model, including the uniformity of temperature fluctuations in the CMB, were problematic but are addressed by an understanding of the inflationary epoch in the very early Universe. These temperature fluctuations in the CMB have been mapped to ever-greater spatial resolutions by several satellites, including COBE, WMAP, and (most recently) Planck. Temperature fluctuations represent differences in density, which themselves represent the seeds of the larger structures we see today. The relationship between the expansion of the Universe and the density of matter-energy is key to understanding the future evolution of the Universe. On its own, the gravity of "normal" cosmic matter and energy might have eventually slowed or reversed the expansion. This ratio of current density to the density needed to make the Universe stop expanding is called omega (Ω). When experiments indicated an acceleration of the expansion of space, astronomers were led to the conclusion that there is a repulsive force powered by dark energy. The dark energy makes up 70 percent of the Universe, while luminous and dark matter make up the other 30 percent. Together, they support the conclusion that $\Omega = 1$.

18.5 Questions about "Before" and "Everything"

To understand what may have preceded the Big Bang, astronomers are still working to formulate a theory of quantized space-time, called quantum gravity, which describes how space-time would behave at the ultrahot, ultradense conditions of the singularity.

●→ questions and problems

Narrow It Down: Multiple-Choice Questions

1. Which of the following is/are *not* considered direct evidence for the Big Bang? Choose all that apply.
   a. the existence of black holes
   b. the expansion of the Universe
   c. the primordial abundance of elements
   d. the presence of life in the Universe
   e. cosmic microwave background radiation

2. Which of the following statement(s) accurately state(s) ways in which Big Bang nucleosynthesis differed from nuclear fusion in stars? Choose all that apply.
   a. Its duration was a tiny fraction of time over which nuclear fusion has occurred in stars.
   b. It produced fewer elements and isotopes.
   c. It was incapable of producing carbon.
   d. It resulted in the production of hydrogen.
   e. The temperature was falling as nuclear fusion began.

3. Which of the following statements about space-time in our Universe is *not* correct?
   a. Evidence suggests that it is finite and bounded.
   b. It constitutes a four-dimensional continuum.
   c. It was first recognized as a unified quantity within special relativity.
   d. There is a limit to its expansion.
   e. There is strong evidence to suggest its current geometry.

4. Which of the following was/were "steady" according to Hoyle's steady-state model?
   a. the amount of matter
   b. the amount of space
   c. the average cosmic density
   d. the distance between any two galaxies
   e. all quantities and measurements

5. Which particle carries the strong nuclear force?
   a. photon
   b. quark
   c. gluon
   d. newton
   e. graviton

6. Which of the following statements is false?
   a. The tau lepton carries the electromagnetic force.
   b. The photon is a force boson.
   c. W and Z bosons carry the weak nuclear force.
   d. The muon is a lepton.
   e. Gravitons are hypothesized to carry the gravitational force.

7. Which of the following does *not* describe conditions since shortly after the Big Bang?
   a. Temperature and density have decreased.
   b. The volume of space-time has increased.
   c. The overall size of the Universe has increased.
   d. Inflation occurred and ended.
   e. Initial perturbations have been mostly smoothed away.

8. *Cosmic inflation* refers to
   a. the Universe filling up with hydrogen gas.
   b. the ongoing expansion of the Universe.
   c. a brief early period of hyperrapid expansion of space-time.
   d. the current period of cosmic acceleration.
   e. stellar winds creating bubbles of gas.

9. How long after the Big Bang was the primordial abundance of elements established?
   a. It is still changing today.
   b. 10^{-33} second
   c. 0.001 second
   d. 3 minutes
   e. 380,000 years

10. What temperature can the CMB reasonably be expected to have 5 billion years in our future?
    a. 0 K
    b. less than 2.7 K
    c. 2.7 K
    d. There is no way to know.
    e. The CMB will not exist.

11. Which of the following statements about the CMB is/are *not* true? Choose all that apply.
    a. Its photons originated at the moment of the Big Bang.
    b. Its photons have been redshifted dramatically since their creation.
    c. Its current temperature is 2.7 K.
    d. It is a blackbody.
    e. It is completely homogeneous in temperature.

12. Inflation theory helps explain many current properties of the Universe. Which of the following *cannot* be attributed to it? Choose all that apply.
    a. It drives the value of Ω toward 1.
    b. It accounts for the size and strength of perturbations.
    c. It explains why the Big Bang occurred.
    d. It explains the primordial abundance of elements.
    e. It explains why different sides of the sky show the same CMB fluctuations.

13. Dark energy accounts for
    a. 30 percent of the matter and energy in the Universe.
    b. a percentage of the matter-energy in the Universe that is less than dark matter but more than luminous matter.
    c. a percentage of the matter-energy in the Universe that is more than dark matter but less than luminous matter.
    d. a percentage of the matter-energy in the Universe that is more than dark matter and luminous matter combined.
    e. inflation.

14. Ignoring the effect of dark energy, a value of Ω greater than 1
    a. would be impossible.
    b. would mean the Universe will eventually contract.
    c. would mean there is more luminous matter than dark matter.
    d. would mean the Universe will expand at an ever-increasing rate forever.
    e. is what astronomers believe is the actual value of Ω.

15. Which of the following is true about the era of recombination?
    a. It was the period when photons could not escape.
    b. It occurred within seconds of the Big Bang.
    c. It was marked by the re-creation of H atoms from component parts that had been separated long before.
    d. It marked the release of the CMB.
    e. It smoothed out perturbations in the dense matter-energy.

16. Which of the following statements about dark matter and dark energy is/are true? Choose all that apply.
    a. While the nature of dark matter has been determined, the nature of dark energy has not.
    b. The effect of dark energy is observed only on cosmological scales, while the effect of dark matter has been observed on the scale of galaxies.
    c. Both far exceed the matter-energy equivalent of luminous matter.
    d. Evidence for the existence of dark energy was identified first.
    e. Dark matter has only an attractive gravitational effect, while dark energy is repulsive.

17. Which of the following describes commonalities between the periods of Big Bang nucleosynthesis and recombination?
    a. average temperature
    b. duration of the period
    c. size of the Universe
    d. initial formation of a new configuration of matter
    e. release of photons

18. True/False: The Universe is continuing to cool today.

19. True/False: The primal abundance of elements provides evidence for the Big Bang by indicating that the Universe was briefly hot enough for nucleosynthesis to occur.

20. True/False: A fully developed theory uniting all four fundamental forces has yet to be defined.

To the Point: Qualitative Questions

21. How did the term *Big Bang* come into being?

22. What did Gamow and Alpher contribute to the search for a theory of cosmic evolution?

23. In what sense is the Big Bang not a theory of origin?

24. Describe what is meant by positive, negative, and flat curvature.

25. What are the two classes of fundamental matter particles? What distinguishes the two classes?

26. Which force boson carries the electromagnetic force? the weak nuclear force? the strong nuclear force?

27. What is the difference between the "stage" and the "actors" in Big Bang theory as described in the chapter?

28. By what process did the first particles emerge in the Universe?

29. What are the proportions of hydrogen and helium in the primordial abundance of elements? Why were heavier elements not formed during Big Bang nucleosynthesis?

30. Describe the differences between the steady-state model and the Big Bang model.

31. How did observations of the Universe's expansion point to the presence of dark energy?

32. What role did phase transitions play in leading to the contents of the Universe that we see today?

33. How is dark energy similar to Einstein's cosmological constant?

34. Explain the concept of quantum gravity.

35. Recall your view of the origin, scope, and age of the Universe before you began this course. How has it changed (if at all) during this course?

Going Further: Quantitative Questions

In the questions that follow, ignore any acceleration of the expansion rate due to dark energy.

36. What value would you expect for the peak wavelength of the CMB when the age of the Universe has doubled from its current value?

37. What value would you expect for the peak wavelength of the CMB if the Universe had expanded by a factor of 800 since recombination?

38. By what factor had the Universe expanded since recombination when the peak wavelength of the CMB was 0.00038 meter?

39. By what factor will the Universe have expanded when the current wavelength of the CMB is 0.0076 meter?

40. What do you expect the temperature of the CMB to be when the Universe has expanded to 1.5 times its current size?

41. What was the temperature of the CMB when the Universe had expanded to one-third of its current size?

42. What is the recession velocity, in kilometers per second, of a galaxy that is 420 megaparsecs (Mpc) from Earth? (See Going Further 16.1).

43. The discovery of dark energy was based on the variation of the observed brightness of Type Ia supernovas to their expected brightness. Comparing two such supernovas, A and B, what would be the expected ratio of A's brightness to that of B, if A is 1/7th as far from Earth as B?

44. If the evolution of the Universe were at a point where the temperature of the CMB was 4.5 K, how would the Universe's size compare to its current size?

45. If the evolution of the Universe were at a point where the temperature of the CMB was 1.3 K, how would its size compare to its current size?

 If your instructor assigns homework in **smartwork5**, access it at the Digital Landing Page for *At Play in the Cosmos*: **digital.wwnorton.com/cosmos**

APPENDIX 1

PHYSICAL PROPERTIES OF PLANETS AND DWARF PLANETS

Physical Properties of Planets and Dwarf Planets

	Mass (M_E)	Mass ($\times 10^{24}$ kg)	Radius (km)	Density (kg/m^3)	Average Surface Temperature (°C)	Average Surface Temperature (K)	Escape Velocity (km/s)
Mercury	0.055	0.33	2,440	5,430	167	440	4.3
Venus	0.82	4.87	6,052	5,240	464	737	10.4
Earth	1	5.97	6,378	5,515	15	288	11.2
Mars	0.11	0.642	3,397	3,940	−65	208	5.05
Jupiter	317.82	1,898	71,492	1,330	−110	163	59.5
Saturn	95.16	568	60,268	700	−140	133	35.5
Uranus	14.37	86.8	25,559	1,300	−195	78	21.3
Neptune	17.15	102	24,764	1,760	−200	73	23.5
Vesta (asteroid)	4.47×10^{-5}	0.00026	251	3,456	−93	180	0.36
Ceres (dwarf planet)	1.5×10^{-4}	0.00087	476	2,170	−105	168	0.5
Pluto (dwarf planet)	2.2×10^{-3}	0.013	1,195	1,100	−225	48	1.1
Haumea (dwarf planet)	6.6×10^{-4}	0.004	1,700	2,550	−241	32	0.91
Makemake (dwarf planet)	6.7×10^{-4}	0.004	715	2,300	−239	34	0.8
Eris (dwarf planet)	2.8×10^{-3}	0.017	1,163	2,520	−231	42	1.38

APPENDIX 2
ORBITAL PROPERTIES OF PLANETS AND DWARF PLANETS

Orbital Properties of Planets and Dwarf Planets

	Orbital Inclination to Ecliptic (degrees)	Rotation Period (hours)	Orbital Radius (AU)	Orbital Period (years)	Orbital Speed (km/s)	Eccentricity
Mercury	7	1,407.6	0.39	0.24	47.9	0.205
Venus	3.39	5,832.5	0.72	0.62	35	0.007
Earth	7.16	23.9	1	1	29.8	0.017
Mars	1.85	24.6	1.52	1.88	24.1	0.093
Jupiter	1.31	9.9	5.2	11.86	13.1	0.048
Saturn	2.49	10.7	9.54	29.42	9.7	0.054
Uranus	0.77	17.2	19.19	83.75	6.8	0.047
Neptune	1.77	16.1	30.07	163.72	5.4	0.009
Vesta	7.14	5.4	2.36	3.63	19.34	0.089
Ceres	10.59	9.1	2.77	4.6	17.9	0.076
Pluto	17.15	153.3	39.48	248.02	4.7	0.249
Haumea	28.2	3.9	43.22	284.12	4.5	0.191
Makemake	29	7.77	45.79	309.09	4.4	0.156
Eris	44.04	25.9	67.78	558.04	3.4	0.441

Properties of Major Solar System Moons

Planet	Moon	Radius (km)	Mass (kg)	Density (kg/m^3)	Orbital Radius (10^3 km)	Orbital Period (days)	Eccentricity
Earth	Moon	1,737	7.35×10^{22}	3,341	384.4	27.32	0.055
Mars	Phobos	11.2	1.06×10^{16}	1,876	9.38	0.32	0.015
	Deimos	6.2	2.4×10^{15}	1,471	23.46	1.26	0.00033
Jupiter	Callisto	2,403	1.08×10^{23}	1,834	1,883	16.69	0.0074
	Europa	1,565	4.80×10^{22}	3,013	671	3.55	0.009
	Ganymede	2,634	1.48×10^{23}	1,936	1,070	7.15	0.0013
	Io	1,821	8.93×10^{22}	3,528	422	1.77	0.0041
Saturn	Pan	14.1	4.95×10^{15}	420	133.58	0.58	0
	Prometheus	43	1.60×10^{17}	480	139.35	0.61	0.0022
	Pandora	40.7	1.37×10^{17}	490	141.7	0.63	0.0042
	Mimas	198	3.74×10^{19}	1,148	185.52	0.94	0.0196
	Enceladus	252	1.08×10^{20}	1,610	238.02	1.37	0.0047
	Titan	2,576	1.35×10^{23}	1,881	1,222	15.95	0.0288
Uranus	Miranda	236	6.59×10^{19}	1,200	129.39	1.41	0.0013
	Ariel	579	1.35×10^{21}	1,660	191.02	2.52	0.0012
	Umbriel	585	1.17×10^{21}	1,390	266.3	4.14	0.0039
	Titania	789	3.53×10^{21}	1,711	435.91	8.71	0.0011
	Oberon	761	3.01×10^{21}	1,630	583.52	13.46	0.0014
Neptune	Proteus	210	4.4×10^{19}	1,300	117.65	1.12	0
	Triton	1,353	2.14×10^{22}	2,061	355.76	5.88	0
	Nereid	170	2.70×10^{19}	—	5,513.4	360.13	0.751
Pluto	Charon	593	1.62×10^{21}	1,650	19.6	6.39	0

Properties of the Nearest Stars to Earth

Common Name	Constellation	Scientific Name	Distance (ly)	Distance (pc)	Spectral Type	Apparent Magnitude	Absolute Magnitude	Luminosity (L_{Sun})
Sun	N/A	Sol	0.000016	0.000005	G2	−26.74	4.83	1
Alpha Centauri	Centaurus	Alpha Cen	4.2	1.3	G2	−0.27	4.38	1.50
Barnard's Star	Ophiuchus	Gliese 699	5.9	1.8	M4	9.53	13.22	0.0004
Wolf 359	Leo	Gliese 409	7.8	2.4	M6	13.44	16.55	0.00002
Lalande 21185	Ursa Major	Gliese 411	8.3	2.5	M2	7.47	10.44	0.021
Sirius	Canis Major	Alpha CMa	8.6	2.6	A1	−1.46	1.42	25
Luyten 726	Cetus	Gliese 65b	8.6	2.6	M5.5	12.99	15.85	0.00006
Ross 154	Sagittarius	Gliese 729	9.7	3.0	M3	10.43	13.07	0.0038
Ross 248	Andromeda	Gliese 905	10.3	3.2	M6	12.29	14.79	0.0018
Epsilon Eridani	Eridanus	Gliese 144	10.5	3.2	K2	3.73	6.19	0.34
Lacaille 9352	Piscis Austinus	Gliese 887	10.7	3.3	M2	7.34	9.75	0.011
Luyten 789	Aquarius	Gliese 866	10.9	3.4	M5.5	11.06	13.30	0.00001
Ross 128	Virgo	Gliese 447	10.9	3.4	M4	11.13	13.51	0.00036
Struve 2398	Draco	Gliese 725	11.3	3.5	M3	8.90	11.16	0.039
61 Cygni	Cygnus	Gliese 820	11.4	3.5	K5	5.21	7.49	0.15
Procyon	Canis Minor	Alpha CMi	11.4	3.5	F5	0.34	2.66	6.93
Groombridge 34	Andromeda	Gliese 15	11.7	3.6	M2	8.08	10.32	0.0064
Epsilon Indi	Indus	Gliese 845	11.8	3.6	K5	4.69	6.89	0.15
Tau Ceti	Cetus	Gliese 71.0	11.9	3.7	G8	3.49	5.68	0.52
YZ Ceti	Cetus	Gliese 54.1	12	3.7	M4	12.02	14.17	0.0002
Luyten's Star	Canis Minor	Gliese 273	12.4	3.8	M3.5	9.86	11.97	0.0004

Properties of the Brightest Stars

Common Name	Constellation	Scientific Name	Distance (ly)	Distance (pc)	Spectral Type	Apparent Magnitude	Absolute Magnitude	Luminosity (L_{Sun})
Sun	N/A	Sol	0.000016	0.000005	G2	-26.74	4.83	1
Sirius	Canis Major	Alpha CMa	8.6	2.6	A1	-1.46	1.42	25
Canopus	Carina	Alpha Car	309	95	A9	-0.72	-5.53	15,100
Alpha Centauri	Centaurus	Alpha Cen	4.2	1.3	G2	-0.27	4.38	1.5
Arcturus	Bootes	Alpha Boo	37	11	K0	-0.04	-0.31	170
Vega	Lyra	Alpha Lyr	25	8	A0	0.03	0.58	40
Capella	Auriga	Alpha Aur	43	13	G1	0.08	0.40	79
Rigel	Orion	Beta Ori	862	265	B8	0.12	-7.84	120,000
Procyon	Canis Minor	Alpha CMi	11.4	3.5	F5	0.34	2.66	6.93
Betelgeuse	Orion	Alpha Ori	498	153	M1	0.42	-5.14	150,000
Achernar	Eridanus	Alpha Eri	139	43	B6	0.50	-2.77	3,150
Hadar/Agena	Centaurus	Beta Cen	392	120	B1	0.60	-5.42	41,700
Altair	Aquila	Alpha Aql	17	5	A7	0.77	2.21	11
Acrux	Crux	Alpha Cru	322	99	B1	0.77	-4.14	25,000
Aldebaran	Taurus	Alpha Tau	67	20	K5	0.85	-0.63	518
Spica	Virgo	Alpha Vir	250	77	B1	1.04	-3.55	12,100
Antares	Scorpius	Alpha Sco	553	170	M0.5	1.09	-7.20	57,500
Pollux	Gemini	Beta Gem	34	10	K0	1.15	1.08	43
Fomalhaut	Piscis Austrinus	Alpha PsA	25	8	A4	1.16	1.72	17
Deneb	Cygnus	Alpha Cyg	1,411	433	A2	1.25	-8.38	196,000
Mimosa	Crux	Beta Cru	278	85	B0.5	1.30	-3.92	34,000

APPENDIX 6
PERIODIC TABLE

PERIODIC TABLE OF THE ELEMENTS

Legend:
- 1 — Atomic number
- H — Symbol
- Hydrogen — Name
- 1.0079 — Average atomic mass

- Metals
- Metalloids
- Nonmetals

1 / 1A	2 / 2A	3 / 3B	4 / 4B	5 / 5B	6 / 6B	7 / 7B	8 / 8B	9 / 8B	10 / 8B	11 / 1B	12 / 2B	13 / 3A	14 / 4A	15 / 5A	16 / 6A	17 / 7A	18 / 8A
1 **H** Hydrogen 1.0079																	2 **He** Helium 4.0026
3 **Li** Lithium 6.941	4 **Be** Beryllium 9.0122											5 **B** Boron 10.811	6 **C** Carbon 12.011	7 **N** Nitrogen 14.007	8 **O** Oxygen 15.999	9 **F** Fluorine 18.998	10 **Ne** Neon 20.180
11 **Na** Sodium 22.990	12 **Mg** Magnesium 24.305											13 **Al** Aluminum 26.982	14 **Si** Silicon 28.086	15 **P** Phosphorus 30.974	16 **S** Sulfur 32.065	17 **Cl** Chlorine 35.453	18 **Ar** Argon 39.948
19 **K** Potassium 39.098	20 **Ca** Calcium 40.078	21 **Sc** Scandium 44.956	22 **Ti** Titanium 47.867	23 **V** Vanadium 50.942	24 **Cr** Chromium 51.996	25 **Mn** Manganese 54.938	26 **Fe** Iron 55.845	27 **Co** Cobalt 58.933	28 **Ni** Nickel 58.693	29 **Cu** Copper 63.546	30 **Zn** Zinc 65.38	31 **Ga** Gallium 69.723	32 **Ge** Germanium 72.63	33 **As** Arsenic 74.922	34 **Se** Selenium 78.96	35 **Br** Bromine 79.904	36 **Kr** Krypton 83.798
37 **Rb** Rubidium 85.468	38 **Sr** Strontium 87.62	39 **Y** Yttrium 88.906	40 **Zr** Zirconium 91.224	41 **Nb** Niobium 92.906	42 **Mo** Molybdenum 95.96	43 **Tc** Technetium [98]	44 **Ru** Ruthenium 101.07	45 **Rh** Rhodium 102.91	46 **Pd** Palladium 106.42	47 **Ag** Silver 107.87	48 **Cd** Cadmium 112.41	49 **In** Indium 114.82	50 **Sn** Tin 118.71	51 **Sb** Antimony 121.76	52 **Te** Tellurium 127.60	53 **I** Iodine 126.90	54 **Xe** Xenon 131.29
55 **Cs** Cesium 132.91	56 **Ba** Barium 137.33	57 **La** Lanthanum 138.91	72 **Hf** Hafnium 178.49	73 **Ta** Tantalum 180.95	74 **W** Tungsten 183.84	75 **Re** Rhenium 186.21	76 **Os** Osmium 190.23	77 **Ir** Iridium 192.22	78 **Pt** Platinum 195.08	79 **Au** Gold 196.97	80 **Hg** Mercury 200.59	81 **Tl** Thallium 204.38	82 **Pb** Lead 207.2	83 **Bi** Bismuth 208.98	84 **Po** Polonium [209]	85 **At** Astatine [210]	86 **Rn** Radon [222]
87 **Fr** Francium [223]	88 **Ra** Radium [226]	89 **Ac** Actinium [227]	104 **Rf** Rutherfordium [265]	105 **Db** Dubnium [268]	106 **Sg** Seaborgium [271]	107 **Bh** Bohrium [270]	108 **Hs** Hassium [277]	109 **Mt** Meitnerium [276]	110 **Ds** Darmstadtium [281]	111 **Rg** Roentgenium [280]	112 **Cn** Copernicium [285]	113 **Uut** Ununtrium [284]	114 **Fl** Flerovium [289]	115 **Uup** Ununpentium [288]	116 **Lv** Livermorium [293]	117 **Uus** Ununseptium [294]	118 **Uuo** Ununoctium [294]

6 Lanthanides

58 **Ce** Cerium 140.12	59 **Pr** Praseodymium 140.91	60 **Nd** Neodymium 144.24	61 **Pm** Promethium [145]	62 **Sm** Samarium 150.36	63 **Eu** Europium 151.96	64 **Gd** Gadolinium 157.25	65 **Tb** Terbium 158.93	66 **Dy** Dysprosium 162.50	67 **Ho** Holmium 164.93	68 **Er** Erbium 167.26	69 **Tm** Thulium 168.93	70 **Yb** Ytterbium 173.05	71 **Lu** Lutetium 174.97

7 Actinides

90 **Th** Thorium 232.04	91 **Pa** Protactinium 231.04	92 **U** Uranium 238.03	93 **Np** Neptunium [237]	94 **Pu** Plutonium [244]	95 **Am** Americium [243]	96 **Cm** Curium [247]	97 **Bk** Berkelium [247]	98 **Cf** Californium [251]	99 **Es** Einsteinium [252]	100 **Fm** Fermium [257]	101 **Md** Mendelevium [258]	102 **No** Nobelium [259]	103 **Lr** Lawrencium [262]

We have used the U.S. system as well as the system recommended by the International Union of Pure and Applied Chemistry (IUPAC) to label the groups in this periodic table. The system used in the United States includes a letter and a number (1A, 2A, 3B, 4B, etc.), which is close to the system developed by Mendeleev. The IUPAC system uses numbers 1–18 and has been recommended by the American Chemical Society (ACS). While we show both numbering systems here, we use the IUPAC system exclusively in the book.

APPENDIX 7

STAR MAPS

Northern Hemisphere Night Sky in Spring

1 a.m. on March 1

11 p.m. on April 1

9 p.m. on May 1

(Add 1 hour for daylight saving time)

Northern horizon

Eastern horizon

Western horizon

Southern horizon

Northern Hemisphere Night Sky in Summer

1 a.m. on June 1

11 p.m. on July 1

9 p.m. on August 1

(Add 1 hour for daylight saving time)

Northern horizon

Eastern horizon

Western horizon

Southern horizon

Northern Hemisphere Night Sky in Autumn

1 a.m. on September 1

11 p.m. on October 1

9 p.m. on November 1

(Add 1 hour for daylight saving time)

Northern horizon

Eastern horizon

Western horizon

Southern horizon

Northern Hemisphere Night Sky in Winter

2 a.m. on December 1

Midnight on January 1

10 p.m. on February 1

Northern horizon

Eastern horizon

Western horizon

Southern horizon

SELECTED ANSWERS

CHAPTER 1

Narrow It Down:
Multiple-Choice Questions

3. b

10. b

17. b, d, e

To the Point: Qualitative and Discussion Questions

21. As science continues to evolve, astronomers and other scientists are willing to modify or update their findings to accommodate new data.

27. As a result of technological advancement, the world faces many challenges, including climate change, depletion of resources, exhaustion of many of the world's fisheries, the threat of the use of nuclear weapons, and biological terrorism.

31. A solstice is a point on the celestial sphere where the ecliptic is farthest north (June 21) or south (December 21) of the celestial equator, and the moment at which the Sun returns to that point each year.

Going Further:
Quantitative Questions

37. The rental of a luxury car is four orders of magnitude more costly.

41. The number of powers of 10 between the galaxy and planetary system is larger, by two orders of magnitude.

42. The ratio of time between modern humans and *Homo sapiens* is 1:20.

CHAPTER 2

Narrow It Down:
Multiple-Choice Questions

8. d

12. c, d

20. b

To the Point: Qualitative and Discussion Questions

24. If the Moon crosses the meridian at midnight (the time it sets), the phase must be first quarter. That means that the Moon rose at noon and was highest on the sky at 6 P.M.

28. All year long you can see every part of the celestial sphere from the north celestial pole, where the North Star is located.

32. The seasons would be more extreme.

Going Further:
Quantitative Questions

36. 14,400″ or 1.44×10^4.

37. The horizon is an imaginary line at 0° altitude. The zenith is directly overhead at an altitude of 90°.

45. 0.06″.

CHAPTER 3

Narrow It Down:
Multiple-Choice Questions

5. b, d

8. b

14. b

To the Point: Qualitative and Discussion Questions

23. The current star names came from Arabic astronomers as scientific learning continued from 500 CE onward for the next five centuries.

28. Ptolemy and Copernicus both strongly believed in uniform circular motion and that the circle was the perfect shape for orbital motion.

31. If you increased Earth's radius, the acceleration due to gravity would decrease because radius and acceleration are inversely proportional to one another ($g = GM/r^2$).

Going Further:
Quantitative Questions

37. The comet would be, on average, 13.02 AU from the Sun.

42. Satellite B's velocity is 1.3 times as fast as satellite A's velocity because it is closer to Earth's gravitational pull.

45. The mass of the unknown star is a little more than a quarter of the mass of the Sun. It is probably a red dwarf.

CHAPTER 4

Narrow It Down: Multiple-Choice Questions

5. c

10. a, b, e

18. a

To the Point: Qualitative and Discussion Questions

23. Peak wavelength (color) and temperature.

27. Wien's law tells us that the higher the object's temperature, the shorter the peak wavelength for the light emitted by the object.

35. If all wavelengths of light, especially X-rays, could penetrate Earth's atmosphere, our planet would be sterilized and unable to sustain life.

Going Further: Quantitative Questions

37. If the ratio of the wavelengths is 1:3, the frequency is inversely proportional to the wavelength.

38. 3,776 K.

40. The dimmer star is 4 times more distant than the brighter star.

CHAPTER 5

Narrow It Down: Multiple-Choice Questions

8. b, e

9. d

13. a

To the Point: Qualitative and Discussion Questions

21. The discovery of the Kuiper Belt led to the demotion of Pluto because of its small size, location, and composition of ice and rock. Tens of thousands of icy, rocky bodies reside in this region.

23. Giordano Bruno was persecuted for believing that the Sun, not Earth, was the center of the Solar System.

26. Long-period comets are the greatest potential threat to Earth because of their very large eccentricities. Comets on highly elliptical orbits must, by definition, cross the orbits of some or all of the planets.

Going Further: Quantitative Questions

38. The orbital velocity at perihelion is almost 8 times faster than at aphelion.

43. 9,000 m/s.

45. 375 years.

CHAPTER 6

Narrow It Down: Multiple-Choice Questions

2. a

8. a–d

20. False

To the Point: Qualitative and Discussion Questions

23. Seismographic studies revealed the properties of Earth's interior and its slowly convecting mantle that is the force that drives continental drift. Combining current data with fossil records and geologic studies, scientists have reconstructed plate tectonic movement of the entire history of the planet.

29. One day-night cycle on the Moon is 27.3 days. That's because as it completes its orbit around Earth, the Moon makes only one complete rotation on its axis.

32. Auroras are caused when charged particles collide, driving the atoms and molecules in the atmosphere at the poles into excited states. As they drop back to lower energy levels, the atoms and molecules emit photons, which produce the glowing auroras visible in the polar regions.

Going Further: Quantitative Questions

36. 19.8% stronger.

39. By a factor of 16.

44. Yes, oxygen would be retained, because $V_t < V_e$.

CHAPTER 7

Narrow It Down: Multiple-Choice Questions

2. a–e

8. c, e

14. a, b, d

To the Point: Qualitative and Discussion Questions

25. Mercury, Venus, Earth, Mars.

28. The requirements for the formation of a planetary magnetic field are (1) an interior region of an electrically conducting fluid (molten metal), (2) convection in that layer, and (3) a moderately rapid rotation.

33. The Martian day is 24 hours and 37 minutes, versus Earth's day of 24 hours. Engineers and scientists working with orbiters and rovers on Mars must make only a slight adjustment in the amount of time it takes to complete a task.

Going Further: Quantitative Questions

36. 1.1 times as great.
38. 5.05 km/s.
43. 206 K.

CHAPTER 8

Narrow It Down: Multiple-Choice Questions

1. a, b
9. a–e
17. a, c, d

To the Point: Qualitative and Discussion Questions

22. The more extreme the axis of tilt for a planet, the more dramatic the change in seasons.
25. The terrestrial worlds are almost twice as dense as the giant planets. This difference tells us where they formed in the Solar System.
28. Jupiter acts like a kind of planetary vacuum cleaner, sweeping away wandering debris that could pose a threat to life-bearing planets like Earth.

Going Further: Quantitative Questions

36. 88.82 K.
37. 12.5 AU.
40. 6.91×10^{29} kg.

CHAPTER 9

Narrow It Down: Multiple-Choice Questions

2. a
13. a, d
14. a

To the Point: Qualitative and Discussion Questions

23. Hot stars are not good candidates for highly evolved life because they have very short lifetimes, as short as a few million years. Complex life would not have time to form.
26. Our understanding of the early Earth changed when we realized that volcanism could have provided a lot more CO_2 and N_2 in the atmosphere, yielding a set of chemical reactions very different from those that Miller and Urey's experiment showed.
31. The Fermi paradox states, "There are no other intelligent species out there in the galaxy. If extraterrestrials exist, why aren't they here already?"

Going Further: Quantitative Questions

38. 46 years.
39. 4.97×10^{86}.
40. 2.43×10^{107}.

CHAPTER 10

Narrow It Down: Multiple-Choice Questions

1. b, d
5. b, c
14. a

To the Point: Qualitative and Discussion Questions

21. The P-P chain is a type of fusion reaction that the Sun uses to convert hydrogen into helium and release subatomic particles and energy in the form of gamma rays.
26. The photosphere is the layer of the Sun that we can see with our eyes.
31. Groups of sunspots appear at middle latitudes, and then, as the cycle progresses, the spots increase in number and make their appearance closer and closer to the solar equator.

Going Further: Quantitative Questions

37. 285.
38. Over 21 years.
42. The hotter light source is brighter by a factor of 33.18.

CHAPTER 11

Narrow It Down: Multiple-Choice Questions

2. True
8. c
16. a

To the Point: Qualitative and Discussion Questions

22. Luminosity, distance, temperature, size, mass, and composition.
25. Every element and molecule has a unique set of spectral lines. Thus, the different types of spectral lines found in a star's spectrum can indicate its chemical composition.
32. The cluster turnoff mass is determined from an HR diagram. The mass of the most luminous stars on the main sequence is the turnoff mass and can be used to determine the age of the cluster.

Going Further: Quantitative Questions

37. 446 nm.
42. 0.5 pc. The closest star other than the Sun (Proxima Centauri) is at a distance of 1.3 pc, so it would be a big surprise to learn of a closer star!
43. 2.33 pc.

CHAPTER 12

Narrow It Down: Multiple-Choice Questions

1. a
8. d
11. e

To the Point: Qualitative and Discussion Questions

22. The WISM's warmth comes from ionization caused by weak, diffuse background radiation. This background flux of ionizing photons is the sum of all the sources in the galaxy.
25. Dust temperatures can be determined by measuring their blackbody radiation, and spectral lines can indicate the composition of the dust.
27. Molecular clouds have limited lifetimes because there are competing forces acting on them. If the force of gravity overwhelms gas pressure, the cloud will collapse on itself and ultimately form a protostar.

Going Further: Quantitative Questions

36. The pressure in cloud B is 1.5 times the pressure in cloud A.
37. The density in region 1 is 0.4 times that of region 2 (it is less dense).
41. The free-fall time would change by a factor of 0.56: it would be slightly more than ½ of what it was originally (the cloud would collapse faster).

CHAPTER 13

Narrow It Down: Multiple-Choice Questions

3. e
4. b
14. c

To the Point: Qualitative and Discussion Questions

21. The more mass a star has, the greater the gravitational weight crushing down on the core, which determines the temperature in the core. The core temperature determines the rate at which stars burn hydrogen fuel. Massive stars have higher core temperatures, burn their fuel faster, and therefore have shorter lives.
23. The majority of a star's life is spent on the main sequence. Its death process begins when it runs out of hydrogen in the core and begins fusing heavier elements.
31. Massive stars cannot burn elements heavier than iron because iron fusion requires more energy to be put in than it gives off.

Going Further: Quantitative Questions

36. The star's luminosity is 64 times larger than that of the Sun. Doubling the temperature has a greater effect, since temperature is raised to the fourth power while radius is raised to only the second power.
38. The star is 377 times more luminous than the Sun; that is, its luminosity is 377 L_{Sun}. This makes sense because although the red giant is much cooler, its radius is vastly larger.
42. Since the peak wavelength, 6.7×10^{-7}, is equivalent to 670 nm and is in the visible range, this star has a fairly cool temperature of 4,300 K.

CHAPTER 14

Narrow It Down: Multiple-Choice Questions

2. c
7. a
19. True

To the Point: Qualitative and Discussion Questions

24. The luminiferous aether was believed to be the material through which light waves propagated. Scientists thought all waves needed a material to propagate in.
32. The event horizon of a black hole is defined by the radius at which the escape velocity equals the speed of light.
35. In TV and movies, black holes are depicted as portals to other parts of space. Real black holes, though, would tear any spacecraft to pieces with tidal forces and destroy any spacecraft with radiation from an accretion disk or jets.

Going Further: Quantitative Questions

42. Since the spacecraft is traveling at 90% the speed of light, time is dilated and a 35-minute dinner on the spacecraft takes 80.3 minutes for scientists on Earth.

GLOSSARY

A

abiotic synthesis The creation of life from nonlife.

absorption The capture of a photon of a specific wavelength, causing an electron to jump from a lower to a higher energy level.

absorption line A dark line in a continuous spectrum. Specific patterns of absorption lines indicate the presence of a specific element in the source.

acceleration The velocity change of distance per unit time.

active galactic nucleus (AGN) A highly luminous, energetic galactic nucleus powered by a supermassive black hole.

adaptive optics A telescope technology in which mirror segments are rapidly adjusted to account for atmospheric turbulence and produce higher-resolution images.

AGB See *asymptotic giant branch*.

AGN See *active galactic nucleus*.

AGN outflow Material driven away from the central regions of an active galactic nucleus.

albedo The reflecting power of an object, measured as the ratio of light reflected to light received.

altitude The angular distance from a point on the sky to a point on the horizon directly below it.

Amazonian period The evolutionary period of Mars that began about 3 billion years ago.

amplitude A measure of a wave's strength.

anaerobic Describing an organism that requires the absence of free oxygen to survive.

analemma The figure-eight pattern formed by charting the daily position of the Sun at noon over the course of the year.

angular resolution A measure of the ability of a telescope to separate or distinguish features in a distant object.

angular size Angular distance measured on a sphere (such as Earth or the celestial sphere) in degrees, arcminutes, and arcseconds.

annular eclipse A solar eclipse that occurs when the positions of Moon and Sun are such that the lunar disk is not able to fully block the solar disk.

antimatter A form of matter that annihilates on contact with normal matter. Particles and antiparticles of the same type have the same mass but opposite charge.

aperture The diameter of a telescope's main light-collecting lens or mirror.

aphelion (pl. aphelia) The distance of farthest approach for an object orbiting the Sun.

apogee The distance of farthest approach for an object orbiting Earth.

arcminute A measure of angular size. There are 60 arcminutes in a degree.

asteroid A rocky object that orbits the Sun and has a diameter from hundreds of meters to 1,000 km.

asteroid belt The disk-shaped region from 2.1 to 3.5 AU (as measured from the Sun) where many asteroids orbit.

astrobiology The interdisciplinary study of life in its astronomical context.

astrology The belief that there is a relationship between astronomical phenomena and events in the human world.

astrometric binary or **visual binary** Co-orbiting stars that have been detected by the proper motions of the components.

astronomical unit (AU) The average distance from Earth to the Sun: 1.5×10^{11} meters; used as a standard of measure for Solar System objects.

astronomy The study of celestial objects such as moons, planets, stars, nebulas, and galaxies.

asymptotic giant branch (AGB) The region of the HR diagram to which a low- or intermediate-mass star evolves when it has burned through the hydrogen and helium in its core. The AGB's external characteristics are a cool surface, large radius, and high luminosity.

atom The fundamental building block of matter, consisting of a dense central nucleus surrounded by a cloud of negatively charged electrons.

AU See *astronomical unit*.

aurora The emission of light in the upper atmosphere caused by charged particles from the solar wind flowing along Earth's magnetosphere. Auroras tend be seen at upper latitudes.

axial tilt See *obliquity*.

B

Balmer series Emission or absorption lines of the hydrogen atom as an electron moves between the first excited state and higher energy states.

bar A thick, rectangular region of stars extending across the center of some spiral galaxies.

barred spiral galaxy A spiral galaxy with a central rectangular bar.

baryogenesis A process that produced an asymmetry between matter and antimatter in the early Universe.

basalt A type of volcanic rock that forms at the ocean floor.

BBN See *Big Bang nucleosynthesis*.

belt A rising, high-pressure cloud region that forms a horizontal band in Jupiter's atmosphere because of convection and the planet's rapid rotation.

Big Bang The event driving cosmic expansion.

Big Bang nucleosynthesis (BBN) Nuclear fusion of light elements in the first 3 minutes after the Big Bang.

Big Crunch A theoretical contraction of the Universe that would occur if the combined matter-energy exceeded critical density.

binary accretion The process by which the collision of two objects results in the formation of one larger object.

binary-accretion-disk-magnetic-field model A theory of planetary nebula formation that involves a binary system and accretion disks. Jets from the disk sculpt the shape of the planetary nebula.

binary star Two stars in orbit around their common center of mass. Binary stars are born together.

binding energy The energy needed to break up an object being held together by a given force. The same amount of energy will be released in the formation of the object.

birth line The region on the HR diagram where protostars begin their descent onto the main sequence.

blackbody radiation Electromagnetic radiation from a dense, opaque body, whose spectrum depends only on the temperature of the body.

black hole A region of space-time whose strong gravitational distortion prevents anything, including light, from escaping.

blueshift A decrease in the wavelength of light that results when the source of the light moves toward the observer or the observer moves toward the source.

Bohr atom A model of the atom, devised by Niels Bohr, in which electrons in discrete orbits surround a positively charged nucleus. Electron "jumps" between the discrete orbits determine the emission or absorption of light.

Bok globule See *dark globule*.

bow shock The boundary where a strong flow runs into an obstacle.

brown dwarf A failed star that has too little mass to drive core temperatures high enough to initiate hydrogen-burning nuclear fusion.

bulge The roughly spherical distribution of stars and interstellar material in the central region of a galaxy.

C

caldera The depression at the top of a volcano.

calibration The use of a known parameter value obtained by a proven method to validate a measurement of that parameter by an unproven method.

Cambrian explosion The period 540 million years ago when the diversity of life expanded at an accelerated rate.

capture hypothesis A hypothesis of the Moon's origin stating that the Moon formed elsewhere and was captured intact by Earth's gravity.

carbon-nitrogen-oxygen cycle See *CNO cycle*.

Cassini Division A 4,800-kilometer-wide region between the A and B rings of Saturn's rings that has been shaped by Saturn's moon Mimas.

cD galaxy The largest, most massive example of an elliptical galaxy.

CDM See *cold dark matter*.

celestial equator The extension of Earth's equator onto the celestial sphere.

celestial sphere An imaginary transparent sphere surrounding Earth, on which positions of stars, planets, and other celestial bodies are projected from extensions of Earth's coordinate system.

center of mass The point around which two orbiting bodies move. It lies closer to the more massive of the two objects.

centrifugal force The illusion of outward force when an object is moving on a curved path.

centripetal acceleration Acceleration involving a force that keeps an object moving on a curved path. The acceleration is directed toward the center of curvature of the path.

Chandrasekhar limit The maximum mass ($1.4\ M_{Sun}$) of a white dwarf; the mass that can be supported by electron degeneracy pressure.

chromatic aberration The failure of a lens to focus all colors to the same point.

chromosphere The lower layer of the Sun's atmosphere.

circumpolar constellation A constellation that, from the viewer's perspective, never rises or sets.

climate change A significant, long-term change in the statistical distribution of weather patterns.

CMB See *cosmic microwave background*.

CME See *coronal mass ejection*.

CNO cycle A chain of nuclear fusion reactions, involving carbon, nitrogen, and oxygen, that produces helium. It is the mechanism for powering stars of more than 1.3 solar masses.

cold dark matter (CDM) Nonluminous matter moving at slower-than-relativistic speed.

coma The nebulous envelope around the nucleus of a comet, formed when the comet passes close to the Sun.

comet A body of rock and ice ranging in size from hundreds of meters to 1,000 km and typically orbiting the Sun on a highly elliptical orbit.

comparative planetology The study of planets' characteristics based on comparison with other known planets.

compression wave A wave of increased density that passes through background material. The density returns to normal after the wave passes.

condensation theory A theory of planetary formation in which planets are created from a disk of gas and dust around a newly formed star.

conduction The transfer of thermal energy from a hotter region to a cooler one via the collisions of atoms or molecules.

conservation of angular momentum The principle whereby a spinning or orbiting object increases its rotational speed as the radius (of either the spinning object or the object's orbit) decreases, and vice versa.

constant of nature A term that does not change from one situation or time to another. The value of a constant of nature must be measured to be determined.

constellation A group of stars that form a pattern, and the designated region of the sky surrounding them.

convection The transfer of heat from one place to another by fluid motion.

convective zone The region of the Sun above the radiative zone, where energy is transported outward by means of the motion of gas (convection).

core 1. Earth's innermost layer, composed primarily of iron. The inner core is solid; the outer core is in a fluid state. 2. The central region of a star where high temperatures and high densities allow thermonuclear fusion reactions to occur, releasing energy.

core accretion model A model of giant gas planet formation in which an icy terrestrial planet grows by binary accretion up to a critical point and then rapidly pulls gas in from the surrounding disk.

corona 1. A large oval structure surrounding volcanoes on some Solar System bodies, formed by upwelling of material. 2. The extended region of extremely hot, low-density gas surrounding the Sun.

coronal gas The phase of the interstellar medium that consists of extremely hot, low-density gas resulting from explosions of high-mass stars (supernovas).

coronal hole A region of low density where the solar wind is generated.

coronal mass ejection (CME) A large-scale ejection of mass and magnetic energy from the Sun's corona.

cosmic censorship The conjecture that the Universe will not allow the existence of a singularity without an event horizon.

cosmic distance ladder The sequence of methods for determining distance to astronomical objects; arranged by increasing distance over which each method is effective.

cosmic expansion The expansion of space-time since the Big Bang.

cosmic microwave background (CMB) Blackbody radiation released during recombination, with a peak wavelength in the red part of the electromagnetic spectrum. The expansion of the Universe now puts the peak of CMB in the microwave region of the electromagnetic spectrum.

cosmic ray A high-speed subatomic particle (atomic nucleus) from space.

cosmic singularity The initial condition of the expanding Universe as a state of infinite density, temperature, and curvature from which the Big Bang was initiated. The presence of the infinities implies that a yet-to-be-discovered theory will provide a more complete account of initial condition.

cosmic web Structure on the largest cosmic scales appearing as a network of interconnected filaments, surrounding voids, joined at nodes where dense clusters are found.

cosmological look-back time A measure of time elapsed from the moment photons are detected on Earth back to the moment they were emitted by a distant object.

cosmological principle The idea that, on average, the conditions in one part of the Universe do not differ from those in any other part.

cosmology The field of astronomy that studies the Universe as a whole.

crater An approximately circular depression in the surface of an object formed by a high-velocity impact of a smaller body.

cratering The change in surface features of a planet due to impacts with comets, asteroids, and meteoroids.

critical density The matter-energy per unit volume that is just sufficient to slow the expansion of the Universe such that while the expansion proceeds indefinitely, the expansion speed approaches zero more and more closely as time passes.

crust Earth's solid outer layer.

C-type asteroid An asteroid that is carbonaceous (rich in carbon) and is typically dark.

D

Dark Ages The early era of cosmic evolution, when neutral hydrogen filled space and absorbed all ultraviolet and optical photons.

dark energy Energy driving the acceleration of the expansion of space-time.

dark globule or **Bok globule** A small molecular cloud containing a few solar masses' worth of material whose density and dust content render it opaque to light in visible wavelengths.

dark matter Nonluminous mass that is evident only by its gravitational influence.

declination The position of an object on the celestial sphere, measured similarly to latitude on Earth.

degeneracy pressure Pressure exerted because of quantum mechanical motions of particles resulting from the associated confinement to a small volume. See also *Heisenberg uncertainty principle*.

density perturbation A small region where the density of hydrogen gas is slightly higher or lower than that of the surrounding regions.

deoxyribonucleic acid (DNA) A molecule that encodes genetic information. It consists of two polymer strands wound around each other in a double-helix shape. DNA is the biological basis for life on Earth.

deuteron An isotope of hydrogen with one proton and one neutron.

differential force of gravity The different strengths of the gravitational force from a massive body felt on the different parts of an extended object because the force decreases as the inverse square of the distance.

differential rotation The rotation of different regions of an object (different latitudes or depths) at different rates.

differentiation The process in planet formation whereby, during a planet's molten stage, the denser materials sink to the center while less dense materials rise to the surface, forming distinct layers.

dipole field The magnetic-field configuration consisting of a north and south pole and magnetic-field lines connecting them.

DNA See *deoxyribonucleic acid*.

Doppler shift A change in wavelength that results because either a wave source or an observer moves relative to the other.

droplet condensation The transformation of molecules from a gaseous to a liquid state in the form of small droplets in an atmosphere. It generally occurs where temperature is decreasing.

dust The smallest bits of solid matter in the Universe, composed of atoms such as carbon and silicon.

dust tail The tail of a comet that is composed of dust driven off the coma by solar radiation pressure; its particles then move on their own orbits. It always points away from the Sun.

dwarf planet A body that orbits a star and has enough mass to become spherical by self-gravity but that has not cleared its neighboring region of small objects.

E

Earth-crossing asteroid (ECA) An asteroid that, because its highly elliptical orbit crosses Earth's orbit, is a potential risk for collision with Earth.

Earth mass The mass of Earth, 5.97×10^{24} kg, used as a unit of measure for other objects.

ECA See *Earth-crossing asteroid*.

eccentricity A measure of the roundness of an ellipse, calculated as the ratio of the distance from the ellipse's center to its foci, divided by the length of the semimajor axis.

eclipse The passing of one celestial body through the shadow of another.

eclipsing binary Co-orbiting stars in which one star is observed passing in front of another.

ecliptic The path that the Sun appears to follow against the background stars, as defined by Earth's orbit around the Sun.

Einstein ring The ring-shaped appearance of a distant background object that is caused by a gravitational lens formed via symmetric distribution of dark matter in the foreground.

ejecta blanket A layer of pulverized material that is blown out of an impact site (a crater).

electromagnetic radiation Oscillating electric and magnetic fields traveling through space.

electromagnetic spectrum The full range of electromagnetic radiation at different wavelengths running from gamma rays (very short) to radio waves (very long).

electron A subatomic particle with a negative electric charge.

elephant trunk A linear feature at the edge of an H II region composed of molecular gas and pointing toward the ionizing star or stars.

ellipse A geometric form in the shape of an oval or circle, in which the sum of the distances from two points (foci) to every point on the ellipse is constant.

elliptical galaxy A galaxy that appears elliptical on the sky. Elliptical galaxies are generally triaxial, meaning that the width is different in each dimension.

emission The release of a photon of a specific wavelength when an electron jumps from a higher to a lower energy level.

emission line A bright line at a particular wavelength in a spectrum. Specific patterns of emission lines indicate the presence of a specific element in the source.

Encke Division A 325-kilometer-wide gap within Saturn's A ring that has been shaped by Saturn's moon Pan.

energy The ability to perform work. Energy comes in many forms, such as kinetic, thermal, gravitational, and electromagnetic.

energy level Any one of the certain discrete values of orbital energy possible for an atom.

envelope The region surrounding a protostar that is composed of free-falling gas.

epicycle A secondary orbit whose center point orbits Earth; devised by Ptolemy to explain retrograde motion.

equant A point displaced from the center of a planet's orbit around which the planet displays uniform motion.

equinox A day when the hours of daylight and darkness are equal. The autumn equinox occurs on approximately September 21; the spring equinox, approximately March 21.

erosion or **weathering** The process by which surface features are worn away by the action of wind and water (including ice).

escape velocity The velocity required to escape the pull of gravity from a body.

eukaryotic cell A cell that stores its DNA in a cell nucleus.

event horizon The boundary of a black hole, located one Schwarzschild radius out from the singularity.

excited state Any energy state within an atom that is higher than the ground state.

extrasolar planet or **exoplanet** A planet orbiting any star other than the Sun.

extremophile An organism that thrives under extremes of temperature, pressure, and other conditions.

eyepiece The small lens of a telescope, through which the observer looks.

F

Fermi paradox The conclusion that no intelligent life exists in our galaxy beyond Earth, since the billions of years following the origin of the galaxy would have allowed advanced civilizations to have spread across interstellar space and reached Earth.

field reversal A reorientation of north and south poles within a magnetic field such that the two poles swap positions.

first law of planetary motion The principle, advanced by Johannes Kepler, stating that planets move on elliptical orbits with the Sun at one focus.

fission hypothesis A hypothesis of the Moon's origin stating that the Moon formed from material pulled into orbit from the still-molten Earth by a close encounter with another body.

flare An eruption that launches material from the solar surface.

flocculent spiral A spiral galaxy with poorly defined spiral arms.

flux tube A bundle of magnetic-field lines.

focus (pl. foci) Either of two points interior to an ellipse that are used to define its shape. The Sun is always at one focus of a planet's elliptical orbit.

force An interaction between two bodies that, if unbalanced, can change their state of motion.

force boson A particle that carries one of the four fundamental forces.

force law The principle, advanced by Isaac Newton, stating that the change in an object's acceleration due to an applied net force is in the same direction as the force and directly proportional to it, but inversely proportional to the object's mass.

fossil magnetism Magnetic fields imprinted in a planet that were created by an earlier active dynamo phase.

fragmentation The shattering of solid bodies, such as planetesimals, caused by collisions.

frame dragging The process by which a massive, spinning object pulls the surrounding space-time around with it.

frame of reference An individual perspective from which observations are made.

free-fall time The length of time required for a cloud to collapse under the force of its own gravity.

frequency The number of peaks (or troughs) of a wave that pass a point in space in a single second.

G

galactic disk A flattened distribution of stars and interstellar material within a galaxy.

galaxy Millions to billions of gravitationally bound stars. Galaxies also include gas, dust, and dark matter.

galaxy cluster A collection of hundreds or thousands of galaxies bound together by gravity.

galaxy group A gravitationally bound collection of galaxies, usually containing 50 or more members.

galaxy supercluster See *supercluster*.

Galilean moons The four largest moons of Jupiter: Io, Ganymede, Callisto, and Europa.

gamma ray A high-energy electromagnetic wave with a wavelength less than 10^{-11} meter.

gas giant A planet consisting mostly of hydrogen and helium that has a mass many times that of Earth.

gene A section of a DNA or RNA molecule that is the code for a particular trait or function.

general theory of relativity Albert Einstein's second theory of relativity, which extends the special theory of relativity by incorporating gravity and states that matter and energy tell space-time how to curve while space-time tells matter and energy how to move.

genetic drift Changes in genes due to random events other than gene mutation.

geocentric model A model stating that Earth is the body around which all other Solar System objects orbit.

giant molecular cloud (GMC) A large-scale region of gas in molecular form spanning tens to hundreds of parsecs and containing up to 10 million solar masses.

globular cluster A gravitationally bound cluster of 10,000 or more stars all born at the same time.

GMC See *giant molecular cloud*.

GOE See *Great Oxygenation Event*.

grand design spiral A spiral galaxy with well-defined spiral arms that may be traced all the way from the bulge/bar to the outer regions of the disk.

grand unified theory (GUT) A yet-to-be-discovered theory that would explain how the strong and weak nuclear forces along with the electromagnetic force emerged from a unified force in the early instants following the Big Bang.

gravitational contraction The shrinking of an object's radius and accompanying increase in its density that is due to the force of its own gravity.

gravitational lensing A change in the path of light passing a massive object produced by the curvature of space-time.

gravitational potential energy Energy that an object possesses because of its position in a gravitational field.

gravity An attractive force between any two massive bodies that depends on the product of the bodies' masses and the inverse square of the distance between them.

Great Attractor An unseen concentration of mass whose gravity affects the motion of galaxies in a large volume of space that includes the Milky Way.

Great Oxygenation Event (GOE) The period in Earth's history during which oxygen built up relatively rapidly in the planet's atmosphere.

greenhouse effect The phenomenon by which the fraction of incoming solar energy that would have been radiated back into space is trapped by greenhouse gases, leading to an increase in the average temperature of the planet.

greenhouse forcing An increase in greenhouse gas concentrations that leads to a change in a planet's climatic state.

greenhouse gas An atmospheric gas that absorbs infrared radiation and keeps it from being reemitted into space.

ground state The lowest energy state within an atom.

GUT See *grand unified theory*.

H

H I clouds The phase of the interstellar medium that is composed of relatively low-temperature neutral hydrogen gas.

H II region The phase of the interstellar medium that contains ionized hydrogen gas.

habitable zone The region around a star where liquid water can exist on the surface of a planet.

half-life The period of time over which the total number of nuclei of a radioactive isotope will drop by one-half.

heat energy See *thermal energy*.

Heisenberg uncertainty principle The quantum mechanical principle stating that as the position of a particle is specified to higher accuracy, its velocity becomes increasingly uncertain, and vice versa.

heliocentric model A model stating that the Sun is the body around which all other Solar System objects orbit.

heliopause The outer boundary of the heliosphere.

helioseismology The study of oscillations in the Sun and their propagation. Helioseismological studies are used to develop precise models of solar structure.

heliosphere A cavity, created by the solar wind, in the interstellar gas surrounding the Solar System.

helium flash Rapid internal ignition of helium fusion in the core of an evolved star.

Herbig-Haro (HH) object A protostellar jet that is distinguished by knots in the jet beam.

Hesperian period The evolutionary period of Mars from about 3.5 to 3 billion years ago.

HH object See *Herbig-Haro object*.

hierarchical formation An ordered sequence of formation, from smallest to largest structures.

homogeneous Having the matter distribution property of being uniform throughout space. A homogeneous distribution has the same average properties at every point in space.

horizon The horizontal line defining the lower edge of the sky.

hot dark matter Nonluminous matter moving at relativistic speed.

hot Jupiter A Jupiter-sized planet orbiting very close to its star, usually at a fraction of an astronomical unit.

HR diagram A plot of luminosity versus temperature (or spectral class) for a population of stars.

Hubble Deep Field Data resulting from a week-long exposure in 1995 by the Hubble Space Telescope of a tiny, dark patch of sky revealing thousands of galaxies at great distances. Since the original, other deep-exposure projects have been carried out.

Hubble flow The pattern of large-scale galaxy motion associated with the expansion of space-time.

Hubble's constant The slope of Hubble's law relating recession velocity and distance, which is expressed in kilometers per second per megaparsec (km/s/Mpc).

Hubble's law The linear relationship between a galaxy's recession velocity and its distance.

Hubble tuning-fork diagram A scheme for representing galaxy types developed by Edwin Hubble that shows a progression from highly elliptical to spiral or barred spiral.

hydrated Containing water.

hydrocarbon An organic compound consisting entirely of hydrogen and carbon.

hydrodynamic instability model A model of giant gas planet formation in which small regions collapse to form planets within a gravitationally unstable disk.

hydrogen-burning shell The nuclear burning layer outside the core of a star where hydrogen continues to be fused into helium after such reactions have ended in the core.

hydrostatic equilibrium The condition in which gravity and gas pressure are balanced.

hypothesis (pl. hypotheses) A proposed explanation of a phenomenon that must be rigorously tested by experiments.

I

ice giant A planet consisting mostly of ices (water, methane, carbon dioxide, and other compounds) that has a mass many times that of Earth.

IMF See *initial mass function*.

impact hypothesis The currently accepted hypothesis of the Moon's origin, which states that a large, Mars-sized body striking Earth ejected mantle material into orbit, which gravitationally coalesced into the Moon.

inertial law The principle, advanced by Isaac Newton, stating that objects in motion at constant velocity along a straight line continue in that way unless acted on by a net force.

inflation A hyperrapid expansion of space-time that occurred extremely early after the Big Bang.

infrared wave An electromagnetic wave with a wavelength of 700 nanometers (1×10^{-9} meter) to 1 millimeter, just longer than that of visible light.

initial mass function (IMF) The mathematical relation describing the relative abundance of stars of different masses.

instability A complex pattern of gas flow resulting from a disruption in an originally smooth flow.

interferometry The technique of combining the signal from many smaller telescopes to achieve the resolving power of a larger one.

interstellar dust grains Tiny bits of solid matter found in diffuse clouds in the regions between the stars.

interstellar gas Ionic, atomic, or molecular gas found in diffuse clouds in the regions between the stars.

interstellar medium (ISM) The gas and dust spread between the stars in different phases that depend on temperature and density conditions.

intracluster medium Thin gas in the space between galaxies in a galaxy cluster.

inverse square law A relationship whereby a quantity (such as gravity) decreases in proportion to the square of a variable (such as distance) as the latter increases.

ionosphere or **thermosphere** The fourth (and highest) level of Earth's atmosphere, extending up to 1,000 km, where most of the atoms are ionized.

ion tail The tail of a comet that is composed of gas and reacts to the solar wind and thus always points directly away from the Sun.

ion torus (pl. tori) A doughnut-shaped region of emission where ions spiral around the magnetic field.

irregular galaxy An amorphous galaxy lacking common structures.

irregular satellite A satellite orbiting in the opposite direction from its central object or on a highly inclined orbit in relation to the planet's orbital plane.

ISM See *interstellar medium*.

isotopes Forms of an element with different numbers of neutrons in the nucleus.

isotropic Having the matter distribution property of being uniform in terms of its appearance when viewed from different angles. An isotropic distribution looks the same no matter which direction a viewer chooses to observe.

J

jet A stream of a liquid, gas, or small solid particles shot outward from a comet nucleus or other object in a focused beam.

K

KBO See *Kuiper Belt object*.

kinetic energy Energy that is due to an object's bulk motion.

Kuiper Belt The disk-shaped Solar System zone outside Neptune's orbit, from 30 to 50 AU, containing small bodies made of ice and rock.

Kuiper Belt object (KBO) A small body made of ice and rock that orbits beyond Neptune's orbit, in the Kuiper Belt.

L

large-scale structure of the Universe The hierarchy of cosmic structures on ever-larger scales from galaxy groups up to and beyond superclusters and voids.

Late Heavy Bombardment A hypothetical epoch, from 4.1 to 3.8 billion years ago, during which a large number of asteroids collided with bodies in the inner Solar System.

latitude A measure of location specifying north–south position on Earth running from 0° at the equator to 90° north (North Pole) and 90° south (South Pole).

lenticular galaxy A galaxy with a bulge and limited evidence of a disk; it has no spiral arms.

lepton A class of fundamental particles of matter that includes electron, muon, and tau particles.

life expectancy–mass relationship The mathematical connection between the life expectancy and mass of a star: the greater the star's mass, the shorter its life expectancy.

light curve A graph of the light output versus time of an astronomical object.

light-year (ly) The distance that light travels in a year: 9.5×10^{15} meters; used as a standard of measure for large distances.

line of nodes The line defined by the intersection of the Moon's orbital plane and Earth's orbital plane around the Sun.

lithosphere Earth's solid outer layers, consisting of the crust and the uppermost regions of the mantle.

Local Group A gravitationally bound collection of galaxies within which the Milky Way resides.

longitude A measure of location specifying east–west position on Earth.

long-period comet A comet that originated in the Oort Cloud and has a highly eccentric orbit and an orbital period from 200 years to millions of years.

luminosity The total energy output per time of a light source.

lunar highlands Older, lighter-colored, heavily cratered portions of the Moon's surface.

ly See *light-year*.

Lyman series Emission or absorption lines of the hydrogen atom as an electron moves between the ground state and excited states.

M

M-sigma relation The correlation between the average rotational velocity of stars in the outer region of a galaxy and the mass of the central black hole.

MACHO See *massive compact halo object*.

macromolecule See *polymer*.

magma flooding The creation of smooth "lowlands" via lava flows that fill in craters.

magnetic dynamo The flow of electrically charged fluid that can generate a large-scale magnetic field in a planet, a star, or any other astronomical object.

magnetosphere A region of space around a planet where the planet's magnetic field exerts a strong influence.

magnification The ability of a telescope to enlarge appearances in an image.

magnitude scale A categorization of stellar brightness in which faint objects are given higher numbers than brighter objects.

main sequence The diagonal band on the HR diagram representing core hydrogen-burning stars. High-mass stars appear on the upper left; low-mass stars, on the lower right.

mantle The layer between Earth's core and crust, composed of silicate rocks. High pressures found in the mantle cause material to deform and flow.

maria (sing. mare) Large, dark plains on the Moon, formed by ancient volcanic eruptions.

Martian spherule Any of the tiny pebble-like structures ("blueberries") found in Martian soil, which may have crystallized out of a water-rich solution.

mass A property of matter that is related to its resistance to changes in motion or inertia. Mass is measured in units of kilograms.

massive compact halo object (MACHO) A proposed form for dark matter that includes objects such as black holes and extremely faint neutron stars.

matter Any substance made up of particles with mass that occupies space.

Maunder Minimum The period in the late 1600s when the number of sunspots remained extremely low from one sunspot cycle to another.

megabar A unit of pressure equal to a million times Earth's surface pressure.

megalith A large stone used as a monument or part of a monument.

megaparsec (Mpc) A unit of distance equal to 1 million parsecs.

meridian A circular arc crossing the celestial sphere and passing through the local zenith and celestial pole.

mesosphere The third level of Earth's atmosphere, extending up to 85 km.

metallic hydrogen Hydrogen at a pressure and temperature high enough to have atoms closely spaced but with electrons able to move freely as in a metal.

metallicity The abundance of elements heavier than helium in an object.

meteor A meteoroid that has entered Earth's atmosphere. Heat from friction causes radiation that may be seen as a very brief streak of light or as a fireball.

meteor shower The appearance of frequent meteors, over a period of a few hours or several days. The meteors seem to emanate from a common location on the sky, which is caused by Earth's passing through the debris deposited along the orbit of a comet.

meteorite A rock that was a meteoroid, then a meteor, and has reached Earth's surface.

meteoroid A rock ranging in size from a grain of sand to a boulder and orbiting the Sun.

microwave The shortest-wavelength radio wave, with a wavelength of 1 millimeter (1×10^{-3} meter) to about 0.1 meter.

Milky Way The spiral galaxy in which our Sun is located.

mini-Neptune A planet with a mass less than 15 Earth masses but substantially greater than Earth's mass.

molecular cloud The low-temperature, high-density phase of the interstellar medium where molecules can form. Molecular clouds are the site of star formation.

molecular disk See *molecular ring*.

molecular outflow The bipolar swept-up shells of molecular material, which may be driven by protostellar jets.

molecular ring or **molecular disk** A dense ring of molecular material orbiting the galactic center.

momentum The tendency of a moving object to continue in motion. Only the application of a force can change an object's momentum.

monomer A small molecular unit that can combine to form larger, more complex polymers.

moon A natural satellite of a planet.

Mpc See *megaparsec*.

muon An elementary, extremely short-lived particle that is heavier than an electron.

mythology A collection of stories used to explain phenomena whose physical basis is not understood.

N

naked singularity A singularity that lacks an event horizon.

natural selection The process by which an organism that is better adapted to its environment is preferentially selected to survive and reproduce over less adapted organisms.

neap tide A tide that is less than average because the Sun's gravity partially cancels out that of the Moon at the Moon's first and third quarter phases.

nebula An interstellar cloud of dust, hydrogen, helium, and other ionized gases.

nebulosity Clouds in space.

Neolithic period The period extending from about 10,000 BCE to between 4500 and 2000 BCE, when farming was introduced.

neutrino An electrically neutral, weakly interacting elementary subatomic particle often created in nuclear reactions.

neutron A subatomic particle with no net electric charge within an atomic nucleus.

neutron star A stellar remnant supported by neutron degeneracy pressure that results from the gravitational collapse of a massive star.

Noachian period The evolutionary period of Mars from about 4 to 3.5 billion years ago.

no-hair theorem The description of a black hole as having only three properties: mass, spin, and electric charge.

north celestial pole The extension of Earth's North Pole onto the celestial sphere.

north magnetic pole The point on Earth's Northern Hemisphere at which the planet's magnetic field points vertically downward.

North Star See *Polaris*.

nova Nonterminal explosive nuclear fusion occurring on the surface of a white dwarf after it has accreted matter from a companion star.

nuclear fission The splitting apart of a nucleus into two or more nuclei of lower atomic weight.

nucleobase A polymer that links nucleotides in DNA and RNA.

nucleotide A polymer built of sugar molecules and phosphate groups that link to compose the backbones of DNA and RNA.

nucleus (pl. nuclei) 1. The dense region at the center of an atom consisting or protons and neutrons. 2. The main body of a comet, composed of ice and rock.

O

O and B association A group of massive stars of O and B spectral types found tracing out a spiral arm.

objective In a refractor telescope, the large lens that gathers light from the object being observed and focuses the light rays to form an image.

oblate Roughly spherical, showing flattening in the equatorial plane.

obliquity or **axial tilt** The angle between a planet's spin axis and a line perpendicular to the planet's orbital plane.

Occam's razor A principle stating that among competing hypotheses, the simplest one should be selected.

omega (Ω) The ratio of the observed density of matter-energy in the Universe to critical density.

Oort Cloud A spherical region at the outer limits of the Solar System, with a radius of about 50,000 AU, that is believed to contain potentially trillions of comets.

opacity The degree to which light is absorbed when passing through a material.

open star cluster A loose distributed group of stars that were born from the same cloud at the same time and remain bound for some period before dispersing.

orbit The cyclical path in space that one object makes around another object.

orbital period The time an object takes to complete one revolution of an orbit.

orbital resonance The synchronized gravitational accelerations that arise when objects orbiting a body have periods that are whole-number ratios of each other.

orbital velocity The velocity required to keep a body in orbit around another body.

order of magnitude The size of a physical quantity in powers of 10.

outflow channel An extended region of scoured ground on Mars that includes features indicating high rates of ground fluid flow.

ozone A molecule composed of three oxygen atoms: O_3.

ozone hole An atmospheric region near either of Earth's poles and within the stratosphere that shows a substantial decrease in ozone abundance.

P

P-P chain See *proton-proton chain*.

P-wave or **pressure wave** A seismic wave in which atoms and molecules oscillate back and forth in the same direction as the wave's propagation.

Paleolithic period The period extending from 2.6 million years ago (characterized by the earliest known use of stone tools) until 10,000 years ago.

parallax The displacement or difference in the apparent position of an object viewed along two different lines of sight.

parsec (pc) The preferred unit for large distances, equal to 3.26 light-years.

partial lunar eclipse An eclipse that occurs when the Moon passes through Earth's penumbral shadow.

partial solar eclipse An eclipse that occurs when a region of Earth's surface passes under the Moon's penumbral shadow.

particle accelerator An experimental apparatus used by physicists to study the structure of matter by colliding atoms and subatomic particles together at very high velocities (high energies).

particle era A period in cosmic evolution when the particles described by the standard model of particle physics emerged.

pc See *parsec*.

peculiar motion The motions of stars in the Milky Way that depart from pure revolution around the galactic center.

peculiar velocity The motion of a galaxy through space that is due to gravitational attraction, distinct from motion due to the expansion of space-time.

penumbra The outer region of a shadow cast by an extended object.

perigee The distance of closest approach for an object orbiting Earth.

perihelion (pl. perihelia) The distance of closest approach for an object orbiting the Sun.

period-luminosity relationship The correlation between the average luminosity and period of a variable star, allowing a star to act as a standard candle for determining distance.

phase (of the Moon) The appearance of the illuminated portion of the Moon as seen by an observer on Earth, which changes cyclically as the Moon orbits Earth.

phase transition A change in the state of matter based on changes in temperature (or other quantity).

photoevaporation The process by which gas is launched from the surface of an accretion disk or molecular cloud through heating by ionizing radiation from a young, massive star.

photoionization The stripping of an electron from an atom or ion via the absorption of a photon.

photon A particle of light.

photosphere The thin, "surface" region of the Sun, from which radiation escapes into space.

phylogeny Relationships among species.

Planck length The size scale at which space-time should begin showing its quantum nature, about 10^{-44} meter.

planet A body that orbits a star and is massive enough for self-gravity to have pulled it into a spherical shape and to have cleared smaller objects from the neighborhood of its orbit.

planet migration A change in a planet's orbit after the planet forms.

planetary cooling The process by which a planet loses the heat acquired during its formation.

planetary differentiation The separation by gravity of a planet into different layers running from densest at the core to less dense in the outer regions.

planetary nebula An evolutionary stage for low- and intermediate-mass stars that comes between the AGB and white dwarf phases, in which stellar wind material is driven off the hot star and ionized.

planetary system A star and its orbiting planets; the Solar System is an example of a planetary system.

planetesimal An irregular rocky object, typical of those from which planets are believed to have formed by accretion.

plasma Electrically charged gas.

plate A large segment of Earth's crust that floats on the mantle.

plate tectonics The movement of Earth's crustal plates due to underlying mantle flow.

plume Molten rock from a planet's mantle that breaks through the crust.

Polaris The star that is found very near Earth's celestial north pole (hence called the North Star) and around which the sky of the Northern Hemisphere appears to turn.

polymer or **macromolecule** A large molecule containing many atoms, often with a chain structure.

Population I Stars found primarily in the disk and bulge, which have higher metallicity and therefore must have been born relatively recently in cosmic history.

Population II Stars found primarily in the stellar halo, which have lower metallicity and therefore must have been born relatively early in cosmic history.

positron The antimatter version of the electron, having the mass of the electron but a positive charge.

precession The rotation of a planet's spin axis, similar to that of a wobbling top.

pressure wave See *P-wave*.

primordial abundance of light elements The proportions of hydrogen, helium, deuterium, and other light elements created in the first 3 minutes after the Big Bang.

prominence A giant arc of magnetism anchored in the photosphere.

proper motion An object's motion that is on the plane of the sky and can be detected by observations taken over time.

proplyd A teardrop-shaped knot of gas formed by the interaction of stellar winds and the photoevaporation flow from a protostar's accretion disk.

protein A polymer made of amino acids that assists in cellular functions.

proton A subatomic particle with a positive electric charge.

proton-proton chain or **P-P chain** The sequence of reactions required to convert hydrogen nuclei into helium nuclei.

protoplanetary disk The disk that surrounds a young star, from which planets may form.

protostar A dense central object, plus an accretion disk and envelope, that forms after cloud collapse but before nuclear fusion begins.

pseudoscience A claim, belief, or practice that is presented as scientific but does not adhere to a valid scientific method, lacks supporting evidence or plausibility, and/or cannot be reliably tested.

pulsar A rotating, magnetized neutron star whose radio waves periodically sweep across Earth, producing emission patterns that are observed as bursts of energy.

Q

quantum field The distribution of energy that fills space, within which fundamental particles appear as excited states.

quantum gravity A theory (yet to be fully worked out) that unifies general relativity and quantum mechanics, describing how space-time becomes granular ("quantized") at the smallest scales.

quantum mechanics or **quantum physics** A branch of physics dealing with physical phenomena at the level of molecular, atomic, and subatomic scales.

quark A class of fundamental particles of matter that make up other particles, such as protons and neutrons.

quasar Quasi-stellar radio source; an extremely luminous and distant active galactic nucleus.

R

radial velocity An object's motion that is perpendicular to the plane of the sky along the observer's line of sight and can be detected via Doppler shift measurements.

radial velocity method An exoplanet detection method that measures changes in a star's velocity as it orbits a common center of mass with one or more planets.

radiative energy transport The transfer of thermal energy via electromagnetic radiation.

radiative zone The region of the Sun above the core, where energy is transported outward by means of electromagnetic radiation (photons scattering off of particles of mass).

radioactive decay The process by which some isotopes of an element transform over time into other elements, emitting subatomic particles and energy in the process.

radio galaxy A galaxy whose nucleus is the origin of relativistic bipolar jets.

radio wave An electromagnetic wave with a wavelength greater than 0.1 meter.

radiometric dating A technique used to date a sample of material by comparing relative amounts of radioactive atoms with the "daughter" atoms they decay into.

reaction law The principle, advanced by Isaac Newton, stating that for every applied force, there is an equal and opposite force.

recombination era The era approximately 380,000 years after the Big Bang in which the first electrons were captured by protons to form hydrogen atoms, thus decoupling matter from the cosmic blackbody radiation.

reconnection The breaking and simultaneous rejoining of magnetic-field lines.

red giant A cool star with a large radius that has evolved off the main sequence because hydrogen fuel in its core is depleted.

redshift An increase in the wavelength of light that results when the source of the light moves away from the observer or the observer moves away from the source.

redshift survey Measurements of redshift (recessional velocity) for many galaxies across a region of the sky. Redshift surveys produce maps of galaxies across all three dimensions.

reflection nebula A dense, dusty gas cloud made visible by reflected starlight.

reflector An optical telescope that uses curved mirrors to collect light and form an image.

reflex motion The movement of a star in response to the gravity of an orbiting planet.

refraction The bending of light as it passes between two media, such as water and air.

refractor An optical telescope that uses a lens to collect light and form an image.

regular rich cluster A spherical, centrally condensed galaxy cluster with more than 1,000 galaxies.

regular satellite A satellite orbiting in the same direction as its planet's movement around the Sun and with a low inclination to the plane of the planet's orbit.

Renaissance A cultural movement spanning the 14th to 17th centuries that encompassed a flowering of education and art, much of which was based on principles established by the ancient Greeks.

residual ice cap A polar region of frozen material such as CO_2 and water that stays constant in size with seasonal changes.

retrograde motion Apparent motion of a planet in a direction opposite to its normal motion.

ribonucleic acid (RNA) Single-stranded polymer that transcribes the code of DNA, placing amino acids in the correct sequence to create proteins.

RNA See *ribonucleic acid*.

Roche limit The closest that a satellite can get to its central object before tidal forces from that object pull the satellite apart (because the satellite can no longer be held together by its own gravity).

Roche lobe The region around a star within which orbiting material is gravitationally bound to that star.

rotation The spin of an object around an internal axis.

rotation curve The change in rotation speed as a function of distance for objects orbiting around a common center.

runoff channel A long, meandering feature on Mars that resembles a river system on Earth.

S

S-type asteroid An asteroid that contains mostly silicate rock and is typically more reflective than C-type asteroids are.

S-wave or **shear wave** A seismic wave in which atoms and molecules oscillate perpendicular to the direction of the wave's propagation.

scatter To change a photon's direction of propagation.

scattered disk The disk-shaped region beyond the Kuiper Belt that contains many small bodies made of ice and rock. Their highly elliptical orbits may reach perihelion as low as 30 AU and aphelion as high as 100 AU.

scattering The rearrangement of the orbits of planets due to mutual gravitational interactions.

Schwarzschild radius The distance from a black-hole singularity where the escape velocity equals the speed of light.

scientific method A body of techniques for acquiring new knowledge that includes systematic observation, experimentation, and theorizing.

scientific model An idea or set of ideas used to create testable explanations.

scientific notation A way to write very large or very small numbers using the form $a \times 10^b$, where the coefficient a is any real number and the exponent b is an integer.

seasonal ice cap A polar region of frozen material such as CO_2 and water that increases and decreases in area and thickness with the seasons.

second law of planetary motion The principle, advanced by Johannes Kepler, stating that the line connecting a planet to the Sun will sweep out equal areas of its orbit in equal times.

seismic wave A propagating wave in Earth's interior caused by an earthquake or a volcano.

semimajor axis Half of the long axis of an ellipse.

semiminor axis Half of the short axis of an ellipse.

Seyfert galaxy A galaxy whose nucleus shows broad emission lines, indicating highly ionized material moving at great speed.

shear wave See *S-wave*.

shepherd moon A planetary satellite whose gravity confines the ring to a narrow band.

shield volcano A volcano formed from a mantle plume that tends to have shallow, sloping sides.

short-period comet A comet that originated in the Kuiper Belt or scattered disk and has an orbital period of 200 years or less.

sidereal day The time taken by Earth to rotate on its axis relative to the stars.

sidereal month The amount of time it takes the Moon to make one complete orbit in relation to the background stars.

silicate A mineral compound containing silicon and oxygen.

singularity The central point of a black hole, where density and space-time curvature is infinite.

small-angle formula A mathematical relation between the angular size of an object as measured on the sky and the object's distance and physical size.

snow line The distance from the Sun or other star beyond which it is cold enough for water to exist as ice.

solar day The time between successive crossings of the same meridian on the sky by the Sun.

solar granulation The pattern of cellular features on the solar surface, showing bright interiors and dark edges.

solar magnetic cycle A large-scale reorientation of the Sun's magnetic field occurring every 22 years.

solar max The period of peak activity in the sunspot cycle.

solar neutrino problem The apparent discrepancy between the expected number of neutrinos being emitted from the Sun and the number measured experimentally, which has been resolved with the discovery that neutrino species can change form.

solar wind A stream of energetic particles flowing off the Sun.

south celestial pole The extension of Earth's South Pole onto the celestial sphere.

south magnetic pole The point on Earth's Southern Hemisphere at which the planet's magnetic field points vertically upward.

space-time A unified, four-dimensional geometry of the Universe that incorporates both space and time.

space weather Coronal mass ejections sweeping through interplanetary space, which produce auroras when they hit Earth and may pose threats to astronauts and technological systems.

special theory of relativity Albert Einstein's first theory of relativity, stating that the laws of physics are the same for all observers that are in uniform motion relative to one another, and that the speed of light in a vacuum is the same for all observers, regardless of their relative motion.

spectral classification A fundamental grouping of stellar types, based on observed spectral lines, that specifies seven categories denoted by the letters O, B, A, F, G, K, and M.

spectrograph A device for separating light from a source into its component wavelengths and measuring the energy at each wavelength.

spectroscopic binary Co-orbiting stars that have been detected by the Doppler shift of absorption lines.

spectroscopic parallax A method for determining distance to an astronomical object that is based on obtaining the width of absorption lines in the object's spectrum.

spectrum (pl. spectra) The distribution of light intensity versus wavelength (or frequency).

spectrum of perturbations The range and strength of variations in density imposed on cosmic matter during inflation.

speed The change in distance per unit of time.

spiral arm A spiral-shaped distribution of stars and interstellar material, often extending from the bulge outward.

spiral density wave A wavelike pattern of compression that moves through the physical material of a galaxy's disk. Spiral density waves are seen in spiral arms.

spiral galaxy A pinwheel-shaped galaxy that may or may not have spiral arms.

spring tide A tide that is greater than average because it is reinforced by the linear alignment of Sun, Earth, and Moon at the new and full Moon phases.

standard candle An object with known luminosity that can be used to determine distance.

standard model of particle physics A description of subatomic particles and forces.

star A sphere of gas, such as the Sun, that produces energy via nuclear fusion in its core.

starburst A rapid increase in star formation in which an appreciable fraction of the molecular gas is quickly converted into stars.

steady-state model A theory, advanced by astronomer Fred Hoyle in the 1950s, stating that although the Universe is expanding, it always appears the same because new matter is slowly created between galaxies to maintain the average density.

Stefan-Boltzmann law The mathematical relationship between a blackbody's luminosity and its surface area and temperature.

stellar ejecta Material expelled in a stellar explosion such as a supernova.

stellar evolutionary track A path on the HR diagram representing the evolution of a star through time.

stellar halo The spherical distribution of Population II stars that are found mainly in globular clusters.

stellar occultation The blocking of a star's light by an object between the star and an observer's view.

stratosphere The second level of Earth's atmosphere, extending about 20 km to about 50 km.

strong nuclear force One the four fundamental forces of nature. It binds particles together in the nucleus of an atom.

subatomic particle Any particle smaller than a composite atom. Subatomic particles include protons, neutrons, and electrons.

subduction A geologic process in which one edge of a crustal plate is forced below the edge of another.

summer solstice The day (approximately June 21 in the Northern Hemisphere) with the most hours of sunlight, when the Sun appears highest in the sky.

sunspot A large, strongly magnetized region of the Sun's surface that appears darker because it has a lower temperature than its surroundings have.

supercluster A collection of galaxy clusters; the largest type of object in the Universe.

super-Earth A planet with a mass greater than Earth's but substantially less than 15 Earth masses.

superluminal motion The illusion that an object's speed exceeds the speed of light, which is created by the object's motion toward the observer.

supermassive black hole A black hole of millions or billions of solar masses that is found in the center of a galaxy.

supernova The highly energetic explosion of a massive star.

supernova remnant An expanding shock wave that contains ejected material and swept-up interstellar material resulting from the explosion of a star.

synodic month The amount of time it takes the Moon to cycle through its phases.

T

T Tauri star A variable star of less than 3 solar masses that is in a later stage of star formation and exhibits bright X-ray flares.

terracing The steplike appearance of steep walls in an impact crater created by slumping of the walls due to gravity.

terrestrial planet A planet consisting primarily of solid material.

theory An explanation for a phenomenon that has undergone extensive testing and is generally accepted to be accurate.

thermal energy or **heat energy** Energy that is due to random motions of atoms.

thermal history The evolutionary pattern of heating and cooling for a planet.

thermal inversion An increase in temperature that occurs with increasing altitude (whereas temperature usually decreases with increasing altitude).

thermal velocity The average speed of particles in a gas, which depends on the temperature of the gas.

thermonuclear fusion The fusion of lighter atomic particles into heavier ones, requiring high temperatures.

thermonuclear runaway Nuclear burning that ignites and spreads catastrophically, consuming all available fuel.

thermosphere See *ionosphere*.

thin disk An accretion disk surrounding a supermassive black hole; an element of the unified model for AGNs.

third law of planetary motion The principle, advanced by Johannes Kepler, stating that the square of a planet's orbital period equals its average orbital radius cubed.

tidal bulge A distortion from spherical symmetry that is caused by tidal force.

tidal force Force acting on an object that is due to differential gravity from a second object.

tidal locking The one-to-one relationship between the rotation and revolution periods of an orbiting body that is created by tidal effects, such that the orbiting object always shows the same side to the object being orbited.

tidal synchronization The modification of the spin rate of a satellite (such as a planet or a moon) by the gravity of the body it orbits such that the satellite's rotation and orbital periods are synchronized (for example, the satellite completes two rotations for every three of its orbits).

tide The effect of differential gravity on an extended object. Tides lead to distortion of spherical mass distributions, among other effects.

time dilation The effect of relativity in which time runs slower for objects near a strong gravitational source or for objects that are moving at relativistic speeds in the reference frame of distant observers.

torus (pl. tori) A doughnut-shaped ring surrounding an accretion disk and supermassive black hole; an element of the unified model for AGNs.

total lunar eclipse An eclipse that occurs when the Moon passes through Earth's umbral shadow.

total solar eclipse An eclipse that occurs when a region of Earth's surface passes under the Moon's umbral shadow and the Sun's disk is fully blocked.

totality The duration of total obscuration of the Sun or Moon during an eclipse.

transit method An exoplanet detection method that measures changes in the brightness of a star as a planet passes in front of it.

transmission The passing of light through matter without being absorbed.

triggered star formation Star formation that is initiated by the compression of a molecular cloud by a supernova blast wave.

Trojan asteroid An asteroid co-orbiting with Jupiter in one of two groups: one group preceding Jupiter at a specific distance and the other group following Jupiter at the same distance.

troposphere The lowest level of Earth's atmosphere, reaching a maximum altitude of 17 km; the location of most of Earth's weather and clouds.

Tully-Fisher law A relationship of galactic rotation speed to galaxy mass that is used to infer the intrinsic brightness of a galaxy.

turbulence The random jostling motions of moving air.

21-cm emission line A wavelength of radio light emitted by hydrogen and considered by some to be a natural communication channel between Earth and other civilizations.

Type Ia supernova The thermonuclear explosion of a white dwarf that has accreted mass from a binary companion, causing it to exceed the Chandrasekhar limit and experience thermonuclear runaway.

Type II supernova The rapid collapse and violent explosion of a massive star.

U

ultraviolet light An electromagnetic wave with a wavelength of 10–400 nanometers, just shorter than that of visible light.

umbra The inner region of a shadow cast by an extended object.

uncompressed density The density an object would have if the effect of gravity were excluded.

unified model of AGNs A consolidated depiction of the morphology of the various active galactic nuclei.

Universe The total of space, time, matter, and energy.

V

variable star A star with regular periodic changes in its temperature and luminosity. Periods tend to be on the order of years.

velocity The change of distance and the direction in which it is changing per unit of time.

velocity dispersion The characteristic range of velocities of bodies in an astronomical object, such as stars in an elliptical galaxy.

visible light The form of light to which the human eye responds, ranging from 400 nm to 700 nm.

visual binary See *astrometric binary*.

void A region between superclusters, spanning tens to hundreds of megaparsecs, where galaxy density is much lower than average.

volatile Easily vaporized (transformed into a gaseous state) at relatively low temperatures; or, a volatile substance.

W

warm interstellar medium (WISM) The phase of the interstellar medium that is warmed by the galaxy's diffuse ionizing radiation (the sum of all its ionizing stellar sources).

wavelength The distance between successive peaks or troughs of a wave.

weakly interacting massive particle (WIMP) A proposed form for dark matter consisting of particles that interact only via gravity and the weak nuclear force.

weathering See *erosion*.

weight The force on an object that is produced by gravity.

white dwarf The compact remnant of a low- or intermediate-mass star supported by electron degeneracy pressure.

Wien's law The mathematical relation between a blackbody's temperature and the wavelength at which it radiates with peak intensity.

WIMP See *weakly interacting massive particle*.

winter solstice The day (approximately December 21 in the Northern Hemisphere) with the fewest hours of daylight, when the Sun appears lowest in the sky.

WISM See *warm interstellar medium*.

X

X-ray An electromagnetic wave with a wavelength between 10^{-11} and 10^{-8} meter.

X-ray burster A binary system that shows intense X-ray outbursts caused by the pileup of material on the surface of a neutron star.

X-ray halo A large, quasi-spherical region of hot gas surrounding a galaxy or galaxy cluster.

Z

zenith The point on the sky directly overhead from an observer.

zodiac The 13 constellations that lie close to the plane of the ecliptic, which the Sun appears to pass through over the course of the year.

zone A sinking, low-pressure cloud region that forms a horizontal band in Jupiter's atmosphere because of convection and the planet's rapid rotation.

CREDITS

FRONT MATTER

Page xx: Brandon Vick/University of Rochester.

CHAPTER 1

Pages 2–3: Al Russell, Astronomy Section, Rochester Academy of Science; p. 3 (from top down): Gary Weathers/Tetra Images/Corbis; NASA, ESA, and J. Lotz, M. Mountain, A. Koekemoer, and the HFF Team (STScI); Robert Gendler/Stocktrek Images/Corbis; Michael Wald/Alamy; Robert Harding World Imagery/Alamy; p. 4: Design Pics Inc./Alamy; p. 5: Andrew Clegg, NSF; (bottom, from left to right): NASA; Guillem Lopez/Alamy; Gary Weathers/Tetra Images/Corbis; p. 7 (left, right and down): Wikimedia Commons, pd; Shutterstock; p. David Franklin/Shutterstock; Wikimedia Commons, pd; (11): Shutterstock ; p. 9: NASA/NOAA/GSFC/Suomi NPP/VIIRS/Norman Kuring; p. 10 (top): NASA/SDO; (bottom): Robert Gendler/Stocktrek Images/Corbis; p. 11 (top left): NASA, ESA, and J. Lotz, M. Mountain, A. Koekemoer, and the HFF Team (STScI); (top right): Richard Powell www.atlasoftheuniverse.com/supercls.gif, http://creativecommons.org/licenses/by-sa/2.5; (right, center): Wikipedia, pd; p. 13 (counterclockwise from top): Steffen Hauser/botanikfoto/Alamy; Michael Wald/Alamy; Tony Gale/Alamy; p. 14 Everett Collection/Alamy; p. 15: Alberto Morante/EPA/Newscom; p. 17: Corbis; p. 18: Robert Harding World Imagery/Alamy.

CHAPTER 2

Pages 22–23: Photo by Nicholas M. Lamendola, Astronomy Section, Rochester Academy of Science; p. 23 (from top down): NASA; © David Malin; Radius Images/Corbis; Cenk E. Tezel & Tunc Tezel; nagelestock.com/Alamy; p. 24 (a): NASA; (b): NASA and The Hubble Heritage Team (AURA/STScI); p. 24 (c): Michael Rich, Kenneth Mighell, and James D. Neill (Columbia University), Wendy Freedman (OCIW) and NASA/ESA; p. 25 (a): woodleywonderworks/Foter/Creative Commons Attribution 2.0 Generic (CC BY 2.0); (b): Chris Cook/Science Source; p. 27 (both): © David Malin Images; p. 29 (a): Frank Zullo/Science Source; (b): AA/USNO; p. 32 (a): A. Morton/Science Source; (b): Babak Tafreshi/National Geographic Creative; p. 33: Radius Images/Corbis; p. 36 : Alberto Levy; p. 37: Credit & Copyright: Fred Espenak (NASA /GSFC); p. 38: Photo by Mrpulley; http://creativecommons.org/licenses/by-sa/3.0/deed.en; p. 39 (a): Bob King/Stellarium; (b): Alan Dyer/ Visuals Unlimited/Corbis; p. 40 (right): Cenk E. Tezel & Tunc Tezel; p. 41 (top a): Alfredo Dagli Orti/The Art Archive/Corbis; (top b): Ivy Close Images/Alamy; (bottom a): nagelestock.com/Alamy; (bottom b): 2/Steve Allen/Ocean/Corbis; p. 42 (top): Trustees of Dartmouth College, all rights reserved; (bottom): Arterra Picture Library/Alamy; p. 43: Gianni Dagli Orti/Corbis.

CHAPTER 3

Pages 50–51: James Canning, Astronomy Section, Rochester Academy of Science; p. 51 (from top down): Christophe Boisvieux/Corbis; World History Archive/Alamy; GL Archive/Alamy FineArt/Alamy; Lebrecht Music & Arts/Corbis; p. 52: Sonia Halliday Photographs/Alamy; p. 53: Christophe Boisvieux / Corbis; p. 55: World History Archive/Alamy; p. 56 (top): Photo by Bachrach; (bottom): GL Archive/Alamy; p. 60 (from left to right): FineArt/Alamy; Gianni Tortoli/Science Source; Jim Sugar/Corbis; p. 61: Sean Sexton Collection /Corbis; p. 62 (a): Chronicle/Alamy; (b): Granger Collection; (right): Gary Weathers/Tetra Images/Corbis; p. 63: Leemage/Corbis; p. 65 (a): Lebrecht Music & Arts/Corbis; (b): Bettmann/Corbis; p. 67 (top): Liliya Williams; (bottom): Corbis.

CHAPTER 4

Pages 76–77: Joel Schmid, Astronomy Section, Rochester Academy of Science; p. 77 (3): *New Scientist*, "Smile, Hydrogen Atom, You're on Quantum Camera," by Lisa Grossman, 30 May 2013, image by Aneta Stodolna of the FOM Institute AMOLF in Amsterdam; (4): NASA, ESA, Andrew Fruchter (STScI), and the ERO team (STScI + ST-ECF); (5): NASA, JPL-Caltech, SINGS Team (SSC); p. 78: H. Schweiker/WIYN and NOAO/AURA/NSF; p. 79 (a): Science Source; (b): Royal Astronomical Society/Science Source; (c): S. Beckwith (STScI) Hubble Heritage Team, (STScI/AURA), ESA, NASA; p. 84 (a): Atlantide Phototravel/Corbis; (b): Cultura/Alamy Stock Photo; (right): Hubble Heritage Team (AURA/ STScI/ NASA); p. 85: Courtesy of Wabash Instrument Corporation; p. 86 (top left): Alyssa Goodman; (top right): NASA, ESA, and K. Noll (STScI); (bottom right): NOAO/Science Source; p. 88 (b): *New Scientist*, "Smile, Hydrogen Atom, You're on Quantum Camera," by Lisa Grossman, 30 May 2013, image by Aneta Stodolna of the FOM Institute AMOLF in Amsterdam; p. 90 (a): © John Chumack/www.galacticimages.com; (b): Phil Degginger/Science Source; Pat Hartigan; p. 92 (left): Luc Viatour/www.Lucnix.be; (right): Photo: Thomas Bresson; p. 94 (a): Eskimo Nebula Credit: Maximilian-Vlad Teodorescu; (b): NASA, ESA, Andrew Fruchter (STScI), and the ERO team (STScI + ST-ECF); p. 95: Enrico Sacchetti/Science Source; p. 97 (a): Luc Novovitch/Alamy; (b): ALMA (ESO/NAOJ/NRAO); p. 98 (a): RGB Ventures/SuperStock/Alamy; (b): NASA, JPL-Caltech, SINGS Team (SSC); p. 99 (a): NASA; (b): NASA, ESA, G. Illingworth, D. Magee, and P. Oesch (University of California, Santa Cruz), R. Bouwens (Leiden University), and the HUDF09 Team; p. 100 (a): Northrup Grumman Space Technologies (NGST) (Public Domain); (b): ESO/WFI (Optical); MPIfR/ESO/APEX/ A.Weiss et al. (Submillimetre); NASA/CXC/CfA/R.Kraft et al. (X-ray).

CHAPTER 5

Pages 104–105: ESA/NASA, ESO and Danny LaCrue; p. 105: (1): ESO: (3): Al Russell, Astronomy Section, Rochester Academy of Science; (4): STSci; (5): NASA and The Hubble Heritage Team (STScI/AURA); p. 106: ESO: p. 107: Wikimedia, pd p. 110 (all): JPL/NASA; p. 111: NEAR Project, Galileo Project, NASA; p. 113: Tunguska, Siberia, 1908, Wikimedia, pd; p. 114 (a): Al Russell, Astronomy Section, Rochester Academy of Science; (b): Wikimedia, pd; p 115 (a): NASA/JPL/Caltech/Ames Research Center/University of Arizona; (b): Halley Multicolor Camera Team, Giotto Project, ESA; (c) : NASA/JPL-Caltech/UMD; p. 117 (top a): NASA/ESA; (bottom a): ESA/Rosetta/NAVCAM/Marco Di Lorenzo/Ken Kremer—www.kenkremer.com; (top b): Hubble Space Telescope Comet Team; (bottom b): ESA/Rosetta/MPS for Osiris Team; p. 118 (center left): Al Russell, ASRAS; (bottom left): AP Photo/AP Video; (right down from top): NASA/JPL/Cornell; (2): NASA/JPL; p. 119 (top): NASA/MSFC/D. Moser, NASA's Meteoroid Environment Office; (bottom): NASA Earth Observatory; p. 120: Courtesy David Jewitt; p. 124 (a): ESO (b): NASA/JPL-Caltech/Palomar Observatory; (c): Gemini Observatory/AURA; p. 125: PHL/UPR Arecibo; 126: STSci; p. 128: Courtesy Heather Knutson; p. 131: NASA and The Hubble Heritage Team (STScI/AURA).

CHAPTER 6

Pages 138–139: NASA/JPL/Caltech; p. 139 (1): All Canada Photos/Alamy; (2): Accent Alaska.com/Alamy; (3): United States Air Force photo by Senior Airman Joshua Strang; (4): Lunar and Planetary Institute; p. 140 (top): NASA; (bottom): All Canada Photos/Alamy; p.145 (a): Imaginechina via AP Images; (b): Accent Alaska.com/Alamy; p. 146 (a): Archive of Alfred Wegener Institute; Wikimedia; public domain; (b): Gary Hincks/Science Source; p. 149: NASA; p. 152: Les Mentel; p. 155 (top): United States Air Force photo by Senior Airman Joshua Strang; (bottom, both): Bill Brooks/Alamy; p. 156: Courtesy of Cornell University; p. 158 (a): James Stuby based on NASA image, Public Domain; (b): LRO/NASA; (c): Steve Mandel, Hidden Valley Observatory, apod010809; p. 160 (top): Lunar and Planetary Institute; (bottom both): NASA; p. 161: NASA.

CHAPTER 7

Pages 166–167: Mars Exploration Rover Mission, Cornell, JPL, NASA; p. 167 (2): NASA/Johns Hopkins University Applied Physics Laboratory/Carnegie Institution of Washington; (3): NASA; (4): Images courtesy of Phil James (Univ. Toledo), Todd Clancy (Space Science Inst., Boulder, CO), Steve Lee (Univ. Colorado), and NASA [North

Pole image]; and NASA, J. Bell (Cornell U.) and M. Wolff (SSI) [South Pole image]; p. 168: (a): NASA/JPL-Caltech/MSSS; (b): JPL/NASA; p. 171: NASA/Johns Hopkins University Applied Physics Laboratory/Carnegie Institution of Washington; p. 172 (a): NASA; (b): Copyright Calvin J. Hamilton. http://www.solarviews.com; p. 173 (top): Photo by Carolyn Russo/NASM; (bottom): NASA/Johns Hopkins University Applied Physics Laboratory/ Carnegie Institution of Washington; p. 174: ESA/ MPS/DLR/IDA; p. 176 (top left): NASA; (top right): Processed Image Copyright Ted Stryk, Data Courtesy the Russian Academy of Sciences; (bottom a): NASA/JPL/USGA; (bottom b): E. De Jong et al. (JPL), MIPL, Magellan Team, NASA; (bottom c): NASA; p. 177 (top left): NASA; (left, down): NASA/JPL/USGS; NASA/USGS; NASA/JPL; p. 181 (top both): Wikimedia, pd; (bottom): NASA/ USGS; p. 182:Bob Paz/Keck Institute for Space Studies; p. 183 (top, from left to right): NASA/ JPL-Caltech/University of Arizona; Malin Space Science Systems/NASA; NASA/MOLA Science Team/O. de Goursac, Adrian Lark; (bottom a): NASA/JPL-Caltech/Arizona State University; (bottom b): © Guillermo Abramson; p. 184 (a): Images courtesy of Phil James (Univ. Toledo), Todd Clancy (Space Science Inst., Boulder, CO), Steve Lee (Univ. Colorado), and NASA [North Pole image]; and NASA, J. Bell (Cornell U.) and M. Wolff (SSI) [South Pole image]; (b): NASA/JPL/Malin Space Science Systems; p. 186 (a): HiRISE, MRO, LPL (U. Arizona), NASA; (b): NASA; p. 187 (top): AP Photo/NASA/JPL-Caltech/Cornell University; (bottom): NASA/JPL-Caltech/University of Arizona/Texas A&M University; p. 188: NASA/ JPL-Caltech/Univ. of Arizona.

CHAPTER 8

Pages 194–195: NASA/JPL/University of Arizona; p. 195 (1): NASA/JPL; (2): NASA/JPL/University of Arizona; (3): Time Life Pictures/JPL/NASA/The LIFE Picture Collection/Getty Images; (4): NASA/ JPL-Caltech/Space Science Institute; (5): Lawrence A. Sromovsky (University of Wisconsin–Madison) and NASA/ESA; p. 196 (a both): NASA; (b): NASA/ JPL; p. 198 (from left to right): NASA/JPL; NASA/ JPL/Space Science Institute; Lawrence Sromovsky, (Univ. Wisconsin–Madison), Keck Observatory; (2): NASA; p. 202: Mark Halper; p. 204: NASA/ ESA; p. 205 (top): NASA/JPL Illustration Credit: NASA; (center): NASA Planetary Photojournal; (bottom): NASA; p. 206 (a): Time Life Pictures /JPL /NASA/The LIFE Picture Collection/Getty Images; (b–c): NASA/JPL; p. 207 (top a): NASA; (top, b): World History Archive/Alamy; (top, c): NASA/ Science Source; (top right): NASA/JPL/University of Arizona/University of Colorado; (bottom right): NASA /JPL /University of Arizona; (top left): Galileo Project/JPL/NASA; (bottom left): NASA/ JPL/DLR; p. 208 (a): NASA/Galileo; (b): NASA/ JPL/USGS; (c): NASA, Voyager Project, Copyright Calvin J. Hamilton; p. 213: (top a): NASA/JPL-Caltech/Space Science Institute; (top b): NASA/ JPL/University of Arizona; (top c): Reta Beebe (New Mexico State University), D. Gilmore, L. Bergeron (STScI), and NASA; (bottom a): NASA /JPL/Space

Science Institute—NASA planetary photojournal, prepared by Alfred McEwen; (bottom b): Photo by David Monniaux, 2005; https://creativecommons. org/licenses/by-sa/3.0/deed.en; (bottom c): NASA/ JPL-Caltech/ASI/USGS; (bottom d): NASA/JPL/ ESA/University of Arizona; p. 214 (a): Cassini Imaging Team, SSI, JPL, ESA, NASA; (b): NASA/ JPL/Space Science Institute; (c): NASA /JPL/Space Science Institute; p. 215 (top): NASA/JPL/Space Science Institute;(center): Image copyright by Joe Bergeron; (a–b): NASA/JPL/Space Science Institute; (c): NASA/JPL; p. 216: Bill Youngblood; p. 217 (a): NASA/JPL-CalTech; (b): NASA/JPL; p. 218 (a): Lawrence Sromovsky, Space Science and Engineering Center, University of Wisconsin–Madison; (b): Lawrence A. Sromovsky (University of Wisconsin–Madison) and NASA/ESA; p. 219 (top): NASA/ JPL-Caltech; (bottom, from left to right): NASA/ JPL/USGS; NASA, Voyager 2; NASA/JPL; p. 220 (a): Erich Karkoschka (University of Arizona) and NASA; (b): NASA/JPL.

CHAPTER 9

Pages 224–225: NASA/JPL-Caltech/T. Pyle; p. 225 (1): Copyright Dale Andersen, All Rights Reserved; (3): Cultura RM/Alamy; (4): NASA/JPL; p. 226: Copyright Dale Andersen, All Rights Reserved; p. 230 (a): NOAA; (b): University of Delaware College of Earth, Ocean, and Environment; p. 232 (a): Bettmann/Corbis; (b): Omikron/Science Source; p. 233: NASA; p. 235 (a, c): Shutterstock; (b): David Shale/Nature Picture Library/Corbis; (d): blickwinkel/ Alamy; p. 236: Pictorial Press Ltd/Alamy; p. 237 (all): Shutterstock; p. 239 (a): P. Carrara, NPS—National Park Service/USGS Photographic Library; (b): Cultura RM/Alamy; p. 240 (both): Shutterstock; p. 241 (a): NRAO/NSF/AUI; (b): NASA; (c): NASA/JPL; p. 244: Woody Sullivan; p. 245 (left): Arne Nordmann, 2005 http://creativecommons. org/licenses/by-sa/3.0/deed.en; (right): Namazu-tron, 2008 http://commons.wikimedia.org/wiki/ File:SETI@home_Multi-Beam_screensaver.png; p.246 (a): Moviestore collection Ltd/Alamy; (b): United Archives GmbH/Alamy.

CHAPTER 10

Pages 250–251: Bob McGiovern, Astronomy Section, Rochester Academy of Science; p. 251 (from top down): ESA & NASA; © Kamioka Observatory, ICRR (Institute for Cosmic Ray Research), The University of Tokyo; New Jersey Institute of Technology's Big Bear Solar Observatory; Courtesy of SOHO/[instrument] consortium. SOHO is a project of international cooperation between ESA and NASA; p. 252: ESA & NASA; p. 253: NSO/ AURA/NSF; p. 256: Wikimedia, pd; p. 260 (both): © Kamioka Observatory, ICRR (Institute for Cosmic Ray Research), The University of Tokyo; p. 263: New Jersey Institute of Technology's Big Bear Solar Observatory; p. 264: Courtesy: Friedrich Woeger, KIS, and Chris Berst and Mark Komsa, NSO/AURA/NSF; p. 265: NASA/NSSTC/ HATHAWAY 2006/03; p. 268 (top): SOHO-EIT Consortium ESA, NASA; (bottom): Solar Dynamics Observatory/NASA; p. 269 (top): Courtesy of

SOHO/[instrument] consortium. SOHO is a project of international cooperation between ESA and NASA; (bottom a): Dynamics Explorer mission, The University of Iowa and NASA Goddard Space Flight Center; (bottom b): NASA's Earth Observatory; p. 270: NASA/AIA.

CHAPTER 11

Pages 274–275: Marco Burali, Tiziano Capecchi, Marco Mancini (Osservatorio MTM); p. 275 (from top down): www.Wikisky.org; Martin S. Mitchell of Nottingham, England; Hubble Heritage Team (AURA/ STScI/ NASA); p. 276 (a): NASA, ESA, H. Bond (STScI), and M. Barstow (University of Leicester); (b): www.Wikisky.org; (c): Palomar Observatory STScI/WikiSky; p. 280: NASA, ESA, and the Hubble Heritage Team (STScI/AURA); p. 281 (left): Royal Astronomical Society/Science Source; (a): Martin S. Mitchell of Nottingham, England; (b): Professor Greg Parker; p. 282: Andrea Dupree (Harvard-Smithsonian CfA), Ronald Gilliland (STScI), NASA and ESA; p. 284: University of Illinois; p. 285: Smithsonian Institution Archives; p. 286 (a): Interfoto/Alamy; (b): Bettmann/Corbis; p. 291: Lara Eakins, Department of Astronomy, University of Texas at Austin; p. 292: Hubble Heritage Team (AURA/ STScI/ NASA).

CHAPTER 12

Pages 296–297: NASA/JPL-Caltech/J. Rho (SSC/ Caltech); p. 297 (from top down): CalTech/JPL/ NASA; NASA, ESA, and the Hubble Heritage Team (STScI/AURA) Acknowledgment: J. Hughes (Rutgers University); Gary L. Stevens/NASA/ JPL; NASA/JPL-Caltech/UCLA; NASA,ESA, M. Robberto (Space Telescope Science Institute/ ESA) and the Hubble Space Telescope Orion Treasury Project Team; p. 298 (both): CalTech/ JPL/NASA; p. 299 (top): LAMBDA/NASA; (bottom): NASA, ESA, and the Hubble Heritage Team (STScI/AURA) Acknowledgment: J. Hughes (Rutgers University); p. 300: Copyright: Tony Hallas; p. 301: Angel R. Lopez-Sanchez & Barbel Koribalski CSIRO/Australia Telescope National Facility; p. 305: Trippy Photography; p. 307 (top a): Gary L. Stevens/NASA/JPL; (top b): NASA, ESA, and The Hubble Heritage Team (STScI/ AURA), P. McCullough (STScI/AURA); (bottom a): NRAO; (bottom b): Richard Klein, Lawrence Livermore National Laboratory; Pak Shing Li, University of California, Berkeley; Tim Sandstrom, NASA Ames Research Center; p. 311: Photo by Andrea Cipriani Mecchi; p. 313 (a): NASA and A. Watson (Institut de Astronomia, UNAM, Mexico (STScI-PRC00-32b); (b): NASA, ESA, M. Livio and the Hubble 20th Anniversary Team (STScI); p. 314: ESO /ALMA/NAOJ/NRAO/H. Arce. Acknowledgments: Bo Reipurth; p. 316: NASA/ JPL-Caltech/UCLA; p. 317: Al Russell, Astronomy Section, Rochester Academy of Science; p. 318 (a): NASA, ESA, N. Smith (U. California, Berkeley) et al., and The Hubble Heritage Team (STScI/AURA); (b): NASA/JPL-Caltech/N. Smith (University of Colorado at Boulder); p. 319 (both) NASA and ESA, STScI-PRC15-01c; p. 320 (a): NASA,ESA, M.

INDEX

Page numbers in *italics* refer to illustrations and tables.